ELECTRONIC COMMUNICATIONS SYSTEMS
Fundamentals Through Advanced
Second Edition

Wayne Tomasi

Mesa Community College

PRENTICE HALL CAREER & TECHNOLOGY, Englewood Cliffs, New Jersey 07632

Library of Congress Cataloging-in-Publication Data

TOMASI, WAYNE.
    Electronic communications systems : fundamentals through advanced
/ Wayne Tomasi. — 2nd ed.

p.   cm.
    Includes bibliographical references and index.
ISBN 0-13-220021-X
    1. Telecommunication systems.   I. Title.
TK5101.T625 1994   93-5577
621.382—dc20   CIP

Chapters 1–11 are published as
Fundamentals of Electronic Communications Systems
by Wayne Tomasi (© 1994, 1988);
chapters 12–20 are published as
Advanced Electronic Communications Systems
by Wayne Tomasi (© 1994, 1992, 1987).

Editorial/production supervision,
    and interior design: *Barbara Marttine*
Cover design: *Ruta Kysilewskyj*
Cover photograph: *Joshua Sheldon*
Acquisitions editor: *Holly Hodder*
Editorial assistant: *Melissa Steffens*
Manufacturing buyers: *Ed O'Dougherty/Ilene Sanford*

Prentice Hall Career & Technology
© 1994, 1988 by Prentice-Hall, Inc.
A Paramount Communications Company
Englewood Cliffs, NJ 07632

Printed in the United States of America

10 9 8 7 6 5 4 3 2 1

ISBN 0-13-220021-X
          0-13-220039-2 (IE)

Prentice-Hall International (UK) Limited, *London*
Prentice-Hall of Australia Pty. Limited, *Sydney*
Prentice-Hall Canada Inc., *Toronto*
Prentice-Hall Hispanoamericana, S.A., *Mexico*
Prentice-Hall of India Private Limited, *New Delhi*
Prentice-Hall of Japan, Inc., *Tokyo*
Simon & Schuster Asia Pte. Ltd., *Singapore*
Editora Prentice-Hall do Brasil, Ltda., *Rio de Janeiro*

To my loving
and very patient wife, Cheryl

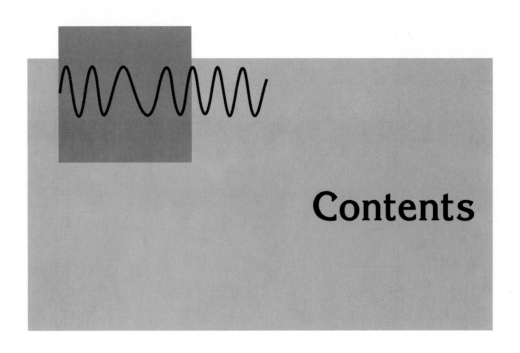

Contents

8   TRANSMISSION LINES   **318**

9   WAVE PROPAGATION   **355**

10   ANTENNAS AND WAVEGUIDES   **377**

Contents   xi

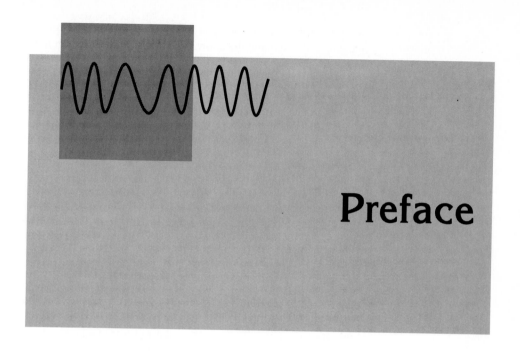

Preface

The second edition of *Electronic Communications Systems: Fundamentals Through Advanced* provides a modern, comprehensive coverage of the field of electronic communications. The major new topics and extended coverage of existing topics incorporated into this edition are:

1. A more detailed explanation of cellular radio
2. Transmission modes
3. Circuit arrangements
4. Special purpose antennas
5. UHF and Microwave antennas
6. Waveguides
7. Time-domain reflectometry
8. Phase-locked loops and frequency synthesizers have been rewritten and are now included in the discussion of signal generation
9. Bit-error-rate and probability of error
10. Standards organizations for data communications
11. Open Systems Interconnection (OSI)
12. ETHERNET
13. Pulse transmission and eye patterns
14. Radio-wave path characteristics
15. ANIK-D communications satellite system
16. Introduction to laser technology
17. Integrated-Services Digital Network (ISDN)

The purpose of this book is to introduce basic electronic communications fundamentals and to expand the knowledge of the reader to more modern digital and data communications systems. It was written so that a reader with previous knowledge in basic electronic principles and an understanding of mathematics through trigonometry will have little trouble understanding the concepts presented. In several sections, a basic understanding of calculus principles (that is, differentiation and integration) would be helpful but is not a prerequisite. Within the text, there are numerous examples that emphasize the most important concepts, and questions and problems are included at the end of each chapter. Also, answers to the odd-numbered problems are given at the end of the book.

Chapter 1 is an introduction to electronic communications. Fundamental communications terms and concepts such as modulation and demodulation, bandwidth and information capacity, signal analysis, and electrical noise are explained. Chapter 2 covers signal generation, phase-locked loops, and frequency synthesizers. The basic requirements for oscillations to occur are outlined and discussed. Standard LC and crystal oscillator configurations are explained as well as an introduction to large-scale integration oscillator chips and how they are used in frequency synthesizers. Chapter 3 defines and explains the basic concepts of amplitude modulation transmission. A detailed analysis is presented on the voltage, power, and bandwidth characteristics of amplitude modulation in both the frequency and time domain. Chapter 4 introduces the basic concepts of AM radio receivers, including a detailed analysis of tuned radio frequency and superheterodyne receivers. Chapter 5 extends the coverage of amplitude modulation given in Chapters 3 and 4 to single sideband transmission. The various types of sideband and suppressed carrier systems are explained and contrasted. Several types of single sideband transmitters and receivers are discussed. Chapter 6 introduces the basic concepts of angle modulation. Frequency and phase modulation are explained and contrasted. The amplitude, power, and frequency characteristics of an angle modulated wave are explained in detail. Both direct and indirect frequency and phase modulation transmitters are shown and discussed. Chapter 7 extends the coverage of angle modulation to receivers and systems. Several types of angle modulation demodulators are introduced and discussed. Frequency modulation stereo transmission, two-way radio communications, mobile radio, and cellular radio are discussed. Chapter 8 explains the characteristics of an electromagnetic wave and wave propagation on metallic transmission lines. Several basic transmission line configurations are discussed and contrasted. Transmission line characteristic impedance and input impedance are also discussed and time domain reflectometry is introduced. Chapter 9 gives a complete explanation of the optical properties of electromagnetic waves: refraction, reflection, diffraction, and interference. Ground and space wave propagation are defined and explained. Chapter 10 introduces the antenna and describes basic antenna operation. Fundamental antenna terms are defined and explained. Explanations are given for antenna configurations from the simplest elementary doublet to complex, highly directional UHF and microwave antennas. Waveguides and waveguide principles are also explained in this chapter. Chapter 11 introduces the basic concepts of television broadcasting. Both monochrome and color television transmission and reception are explained. Chapter 12 introduces the concepts of digital transmission and digital modulation. In this chapter, the most common modulation schemes used in modern digital radio systems are described. The concepts of information capacity and bandwidth efficiency are also explained. Chapter 13 introduces the field of data communications. Detailed explanations are given for numerous data communications concepts, including transmission methods, circuit configurations, topologies, character codes, error control mechanisms, data formats, and data modems. Standards organizations is also included in this chapter. Chapter 14 describes data communications protocols. Synchronous and asynchronous data protocols are first defined, then explicit examples are given for each. This chapter also includes sections on Open System Interconnection (OSI) and Integrated-Services Digital Network (ISDN). Chapter 15 introduces digital transmission techniques. This includes a detailed explanation of pulse code modulation. The concepts of sampling, encoding, and companding are explained. Chapter 16 explains multiplexing. First time division multiplexing is explained and then frequency division multiplexing. The North American Hierarchy for both time and frequency division multiplexing is described. Chapter 17 introduces microwave radio communications and the concept of system gain. A block diagram approach to the operation of a microwave radio system is presented and numerous examples are included. Chapter 18 introduces the basic concepts of satellite systems including orbital patterns, radiation patterns, and geosynchronous and nonsynchronous systems. System parameters and link equations are discussed and a detailed explanation of a satellite link budget is given. Chapter 19 extends the coverage of satellite systems to methods of multiple accessing. The three predominant methods for multiple accessing are explained, and a section on the ANIK-D communications satellite system is described. Chapter 20 covers the basic concepts of an optical fiber communications system. A detailed explanation is given for light-wave propagation through a guided fiber cable. Also, several light sources and detectors are discussed and an introduction of lasers is presented. Appendix A describes the Smith chart and how it is used for transmission line calculations.

Wayne Tomasi

Acknowledgments

I would like to acknowledge the following individuals for their contributions to this book: Holly Hodder, Senior Managing Editor, Electronic Technology, for giving me the opportunity to write this edition and for her outstanding professional guidance throughout this project; Kathryn Pavelec, Production Editor for the first edition; Barbara Marttine, Production Editor for this edition; and the following reviewers — Merle Parmer, DeVry, Phoenix; Robert E. Greenwood, Ryerson Polytechnical Institute, James W. Stewart, DeVry, Woodbridge; Susan A. R. Garrod, Purdue University; James C. Chapel, Lincoln Land Community College; James Fisk, Northern Essex Community College; Allan Shapiro, Northern Virginia Community College; William H. Maxwell, Nashville State Technical Institute; and Ladimer S. Nagurney, University of Hartford. I would also like to give special thanks to my friend and former associate, Merle Parmer, DeVry Institute of Technology — Phoenix, for the many hours he spent reviewing the technical content and accuracy of the final manuscript.

We would also like to acknowledge Terry O'Neil and Frank Tatulli at Paramount Communications, Inc., 15 Columbus Circle, New York, New York, for allowing us access to the roof in order to shoot the cover photograph.

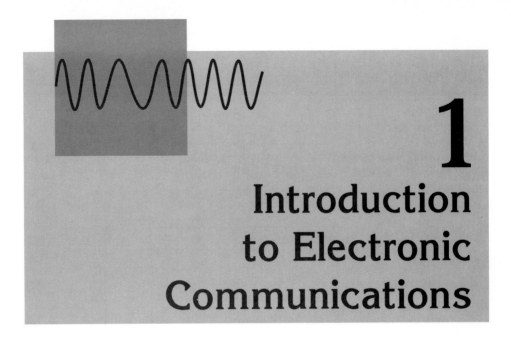

1

Introduction to Electronic Communications

INTRODUCTION

The purpose of this chapter is to introduce the reader to the fundamental concepts of *electronic communications systems* and to explain some of the basic terminology necessary to the understanding of more complex topics discussed later in this book.

In essence, *electronic communications* is the transmission, reception, and processing of *information* using electronic circuits. Information is defined as knowledge, wisdom, or facts and can be in *analog* form (proportional or continuous), such as the human voice, video picture information, or music, or *digital* form (discrete steps), such as binary-coded numbers, alphanumeric codes, graphic symbols, microprocessor op-codes, or database information. All information must be converted to *electromagnetic energy* before it can be propagated through an electronic communications system.

Figure 1-1a is a simplified block diagram of an electronic communications system showing the relationship among the original *source information,* the *transmitter,* the *transmission medium (facility),* the *receiver,* and the *received information* at the *destination.* As the figure shows, an electronic communications system is comprised of three primary sections: a transmitter, a transmission medium, and a receiver. The transmitter converts the original source information to a form more suitable for transmission, the transmission medium provides a means of connecting the transmitter to the receiver (such as a metallic conductor, an optical fiber, or free space), and the receiver converts the received information back to its original form and transfers it to the destination. The original information can originate from a variety of different sources and be in analog or digital form. The communications system shown in Figure 1-1a is capable of conveying

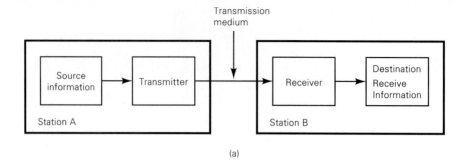

(a)

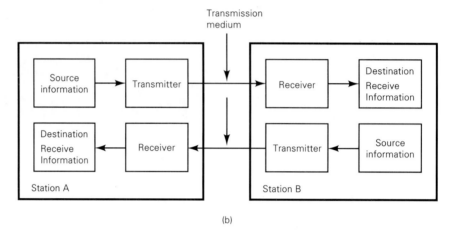

(b)

Figure 1-1 Simplified block diagram of a communications system: (a) one direction only; (b) both directions.

information only in one direction (from station A to station B), while the communications system shown in Figure 1-1b is capable of conveying information in both directions (from station A to station B and from station B to station A).

When transmitting information from more than one source over a common transmission medium, the information must be combined into a single composite information signal. The process of combining information into a composite information signal is called *multiplexing,* and the process of separating the information is called *demultiplexing.* Multiplexing and demultiplexing are covered in detail in Chapter 16.

There are two basic types of electronic communications systems: analog and digital. An *analog communications system* is a system in which electromagnetic energy is transmitted and received in analog form (a continuously varying signal such as a sine wave). Commercial radio systems broadcast analog signals. A *digital communications system* is a system in which electromagnetic energy is transmitted and received in digital form (discrete levels such as +5 V and ground). Binary systems use digital signals that have only two discrete levels (*bi* meaning two). Very often the original source information is in a form unsuitable for transmission and must be converted to a more suitable form prior to transmission. For example, with digital communications systems, analog information is converted to digital form prior to transmission, and with analog communications systems, digital information is converted to analog form prior to transmission.

Analog communications systems were the first to be developed; however, in recent years digital communications systems have become more prevalent. The reasons for the change from analog to digital will become evident as the reader progresses through this book. The beginning chapters of this book deal predominantly with analog communication systems, and the later chapters deal with the more contemporary systems.

History of Electronic Communications

The theory of electronic communications began in the middle of the nineteenth century with the British physicist, James Clerk Maxwell. Maxwell's mathematical research indicated that electricity and light both travel in the form of *electromagnetic waves* and are therefore related to each other. Maxwell predicted that it was possible to propagate electromagnetic waves through *free space* using electrical discharges. However, such wave propagation was not accomplished until 1888 when Heinrich Hertz, a German scientist, was able to radiate electromagnetic energy from a machine he called an *oscillator.* Hertz developed the first radio transmitter and, using these devices, was able to generate radio frequencies between 31 MHz and 1.25 GHz. Hertz also developed the first rudimentary *antenna,* which is still used in modified form today. In 1892, E. Branly of France developed the first *radio detector,* and exactly one year later a Russian experimenter, A. S. Popoff, recorded radio waves emanating from lightning.

The first electronic communications system was developed in 1837 by Samuel Morse. Morse, using *electromagnetic induction,* was able to transmit information in the form of dots, dashes, and spaces across a metallic wire. He called his invention the *telegraph.* In 1876, a Canadian educator and speech therapist named Alexander Graham Bell and his assistant, Thomas A. Watson (a well-known inventor himself), successfully transmitted human conversation over a functional *telephone* system using metallic wires for the transmission medium.

In 1894, Guglielmo Marconi, a young Italian scientist, accomplished the first wireless electronic communications when he transmitted radio signals for three-quarters of a mile through Earth's atmosphere on his father's estate. By 1896, Marconi was transmitting radio signals up to two miles from ships to shore, and in 1899 he sent the first wireless message across the English Channel from France to Dover, England. In 1902, the first transatlantic signals were sent from Poldu, England, to Newfoundland. Lee DeForest invented the *triode vacuum tube* in 1908, which allowed for the first practical amplification of electronic signals. Regular radio broadcasting began in 1920, when AM (amplitude modulation) radio stations WWJ in Detroit, Michigan, and KDKA in Pittsburgh, Pennsylvania, began commercial broadcasting. In 1933, Major Edwin Howard Armstrong invented frequency modulation (FM), and commercial broadcasting of FM signals began in 1936. In 1948, the transistor was invented at Bell Telephone Laboratories by William Shockley, Walter Brattain, and John Bardeen. The transistor lead to the development and refinement of the integrated circuit in the 1960s.

Although the general concepts of electronic communications have not changed much since their inception, the methods by which these concepts are implemented have undergone both dramatic and remarkable changes in the recent past. There are really no limits on the expectations for electronic communications systems of the future.

Modulation and Demodulation

For reasons that are explained in Chapter 10, it is impractical to propagate low-frequency electromagnetic energy through Earth's atmosphere. Therefore, with radio communications, it is necessary to superimpose a relatively low frequency intelligence signal onto a relatively high frequency signal for transmission. In analog electronic communications systems, the source information (intelligence signal) acts on or *modulates* a single-frequency sinusoidal signal. *Modulate* simply means to vary, change, or regulate. Therefore, the relatively low frequency source information is called the *modulating signal,* the relatively high frequency signal that is acted on (modulated) is called the *carrier,* and the resultant signal is called the *modulated wave* or signal. In essence, the source information is transported through the system on the carrier.

With analog communications systems, *modulation* is the process of varying or changing some property of an analog carrier in accordance with the original source information. Conversely, *demodulation* is the process of converting the changes in the analog carrier back to the original source information. Modulation is performed in the transmitter in a circuit called a *modulator,* and demodulation is performed in the receiver in a circuit called a *demodulator.* The information signal that modulates the main carrier is called the *baseband signal* or simply *baseband.* Baseband is an information signal, such as a single telephone channel, and the *composite baseband signal* is the total information signal, such as several hundred telephone channels. Baseband signals are converted from their original frequency band to a band more suitable for transmission through the communications system. Baseband signals are *up-converted* in the transmitter and *down-converted* in the receiver. *Frequency translation* is the process of converting a single frequency or a band of frequencies to another location in the total frequency spectrum.

The term *channel* is often used when referring to a specific band of frequencies allocated to a particular service or transmission. For example, a standard voice-band channel occupies a 3-kHz bandwidth and is used for the transmission of voice-quality signals. An RF channel refers to a band of frequencies used to propagate radio-frequency signals, such as a single commercial FM broadcast channel that occupies approximately a 200-kHz frequency band within the total 88- to 108-MHz band allotted for commercial FM transmission.

Equation 1-1 is the general expression for a time-varying sine wave of voltage such as an analog carrier. Three properties of a sine wave can be varied: the amplitude (*V*), the frequency (*f*), the phase (θ), or any combination of two or more of these properties. If the amplitude of the carrier is varied proportional to the source information, *amplitude modulation (AM)* results. If the frequency of the carrier is varied proportional to the source information, *frequency modulation (FM)* results. If the phase of the carrier is varied proportional to the source information, *phase modulation (PM)* results.

$$v(t) = V \sin(2\pi f t + \theta) \qquad (1\text{-}1)$$

where $v(t)$ = time-varying sine wave of voltage
$\quad V$ = peak amplitude (volts)
$\quad f$ = frequency (hertz)
$\quad \theta$ = phase (radians)

Figure 1-2 is a simplified block diagram of a communications system showing the relationship among the modulating signal (information), the modulated signal (carrier), the modulated wave (resultant), and system noise.

There are two important reasons why modulation is necessary in an electronic communications system. The first is the fact that it is extremely difficult to radiate low frequency signals through Earth's atmosphere in the form of electromagnetic energy. Second, information signals often occupy the same frequency band and, if transmitted in their original form, would interfere with each other. An example of this is the commercial FM broadcast band. All FM stations broadcast voice and music information that occupy the

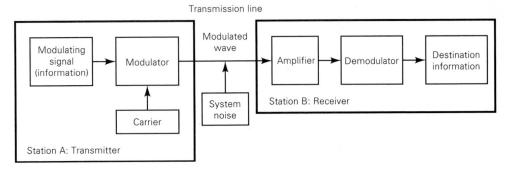

Figure 1-2 Communications system block diagram.

audio-frequency band from 0 to 15 kHz. Each station translates their information to a different frequency band (channel) so that their transmissions will not interfere with everyone else's transmissions. The reasons for modulation and demodulation are explained in detail in Chapters 3 and 4, when their purpose will become more evident.

THE ELECTROMAGNETIC SPECTRUM

The purpose of an electronic communications system is to communicate information between two or more locations (generally called *stations*). This is accomplished by converting the original source information into electromagnetic energy and then transmitting that energy to one or more destinations, where it is converted back to its original form. Electromagnetic energy can propagate in various modes: as a voltage or a current along a metallic wire, as emitted radio waves through free space, or as light waves down an optical fiber.

Electromagnetic energy is distributed throughout an almost infinite range of frequencies. The total *electromagnetic frequency spectrum* showing the approximate locations of various services within the band is shown in Figure 1-3. It can be seen that the frequency spectrum extends from the *subsonic* frequencies (a few hertz) to *cosmic rays* (10^{22} Hz). Each band of frequencies has a unique characteristic that makes it different from the other bands.

When dealing with radio waves, it is common to use the units of wavelength rather than frequency. Wavelength is the length that one cycle of an electromagnetic wave occupies in space (that is, the distance between similar points in a repetitive wave). Wavelength is inversely proportional to the frequency of the wave and directly proportional to the velocity of propagation (the velocity of propagation of electromagnetic energy in free

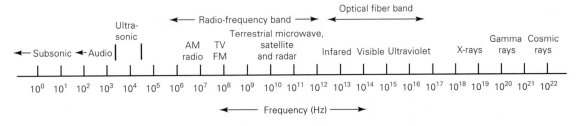

Figure 1-3 Electromagnetic frequency spectrum.

space is assumed to be the speed of light, 3×10^8 m/s). The relationship among frequency, velocity, and wavelength is expressed mathematically as

$$\text{wavelength} = \frac{\text{velocity}}{\text{frequency}}$$

(1-2)

$$\lambda = \frac{c}{f}$$

where λ = wavelength (meters per cycle)
c = velocity of light (300,000,000 m/s)
f = frequency (hertz)

The total *electromagnetic wavelength* spectrum showing the various services within the band is shown in Figure 1-4.

EXAMPLE 1-1

Determine the wavelength for the following frequencies: 1 kHz, 100 kHz, and 10 MHz.
Solution Substituting into Equation 1-2,

$$\lambda = \frac{300,000,000}{1,000} = 300,000 \text{ m}$$

$$\lambda = \frac{300,000,000}{100,000} = 3000 \text{ m}$$

$$\lambda = \frac{300,000,000}{10,000,000} = 30 \text{ m}$$

Transmission Frequencies

The total electromagnetic frequency spectrum is divided into subsections or bands. Each band has a name and boundaries. In the United States, frequency assignments for *free-space radio propagation* are assigned by the *Federal Communications Commission (FCC)*. For example, the commercial FM broadcast band extends from 88 to 108 MHz. The exact frequencies assigned specific transmitters operating in the various classes of services are constantly being updated and altered to meet the nation's communications needs. However, the general division of the total usable frequency spectrum is decided at the *International Telecommunications Conventions,* which are held approximately once every 10 years.

The total usable *radio-frequency (RF)* spectrum is divided into narrower frequency bands, which are given descriptive names and band numbers. The *International Radio Consultative Committee's (CCIR's)* band designations are listed in Table 1-1. Several of

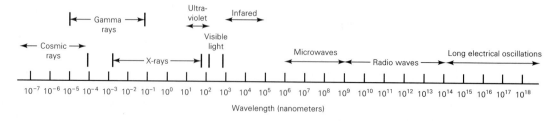

Figure 1-4 Electromagnetic wavelength spectrum.

TABLE 1-1 CCIR BAND DESIGNATIONS

Band Number	Frequency Range[a]	Designations
2	30–300 Hz	ELF (extremely low frequencies)
3	0.3–3 kHz	VF (voice frequencies)
4	3–30 kHz	VLF (very low frequencies)
5	30–300 kHz	LF (low frequencies)
6	0.3–3 MHz	MF (medium frequencies)
7	3–30 MHz	HF (high frequencies)
8	30–300 MHz	VHF (very high frequencies)
9	0.3–3 GHz	UHF (ultrahigh frequencies)
10	3–30 GHz	SHF (superhigh frequencies)
11	30–300 GHz	EHF (extremely high frequencies)
12	0.3–3 THz	Infrared light
13	3–30 THz	Infrared light
14	30–300 THz	Infrared light
15	0.3–3 PHz	Visible light
16	3–30 PHz	Ultraviolet light
17	30–300 PHz	X-rays
18	0.3–3 EHz	Gamma rays
19	3–30 EHz	Cosmic rays

[a]10^0, hertz (Hz); 10^3, kilohertz (kHz); 10^6, megahertz (MHz); 10^9 gigahertz (GHz); 10^{12}, terahertz (THz); 10^{15}, petahertz (PHz); 10^{18}, exahertz (EHz)

these bands are further broken down into various types of services, such as shipboard search, microwave, satellite, mobile land-based search, shipboard navigation, aircraft approach, airport surface detection, airborne weather, mobile telephone, and many more.

Classification of Transmitters

For licensing purposes in the United States, radio transmitters are classified according to their bandwidth, type of modulation, and the type of intelligence information that they carry. The *emission classifications* are identified by a three-symbol code containing a combination of letters and numbers as shown in Table 1-2. The first symbol is a letter that designates the type of modulation of the main carrier (amplitude, frequency, phase, pulse, or no modulation). The second symbol is a number that identifies the type of emission (analog, digital, and so on), and the third symbol is another letter that describes the type of information being transmitted (data, telephony, and so on). For example, the designation A3E describes a double-sideband, full-carrier, amplitude-modulated signal carrying telephony (voice or music) information.

BANDWIDTH AND INFORMATION CAPACITY

The two most significant limitations on communications system performance are *noise* and *bandwidth*. The significance of noise is discussed later in this chapter. The bandwidth of a communications system is the minimum passband (range of frequencies) required to propagate the source information through the system. The bandwidth of a communications system must be sufficiently large (wide) to pass all significant information frequencies.

The *information capacity* of a communications system is a measure of how much source information can be carried through the system in a given period of time. The amount of information that can be propagated through a *transmission system* is a function

TABLE 1-2 FCC EMISSION CLASSIFICATIONS

Symbol	Letter	Type of Modulation
First	Unmodulated	
	N	Unmodulated carrier
	Amplitude Modulation	
	A	Double sideband, full carrier (DSBFC)
	B	Independent sideband, full carrier (ISBFC)
	C	Vestigial sideband, full carrier (VSB)
	H	Single-sideband, full carrier (SSBFC)
	J	Single-sideband, suppressed carrier (SSBSC)
	R	Single-sideband, reduced carrier (SSBRC)
	Angle Modulation	
	F	Frequency modulation (direct FM)
	G	Phase modulation (indirect FM)
	D	AM and FM simultaneously or sequenced
	Pulse Modulation	
	K	Pulse-amplitude modulation (PAM)
	L	Pulse-width modulation (PWM)
	M	Pulse-position modulation (PPM)
	P	Unmodulated pulses (binary data)
	Q	Angle modulated during pulses
	V	Any combination of pulse-modulation category
	W	Any combination of two or more of the above forms of modulation
	X	Cases not otherwise covered
Second	0	No modulating signal
	1	Digitally keyed carrier
	2	Digitally keyed tone
	3	Analog (sound or video)
	7	Two or more digital channels
	8	Two or more analog channels
	9	Analog and digital
Third	A	Telegraphy, manual
	B	Telegraphy, automatic (teletype)
	C	Facsimile
	D	Data, telemetry
	E	Telephony (sound broadcasting)
	F	Television (video broadcasting)
	N	No information transmitted
	W	Any combination of second letter

of the system bandwidth and the transmission time. The relationship among bandwidth, transmission time, and information capacity was developed in 1920 by R. Hartley of Bell Telephone Laboratories. Simply stated, Hartley's law is

$$I \propto B \times t \qquad (1\text{-}3)$$

where
I = information capacity
B = bandwidth (hertz)
t = transmission time (seconds)

Equation 1-3 shows that information capacity is a linear function and directly proportional to both system bandwidth and transmission time. If either bandwidth or transmission time is changed, a directly proportional change in information capacity will occur.

Approximately 3 kHz of bandwidth is required to transmit voice-quality telephone signals. More than 200 kHz of bandwidth is required for commercial FM transmission of high-fidelity music, and almost 6 MHz of bandwidth is required for broadcast-quality television signals (that is, the more information per unit time, the more bandwidth required).

TRANSMISSION MODES

Electronic communications systems can be designed to handle transmission only in one direction, in both directions but in only one at a time, or in both directions at the same time. These are called *transmission modes.* Four transmission modes are possible: *simplex, half-duplex, full-duplex,* and *full/full duplex.*

Simplex (SX)

With simplex operation, transmissions can occur only in one direction. Simplex systems are sometimes called *one-way-only, receive-only,* or *transmit-only* systems. A location may be a transmitter or a receiver, but not both. An example of simplex transmission is commercial radio or television broadcasting; the radio station always transmits and you always receive.

Half-duplex (HDX)

With half-duplex operation, transmissions can occur in both directions, but not at the same time. Half-duplex systems are sometimes called *two-way-alternate, either-way,* or *over-and-out* systems. A location may be a transmitter and a receiver, but not both at the same time. Two-way radio systems that use *push-to-talk (PTT)* buttons to key their transmitters such as citizens-band and police-band radio are examples of half-duplex transmission.

Full-duplex (FDX)

With full-duplex operation, transmissions can occur in both directions at the same time. Full-duplex systems are sometimes called *two-way simultaneous, duplex,* or *both-way* lines. A location can transmit and receive simultaneously; however, the station it is transmitting to must also be the station it is receiving from. A standard telephone system is an example of full-duplex transmission.

Full/full Duplex (F/FDX)

With full/full duplex operation, it is possible to transmit and receive simultaneously, but not necessarily between the same two locations (that is, one station can transmit to a second station and receive from a third station at the same time). Full/full duplex transmissions are used almost exclusively with data communications circuits. The U.S. Postal Service is an example of full/full duplex operation.

CIRCUIT ARRANGEMENTS

Electronic communications circuits can be configured in several different ways. These configurations are called *circuit arrangements* and can include both two- and four-wire transmission.

Two-wire Transmission

As the name implies, *two-wire transmission* involves two wires (one for the signal and one for a reference or ground) or a circuit configuration that is equivalent to only two wires. Two-wire circuits are ideally suited to simplex transmission, although they can be used for half- and full-duplex transmission. The telephone line between your home and the nearest telephone office is a two-wire circuit.

Figure 1-5 shows the block diagrams for two different two-wire circuit configurations. Figure 1-5a shows the simplest two-wire configuration, which is a passive circuit consisting of two wires connecting an information source through a transmitter to a destination at a receiver. The wires themselves are capable of two-way transmission, but the transmitter and receiver are not. To exchange information in the opposite direction, the locations of the transmitter and receiver would have to be switched. Therefore, this configuration is capable of only one-way transmission and provides no gain to the signal. To achieve half-duplex transmission with a two-wire circuit, a transmitter and receiver would be required at each location, and they would have to be connected to the same wire pair in a manner such that they do not interfere with each other.

Figure 1-5b shows an active two-wire circuit (that is, one that provides gain). With this configuration, an amplifier is placed in the circuit between the transmitter and the receiver. The amplifier is a unidirectional device and thus limits transmissions to only one direction.

To achieve half- or full-duplex capabilities with a two-wire circuit, the information traveling in opposite directions would have to be altered in some way or by some method of converting the source to a destination and the destination to a source. Half- and full-duplex transmission can be accomplished with a two-wire circuit by using some form of modulation technique to *multiplex* or combine the two signals in such a way that they do not interfere with each other but can still be separated or converted back to their origi-

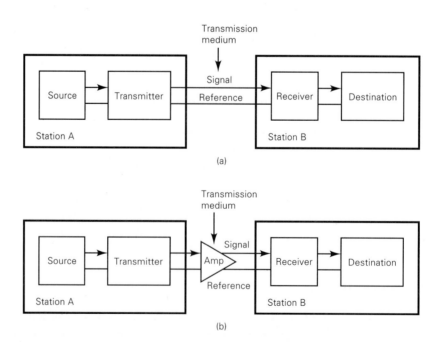

Figure 1-5 Two-wire circuit configurations: (a) passive; (b) active.

nal form at the receiver. Both modulation and multiplexing are described in detail later in this book.

Four-wire Transmission

Four-wire transmission involves four wires (two for each direction, a signal and a reference ground) or a circuit configuration that is equivalent to four wires. Four-wire circuits are ideally suited to full-duplex transmission. Figure 1-6 shows the block diagram of an active four-wire system. As the figure shows, a four-wire circuit is equivalent to two two-wire circuits, one for each direction of transmission. With four-wire operation, the transmitter at one location is connected through a transmission medium to the receiver at the other location, and vice versa. However, the transmitters and receivers at a given location may be operated completely independently of each other.

There are several inherent advantages of four-wire circuits over two-wire circuits. For instance, four-wire circuits are considerably less noisy and provide more isolation between the two directions of transmission when either half- and full-duplex operation is used. However, two-wire circuits require less wire, less circuitry, and thus less money than their four-wire counterparts. The advantages and disadvantages of two- and four-wire circuits will become more apparent as you continue your studies in electronic communications.

Hybrids and Echo Suppressors

When a two-wire circuit is connected to a four-wire circuit as in a long-distance telephone call, an interface circuit called a *hybrid* or *terminating set* is used to affect the interface. The hybrid set is used to match impedances and provide isolation between the two directions of signal flow.

Figure 1-7 shows the block diagram for a two-wire to four-wire hybrid network. The hybrid coil compensates for impedance variations in the two-wire portion of the circuit. The amplifiers and attenuators adjust the signal voltages to required levels, and the equalizers compensate for impairments in the transmission line that affect the frequency response of the transmitted signal, such as line inductance, capacitance, and resistance.

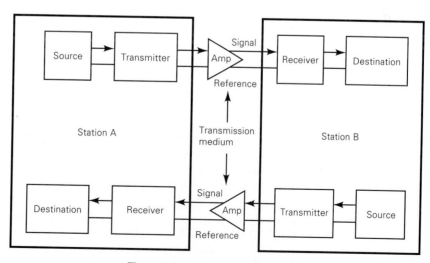

Figure 1-6 Active four-wire circuit.

Circuit Arrangements

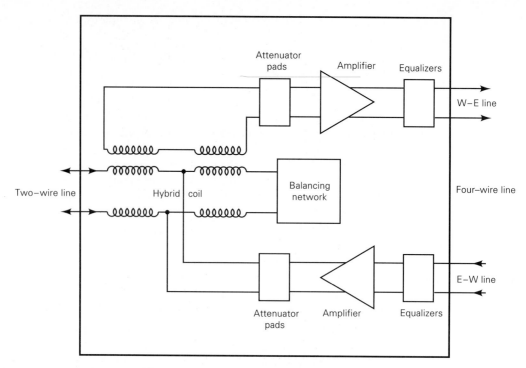

Figure 1-7 Two-wire to four-wire terminating set (hybrid).

Signals traveling west–east (W–E) enter the terminating set from the two-wire line where they are inductively coupled into the west-to-east transmitter section of the four-wire circuit. Signals received from the line are coupled into the east–west (E–W) receiver section of the four-wire circuit where they are applied to the center taps of the hybrid coils. If the impedances of the two-wire line and the balancing network are properly matched, all currents produced in the upper half of the hybrid by the E–W signal will be equal in magnitude but opposite in polarity. Therefore, the voltages induced in the secondaries will be 180 degrees out of phase with each other and thus cancel. This prevents any portion of the received signal from being returned to the sender as an echo.

If the impedances of the two-wire line and the balancing network are not matched, the voltages induced in the secondaries of the hybrid coil will not completely cancel. This imbalance causes a portion of the received signal to be returned to the sender on the W–E portion of the four-wire circuit. The returned portion of the signal is heard as an echo by the talker and, if the round-trip delay of this signal exceeds approximately 45 ms, the echo can become quite annoying. To eliminate this echo, devices called *echo suppressors* are inserted at one end of the four-wire circuit. Figure 1-8 shows a simplified block diagram of an echo suppressor. The speech detector senses the presence and direction of the signal. It then enables the amplifier in the appropriate direction and disables the amplifier in the opposite direction, thus preventing the echo from returning to the speaker. If the conversation is changing direction rapidly, the people listening may be able to hear the echo suppressor turning on and off (every time an echo suppressor detects speech and is activated, the first instant of sound is removed from the message, giving the speech a choppy sound). With an echo suppressor in the circuit, transmissions cannot occur in both directions at the same time, thus limiting the circuit to half-duplex operation. Long-distance common carriers, like AT&T, generally place echo suppressors in four-wire circuits that exceed 1500 electrical miles in length (the longer the circuit, the longer the round-trip time delay).

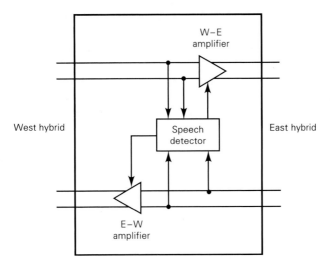

Figure 1-8 Echo suppressor.

SIGNAL ANALYSIS

When designing electronic communications circuits, it is often necessary to analyze and predict the performance of the circuit based on the power distribution and frequency composition of the information signal. This is done with mathematical *signal analysis*. Although all signals in electronic communications are not single-frequency sine or cosine waves, many of them are, and the signals that are not can be represented by a series of sine or cosine functions.

Sinusoidal Signals

In essence, signal analysis is the mathematical analysis of the frequency, bandwidth, and voltage level of a signal. Electrical signals are voltage– or current–time variations that can be represented by a series of sine or cosine waves. Mathematically, a single-frequency voltage or current waveform is

$$v(t) = V \sin (2\pi ft + \theta) \quad \text{or} \quad v(t) = V \cos (2\pi ft + \theta)$$
$$i(t) = I \sin (2\pi ft + \theta) \quad \text{or} \quad i(t) = I \cos (2\pi ft + \theta)$$

where
$v(t)$ = time-varying voltage sine wave
$i(t)$ = time-varying current sine wave
V = peak voltage (volts)
f = frequency (hertz)
θ = phase (radians)
I = peak current (amperes)
$2\pi f = \omega$ angular velocity (radians per second)

Whether a sine or a cosine function is used to represent a signal is purely arbitrary and depends on which is chosen as the reference. However, it should be noted that $\sin \theta = \cos (\theta - 90°)$. Therefore, the following relationships hold true:

$$v(t) = V \sin (2\pi ft + \theta) = V \cos (2\pi ft + \theta - 90°)$$
$$v(t) = V \cos (2\pi ft + \theta) = V \sin (2\pi ft + \theta + 90°)$$

Signal Analysis

The preceding formulas are for a single-frequency, repetitive waveform. Such a waveform is called a *periodic* wave because it repeats at a uniform rate (that is, each successive cycle of the signal takes exactly the same length of time and has exactly the same amplitude variations as every other cycle, each cycle has exactly the same shape). A series of sine, cosine, or square waves are examples of periodic waves. Periodic waves can be analyzed in either the *time* or the *frequency domain.* In fact, it is often necessary when analyzing system performance to switch from the time domain to the frequency domain, and vice versa.

Time domain.   A standard oscilloscope is a time-domain instrument. The display on the cathode ray tube (CRT) is an amplitude-versus-time representation of the input signal and is commonly called a *signal waveform.* Essentially, a signal waveform shows the shape and the instantaneous magnitude of the signal with respect to time, but does not necessarily indicate its frequency content. With an oscilloscope, the vertical deflection is proportional to the amplitude of the total input signal, and the horizontal deflection is a function of time (sweep rate). Figure 1-9 shows the signal waveform for a single-frequency sinusoidal signal with a peak amplitude of V volts and a frequency of f hertz.

Frequency domain.   A spectrum analyzer is a frequency-domain instrument. Essentially, no waveform is displayed on the CRT. Instead, an amplitude-versus-frequency plot is shown (this is called a *frequency spectrum*). With a spectrum analyzer, the horizontal axis represents frequency and the vertical axis amplitude. Therefore, there is a vertical deflection for each frequency present at its input. Effectively, the input waveform is swept with a variable-frequency, high-Q bandpass filter whose center frequency is synchronized to the horizontal sweep rate of the CRT. Each frequency present in the input waveform produces a vertical line on the CRT (these are called *spectral components*). The vertical deflection (height) of each line is proportional to the amplitude of the frequency that it represents. A frequency-domain representation of a wave shows the frequency content, but does not necessarily indicate the shape of the waveform or the combined amplitude of all the input components at any specific time. Figure 1-10 shows the frequency spectrum for a

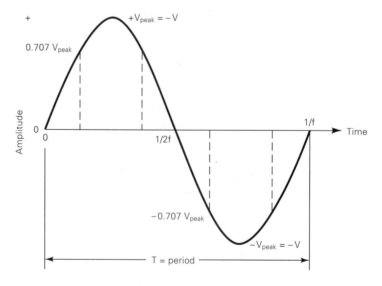

Figure 1-9   Time-domain representation (signal waveform) for a single-frequency sinusoidal wave.

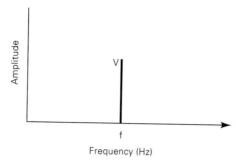

Figure 1-10 Frequency-domain representation (spectrum) for a single-frequency sinusoidal wave.

single-frequency sinusoidal signal with a peak amplitude of V volts and a frequency of f hertz.

Nonsinusoidal Periodic Waves (Complex Waves)

Essentially, any repetitive waveform that comprises more than one sine or cosine wave is a *nonsinusoidal* or *complex periodic wave.* To analyze a complex periodic waveform, it is necessary to use a mathematical series developed in 1826 by the French physicist and mathematician Baron Jean Fourier. This series is appropriately called the *Fourier series.*

The Fourier series. The *Fourier series* is used in signal analysis to represent the sinusoidal components of a nonsinusoidal periodic waveform (that is, to change a time-domain signal to a frequency-domain signal). In general, a Fourier series can be written for any periodic function as a series of terms that include trigonometric functions with the following mathematical expression:

$$f(t) = A_0 + A_1 \cos \alpha + A_2 \cos 2\alpha + A_3 \cos 3\alpha + \cdots + A_n \cos n\alpha \\ + B_1 \sin \beta + B_2 \sin 2\beta + B_3 \sin 3\beta + \cdots + B_n \sin n\beta \qquad (1\text{-}4)$$

where $\alpha = \beta$

Equation 1-4 states that the waveform $f(t)$ comprises an average (dc) value (A_0), a series of cosine functions in which each successive term has a frequency that is an integer multiple of the frequency of the first cosine term in the series, and a series of sine functions in which each successive term has a frequency that is an integer multiple of the frequency of the first sine term in the series. There are no restrictions on the values or relative values of the amplitudes for the sine or cosine terms. Equation 1-4 is stated in words as follows: Any *periodic waveform* is comprised of an average component and a series of harmonically related sine and cosine waves. A *harmonic* is an integral multiple of the fundamental frequency. The *fundamental frequency* is the *first harmonic* and is equal to the frequency (*repetition rate*) of the waveform. The second multiple of the fundamental is called the *second harmonic,* the third multiple is called the *third harmonic,* and so on. The fundamental frequency is the minimum frequency necessary to represent a waveform. Therefore, Equation 1-4 can be rewritten as

$$f(t) = \text{dc} + \text{fundamental} + \text{2nd harmonic} + \text{3rd harmonic} + \cdots n\text{th harmonic}$$

Wave symmetry. Simply stated, *wave symmetry* describes the symmetry of a waveform in the time domain, that is, its relative position with respect to the horizontal (time) and vertical (amplitude) axes.

Even Symmetry. If a periodic voltage waveform is symmetric about the vertical (amplitude) axis, it is said to have *axes* or *mirror symmetry* and is called an *even function*. For all even functions, the *B* coefficients in Equation 1-4 are zero. Therefore, the signal simply contains a dc component and the cosine terms (note that a cosine wave is itself an even function). The sum of a series of even functions is an even function. Even functions satisfy the condition

$$f(t) = f(-t) \qquad (1\text{-}5)$$

Equation 1-5 states that the magnitude and polarity of the function at $+t$ is equal to the magnitude and polarity at $-t$. A waveform that contains only the even functions is shown in Figure 1-11a.

   Odd Symmetry.   If a periodic voltage waveform is symmetric about a line midway between the vertical and the negative horizontal axes (that is, the axes in the second and fourth quadrants) and passing through the coordinate origin, it is said to have *point* or *skew* symmetry and is called an *odd function*. For all odd functions, the *A* coefficients in Equation 1-4 are zero. Therefore, the signal simply contains a dc component and the sine terms (note that a sine wave is itself an odd function). The sum of a series of odd functions

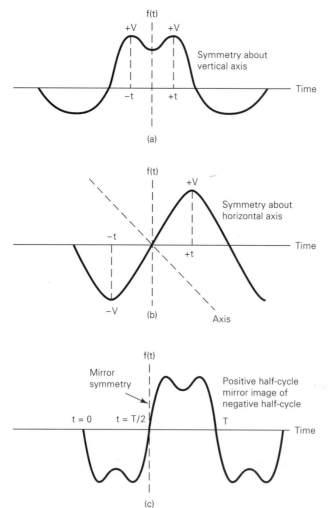

Figure 1-11 Wave symmetries: (a) even symmetry; (b) odd symmetry; (c) half-wave symmetry.

is an odd function. This form must be mirrored first in the Y axis and then in the X axis for superposition. Thus

$$f(t) = -f(-t) \qquad (1\text{-}6)$$

Equation 1-6 states that the magnitude of the function at $+t$ is equal to the negative of the magnitude at $-t$ (that is, equal in magnitude but opposite in sign). A periodic waveform that contains only the odd functions is shown in Figure 1-11b.

Half-wave Symmetry. If a periodic voltage waveform is such that the waveform for the first half cycle ($t = 0$ to $t = T/2$) repeats itself except with the opposite sign for the second half cycle ($t = T/2$ to $t = T$), it is said to have *half-wave symmetry*. For all waveforms with half-wave symmetry, the even harmonics in the series for both the sine and cosine terms are zero. Therefore, half-wave functions satisfy the condition

$$f(t) = -f\left(\frac{T}{2} + t\right) \qquad (1\text{-}7)$$

A periodic waveform that exhibits half-wave symmetry is shown in Figure 1-11c. It should be noted that a waveform can have half-wave as well as either odd or even symmetry at the same time. The coefficients A_0, B_1 to B_n, and A_1 to A_n can be evaluated using the following integral formulas:

$$A_0 = \frac{1}{T} \int_0^T f(t)\, dt \qquad (1\text{-}8)$$

$$A_n = \frac{2}{T} \int_0^T f(t) \cos n\omega t\, dt \qquad (1\text{-}9)$$

$$B_n = \frac{2}{T} \int_0^T f(t) \sin n\omega t\, dt \qquad (1\text{-}10)$$

Solving Equations 1-8, 1-9, and 1-10 requires integral calculus, which is beyond the intent of this book. Therefore, in subsequent discussions, the appropriate solutions are given.

Table 1-3 is a summary of the Fourier series for several of the more common nonsinusoidal periodic waveforms.

EXAMPLE 1-2

For the train of square waves shown in Figure 1-12:

(a) Determine the peak amplitudes and frequencies of the first five odd harmonics.
(b) Draw the frequency spectrum.
(c) Calculate the total instantaneous voltage for several times and sketch the time-domain waveform.

Solution (a) From inspection of the waveform in Figure 1-12, it can be seen that the average dc component is 0 V and the waveform has both odd and half-wave symmetry. Evaluating Equations 1-8, 1-9, and 1-10 yields the following Fourier series for a square wave.

$v(t)$

$$= V_0 + \frac{4V}{\pi}\left[\sin \omega t + \frac{1}{3}\sin 3\omega t + \frac{1}{5}\sin 5\omega t + \frac{1}{7}\sin 7\omega t + \frac{1}{9}\sin 9\omega t + \cdots\right] \qquad (1\text{-}11a)$$

TABLE 1-3 FOURIER SERIES SUMMARY

Waveform	Fourier Series
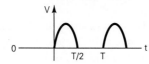	$v(t) = \dfrac{V}{\pi} + \dfrac{V}{2} \sin \omega t - \dfrac{2V}{3\pi} \cos 2\omega t - \dfrac{2V}{15\pi} \cos 4\omega t + \cdots$ $v(t) = \dfrac{V}{\pi} + \dfrac{V}{2} \sin \omega t + \displaystyle\sum_{N=2}^{\infty} \dfrac{V[1 + (-1)^N]}{\pi(1 - N^2)} \cos N\omega t$
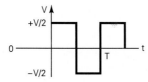	$v(t) = \dfrac{2V}{\pi} + \dfrac{4V}{3\pi} \cos \omega t - \dfrac{4V}{15\pi} \cos 2\omega t + \cdots$ $v(t) = \dfrac{2V}{\pi} + \displaystyle\sum_{N=1}^{\infty} \dfrac{4V(-1)^N}{\pi[1 - (2N)^2]} \cos N\omega t$
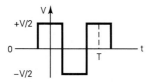	$v(t) = \dfrac{2V}{\pi} \sin \omega t + \dfrac{2V}{3\pi} \sin 3\omega t + \cdots$ $v(t) = \displaystyle\sum_{N=\text{odd}}^{\infty} \dfrac{2V}{N\pi} \sin N\omega t$
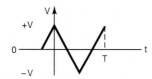	$v(t) = \dfrac{2V}{\pi} \cos \omega t - \dfrac{2V}{3\pi} \cos 3\omega t + \dfrac{2V}{5\pi} \cos 5\omega t + \cdots$ $v(t) = \displaystyle\sum_{N=\text{odd}}^{\infty} \dfrac{V \sin N\pi/2}{N\pi/2} \cos N\omega t$
(pulse waveform)	$v(t) = \dfrac{Vt}{T} + \displaystyle\sum_{N=1}^{\infty} \left(\dfrac{2Vt}{T} \dfrac{\sin N\pi t/T}{N\pi t/T} \right) \cos N\omega t$
(triangle waveform)	$v(t) = \dfrac{4V}{\pi^2} \cos \omega t + \dfrac{4V}{(3\pi)^2} \cos 3\omega t + \dfrac{4V}{(5\pi)^2} \cos 5\omega t + \cdots$ $v(t) = \displaystyle\sum_{N=\text{odd}}^{\infty} \dfrac{4V}{(N\pi)^2} \cos N\omega t$

where $v(t)$ = time-varying voltage
V_0 = average dc voltage (volts)
V = peak amplitude of the square wave (volts)
$\omega = 2\pi f$ (radians per second)
T = period of the square wave (seconds)
f = fundamental frequency of the square wave $(1/T)$ (hertz)

The fundamental frequency of the square wave is

$$f = \frac{1}{T}$$

$$= \frac{1}{1 \text{ ms}}$$

$$= 1 \text{ kHz}$$

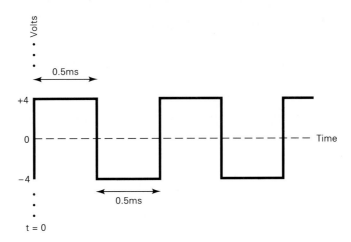

Figure 1-12 Waveform for Example 1-1.

From Equation 1-11a, it can be seen that the frequency and amplitude of the nth odd harmonic can be determined from the following expressions:

$$f_n = n \times f \tag{1-11b}$$

$$V_n = \frac{4V}{n\pi}, \qquad n = \text{odd} \tag{1-11c}$$

where n = nth harmonic (odd harmonics only for a square wave)
        f = fundamental frequency of the square wave (hertz)
       V_n = peak amplitude of the nth harmonic (volts)
       f_n = frequency of the nth harmonic (hertz)
       V = peak amplitude of the square wave (volts)

Substituting $n = 1$ into Equations 1-11b and 1-11c gives

$$V_1 = \frac{4(4)}{\pi} = 5.09 \text{ V}, \qquad f_1 = 1 \times 1000 = 1000 \text{ Hz}$$

Substituting $n = 3, 5, 7,$ and 9 into Equations 1-11b and 1-11c gives

n	Harmonic	Frequency (Hz)	Peak Voltage (Vp)
1	First	1000	5.09
3	Third	3000	1.69
5	Fifth	5000	1.02
7	Seventh	7000	0.73
9	Ninth	9000	0.57

(b) The frequency spectrum is shown in Figure 1-13.
(c) Substituting the results of the previous steps into Equation 1-11a gives

$$v(t) = 5.09 \sin[2\pi 1000t] + 1.69 \sin[2\pi 3000t] + 1.02 \sin[2\pi 5000t]$$

$$+ 0.73 \sin[2\pi 7000t] + 0.57 \sin[2\pi 9000t]$$

Signal Analysis

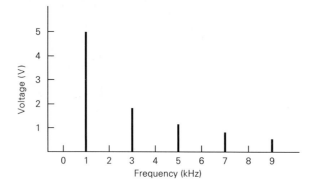

Figure 1-13 Frequency spectrum for Example 1-1.

Solving for $v(t)$ at $t = 62.5$ μs gives

$$v(t) = 5.09 \sin[2\pi\,1000(62.5\ \mu s)] + 1.69 \sin[2\pi3000(62.5\ \mu s)]$$

$$+\ 1.02 \sin[2\pi\,5000(62.5\ \mu s)] + 0.73 \sin[2\pi7000(62.5\ \mu s)]$$

$$+\ 0.57 \sin[2\pi\,9000t(62.5\ \mu s)]$$

$$v(t) = 4.51\ \text{V}$$

Solving for $v(t)$ for several additional values of time gives the following table:

Time (μs)	$v(t)$ (V)
0	0
62.5	4.51
125	3.96
250	4.26
375	3.96
437.5	4.51
500	0
562.5	−4.51
625	−3.96
750	−4.26
875	−3.96
937.5	−4.51
1000	0

The time-domain signal is derived by plotting the times and voltages calculated above on graph paper and is shown in Figure 1-14. Although the waveform shown is not an exact square wave, it does closely resemble one. To achieve a more accurate time-domain waveform, it would be necessary to solve for $v(t)$ for more values of time than are shown in this diagram.

Fourier Series for a Rectangular Waveform

When analyzing electronic communications circuits, it is often necessary to use *rectangular pulses*. A waveform showing a string of rectangular pulses is given in Figure 1-15. The *duty cycle* (*DC*) for the waveform is the ratio of the active time of the pulse to the period of the waveform. Mathematically, duty cycle is

$$DC = \frac{\tau}{T} \tag{1-12a}$$

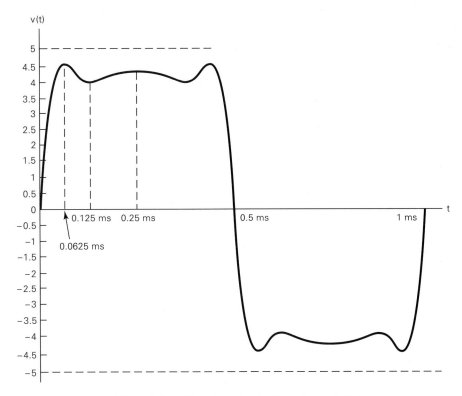

Figure 1-14 Time-domain signal for Example 1-1.

$$DC(\%) = \frac{\tau}{T} \times 100 \qquad (1\text{-}12b)$$

where   DC = duty cycle as a decimal

     DC(%) = duty cycle as a percent

         τ = pulse width of the rectangular wave (seconds)

        T = period of the rectangular wave (seconds)

      Regardless of the duty cycle, a rectangular waveform is made up of a series of harmonically related sine waves. However, the amplitude of the spectral components depends

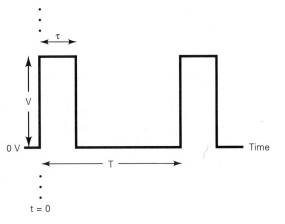

Figure 1-15 Rectangular pulse waveform.

Signal Analysis

on the duty cycle. The Fourier series for a rectangular voltage waveform with even symmetry is

$$v(t)$$

$$= \frac{V\tau}{T} + \frac{2V\tau}{T} \left[\frac{\sin x}{x} (\cos \omega t) + \frac{\sin 2x}{2x} (\cos 2\omega t) + \cdots + \frac{\sin nx}{nx} (\cos n\omega t) \right] \quad (1\text{-}13)$$

where $v(t)$ = time-varying voltage wave
τ = pulse width of the rectangular wave (seconds)
T = period of the rectangular wave (seconds)
$x = \pi(\tau/T)$
n = nth harmonic and can be any positive integer value
V = peak pulse amplitude (volts)

From Equation 1-13, it can be seen that a rectangular waveform has a 0-Hz (dc) component equal to

$$V_0 = V \times \frac{\tau}{T} \quad or \quad V \times DC \quad (1\text{-}14)$$

where V_0 = dc voltage (volts)
DC = duty cycle as a decimal
τ = pulse width of rectangular wave (seconds)
T = period of rectangular wave (seconds)

The narrower the pulse width is, the smaller the dc component. Also, from Equation 1-13, the amplitude of the nth harmonic is

$$V_n = \frac{2V\tau}{T} \times \frac{\sin nx}{nx} \quad (1\text{-}15a)$$

or

$$V_n = \frac{2V\tau}{T} \times \frac{\sin[(n\pi\tau)/T)]}{[(n\pi\tau)/T)]} \quad (1\text{-}15b)$$

where V_n = peak amplitude of the nth harmonic (volts)
n = nth harmonic (any positive integer)
π = 3.14159 radians
V = peak amplitude of the rectangular wave (volts)
τ = pulse width of the rectangular wave (seconds)
T = period of the rectangular wave (seconds)

The *(sin x)/x* function is used to describe repetitive pulse waveforms. Sin *x* is simply a sinusoidal waveform whose instantaneous amplitude depends on *x* and varies both positively and negatively between its peak amplitudes at a sinusoidal rate as *x* increases. With only *x* in the denominator, the denominator increases with *x*. Therefore, a (sin *x*)/*x* function is simply a damped sine wave in which each successive peak is smaller than the preceding one. A (sin *x*)/*x* function is shown in Figure 1-16.

Figure 1-17 shows the frequency spectrum for a rectangular pulse with a pulse width-to-period ratio of 0.1. It can be seen that the amplitudes of the harmonics follow a damped sinusoidal shape. At the frequency whose period equals 1/τ (that is, at frequency 10*f* hertz), there is a 0-V component. A second null occurs at 20*f* hertz (period = 2/τ), a third at 30*f* hertz (period = 3/τ), and so on. All spectrum components between 0 Hz and the first null frequency are considered in the first lobe of the frequency spectrum and are

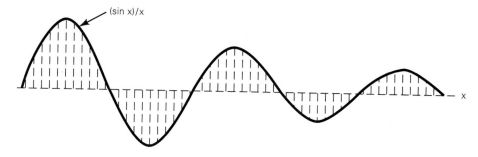

Figure 1-16 (sin x)/x function.

positive. All spectrum components between the first and second null frequencies are in the second lobe and are negative; components between the second and third nulls are in the third lobe and positive; and so on.

The following characteristics are true for all repetitive rectangular waveforms:

1. The dc component is equal to the pulse amplitude times the duty cycle.
2. There are 0-V components at frequency 1/τ hertz and all integer multiples of that frequency providing $T = n\tau$, where $n =$ any odd integer.
3. The amplitude-versus-frequency time envelope of the spectrum components take on the shape of a damped sine wave in which all spectrum components in odd-numbered lobes are positive and all spectrum components in even-numbered lobes are negative.

EXAMPLE 1-3

For the pulse waveform shown in Figure 1-18:

(a) Determine the dc component.
(b) Determine the peak amplitudes of the first 10 harmonics.

(a)

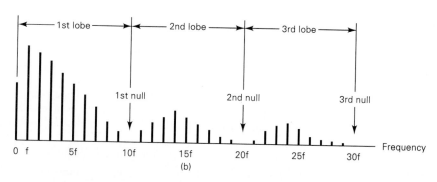

(b)

Figure 1-17 (sin x)/x function: (a) rectangular pulse waveform; (b) frequency spectrum.

Signal Analysis

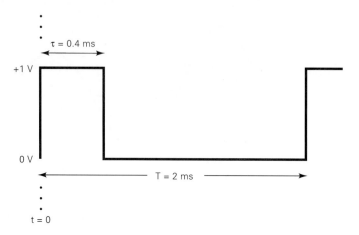

Figure 1-18 Pulse waveform for Example 1-2.

(c) Plot the $(\sin x)/x$ function.
(d) Sketch the frequency spectrum.

Solution (a) From Equation 1-14, the dc component is

$$V_0 = \frac{1(0.4 \text{ ms})}{2 \text{ ms}} = 0.2 \text{ V}$$

(b) The peak amplitudes of the first 10 harmonics are determined by substituting the values for τ, T, V, and n into Equation 1-15b, as follows:

$$Vn = 2(1) \frac{[0.4 \text{ ms}]}{2 \text{ ms}} \frac{\{\sin[(n\pi)(0.4 \text{ ms/2 ms})]\}}{(n\pi)(0.4 \text{ ms/2 ms})}$$

n	Frequency (Hz)	Amplitude (V)
0	0	0.2 V dc
1	500	0.374 Vp
2	1000	0.303 Vp
3	1500	0.202 Vp
4	2000	0.094 Vp
5	2500	0.0 V
6	3000	−0.063 Vp
7	3500	−0.087 Vp
8	4000	−0.076 Vp
9	4500	−0.042 Vp
10	5000	0.0 V

(c) The *(sin x)/x* function is shown in Figure 1-19.

(d) The frequency spectrum is shown in Figure 1-20.
Although the frequency components in the even lobes are negative, it is customary to plot all voltages in the positive direction on the frequency spectrum.

Figure 1-21 shows the effect that reducing the duty cycle (that is, reducing the τ/T ratio) has on the frequency spectrum for a nonsinusoidal waveform. It can be seen that narrowing the pulse width produces a frequency spectrum with a more uniform amplitude. In fact, for infinitely narrow pulses, the frequency spectrum comprises an infinite number

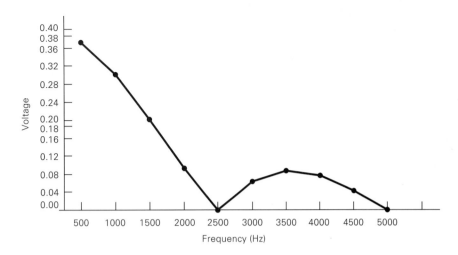

Figure 1-19 (sin x)/x function for Example 1-2.

of harmonically related frequencies of equal amplitude. Such a spectrum is impossible to produce, let alone to propagate, which explains why it is difficult to produce extremely narrow pulses. Increasing the period of a rectangular waveform while keeping the pulse width constant has the same effect on the frequency spectrum.

Power and Energy Spectra

In the previous sections, we used the Fourier series to better understand the frequency- and time-domain representation of a complex signal. Both the frequency and time domain can be used to illustrate the relationship of signal voltages (magnitudes) with respect to either frequency or time for a time-varying signal.

However, there is another important application of the Fourier series. The goal of a communications channel is to transfer electromagnetic energy from a source to a destination. Thus, the relationship between the amount of energy transmitted and the amount received is an important consideration. Therefore, it is important that we examine the relationship between energy and power versus frequency.

Electrical power is the rate at which energy is dissipated, delivered, or used and is a function of the square of the voltage or current ($P = E^2/R$ or $P = I^2 \times R$). For power relationships, in the Fourier equation, $f(t)$ is replaced by $[f(t)]^2$. Figure 1-22 shows the power spectrum for a rectangular waveform with a 25% duty cycle. It resembles its voltage-versus-frequency spectrum except it has more lobes and a much larger primary lobe. Note also that all the lobes are positive, since there is no such thing as negative power.

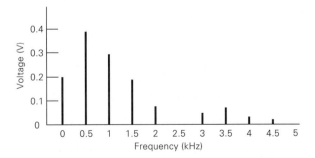

Figure 1-20 Frequency spectrum for Example 1-2.

Signal Analysis

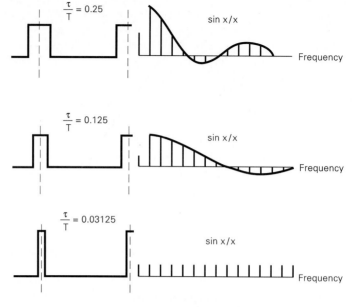

Figure 1-21 Effects of reducing the τ/T ratio (either decreasing τ or increasing T).

From Figure 1-22, it can be seen that the power in a pulse is dispersed throughout a relatively wide frequency spectrum. However, note that most of that power is within the primary lobe. Consequently, if the bandwidth of a communications channel is sufficiently wide to pass only the frequencies within the primary lobe, it will transfer most of the energy contained in the pulse to the receiver.

Discrete and Fast Fourier Transforms

Many waveforms encountered in typical communications systems cannot be satisfactorily defined by mathematical expressions. However, their frequency-domain behavior is of primary interest. Often there is a need to obtain the frequency-domain behavior of signals that are being collected in the time domain (that is, in real time). This is why the *discrete Fourier transform* was developed. With the discrete Fourier transform, a time-domain signal is sampled at discrete times. The samples are fed into a computer where an algorithm computes the transform. However, the computation time is proportional to n^2, where n is

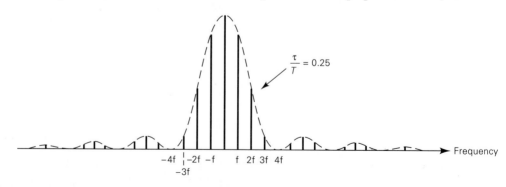

Figure 1-22 Power spectrum of a 25% duty cycle rectangular pulse.

the number of samples. For any reasonable number of samples, the computation time is excessive. Consequently, in 1965 a new algorithm called the *fast Fourier transform* or *FFT* was developed by Cooley and Tukey. With the FFT the computing time is proportional to $n \log 2n$ rather than n^2. The FFT is now available as a subroutine in many scientific subroutine libraries at large computer centers.

Effects of Bandlimiting on Signals

All communications channels have a limited bandwidth and therefore have a limiting effect on signals that are propagated through them. We can consider a communications channel to be equivalent to an ideal *linear-phase filter* with a finite bandwidth. If a nonsinusoidal

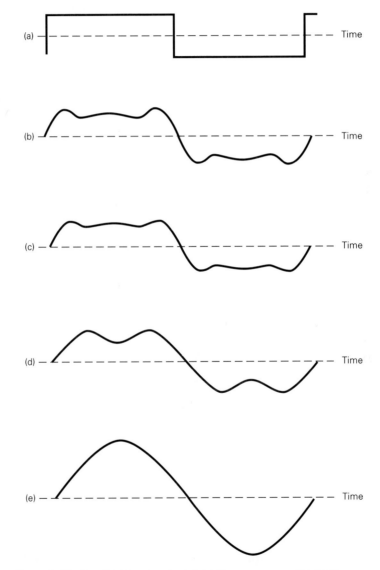

Figure 1-23 Bandlimiting signals: (a) 1-kHz square wave; (b) 1-kHz square wave bandlimited to 8 kHz; (c) 1-kHz square wave bandlimited to 6 kHz; (d) 1-kHz square wave bandlimited to 4 kHz; (e) 1-kHz square wave bandlimited to 2 kHz.

Signal Analysis

repetitive waveform passes through an ideal low-pass filter, the harmonic frequency components that are higher in frequency than the upper cutoff frequency of the filter are removed. Consequently, both the frequency content and shape of the waveform are changed. Figure 1-23a shows the time-domain waveform for the square wave used in Example 1-2. If this waveform is passed through a low-pass filter with an upper cutoff frequency of 8 kHz, frequencies above the eighth harmonic (9 kHz and above) are cut off, and the waveform shown in Figure 1-23b results. Figures 1-23c, d, and e show the waveforms produced when low-pass filters with upper cutoff frequencies of 6, 4, and 2 kHz are used, respectively.

It can be seen from Figure 1-23 that *bandlimiting* a signal changes the frequency content and thus the shape of its waveform and, if sufficient bandlimiting is imposed, the waveform eventually comprises only the fundamental frequency. In a communications system, bandlimiting reduces the information capacity of the system and, if excessive bandlimiting is imposed, a portion of the information signal can be removed from the composite waveform.

MIXING

Mixing is the process of combining two or more signals and is an essential process in electronic communications. In essence, there are two ways in which signals can be combined or mixed: linearly and nonlinearly.

Linear Summing

Linear summing occurs when two or more signals combine in a linear device, such as a passive network or a small-signal amplifier. The signals combine in such a way that no new frequencies are produced, and the combined waveform is simply the linear addition of the individual signals. In the audio recording industry, linear summing is sometimes called linear *mixing*; however, in radio communications, mixing almost always implies a nonlinear process.

Single-input frequency. Figure 1-24a shows the amplification of a single-input frequency by a linear amplifier. The output is simply the original input signal amplified by the gain of the amplifier (A). Figure 1-24b shows the output signal in the time domain, and Figure 1-24c shows the frequency domain. Mathematically, the output is

$$V_{out} = AV_{in} \tag{1-16}$$

where
$$V_{in} = V_a \sin 2\pi f_a t$$

Thus
$$V_{out} = AV_a \sin 2\pi f_a t$$

Multiple-input frequencies. Figure 1-25a shows two input frequencies combining in a small-signal amplifier. Each input signal is amplified by the gain (A). Therefore, the output is expressed mathematically as

$$V_{out} = AV_{in}$$

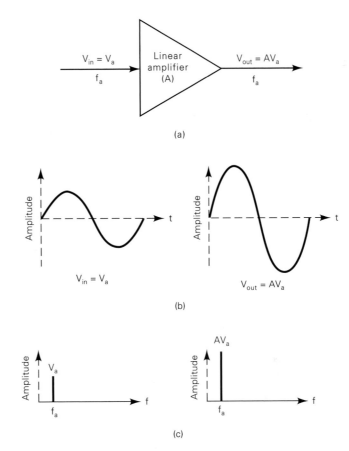

Figure 1-24 Linear amplification of a single-input frequency: (a) linear amplification; (b) time domain; (c) frequency domain.

where
$$V_{in} = V_a \sin 2\pi f_a t + V_b \sin 2\pi f_{bt}$$

Therefore,

$$V_{out} = A(V_a \sin 2\pi f_a t + V_b \sin 2\pi f_{bt}) \qquad (1\text{-}17a)$$

or

$$V_{out} = AV_a \sin 2\pi f_a t + AV_b \sin 2\pi f_{bt}) \qquad (1\text{-}17b)$$

V_{out} is simply a complex waveform containing both input frequencies and is equal to the algebraic sum of V_a and V_b. Figure 1-25b shows the linear summation of V_a and V_b in the time domain, and Figure 1-25c shows the linear summation in the frequency domain. If additional input frequencies are applied to the circuit, they are linearly summed with V_a and V_b. In high-fidelity audio systems, it is important that the output spectrum contain only the original input frequencies; therefore, linear operation is desired. However, in radio communications where modulation is essential, nonlinear mixing is often necessary.

Nonlinear Mixing

Nonlinear mixing occurs when two or more signals are combined in a nonlinear device such as a diode or large signal amplifier. With nonlinear mixing, the input signals combine in a nonlinear fashion and produce additional frequency components.

Mixing

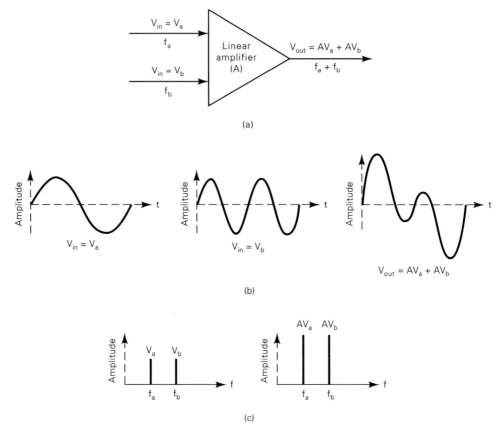

Figure 1-25 Linear mixing: (a) linear amplification; (b) time domain; (c) frequency domain.

Single-input frequency. Figure 1-26a shows the amplification of a single-frequency input signal by a nonlinear amplifier. The output from a nonlinear amplifier with a single-frequency input signal is not a single sine or cosine wave. Mathematically, the output is in the infinite power series

$$V_{out} = AV_{in} + BV_{in}^2 + CV_{in}^3 \qquad (1\text{-}18a)$$

where $$V_{in} = V_a \sin 2\pi f_a t$$

Therefore,

$$V_{out} = A(V_a \sin 2\pi f_a t) + B(V_a \sin 2\pi f_a t)^2 + C(V_a \sin 2\pi f_a t)^3 \qquad (1\text{-}18b)$$

where AV_{in} = linear term or simply the input signal (f_a) amplified by the gain (A)
BV_{in}^2 = quadratic term that generates the second harmonic frequency ($2f_a$)
CV_{in}^3 = cubic term that generates the third harmonic frequency ($3f_a$)

V_{in}^n produces a frequency equal to n times f. For example, BV_{in}^2 generates a frequency equal to $2f_a$. CV_{in}^3 generates a frequency equal to $3f_a$, and so on. Integer multiples of a *base* frequency are called *harmonics*. As stated previously, the original input frequency (f_a) is the first harmonic or the fundamental frequency. $2f_a$ is the second harmonic, $3f_a$ the third, and so on. Figure 1-26b shows the output waveform in the time domain for a nonlinear amplifier with a single-input frequency. It can be seen that the output waveform

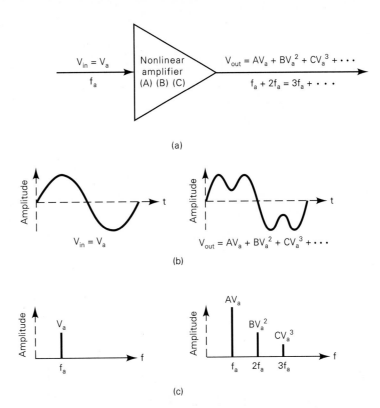

$V_{in} = V_a$

f_a

Nonlinear amplifier (A) (B) (C)

$V_{out} = AV_a + BV_a^2 + CV_a^3 + \cdots$

$f_a + 2f_a = 3f_a + \cdots$

(a)

Amplitude

t

$V_{in} = V_a$

Amplitude

t

$V_{out} = AV_a + BV_a^2 + CV_a^3 + \cdots$

(b)

Amplitude

V_a

f

f_a

Amplitude

AV_a

BV_a^2

CV_a^3

f

f_a $2f_a$ $3f_a$

(c)

Figure 1-26 Nonlinear amplification of a single-input frequency: (a) nonlinear amplification; (b) time domain; (c) frequency domain.

is simply the summation of the input frequency and its higher harmonics (multiples of the fundamental frequency). Figure 1-26c shows the output spectrum in the frequency domain. Note that adjacent harmonics are separated in frequency by a value equal to the fundamental frequency, f_a.

Nonlinear amplification of a single frequency results in the generation of multiples or harmonics of that frequency. If the harmonics are undesired, it is called *harmonic distortion*. If the harmonics are desired, it is called *frequency multiplication*.

A JFET is a special-case nonlinear device that has characteristics that are approximately those of a square-law device. The output from a square-law device is

$$V_{out} = BV_{in}^2 \tag{1-19}$$

The output from a square-law device with a single-input frequency is dc and the second harmonic. No additional harmonics are generated beyond the second. Therefore, less harmonic distortion is produced with a JFET than with a comparable BJT.

Multiple-input frequencies. Figure 1-27 shows the nonlinear amplification of two input frequencies by a large-signal (nonlinear) amplifier. Mathematically, the output of a large-signal amplifier with two input frequencies is

$$V_{out} = AV_{in} + BV_{in}^2 + CV_{in}^2$$

where

$$V_{in} = V_a \sin 2\pi f_a t + V_b \sin 2\pi f_b t$$

Mixing

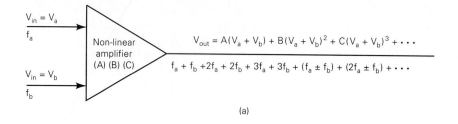

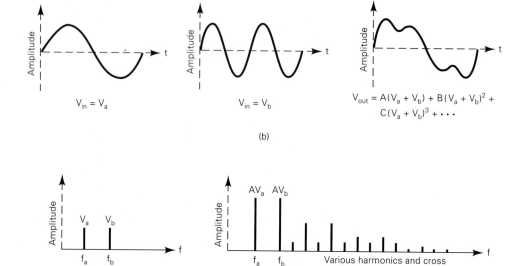

Figure 1-27 Nonlinear amplification of two sine waves: (a) nonlinear amplification; (b) time domain; (c) frequency domain.

Therefore,

$$V_{out} = A(V_a \sin 2\pi f_a t + V_b \sin 2\pi f_b t) + B(V_a \sin 2\pi f_a t + V_b \sin 2\pi f_b t)^2$$
$$+ C(V_a \sin 2\pi f_a t + V_b \sin 2\pi f_b t)^3 + \cdots \quad (1\text{-}20a)$$

The preceding formula is an infinite series and there is no limit to the number of terms it can have. If the binomial theorem is applied to each higher-power term, the formula can be rearranged and written as

$$V_{out} = (AV'_a + BV'^2_a + CV'^3_a + \cdots) + (AV'_b + BV'^2_b + CV'^3_b + \cdots)$$
$$+ (2BV'_a V'_b + 3CV'^2_a V'_b + 3CV'_a V'^2_b + \cdots) \quad (1\text{-}20b)$$

where
$$V_a' = V_a \sin 2\pi f_a t$$
$$V_b' = V_b \sin 2\pi f_b t$$

The terms in the first set of parentheses generate harmonics of f_a ($2f_a$, $3f_a$, and so on). The terms in the second set of parentheses generate harmonics of f_b ($2f_b$, $3f_b$, and so on). The terms in the third set of parentheses generate the *cross products* ($f_a + f_b$, $f_a - f_b$, $2f_a + f_b$, $2f_a - f_b$, and so on). The cross products are produced from *intermodulation* among the two original frequencies and their harmonics. The cross products are the *sum* and *difference* frequencies; they are the sum and difference of the two original frequencies, the sums and differences of their harmonics, and the sums and differences of the original frequencies and all the harmonics. An infinite number of harmonic and cross-product frequencies are produced when two or more frequencies *mix* in a nonlinear device. If the cross products are undesired, it is called *intermodulation distortion*. If the cross products are desired, it is called *modulation*. Mathematically, the sum and difference frequencies are

$$\text{cross products} = mf_a + nf_b \qquad\qquad (1\text{-}21)$$

where m and n are positive integers between one and infinity. Figure 1-28 shows the output spectrum from a nonlinear amplifier with two input frequencies.

Intermodulation distortion is the generation of any unwanted cross-product frequency when two or more frequencies are mixed in a nonlinear device. Consequently, whenever two or more frequencies are amplified in a nonlinear device, both harmonic and intermodulation distortions are present in the output.

EXAMPLE 1-4

For a nonlinear amplifier with two input frequencies, 5 kHz and 7 kHz:

(a) Determine the first three harmonics present in the output for each input frequency.
(b) Determine the cross products produced in the output for values of m and n of 1 and 2.
(c) Draw the output frequency spectrum for the harmonics and cross-product frequencies determined in steps (a) and (b).

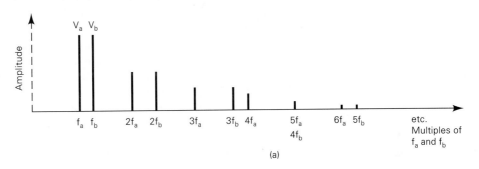

(a)

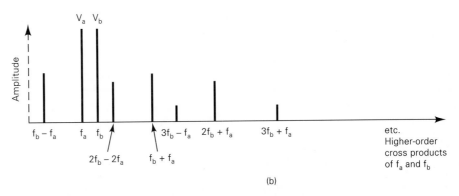

(b)

Figure 1-28 Output spectrum from a nonlinear amplifier with two input frequencies: (a) harmonic distortion; (b) intermodulation distortion.

Mixing   **33**

Solution (a) The first three harmonics include the two original input frequencies of 5 and 7 kHz; two times each of the original input frequencies, 10 and 14 kHz; and three times each of the original input frequencies, 15 and 21 kHz.

(b) The cross products for values of m and n of 1 and 2 are determined from Equation 1-21 and are summarized next.

m	n	Cross-Products
1	1	7 kHz ± 5 kHz = 2 and 12 kHz
1	2	7 kHz ± 10 kHz = 3 and 17 kHz
2	1	14 kHz ± 5 kHz = 9 and 19 kHz
2	2	14 kHz ± 10 kHz = 4 and 24 kHz

(c) The output frequency spectrum is shown in Figure 1-29.

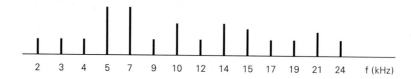

Figure 1-29 Output spectrum for Example 1-4.

ELECTRICAL NOISE

In general, *electrical noise* is defined as any unwanted electrical energy present in the usable *passband* of a communications circuit. For instance, in audio recording any undesired signals that fall into the frequency band from 0 to 15 kHz are audible and will interfere with the audio information. Consequently, for audio circuits, any unwanted electrical energy in the frequency band from 0 to 15 kHz is considered noise.

Figure 1-30 shows the effect noise has on an electrical signal. Figure 1-30a shows a perfect, noiseless signal, and Figure 1-30b shows the same signal except in the presence of noise. As the figures show, the signal that has been contaminated with noise is distorted and obviously contains frequencies other than the original.

Essentially, noise can be divided into two general categories, *correlated* and *uncorrelated*. Correlation implies a relationship between the signal and the noise. *Uncorrelated noise* is present in the absence of any signal. (This is not to say that, when present, the signal has no effect on the magnitude of the noise. It will be shown in later chapters that in

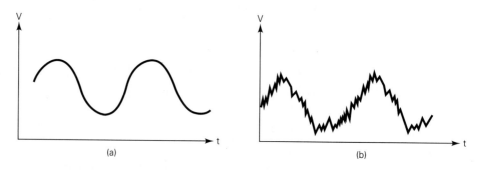

Figure 1-30 Effects of noise on a signal: (a) signal without noise; (b) signal with noise.

Chap. 1   **Introduction to Electronic Communications**

radio receivers using *automatic gain control* the noise magnitude is affected by the magnitude of the signal.) *Correlated noise* is produced directly as a result of the signal.

Uncorrelated Noise

Uncorrelated noise is present regardless of whether there is a signal present or not. Uncorrelated noise can be divided into two general categories: external and internal.

External noise. *External noise* is generated external to a circuit and allowed to enter the circuit. Externally generated signals are considered noise only if their frequencies fall within the passband of the circuit's input filter. There are three primary kinds of external noise: atmospheric, extraterrestrial, and man-made.

Atmospheric noise. *Atmospheric noise* is naturally occurring electrical energy that originates within Earth's atmosphere. Atmospheric noise is commonly called *static electricity*. The source of most static electricity is natural electrical disturbances, such as lightning. Static electricity often comes in the form of impulses that spread their energy throughout a wide range of radio frequencies. The magnitude of these impulses measured from naturally occurring events has been observed to be inversely proportional to frequency. Consequently, at frequencies above approximately 30 MHz, atmospheric noise is insignificant. Also, frequencies above 30 MHz are limited predominantly to line-of-sight propagation, which limits their interfering range to approximately 80 km (50 miles).

Atmospheric noise is the summation of the electrical energy from all external sources both local and distant. Atmospheric noise propagates through Earth's atmosphere in the same manner as radio waves. Therefore, the magnitude of the static noise received depends on the propagation conditions at the time and, in part, on diurnal and seasonal variations. Atmospheric noise is the familiar sputtering, crackling, and so on, heard on a radio receiver predominantly in the absence of a received signal and is relatively insignificant compared to the other sources of noise.

Extraterrestrial noise. *Extraterrestrial noise* originates outside Earth's atmosphere and is therefore sometimes called *deep-space noise*. Extraterrestrial noise originates from the Milky Way, other galaxies, and the sun. Extraterrestrial noise is divided into two categories: solar and cosmic (galactic).

Solar noise is generated directly from the sun's heat. There are two components of solar noise: a *quiet* condition when a relatively constant radiation intensity exists and *high intensity,* sporadic disturbances caused by *sun spot* activity and *solar flare-ups*. The sporadic disturbances come from specific locations on the sun's surface. The magnitude of these disturbances caused from sun spot activity follows a cyclic pattern that repeats every 11 years. Also, these 11-year periods follow a supercycle pattern in which approximately every 99 years a new maximum intensity is realized.

Cosmic noise sources are continuously distributed throughout our galaxy and other galaxies. Distant stars are also suns and therefore have high temperatures associated with them. Consequently, they radiate noise in the same manner as our sun. Because the sources of galactic noise are located much farther away than our sun, their noise intensity is relatively small. Cosmic noise is often called *black-body noise* and is distributed fairly evenly throughout the sky.

Extraterrestrial noise contains frequencies from approximately 8 MHz to 1.5 GHz, although frequencies below 20 MHz seldom penetrate Earth's atmosphere and are therefore generally insignificant.

Electrical Noise

Man-made noise.   *Man-made noise* is simply noise that can be attributed to man. The sources of man-made noise include spark-producing mechanisms such as commutators in electric motors, automobile ignition systems, power switching equipment, and fluorescent lights. Such noise is also impulsive in nature and therefore contains a wide range of frequencies that are propagated through space in the same manner as radio waves. This noise is most intense in more populated metropolitan and industrial areas and is sometimes called *industrial noise.*

Internal noise.   *Internal noise* is electrical interference generated within a device. There are three primary kinds of internally generated noise: thermal, shot, and transit time.

Thermal noise.   *Thermal noise* is associated with *Brownian movement* of electrons within a conductor. In accordance with the *kinetic theory of matter*, electrons within a conductor are in thermal equilibrium with the molecules and in constant random motion. This random movement is accepted as being a confirmation of the kinetic theory of matter and was first noted by the English botanist, Robert Brown (hence the name *Brownian noise*). Brown first observed evidence for the kinetic (moving-particle) nature of matter while observing pollen grains under a microscope. Brown noted an extraordinary agitation of the pollen grains that made them extremely difficult to examine. He later noted that this same phenomenon existed for smoke particles in the air. Brownian movement of electrons was first recognized in 1927 by J. B. Johnson of Bell Telephone Laboratories. In 1928, a quantitative theoretical treatment was furnished by H. Nyquist (also of Bell Telephone Laboratories). Electrons within a conductor carry a unit negative charge, and the mean-square velocity of an electron is proportional to the absolute temperature. Consequently, each flight of an electron between collisions with molecules constitutes a short pulse of current. Because electron movement is totally random and in all directions, the average voltage produced in the substance by their movement is 0 V dc. However, such a random movement gives rise to an ac component. This ac component has several names, which include *thermal noise* (because it is temperature dependent), *Brownian noise* (after its discoverer), *Johnson noise* (after the person who related Brownian particle movement to electron movement), *random noise* (because the direction of electron movement is totally random), *resistance noise* (because the magnitude of its voltage depends on resistance), and *white noise* (because it contains all frequencies). Hence, thermal noise is the random motion of free electrons within a conductor caused by thermal agitation.

The *equipartition law* of Boltzmann and Maxwell combined with the works of Johnson and Nyquist states that the thermal noise power generated within a source for a 1-Hz bandwidth (watts per hertz) is the *noise power density*, which is stated mathematically as

$$N_o = KT \qquad (1\text{-}22)$$

where N_o = noise power density (watts per hertz)
   K = Boltzmann's constant (1.38×10^{-23} J/K)
   T = absolute temperature (kelvin) (room temperature = 17°C or 290 K)*
   *0 K = −273°C

Thus, at room temperature with a 1-Hz bandwidth, the available noise power density is

$$N_o = 1.38 \times 10^{-23} \ J/K \times 290 \ K$$

$$= 4 \times 10^{-21} \ W/Hz$$

Stated in dBm
$$N_o \text{ (dBm)} = 10 \log \frac{KT}{0.001} \tag{1-23}$$

$$= 10 \log \frac{4 \times 10^{-21}}{0.001}$$

$$= -174 \text{ dBm}$$

The total noise power is equal to the product of the bandwidth and the noise power density. Therefore, the total noise power present in bandwidth B is

$$N = KTB \tag{1-24}$$

where
N = total noise power in bandwidth B (watts)
$N_o = KT$ = noise power density (watts per hertz)
B = bandwidth of the device or system (hertz)

and stated in dBm

$$N \text{(dBm)} = 10 \log \frac{KTB}{0.001} \tag{1-25}$$

The result of the equipartition theory is one of a constant power density versus frequency. Equation 1-24 indicates that the available power from a thermal noise source is proportional to bandwidth over any range of frequencies. This has been found to be true for frequencies from 0 Hz to the highest microwave frequencies commonly used today. Thus, if the bandwidth is unlimited, the results of the equipartition theory say that the available power from a thermal noise source is also unlimited. This, of course, is not true and can be proved by applying a few principles of *quantum mechanics* to the problem. In Equation 1-22, if KT is replaced by $hf/[\exp(hf/KT) -1]$, where h is Planck's constant (6.625×10^{-34} J-s) and f is frequency, it can be shown that at arbitrarily high frequencies the thermal noise spectrum eventually drops to zero. For most practical purposes, thermal noise power is directly proportional to the product of the bandwidth of the system and the absolute temperature of the source. Thus, total thermal noise can be expressed as

$$N_{\text{(dBm)}} = -174 + 10 \log B \tag{1-26}$$

Shot noise. *Shot noise* is caused by the random arrival of carriers (holes and electrons) at the output element of an electronic device, such as a diode, field-effect transistor (FET), bipolar transistor (BJT), or vacuum tube. Shot noise was first observed in the anode current of vacuum-tube amplifiers and was described by W. Schottky in 1918. The current carriers (for both ac and dc) are not moving in a continuous, steady flow because the distance they travel varies due to their random paths of motion. Shot noise is randomly varying and is superimposed onto any signal present. Shot noise, when amplified, sounds like a shower of metal pellets falling on a tin roof. Shot noise is sometimes called *transistor noise*. Shot noise is proportional to the charge of an electron (1.6×10^{-19}), direct current, and system bandwidth. Also, shot noise power is additive with thermal noise and other shot noise.

Transit-time noise. Any modification to a stream of carriers as they pass from the input to the output of a device (such as from the emitter to the collector of a transistor) produces an irregular, random variation categorized as *transit noise*. When the time it takes for a carrier to propagate through a device is an appreciable part of the time of one cycle of the signal, the noise becomes noticeable. Transit-time noise in transistors is determined by ion mobility, the bias voltages, and the actual transistor construction. Carriers traveling from the emitter to the collector suffer from emitter delay times, base transit-time delays, and collector recombination and propagation delay times. At high frequencies and if transit delays are excessive, the device may add more noise than amplification to the signal.

Gaussian Distribution

The *Gaussian distribution* is the limiting form for the distribution function of the summation of a large number of independent quantities, which individually may have a variety of different distributions. In statistics, this result is known as the *central limiting theorem*.

Thermal noise is sometimes considered the superposition of an exceedingly large number of random, practically independent electrical noise contributions. Therefore, thermal noise satisfies the theoretical conditions for a Gaussian distribution. The Gaussian probability density function for zero mean is shown in Figure 1-31a and is stated mathematically as

$$p(V) = \frac{1}{\sigma_n \sqrt{(2\pi)}} \exp(-V^2/2\sigma_n^2) \qquad (1\text{-}27a)$$

where $p(V)$ = Gaussian probability density function
V^2 = mean-square voltage
σ_n = Gaussian distributed noise source (standard deviation)
σ_n^2 = variance

The Gaussian distribution function is shown in Figure 1-31b and is expressed mathematically as the integral of Equation 1-27a.

$$P(V) = \frac{1}{\sigma_n \sqrt{(2\pi)}} \int_{-\infty}^{V} \exp\left(\frac{-x^2}{2\sigma_n^2}\right) dx \qquad (1\text{-}27b)$$

where $P(V)$ = Gaussian distribution function.

It can be easily shown that the expected value of V^2 (the mean-square voltage) is equal to the variance, σ_n^2. Thus, the *rms* voltage of a Gaussian distributed noise source is given by the standard deviation σ_n.

The fact that thermal noise is white as well as Gaussian has caused some engineers to carelessly treat white and Gaussian noise as the same thing, which they are not. For

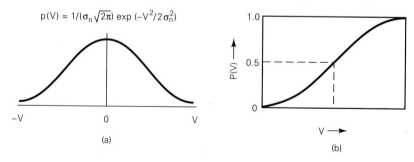

Figure 1-31 Gaussian probability density and distribution functions: (a) density function; (b) distribution function.

example, passing Gaussian noise through a linear network, such as a filter, will leave it Gaussian, but may drastically change its frequency spectrum. Also, a single impulse of noise will not have a Gaussian amplitude distribution, but will have a flat or white frequency spectrum.

Noise Voltage

Figure 1-32 shows the equivalent circuit for an electrical noise source. The internal resistance of the noise source (R_I) is in series with the rms noise voltage (V_N). For the worst-case condition (that is, maximum transfer of power), the load resistance (R) is made equal to R_I. Therefore, the noise voltage dropped across R is equal to $V_N/2$, and the noise power (N) developed across the load resistor is equal to KTB. Therefore, the mathematical expression for V_N is determined as follows:

$$N = KTB = \frac{(V_N/2)^2}{R} = \frac{V_N^2}{4R}$$

and
$$V_N^2 = 4RKTB$$

$$V_N = \sqrt{(4RKTB)} \qquad (1\text{-}28)$$

Thermal noise is equally distributed throughout the frequency spectrum. Because of this property, a thermal noise source is called a *white noise source* (this is an analogy to white light, which contains all visible-light frequencies). Therefore, the noise power measured at any frequency from a noise source is equal to the noise power measured at any other frequency from the same noise source. Similarly, the noise power measured in any given bandwidth is equal to the noise power measured in any other equal bandwidth, regardless of the center frequency. In other words, the thermal noise power present in the band from 1000 to 2000 Hz is equal to the thermal noise power present in the band from 1,001,000 to 1,002,000 Hz.

EXAMPLE 1-5

For a device operating at a temperature of 17°C with a bandwidth of 10 kHz, determine:

(a) The noise power density (N_o).
(b) The total noise power (N).
(c) The rms noise voltage (V_N) for a 100-Ω internal resistance and a 100-Ω load resistor.

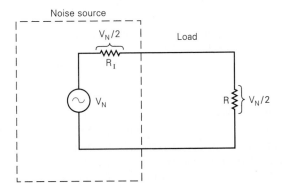

Figure 1-32 Noise source equivalent circuit.

Electrical Noise

Solution (a) By definition, the noise density is the noise measured in any 1-Hz bandwidth with constant temperature. Therefore, substituting into Equation 1-22, noise density equals

$$N_o = KT = 1.38 \times 10^{-23} \times 290 = 4 \times 10^{-21} \text{ W/Hz}$$

(b) Substituting into Equation 1-24, the total thermal noise power is found to be

$$N = KTB$$
$$= 1.38 \times 10^{-23} \times 290 \times 10^4$$
$$= 4 \times 10^{-17} \text{ W}$$

(c) Substituting into Equation 1-28, we find that the *rms* noise voltage is

$$V_N = \sqrt{(4RKTB)}$$
$$= \sqrt{[4 \times 100 \times 1.38 \times 10^{-23} \times 290 \times 10^4]}$$
$$= 0.1265 \text{ } \mu\text{V}$$

For equal-value load and internal resistances, the noise voltage dropped across the load resistor is equal to one-half the noise voltage, or 0.06325 μV. Therefore, the total noise power is

$$N = \frac{E^2}{R} = \frac{(V_N/2)^2}{R} = \frac{(0.1265 \text{ } x \text{ } 10^{-6}/2)^2}{100} = \frac{(0.06325 \text{ } x \text{ } 10^{-6})^2}{100} = 4 \times 10^{-17} \text{ W}$$

Thermal noise is random, continuous, and occurs at all frequencies. Also, thermal noise is present in all devices, is predictable, and is additive. This is why thermal noise is generally the most significant of all the noise sources. In subsequent discussions in this book, thermal noise will be referred to as simply thermal or random white noise.

Correlated Noise

Correlated noise is unwanted electrical energy that is present as a direct result of a signal, such as harmonic and intermodulation distortion. Harmonic and intermodulation distortions are both forms of nonlinear distortion; they are produced from nonlinear amplification. Correlated noise cannot be present in a circuit unless there is an input signal. Simply stated, no signal, no noise! Both harmonic and intermodulation distortions change the shape of the wave in the time domain and the spectral content in the frequency domain.

Harmonic distortion. *Harmonic distortion* is the unwanted multiples of a single-frequency sine wave that are created when the sine wave is amplified in a nonlinear device, such as a large-signal amplifier. *Amplitude distortion* is another name for harmonic distortion. Generally, the term amplitude distortion is used for analyzing a waveform in the time domain, and the term harmonic distortion is used for analyzing a waveform in the frequency domain. The original input frequency is the first harmonic and is called the *fundamental frequency*.

There are various degrees or orders of harmonic distortion. Second order harmonic distortion is the ratio of the rms amplitude of the second harmonic frequency to the rms amplitude of the fundamental frequency. Third order harmonic distortion is the ratio of the rms amplitude of the third harmonic frequency to the rms amplitude of the fundamental frequency, and so on. The ratio of the combined rms amplitudes of the higher harmonics to the rms amplitude of the fundamental frequency is called *total harmonic distortion* (THD). Mathematically, total harmonic distortion is

$$\% \text{ THD} = \frac{V_{\text{higher}}}{V_{\text{fund}}} \times 100 \qquad (1\text{-}29)$$

where $\% \text{ THD}$ = percent total harmonic distortion
         V_{higher} = quadratic sum of the root-mean-square (rms) voltages of
                   the higher harmonics
               $= \sqrt{[V_2^2 + V_3^2 + V_n^2]}$
         V_{fund} = rms voltage of the fundamental frequency

EXAMPLE 1-6

Determine the percent second order, third order, and total harmonic distortion for the output spectrum shown in Figure 1-33.

Solution

$$\% \text{ 2nd order} = \frac{V_2}{V_1} \times 100 = \frac{2}{6} \times 100 = 33\%$$

$$\% \text{ 3rd order} = \frac{V_3}{V_1} \times 100 = \frac{1}{6} \times 100 = 16.7\%$$

$$\% \text{ THD} = \frac{\sqrt{[V_2^2 + V_3^2]}}{V_1} \times 100$$

$$= \frac{\sqrt{[2^2 + 1^2]}}{6} \times 100 = 37.3\%$$

Intermodulation noise.   *Intermodulation noise* is the unwanted cross-product (sums and difference) frequencies created when two or more signals are amplified in a nonlinear device, such as a large-signal amplifier. As with harmonic distortion, there are various degrees of intermodulation distortion. It would be impossible to measure all the intermodulation components produced when two or more frequencies mix in a nonlinear device. Therefore, for comparison purposes, a common method used to measure intermodulation distortion is *percent second order intermodulation distortion.* Second order intermodulation distortion is the ratio of the total rms amplitude of the second order cross products to the combined rms amplitude of the original input frequencies. Generally, to measure second order intermodulation distortion, four test frequencies are used; two designated the A-band (f_{a1} and f_{a2}) and two B-band frequencies (f_{b1} and f_{b2}). The second

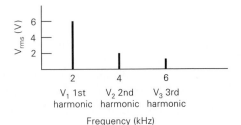

Figure 1-33 Harmonic distortion for Example 1-5.

Electrical Noise

order cross-products (2A − B) are $2f_{a1} - f_{b1}$, $2f_{a1} - f_{b2}$, $2f_{a2} - f_{b1}$, $2f_{a2} - f_{b2}$, $(f_{a1} + f_{a2}) - f_{b1}$, and $(f_{a1} + f_{a2}) - f_{b2}$. Mathematically, the percent second order intermodulation distortion (% 2nd order IMD) is

$$\% \text{ 2nd order IMD} = \frac{V_{\text{2nd order cross products}}}{V_{\text{original}}} \times 100 \qquad (1\text{-}30)$$

where $V_{\text{2nd order}}$ = quadratic sum of the rms amplitudes of the second order cross-product frequencies

             V_{original} = quadratic sum of the rms amplitudes of the input frequencies

EXAMPLE 1-7

Determine the percent second order intermodulation distortion for the A-band, B-band, and second order intermodulation components shown in Figure 1-34.

Solution   Substituting into Equation 1-30

$$\% \text{ 2nd order IMD} = \frac{\sqrt{[2^2 + 2^2 + 2^2 + 2^2 + 1^2 + 1^2]}}{\sqrt{[6^2 + 6^2 + 6^2 + 6^2]}} \times 100$$

$$= 35.5\%$$

Both harmonic and intermodulation distortion are caused by the same thing, nonlinear amplification. Essentially, the only difference between the two is that harmonic distortion can occur when there is a single-input frequency, and intermodulation distortion can only occur when there are two or more input frequencies.

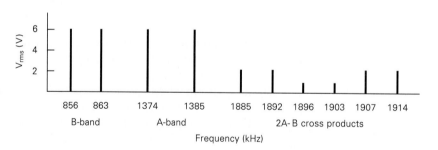

Figure 1-34   Intermodulation distortion for Example 1-7.

MISCELLANEOUS TYPES OF NOISE

Excess noise.   *Excess noise* is a form of uncorrelated internal noise that is not totally understood. It is found in transistors and is directly proportional to the emitter current and junction temperature and inversely proportional to frequency. Excess noise is also called *low-frequency noise, flicker noise, 1/f noise,* and *modulation noise.* It is believed to be caused by or at least associated with *carrier traps* in the emitter depletion layer of transistors. These traps capture and release holes and electrons at different rates, but with energy levels that vary inversely with frequency. Excess noise is insignificant above approximately 1 kHz.

Resistance noise. *Resistance noise* is a form of thermal noise that is associated with the internal resistance of the base, emitter, and collector of a transistor. Resistance noise is fairly constant from about 500 Hz up and may, therefore, be lumped with thermal and shot noise.

Precipitation noise. *Precipitation noise* is a type of static noise caused when an airplane passes through snow or rain. The airplane becomes electrically charged to a potential high enough with respect to the surrounding space that a *corona discharge* occurs at a sharp point on the airplane. The interference from precipitation static is most annoying at shortwave frequencies and lower.

Signal-to-Noise Ratio

Signal-to-noise ratio (S/N) is a simple mathematical relationship of the signal level with respect to the noise level at a given point in a circuit, amplifier, or system. The signal-to-noise ratio can be expressed as both a voltage ratio and a power ratio. Mathematically, S/N is

$$\frac{S}{N} = \left[\frac{\text{signal voltage}}{\text{noise voltage}} \right]^2 = \left(\frac{V_s}{V_n} \right)^2 \qquad \text{as a voltage ratio} \qquad (1\text{-}31)$$

$$\frac{S}{N} = \frac{\text{signal power}}{\text{noise power}} = \frac{P_s}{P_n} \qquad \text{as a power ratio} \qquad (1\text{-}32)$$

The signal-to-noise ratio is often expressed as a logarithmic function with the decibel unit.

For voltage ratios,
$$\frac{S}{N} \text{ (dB)} = 20 \log \frac{V_s}{V_n} \qquad (1\text{-}33)$$

For power ratios,
$$\frac{S}{N} \text{ (dB)} = 10 \log \frac{P_s}{P_n} \qquad (1\text{-}34)$$

If the input and output resistances of the amplifier, receiver, or network being evaluated are equal, the signal power-to-noise power ratio will equal the signal voltage-to-noise voltage ratio squared.

Signal-to-noise ratio is probably the most important and most often used parameter for evaluating the performance of an amplifier or an entire radio communications system or comparing the performance of one amplifier or system to another. The higher the signal-to-noise ratio is, the better the system performance. From the signal-to-noise ratio, the general quality of a system can be determined.

Noise Factor and Noise Figure

Noise factor (F) and *noise figure (NF)* are figures of merit that indicate the degradation in the signal-to-noise ratio as a signal propagates through a single amplifier, a series of amplifiers, or a communications system. Noise factor is the ratio of the input signal power-to-noise power ratio to the output signal power-to-noise power ratio. Therefore, noise factor is a ratio of ratios. Mathematically, noise factor is

$$F = \frac{\text{input signal-to-noise ratio}}{\text{output signal-to-noise ratio}} \qquad (1\text{-}35)$$

Noise figure is the noise factor expressed in logarithmic form. Mathematically, noise figure is

$$\text{NF(dB)} = 10 \log \frac{\text{input signal-to-noise ratio}}{\text{output signal-to-noise ratio}} \qquad (1\text{-}36)$$

or
$$= 10 \log F \qquad (1\text{-}37)$$

An amplifier will amplify equally all signals and noise present at its input that fall within its passband. Therefore, if the amplifier is ideal and noiseless, the signal and noise are amplified by the same factor, and the signal-to-noise ratio at the output of the amplifier will equal the signal-to-noise ratio at the input. However, in reality, amplifiers are not ideal, noiseless devices. Therefore, although the input signal and noise are amplified equally, the device adds internally generated noise to the waveform, thus reducing the overall signal-to-noise ratio. The most predominant form of electrical noise is thermal noise, which is generated in all electrical components. Therefore, all networks, amplifiers, and systems add noise to the signal and thus reduce the overall signal-to-noise ratio as the signal passes through them.

Figure 1-35a shows an ideal noiseless amplifier with a power gain (A_p), an input signal level (S_i), and an input noise level (N_i). It can be seen that the output S/N ratio is the same as the input S/N ratio. In Figure 1-35b, the same amplifier is shown except that, instead of being ideal and noiseless, it adds internally generated noise (N_d) to the waveform.

From Figure 1-35b, it can also be seen that the output S/N ratio is less than the input S/N ratio by an amount proportional to N_d. Also, it can be seen that the noise figure of a perfect noiseless device is 0 dB.

EXAMPLE 1-8

For the amplifier shown in Figure 1-35 and the following parameters, determine:

(a) Input S/N ratio for both voltage and power
(b) Output S/N ratio for both voltage and power
(c) Noise factor and noise figure

$$\text{input signal voltage} = 0.1 \times 10^{-3} \text{ V}$$

$$\text{input signal power} = 2 \times 10^{-10} \text{ W}$$

$$\text{input noise voltage} = 0.01 \times 10^{-6} \text{ V}$$

$$\text{input noise power} = 2 \times 10^{-18} \text{ W}$$

$$\text{voltage gain} = 1000$$

$$\text{power gain} = 1,000,000$$

$$\text{amplifier internal noise voltage} = 1 \times 10^{-5} \text{ V}$$

$$\text{amplifier internal noise power} = 6 \times 10^{-12} \text{ W}$$

Solution (a) For the input signal and noise voltages given and substituting into Equation 1-31, we have

$$\frac{V_s}{V_n} = \frac{0.1 \text{ mV}}{0.01 \text{ }\mu\text{V}} = 10,000$$

Substituting into Equation 1-33 yields

$$\frac{V_s}{V_n} \text{ (dB)} = 20 \log 10,000 = 80 \text{ dB}$$

(a)

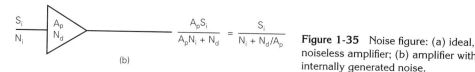

(b)

Figure 1-35 Noise figure: (a) ideal, noiseless amplifier; (b) amplifier with internally generated noise.

For the input signal and noise powers given and substituting into Equation 1-32, we have

$$\frac{P_s}{P_n} = \frac{2 \times 10^{-10}\text{ W}}{2 \times 10^{-18}\text{ W}} = 100{,}000{,}000$$

Substituting into Equation 1-34 gives us

$$\frac{P_s}{P_n}\text{ (dB)} = 10 \log 100{,}000{,}000 = 80\text{ dB}$$

(b) For the calculated output signal and noise voltages and substituting into Equation 1-31, we obtain

$$\text{output } \frac{V_s}{V_n} = \frac{(1000)(0.1\text{ mV})}{(1000)(0.01\ \mu\text{V}) + 10\ \mu\text{V}} = 5000$$

Substituting into Equation 1-33 yields

$$\text{output } \frac{V_s}{V_n}\text{(dB)} = 20 \log 5000 = 74\text{ dB}$$

For the calculated output signal and noise powers and substituting into Equation 1-32, we have

$$\text{output } \frac{P_s}{P_n} = \frac{(1 \times 10^6)(2 \times 10^{-10})}{(1 \times 10^6)(2 \times 10^{-18}) + 6 \times 10^{-12}} = 25 \times 10^6$$

Substituting into Equation 1-34 gives us

$$\text{output } \frac{P_s}{P_n} = 10 \log 25 \times 10^6 = 74\text{ dB}$$

(c) From the results of parts (a) and (b) and using Equations 1-35, 1-36, and 1-37, we obtain the power noise factor:

$$F = \frac{\text{input } P_s/P_n}{\text{output } P_s/P_n} = \frac{100{,}000{,}000}{25{,}000{,}000} = 4$$

And power noise figure is

$$NF = 10 \log 4 = 6\text{ dB}$$

A noise figure of 6 dB indicates that the signal power-to-noise power ratio decreased by a factor of 4 as the signal propagated from the input to the output of the amplifier.

When two or more amplifiers or devices are cascaded, the total noise factor is the accumulation of the individual noise factors. Mathematically, the total noise factor is expressed as

$$F_T = F_1 + \frac{F_2 - 1}{A_1} + \frac{F_3 - 1}{A_1 A_2} + \frac{F_N - 1}{A_1 A_2 A_3} \ldots \tag{1-38}$$

Miscellaneous Types of Noise

where F_T = total noise factor for N cascaded amplifiers
$\quad\quad F_1$ = noise factor, amplifier 1
$\quad\quad F_2$ = noise factor, amplifier 2
$\quad\quad F_3$ = noise factor, amplifier 3
$\quad\quad F_N$ = noise factor, amplifier N
$\quad\quad A_1$ = power gain, amplifier 1
$\quad\quad A_2$ = power gain, amplifier 2
$\quad\quad A_3$ = power gain, amplifier 3

and the total noise figure is simply

$$NF_T \,(\text{dB}) = 10 \log F_T \quad\quad\quad\quad (1\text{-}39)$$

where $NF_T\,(\text{dB})$ = total noise figure of N cascaded amplifiers.

It can be seen in Equation 1-39 that the noise factor of the first amplifier (F_1) contributes the most toward the overall noise figure (F_T). This is because the noise introduced in the first stage is amplified by each succeeding stage. Therefore, when compared to the noise introduced in the first stage, the noise added by each succeeding stage is effectively reduced by a factor equal to the product of the gains of the preceding stages.

EXAMPLE 1-9

For three cascaded amplifier stages, each with noise figures of 3 dB and power gains of 10 dB, determine the total noise figure.

Solution The power noise figures must be converted to power noise factors and then substituted into Equation 1-38 giving us a total power noise factor of

$$F_T = F_1 + \frac{F2 - 1}{A_1} + \frac{F3 - 1}{A_1 A_2}$$

$$= 2 + \frac{2 - 1}{10} + \frac{2 - 1}{10^2}$$

$$= 2.11$$

Thus the total noise figure is

$$NF_T = 10 \log 2.11$$

$$= 3.24 \text{ dB}$$

An overall noise figure of 3.24 dB indicates that the S/N ratio at the output of A_3 is 3.24 dB less than the S/N ratio at the input of A_1.

QUESTIONS

1-1. Define *electronic communications*.

1-2. What three primary components make up a communications system?

1-3. Define *modulation*.

1-4. Define *demodulation*.

1-5. Define *carrier signal*.

1-6. Explain the relationships among the source information, the carrier, and the modulated wave.

1-7. What are the three properties of an analog carrier that can be varied?

1-8. What organization assigns frequencies for free-space radio propagation in the United States?

1-9. Briefly describe the significance of Hartley's law and give the relationship between information capacity and bandwidth; information capacity and transmission time.

1-10. What are the two primary limitations on the performance of a communications system?

1-11. Describe signal analysis as it pertains to electronic communications.

1-12. Describe a time-domain display of a signal waveform; a frequency-domain display.

1-13. What is meant by the term *even symmetry*? What is another name for even symmetry?

1-14. What is meant by the term *odd symmetry*? What is another name for odd symmetry?

1-15. What is meant by the term *half-wave symmetry*?

1-16. Describe the term *duty cycle*.

1-17. Describe a *(sin x)/x* function.

1-18. Define *linear summing*.

1-19. Define *nonlinear mixing*.

1-20. Contrast the input and output spectra for a linear amplifier.

1-21. When will harmonic and intermodulation distortion occur?

1-22. Define *electrical noise*.

1-23. What is meant by the term *correlated noise*? List and describe two common forms of correlated noise.

1-24. What is meant by the term *uncorrelated noise*? List several types of uncorrelated noise and state their sources.

1-25. Briefly describe *thermal noise*.

1-26. What are four alternative names for thermal noise?

1-27. Describe the relationship between thermal noise and temperature; thermal noise and bandwidth.

1-28. Define signal-to-noise ratio. What does a signal-to-noise ratio of 100 indicate? 100 dB?

1-29. Define noise figure and noise factor. An amplifier has a noise figure of 10 dB. What does this mean?

1-30. What is the noise figure for a totally noiseless device?

PROBLEMS

1-1. For the train of square waves shown:
(a) Determine the amplitudes of the first five harmonics.
(b) Draw the frequency spectrum.
(c) Sketch the time-domain signal for frequency components up to the first five harmonics.

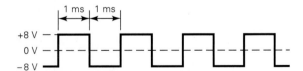

1-2. For the pulse waveform shown:
(a) Determine the dc component.
(b) Determine the peak amplitudes of the first five harmonics.
(c) Plot the *(sin x)/x* function.
(d) Sketch the frequency spectrum.

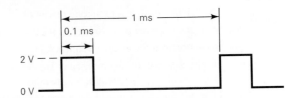

1-3. Describe the spectrum shown. Determine the type of amplifier (linear or nonlinear) and the frequency content of the input signal.

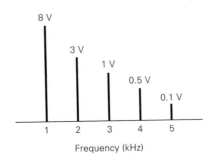

Frequency (kHz)

1-4. Repeat Problem 1-3 for the spectrum shown next.

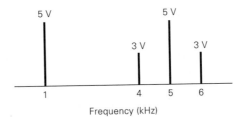

Frequency (kHz)

1-5. For a nonlinear amplifier with two input frequencies of 7 kHz and 4 kHz:
   (a) Determine the first three harmonics present in the output for each frequency.
   (b) Determine the cross-product frequencies produced in the output for values of m and n of 1 and 2.
   (c) Draw the output spectrum for the harmonics and cross-product frequencies determined in steps (a) and (b).

1-6. Determine the percent second order, third order, and total harmonic distortion for the output spectrum shown.

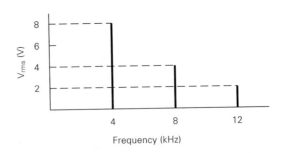

Frequency (kHz)

1-7. Determine the percent second order intermodulation distortion for the A-band, B-band, and second order intermodulation components shown.

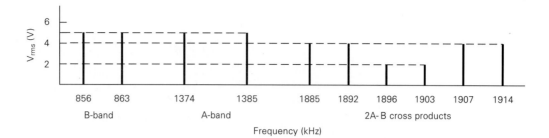

1-8. Determine the second order cross-product frequencies for the following A-band and B-band frequencies: B = 822 and 829, A = 1356 and 1365.

1-9. For an amplifier operating at a temperature of 27°C with a bandwidth of 20 kHz, determine:
   (a) The noise power density (N_o) in watts and dBm.
   (b) The total noise power (N) in watts per hertz and dBm.
   (c) The rms noise voltage (V_N) for a 50-Ω internal resistance and a 50-Ω load resistor.

1-10. **(a)** Determine the noise power (N) in watts and dBm for an amplifier operating at a temperature of 400°C with a 1-MHz bandwidth.
   (b) Determine the decrease in noise power in decibels if the temperature decreased to 100°C.
   (c) Determine the increase in noise power in decibels if the bandwidth doubled.

1-11. Determine the overall power noise figure and noise factor for three cascaded amplifiers, each with individual noise figures of 3 dB and power gains of 20 dB.

1-12. Determine the duty cycle for the pulse waveform shown.

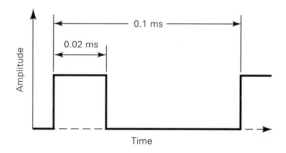

1-13. Determine the % THD for a fundamental frequency voltage V_{fund} = 12 V rms and a total voltage for the higher harmonics V_{higher} = 1.2 V rms.

1-14. Determine the percent second order IMD for a quadratic sum of the A- and B-band components of V_{original} = 2.6 V rms and a quadratic sum of the second order cross-products of $V_{\text{2nd order}}$ = 0.02 V rms.

1-15. If an amplifier has a bandwidth B = 20 kHz and a total noise power $N = 2 \times 10^{-17}$ W, determine:
   (a) Noise density.
   (b) Total noise if the bandwidth is increased to 40 kHz.
   (c) Noise density if the bandwidth is increased to 30 kHz.

Problems

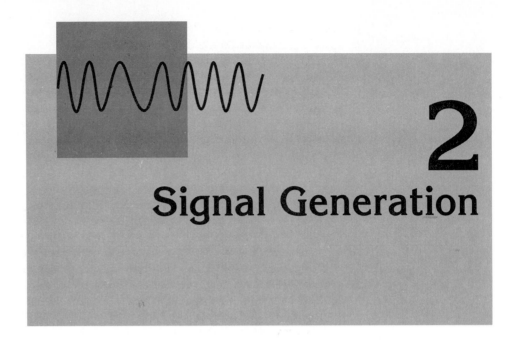

2

Signal Generation

INTRODUCTION

Modern electronic communications systems have many applications that require stable, repetitive waveforms (both sinusoidal and nonsinusoidal). In many of these applications, more than one frequency is required, and very often these frequencies must be synchronized to each other. Therefore, *signal generation*, *frequency synchronization*, and *frequency synthesis* are essential parts of an electronic communications system. The purpose of this chapter is to introduce the reader to the basic operation of oscillators, phase-locked loops, and frequency synthesizers and to show how these circuits are used for signal generation.

OSCILLATORS

The definition of *oscillate* is to fluctuate between two states or conditions. Therefore, to *oscillate* is to vibrate or change, and *oscillating* is the act of fluctuating from one state to another. An *oscillator* is a device that produces oscillations (that is, generates a repetitive waveform). There are many applications for oscillators in electronic communications, such as high-frequency carrier supplies, pilot supplies, clocks, and timing circuits.

In electronic applications, an oscillator is a device or circuit that produces electrical oscillations. An electrical oscillation is a repetitive change in a voltage or current waveform. If an oscillator is *self-sustaining*, the changes in the waveform are *continuous* and *repetitive*; they occur at a periodic rate. A self-sustaining oscillator is also called a *free-running* oscillator. Oscillators that are not self-sustaining require an external input signal or *trigger* to produce a change in the output waveform. Oscillators that are not self-sustaining are called *triggered* or *one-shot* oscillators. The remainder of this chapter is restricted to explaining self-sustaining oscillators, which require no external input other than a dc supply voltage. Essentially, an oscillator converts a dc input voltage to an ac output voltage. The shape of the output waveform can be a sine wave, a square wave, a sawtooth wave, or any other waveform shape as long as it repeats at periodic intervals.

Feedback Oscillators

A *feedback oscillator* is an amplifier with a *feedback loop* (that is, a path for energy to propagate from the output back to the input). Free-running oscillators are feedback oscillators. Once started, a feedback oscillator generates an ac output signal of which a small portion is fed back to the input, where it is amplified. The amplified input signal appears at the output and the process repeats; a *regenerative* process occurs in which the output is dependent on the input, and vice versa.

According to the *Barkhausen criterion*, for a feedback circuit to sustain oscillations, the net voltage gain around the feedback loop must be unity or greater, and the net phase shift around the loop must be a positive integer multiple of 360°.

There are four requirements for a feedback oscillator to work: *amplification*, *positive feedback*, *frequency dependency*, and a *source* of electrical power.

1. *Amplification.* An oscillator circuit must include at least one active device and be capable of voltage amplification. In fact, at times it may be required to provide an infinite gain.

2. *Positive feedback.* An oscillator circuit must have a complete path for a portion of the output signal to be returned to the input. The feedback signal must be *regenerative*, which means it must have the correct phase and amplitude necessary to sustain oscillations. If the phase is incorrect or if the amplitude is insufficient, oscillations will cease. If the amplitude is excessive, the amplifier will saturate. *Regenerative feedback* is called *positive feedback*, where "positive" simply means that its phase aids the oscillation process and does not necessarily indicate a positive (+) or negative (−) polarity. *Degenerative feedback* is called *negative feedback* and supplies a feedback signal that inhibits oscillations from occurring.

3. *Frequency determining components.* An oscillator must have frequency determining components such as resistors, capacitors, inductors, or crystals to allow the frequency of operation to be set or changed.

4. *Power source.* An oscillator must have a source of electrical energy, such as a dc power supply.

Figure 2-1 shows an electrical model for a *feedback oscillator* circuit (that is, a voltage amplifier with regenerative feedback). A feedback oscillator is a *closed-loop* circuit comprised of a voltage amplifier with an *open-loop voltage gain* (A_{ol}), a frequency-determining regenerative feedback path with a *feedback ratio* (β), and either a summer or a subtractor circuit. The open-loop voltage gain is the voltage gain of the amplifier with the feedback path open circuited. The *closed-loop voltage gain* (A_{cl}) is the overall voltage gain of the complete circuit with the feedback loop closed and is always less than the open-loop voltage gain. The feedback ratio is simply the transfer function of the feedback network (that is, the ratio of its output to its input voltage). For a passive feedback network, the feedback ratio is always less than 1.

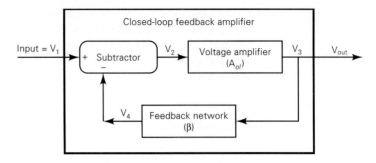

Figure 2-1 Model of an amplifier with feedback.

From Figure 2-1, the following mathematical relationships are derived:

$$\frac{V_{\text{out}}}{V_{\text{in}}} = \frac{V_3}{V_1}$$

$$V_2 = V_1 - V_4$$

$$V_3 = A_{ol}V_2$$

$$A_{ol} = \frac{V_3}{V_2}$$

$$V_4 = \beta V_3$$

$$\beta = \frac{V_4}{V_3}$$

where V_1 = external input voltage
V_2 = input voltage to the amplifier
V_3 = output voltage
V_4 = feedback voltage
A_{ol} = open loop voltage gain
β = feedback ratio of the feedback network

Substituting for V_4 gives us $\qquad V_2 = V_1 - \beta V_3$

Thus $\qquad\qquad\qquad\qquad V_3 = (V_1 - \beta V_3)A_{ol}$

and $\qquad\qquad\qquad\qquad V_3 = V_1 A_{ol} - V_3 \beta A_{ol}$

Rearranging and factoring yields

$$V_3 + V_3 \beta A_{ol} = V_1 A_{ol}$$

Thus $\qquad\qquad\qquad\qquad V_3(1 + \beta A_{ol}) = V_1 A_{ol}$

and $\qquad\qquad \dfrac{V_{\text{out}}}{V_{\text{in}}} = \dfrac{V_3}{V_1} = \dfrac{A_{ol}}{1 + \beta A_{ol}} = A_{cl}$ $\qquad$ (2-1)

where A_{cl} = closed-loop voltage gain.

$A_{ol}/(1 + \beta A_{ol})$ is the standard formula used for the closed-loop voltage gain of an amplifier with feedback. If at any frequency, βA_{ol} goes to $+1$, the denominator in Equation 2-1 goes to zero and $V_{\text{out}}/V_{\text{in}}$ is infinity. When this happens, the circuit will oscillate and the external input may be removed.

For self-sustained oscillations to occur, a circuit must fulfill the four basic requirements for oscillation outlined previously, meet the criterion of Equation 2-1, and fit the basic feedback circuit model shown in Figure 2-1. Although oscillator action can be accomplished in many different ways, the most common configurations use *RC* phase-shift networks, *LC* tank circuits, quartz crystals, or integrated-circuit chips. The type of oscillator used for a particular application depends on the following criteria:

1. Desired frequency of operation
2. Required frequency stability
3. Variable or fixed frequency operation
4. Distortion requirements or limitations
5. Desired output power
6. Physical size
7. Application (that is, digital or analog)
8. Cost
9. Reliability and durability
10. Desired accuracy

Wien-bridge Oscillator

The Wien-bridge oscillator is an *RC* phase-shift oscillator that uses both positive and negative feedback. It is a relatively stable, low-frequency oscillator circuit that is easily tuned and commonly used in signal generators to produce frequencies between 5 Hz and 1 MHz. The Wien-bridge oscillator is the circuit that Hewlett and Packard used in their original signal generator design.

Figure 2-2a shows a simple lead–lag network. At the frequency of oscillation (f_o), $R = X_C$ and the signal undergoes a $-45°$ phase shift across Z_1 and a $+45°$ phase shift across Z_2. Consequently, at f_o, the total phase shift across the lead–lag network is exactly $0°$. At frequencies below the frequency of oscillation, the phase shift across the network leads and for frequencies above the phase shift lags. At extreme low frequencies, C_1 looks like an open circuit and there is no output. At extreme high frequencies, C_2 looks like a short circuit and there is no output.

A lead–lag network is a reactive voltage divider in which the input voltage is divided between Z_1 (the series combination of R_1 and C_1) and Z_2 (the parallel combination

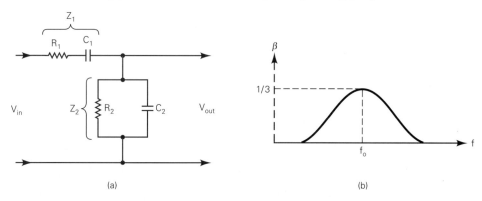

Figure 2-2 Lead–lag network: (a) circuit configuration; (b) input-versus-output transfer curve (β).

Oscillators 53

of R_2 and C_2). Therefore, the lead–lag network is frequency selective and the output voltage is maximum at f_o. The transfer function for the feedback network (β) equals $Z_2/(Z_1 + Z_2)$ and is maximum and equal to 1/3 at f_o. Figure 2-2b shows a plot of β versus frequency when $R_1 = R_2$ and $C_1 = C_2$. Thus, f_o is determined from the following expression:

$$f_o = \frac{1}{2\pi RC}$$

where $R = R_1 = R_2$
$C = C_1 = C_2$

Figure 2-3 shows a Wien-bridge oscillator. The lead–lag network and the resistive voltage divider make up a Wien bridge (hence the name *Wien-bridge oscillator*). When the bridge is balanced, the difference voltage equals zero. The voltage divider provides negative or degenerative feedback that offsets the positive or regenerative feedback from the lead–lag network. The ratio of the resistors in the voltage divider is 2:1, which sets the noninverting voltage gain of amplifier A_1 to $R_f/R_i + 1 = 3$. Thus, at f_o, the signal at the output of A_1 is reduced by a factor of 3 as it passes through the lead–lag network ($\beta = 1/3$) and then amplified by 3 in amplifier A_1. Thus, at f_o the loop voltage gain is equal to $A_{ol}\beta$ or $3 \times 1/3 = 1$.

To compensate for imbalances in the bridge and variations in component values due to heat, *automatic gain control (AGC)* is added to the circuit. A simple way of providing automatic gain is to replace R_i in Figure 2-3 with a variable resistance device such as a FET. The resistance of the FET is made inversely proportional to V_{out}. The circuit is designed such that, when V_{out} increases in amplitude, the resistance of the FET increases, and when V_{out} decreases in amplitude, the resistance of the FET decreases. Therefore, the voltage gain of the amplifier automatically compensates for changes in amplitude in the output signal.

The operation of the circuit shown in Figure 2-3 is as follows. On initial power-up, noise (at all frequencies) appears at V_{out} and is fed back through the lead–lag network. Only noise at f_o passes through the lead–lag network with a 0° phase shift and a transfer ratio of 1/3. Consequently, only a single frequency (f_o) is fed back in phase, undergoes a loop voltage gain of 1, and produces self-sustained oscillations.

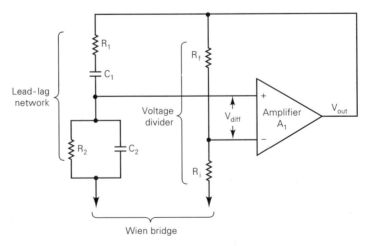

Figure 2-3   Wien-bridge oscillator.

LC Oscillators

LC oscillators are oscillator circuits that utilize *LC tank circuits* for the frequency-determining components. Tank-circuit operation involves an exchange of energy between *kinetic* and *potential*. Figure 2-4 illustrates *LC* tank circuit operation. As shown in Figure 2-4a, once current is injected into the circuit (time t_1), energy is exchanged between the inductor and capacitor, producing a corresponding ac output voltage (times t_2 to t_4). The

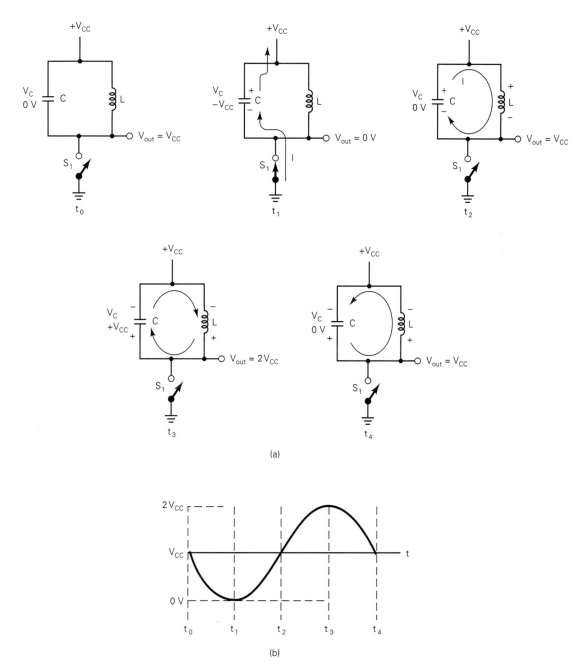

Figure 2-4 *LC* tank circuit: (a) oscillator action and flywheel effect; (b) output waveform.

Oscillators

55

output voltage waveform is shown in Figure 2-4b. The frequency of operation of an LC tank circuit is simply the resonant frequency of the parallel LC network, and the bandwidth is a function of the circuit Q. Mathematically, the resonant frequency of an LC tank circuit with a $Q \geq 10$ is closely approximated by

$$f_o = \frac{1}{2\pi \sqrt{(LC)}} \tag{2-2}$$

LC oscillators include the Hartley and Colpitts oscillators.

Hartley oscillator.   Figure 2-5a shows the schematic diagram of a *Hartley oscillator*. The transistor amplifier (Q_1) provides the amplification necessary for a loop voltage gain of unity at the resonant frequency. The coupling capacitor (C_C) provides the path for regenerative feedback. L_1 and C_1 are the frequency-determining components, and V_{CC} is the dc supply voltage.

Figure 2-5b shows the dc equivalent circuit for the Hartley oscillator. C_C is a blocking capacitor that isolates the dc base bias voltage and prevents it from being shorted to ground through L_{1b}. C_2 is also a blocking capacitor that prevents the collector supply voltage from being shorted to ground through L_{1a}. The *radio-frequency choke* (RFC) is a dc short.

Figure 2-5c shows the ac equivalent circuit for the Hartley oscillator. C_C is a coupling capacitor for ac and provides a path for regenerative feedback from the tank circuit to the base of Q_1. C_2 couples ac signals from the collector of Q_1 to the tank circuit. The RFC looks open to ac, consequently isolating the dc power supply from ac oscillations.

The Hartley oscillator operates as follows. On initial power-up, a multitude of frequencies appear at the collector of Q_1 and are coupled through C_2 into the tank circuit. The initial noise provides the energy necessary to charge C_1. Once C_1 is partially charged, oscillator action begins. The tank circuit will only oscillate efficiently at its resonant frequency. A portion of the oscillating tank circuit voltage is dropped across L_{1b} and fed back to the base of Q_1, where it is amplified. The amplified signal appears at the collector 180° out of phase with the base signal. An additional 180° of phase shift is realized across L_1; consequently, the signal fed back to the base of Q_1 is amplified and shifted in phase 360°. Thus, the circuit is regenerative and will sustain oscillations with no external input signal.

The proportion of oscillating energy that is fed back to the base of Q_1 is determined by the ratio of L_{1b} to the total inductance ($L_{1a} + L_{1b}$). If insufficient energy is fed back, oscillations are damped. If excessive energy is fed back, the transistor saturates. Therefore, the position of the wiper on L_1 is adjusted until the amount of feedback energy is exactly what is required for a unity loop voltage gain and oscillations to continue.

The frequency of oscillation for the Hartley oscillator is closely approximated by the following formula:

$$f_o = \frac{1}{2\pi \sqrt{(LC)}} \tag{2-3}$$

where $L = L_{1a} + L_{1b}$
$\qquad C = C_1$

Colpitts oscillator.   Figure 2-6a shows the schematic diagram of a *Colpitts oscillator*. The operation of a Colpitts oscillator is very similar to that of the Hartley except that a capacitive divider is used instead of a tapped coil. Q_1 provides the amplification, C_C provides the regenerative feedback path, L_1, C_{1a}, and C_{1b} are the frequency-determining components, and V_{CC} is the dc supply voltage.

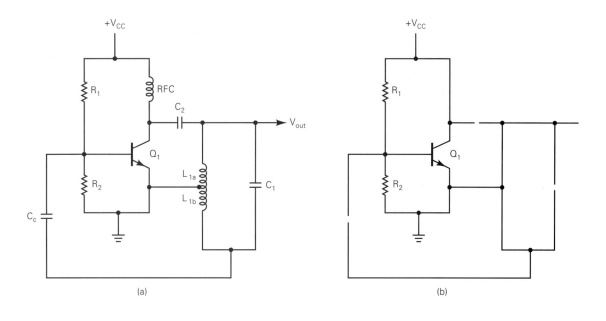

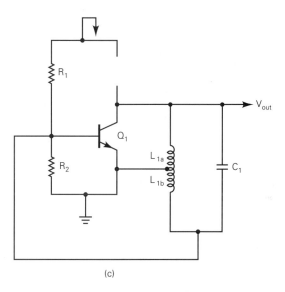

Figure 2-5 Hartley oscillator: (a) schematic diagram; (b) dc equivalent circuit; (c) ac equivalent circuit.

Figure 2-6b shows the dc equivalent circuit for the Colpitts oscillator. C_2 is a blocking capacitor that prevents the collector supply voltage from appearing at the output. The RFC is again a dc short.

Figure 2-6c shows the ac equivalent circuit for the Colpitts oscillator. C_C is a coupling capacitor for ac and provides the feedback path for regenerative feedback from the tank circuit to the base of Q_1. The RFC is open to ac and decouples oscillations from the dc power supply.

The operation of the Colpitts oscillator is almost identical to that of the Hartley oscillator. On initial power-up, noise appears at the collector of Q_1 and supplies energy to the tank circuit, causing it to begin oscillating. C_{1a} and C_{1b} make up an ac voltage divider.

Oscillators

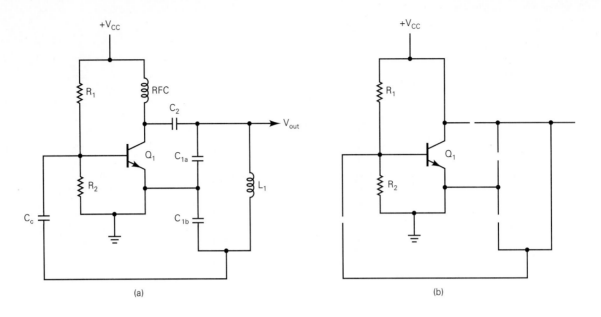

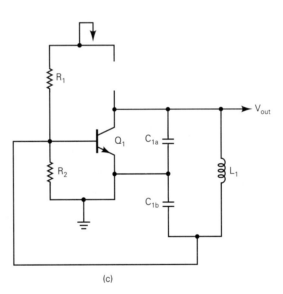

Figure 2-6 Colpitts oscillator: (a) schematic diagram; (b) dc equivalent circuit; (c) ac equivalent circuit.

The voltage dropped across C_{1b} is fed back to the base of Q_1 through C_C. There is a 180° phase shift from the base to the collector of Q_1 and an additional 180° phase shift across C_1. Consequently, the total phase shift is 360° and the feedback signal is regenerative. The ratio of C_{1a} to $C_{1a} + C_{1b}$ determines the amplitude of the feedback signal.

The frequency of oscillation of the Colpitts oscillator is closely approximated by the following formula:

$$f_o = \frac{1}{2\pi \sqrt{(LC)}}$$

(2-4)

where $L = L_1$

$$C = \frac{C_{1a}C_{1b}}{C_{1a} + C_{1b}}$$

Frequency Stability

Frequency stability is the ability of an oscillator to remain at a fixed frequency and is of primary importance in communications systems. Frequency stability is often stated as either short or long term. *Short-term stability* is affected predominantly by fluctuations in dc operating voltages, while *long-term stability* is a function of component aging and changes in the ambient temperature and humidity. In the *LC* tank-circuit and *RC* phase-shift oscillators discussed previously, the frequency stability is inadequate for most radio communications applications. This is because *RC* phase-shift oscillators are susceptible to both short- and long-term variations. In addition, the *Q*-factors of the *LC* tank circuits are relatively low, allowing the resonant tank circuit to oscillate over a wide range of frequencies.

Frequency stability is generally given as a percentage of change in frequency (tolerance) from the desired value. For example, an oscillator operating at 100 kHz with a ± 5% stability will operate at a frequency of 100 kHz ± 5 kHz or between 95 and 105 kHz. Commercial FM broadcast stations must maintain their carrier frequencies to within ± 2 kHz of their assigned frequency, which is approximately a 0.002% tolerance. In commercial AM broadcasting, the maximum allowable shift in the carrier frequency is only ± 20 Hz.

Several factors affect the stability of an oscillator. The most obvious are those that directly affect the value of the frequency-determining components. These include changes in inductance, capacitance, and resistance values due to environmental variations in temperature and humidity and changes in the quiescent operating point of transistors and field-effect transistors. Stability is also affected by ac ripple in dc power supplies. The frequency stability of *RC* or *LC* oscillators can be greatly improved by regulating the dc power supply and minimizing the environmental variations. Also, special temperature-independent components can be used.

The FCC has established stringent regulations concerning the tolerances of radio-frequency carriers. Whenever the airway (free-space radio propagation) is used as the transmission medium, it is possible that transmissions from one source could interfere with transmissions from other sources if their transmit frequency or transmission bandwidths overlap. Therefore, it is important that all sources maintain their frequency of operation within a specified tolerance.

Crystal Oscillators

Crystal oscillators are feedback oscillator circuits in which the *LC* tank circuit is replaced with a crystal for the frequency-determining component. The crystal acts in a manner similar to the *LC* tank, except with several inherent advantages. Crystals are sometimes called crystal resonators and they are capable of producing precise, stable frequencies for frequency counters, electronic navigation systems, radio transmitters and receivers, televisions, video cassette recorders (VCRs), computer system clocks, and many other applications too numerous to list.

Crystallography is the study of the form, structure, properties, and classifications of crystals. Crystallography deals with lattices, bonding, and the behavior of slices of crystal material that have been cut at various angles with respect to the crystal's axes. The mechanical properties of crystal lattices allow them to exhibit the piezoelectric effect. Sections of crystals that have been cut and polished vibrate when alternating voltages are applied across their faces. The physical dimensions of a crystal, particularly its thickness and where and how it was cut, determine its electrical and mechanical properties.

Piezoelectric effect. Simply stated, the *piezoelectric effect* occurs when oscillating mechanical stresses applied across a *crystal lattice structure* generate electrical oscillations, and vice versa. The stress can be in the form of squeezing (compression), stretching (tension), twisting (torsion), or shearing. If the stress is applied periodically, the output voltage will alternate. Conversely, when an alternating voltage is applied across a crystal at or near the natural resonant frequency of the crystal, the crystal will break into mechanical oscillations. This is called *exciting* a crystal into *mechanical vibrations*. The mechanical vibrations are called *bulk acoustic waves* (*BAWs*) and are directly proportional to the amplitude of the applied voltage.

A number of natural crystal substances exhibit piezoelectric properties: *quartz*, *Rochelle salt*, and *tourmaline* and several manufactured substances such as *ADP*, *EDT*, and *DKT*. The piezoelectric effect is most pronounced in Rochelle salt, which is why it is the substance commonly used in crystal microphones. Synthetic quartz, however, is used more often for frequency control in oscillators because of its *permanence*, low *temperature coefficient*, and high *mechanical Q*.

Crystal cuts. In nature, complete quartz crystals have a hexagonal cross section with pointed ends, as shown in Figure 2-7a. Three sets of axes are associated with a crystal: *optical, electrical,* and *mechanical.* The longitudinal axis joining points at the ends of the crystal is called the *optical* or *Z axis.* Electrical stresses applied to the optical axis do not produce the piezoelectric effect. The *electrical* or *X axis* passes diagonally through opposite corners of the hexagon. The axis that is perpendicular to the faces of the crystal is the *Y* or *mechanical axis.* Figure 2-7b shows the axes and the basic behavior of a quartz crystal.

If a thin flat section is cut from a crystal such that the flat sides are perpendicular to an electrical axis, mechanical stresses along the *Y* axis will produce electrical charges on the flat sides. As the stress changes from compression to tension, and vice versa, the polarity of the charge is reversed. Conversely, if an alternating electrical charge is placed on the flat sides, a mechanical vibration is produced along the *Y* axis. This is the piezoelectric effect and is also exhibited when mechanical forces are applied across the faces of a crystal cut with its flat sides perpendicular to the *Y* axis. When a crystal wafer is cut parallel to the *Z* axis with its faces perpendicular to the *X* axis, it is called an *X*-cut crystal. When the faces are perpendicular to the *Y* axis, it is called a *Y*-cut crystal. A variety of cuts can be obtained by rotating the plane of the cut around one or more axes. If the *Y* cut is made at a 35°20′ angle from the vertical axis (Figure 2-7c), an *AT* cut is obtained. Other types of crystal cuts include the *BT, CT, DT, ET, AC, GT, MT, NT,* and *JT* cuts. The *AT* cut is the most popular for high-frequency and very high frequency crystal resonators. The type, length, and thickness of a cut and the mode of vibration determine the natural resonant frequency of the crystal. Resonant frequencies for *AT*-cut crystals range from approximately 800 kHz up to approximately 30 MHz. *CT* and *DT* cuts exhibit low-frequency shear and are most useful in the 100- to 500-kHz range. The *MT* cut vibrates longitudinally and is useful in the 50- to 100-kHz range while the *NT* cut has a useful range under 50 kHz.

Crystal *wafers* are generally mounted in *crystal holders*, which include the mounting and housing assemblies. A *crystal unit* refers to the holder and the crystal itself. Figure 2-7d shows a common crystal mounting. Because a crystal's stability is somewhat temperature dependent, a crystal unit may be mounted in an oven to maintain a constant operating temperature.

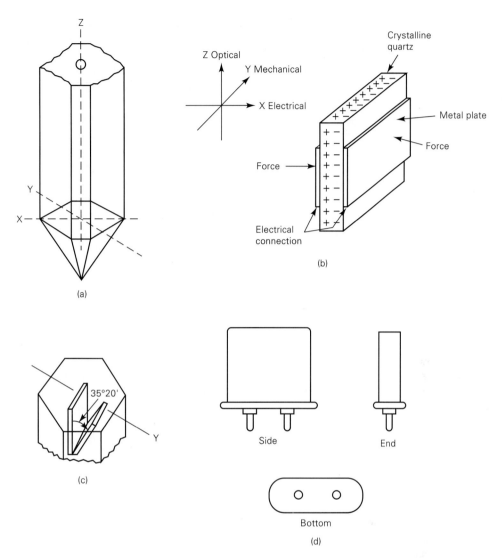

Figure 2-7 Quartz crystal: (a) basic crystal structure; (b) crystal axes; (c) crystal cuts; (d) crystal mountings.

The relationship between a crystal's operating frequency and its thickness is expressed mathematically as

$$h = \frac{65.5}{f_n}$$

where h = crystal thickness in inches
f_n = crystal natural resonant frequency in hertz

This formula indicates that for high-frequency oscillations the quartz wafer must be very thin. This makes it difficult to manufacture crystal oscillators with fundamental frequencies above approximately 30 MHz because the wafer becomes so thin that it is exceptionally fragile, and conventional cutting and polishing can only be accomplished at extreme

Oscillators 61

costs. This problem can be alleviated by using chemical etching to achieve thinner slices. With this process, crystals with fundamental frequencies up to 350 MHz are possible.

Overtone crystal oscillator.   As previously stated, to increase the frequency of vibration of a quartz crystal, the quartz wafer is sliced thinner. This imposes an obvious physical limitation; the thinner the wafer, the more susceptible it is to damage and the less useful it becomes. Although the practical limit for fundamental-mode crystal oscillators is approximately 30 MHz, it is possible to operate the crystal in an overtone mode. In the overtone mode, harmonically related vibrations that occur simultaneously with the fundamental vibration are used. In the overtone mode, the oscillator is tuned to operate at the third, fifth, seventh, or even the ninth harmonic of the crystal's fundamental frequency. The harmonics are called overtones because they are not true harmonics. Manufacturers can process crystals such that one overtone is enhanced more than the others. Using an overtone mode increases the usable limit of standard crystal oscillators to approximately 200 MHz.

Temperature coefficient.   The natural resonant frequency of a crystal is influenced somewhat by its operating temperature. The ratio of the magnitude of frequency change (Δf) to a change in temperature (ΔC) is expressed in hertz change per megahertz of crystal operating frequency per degree Celsius (Hz/MHz/°C). The fractional change in frequency is often given in parts per million (ppm) per °C. For example, a temperature coefficient of +20 Hz/MHz/°C is the same as +20 ppm/°C. If the direction of the frequency change is the same as the temperature change (that is, an increase in temperature causes an increase in frequency and a decrease in temperature causes a decrease in frequency), it is called a *positive temperature coefficient*. If the change in frequency is in the direction opposite to the temperature change (that is, an increase in temperature causes a decrease in frequency and a decrease in temperature causes an increase in frequency), it is called a *negative temperature coefficient*. Mathematically, the relationship of the change in frequency of a crystal to a change in temperature is

$$\Delta f = k(f_n \times \Delta C) \tag{2-5}$$

where Δf = change in frequency (hertz)
$\qquad k$ = temperature coefficient (Hz/MHz/°C)
$\qquad f_n$ = natural crystal frequency in megahertz
$\qquad \Delta C$ = change in temperature (degrees Celsius)

and
$$f_o = f_n + \Delta f \tag{2-6}$$

where f_o = frequency of operation.

The temperature coefficient (k) of a crystal varies depending on the type of crystal cut and its operating temperature. For a range of temperatures from approximately +20° to +50°C, both X- and Y-cut crystals have a temperature coefficient that is nearly constant. X-cut crystals are approximately 10 times more stable than Y-cut crystals. Typically, X-cut crystals have a temperature coefficient that ranges from −10 to −25 Hz/MHz/°C. Y-cut crystals have a temperature coefficient that ranges from approximately −25 to +100 Hz/MHz/°C.

Today, zero-coefficient (GT-cut) crystals are available that have temperature coefficients as low as −1 to +1 Hz/MHz/°C. The GT-cut crystal is almost a perfect zero-coefficient crystal from freezing to boiling, but is useful only at frequencies below a few hundred kilohertz.

EXAMPLE 2-1

For a 10-MHz crystal with a temperature coefficient $k = +10$ Hz/MHz/°C, determine the frequency of operation if the temperature:

(a) Increases 10°C.
(b) Decreases 5°C.

Solution (a) Substituting into Equations 2-5 and 2-6 gives us

$$\Delta f = k(f_n \times \Delta C)$$

$$= +10(10 \times 10) = 1 \text{ kHz}$$

$$f_o = f_n + \Delta f$$

$$= 10 \text{ MHz} + 1 \text{ kHz} = 10.001 \text{ MHz}$$

(b) Again, substituting into Equations 2-5 and 2-6 yields

$$\Delta f = -10[10 \times (-5)] = -500 \text{ Hz}$$

$$f_o = 10 \text{ MHz} + (-500 \text{ Hz})$$

$$= 9.9995 \text{ MHz}$$

Crystal equivalent circuit. Figure 2-8a shows the electrical equivalent circuit for a crystal. Each electrical component is equivalent to a mechanical property of the crystal. C_2 is the actual capacitance formed between the electrodes of the crystal, with the crystal itself being the dielectric. C_1 is equivalent to the mechanical compliance of the crystal (also called the resilience or elasticity). L_1 is equivalent to the mass of the crystal in vibration, and R is the mechanical friction loss. In a crystal, the mechanical *mass-to-friction ratio* (L/R) is quite high. Typical values of L range from 0.1 H to well over 100 H; consequently, Q factors are quite high for crystals. Q factors in the range from 10,000 to 100,000 and higher are not uncommon (as compared to Q factors of 100 to 1000 for the discrete inductors used in LC tank circuits). This provides the high stability of crystal oscillators as compared to discrete LC tank-circuit oscillators. Values for C_1 are typically less than 1 pF, while values for C_2 range between 4 and 40 pF.

Because there is a series and a parallel equivalent circuit for a crystal, there are also two equivalent impedances and two resonant frequencies: a series and a parallel. The series impedance is the combination of R, L, and C_1 (that is, $Z_s = R \pm jX$, where $X = |X_L - X_C|$). The parallel impedance is approximately the impedance of L and C_2 [that is, $Z_p = (X_L \times X_{C2})/(X_L + X_{C2})$]. At extreme low frequencies, the series impedance of L, C_1, and R is very high and capacitive $(-)$. This is shown in Figure 2-8c. As the frequency is increased, a point is reached where $X_L = X_{C1}$. At this frequency (f_1), the series impedance is minimum, resistive, and equal to R. As the frequency is increased even further (f_2), the series impedance becomes high and inductive $(+)$. The parallel combination of L and C_2 causes the crystal to act like a parallel resonant circuit (maximum impedance at resonance). The difference between f_1 and f_2 is usually quite small (typically about 1% of the crystal's natural frequency). A crystal can operate at either its series or parallel resonant frequency, depending on the circuit configuration it is used in. The relative steepness of the impedance curve shown in Figure 2-8b also attributes to the stability and accuracy of a crystal. The series resonant frequency of a quartz crystal is simply

$$f_1 = \frac{1}{2\pi \sqrt{(LC_1)}}$$

Oscillators   **63**

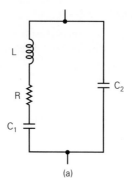

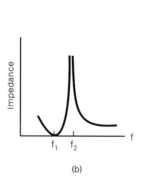

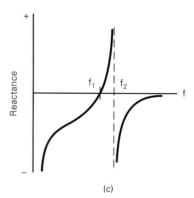

Figure 2-8 Crystal equivalent circuit: (a) equivalent circuit; (b) impedance curve; (c) reactance curve.

and the parallel resonant frequency is

$$f_2 = \frac{1}{2\pi \sqrt{(LC)}}$$

where C = parallel combination of C_1 and C_2.

Crystal oscillator circuits. Although there are many different crystal-based oscillator configurations, the most common are the discrete and integrated-circuit Pierce and the *RLC* half-bridge. If you need very good frequency stability and reasonably simple circuitry, the discrete Pierce is a good choice. If low cost and simple digital interfacing capabilities are of primary concern, an IC-based Pierce oscillator will suffice. However, for the best frequency stability, the *RLC* half-bridge is the best choice.

Discrete pierce oscillator. The discrete Pierce crystal oscillator has many advantages. Its operating frequency spans the full fundamental crystal range (1 kHz to approximately 30 MHz). It uses relatively simple circuitry requiring few components (most medium-frequency versions require only one transistor). The Pierce oscillator design develops a high output signal power while dissipating very little power in the crystal itself. Finally, the short-term frequency stability of the Pierce crystal oscillator is excellent (this is because the in-circuit loaded *Q* is almost as high as the crystal's internal *Q*). The only drawback to the Pierce oscillator is that it requires a high-gain amplifier (approximately

70). Consequently, you must use a single high-gain transistor or possibly even a multiple-stage amplifier.

Figure 2-9 shows a discrete 1-MHz Pierce oscillator circuit. Q_1 provides all the gain necessary for self-sustained oscillations to occur. R_1 and C_1 provide a 65° phase lag to the feedback signal. The crystal impedance is basically resistive with a small inductive component. This impedance combined with the reactance of C_2 provides an additional 115° of phase lag. The transistor inverts the signal (180° phase shift), giving the circuit the necessary 360° of total phase shift. Because the crystal's load is primarily nonresistive (mostly the series combination of C_1 and C_2), this type of oscillator provides very good short-term frequency stability. Unfortunately, C_1 and C_2 introduce substantial losses and, consequently, the transistor must have a relatively high voltage gain; this is an obvious drawback.

Integrated-circuit pierce oscillator.   Figure 2-10 shows an IC-based Pierce crystal oscillator. Although it provides less frequency stability, it can be implemented using simple digital IC design and reduces costs substantially over conventional discrete designs.

To ensure that oscillations begin, RFB dc biases inverting amplifier A_1's input and output for class A operation. A_2 converts the output of A_1 to a full rail-to-rail swing (cutoff to saturation), reducing the rise and fall times and buffering A_1's output. The output resistance of A_1 combines with C_1 to provide the *RC* phase lag needed. CMOS (complementary metal-oxide semiconductor) versions operate up to approximately 2 MHz, and ECL (emitter-coupled logic) versions operate as high as 20 MHz.

RLC Half-bridge crystal oscillator.   Figure 2-11 shows the Meacham version of the *RLC* half-bridge crystal oscillator. The original Meacham oscillator was developed in the 1940s and used a full four-arm bridge and a negative-temperature-coefficient tungsten lamp. The circuit configuration shown in Figure 2-11 uses only a two-arm bridge and employs a negative-temperature-coefficient thermistor. Q_1 serves as a phase splitter and provides two 180° out-of-phase signals. The crystal must operate at its series resonant frequency, so its internal impedance is resistive and quite small. When oscillations begin, the

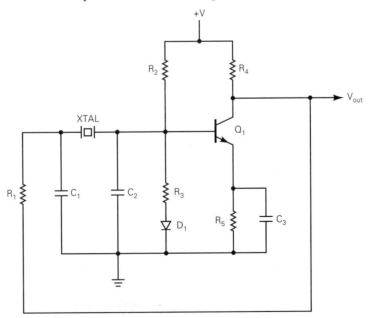

Figure 2-9 Discrete Pierce crystal oscillator.

Oscillators

65

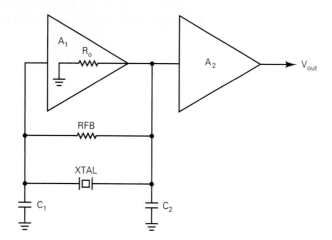

Figure 2-10 IC Pierce crystal oscillator.

signal amplitude increases gradually, decreasing the thermistor resistance until the bridge almost nulls. The amplitude of the oscillations stabilizes and determines the final thermistor resistance. The *LC* tank circuit at the output is tuned to the crystal's series resonant frequency.

Crystal oscillator module. A *crystal oscillator module* consists of a crystal-controlled oscillator and a voltage-variable component such as a *varactor diode*. The entire oscillator circuit is contained in a single *metal can*. A simplified schematic diagram for a Colpitts crystal oscillator module is shown in Figure 2-12a. X_1 is a crystal itself and Q_1 is the active component for the amplifier. C_1 is a shunt capacitor that allows the crystal oscillator frequency to be varied over a narrow range of operating frequencies. VC_1 is a voltage-variable capacitor (*varicap* or *varactor diode*). A varactor diode is a specially constructed diode whose internal capacitance is enhanced when reverse biased, and by varying the reverse-bias voltage, the capacitance of the diode can be adjusted. A varactor diode has a special depletion layer between the *p*- and *n*-type materials that is constructed with various degrees and types of doping material (the term *graded junction* is often used when describing varactor-diode fabrication). Figure 2-13 shows the capacitance versus

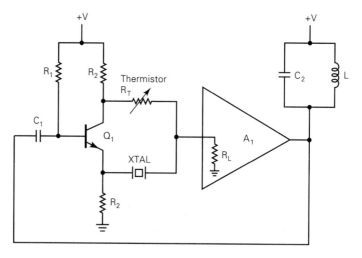

Figure 2-11 *RLC* half-bridge crystal oscillator.

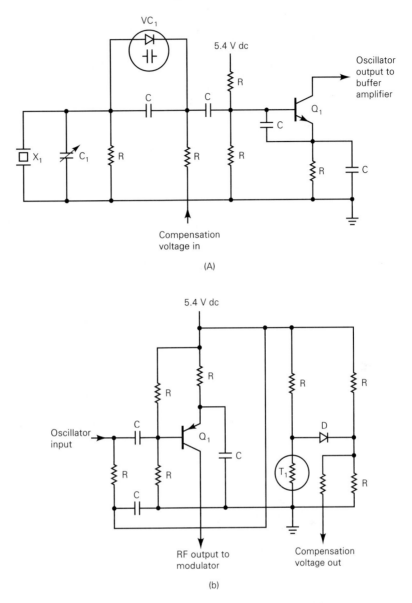

Figure 2-12 Crystal oscillator module: (a) schematic diagram; (b) compensation circuit.

reverse-bias voltage curves for a typical varactor diode. The capacitance of a varactor diode is approximated as

$$C_d = \frac{C}{\sqrt{(1 + 2|\,V_r\,|)}} \tag{2-7}$$

where C = diode capacitance with 0-V reverse bias (farads)
   $|V_r|$ = magnitude of diode reverse-bias voltage (volts)
   C_d = reverse-biased diode capacitance (farads)

The frequency at which the crystal oscillates can be adjusted slightly by changing

Oscillators

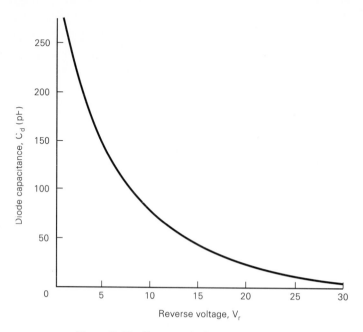

Figure 2-13 Varactor diode characteristics.

the capacitance of VC_1 (that is, changing the value of the reverse-bias voltage). The varactor diode, in conjunction with a temperature-compensating module, provides instant frequency compensation for variations caused by changes in temperature. The schematic diagram of a temperature-compensating module is shown in Figure 2-12b. The compensation module includes a buffer amplifier (Q_1) and a temperature-compensating network (T_1). T_1 is a negative-temperature-coefficient thermistor. When the temperature falls below the threshold value of the thermistor, the compensation voltage increases. The compensation voltage is applied to the oscillator module, where it controls the capacitance of the varactor diode. Compensation modules are available that can compensate for a frequency stability of 0.0005% from $-30°$ to $+80°$C.

LARGE-SCALE INTEGRATION OSCILLATORS

In recent years the use of *large-scale integration* (*LSI*) integrated circuits for frequency and waveform generation has increased at a tremendous rate because integrated-circuit oscillators have excellent frequency stability and a wide tuning range and are easy to use. *Waveform* and *function generators* are used extensively in communications and telemetry equipment, as well as in laboratories for test and calibration equipment. In many of these applications, commercial monolithic integrated-circuit oscillators and function generators are available that provide the circuit designer with a low-cost alternative to their noninte-grated-circuit counterparts.

The basic operations required for waveform generation and shaping are well suited to monolithic integrated-circuit technology. In fact, *monolithic linear integrated circuits* (*LICs*) have several inherent advantages over discrete circuits, such as the availability of a large number of active devices on a single chip and close matching and thermal tracking of component values. It is now possible to fabricate integrated-circuit waveform generators that provide a performance comparable to that of complex discrete generators at only a fraction of the cost.

LSI waveform generators currently available include function generators, timers, programmable timers, voltage-controlled oscillators, precision oscillators, and waveform generators.

Integrated-circuit Waveform Generation

In its simplest form, a waveform generator is an oscillator circuit that generates well-defined, stable waveforms that can be externally modulated or swept over a given frequency range. A typical waveform generator consists of four basic sections: (1) an oscillator to generate the basic periodic waveform, (2) a waveshaper, (3) an optional AM modulator, and (4) an output buffer amplifier to isolate the oscillator from the load and provide the necessary drive current.

Figure 2-14 shows a simplified block diagram of an integrated-circuit waveform generator circuit showing the relationship among the four sections. Each section has been built separately in monolithic form for several years; therefore, fabrication of all four sections onto a single monolithic chip was a natural extension of a preexisting technology. The oscillator section generates the basic oscillator frequency and the waveshaper circuit converts the output from the oscillator to either a sine-, square-, triangular-, or ramp-shaped waveform. The modulator, when used, allows the circuit to produce amplitude-modulated signals, and the output buffer amplifier isolates the oscillator from its load and provides a convenient place to add dc levels to the output waveform. The sync output can be used either as a square-wave source or as a synchronizing pulse for external timing circuitry.

A typical IC oscillator circuit utilizes the constant-current charging and discharging of external timing capacitors. Figure 2-15a shows the simplified schematic diagram for such a waveform generator that uses an emitter-coupled multivibrator, which is capable of generating square waves as well as triangle and linear ramp waveforms. The circuit operates as follow. When transistor Q_1 and diode D_1 are conducting, transistor Q_2 and diode D_2 are off, and vice versa. This action alternately charges and discharges capacitor C_o from constant current source I_1. The voltage across D_1 and D_2 is a symmetrical square wave with a peak-to-peak amplitude of $2V_{BE}$. V_A is constant when Q_1 is on, but becomes a linear ramp with a slope equal to $-I_1/C_o$ when Q_1 goes off. Output $V_B(t)$ is identical to

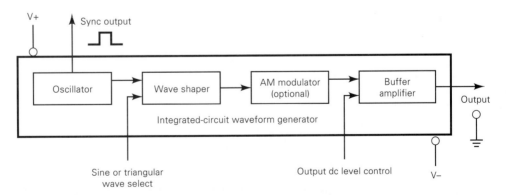

Figure 2-14 Integrated-circuit waveform generator.

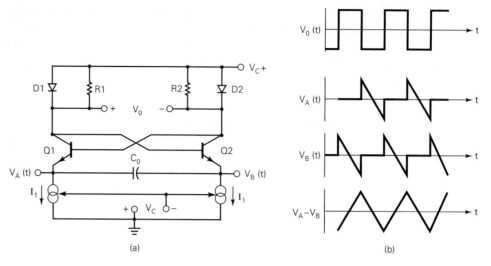

Figure 2-15 Simplified integrated-circuit waveform generator: (a) schematic diagram; (b) waveforms.

$V_A(t)$, except it is delayed by a half-cycle. Differential output, $V_A(t) - V_B(t)$ is a triangle wave. Figure 2-15b shows the output voltage waveforms typically available.

Monolithic function generators. The XR-2206 is a monolithic function generator integrated circuit manufactured by EXAR Corporation that is capable of producing high-quality sine, square, triangle, ramp, and pulse waveforms with both a high degree of stability and accuracy. The output waveforms from the XR-2206 can be both amplitude and frequency modulated by an external modulating signal, and the frequency of operation can be selected externally over a range from 0.01 Hz to more than 1 MHz. The XR-2206 is ideally suited to communications, instrumentation, and function generator applications requiring sinusoidal tone, AM, or FM generation. The XR-2206 has a typical frequency stability of 20 ppm/°C and can be linearly swept over a 2000 : 1 frequency range with an external control voltage.

The block diagram for the XR-2206 is shown in Figure 2-16. The function generator is comprised of four functional blocks: a voltage-controlled oscillator (VCO), an analog

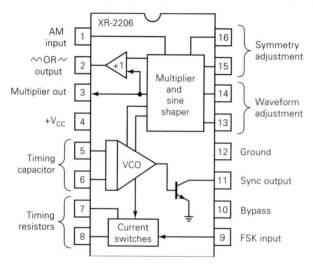

Figure 2-16 Block diagram for the XR-2206 monolithic function generator.

multiplier and sineshaper, a unity-gain buffer amplifier, and a set of input current switches. A *voltage-controlled oscillator* is a free-running oscillator with a stable frequency of oscillation that depends on an external timing capacitance, timing resistance, and control voltage. The output from a VCO is a frequency, and its input is a bias or control signal that can be either a dc or an ac voltage. The VCO actually produces an output frequency that is proportional to an input current that is produced by a resistor from the timing terminals (either pin 7 or 8) to ground. The current switches route the current from one of the timing pins to the VCO. The current selected depends on the voltage level on the frequency shift keying input pin (pin 9). Therefore, two discrete output frequencies can be independently produced. If pin 9 is open circuited or connected to a bias voltage ≥ 2 V, the current passing through the resistor connected to pin 7 is selected. Similarly, if the voltage level at pin 9 is ≤ 1 V, the current passing through the resistor connected to pin 8 is selected. Thus, the output frequency can be keyed between f_1 and f_2 by simply changing the voltage on pin 9. The formulas for determining the two frequencies of operation are

$$f_1 = \frac{1}{R_1 C} \qquad f_2 = \frac{1}{R_2 C}$$

where R_1 = resistor connected to pin 7
        R_2 = resistor connected to pin 8

The frequency of oscillation is proportional to the total timing current on either pin 7 or 8. Frequency varies linearly with current over a range of current values between 1 μA to 3 mA. The frequency can be controlled by applying a control voltage, V_C, to the selected timing pin, as shown in Figure 2-17. The frequency of oscillation is related to V_C by

$$f = \frac{1}{RC} \left[1 + \frac{R}{R_C} \frac{(1-V_C)}{3} \right] \quad \text{Hz} \tag{2-8}$$

The voltage-to-frequency conversion gain K is given as

$$K = \frac{\Delta f}{\Delta V_C} = \frac{-0.32}{R_C C} \quad \text{Hz/V} \tag{2-9}$$

Monolithic voltage-controlled oscillators.   The XR-2207 is a monolithic voltage-controlled oscillator (VCO) integrated circuit featuring excellent frequency stability and a wide tuning range. The circuit provides simultaneous triangle- and square-wave outputs over a frequency range of from 0.01 Hz to 1 MHz. The XR-2207 is ideally suited for FM, FSK, and sweep or tone generation, as well as for phase-locked-loop applications. The XR-2207 has a typical frequency stability of 20 ppm/°C and can be linearly swept over a

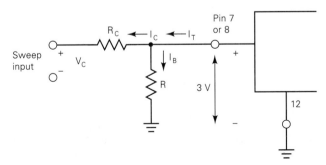

Figure 2-17   Circuit connection for control voltage frequency sweep of the XR-2206.

Large-scale Integration Oscillators

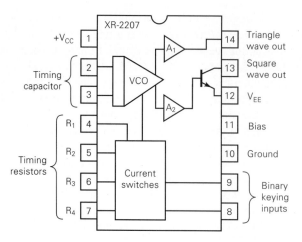

Figure 2-18 Block diagram for the XR-2207 monolithic voltage-controlled oscillator (VCO).

1000 : 1 frequency range with an external control voltage. The duty cycle of the triangular- and square-wave outputs can be varied from 0.1% to 99.9%, generating stable pulse and sawtooth waveforms.

The block diagram for the XR-2207 is shown in Figure 2-18. The circuit is a modified emitter-coupled multivibrator that utilizes four main functional blocks for frequency generation: a voltage-controlled oscillator (VCO), four current switches that are activated by binary keying inputs, and two buffer amplifiers. Two binary input pins (pins 8 and 9) determine which of the four timing currents are channeled to the VCO. These currents are set by resistors to ground from each of the four timing input terminals (pins 4 through 7). The triangular output buffer provides a low-impedance output (10 Ω typical), while the square-wave output is open collector.

Monolithic precision oscillators.   The XR-2209 is a monolithic variable-frequency oscillator circuit featuring excellent temperature stability and a wide linear sweep range. The circuit provides simultaneous triangle- and square-wave outputs, and the frequency is set by an external RC product. The XR-2209 is ideally suited for frequency modulation, voltage-to-frequency conversion and sweep or tone generation, as well as for phase-locked-loop applications when used in conjunction with an appropriate phase comparator.

The block diagram for the XR-2209 precision oscillator is shown in Figure 2-19. The oscillator is comprised of three functional blocks: a variable-frequency oscillator that generates the basic periodic waveforms and two buffer amplifiers for the triangular- and square-wave outputs. The oscillator frequency is set by an external capacitor and timing resistor. The XR-2209 is capable of operating over eight frequency decades from 0.01 Hz to 1 MHz. With no external sweep signal or bias voltage, the frequency of oscillation is simply equal to $1/RC$.

The frequency of operation for the XR-2209 is proportional to the timing current

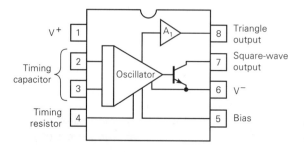

Figure 2-19 Block diagram for the XR-2209 monolithic precision oscillator.

Chap. 2   **Signal Generation**

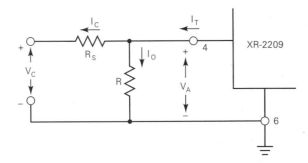

Figure 2-20 Circuit connection for control voltage frequency sweep of the XR-2209.

drawn from the timing pin. This current can be modulated by applying a control voltage, V_C, to the timing pin through series resistor R_S as shown in Figure 2-20. If V_C is negative with respect to the voltage on pin 4, an additional current, I_O, is drawn from the timing pin, causing the total input current to increase, thus increasing the frequency of oscillation. Conversely, if V_C is higher than the voltage on pin 4, the frequency of oscillation is decreased.

PHASE-LOCKED LOOPS

The *phase-locked loop (PLL)* is used extensively in electronic communications for performing modulation, demodulation, frequency generation, and frequency synthesis. PLLs are used in both transmitters and receivers with both analog and digital modulation and with the transmission of digital pulses. Phase-locked loops were first used in 1932 for synchronous detection of radio signals, instrumentation circuits, and space telemetry systems. However, for many years the use of PLLs was avoided because of their large size, necessary complexity, narrow bandwidth, and expense. With the advent of large-scale integration, PLLs now take up little space, are easy to use, and are more reliable. Therefore, PLLs have changed from a specialized design technique to a general-purpose, universal building block with numerous applications. Today, over a dozen different integrated-circuit PLL products are available from several IC manufacturers. Some of these are designated as general-purpose circuits suitable for a multitude of uses, while others are intended or optimized for special applications such as tone detection, stereo decoding, and frequency synthesis.

Essentially, a PLL is a closed-loop feedback control system in which the feedback signal is a frequency rather than simply a voltage. The PLL provides frequency selective tuning and filtering without the need for coils or inductors. The basic phase-locked-loop circuit is shown in Figure 2-21 and consists of four primary blocks: a phase comparator (multiplier), a low-pass filter, a low-gain amplifier (op-amp), and a voltage-controlled oscillator (VCO). With no external input signal, the output voltage, V_{out}, is equal to zero. The VCO operates at a set frequency called its *natural* or *free-running frequency* (f_n), which is set by external resistor (R_t) and capacitor (C_t). If an input signal is applied to the system, the phase comparator compares the phase and frequency of the input signal with the VCO natural frequency and generates an error voltage, $V_d(t)$, that is related to the phase and frequency difference between the two signals. This error voltage is then filtered, amplified, and applied to the input terminal of the VCO. If the input frequency, f_i, is sufficiently close to the VCO natural frequency, f_n, the feedback nature of the PLL causes the VCO to synchronize, or lock, to the incoming signal. Once in lock, the VCO frequency is

Phase-locked Loops   **73**

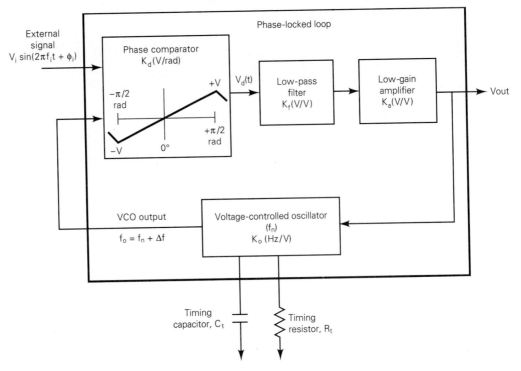

Figure 2-21 Block diagram for phase-locked loop.

identical to the input signal, except for a finite phase difference that is equal to the phase of the incoming signal minus the phase of the VCO output signal.

Lock and Capture Range

Two key parameters of PLLs that indicate their useful frequency range are lock and capture range.

Lock range. *Lock range* is defined as the range of frequencies in the vicinity of the VCO's natural frequency (f_n) over which the PLL can maintain lock with an input signal. This presumes that the PLL was initially locked onto the input signal. Lock range is also known as *tracking range*. It is the range of frequencies over which the PLL will accurately track or follow the input frequency. Lock range increases as the overall loop gain of the PLL is increased (loop gain is discussed in a later section of this chapter). *Hold-in range* is equal to half the lock range (that is, lock range = 2 × hold-in range). The relationship between lock and hold-in range is shown in frequency diagram form in Figure 2-22. The lowest frequency that the PLL will track is called the *lower lock limit* (f_{ll}), and the highest frequency that the PLL will track is called the *upper lock limit* (f_{lu}). The lock range depends on the transfer functions (gains) of the phase comparator, low-gain amplifier, and VCO.

Capture range. *Capture range* is defined as the band of frequencies in the vicinity of f_n where the PLL can establish or acquire lock with an input signal. The capture range is generally between 1.1 and 1.7 times the natural frequency of the VCO. Capture range is also known as *acquisition range*. Capture range is related to the bandwidth of the

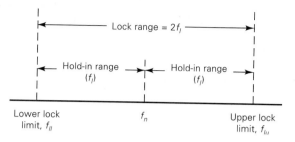

Figure 2-22 PLL lock range.

low-pass filter. The capture range of a PLL decreases as the bandwidth of the filter is reduced. *Pull-in range* is the peak capture range (that is, capture range = 2 × pull-in range). Capture and pull-in range are shown in frequency diagram form in Figure 2-23. The lowest frequency the PLL can lock onto is called the *lower capture limit* (f_{cl}), and the highest frequency the PLL can lock onto is called the *upper capture limit* (f_{cu}).

The capture range is never greater than and is almost always less than the lock range. The relationship among capture, lock, hold-in, and pull-in range is shown in frequency diagram form in Figure 2-24. Note that lock range ≥ capture range and hold-in range ≥ pull-in range.

Voltage-controlled Oscillator

A *voltage-controlled oscillator* (VCO) is an oscillator (more specifically, a free-running multivibrator) with a stable frequency of oscillation that depends on an external bias voltage. The output from a VCO is a frequency, and its input is a bias or control signal that may be a dc or ac voltage. When a dc or slowly changing ac voltage is applied to the VCO input, the output frequency changes or deviates proportionally. Figure 2-25 shows a transfer curve (output frequency-versus-input bias voltage characteristics) for a typical VCO. The output frequency (f_o) with 0-V input bias is the VCO's natural frequency (f_n), which is determined by an external *RC* network, and the change in the output frequency caused by a change in the input voltage is called frequency deviation (Δf). Consequently, $f_o = f_n + \Delta f$,

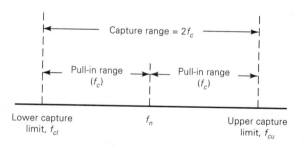

Figure 2-23 PLL capture range.

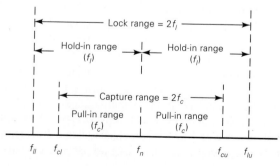

Figure 2-24 PLL capture and lock ranges.

Phase-locked Loops

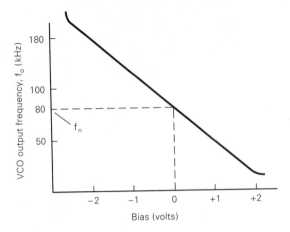

Figure 2-25 Voltage-controlled oscillator input bias voltage-versus-output frequency characteristics.

where f_o = VCO output frequency. For a symmetrical Δf, the natural frequency of the VCO should be centered within the linear portion of the input-versus-output curve. The transfer function for a VCO is

$$K_o = \frac{\Delta f}{\Delta V} \tag{2-10}$$

where K_o = input-versus-output transfer function (hertz-per-volt)
    ΔV = change in the input control voltage (volts)
    Δf = change in the output frequency (hertz)

Phase Comparator

A phase comparator, sometimes called a *phase detector*, is a nonlinear device with two input signals: an externally generated frequency (f_i) and the VCO output frequency (f_o). The output from a phase comparator is the product of the two signals of frequencies f_i and f_o and, therefore, contains their sum and difference frequencies ($f_i \pm f_o$). This is analyzed in more detail later in this chapter. Figure 2-26a shows the schematic diagram for a simple phase comparator. v_o is applied simultaneously to the two halves of input transformer T_1. D_1, R_1, and C_1 make up a half-wave rectifier, as do D_2, R_2, and C_2 (note that $C_1 = C_2$ and $R_1 = R_2$). During the positive alternation of v_0, D_1 and D_2 are forward biased and *on*, charging C_1 and C_2 to equal values but with opposite polarities. Therefore, the average output voltage is $V_{\text{out}} = V_{C1} + (-V_{C2}) = 0$ V. This is shown in Figure 2-26b. During the negative half-cycle of v_o, D_1 and D_2 are reverse biased and *off*. Therefore, C_1 and C_2 discharge equally through R_1 and R_2, respectively, keeping the output voltage equal to 0 V. This is shown in Figure 2-26c. The two half-wave rectifiers produce equal-magnitude, opposite-polarity output voltages. Therefore, the output voltage due to v_o is constant and equal to 0 V. The corresponding input and output waveforms for a square-wave VCO signal are shown in Figure 2-26d.

Circuit operation.   When an external input signal [$v_{\text{in}} = V_i \sin(2\pi f_i t)$] is applied to the phase comparator, its voltage adds to v_o, causing C_1 and C_2 to charge and discharge, producing a proportional change in the output voltage. Figure 2-27a shows the unfiltered output waveform shaded when $f_o = f_i$ and v_o leads v_i by 90°. For the phase comparator to operate properly, v_o must be much larger than v_i. Therefore, D_1 and D_2 are switched *on* only during the positive alternation of v_o and are *off* during the negative alternation. Dur-

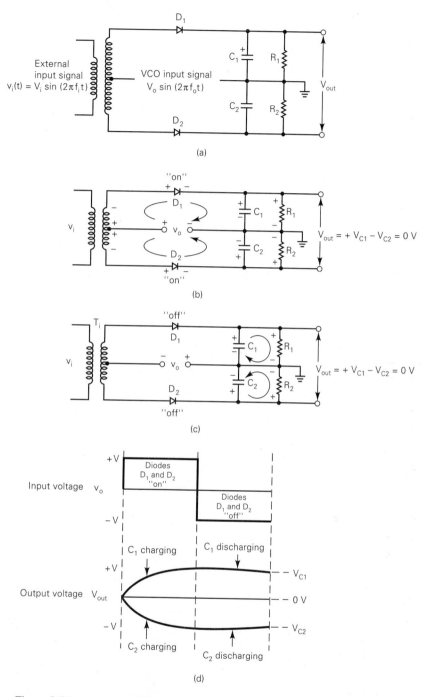

Figure 2-26 Phase comparator: (a) schematic diagram; (b) output voltage due to positive half-cycle of v_o; (c) output voltage do to negative half-cycle of v_o; (d) input and output voltage waveforms.

ing the first half of the *on* time, the voltage applied to $D_1 = v_o - v_i$, and the voltage applied to $D_2 = v_o + v_i$. Therefore, C_1 is discharging while C_2 is charging. During the second half of the *on* time, the voltage applied to $D_1 = v_o + v_i$, the voltage applied to D_2

Phase-locked Loops

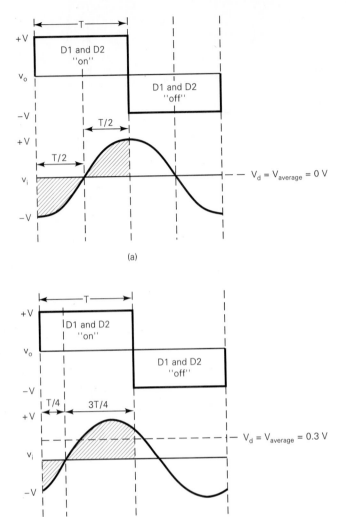

Figure 2-27 Phase comparator output voltage waveforms: (a) v_o leads v_i by 90°; (b) v_o leads v_i by 45°; *(Continued on next page.)*

$= v_o - v_i$, and C_1 is charging while C_2 is discharging. During the *off* time, C_1 and C_2 are neither charging nor discharging. For each complete cycle of v_o, C_1 and C_2 charge and discharge equally and the average output voltage remains at 0 V. Thus, the average value of V_{out} is 0 V when the input and VCO output signals are equal in frequency and 90° out of phase.

Figure 2-27b shows the unfiltered output voltage waveform shaded when v_o leads v_i by 45°. v_i is positive for 75% of the *on* time and negative for the remaining 25%. As a result, the average output voltage for one cycle of v_o is positive and approximately equal to 0.3 V, where V is the peak input voltage. Figure 2-27c shows the unfiltered output waveform when v_o and v_i are in phase. During the entire *on* time, v_1 is positive. Consequently, the output voltage is positive and approximately equal to 0.636 V. Figures 2-27d and e show the unfiltered output waveform when v_o leads v_i by 135° and 180°, respectively. It can be seen that the output voltage goes negative when v_o leads v_i by more than 90° and reaches its maximum value when v_o leads v_i by 180°. In essence, a phase comparator rectifies v_i and integrates it to produce an output voltage that is proportional to the difference in phase between v_o and v_i.

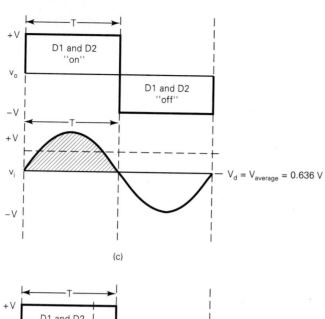

(c)

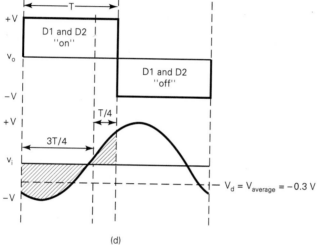

(d)

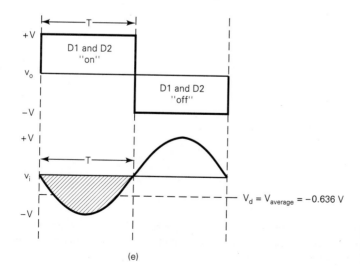

(e)

Figure 2-27 *(continued)* (c) v_o and v_i in phase; (d) v_o leads v_i by 135°; (e) v_o leads v_i by 180°.

Phase-locked Loops

Figure 2-28 shows the output voltage-versus-input phase difference characteristics for the phase comparator shown in Figure 2-26a. Figure 2-28a shows the curve for a square-wave phase comparator. The curve has a triangular shape with a negative slope from 0° to 180°. V_{out} is maximum positive when v_o and v_i are in phase, 0 V when v_o leads v_i by 90°, and maximum negative when v_o leads v_i by 180°. If v_o advances more than 180°, the output voltage become less negative, and if v_o lags behind v_i, the output voltage become less positive. Therefore, the maximum phase difference that the comparator can track is 90° ± 90° or from 0° to 180°. The phase comparator produces an output voltage that is proportional to the difference in phase between v_o and v_i. This phase difference is called the *phase error*. The phase error is expressed mathematically as

$$\theta e = \theta o - \theta i \qquad (2\text{-}11)$$

where θe = phase error (radians)
θo = phase of the VCO output signal voltage (radians)
θi = phase of the external input signal voltage (radians)

The output voltage from the phase comparator is linear for phase errors between 0° and 180° (0 to π radians). Therefore, the transfer function for a square-wave phase comparator for phase errors between 0° and 180° is given as

$$K_d = \frac{V_{out}}{\theta_e} = \frac{2\,v_i}{\pi} \qquad (2\text{-}12)$$

where Kd = transfer function or gain (volts per radian)
V_{out} = phase comparator output voltage (volts)
θe = phase error $(\theta_o - \theta_i)$ (radians)

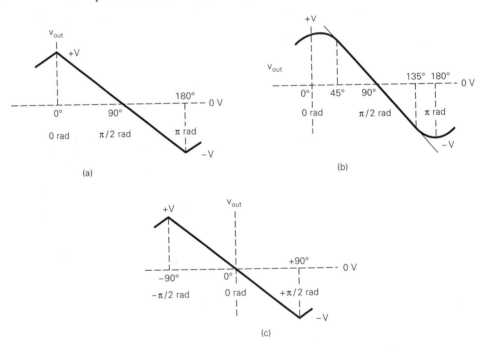

Figure 2-28 Phase comparator output voltage (V_d) versus phase difference ($\emptyset_e$) characteristics: (a) square-wave inputs; (b) sinusoidal inputs; (c) square-wave inputs, phase bias reference.

Chap. 2   **Signal Generation**

$$\pi = 3.14 \text{ radians}$$
$$v_i = \text{peak input signal voltage (volts)}$$

Figure 2-28b shows the output voltage-versus-input phase difference curve for an analog phase comparator with sinusoidal characteristics. The phase error versus output is nearly linear only from 45° to 135°. Therefore, the transfer function is given as

$$K_d = \frac{V_{\text{out}}}{\theta_e} \text{ volts per radian} \tag{2-13}$$

where Kd = transfer function or gain (volts per radian)
   θe = phase error $(\theta_o - \theta_i)$ (radians)
   V_{out} = phase comparator output voltage (volts)

From Figures 2-28a and b, it can be seen that the phase comparator output voltage $V_{\text{out}} = 0$ V when $f_o = f_i$ and v_o and v_i are 90° out of phase. Therefore, if the input frequency (f_i) is initially equal to the VCO's natural frequency (f_n), a 90° phase difference is required to keep the phase comparator output voltage at 0 V and the VCO output frequency equal to its natural frequency $(f_o = f_n)$. This 90° phase difference is equivalent to a bias or offset phase. Generally, the phase bias is considered as the reference phase, which can be deviated $\pm \pi/2$ radians ($\pm$ 90°). Therefore, V_{out} goes from its maximum positive value at $-\pi/2$ radians ($-90°$) and to its maximum negative value at $+\pi/2$ radians ($+90°$). Figure 2-28c shows the phase comparator output voltage-versus-phase error characteristics for square wave inputs with the 90° phase bias as the reference

Figure 2-29a shows the unfiltered output voltage waveform when v_i leads v_o by 90°. Note that the average values 0 V (the same as when v_o lead v_i by 90°). When frequency lock occurs, it is uncertain whether the VCO will lock onto the input frequency with a $+$ or $-90°$ phase difference. Therefore, there is a 180° phase ambiguity in the phase of VCO output frequency. Figure 2-29b shows the output voltage-versus-phase difference characteristics for square-wave inputs when the VCO output frequency equals its natural frequency and it has locked onto the input signal with a $-90°$ phase difference. Note that the opposite voltages occur for the opposite direction phase error, and the slope is positive rather than negative from $-\pi/2$ to $+\pi/2$ radians. When frequency lock occurs, the PLL produces a coherent frequency $(f_o = f_i)$, but the phase of the incoming signal is uncertain (either f_o leads f_i by 90° $\pm$ θ_e, or vice versa.

Loop Operation

For the following explanations, refer to Figure 2-30.

Loop acquisition.   An external input signal $[(V_i \sin(2\pi f_i t + \theta_i)]$ enters the phase comparator and mixes with the VCO output signal (a square wave with fundamental frequency f_o). Initially, the two frequencies are not equal $(f_o \neq f_i)$ and the loop is *unlocked*. Because the phase comparator is a nonlinear device, the input and VCO signals mix and generate cross-product frequencies (that is, sum and difference frequencies). Therefore, the primary output frequencies from the phase comparator are the external input frequency (f_i), the VCO output frequency (f_o), and their sum $(f_i + f_o)$ and difference $(f_i - f_o)$ frequencies.

The low-pass filter (LPF) blocks the two original input frequencies and the sum frequency; thus, the input to the amplifier is simply the difference frequency $(f_i - f_o)$, sometimes called the beat frequency). The beat frequency is amplified and then applied to the

Phase-locked Loops

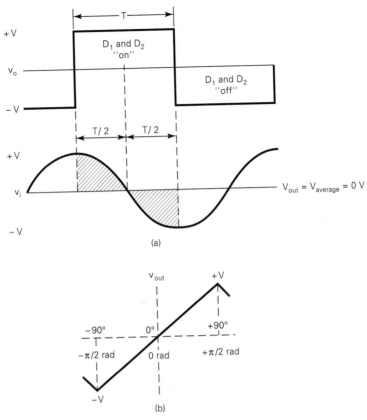

Figure 2-29 Phase comparator output voltage: (a) unfiltered output voltage waveform when v_i leads v_o by 90°; (b) output voltage-versus-phase difference characteristics.

input of the voltage-controlled oscillator, where it deviates the VCO by an amount proportional to its polarity and amplitude. As the VCO output frequency changes, the amplitude and frequency of the beat frequency changes proportionately. Figure 2-30b shows the beat frequency produced when the VCO is swept by the difference frequency (f_d). After several cycles around the loop, the VCO output frequency equals the external input frequency and the loop is said to be locked. Once lock has occurred, the beat frequency at the output of the LPF is 0 Hz (a dc voltage), which is necessary to bias the VCO and keep it locked to the external input frequency. In essence, the phase comparator is a frequency comparator until frequency acquisition (zero beat) is achieved, then it becomes a phase comparator. Once the loop is locked, the difference in phase between the external input and VCO output frequencies is converted to a bias voltage (V_d) in the phase comparator, amplified, and then fed back to the VCO to hold lock. Therefore, it is necessary that a phase error be maintained between the external input signal and the VCO output signal. The change in the VCO frequency required to achieve lock and the time required to achieve lock (*acquisition* or *pull-in time*) for a PLL with no loop filter (loop filters are explained later in this chapter) is approximately equal to $1/K_v$ seconds, where K_v is the open-loop gain of the PLL. Once the loop is locked, any change in the input frequency is seen as a phase error, and the comparator produces a corresponding change in its output voltage, V_d. The change in voltage is amplified and fed back to the VCO to reestablish lock. Thus, the loop dynamically adjusts itself to follow input frequency changes.

Chap. 2   **Signal Generation**

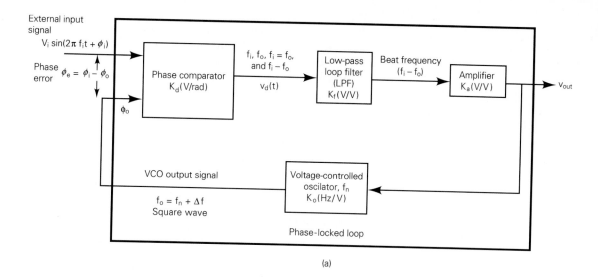

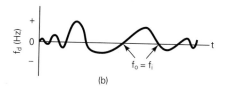

(b)

Figure 2-30 PLL operation: (a) block diagram; (b) beat frequency.

Mathematically, the output from the phase comparator is (considering only the fundamental frequency for V_o and excluding the 90° phase bias)

$$V_d = [V \sin(2\pi f_o t + \theta_o) \times V \sin (2\pi f_i t + \theta_i)]$$

$$= \frac{V}{2} \sin (2\pi f_o t + \theta_o - 2\pi f_i t - \theta_i) - \frac{V}{2} \sin (2\pi f_o t + \theta_o + 2\pi f_i t - \theta_i)$$

where V_d = the phase detector output voltage (volts)
  $V = V_o V_i$ (peak volts)

When $f_o = f_i$,

$$V_d = \frac{V}{2} \sin (\theta_o - \theta_i) \tag{2-14}$$

$$= \frac{V}{2} \sin \theta_e$$

where $\theta_o - \theta_i = \theta_e$ (phase error). θ_e is the phase error required to change the VCO output frequency from f_n to f_i (a change $= \Delta f$) and is often called the *static phase error*.

Loop gain. The *loop gain* for a PLL is simply the product of the individual gains or transfer functions around the loop. In Figure 2-30, the open-loop gain is the product of the phase comparator gain, the low-pass filter gain, the amplifier gain, and the VCO gain. Mathematically, open-loop gain is

$$K_v = K_d K_f K_a K_o \tag{2-15}$$

Phase-locked Loops

where K_v = PLL open-loop gain (hertz per radian or s^{-1})

K_d = phase comparator gain (volts per radian)

K_f = low-pass filter gain (volts per volt)

K_a = amplifier gain (volts per volt)

K_o = VCO gain (hertz per volt)

and K_v = open-loop gain

$$= \frac{(\text{volt})(\text{volt})(\text{volt}) \, (\text{hertz})}{(\text{rad}) \, (\text{volt})(\text{volt}) \, (\text{volts})} = \frac{\text{hertz}}{\text{rad}}$$

or $$= \frac{\text{cycles/s}}{\text{rad}} = \frac{\text{cycles}}{\text{rad-s}} \times \frac{2\pi \, \text{rad}}{\text{cycle}} = 2\pi \quad s^{-1}$$

Expressed in decibels, this gives us

$$K_{v \, (\text{dB})} = 20 \log K_v \qquad (2\text{-}16)$$

From Equations 2-10, 2-13, and 2-16, the following relationships are derived:

$$V_d = (\theta_e)(K_d) \, (\text{volts}) \qquad (2\text{-}17)$$

$$V_{\text{out}} = (V_d)(K_f)(K_a) \, (\text{volts}) \qquad (2\text{-}18)$$

$$\Delta f = (V_{\text{out}})(K_o) \, (\text{hertz}) \qquad (2\text{-}19)$$

As previously stated, the hold-in range for a PLL is the range of input frequencies over which the PLL will remain locked. This presumes that the PLL was initially locked. The hold-in range is limited by the peak-to-peak swing in the phase comparator output voltage (ΔV_d) and depends on the phase comparator, amplifier, and VCO transfer functions. From Figure 2-28c it can be seen that the phase comparator output voltage (V_d) is corrective for $\pm\pi/2$ radians ($\pm90°$). Beyond these limits, the polarity of V_d reverses and actually chases the VCO frequency away from the external input frequency. Therefore, the maximum phase error (θ_e) that is allowed is $\pm\pi/2$ radians and the maximum phase comparator output voltage is

$$\pm V_{d(\text{max})} = [\theta_{e(\text{max})}](K_d) \qquad (2\text{-}20a)$$

$$= \pm \frac{\pi}{2} \, \text{rad}(K_d) \qquad (2\text{-}20b)$$

where $\pm V_{d(\text{max})}$ = maximum peak change at the phase comparator output voltage

K_d = phase comparator transfer function

Consequently, the maximum change in the VCO output frequency is

$$\pm \Delta f_{\text{max}} = \pm \left(\frac{\pi}{2} \, \text{rad} \right)(K_d)(K_f)(K_a)(K_o) \qquad (2\text{-}21a)$$

where $\pm \Delta f_{\text{max}}$ = hold-in range (maximum peak change in VCO output frequency)

and $$\pm \Delta f_{\text{max}} = V_{d(\text{max})}K_o \qquad (2\text{-}21b)$$

where K_o = VCO transfer function. Substituting into Equation 2-15,

$$\pm \Delta f_{\text{max}} = \pm \left(\frac{\pi}{2} \, \text{rad} \right)(K_v) \qquad (2\text{-}22)$$

EXAMPLE 2-2

For the PLL shown in Figure 2-30, a VCO natural frequency $f_n = 200$ kHz, an external input frequency $f_i = 210$ kHz, and the transfer functions $K_d = 0.2$ V/rad, $K_f = 1$, $K_a = 5$, and $K_o = 20$ kHz/V; determine:

(a) PLL open-loop gain
(b) Change in VCO frequency necessary to achieve lock (Δf)
(c) PLL output voltage (V_{out})
(d) Phase detector output voltage (V_d)
(e) Static phase error (θ_e)
(f) Hold-in range (Δf_{max})

Solution (a) From Equation 2-15,

$$K_v = \frac{0.2 \text{ V}}{\text{rad}} \frac{1 \text{ V}}{\text{V}} \frac{5 \text{ V}}{\text{V}} \frac{20 \text{ kHz}}{\text{V}} = \frac{20 \text{ kHz}}{\text{rad}}$$

$$\frac{20 \text{ kHz}}{\text{rad}} = \frac{20 \text{ kilocycles}}{\text{rad-s}} \times \frac{2\pi \text{ rads}}{\text{cycle}} = 125,600 \text{ s}^{-1}$$

$$K_v(\text{dB}) = 20 \log 125.6 \text{ ks}^{-1} = 102 \text{ dB}$$

(b) $\Delta f = f_i - f_n = 210$ kHz $- 200$ kHz $= 10$ kHz
(c) Rearranging Equation 2-10 gives us

$$V_{out} = \frac{\Delta f}{K_o} = \frac{10 \text{ kHz}}{20 \text{ kHz/V}} = 0.5 \text{ V}$$

(d) $V_d = \dfrac{V_{out}}{(K_f)(K_a)} = \dfrac{0.5}{(1)(5)} = 0.1$ V

(e) Rearranging Equation 2-13 gives us

$$\theta_e = \frac{V_d}{K_d} = \frac{0.1 \text{ V}}{0.2 \text{ V/rad}} = 0.5 \text{ rad} \quad \text{or} \quad 28.65°$$

(f) Substituting into Equation 2-22 yields

$$\Delta f_{max} = \frac{(\pm \pi/2 \text{ rad})(20 \text{ kHz})}{\text{rad}} = \pm 31.4 \text{ kHz}$$

Lock range is the range of frequencies over which the loop will stay locked onto the external input signal once lock has been established. Lock range is expressed in rad/s and is related to the open-loop gain K_v as

$$\text{lock range} = 2\Delta f_{max} = \pi K_v$$

where $K_v = (K_d)(K_f)(K_o)$ for a simple loop with a LPF, phase comparator, and VCO

or $\quad K_v = (K_d)(K_f)(K_a)(K_o)$ for a loop with an amplifier

The lock range in radians per second is π times the dc loop voltage gain and is independent of the LPF response. The capture range depends on the lock range and on the LPF response, so it changes with the type of filter used and with the filter cutoff frequency. For a simple single-pole RC LPF, it is given by

$$\text{capture range} = \frac{2 \sqrt{(\Delta f_{min})}}{RC}$$

Phase-locked Loops

Closed-loop frequency response. The *closed-loop frequency response* for an *uncompensated* (unfiltered) PLL is shown in Figure 2-31. The open-loop gain of a PLL for a frequency of 1 rad/sec $= K_v$. The frequency response shown in Figure 2-31 is for the circuit and PLL parameters given in Example 2-2. It can be seen that the open-loop gain (K_v) at 1 rad/sec $= 102$ dB, and the open-loop gain equals 0 dB at the loop cutoff frequency (ω_v). Also, the closed-loop gain is unity up to ω_v, where it drops to -3 dB and continues to roll off at 6 dB/octave (20 dB/decade). Also, $\omega_v = K_v = 125.6$ krad/s, which is the single-sided bandwidth of the uncompensated closed loop.

From Figure 2-31 it can be seen that the frequency response for an uncompensated PLL is identical to that of a single-pole (first-order) low-pass filter with a break frequency of $\omega_c = 1$ rad/s. In essence, a PLL is a low-pass tracking filter that follows input frequency changes that fall within a bandwidth equal to $\pm K_v$.

If additional bandlimiting is required, a low-pass filter can be added between the phase comparator and amplifier as shown in Figure 2-30. This filter can be either a single- or multiple-pole filter. Figure 2-32 shows the loop frequency response for a simple single-pole *RC* filter with a cutoff frequency of $\omega_c = 100$ rad/s. The frequency response follows that of Figure 2-31 up to the loop filter break frequency; then the response rolls off at 12 dB/octave (40 dB/decade). As a result, the compensated unity-gain frequency (ω'_c) is reduced to approximately ± 3.5 krad/s.

EXAMPLE 2-3

Plot the frequency response for a PLL with a loop gain of $K_v = 15$ kHz/rad ($\omega_v = 94.2$ krad/s). On the same log paper, plot the response with the addition of a single-pole loop filter with a cutoff frequency $\omega_c = 1.59$ Hz/rad (10 rad/s) and a two-pole loop filter with the same cutoff frequency.

Solution The specified frequency response curves are shown in Figure 2-33. It can be seen that with the single-pole filter the compensated loop response $= \omega'_v = 1$ krad/s and with the two-pole filter, $\omega''_v = 200$ rad/s.

The bandwidth of the loop filter (or for that matter, whether a loop filter is needed) depends on the specific application.

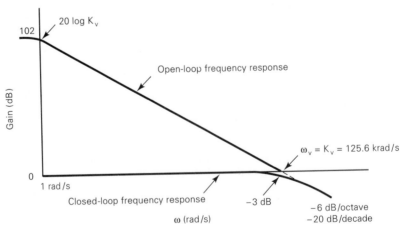

Figure 2-31 Frequency response for an uncompensated phase-locked loop.

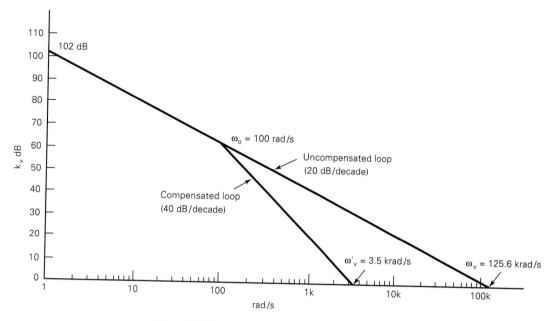

Figure 2-32 PLL frequency response for a single-pole RC filter.

Integrated-circuit Precision Phase-locked Loop

The XR-215 is an ultrastable monolithic phase-locked-loop system designed by EXAR Corporation for a wide variety of applications in both analog and digital communications systems. It is especially well suited for FM or FSK demodulation, frequency synthesis, and tracking filter applications. The XR-215 can operate over a relatively wide frequency range from 0.5 Hz to 35 MHz and can accommodate analog input voltages between

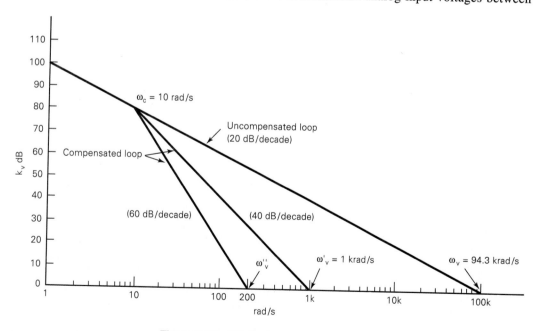

Figure 2-33 PLL frequency response for Example 2-3.

300 μV and 3 V. The XR-215 can interface with conventional DTL, TTL, and ECL logic families.

The block diagram for the XR-215 is shown in Figure 2-34 and consists of three main sections: a balanced phase comparator, a highly stable voltage-controlled oscillator (VCO), and a high-speed operational amplifier (op-amp). The phase comparator outputs are internally connected to the VCO inputs and to the noninverting amplifier of the op-amp. A self-contained PLL system is formed by simply ac coupling the VCO output to either of the phase comparator inputs and adding a low-pass filter to the phase comparator output terminals.

The VCO section has frequency sweep, on-off keying, sync, and digital programming capabilities. Its frequency is highly stable and determined by a single external capacitor. The op-amp can be used for audio preamplification in FM detector applications or as a high-speed sense amplifier (or comparator) in FSK demodulation.

Phase comparator. One input to the phase comparator (pin 4) is connected to the external input signal, and the second input (pin 6) is ac coupled to the VCO output pin. The low-frequency ac (or dc) voltage across the phase comparator output pins (pins 2 and 3) is proportional to the phase difference between the two signals at the phase comparator inputs. The phase comparator outputs are internally connected to the VCO control terminals. One output (pin 3) is internally connected to the operational amplifier. The low-pass filter is achieved by connecting an RC network to the phase comparator outputs as shown in Figure 2-35. A typical transfer function (conversion gain) for the phase detector is 2 V/rad for input voltages ≥50 mV.

Voltage-controlled oscillator (VCO). The VCO free-running or natural frequency (f_n) is inversely proportional to the capacitance of a timing capacitor (C_o) connected between pins 13 and 14. The VCO produces an output signal with approximately 2.5 V_{p-p} amplitude at pin 15 with a dc output level of approximately 2 V. The VCO can be swept over a broad range of output frequencies by applying an analog sweep voltage (V_s) to pin 12 as shown in Figure 2-36. Typical sweep characteristics are also shown. The frequency range of the XR-215 can be extended by connecting an external resistor between pins 9 and 10. The VCO output frequency is proportional to the sum of currents I_1 and I_2

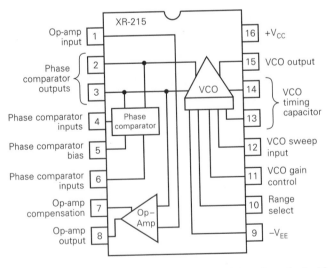

Figure 2-34 Block diagram for the XR-215 monolithic phase-locked loop.

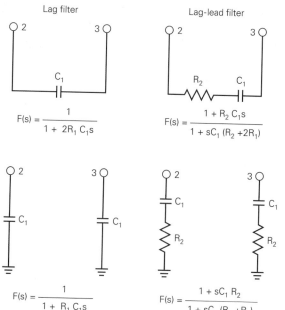

Lag filter

$$F(s) = \frac{1}{1 + 2R_1 C_1 s}$$

Lag-lead filter

$$F(s) = \frac{1 + R_2 C_1 s}{1 + sC_1 (R_2 + 2R_1)}$$

$$F(s) = \frac{1}{1 + R_1 C_1 s}$$

$$F(s) = \frac{1 + sC_1 R_2}{1 + sC_1 (R_1 + R_2)}$$

Figure 2-35 XR-215 low-pass filter connections.

flowing through two internal transistors. Current I_1 is set internally, whereas I_2 is set by an external resistor, R_x. Thus, for any value of C_o, the VCO free-running frequency can be expressed as

$$f_n = f\left(1 + \frac{0.6}{R_x}\right) \tag{2-24}$$

where f_n = VCO free-running frequency (hertz)
$\quad\quad f$ = VCO output frequency with pin 10 open circuited (hertz)
$\quad\quad R_x$ = external resistance (kilohms)

or

$$f_n = \frac{200}{C_o}\left(1 + \frac{0.6}{R_x}\right) \tag{2-25}$$

where C_o = external timing capacitor (microfarads)
$\quad\quad R_x$ = external resistance (kilohms)

The VCO voltage-to-frequency conversion gain (transfer function) is determined by the choice of timing capacitor C_o and gain control resistor R_o connected externally across pins 11 and 12. Mathematically, the transfer function is expressed as

$$K_o = \frac{700}{C_o R_o} \quad \text{(radians/second)/volt} \tag{2-26}$$

where K_o = VCO conversion gain (radians per second per volt)
$\quad\quad C_o$ = capacitance (microfarads)
$\quad\quad R_o$ = resistance (kilohms)

Operational amplifier.   Pin 1 is the external connection to the inverting input of the operational amplifier section and is normally connected to pin 2 through a 10-kΩ resistor. The noninverting input is internally connected to one of the phase detector outputs.

Phase-locked Loops

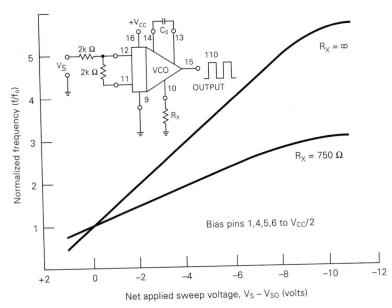

Figure 2-36 Typical frequency sweep characteristics as a function of applied sweep voltage.

Pin 8 is used for the output terminal for FM or FSK demodulation. The amplifier voltage gain is determined by the resistance of feedback resistor R_f connected between pins 1 and 8. Typical frequency response characteristics for the amplifier are shown in Figure 2-37.

The voltage gain of the op-amp section is determined by feedback resistors R_f and R_p between pins (8 and 1) and (2 and 1), respectively, and stated mathematically as

$$A_v = \frac{-R_f}{R_1 + R_p} \qquad (2\text{-}27)$$

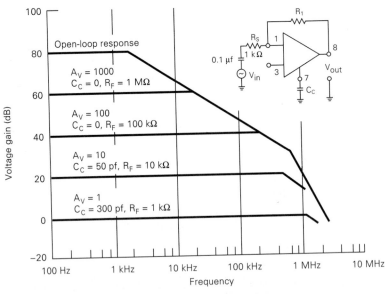

Figure 2-37 XR-215 Operational-amplifier frequency response.

     Chap. 2   **Signal Generation**

where A_v = voltage gain (volts per volt)
R_f = feedback resistor (ohms)
R_p = external resistor between pins 1 and 2 (ohms)
R_1 = internal 6-kΩ impedance at pin 2

Lock range. Lock range is the range of frequencies in the vicinity of the VCO's natural frequency over which the PLL can maintain lock with an external input signal. For the XR-215, if saturation or limiting does not occur, the lock range is equal to the open-loop gain or

$$\Delta\omega_L = K_v = (K_d)(K_o) \tag{2-28}$$

where $\Delta\omega_L$ = lock range (radians per second)
K_v = open-loop gain (second^{-1})
K_d = phase detector conversion gain (volts per radian)
K_o = VCO conversion gain (radians per second per volt)

Capture range. Capture range is the range of frequencies in the vicinity of the VCO's natural frequency where the PLL can establish or acquire lock with an input signal. For the XR-215, it can be approximated by a parametric equation of the form

$$\Delta\omega_C = \Delta\omega_L|F(j\Delta\omega_C)| \tag{2-29}$$

where $\Delta\omega_C$ = capture range (radians/second)
$\Delta\omega_L$ = lock range (radians/second)
$|F(j\Delta\omega_C)|$ = low-pass filter magnitude response at $\omega = \Delta\omega_C$

FREQUENCY SYNTHESIZERS

Synthesize means to form an entity by combining parts or elements. A *frequency synthesizer* is used to generate many output frequencies through the addition, subtraction, multiplication, and division of a smaller number of fixed frequency sources. Simply stated, a frequency synthesizer is a crystal-controlled variable-frequency generator. The objective of a synthesizer is twofold. It should produce as many frequencies as possible from a minimum number of sources, and each frequency should be as accurate and stable as every other frequency. The ideal frequency synthesizer can generate hundreds or even thousands of different frequencies from a single-crystal oscillator. A frequency synthesizer may be capable of simultaneously generating more than one output frequency, with each frequency being synchronous to a single reference or master oscillator frequency. Frequency synthesizers are used extensively in test and measurement equipment (audio and RF signal generators), tone-generating equipment (Touch-Tone), remote-control units (electronic tuners), multichannel communications systems (telephony), and music synthesizers.

Essentially, there are two methods of frequency synthesis: direct and indirect. With *direct frequency synthesis*, multiple output frequencies are generated by mixing the outputs from two or more crystal-controlled frequency sources or by dividing or multiplying the output frequency from a single-crystal oscillator. With *indirect frequency synthesis*, a feedback-controlled divider/multiplier (such as a PLL) is used to generate multiple output frequencies. Indirect frequency synthesis is slower and more susceptible to noise; however, it is less expensive and requires fewer and less complicated filters than direct frequency synthesis.

Direct Frequency Synthesizers

Multiple-crystal frequency synthesizer. Figure 2-38 shows a block diagram for a *multiple-crystal frequency synthesizer* that uses nonlinear mixing (heterodyning) and filtering to produce 128 different frequencies from 20 crystals and two oscillator modules. For the crystal values shown, a range of frequencies from 510 to 1790 kHz in 10-kHz steps is synthesized. A synthesizer such as this can be used to generate the carrier frequencies for the 106 AM broadcast-band stations (540 to 1600 kHz). For the switch positions shown, the 160- and 700-kHz oscillators are selected, and the outputs from the balanced mixer are their sum and difference frequencies (700 kHz ± 160 kHz = 540 and 860 kHz). The output filter is tuned to 540 kHz, which is the carrier frequency for channel 1. To generate the carrier frequency for channel 106, the 100-kHz crystal is selected with either the 1700-kHz (difference) or 1500-kHz (sum) crystal. The minimum frequency separation between output frequencies for a synthesizer is called *resolution*. The resolution for the synthesizer shown in Figure 2-38 is 10 kHz.

Single-crystal frequency synthesizer. Figure 2-39 shows a block diagram for a *single-crystal frequency synthesizer* that again uses frequency addition, subtraction, multiplication, and division to generate frequencies (in 1-Hz steps) from 1 to 999,999 Hz. A 100-kHz crystal is the source for the master oscillator from which all frequencies are derived.

The master oscillator frequency is a base frequency that is repeatedly divided by 10 to generate five additional subbase frequencies (10 kHz, 1 kHz, 100 Hz, 10 Hz, and 1 Hz). Each subbase frequency is fed to a separate harmonic generator (frequency multiplier), which consists of a nonlinear amplifier with a tunable filter. The filter is tunable to each of

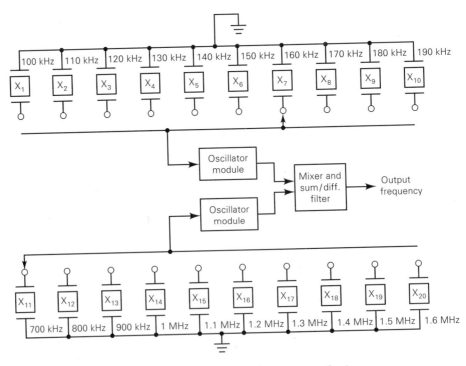

Figure 2-38 Multiple-crystal frequency synthesizer.

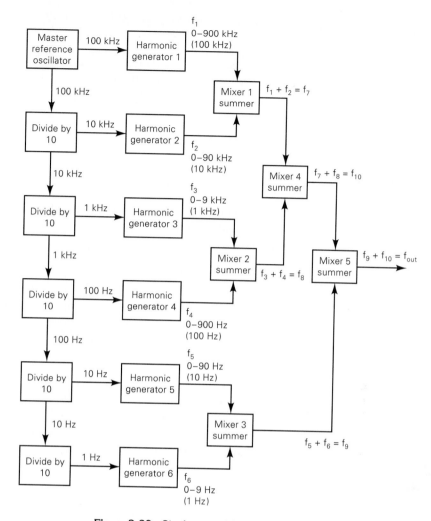

Figure 2-39 Single-crystal frequency synthesizer.

the first nine harmonics of its base frequency. Therefore, the possible output frequencies for harmonic generator 1 are 100 to 900 kHz in 100-kHz steps; for harmonic generator 2, 10 to 90 kHz in 10-kHz steps; and so on. The resolution for the synthesizer is determined by how many times the master crystal oscillator frequency is divided. For the synthesizer shown in Figure 2-39, the resolution is 1 Hz. The mixers used are balanced modulators with output filters that are tuned to the sum of the two input frequencies. For example, the harmonics shown selected in Table 2-1 produce a 246,313-Hz output frequency. Table 2-1 lists the selector switch positions for each harmonic generator and the input and output frequencies from each mixer. It can be seen that the five mixers simply sum the output frequencies from the six harmonic generators with three levels of mixing (adding).

Indirect Frequency Synthesizers

Phase-locked-loop frequency synthesizers. In recent years, PLL frequency synthesizers have rapidly become the most popular method for frequency synthesis. Figure 2-40 shows a block diagram for a simple *single-loop* PLL frequency synthesizer. The sta-

TABLE 2-1 SWITCH POSITIONS AND HARMONICS

Harmonic Generator	Output Frequency	Mixer	Output Frequency
1	200 kHz	1	240 kHz
2	40 kHz		
3	6 kHz	2	6.3 kHz
4	300 kHz		
5	10 kHz	3	13 Hz
6	3 Hz		
		4	246.3 kHz
		5	246.313 kHz

ble frequency reference is a crystal-controlled oscillator. The range of frequencies generated and the resolution depend on the divider network and the open-loop gain. The frequency divider is a divide-by-n circuit, where n is any integer number. The simplest form of divider circuit is a programmable digital *up–down counter* with an output frequency of $f_c = f_o/n$, where f_o = the VCO output frequency. With this arrangement, once lock has occurred, $f_c = f_{ref}$, and the VCO and synthesizer output frequency $f_o = nf_{ref}$. Thus, the synthesizer is essentially a times-n frequency multiplier. The frequency divider reduces the open-loop gain by a factor of n. Consequently, the other circuits around the loop must have relatively high gains. The open-loop gain for the frequency synthesizer shown in Figure 2-40 is

$$K_v = \frac{(K_d)(K_a)(K_o)}{n} \qquad (2\text{-}30a)$$

From Equation 2-30a, it can be seen that as n changes, the open-loop gain changes inversely proportionally. A way to remedy this problem is to program the amplifier gain as well as the divider ratio. Thus, the open-loop gain is

$$K_v = \frac{n(K_d)(K_a)(K_o)}{n} = (K_d)(K_a)(K_o) \qquad (2\text{-}30b)$$

For the reference frequency and divider circuit shown in Figure 2-40, the range of output frequencies is

$$f_o = nf_{ref}$$
$$= f_{ref} \text{ to } 10f_{ref}$$
$$= 1 \text{ to } 10 \text{ MHz}$$

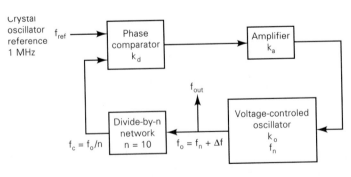

Figure 2-40 Single-loop PLL frequency synthesizer.

Figure 2-41 shows a block diagram for a CB (citizens band) *transceiver* (*transmitter–receiver*) that uses three crystal oscillators and a PLL frequency synthesizer to generate the 23 RF (radio frequency) and two IF (intermediate frequency) carrier frequencies. The divide-by-1024 network is used to improve the resolution of the PLL to 10 kHz and, at the same time, provide the 10.24-MHz beat frequency required to generate the 455-kHz second IF. The first IF is the same for all 23 channels (10.695 MHz). The programmable divider is controlled by a binary code generated in the channel selector switch. The VCO output frequency (f_o) varies from 37.66 MHz (channel 1) to 37.94 MHz (channel 23) and mixes with the receive crystal oscillator frequency (36.38 MHz) to generate a difference frequency $f_1 = 1.28$ to 1.56 MHz, which is the input frequency to the programmable divider. The programmable divider is programmed for values of n from 128 for channel 1 to 156 for channel 23. Consequently, $f_2 = 10$ kHz for all 23 channels. f_2 and the crystal reference frequency (f_3) are compared in the phase comparator. The phase comparator output voltage is used to tune the VCO and lock it onto the crystal reference frequency. The VCO output frequency is also mixed with the received RF carrier to generate a 10.695-MHz difference frequency, which is the first intermediate frequency (1st IF). To generate the 23 transmit RF carrier frequencies, the VCO output frequency is mixed with the output frequency from crystal oscillator 3 (10.695 MHz). The output of the transmit mixer is tuned to their difference frequency. Table 2-2 lists the 23 channel carrier frequencies and their corresponding values for n and f_1.

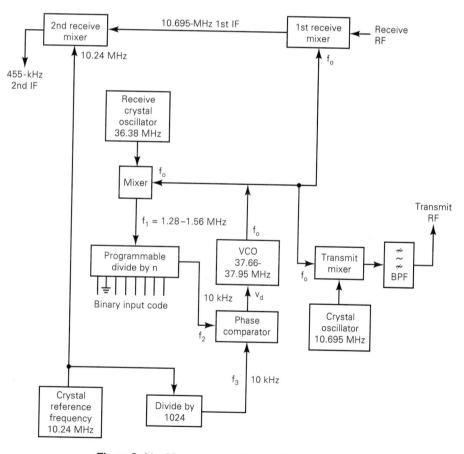

Figure 2-41 23-channel synthesized CB transceiver.

Frequency Synthesizers

TABLE 2-2

Channel	Frequency (MHz)	VCO Frequency (MHz)	n	f_1 (MHz)
1	26.965	37.66	128	1.28
2	26.975	37.67	129	1.29
3	26.985	37.68	130	1.30
4	27.005	37.70	132	1.32
5	27.015	37.71	133	1.33
6	27.025	37.72	134	1.34
7	27.035	37.73	135	1.35
8	27.055	37.75	137	1.37
9	27.065	37.76	138	1.38
10	27.075	37.77	139	1.39
11	27.085	37.78	140	1.40
12	27.105	37.80	142	1.42
13	27.115	37.81	143	1.43
14	27.125	37.82	144	1.44
15	27.135	37.83	145	1.45
16	27.155	37.85	147	1.47
17	27.165	37.86	148	1.48
18	27.175	37.87	149	1.49
19	27.185	37.88	150	1.50
20	27.205	37.90	152	1.52
21	27.215	37.91	153	1.53
22	27.225	37.92	154	1.54
23	27.245	37.94	156	1.56

Prescaled frequency synthesizer.   Figure 2-42 shows the block diagram for a frequency synthesizer that uses a phase-locked loop and a *prescaler* to achieve fractional division. Prescaling is also necessary for generating frequencies greater than 100 MHz because programmable counters are not available that operate efficiently at such high frequencies. The synthesizer shown in Figure 2-42 uses a *two-modulus* prescaler. The

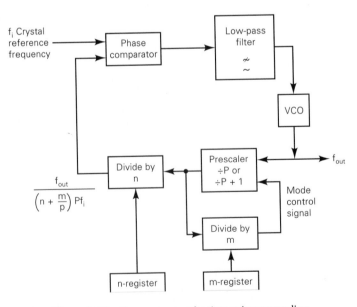

Figure 2-42 Frequency synthesizer using prescaling.

Chap. 2   **Signal Generation**

prescaler has two modes of operation. One mode provides an output for every input pulse (P), and the other mode provides an output for every $P + 1$ input pulse. Whenever the m register contains a nonzero number, the prescaler counts in the $P + 1$ mode. Consequently, once the m and n registers have been initially loaded, the prescaler will count down $(P + 1)m$ times until the m counter goes to zero, the prescaler operates in the P mode, and the n counter counts down $(n - m)$ times. At this time, both the m and n counters are reset to their initial values, which have been stored in the m and n registers, respectively, and the process repeats. Mathematically, the synthesizer output frequency f_o is

$$f_o = \left(n + \frac{m}{P} \right) P f_i \tag{2-31}$$

Integrated-circuit prescalers.   Advanced ECL (emitter-coupled logic) integrated-circuit dual (divide by 128/129 or 64/65) and triple (divide by 64/65/72) modulus prescalers are now available that operate at frequencies from 1 Hz to 1.3 GHz. These integrated-circuit prescalers feature small size, low-voltage operation, low current consumption, and simplicity. Integrated-circuit prescalers are ideally suited for cellular and cordless telephones, RF LANs (local area networks), test and measurement equipment, military radio systems, VHF/UHF mobile radios, and VHF/UHF hand-held radios.

Figure 2-43 shows the block diagram for the NE/SA701 prescaler manufactured by Signetics Company. The NE701 is an advanced dual-modulus (divide by 128/129 or 64/65) low-power, ECL prescaler. It will operate with a minimum supply voltage of 2.5 V and has a maximum current drain of 2.8 mA, allowing application in battery-operated low-power equipment. The maximum input signal frequency is 1.2 GHz for cellular and other land mobile applications. The circuit is implemented in ECL technology on the HS4+ process. The circuit is available in an 8-pin SO package.

The NE701 comprises a frequency divider implemented by using a divide by 4 or 5 synchronous prescaler followed by a fixed five-stage synchronous counter. The normal operating mode is for the SW (modulus set switch) input to be low and the MC (modulus control) input to be high, in which case it functions as a divide-by-128 counter. For divide-by-129 operation, the MC input is forced low, causing the prescaler to switch into divide-by-5

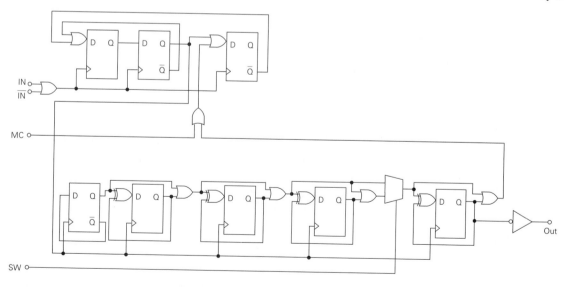

Figure 2-43   Block diagram of the NE/SA701 ECL prescaler.

Frequency Synthesizers

operation for the last cycle of the synchronous counter. Similarly, for divide-by-64 and 65, the NE701 will generate those respective moduli with the SW signal forced high, in which the fourth stage of the synchronous divider is bypassed. With SW open circuited, the divide-by-128/129 mode is selected, and with SW connected to V_{CC}, divide by 64/65 is selected.

Integrated-circuit radio-tuning PLL frequency synthesizer. Figure 2-44 shows the block diagram for the Signetics TSA6057/T radio-tuning PLL frequency synthesizer. The TSA6057 is a bipolar single-chip frequency synthesizer manufactured in SUBILO-N technology (components laterally separated by oxide). It performs all the tuning functions of a PLL radio-tuning system. The IC is designed for applications in all types of radio receivers and has the following features:

1. Separate input amplifiers for the AM and FM VCO signals.
2. On-chip, high input sensitivity AM (3:4) and FM (15:16) prescalers.
3. High-speed tuning due to a powerful digital memory phase detector.
4. On-chip high-performance one-input (two-output) tuning voltage amplifier. One output is connected to the external AM loop filter and the other output to the external FM loop filter.
5. On-chip two-level current amplifier that consists of a 5- and 450-μA current source. This allows adjustment of the loop gain, thus providing high-current, high-speed tuning, and low-current stable tuning.
6. One reference oscillator (4 MHz) for both AM and FM followed by a reference counter. The reference frequency can be 1, 10, or 25 kHz and is applied to the digital memory phase detector. The reference counter also outputs a 40-kHz reference frequency to pin 9 for co-operation with the FM/IF system.
7. Oscillator frequency ranges of 512 kHz to 30 MHz and 30 MHz to 150 MHz.

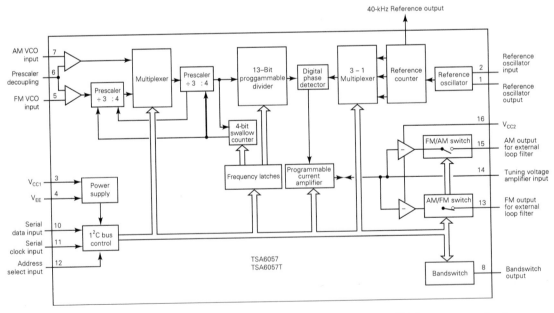

Figure 2-44 Block diagram of the TSA6057 radio tuning PLL frequency synthesizer.

QUESTIONS

2-1. Define *oscillate* and *oscillator*.

2-2. Describe the following terms: *self-sustaining*; *repetitive*; *free-running*, *one-shot*.

2-3. Describe the regenerative process necessary for self-sustained oscillations to occur.

2-4. List and describe the four requirements for a *feedback oscillator* to work.

2-5. What is meant by the terms *positive* and *negative feedback*?

2-6. Define *open-* and *closed-loop gain*.

2-7. List the four most common oscillator configurations.

2-8. Describe the operation of a Wien-bridge oscillator.

2-9. Describe oscillator action for an *LC* tank circuit.

2-10. What is meant by a *damped oscillation*? What causes it to occur?

2-11. Describe the operation of a Hartley oscillator; a Colpitts oscillator.

2-12. Define *frequency stability*.

2-13. List several factors that affect the frequency stability of an oscillator.

2-14. Describe the *piezoelectric effect*.

2-15. What is meant by the term *crystal cut*? List and describe several crystal cuts and contrast their stabilities.

2-16. Describe how an overtone crystal oscillator works.

2-17. What is the advantage of an overtone crystal oscillator over a conventional crystal oscillator?

2-18. What is meant by *positive temperature coefficient*; *negative temperature coefficient*?

2-19. What is meant by a *zero coefficient* crystal?

2-20. Sketch the electrical equivalent circuit for a crystal and describe the various components and their mechanical counterparts.

2-21. Which crystal oscillator configuration has the best stability?

2-22. Which crystal oscillator configuration is the least expensive and most adaptable to digital interfacing?

2-23. Describe a crystal oscillator module.

2-24. What is the predominant advantage of crystal oscillators over *LC* tank-circuit oscillators?

2-25. Describe the operation of a varactor diode.

2-26. Describe a phase-locked loop.

2-27. What types of LSI waveform generators are available?

2-28. Describe the basic operation of an integrated-circuit waveform generator.

2-29. List the advantages of a monolithic function generator.

2-30. List the advantages of a monolithic voltage-controlled oscillator.

2-31. Briefly describe the operation of a monolithic precision oscillator.

2-32. List the advantages of an integrated-circuit PLL over a discrete PLL.

2-33. Describe the operation of a voltage-controlled oscillator.

2-34. Describe the operation of a phase detector.

2-35. Describe how loop acquisition is accomplished with a PLL from an initial unlocked condition until frequency lock is achieved.

2-36. Define the following terms: *beat frequency*; *zero beat*; *acquisition time*; *open-loop gain*.

2-37. Contrast the following terms and show how they relate to each other: *capture range*; *pull-in range*; *closed-loop gain*; *hold-in range*; *tracking range*; *lock range*.

2-38. Define the following terms: *uncompensated PLL*; *loop cutoff frequency*; *tracking filter*.

2-39. Define synthesize. What is a frequency synthesizer?

2-40. Describe direct and indirect frequency synthesis.

2-41. What is meant by the resolution of a frequency synthesizer?

2-42. What are some advantages of integrated-circuit prescalers and frequency synthesizers over conventional nonintegrated-circuit equivalents?

PROBLEMS

2-1. For a 20-MHz crystal with a negative temperature coefficient of $k = -8$ Hz/MHz/°C, determine the frequency of operation for the following temperature changes:
(a) Increase of 10°C. (b) Increase of 20°C. (c) Decrease of 20°C.

2-2. For the Wien-bridge oscillator shown in Figure 2-3 and the following component values, determine the frequency of oscillation: $R_1 = R_2 = 1$ kΩ; $C_1 = C_2 = 100$ pF.

2-3. For the Hartley oscillator shown in Figure 2-5a and the following component values, determine the frequency of oscillation: $L_{1a} = L_{1b} = 50$ μH; $C_1 = 0.01$ μF.

2-4. For the Colpitts oscillator shown in Figure 2-6a, and the following component values, determine the frequency of oscillation: $C_{1a} = C_{1b} = 0.01$ μF; $L_1 = 100$ μH.

2-5. Determine the capacitance for a varactor diode with the following values: $C = 0.005$ μF; $V_r = -2$ V.

2-6. For the VCO input-versus-output characteristic curve shown, determine:
(a) Frequency of operation for a −2-V input signal.
(b) Frequency deviation for a ±2-Vp input signal.
(c) Transfer function, K_o, for the linear portion of the curve (−3 to +3 V).

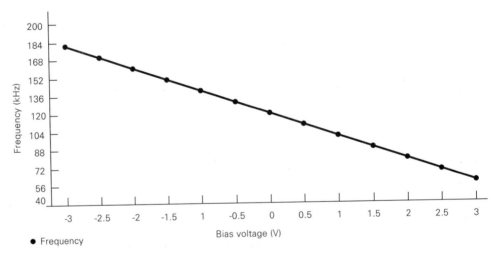

● Frequency

2-7. For the output voltage-versus-phase difference (θ_e) characteristic curve shown, determine:
(a) Output voltage for a −45° phase difference.
(b) Output voltage for a +60° phase difference.
(c) Maximum peak output voltage.
(d) Transfer function, K_d.

Chap. 2   Signal Generation

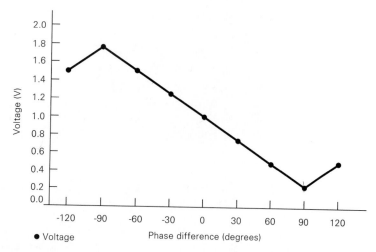

● Voltage   Phase difference (degrees)

2-8. For the PLL shown in Figure 2-30, a VCO natural frequency of $f_n = 150$ kHz, an input frequency of $f_i = 160$ kHz, and the circuit gains $K_d = 0.2$ V/rad, $K_f = 1$, $K_a = 4$, and $K_o = 15$ kHz/V, determine:

(a) Open-loop gain, K_v. (b) Δf. (c) V_{out}.
(d) V_d. (e) θ_e. (f) Hold-in range, Δf_{max}.

2-9. Plot the frequency response for a PLL with an open-loop gain of $K_v = 20$ kHz/rad. On the same log paper, plot the response with a single-pole loop filter with a cutoff frequency of $\omega_c = 100$ rad/s and a two-pole filter with the same cutoff frequency.

2-10. Determine the change in frequency (Δf) for a VCO with a transfer function, of $K_o = 2.5$ kHz/V and a dc input voltage change of $\Delta V = 0.8$ V.

2-11. Determine the voltage at the output of a phase comparator with a transfer function of $K_d = 0.5$ V/rad and a phase error of $\theta_e = 0.75$ rad.

2-12. Determine the hold-in range (Δf_{max}) for a PLL with an open-loop gain of $K_v = 20$ kHz/rad.

2-13. Determine the phase error necessary to produce a VCO frequency shift of $\Delta f = 10$ kHz for an open-loop gain of $K_v = 40$ kHz/rad.

2-14. Determine the output frequency from the multiple-crystal frequency synthesizer shown in Figure 2-38 if crystals X8 and X18 are selected.

2-15. Determine the output frequency from the single-crystal frequency synthesizer shown in Figure 2-39 for the following harmonics.

Harmonic Generator	Harmonic	Harmonic Generator	Harmonic
1	6	4	1
2	4	5	2
3		6	6

2-16. Determine f_c for the PLL shown in Figure 2-40 for a natural frequency of $f_n = 200$ kHz, $\Delta f = 0$ Hz, and $n = 20$.

Problems

3

Amplitude Modulation Transmission

INTRODUCTION

Information signals must be carried between a transmitter and a receiver over some form of transmission medium. However, the information signals are seldom in a form suitable for transmission. *Modulation* is defined as the process of transforming information from its original form to a form that is more suitable for transmission. *Demodulation* is the reverse process (that is, the modulated wave is converted back to its original form). Modulation takes place in the transmitter in a circuit called a *modulator,* and demodulation takes place in the receiver in a circuit called a *demodulator.* The purpose of this chapter is to introduce the reader to the fundamental concepts of AM transmission, to describe some of the circuits used in AM modulators, and to describe two different types of AM transmitters.

AMPLITUDE MODULATION

Amplitude modulation (AM) is the process of changing the amplitude of a relatively high frequency carrier in accordance with the amplitude of the modulating signal (information). Frequencies that are high enough to be efficiently radiated by an antenna and propagated through free space are commonly called *radio frequencies* or simply *RF*. With amplitude modulation, the information is impressed onto the carrier in the form of amplitude changes. Amplitude modulation is a relatively inexpensive, low-quality form of modulation that is used for commercial broadcasting of both audio and video signals. The commercial AM broadcast band extends from 535 to 1605 kHz. Commercial television broadcasting is divided into three bands (two VHF and one UHF). The low-band VHF channels are 2 to 6 (54 to 88 MHz), the high-band VHF channels are 7 to 13 (174 to 216

MHz), and the UHF channels are 14 to 83 (470 to 890 MHz). Amplitude modulation is also used for two-way mobile radio communications such as citizens band (CB) radio (26.965 to 27.405 MHz).

An AM modulator is a nonlinear device with two input signals: a single-frequency, constant-amplitude carrier signal and the information signal. The information acts on or modulates the carrier and may be a single frequency or a complex waveform made up of many frequencies that originated from one or more sources. Because the information acts on the carrier, it is called the *modulating signal*. The resultant is called the *modulated wave* or modulated signal.

The AM Envelope

Several forms or variations of amplitude modulation are possible. Although mathematically it is not the simplest form, AM *double-sideband full carrier* (AM DSBFC) will be discussed first because it is probably the most often used form of amplitude modulation. AM DSBFC is sometimes called *conventional* AM. Several other forms of AM are examined in Chapter 5.

Figure 3-1a shows a simplified AM DSBFC modulator that illustrates the relationship among the carrier [$V_c \sin (2\pi f_c t)$], the input (modulating) signal [$V_m \sin (2\pi f_m t)$], and the modulated wave [$v_{am}(t)$]. Figure 3-1b shows in the time domain how an AM wave is produced from a single-frequency modulating signal. The output modulated wave contains all the frequencies that make up the AM signal and is used to carry the information through the system. Therefore, the shape of the modulated wave is called the *envelope*. With no modulating signal, the output wave is simply the amplified carrier signal. When a modulating signal is applied, the amplitude of the output wave is varied in accordance with the modulating signal. Note that the shape of the AM envelope is identical to the shape of the modulating signal. Also, the time of one cycle of the envelope is the same as the period of the modulating signal. Consequently, the repetition rate of the envelope is equal to the frequency of the modulating signal.

AM Frequency Spectrum and Bandwidth

As stated previously, an AM modulator is a nonlinear device. Therefore, nonlinear mixing occurs and the output envelope is a complex wave made up of a dc voltage, the carrier frequency, and the sum ($f_c + f_m$) and difference ($f_c - f_m$) frequencies (that is, the cross products). The sum and difference frequencies are displaced from the carrier frequency by an amount equal to the modulating signal frequency. Therefore, an AM envelope contains frequency components spaced f_m Hz on either side of the carrier. However, it should be noted that the modulated wave does not contain a frequency component that is equal to the modulating signal frequency. The effect of modulation is to translate the modulating signal in the frequency domain so that it is reflected symmetrically about the carrier frequency.

Figure 3-2 shows the frequency spectrum for an AM wave. The AM spectrum extends from $f_c - f_{m(\max)}$ to $f_c + f_{m(\max)}$, where f_c is the carrier frequency and $f_{m(\max)}$ is the highest modulating signal frequency. The band of frequencies between $f_c - f_{m(\max)}$ and f_c is called the *lower sideband* (*LSB*), and any frequency within this band is called a *lower side frequency* (*LSF*). The band of frequencies between f_c and $f_c + f_{m(\max)}$ is called the *upper sideband* (*USB*), and any frequency within this band is called an *upper side frequency* (*USF*). Therefore, the bandwidth (B) of an AM DSBFC wave is equal to the differ-

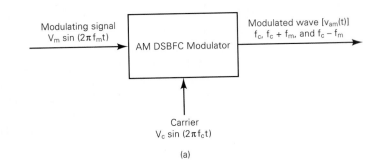

(a)

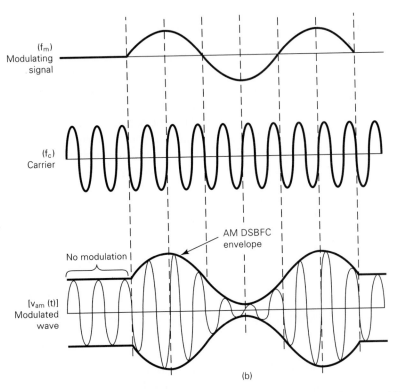

(b)

Figure 3-1 AM generation: (a) AM DSBFC modulator; (b) producing an AM DSBFC envelope—time domain.

ence between the highest upper side frequency and the lowest lower side frequency, or two times the highest modulating signal frequency (that is, $B = 2f_{m(max)}$). For radio wave propagation, the carrier and all the frequencies within the upper and lower sidebands must be high enough to be sufficiently propagated through Earth's atmosphere.

EXAMPLE 3-1

For an AM DSBFC modulator with a carrier frequency $f_c = 100$ kHz and a maximum modulating signal frequency $f_{m(max)} = 5$ kHz, determine:

(a) Frequency limits for the upper and lower sidebands.

(b) Bandwidth.

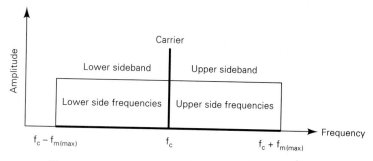

Figure 3-2 Frequency spectrum of an AM DSBFC wave.

(c) Upper and lower side frequencies produced when the modulating signal is a single-frequency 3-kHz tone.

(d) Draw the output frequency spectrum.

Solution (a) The lower sideband extends from the lowest possible lower side frequency to the carrier frequency or

$$LSB = [f_c - f_{m(\max)}] \quad \text{to} \quad f_c$$

$$= (100 - 5)\,\text{kHz} \quad \text{to} \quad 100\,\text{kHz} = 95 \quad \text{to} \quad 100\,\text{kHz}$$

The upper sideband extends from the carrier frequency to the highest possible upper side frequency or

$$USB = f_c \quad \text{to} \quad [f_c + f_{m(\max)}]$$

$$= 100\,\text{kHz} \quad \text{to} \quad (100 + 5)\,\text{kHz} = 100 \quad \text{to} \quad 105\,\text{kHz}$$

(b) The bandwidth is equal to the difference between the maximum upper side frequency and the minimum lower side frequency or

$$B = 2f_{m(\max)}$$

$$= 2(5\,\text{kHz}) = 10\,\text{kHz}$$

(c) The upper side frequency is the sum of the carrier and modulating frequency or

$$f_{\text{usf}} = f_c + f_m = 100\,\text{kHz} + 3\,\text{kHz} = 103\,\text{kHz}$$

The lower side frequency is the difference between the carrier and the modulating frequency or

$$f_{\text{lsf}} = f_c - f_m = 100\,\text{kHz} - 3\,\text{kHz} = 97\,\text{kHz}$$

(d) The output frequency spectrum is shown in Figure 3-3.

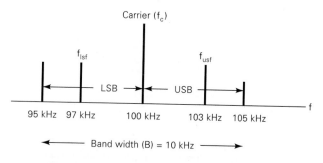

Figure 3-3 Output spectrum for Example 3-1.

Amplitude Modulation

Phasor Representation of an Amplitude-modulated Wave

For a single-frequency modulating signal, an AM envelope is produced from the vector addition of the carrier and the upper and lower side frequencies. The two side frequencies combine and produce a resultant component that combines with the carrier vector. Figure 3-4a shows this phasor addition. The phasors for the carrier and the upper and lower side frequencies all rotate in a counterclockwise direction. However, the upper side frequency rotates faster than the carrier ($\omega_{usf} > \omega_c$), and the lower side frequency rotates slower ($\omega_{lsf} < \omega_c$). Consequently, if the phasor for the carrier is held stationary, the phasor for the upper side frequency will continue to rotate in a counterclockwise direction relative to the carrier, and the phasor for the lower side frequency will rotate in a clockwise direction. The phasors for the carrier and the upper and lower side frequencies combine, sometimes

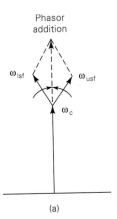

V_{usf} = voltage of the upper side frequency

V_{lsf} = voltage of the lower side frequency

V_c = voltage of the carrier

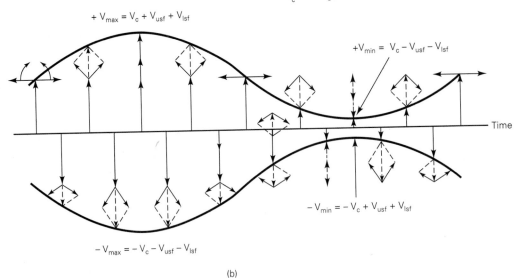

Figure 3-4 Phasor addition in an AM DSBFC envelope: (a) phasor addition of the carrier and the upper and lower side frequencies; (b) phasor addition producing an AM envelope.

Chap. 3 Amplitude Modulation Transmission

in phase (adding) and sometimes out of phase (subtracting). For the waveform shown in Figure 3-4b, the maximum positive amplitude of the envelope occurs when the carrier and the upper and lower side frequencies are at their maximum positive values at the same time ($+V_{max} = V_c + V_{usf} + V_{lsf}$). The minimum positive amplitude of the envelope occurs when the carrier is at its maximum positive value at the same time that the upper and lower side frequencies are at their maximum negative values ($+V_{min} = V_c - V_{usf} - V_{lsf}$). The maximum negative amplitude occurs when the carrier and the upper and lower side frequencies are at their maximum negative values at the same time ($-V_{max} = -V_c - V_{usf} - V_{lsf}$). The minimum negative amplitude occurs when the carrier is at its maximum negative value at the same time that the upper and lower side frequencies are at their maximum positive values ($-V_{min} = -V_c + V_{usf} + V_{lsf}$).

Coefficient of Modulation and Percent Modulation

Coefficient of modulation is a term that is used to describe the amount of amplitude change (modulation) present in an AM waveform. *Percent modulation* is simply the coefficient of modulation stated as a percentage. More specifically, percent modulation gives the percentage change in the amplitude of the output wave when the carrier is acted on by a modulating signal. Mathematically, the modulation coefficient is

$$m = \frac{E_m}{E_c} \qquad (3\text{-}1)$$

where m = modulation coefficient (unitless)
$\quad E_m$ = peak change in the amplitude of the output waveform voltage (volts)
$\quad E_c$ = peak amplitude of the unmodulated carrier voltage (volts)

Equation 3-1 can be rearranged to solve for E_m and E_c as

$$E_m = mE_c \qquad (3\text{-}2)$$

and

$$E_c = \frac{E_m}{m} \qquad (3\text{-}3)$$

and percent modulation (M) is

$$M = \frac{Em}{E_c} \times 100 \quad \text{or simply} \quad m \times 100 \qquad (3\text{-}4)$$

The relationship among m, E_m, and E_c is shown in Figure 3-5.

If the modulating signal is a pure, single-frequency sine wave and the modulation process is symmetrical (that is, the positive and negative excursions of the envelope's amplitude are equal), percent modulation can be derived as follows (refer to Figure 3-5 for the following derivation):

$$E_m = \frac{1}{2}(V_{max} - V_{min}) \qquad (3\text{-}5)$$

and

$$E_c = \frac{1}{2}(V_{max} + V_{min}) \qquad (3\text{-}6)$$

Therefore

$$M = \frac{1/2(V_{max} - V_{min})}{1/2(V_{max} + V_{min})} \times 100$$

Amplitude Modulation

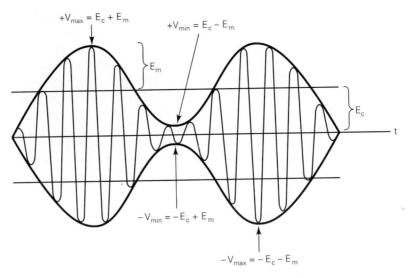

$+V_{max} = E_c + E_m$

$+V_{min} = E_c - E_m$

E_m

E_c

t

$-V_{min} = -E_c + E_m$

$-V_{max} = -E_c - E_m$

Figure 3-5 Modulation coefficient, E_m, and E_c.

$$= \frac{(V_{max} - V_{min})}{(V_{max} + V_{min})} \times 100 \qquad (3\text{-}7)$$

where $\quad V_{max} = E_c + E_m$
$\quad\quad\quad V_{min} = E_c - E_m$

The peak change in the amplitude of the output wave (E_m) is the sum of the voltages from the upper and lower side frequencies. Therefore, since $E_m = E_{usf} + E_{lsf}$ and $E_{usf} = E_{lsf}$, then

$$E_{usf} = E_{lsf} = \frac{E_m}{2} = \frac{1/2(V_{max} - V_{min})}{2} = \frac{1}{4}(V_{max} - V_{min}) \qquad (3\text{-}8)$$

where $\quad E_{usf}$ = peak amplitude of the upper side frequency (volts)
$\quad\quad\quad E_{lsf}$ = peak amplitude of the lower side frequency (volts)

From Equation 3-1 it can be seen that the percent modulation goes to 100% when $E_m = E_c$. This condition is shown in Figure 3-6d. It can also be seen that at 100% modulation, the minimum amplitude of the envelope $V_{min} = 0$ V. Figure 3-6c shows a 50% modulated envelope; the peak change in the amplitude of the envelope is equal to one-half the amplitude of the unmodulated wave. The maximum percent modulation that can be imposed without causing excessive distortion is 100%. Sometimes percent modulation is expressed as the peak change in the voltage of the modulated wave with respect to the peak amplitude of the unmodulated carrier (that is, percent change = $\Delta E_c/E_c \times 100$).

EXAMPLE 3-2

For the AM waveform shown in Figure 3-7, determine:

(a) Peak amplitude of the upper and lower side frequencies.
(b) Peak amplitude of the unmodulated carrier.
(c) Peak change in the amplitude of the envelope.

Chap. 3   Amplitude Modulation Transmission

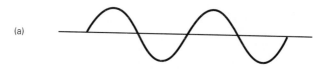

(a)

(b) $\left.\right\} E_c$

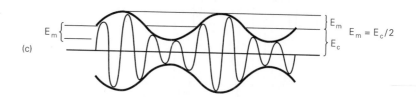

(c) $E_m\left\{\right.$ ... $\left.\right\} E_m$ $E_m = E_c/2$ $\left.\right\} E_c$

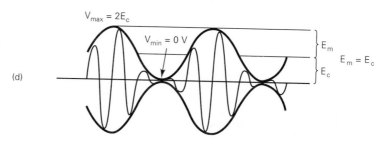

$V_{max} = 2E_c$

$V_{min} = 0\ V$

(d) $\left.\right\} E_m$ $E_m = E_c$ $\left.\right\} E_c$

Figure 3-6 Percent modulation of an AM DSBFC envelope: (a) modulating signal; (b) unmodulated carrier; (c) 50% modulated wave; (d) 100% modulated wave.

(d) Coefficient of modulation.
(e) Percent modulation.

Solution (a) From Equation 3-8,

$$E_{usf} = E_{lsf} = \frac{1}{4}(18 - 2) = 4\ V$$

(b) From Equation 3-6,

$$E_c = \frac{1}{2}(18 + 2) = 10\ V$$

(c) From Equation 3-5,

$$E_m = \frac{1}{2}(18 - 2) = 8\ V$$

(d) From Equation 3-1,

$$m = \frac{8}{10} = 0.8$$

Amplitude Modulation

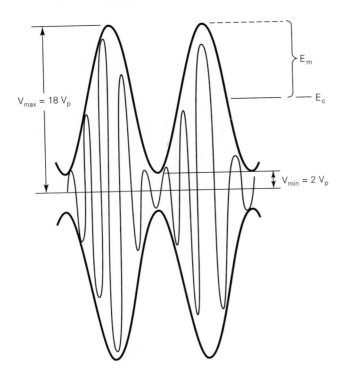

$V_{max} = 18\ V_p$

E_m

E_c

$V_{min} = 2\ V_p$

Figure 3-7 AM envelope for Example 3-2.

(e) From Equation 3-4,

$$M = 0.8 \times 100 = 80\%$$

and from Equation 3-7;

$$M = \frac{18 - 2}{18 + 2} \times 100 = 80\%$$

AM Voltage Distribution

An unmodulated carrier can be described mathematically as

$$v_c(t) = E_c \sin\ (2\pi f_c t)$$

where $v_c(t)$ = time-varying voltage waveform for the carrier
        E_c = peak carrier amplitude (volts)
        f_c = carrier frequency (hertz)

In a previous section it was pointed out that the repetition rate of an AM envelope is equal to the frequency of the modulating signal, the amplitude of the AM wave varies proportional to the amplitude of the modulating signal, and the maximum amplitude of the modulated wave is equal to $E_c + E_m$. Therefore, the instantaneous amplitude of the modulated wave can be expressed as

$$v_{am}(t) = [E_c + E_m \sin\ (2\pi f_m t)][\sin\ (2\pi f_c t)] \qquad (3\text{-}9\text{a})$$

where $[E_c + E_m \sin\ (2\pi f_m t)]$ = amplitude of the modulated wave
                     E_m = peak change in the amplitude of the envelope (volts)
                     f_m = frequency of the modulating signal (hertz)

Chap. 3   Amplitude Modulation Transmission

If mE_c is substituted for E_m,

$$v_{am}(t) = [(E_c + mE_c \sin (2\pi f_m t)][\sin (2\pi f_c t)] \qquad (3\text{-}9\text{b})$$

where $[E_c + mE_c \sin (2\pi f_m t)]$ = amplitude of the modulated wave.

Factoring E_c from Equation 3-9b and rearranging gives

$$v_{am}(t) = [1 + m \sin (2\pi f_m t)][E_c \sin (2\pi f_c t)] \qquad (3\text{-}9\text{c})$$

where $[1 + m \sin (2\pi f_m t)]$ = constant + modulating signal
$\qquad\quad [E_c \sin (2\pi f_c t)]$ = unmodulated carrier

In Equation 3-9c, it can be seen that the modulating signal contains a constant component (1) and a sinusoidal component at the modulating signal frequency $[m \sin (2\pi f_m t)]$. The following analysis will show how the constant component produces the carrier component in the modulated wave and the sinusoidal component produces the side frequencies. Multiplying out Equation 3-9b or c yields

$$v_{am}(t) = E_c \sin (2\pi f_c t) + [mE_c(\sin 2\pi f_m t)][\sin (2\pi f_c t)]$$

The trigonometric identity for the product of two sines with different frequencies is

$$(\sin A)(\sin B) = -\frac{1}{2} \cos(A + B) + \frac{1}{2} \cos(A - B)$$

Therefore,

$$v_{am}(t) = E_c \sin (2\pi f_c t) - \frac{mE_c}{2} \cos [2\pi(f_c + f_m)t] + \frac{mE_c}{2} \cos [2\pi(f_c - f_m)t] \qquad (3\text{-}10)$$

where $\qquad\qquad E_c \sin (2\pi f_c t)$ = carrier signal (volts)

$$-\frac{mE_c}{2} \cos [2\pi(f_c + f_m)t] = \text{upper side frequency signal (volts)}$$

$$+\frac{mE_c}{2} \cos[2\pi(f_c - f_m)t] = \text{lower side frequency signal (volts)}$$

Several interesting characteristics about double-sideband full-carrier amplitude modulation can be pointed out from Equation 3-10. First, note that the amplitude of the carrier after modulation is the same as it was before modulation (E_c). Therefore, the amplitude of the carrier is unaffected by the modulation process. Second, the amplitude of the upper and lower side frequencies depend on both the carrier amplitude and the coefficient of modulation. For 100% modulation, $m = 1$ and the amplitudes of the upper and lower side frequencies are each equal to one-half the amplitude of the carrier ($E_c/2$). Therefore, at 100% modulation,

$$V_{(\text{max})} = E_c + \frac{E_c}{2} + \frac{E_c}{2} = 2E_c$$

and

$$V_{(\text{min})} = E_c - \frac{E_c}{2} - \frac{E_c}{2} = 0 \text{ V}$$

From the relationships shown above and using Equation 3-10, it is evident that, as long as we do not exceed 100% modulation, the maximum peak amplitude of an AM

Amplitude Modulation

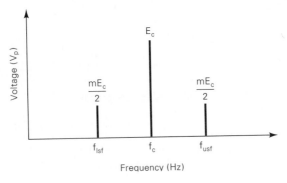

Figure 3-8 Voltage spectrum for an AM DSBFC wave.

envelope $V_{(max)} = 2E_c$, and the minimum peak amplitude of an AM envelope $V_{(min)} = 0$ V. This relationship was shown in Figure 3-6d. Figure 3-8 shows the voltage spectrum for an AM DSBFC wave (note that all the voltages are given in peak values).

Also, from Equation 3-10, the relative phase relationship between the carrier and the upper and lower side frequencies is evident. The carrier component is a + sine function, the upper side frequency a − cosine function, and the lower side frequency a + cosine function. Also, the envelope is a repetitive waveform. Thus, at the beginning of each cycle of the envelope, the carrier is 90° out of phase with both the upper and lower side frequencies, and the upper and lower side frequencies are 180° out of phase with each other. This phase relationship can be seen in Figure 3-9 for $f_c = 25$ Hz and $f_m = 5$ Hz.

EXAMPLE 3-3

One input to a conventional AM modulator is a 500-kHz carrier with an amplitude of 20 Vp. The second input is a 10-kHz modulating signal that is of sufficient amplitude to cause a change in the output wave of ±7.5 Vp. Determine:

(a) Upper and lower side frequencies.
(b) Modulation coefficient and percent modulation.
(c) Peak amplitude of the modulated carrier and the upper and lower side frequency voltages.
(d) Maximum and minimum amplitudes of the envelope.
(e) Expression for the modulated wave.
(f) Draw the output spectrum.
(g) Sketch the output envelope.

Solution (a) The upper and lower side frequencies are simply the sum and difference frequencies, respectively.

$$f_{usf} = 500 \text{ kHz} + 10 \text{ kHz} = 510 \text{ kHz}$$

$$f_{lsf} = 500 \text{ kHz} - 10 \text{ kHz} = 490 \text{ kHz}$$

(b) The modulation coefficient is determined from Equation 3-1:

$$m = \frac{7.5}{20} = 0.375$$

Percent modulation is determined from Equation 3-4:

$$M = 100 \times 0.375 = 37.5\%$$

(c) The peak amplitude of the modulated carrier and the upper and lower side frequencies is

$$E_c(\text{modulated}) = E_c(\text{unmodulated}) = 20 \text{ Vp}$$

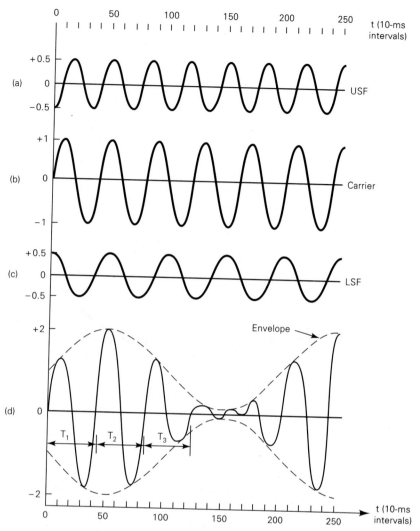

Figure 3-9 Generation of an AM DSBFC envelope shown in the time domain: (a) $-\frac{1}{2}\cos 2\pi 30t$; (b) $\sin 2\pi 25t$; (c) $+\frac{1}{2}\cos 2\pi 20t$; (d) summation of (a), (b), and (c).

$$E_{\text{usf}} = E_{\text{lsf}} = \frac{mE_c}{2} = \frac{(0.375)(20)}{2} = 3.75 \text{ Vp}$$

(d) The maximum and minimum amplitudes of the envelope are determined as follows:

$$V_{\text{(max)}} = E_c + E_m = 20 + 7.5 = 27.5 \text{ Vp}$$

$$V_{\text{(min)}} = E_c - E_m = 20 - 7.5 = 12.5 \text{ Vp}$$

(e) The expression for the modulated wave follows the format of Equation 3-10.

$$v_{am}(t) = 20 \sin (2\pi 500kt) - 3.75 \cos (2\pi 510kt) + 3.75 \cos (2\pi 490kt)$$

(f) The output spectrum is shown in Figure 3-10.

(g) The modulated envelope is shown in Figure 3-11.

Amplitude Modulation

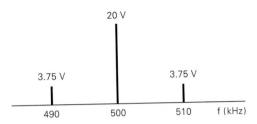

Figure 3-10 Output spectrum for Example 3-3.

AM Time-domain Analysis

Figure 3-9 shows how an AM DSBFC envelope is produced from the algebraic addition of the waveforms for the carrier and the upper and lower side frequencies. For simplicity, the following waveforms are used for the modulating and carrier input signals:

$$\text{carrier} = v_c(t) = E_c \sin (2\pi 25t) \tag{3-11}$$

$$\text{modulating signal} = v_m(t) = E_m \sin (2\pi 5t) \tag{3-12}$$

Substituting Equations 3-11 and 3-12 into Equation 3-10, the expression for the modulated wave is

$$v_{am}(t) = E_c \sin (2\pi 25t) - \frac{mE_c}{2} \cos(2\pi 30t) + \frac{mE_c}{2} \cos(2\pi 20t) \tag{3-13}$$

where $E_c \sin (2\pi 25t) = \text{carrier (volts)}$

$-\dfrac{mE_c}{2} \cos (2\pi 30t) = \text{upper side frequency (volts)}$

$+\dfrac{mE_c}{2} \cos (2\pi 20t) = \text{lower side frequency (volts)}$

Table 3-1 lists the values for the instantaneous voltages of the carrier, the upper and lower side frequency voltages, and the total modulated wave when values of t from 0 to 250 ms, in 10-ms intervals, are substituted into Equation 3-13. The unmodulated carrier voltage $E_c = 1$ Vp, and 100% modulation is achieved. The corresponding waveforms are shown in Figure 3-9. Note that the maximum envelope voltage is 2 V ($2E_c$) and the minimum envelope voltage is 0 V.

In Figure 3-9, note that the time between similar zero crossings within the envelope is constant (that is, $T_1 = T_2 = T_3$, and so on). Also note that the amplitudes of successive peaks within the envelope are not equal. This indicates that a cycle within the envelope is not a pure sine wave and, thus, the modulated wave must be comprised of more than one frequency: the summation of the carrier and the upper and lower side frequencies. Figure 3-9 also shows that the amplitude of the carrier does not vary, but rather, the amplitude of the envelope varies in accordance with the modulating signal. This is accomplished by the addition of the upper and lower side frequencies to the carrier waveform.

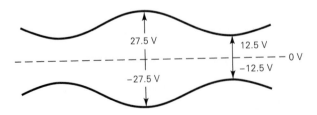

Figure 3-11 AM envelope for Example 3.3.

TABLE 3-1 INSTANTANEOUS VOLTAGES

USF, $-\frac{1}{2}\cos 2\pi 30t$	Carrier, $\sin 2\pi 25t$	LSF, $+\frac{1}{2}\cos 2\pi 20t$	Envelope, $v_{am}(t)$	Time, t (ms)
−0.5	0	+0.5	0	0
+0.155	+1	+0.155	+1.31	10
+0.405	0	−0.405	0	20
−0.405	−1	−0.405	−1.81	30
−0.155	0	+0.155	0	40
+0.5	+1	+0.5	2	50
−0.155	0	+0.155	0	60
−0.405	−1	−0.405	−1.81	70
+0.405	0	−0.405	0	80
+0.155	+1	+0.155	+1.31	90
−0.5	0	+0.5	0	100
+0.155	−1	+0.155	−0.69	110
+0.405	0	−0.405	0	120
−0.405	+1	−0.405	+0.19	130
−0.155	0	+0.155	0	140
+0.5	−1	+0.5	0	150
−0.155	0	+0.155	0	160
−0.405	+1	−0.405	+0.19	170
+0.405	0	−0.405	0	180
+0.155	−1	+0.155	−0.69	190
−0.5	0	+0.5	0	200
+0.155	+1	+0.155	+1.31	210
+0.405	0	−0.405	0	220
−0.405	−1	−0.405	−1.81	230
+0.405	0	−0.405	0	240
+0.155	+1	+0.155	+1.31	250

AM Power Distribution

In any electrical circuit, the power dissipated is equal to the voltage squared, divided by the resistance (that is, $P = E^2/R$). Thus, the power developed across a load by an unmodulated carrier is equal to the carrier voltage squared, divided by the load resistance. Mathematically, the unmodulated carrier power is expressed as

$$P_c = \frac{(0.707E_c)^2}{R}$$

$$= \frac{(E_c)^2}{2R} \tag{3-14}$$

where P_c = carrier power (watts)
   E_c = peak carrier voltage (volts)
   R = load resistance (ohms)

The upper and lower sideband powers are expressed mathematically as

$$P_{usb} = P_{lsb} = \frac{(mE_c/2)^2}{2R}$$

where $mE_c/2$ = peak voltage of the upper and lower side frequencies

Amplitude Modulation

rearranging yields
$$= \frac{m^2 E_c^2}{8R}$$
(3-15)

where P_{usb} = upper sideband power (watts)
           P_{lsb} = lower sideband power (watts)
        $mE_c/2$ = upper and lower sideband peak voltages

Substituting Equation 3-14 into Equation 3-15 gives
$$P_{usb} = P_{lsb} = \frac{m^2 P_c}{4}$$
(3-16)

    It is evident from Equation 3-16 that for a modulation coefficient $m = 0$ the power in the upper and lower sidebands is zero and the total transmitted power is simply the carrier power.

    The total power in an amplitude-modulated wave is equal to the sum of the powers of the carrier, the upper sideband, and the lower sideband. Mathematically, the total power in an AM DSBFC envelope is
$$P_t = P_c + P_{usb} + P_{lsb}$$
(3-17)

where P_t = total power of an AM DSBFC envelope (watts)
          P_c = carrier power (watts)
      P_{usb} = upper sideband power (watts)
      P_{lsb} = lower sideband power (watts)

Substituting Equation 3-16 into Equation 3-17 yields
$$P_t = P_c + \frac{m^2 P_c}{4} + \frac{m^2 P_c}{4}$$
(3-18)

Combining terms gives
$$P_t = P_c + \frac{m^2 P_c}{2}$$
(3-19)

where $\dfrac{m^2 P_c}{2}$ = total sideband power

Factoring P_c gives us
$$P_t = P_c \left(1 + \frac{m^2}{2} \right)$$
(3-20)

    From the preceding analysis, it can be seen that the carrier power in the modulated wave is the same as the carrier power in the unmodulated wave. Thus, it is evident that the power of the carrier is unaffected by the modulation process. Also, because the total power in the AM wave is the sum of the carrier and sideband powers, the total power in an AM envelope increases with modulation (that is, as m increases, P_t increases).

    Figure 3-12 shows the power spectrum for an AM DSBFC wave. Note that with 100% modulation the maximum power in the upper or lower sideband is equal to only one-fourth the power in the carrier. Thus, the maximum total sideband power is equal to one-half the carrier power. One of the most significant disadvantages of AM DSBFC transmission is the fact that the information is contained in the sidebands although most of

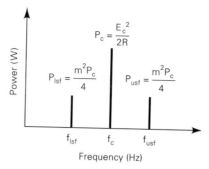

Figure 3-12 Power spectrum for an AM DSBFC wave.

the power is wasted in the carrier. Actually, the power in the carrier is not totally wasted because it does allow for the use of relatively simple, inexpensive demodulator circuits in the receiver, which is the predominant advantage of AM DSBFC.

EXAMPLE 3-4

For an AM DSBFC wave with a peak unmodulated carrier voltage $V_c = 10$ Vp, a load resistance $R_L = 10\Omega$, and a modulation coefficient $m = 1$, determine:

(a) Powers of the carrier and the upper and lower sidebands.
(b) Total sideband power.
(c) Total power of the modulated wave.
(d) Draw the power spectrum.
(e) Repeat steps a through d for a modulation index $m = 0.5$.

Solution (a) The carrier power is found by substituting into Equation 3-14:

$$P_c = \frac{10^2}{2(10)} = \frac{100}{20} = 5 \text{ W}$$

The upper and lower sideband power is found by substituting into Equation 3-16:

$$P_{usb} = P_{lsb} = \frac{(1^2)(5)}{4} = 1.25 \text{ W}$$

(b) The total sideband power is

$$P_{sbt} = \frac{m^2 P_c}{2} = \frac{(1.0^2)(5)}{2} = 2.5 \text{ W}$$

(c) The total power in the modulated wave is found by substituting into Equation 3-20

$$P_t = 5\left[1 + \frac{(1)^2}{2}\right] = 7.5 \text{ W}$$

(d) The power spectrum is shown in Figure 3-13.

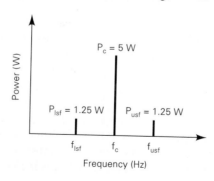

Figure 3-13 Power spectrum for Example 3-4d.

Amplitude Modulation

(e) The carrier power is found by substituting into Equation 3-14:

$$P_c = \frac{10^2}{2(10)} = \frac{100}{20} = 5 \text{ W}$$

The upper and lower sideband power is found by substituting into Equation 3-16:

$$P_{usb} = P_{lsb} = \frac{(0.5)^2(5)}{4} = 0.3125 \text{ W}$$

The total sideband power is

$$P_{sbt} = \frac{m^2 P_c}{2} = \frac{(0.5^2)(5)}{2} = 0.625 \text{ W}$$

The total power of the modulated wave is found by substituting into Equation 3-20

$$P_t = 5\left[1 + \frac{(0.5)^2}{2}\right] = 5.625 \text{ W}$$

The power spectrum is shown in Figure 3-14.

From Example 3-4, it can be seen why it is important to use as high a percentage of modulation as possible while still being sure not to overmodulate. As the example shows, the carrier power remains the same as m changes. However, the sideband power was reduced dramatically when m decreased from 1 to 0.5. Because sideband power is proportional to the square of the modulation coefficient, a reduction in m of one-half results in a reduction in the sideband power of one-fourth (that is, $0.5^2 = 0.25$). The relationship between modulation coefficient and power can sometimes be deceiving because the total transmitted power consists primarily of carrier power and is, therefore, not dramatically affected by changes in m. However, it should be noted that the power in the intelligence-carrying portion of the transmitted signal (that is, the sidebands) is affected dramatically by changes in m. For this reason, AM DSBFC systems try to maintain a modulation coefficient between 0.9 and 0.95 (90% to 95% modulation) for the highest-amplitude intelligence signals.

AM Current Calculations

With amplitude modulation, it is very often necessary and sometimes desirable to measure the current of the carrier and modulated wave and then calculate the modulation index from these measurements. The measurements are made by simply metering the transmit antenna current with and without the presence of a modulating signal. The relationship between carrier current and the current of the modulated wave is

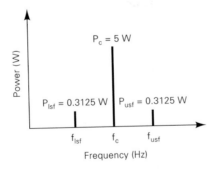

Figure 3-14 Power spectrum for Example 3-4e.

Chap. 3   Amplitude Modulation Transmission

$$\frac{P_t}{P_c} = \frac{I_t^2 R}{I_c^2 R} = \frac{I_t^2}{I_c^2} = 1 + \frac{m^2}{2}$$

where P_t = total transmit power (watts)
P_c = carrier power (watts)
I_t = total transmit current (ampere)
I_c = carrier current (ampere)
R = antenna resistance (ohms)

and

$$\frac{I_t}{I_c} = \sqrt{1 + \frac{m^2}{2}} \tag{3-21a}$$

Thus

$$I_t = I_c \sqrt{\left(1 + \frac{m^2}{2}\right)} \tag{3-21b}$$

Modulation by a Complex Information Signal

In the previous sections of this chapter, frequency spectrum, bandwidth, coefficient of modulation, and voltage and power distribution for double-sideband full carrier AM were analyzed for a single-frequency modulating signal. In practice, however, the modulating signal is very often a complex waveform made up of many sine waves with different amplitudes and frequencies. Consequently, a brief analysis will be given of the effects such a complex modulating signal would have on an AM waveform.

If a modulating signal contains two frequencies (f_{m1} and f_{m2}), the modulated wave will contain the carrier and two sets of side frequencies spaced symmetrically about the carrier. Such a wave can be written as

$$v_{am}(t) = \sin(2\pi f_c t) + \frac{1}{2} \cos[2\pi(f_c - f_{m1})t] - \frac{1}{2} \cos[2\pi(f_c + f_{m1})t]$$

$$+ \frac{1}{2} \cos[2\pi(f_c - f_{m2})t] - \frac{1}{2} \cos[2\pi(f_c + f_{m2})t]$$

When several frequencies simultaneously amplitude modulate a carrier, the combined coefficient of modulation is the square root of the quadratic sum of the individual modulation indexes as follows:

$$m_t = \sqrt{m_1^2 + m_2^2 + m_3^2 + m_n^2} \tag{3-22}$$

where m_t = total coefficient of modulation
$m_1, m_2, m_3,$ and m_n = coefficients of modulation for input signals 1, 2, 3, and n

The combined coefficient of modulation can be used to determine the total sideband and transmit powers as follows:

$$P_{usbt} = P_{lsbt} = \frac{P_c m_t^2}{4} \tag{3-23}$$

and

$$P_{sbt} = \frac{P_c m_t^2}{2} \tag{3-24}$$

Amplitude Modulation

Thus
$$P_t = P_c\left(1 + \frac{m_t^2}{2}\right)$$
(3-25)

where P_{usbt} = total upper sideband power (watts)
P_{lsbt} = total lower sideband power (watts)
P_{sbt} = total sideband power (watts)
P_t = total transmitted power (watts)

In an AM transmitter, care must be taken to ensure that the combined voltages of all the modulating signals do not overmodulate the carrier.

EXAMPLE 3-5

For an AM DSBFC transmitter with an unmodulated carrier power $P_c = 100$ W that is modulated simultaneously by three modulating signals with coefficients of modulation $m_1 = 0.2$, $m_2 = 0.4$, and $m_3 = 0.5$, determine:

(a) Total coefficient of modulation.
(b) Upper and lower sideband power.
(c) Total transmitted power.

Solution (a) The total coefficient of modulation is found by substituting into Equation 3-22.

$$m_t = \sqrt{0.2^2 + 0.4^2 + 0.5^2}$$
$$= \sqrt{0.04 + 0.16 + 0.25}$$
$$= 0.67$$

(b) The total sideband power is found by substituting the results of Step a into Equation 3-24.

$$P_{sbt} = \frac{(0.67^2)100}{2}$$
$$= 22.445 \text{ W}$$

(c) The total transmitted power is found by substituting into Equation 3-25.

$$P_t = 100\left(1 + \frac{0.67^2}{2}\right)$$
$$= 122.445 \text{ W}$$

AM MODULATOR CIRCUITS

The location in a transmitter where modulation occurs determines whether the circuit is a *low-* or *high-level transmitter*. With low-level modulation, the modulation takes place prior to the output element of the final stage of the transmitter, in other words, prior to the collector of the output transistor in a transistorized transmitter, prior to the drain of the output FET in a FET transmitter, or prior to the plate of the output tube in a vacuum-tube transmitter.

An advantage of low-level modulation is that less modulating signal power is required to achieve a high percentage of modulation. In high-level modulators, the modulation takes place in the final element of the final stage where the carrier signal is at its maximum amplitude and, thus, requires a much higher amplitude modulating signal to achieve a reasonable percent modulation. With high-level modulation, the final modulat-

Chap. 3 Amplitude Modulation Transmission

ing signal amplifier must supply all the sideband power, which could be as much as 33% of the total transmit power. An obvious disadvantage of low-level modulation is in high-power applications when all the amplifiers that follow the modulator stage must be linear amplifiers, which is extremely inefficient.

Low-level AM Modulator

Figure 3-15 shows the schematic diagram for a simple low-level modulator with a single active component (Q_1). The carrier [$0.01 \sin (2\pi 5 \times 10^5 t)$] is applied to the base of the transistor, and the modulating signal [$6 \sin 2\pi(1000t)$] to the emitter. Therefore, this method of low-level modulation is called *emitter modulation*. With emitter modulation, it is important that the transistor be biased *class A* with a centered *Q-point*. The modulating signal is multiplied by the carrier producing modulation.

Circuit operation. With emitter modulation, the peak amplitude of the carrier (10 mV) is appreciably smaller than the peak amplitude of the modulating signal (6 V). If the modulating signal is removed or held constant at 0 V, Q_1 operates as a linear amplifier. The base input signal is simply amplified and inverted 180° at the collector. The amplification in Q_1 is determined by the ratio of the ac collector resistance (r_c) to the ac emitter resistance (r'_e) (that is, $A_v = r_c/r'_e$). For the component values shown, r_c and r'_e are determined as follows:

$$r_c = \text{parallel combination of } R_C \text{ and } R_L$$

$$= \frac{(10,000)(2000)}{12,000} = 1667 \ \Omega$$

$$r'_e = \frac{25 \text{ mV}}{I_E}$$

where

$$I_E = \frac{V_{th} - V_{be}}{(R_{th}/\beta) + R_E}$$

$$V_{th} = \frac{V_{CC}R_1}{R_1 + R_2} = \frac{30(10,000)}{30,000} = 10 \text{ V}$$

$$R_{th} = \frac{R_1 R_2}{R_1 + R_2} = 6667 \ \Omega$$

Thus, for a typical $\beta = 100$,

$$I_E = \frac{10 - 0.7}{(6667/100) + 10,000} = 0.924 \text{ mA}$$

and

$$r'_e = \frac{25 \text{ mV}}{0.924 \text{ mA}} = 27 \ \Omega$$

Therefore,

$$A_q = \frac{r_c}{r'_e} = \frac{1667}{27} = 61.7$$

where A_q = quiescent voltage gain.

AM Modulator Circuits

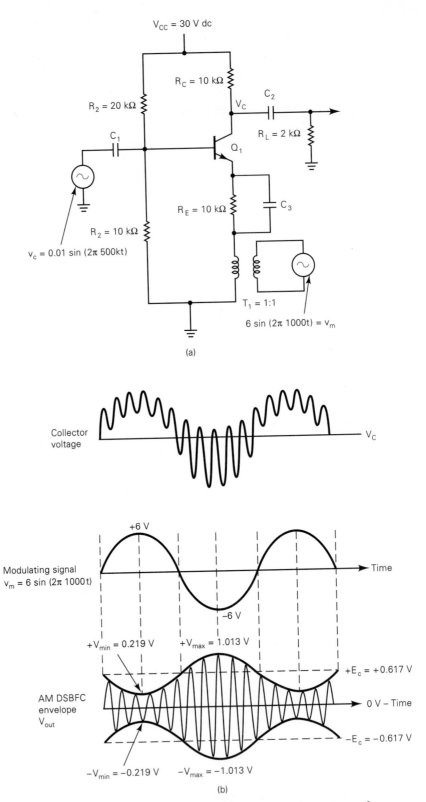

Figure 3-15 (a) Single transistor, emitter modulator; (b) output waveforms.

Chap. 3   **Amplitude Modulation Transmission**

With no modulating input signal, Q_1 is a linear amplifier with a quiescent voltage gain $A_q = 61.7$. With a carrier input voltage $V_c = 10$ mV, V_{out} is

$$V_{out} = A_q V_{in} = 61.7(0.01 \text{ V}) = 0.617 \text{ V}$$

When the modulating signal $[v_m(t)]$ is applied to the circuit, its voltage combines with the dc Thévenin voltage. The result is a bias voltage that has a constant term and a term that varies at a low-frequency sinusoidal rate equal to the frequency of the modulating signal. Thus

$$V_{bias} = V_{th} + V_{in}$$

$$= V_{th} + V_m \sin(2\pi f_m t)$$

For this example,

$$V_{bias} = 10 + 6 \sin(2\pi 1000 t)$$

To analyze the operation of this circuit, it is not necessary to consider every possible value of V_{bias}. Instead, the circuit is analyzed using several key values for V_{in} and the other points are interpolated from them. The three most significant values for V_{bias} are when the input signal is 0 V, maximum positive, and maximum negative. When $V_m = 0$ V, the bias voltage is equal to the Thévenin voltage and the voltage gain is the quiescent value, $A_v = A_q = 61.7$.

When the modulating signal is maximum and negative,

$$V_{in} = -6$$

$$V_{bias} = V_{th} - V_{in} = 10 - (-6) = 16 \text{ V}$$

$$I_E = \frac{16 - 0.7}{(6667/100) + 10,000} = 1.52 \text{ mA}$$

$$r'_e = \frac{25 \text{ mV}}{1.52 \text{ mA}} = 16.45 \text{ } \Omega$$

$$A_v = A_{max} = \frac{1667}{16.45} = 101.3$$

$$V_{out} = A_{max} V_{in}$$

$$= (0.01)(101.3) = 1.013 \text{ V}$$

When the modulating signal is maximum and positive,

$$V_{in} = +6$$

$$V_{bias} = V_{th} - V_{in} = 10 - (+6) = 4 \text{ V}$$

$$I_E = \frac{4 - 0.7}{(6667/100) + 10,000} = 0.328 \text{ mA}$$

$$r'_e = \frac{25 \text{ mV}}{0.328 \text{ mA}} = 76.3 \text{ } \Omega$$

$$A_v = A_{min} = \frac{1667}{76.3} = 21.9$$

$$V_{out} = A_{min} V_{in}$$

$$= (0.01)(21.9) = 0.219 \text{ V}$$

AM Modulator Circuits

In the preceding example, the voltage gain varied at a sinusoidal rate equal to the modulating signal frequency (1000 Hz) from a quiescent value $A_q = 61.7$ to a maximum value $A_{max} = 101.3$ and then to a minimum value $A_{min} = 21.9$. Consequently, the ac output voltage varies from a minimum value of 0.219 V to a maximum value of 1.103 V. The dc collector voltage V_C and ac output voltage waveform are shown in Figure 3-15b. The waveform for $v_{out}(t)$ is a varying amplitude carrier signal riding on top of the low-frequency modulating signal. The average voltage is equal to the quiescent collector dc voltage V_C. The dc voltage and the low-frequency modulating signal are removed from the waveform by coupling capacitor C_2. Consequently, the output waveform developed across R_L is the modulated carrier with an average voltage equal to 0 V, an unmodulated carrier amplitude of 0.617 V, a maximum positive or negative amplitude of ± 1.013 V, and a minimum positive or negative amplitude of ± 0.219 V. It is interesting to note that with emitter modulation the maximum amplitude of the envelope occurs when the modulating signal is maximum negative, and the minimum envelope amplitude occurs when the modulating signal is maximum positive.

From the preceding example, it can be seen that the voltage gain varies at a sinusoidal rate equal to the modulating signal frequency f_m. Therefore, the voltage gain can be expressed mathematically as

$$A_v = A_q[1 + m \sin (2\pi f_m t)]$$

Sin $(2\pi f_m t)$ goes from a maximum value of $+1$ to a minimum value of -1. Thus

$$A_v = A_q (1 \pm m) \tag{3-26}$$

where m = modulation coefficient.

Therefore, at 100% modulation, $m = 1$ and

$$A_{max} = A_q(1 + 1) = 2A_q$$

$$A_{min} = A_q(1 - 1) = 0$$

and the maximum and minimum amplitudes for V_{out} are

$$V_{out(max)} = 2A_q V_{in}$$

$$V_{out(min)} = 0$$

For the preceding example, the modulation coefficient is

$$m = \frac{V_{max} - V_{min}}{V_{max} + V_{min}} = \frac{1.013 - 0.2195}{1.013 + 0.2195} = 0.645$$

Substituting into Equation 3-26 gives

$$A_{max} = 61.7(1 + 0.64) = 101.5$$

$$A_{min} = 61.7(1 - 0.645) = 21.9$$

In the preceding example, the voltage gain changed symmetrically with modulation. In other words, the increase in gain during the negative half-cycle of the modulating signal is equal to the decrease in gain during the positive half-cycle. The voltage gain is approximately $A_q \pm 40$. This is called *linear* or *symmetrical modulation* and is desired. Essentially, a conventional AM DSBFC receiver reproduces the original modulating signal from the shape of the envelope. If modulation is not symmetrical, the envelope does not accurately represent the shape of the modulating signal, and the demodulator will produce a distorted output waveform.

EXAMPLE 3-6

For a low-level AM modulator similar to the one shown in Figure 3-15 with a modulation coefficient $m = 0.8$, a quiescent voltage gain $A_q = 100$, an input carrier frequency $f_c = 500$ kHz with an amplitude $V_c = 5$ mV, and a 1000-Hz modulating signal, determine:

(a) Maximum and minimum voltage gains.
(b) Maximum and minimum amplitudes for V_{out}.
(c) Sketch the output AM envelope.

Solution (a) Substituting into Equation 3-26,

$$A_{max} = 100(1 + 0.8) = 180$$

$$A_{min} = 100(1 - 0.8) = 20$$

(b)
$$V_{out(max)} = 180(0.005) = 0.9 \text{ V}$$

$$V_{out(min)} = 20(0.005) = 0.1 \text{ V}$$

(c) The AM envelope is shown in Figure 3-16.

With no modulating signal, the low-level modulator shown in Figure 3-15 is a linear amplifier. However, when a modulating signal is applied, the Q-point of the amplifier is driven first toward saturation and then toward cutoff (that is, the transistor is forced to operate over a nonlinear portion of its operating curve).

The transistor modulator shown in Figure 3-15 is adequate for low-power applications but is not a practical circuit when high output powers are required. This is because the transistor is biased for class A operation, which is extremely inefficient. In addition, because the transistor characteristics are not the same when driven to cutoff as when driven to saturation, the output envelope is not symmetrical. In addition, the output wave contains harmonic and cross-product components of the modulating signal, the carrier, and their harmonic frequencies. Although most of the unwanted frequencies can be removed with filtering, any amplifiers that follow an AM modulator must be linear. If they are not, intermodulation between the upper and lower side frequencies and the carrier will generate additional cross-product frequencies that could interfere with signals from other transmitters. High-power linear amplifiers are highly undesirable because of their poor efficiency. The modulator shown in Figure 3-15 is a low-level modulator regardless of whether it is the final stage or not because modulation takes place in the emitter, which is not the output element.

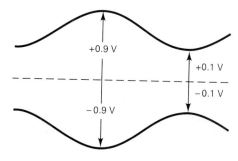

Figure 3-16 AM envelope for Example 3-6.

AM Modulator Circuits

Medium-power AM Modulator

Early medium- and high-power AM transmitters were limited to those that used vacuum tubes for the active devices. However, since the mid 1970s, solid-state transmitters have been available with output powers as high as several thousand watts. This is accomplished by placing several final power amplifiers in parallel such that their output signals combine in phase and are thus additive.

Figure 3-17a shows the schematic diagram for a single-transistor medium-power AM modulator. The modulation takes place in the collector, which is the output element of the transistor. Therefore, if this is the final active stage of the transmitter (that is, there are no amplifiers between it and the antenna), it is a high-level modulator.

To achieve high power efficiency, medium- and high-power AM modulators generally operate class C. Therefore, a practical efficiency of as high as 80% is possible. The circuit shown in Figure 3-17a is a class C amplifier with two inputs: a carrier (v_c) and a single-frequency modulating signal (v_m). Because the transistor is biased class C, it operates nonlinear and is capable of nonlinear mixing (modulation). This circuit is called a *collector modulator* because the modulating signal is applied directly to the collector. The RFC is a radio-frequency choke that acts like a short to dc and an open to high frequencies. Therefore, the RFC isolates the dc power supply from the high-frequency carrier and side frequencies, while still allowing the low-frequency intelligence signals to modulate the collector of Q_1.

Circuit operation. For the following explanation, refer to the circuit shown in Figure 3-17a and the waveforms shown in Figure 3-17b. When the amplitude of the carrier exceeds the barrier potential of the base–emitter junction (approximately 0.7 V for a silicon transistor), Q_1 turns on and collector current flows. When the amplitude of the carrier drops below 0.7 V, Q_1 turns off and collector current ceases. Consequently, Q_1 switches between saturation and cutoff controlled by the carrier signal, collector current flows for less than 180° of each carrier cycle, and class C operation is achieved. Each successive cycle of the carrier turns Q_1 on for an instant and allows current to flow for a short time, producing a negative-going waveform at the collector. The collector current and voltage waveforms are shown in Figure 3-17b. The collector voltage waveform resembles a repetitive half-wave rectified signal with a fundamental frequency equal to f_c.

When a modulating signal is applied to the collector in series with the dc supply voltage, it adds to and subtracts from V_{CC}. The waveforms shown in Figure 3-17c are produced when the maximum peak modulating signal amplitude equals V_{CC}. It can be seen that the output voltage waveform swings from a maximum value of $2V_{CC}$ to approximately 0 V [$V_{CE(sat)}$]. The peak change in collector voltage is equal to V_{CC}. Again, the waveform resembles a half-wave rectified carrier superimposed onto a low-frequency ac intelligence signal.

Because Q_1 is operating nonlinear, the collector waveform contains the two original input frequencies (f_c and f_m) and their sum and difference frequencies ($f_c \pm f_m$). Because the output waveform also contains the higher-order harmonics and intermodulation components, it must be bandlimited to $f_c \pm f_m$ before being transmitted.

A more practical circuit for producing a medium-power AM DSBFC signal is shown in Figure 3-18a, with corresponding waveforms shown in Figure 3-18b. This circuit is also a collector modulator with a maximum peak modulating signal amplitude $V_{m\,(max)}$ = V_{CC}. Operation of this circuit is almost identical to the circuit shown in Figure 3-17a

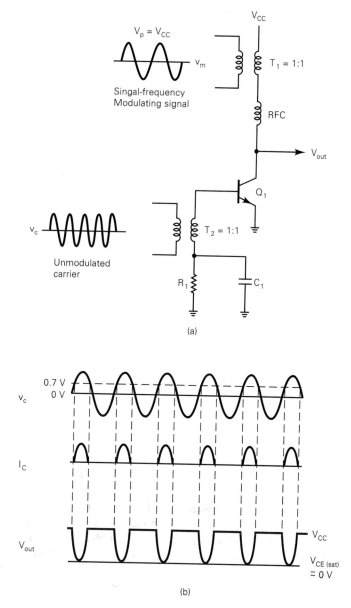

Figure 3-17 Simplified medium-power transistor AM DSBFC modulator:
(a) schematic diagram; (b) collector waveforms with no modulating signal;
(Continued on next page.)

except for the addition of a tank circuit (C_1 and L_1) in the collector of Q_1. Because the transistor is operating between saturation and cutoff, collector current is not dependent on base drive voltage. The voltage developed across the tank circuit is determined by the ac component of the collector current and the impedance of the tank circuit at resonance, which depends on the quality factor (Q) of the coil. The waveforms for the modulating signal, carrier, and collector current are identical to those of the previous example. The output voltage is a symmetrical AM DSBFC signal with an average voltage of 0 V, a maximum positive peak amplitude equal to $2V_{CC}$, and a maximum negative peak amplitude

AM Modulator Circuits

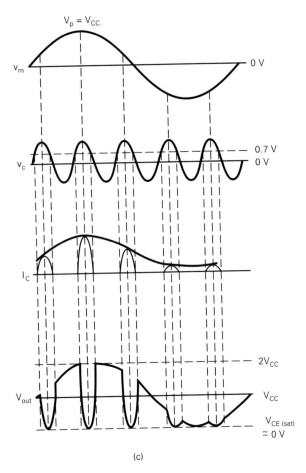

Figure 3-17 *(continued)* (c) collector waveforms with modulating signal.

(c)

equal to $-2V_{CC}$. The positive half-cycle of the output waveform is produced in the tank circuit by the *flywheel effect*. When Q_1 is conducting, C_1 charges to $V_{CC} + V_m$ (a maximum value of $2V_{CC}$ and, when Q_1 is off, C_1 discharges through L_1). When L_1 discharges, C_1 charges to a minimum value of $-2V_{CC}$. This produces the positive half-cycle of the AM envelope. The resonant frequency of the tank circuit is equal to the carrier frequency, and the bandwidth extends from $f_c - f_m$ to $f_c + f_m$. Consequently, the modulating signal, the harmonics, and all the higher-order cross products are removed from the waveform, leaving a symmetrical AM DSBFC wave. One hundred percent modulation occurs when the peak amplitude of the modulating signal equals V_{CC}.

Several components shown in Figure 3-18a have not been explained. R_1 is the bias resistor for Q_1. R_1 and C_2 form a clamper circuit that produce a reverse "self" bias and, in conjunction with the barrier potential of the transistor, determine the turn-on voltage for Q_1. Consequently, Q_1 can be biased to turn on only during the most positive peaks of the carrier voltage. This produces a narrow collector current waveform and enhances class C efficiency.

C_3 is a bypass capacitor that looks like a short to the modulating signal frequencies, preventing the information signals from entering the dc power supply. C_1 is the base-to-collector junction capacitance of Q_1. At radio frequencies, the relatively small junction capacitances within the transistor are insignificant. If the capacitive reactance of C_1 is significant, the collector signal may be returned to the base with sufficient amplitude to cause

Chap. 3 **Amplitude Modulation Transmission**

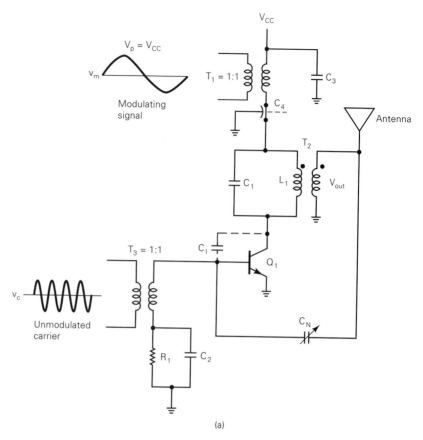

Figure 3-18 Medium-power transistor AM DSBFC modulator: (a) schematic diagram; *(Continued on next page.)*

Q_1 to begin oscillating. Therefore, a signal of equal amplitude and frequency and 180° out of phase must be fed back to the base to cancel or *neutralize* the *intercapacitive feedback*. C_N is a *neutralizing capacitor*. Its purpose is to provide a feedback path for a signal that is equal in amplitude and frequency but 180° out of phase with the signal fed back through C_1. C_4 is an RF bypass capacitor. Its purpose is to isolate the dc power supply from radio frequencies. Its operation is quite similar: at the carrier frequency, C_4 looks like a short circuit, preventing the carrier from *leaking* into the power supply or the modulating signal circuitry and being distributed throughout the transmitter.

Simultaneous Base and Collector Modulation

Collector modulators produce a more symmetrical envelope than low-power emitter modulators, and collector modulators are more power efficient. However, collector modulators require a higher amplitude-modulating signal, and they cannot achieve a full saturation-to-cutoff output voltage swing, thus preventing 100% modulation from occurring. Therefore, to achieve symmetrical modulation, operate at maximum efficiency, develop a high output power, and require as little modulating signal drive power as possible, emitter and collector modulations are sometimes used simultaneously.

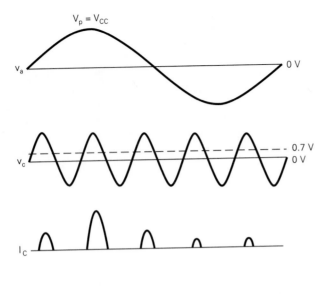

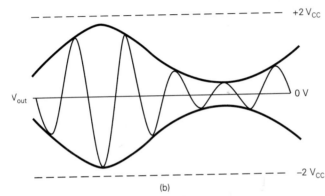

Figure 3-18 *(continued)*
(b) collector and output waveforms.

Circuit operation. Figure 3-19 shows an AM modulator that uses a combination of both emitter and collector modulations. The modulating signal is simultaneously fed into the collectors of the push-pull modulators (Q_2 and Q_3) and to the collector of the driver amplifier (Q_1). Collector modulation occurs in Q_1; thus, the carrier signal on the base of Q_2 and Q_3 has already been partially modulated and the modulating signal power can be reduced. Also, the modulators are not required to operate over their entire operating curve to achieve 100% modulation.

Linear Integrated-circuit AM Modulators

Linear integrated-circuit function generators use a unique arrangement of transistors and FETs to perform signal multiplication. This characteristic makes them ideally suited for generating AM waveforms. Integrated circuits, unlike their discrete counterparts, can precisely match current flow, amplifier voltage gain, and temperature variations. Linear integrated-circuit AM modulators also offer excellent frequency stability, symmetrical modulation characteristics, circuit miniaturization, fewer components, temperature immunity, and simplicity of design. Their disadvantages include low output power, a relatively low usable frequency range, and susceptibility to fluctuations in the dc supply voltage.

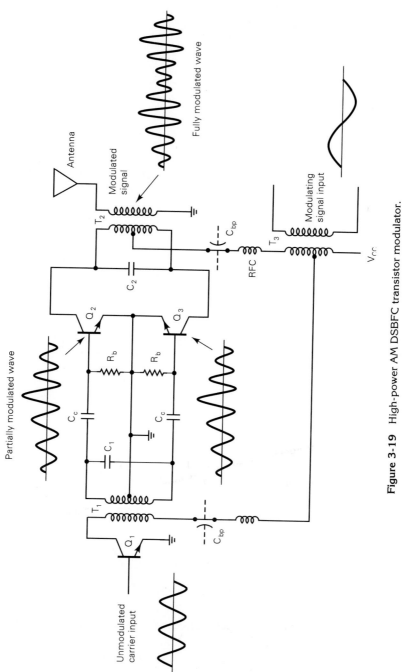

Figure 3-19 High-power AM DSBFC transistor modulator.

Circuit operation. The XR-2206 monolithic function generator is ideally suited for performing amplitude modulation. Figure 3-20a shows the schematic diagram for an integrated-circuit AM modulator that uses the XR-2206. The free-running frequency of the VCO in the XR-2206 function generator is the carrier, and its frequency is determined by external timing capacitor C_1 and resistor R_1. The modulating signal is applied to pin 1, and the output AM envelope appears at pin 2. Figure 3-20b shows the normalized output amplitude-versus-input bias voltage characteristics for the VCO. The amplitude of the out-

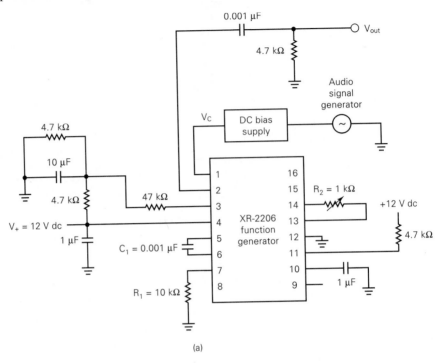

(a)

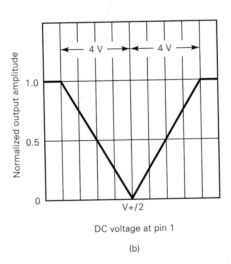

(b)

Figure 3-20 Linear integrated-circuit AM modulator: (a) block diagram of an XR-2206 AM modulator; (b) output voltage versus input voltage curve.

put voltage varies linearly with applied voltage for values between ± 4 V of $V^+/2$. Because the output amplitude is proportional to supply voltage V^+, a well-regulated dc supply must be used with this configuration.

The dc output level on pin 2 is approximately the same as the dc voltage on pin 3, which is generally biased midway between V^+ and ground to allow for a maximum symmetrical ac output signal. The output wave is a symmetrical AM DSBFC signal containing f_c and $f_c \pm f_m$.

Vacuum-tube AM Modulators

With the advent of solid-state medium- and high-power AM modulators, *vacuum-tube circuits* have somewhat gone by the wayside except for specialized applications requiring an exceptionally high transmit power. Vacuum-tube modulators have several disadvantages compared to solid-state modulators. Vacuum tubes require higher dc supply voltages (often, both positive and negative polarities for a single circuit), a separate *filament* (*heater*) supply voltage, and very high modulating signal amplitudes, and vacuum tubes take up considerably more space and weigh much more than their solid-state counterparts.

Consequently, a vacuum-tube modulator (or, for that matter, any other vacuum-tube circuit) is seldom used today. Essentially, vacuum-tube modulators are used in applications where extreme high powers are required, such as in commercial AM and television broadcast-band transmitters. For the most common low- and medium-power applications, such as for two-way radio communications, solid-state transmitters dominate the marketplace.

For historical perspective (that is, a look back into the *cavetronic* era), several vacuum-tube AM modulators are shown in Figure 3-21. Assuming that the reader is familiar with vacuum-tube terminology (plate, grid, cathode, filaments, triode, pentode, and the like), the operation of vacuum-tube modulators is quite similar to comparable transistor modulators. However, because there are more elements in a vacuum tube, more modulator configurations are possible with vacuum tubes than with either transistors or integrated circuits (for example, suppressor- and screen-grid modulators).

High-level vacuum-tube *triode modulators* are class C biased and use place modulation (Figure 3-21a). A high-power modulating signal is transformer coupled into the plate circuit in series with the dc supply voltage. Consequently, the modulating signal adds to and subtracts from E_{BB}. A relatively low power carrier signal is coupled into the grid circuit, where it biases the tube into conduction only during its most positive peaks. Therefore, as with transistor collector modulators, narrow pulses of plate current flow and supply the energy necessary to sustain oscillations in the plate tank circuit.

Figure 3-21b shows a *multigrid* vacuum-tube plate modulator. The vacuum tube is a *pentode*. It has five elements: plate, cathode, control grid, suppressor grid, and screen grid. Pentode vacuum-tube modulators are capable of high output powers and high efficiencies. However, to achieve 100% modulation, it is necessary to modulate the screen grid as well as the plate.

Figure 3-21c shows a vacuum-tube *grid-bias modulator*. The triode can be replaced with a *tetrode* or pentode. The grid-bias modulator requires much less modulating signal power than the triode or pentode plate modulators, but it has poorer modulation symmetry, more distortion, and lower plate efficiency and produces a lower output power.

Figures 3-21d and e show *screen-grid* and *suppressor-grid* modulators, respectively. These two modulator configurations fall somewhere between the grid-bias and plate modulators discussed previously.

AM Modulator Circuits

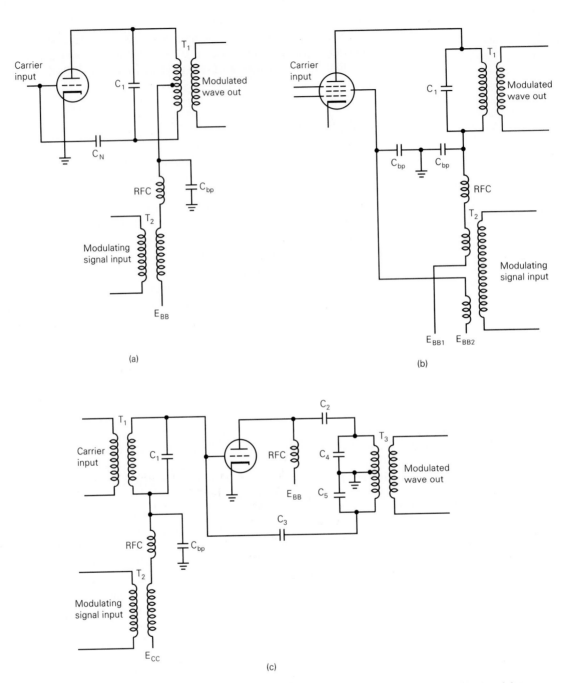

Figure 3-21 Vacuum-tube AM DSBFC modulator circuits: (a) triode plate modulator; (b) multi-grid plate modulator; (c) grid-bias modulator; *(Continued on next page.)*

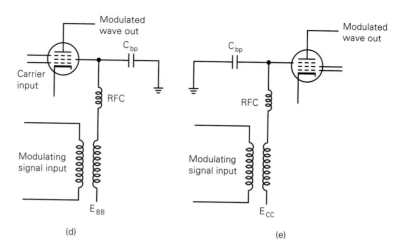

(d)   (e)

Figure 3-21 *(continued)*   (d) screen-grid modulator; (e) suppressor-grid modulator.

AM TRANSMITTERS

Low-level Transmitters

Figure 3-22 shows a block diagram for a low-level AM DSBFC transmitter. For voice or music transmission, the source of the modulating signal is generally an acoustical transducer, such as a microphone, a magnetic tape, a CD disk, or a phonograph record. The *preamplifier* is typically a sensitive, class A linear voltage amplifier with a high input impedance. The function of the preamplifier is to raise the amplitude of the source signal to a usable level while producing minimum nonlinear distortion and adding as little thermal noise as possible. The driver for the modulating signal is also a linear amplifier that simply amplifies the information signal to an adequate level to sufficiently drive the modulator. More than one drive amplifier may be required.

The RF *carrier oscillator* can be any of the oscillator configurations discussed in Chapter 2. The FCC has stringent requirements on transmitter accuracy and stability; therefore, crystal-controlled oscillators are the most common circuits used. The *buffer amplifier* is a low-gain, high-input impedance linear amplifier. Its function is to isolate the oscillator from the high-power amplifiers. The buffer provides a relatively constant load to the oscillator, which helps to reduce the occurrence and magnitude of short-term frequency

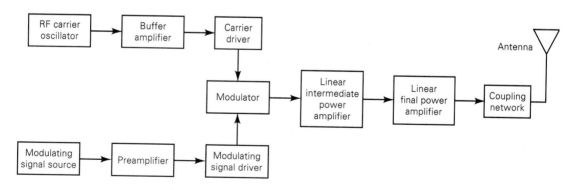

Figure 3-22 Block diagram of a low-level AM DSBFC transmitter.

variations. Emitter followers or integrated-circuit op-amps are often used for the buffer. The modulator can use either emitter or collector modulation. The intermediate and final power amplifiers are either linear class A or class B push-pull. This is required with low-level transmitters to maintain symmetry in the AM envelope. The antenna coupling network matches the output impedance of the final power amplifier to the transmission line and antenna.

Low-level transmitters like the one shown in Figure 3-22 are used predominantly for low-power, low-capacity systems such as wireless intercoms, remote-control units, pagers, and short-range walkie-talkies.

High-level Transmitters

Figure 3-23 shows the block diagram for a high-level AM DSBFC transmitter. The modulating signal is processed in the same manner as in the low-level transmitter except for the addition of a power amplifier. With high-level transmitters, the power of the modulating signal must be considerably higher than is necessary with low-level transmitters. This is because the carrier is at full power at the point in the transmitter where modulation occurs and, consequently, requires a high-amplitude modulating signal to produce 100% modulation.

The RF carrier oscillator, its associated buffer, and the carrier driver are also essentially the same circuits used in low-level transmitters. However, with high-level transmitters, the RF carrier undergoes additional power amplification prior to the modulator stage, and the final power amplifier is also the modulator. Consequently, the modulator is generally a drain-, plate-, or collector-modulated class C amplifier.

With high-level transmitters, the modulator circuit has three primary functions. It provides the circuitry necessary for modulation to occur (that is, nonlinearity), it is the final power amplifier (class C for efficiency), and it is a frequency up-converter. An up-converter simply translates the low-frequency intelligence signals to radio-frequency signals that can be efficiently radiated from an antenna and propagated through free space.

Trapezoidal Patterns

Trapezoidal patterns are used for observing the modulation characteristics of AM transmitters (that is, coefficient of modulation and modulation symmetry). Although the modulation characteristics can be examined with an oscilloscope, a trapezoidal pattern is more easily and accurately interpreted. Figure 3-24 shows the basic test setup for producing a

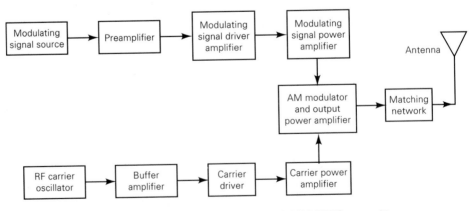

Figure 3-23 Block diagram of a high-level AM DSBFC transmitter.

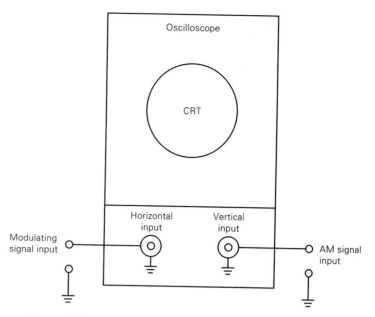

Figure 3-24 Test setup for displaying a trapezoidal pattern on an oscilloscope.

trapezoidal pattern on the CRT of a standard oscilloscope. The AM wave is applied to the vertical input of the oscilloscope, and the modulating signal is applied to the external horizontal input with the internal horizontal sweep disabled. Therefore, the horizontal sweep rate is determined by the modulating signal frequency, and the magnitude of the horizontal deflection is proportional to the amplitude of the modulating signal. The vertical deflection is totally dependent on the amplitude and rate of change of the modulated signal. In essence, the electron beam emitted from the cathode of the CRT is acted on simultaneously in both the horizontal and vertical planes.

Figure 3-25 shows how the modulated signal and the modulating signal produce a trapezoidal pattern. With an oscilloscope, when 0 V is applied to the external horizontal input, the electron beam is centered horizontally on the CRT. When a voltage other than 0 V is applied to the vertical or horizontal inputs, the beam will deflect vertically and horizontally, respectively. If we begin with both the modulated wave and the modulating signal at 0 V (t_0), the electron beam is located in the center of the CRT. As the modulating signal goes positive, the beam deflects to the right. At the same time the modulated signal is going positive, deflecting the beam upward. The beam continues to deflect to the right until the modulating signal reaches its maximum positive value (t_1). While the beam moves toward the right, it is also deflected up and down as the modulated signal alternately swings positive and negative. Notice that on each successive alternation the modulated signal reaches a higher magnitude than the previous alternation. Therefore, as the CRT beam is deflected to the right, its peak-to-peak vertical deflection increases with each successive cycle of the modulated signal. As the modulating signal becomes less positive, the beam is deflected to the left (toward the center of the CRT). At the same time, the modulated signal alternately swings positive and negative, deflecting the beam up and down, except now each successive alternation is lower in amplitude than the previous alternation. Consequently, as the beam moves horizontally toward the center of the CRT, the vertical deflection decreases. The modulating signal and the modulated signal pass through 0 V at the same time, and the beam is again in the center of the CRT (t_2). As the

AM Transmitters

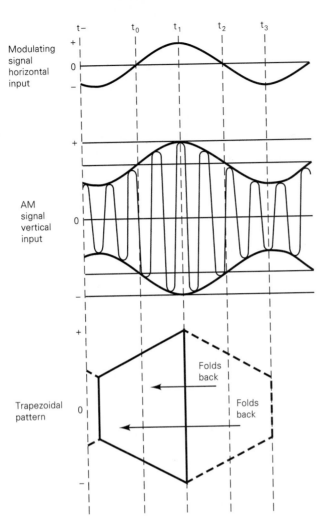

Figure 3-25 Producing a trapezoidal pattern.

modulating signal goes negative, the beam is deflected to the left side of the CRT. At the same time, the modulated signal is decreasing in amplitude on each successive alternation. The modulating signal reaches its maximum negative value at the same time as the modulated signal reaches its minimum amplitude (t_3). The trapezoidal pattern shown between times t_1 and t_3 folds back on top of the pattern displayed during times $t-$ and t_1. Thus, a complete trapezoidal pattern is displayed on the screen after both the left-to-right and right-to-left horizontal sweeps are complete.

If the modulation is symmetrical, the top half of the modulated signal is a mirror image of the bottom half, and a trapezoidal pattern like the one shown in Figure 3-26a is produced. At 100% modulation, the minimum amplitude of the modulated signal is zero, and the trapezoidal pattern comes to a point at one end as shown in Figure 3-26b. If the modulation exceeds 100%, the pattern shown in Figure 3-26c is produced. The pattern shown in Figure 3-26a is a 50% modulated wave. If the modulating signal and the modulated signal are out of phase, a pattern similar to the one shown in Figure 3-26d is produced. If the magnitude of the positive and negative alternations of the modulated signal are not equal, the pattern shown in Figure 3-26e results. If the phase of the modulating signal is shifted 180° (inverted), the trapezoidal patterns would simply point in the opposite direction. As you can see, percent modulation and modulation symmetry are more easily

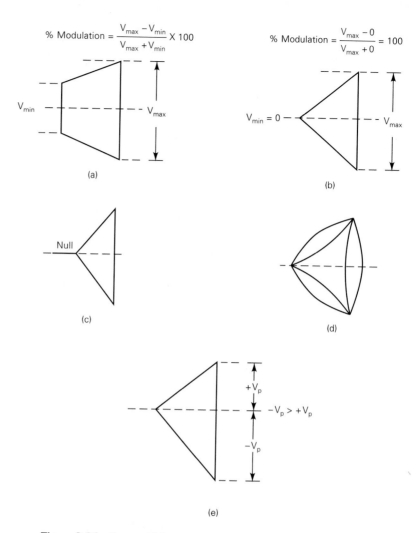

$$\% \text{ Modulation} = \frac{V_{max} - V_{min}}{V_{max} + V_{min}} \times 100$$

$$\% \text{ Modulation} = \frac{V_{max} - 0}{V_{max} + 0} = 100$$

(a)

(b)

(c)

(d)

(e)

Figure 3-26 Trapezoidal patterns: (a) linear 50% AM modulation; (b) 100% AM modulation; (c) more than 100% AM modulation; (d) improper phase relationship; (e) nonsymmetrical AM envelope.

observed with a trapezoidal pattern than with a standard oscilloscope display of the modulated signal.

Carrier Shift

Carrier shift is a term that is often misunderstood or misinterpreted. Carrier shift is sometimes called *upward* or *downward modulation* and has absolutely nothing to do with the frequency of the carrier. *Carrier shift* is a form of amplitude distortion introduced when the positive and negative alternations in the AM modulated signal are not equal (that is, nonsymmetrical modulation). Carrier shift may be either positive or negative. If the positive alternation of the modulated signal has a larger amplitude than the negative alternation, positive carrier shift results. If the negative alternation is larger than the positive, negative carrier shift occurs.

Carrier shift is an indication of the average voltage of an AM modulated signal. If

AM Transmitters

the positive and negative halves of the modulated signal are equal, the average voltage is 0 V. If the positive half is larger, the average voltage is positive, and if the negative half is larger, the average voltage is negative. Figure 3-27a shows a symmetrical AM envelope (no carrier shift); the average voltage is 0 V. Figures 3-27b and c show positive and negative carrier shifts, respectively.

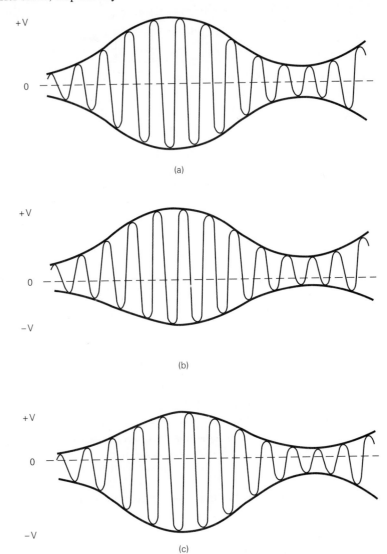

Figure 3-27   Carrier shift: (a) linear modulation; (b) positive carrier shift; (c) negative carrier shift.

QUESTIONS

3-1. Define *amplitude modulation*.

3-2. What is meant by the term *RF*?

3-3. How many inputs are there to an amplitude modulator? What are they?

3-4. In an AM system, what is meant by the following terms: *modulating signal, carrier,* and *modulated wave*?

3-5. Describe an AM DSBFC waveform. Why is the shape of amplitude variations called an envelope?

3-6. Describe upper and lower sidebands and upper and lower side frequencies.

3-7. Define *modulation coefficient.*

3-8. Define *percent modulation.*

3-9. What are the highest coefficient of modulation and percent modulation that can occur without causing excessive distortion?

3-10. Describe the meaning of each term in the following equation

$$v_{am}(t) = E_c \sin (2\pi f_c t) - \frac{mE_c}{2} \cos [2\pi (f_c + f_m)t] + \frac{mE_c}{2} \cos [2\pi (f_c - f_m)t]$$

3-11. Describe the meaning of each term in the following equation:

$$v_{am}(t) = 10 \sin (2\pi 500kt) - 5 \cos [2\pi 515kt] + 5 \cos [2\pi 485kt]$$

3-12. What effect does modulation have on the amplitude of the carrier component of the modulated signal spectrum?

3-13. Describe the significance of the following formula:

$$P_t = P_c \left(1 + \frac{m^2}{2} \right)$$

3-14. What does AM DSBFC stand for?

3-15. Describe the relationship between the carrier and sideband power in an AM DSBFC wave.

3-16. What is the predominant disadvantage of AM double-sideband full carrier transmission?

3-17. What is the predominant advantage of AM double-sideband full carrier transmission?

3-18. What is the maximum ratio of sideband-to-total transmitted power that can be achieved with AM DSBFC?

3-19. Why do any amplifiers that follow the modulator in an AM DSBFC system have to be linear?

3-20. What is the primary disadvantage of a low-power class A transistor modulator?

3-21. Describe the difference between a low- and a high-level modulator.

3-22. List the advantages of low-level modulation; high-level modulation.

3-23. What is the advantage of using a trapezoidal pattern to evaluate an AM envelope?

PROBLEMS

3-1. If a 20 V modulated wave changes in amplitude ±5 V, determine the modulation coefficient and percent modulation.

3-2. For a maximum positive envelope voltage of 12 V and a minimum positive envelope amplitude of 4 V, determine the modulation coefficient and percent modulation.

3-3. For an envelope with $+V_{max}$ = 40 V and $+V_{min}$ = 10 V, determine:
   (a) Unmodulated carrier amplitude.
   (b) Peak change in amplitude of the modulated wave.
   (c) Coefficient of modulation and percent modulation.

3-4. For a unmodulated carrier amplitude of 16 V and a modulation coefficient m = 0.4, determine the amplitudes of the modulated carrier and side frequencies.

3-5. Sketch the envelope for problem 3-4 (label all pertinent voltages).

3-6. For the AM envelope shown below, determine:
  (a) Peak amplitude of the upper and lower side frequencies.
  (b) Peak amplitude of the carrier.
  (c) Peak change in the amplitude of the envelope.
  (d) Modulation coefficient.
  (e) Percent modulation.

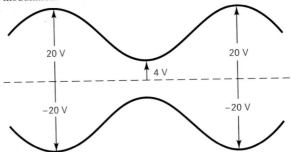

3-7. One input to an AM DSBFC modulator is an 800-kHz carrier with a amplitude of 40 V. The second input is a 25-kHz modulating signal whose amplitude is sufficient to produce a ± 10 V change in the amplitude of the envelope. Determine:
  (a) Upper and lower side frequencies.
  (b) Modulation coefficient and percent modulation.
  (c) Maximum and minimum positive peak amplitudes of the envelope.
  (d) Draw the output spectrum.
  (e) Sketch the envelope (label all pertinent voltages).

3-8. For a modulation coefficient $m = 0.2$ and a carrier power $P_c = 1000$ W, determine:
  (a) Sideband power.
  (b) Total transmitted power.

3-9. For an AM DSBFC wave with an unmodulated carrier voltage of 25 V and a load resistance of 50 Ω, determine:
  (a) Power of the unmodulated carrier.
  (b) Power of the modulated carrier and the upper and lower side frequencies for a modulation coefficient m = 0.6.

3-10. Determine the quiescent, maximum, and minimum voltage gains for the emitter modulator shown below with the indicated carrier and modulating signal amplitudes.

3-11. Sketch the output envelope and draw the output frequency spectrum for the circuit shown in Problem 3-10.

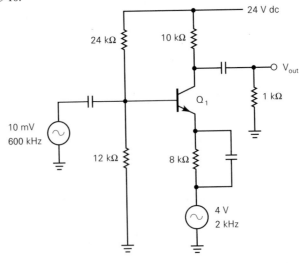

Chap. 3   Amplitude Modulation Transmission

3-12. For a low-power transistor modulator with a modulation coefficient $m = 0.4$, a quiescent gain $A_q = 80$, and an input carrier amplitude of 0.002 V, determine:

    (a) Maximum and minimum voltage gains.

    (b) Maximum and minimum voltages for V_{out}.

    (c) Sketch the envelope.

3-13. For the trapezoidal pattern shown below, determine:

    (a) Modulation coefficient.

    (b) Percent modulation.

    (c) Carrier amplitude.

    (d) Upper and lower side frequency amplitudes.

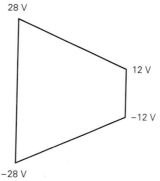

3-14. For an AM modulator with a carrier frequency $f_c = 200$ kHz and a maximum modulating signal frequency $f_{m(max)} = 10$ kHz, determine:

    (a) Frequency limits for the upper and lower sidebands.

    (b) Upper and lower side frequencies produced when the modulating signal is a 7-kHz tone.

    (c) Bandwidth for the maximum modulating signal frequency.

    (d) Draw the output spectrum.

Problems

4

Amplitude Modulation Reception

AM *reception* is the reverse process of AM transmission. A conventional AM receiver simply converts an amplitude-modulated wave back to the original source information (that is, demodulates the AM wave). When an AM wave is demodulated, the carrier and the information-carrying portion of the envelope (that is, the sidebands) are *down-con-verted* or translated from the radio-frequency spectrum to the original source information. The purpose of this chapter is to describe the process of AM demodulation and to show several receiver configurations that accomplish this process.

A receiver must be capable of receiving, amplifying, and demodulating an RF signal. A receiver must also be capable of *bandlimiting* the total radio-frequency spectrum to a specific band of frequencies. In many applications the receiver must be capable of changing the range (band) of frequencies that it is capable of receiving. This process is called *tuning* the receiver. Once an RF signal has been received, amplified, and bandlimited, it must be converted to the original source information. This process is called *demodulation*. Once demodulated, the information may require further bandlimiting and amplification before it is considered usable.

To completely understand the demodulation process, it is necessary first to have a basic understanding of the terminology used to describe the characteristics of receivers and receiver circuits.

Figure 4-1 shows a simplified block diagram of a typical AM receiver. The RF section is the first stage and is therefore often called the receiver *front end*. The primary functions of the RF section are detecting, bandlimiting, and amplifying the received RF signals. In essence, the RF section establishes the *receiver threshold* (that is, the minimum RF signal level that the receiver can detect and demodulate to a usable information signal).

144

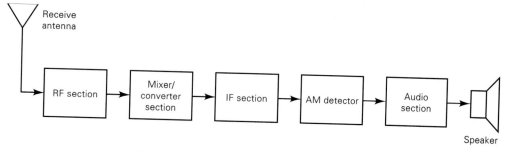

Figure 4-1 Simplified block diagram of an AM receiver.

The RF section comprises one or more of the following circuits: *antenna, antenna coupling network, filter (preselector)*, and one or more *RF amplifiers*. The *mixer/converter* section down-converts the received RF frequencies to *intermediate frequencies* (IF). The IF section generally includes several cascaded amplifiers and bandpass filters. The primary functions of the IF section are amplification and selectivity. The *AM detector* demodulates the AM wave and recovers the original source information. The audio section simply amplifies the recovered information to a usable level.

Receiver Parameters

Several parameters are used to evaluate the ability of a receiver to successfully demodulate an RF signal. These parameters include *selectivity, bandwidth improvement, sensitivity, dynamic range, fidelity, insertion loss, noise temperature,* and *equivalent noise temperature.*

Selectivity. Selectivity is the measure of the ability of a receiver to accept a given band of frequencies and reject all others. For example, with the commercial AM broadcast band, each station's transmitter is allocated a 10-kHz bandwidth (the carrier ±5 kHz). Therefore, for a receiver to select only those frequencies associated with a single channel, the input to the demodulator must be bandlimited to the desired 10-kHz passband. If the passband of the receiver is greater than 10 kHz, more than one channel may be received and demodulated simultaneously. If the passband of the receiver is less than 10 kHz, a portion of the source information for that channel is rejected or blocked from entering the demodulator and, consequently, lost.

Selectivity is defined as the measure of the extent to which a receiver is capable of differentiating between the desired information signals and disturbances or information signals at other frequencies. It may be expressed quantitatively as both the bandwidth and the ratio of the bandwidth of the receiver at some predetermined *attenuation factor* (commonly −60 dB) to the bandwidth at the −3 dB (half-power) points. This ratio is often called the *shape factor (SF)* and is determined by the number of *poles* and the *Q-factors* of the receiver's input filters. The shape factor defines the shape of the gain-versus-frequency plot for a filter and is expressed mathematically as

$$SF = \frac{B(-60 \text{ dB})}{B(-3 \text{ dB})} \tag{4-1a}$$

For perfect filtering, the attenuation factor is infinite and the bandwidth at the −3-dB frequencies is equal to the bandwidth at the −60-dB frequencies. Therefore, the

Introduction

145

ideal shape factor is unity. Selectivity is often given as a percentage and expressed mathematically as

$$\% \text{ Selectivity} = \text{SF} \times 100 \qquad (4\text{-}1b)$$

EXAMPLE 4-1

Determine the shape factor and percent selectivity for the gain-versus-frequency plot shown in Figure 4-2.

Solution The shape factor is determined from Equation 4-1a

$$\text{SF} = \frac{100 \text{ kHz}}{10 \text{ kHz}} = 10$$

and percent selectivity from Equation 4-1b as

$$\% \text{ Selectivity} = 100 \times 10 = 1000\%$$

Selectivity is sometimes stated as simply the ratio of the actual bandwidth for a particular system to the minimum bandwidth required to propagate the information signals through the system. Mathematically, this is stated as

$$\% \text{ Selectivity} = \frac{B_{\text{actual}}}{B_{\text{minimum}}} \times 100 \qquad (4\text{-}1c)$$

Bandwidth improvement. As stated in Chapter 1 and given in Equation 1-24, thermal noise is directly proportional to bandwidth. Therefore, if the bandwidth is reduced, noise is also reduced by the same proportion. The noise reduction ratio achieved by reducing the bandwidth is called *bandwidth improvement (BI)*. As a signal propagates from the antenna through the RF section, the mixer/converter section, and the IF section, the bandwidth is reduced. Effectively, this is equivalent to reducing (improving) the noise figure of the receiver. The bandwidth improvement factor is the ratio of the RF bandwidth to the IF bandwidth. Mathematically, bandwidth improvement is

$$\text{BI} = \frac{B_{\text{RF}}}{B_{\text{IF}}} \qquad (4\text{-}2a)$$

where
$$\text{BI} = \text{bandwidth improvement}$$
$$B_{\text{RF}} = \text{RF bandwidth (hertz)}$$
$$B_{\text{IF}} = \text{IF bandwidth (hertz)}$$

The corresponding reduction in the noise figure due to the reduction in bandwidth is called *noise figure improvement* and is expressed mathematically as

$$\text{NF}_{\text{improvement}} = 10 \log \text{BI} \qquad (4\text{-}2b)$$

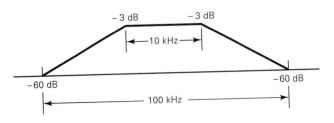

Figure 4-2 Gain-frequency plot for Example 4-1.

EXAMPLE 4-2

Determine the improvement in the noise figure for a receiver with an RF bandwidth equal to 200 kHz and an IF bandwidth equal to 10 kHz.

Solution Bandwidth improvement is found by substituting into Equation 4-2a:

$$BI = \frac{200 \text{ kHz}}{10 \text{ kHz}} = 20$$

and noise figure improvement is found by substituting into Equation 4-2b:

$$NF_{improvement} = 10 \log 20 = 13 \text{ dB}$$

Sensitivity. The *sensitivity* of a receiver is the minimum RF signal level that can be detected at the input to the receiver and still produce a usable demodulated information signal. What constitutes a usable information signal is somewhat arbitrary. Generally, the signal-to-noise ratio and the power of the signal at the output of the audio section are used to determine the quality of a received signal and whether it is usable or not. For commercial AM broadcast band receivers, a 10-dB or more signal-to-noise ratio with 1/2 W (27 dBm) of power at the output of the audio section is considered to be usable. However, for broadband microwave receivers, a 40 dB or more signal-to-noise ratio with approximately 5 mW (7 dBm) of signal power is the minimum acceptable value. The sensitivity of a receiver is usually stated in microvolts of received signal. For example, a typical sensitivity for a commercial broadcast band AM receiver is 50 μV, and a two-way mobile radio receiver generally has a sensitivity between 0.1 and 10 μV. Receiver sensitivity is also called receiver *threshold*. The sensitivity of an AM receiver depends on the noise power present at the input to the receiver, the receiver's noise figure (an indication of the noise generated in the front end of the receiver), the sensitivity of the AM detector, and the bandwidth improvement factor of the receiver. The best way to improve the sensitivity of a receiver is to reduce the noise level. This can be accomplished by reducing either the temperature or the bandwidth of the receiver or improving the receiver's noise figure.

Dynamic range. The *dynamic range* of a receiver is defined as the difference in decibels between the minimum input level necessary to discern a signal and the input level that will overdrive the receiver and produce distortion. In simple terms, dynamic range is the input power range over which the receiver is useful. The minimum receive level is a function of front-end noise, noise figure, and the desired signal quality. The input signal level that will produce overload distortion is a function of the net gain of the receiver (the total gain of all the stages in the receiver). The high-power limit of a receiver depends on whether it will operate with a single- or multiple-frequency input signal. If single-frequency operation is used, the *1-dB compression point is* generally used for the upper limit of usefulness. The 1-dB compression point is defined as the output power when the RF amplifier response is 1 dB less than the ideal linear-gain response. Figure 4-3 shows the linear gain and 1-dB compression point for a typical amplifier where the linear gain drops off just prior to saturation. The 1-dB compression point is often measured directly as the point where a 10-dB increase in input power results in a 9-dB increase in output power.

A dynamic range of 100 dB is considered about the highest possible. A low dynamic range can cause a desensitizing of the RF amplifiers and result in severe intermodulation distortion of the weaker input signals.

Introduction

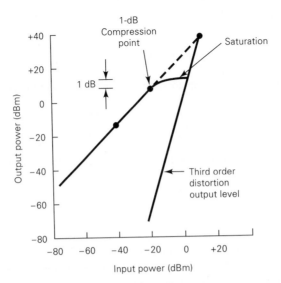

Figure 4-3 Linear gain, 1-dB compression point, and third-order intercept distortion for a typical amplifier.

Fidelity.   *Fidelity* is a measure of the ability of a communications system to produce, at the output of the receiver, an exact replica of the original source information. Any frequency, phase, or amplitude variations that are present in the demodulated waveform that were not in the original information signal are considered distortion.

Essentially, there are three forms of distortion that can deteriorate the fidelity of a communications system: *amplitude, frequency,* and *phase.* Phase distortion is not particularly important for voice transmission because the human ear is relatively insensitive to phase variations. However, phase distortion can be devastating to data transmission. The predominant cause of phase distortion is filtering (both wanted and unwanted). Frequencies at or near the break frequency of a filter undergo varying values of phase shift. Consequently, the cutoff frequency of a filter is often set beyond the minimum value necessary to pass the highest-frequency information signals (typically the upper cutoff frequency of a low-pass filter is approximately 1.3 times the minimum value). *Absolute phase shift* is the total phase shift encountered by a signal and can generally be tolerated as long as all frequencies undergo the same amount of phase delay. *Differential phase shift* occurs when different frequencies undergo different phase shifts and may have a detrimental effect on a complex waveform, especially if the information is encoded into the phase of the carrier as it is with phase-shift keying modulation. If phase shift-versus-frequency is linear, delay is constant with frequency. If all frequencies are not delayed by the same amount of time, the frequency-versus-phase relationship of the received waveform is not consistent with the original source information and the recovered information is distorted.

Amplitude distortion occurs when the amplitude-versus-frequency characteristics of a signal at the output of a receiver differ from those of the original information signal. Amplitude distortion is the result of *nonuniform gain* in amplifiers and filters.

Frequency distortion occurs when frequencies are present in a received signal that were not present in the original source information. Frequency distortion is a result of harmonic and intermodulation distortion and is caused by nonlinear amplification. *Second-order products* ($2f_1$, $2f_2$, $f_1 \pm f_2$, and so on) are usually only a problem in broadband systems because they generally fall outside the bandwidth of a narrowband system. However, *third-order products* often fall within the system bandwidth and produce a distortion called *third-order intercept distortion* . Third-order intercept distortion is a special case of intermodulation distortion and the predominant form of frequency distortion. Third-order intermodulation components are the cross-product frequencies produced when the second harmonic of one signal is added to the fundamental frequency of another signal (that is, $2f_1 \pm f_2$, $2f_2 \pm f_1$, and so on). Frequency distortion can be reduced by using a *square-law device,* such a FET, in the front end of a receiver. Square-law devices have the unique advantage over BJTs in that they produce only second-order harmonic and intermodulation components. Figure 4-3 shows a typical 3rd-order distortion characteristic as a function of amplifier input power and gain.

Insertion loss. *Insertion loss* (*IL*) is a parameter that is associated with the frequencies that fall within the passband of a filter and is generally defined as the ratio of the power transferred to a load with a filter in the circuit to the power transferred to a load without the filter. Because filters are generally constructed from lossy components, such as resistors and imperfect capacitors, even signals that fall within the passband of a filter are attenuated (reduced in magnitude). Typical filter insertion losses are between a few tenths of a decibel to several decibels. In essence, insertion loss is simply the ratio of the output power of a filter to the input power for frequencies that fall within the filter's passband and is stated mathematically in decibels as

$$IL_{(dB)} = 10 \log \frac{P_{out}}{P_{in}} \qquad (4\text{-}3)$$

Noise temperature and equivalent noise temperature. Because thermal noise is directly proportional to temperature, it stands to reason that noise can be expressed in degrees as well as watts or volts. Rearranging Equation 1-24 yields

$$T = \frac{N}{KB} \qquad (4\text{-}4)$$

where T = environmental temperature (kelvin)
   N = noise power (watts)
   K = Boltzmann's constant (1.38×10^{-23} J/K)
   B = bandwidth (hertz)

Equivalent noise temperature (T_e) is a hypothetical value that cannot be directly measured. T_e is a parameter that is often used in low-noise, sophisticated radio receivers rather than noise figure. T_e is an indication of the reduction in the signal-to-noise ratio as a signal propagates through a receiver. The lower the equivalent noise temperature, the better the quality of the receiver. Typical values for T_e range from $20°$ for *cool* receivers to $1000°$ for *noisy* receivers. Mathematically, T_e at the input to a receiver is expressed as

$$T_e = T(F - 1) \qquad (4\text{-}5)$$

where T_e = equivalent noise temperature (kelvin)
   T = environmental temperature (kelvin)
   F = noise factor (unitless)

AM RECEIVERS

There are two basic types of radio receivers: *coherent* and *noncoherent*. With a coherent or *synchronous* receiver, the frequencies generated in the receiver and used for demodulation are synchronized to oscillator frequencies generated in the transmitter (the receiver must have some means of recovering the received carrier and synchronizing to it). With noncoherent or *asynchronous* receivers, either no frequencies are generated in the receiver or the frequencies used for demodulation are completely independent from the transmitter's carrier frequency. *Noncoherent detection* is often called *envelope detection* because the information is recovered from the received waveform by detecting the shape of the modulated envelope. The receivers described in this chapter are noncoherent. Coherent receivers are described in later chapters of this book.

Tuned Radio-frequency Receiver

The *tuned radio-frequency receiver* (*TRF*) was one of the earliest types of AM receivers and was used extensively until the mid-1940s. The TRF replaced earlier crystal and super-regenerative type receivers and is still probably the simplest design available. A block diagram for a TRF is shown in Figure 4-4. A TRF is essentially a three-stage receiver that includes an RF stage, a detector stage, and an audio stage. Generally, two or three RF amplifiers are required to filter and develop sufficient signal amplitude to drive the detector stage. The detector converts RF signals directly to information, and the audio stage amplifies the information signals to a usable level. TRF receivers are advantageous for receivers designed for single-channel operation because of their simplicity and high sensitivity. (A single-channel receiver has a fixed frequency of operation and therefore can only receive a specified band of frequencies unique to a single station's transmissions.)

Tuning a TRF introduces four disadvantages that limit its usefulness to single-station applications. The primary disadvantage of a TRF is that its selectivity (bandwidth) varies when it is tuned over a wide range of input frequencies. The bandwidth of the RF input filter varies with the center frequency of the tuned circuit. This is caused by a phenomenon called *skin effect*. At radio frequencies, current flow is limited to the outermost area of a conductor, and the higher the frequency the smaller the area. Therefore, at radio frequencies, the resistance of the conductor increases with frequency. Consequently, the Q of the tank circuit (X_L/R) remains relatively constant over a wide range of frequencies and, consequently, the bandwidth (f/Q) increases with frequency. As a result, the selectivity of the input filter changes over any appreciable range of input frequencies. If the bandwidth of the input filter is set to the desired value for low-band RF signals, it will be excessive for the high-band signals and possibly cause adjacent channel interference.

The second disadvantage of TRF receivers is instability due to the large number of RF amplifiers all tuned to the same center frequency. When high-gain multistage amplifiers are used, the possibility that a feedback signal will cause the RF stage to break into oscillations is quite high. This problem can be reduced somewhat by tuning each RF amplifier to a different frequency either slightly above or slightly below the center frequency. This technique is called *stagger tuning*. Stagger-tuned RF amplifiers have less gain than center-frequency-tuned amplifiers.

The third disadvantage of TRF receivers is that their gain is not uniform over a very wide frequency range. This is due to the nonuniform L/C ratios of the transformer-coupled tank circuits in the RF amplifiers (that is, the ratio of inductance to capacitance in one tuned amplifier is not the same as that of the other tuned amplifiers).

The fourth disadvantage of the TRF is that it requires *multistage tuning*. To change stations, each RF filter must be tuned simultaneously to the new frequency band, preferably with a single adjustment. This requires exactly the same characteristics for each tuned circuit, which is, of course, impossible to achieve. As you might imagine, this problem is even more severe when stagger tuning is used.

With the development of the *superheterodyne receiver*, TRF receivers are seldom

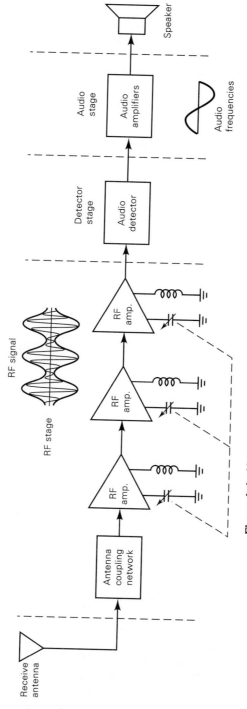

Figure 4-4 Noncoherent tuned radio-frequency receiver block diagram.

used except for special-purpose, single-station receivers and therefore do not warrant further discussion.

EXAMPLE 4-3

For an AM commercial broadcast-band receiver (535 to 1605 kHz) with an input filter Q-factor of 54, determine the bandwidth at the low and high ends of the RF spectrum.

Solution The bandwidth at the low-frequency end of the AM spectrum is centered around a carrier frequency of 540 kHz and is

$$B = \frac{f}{Q} = \frac{540 \text{ kHz}}{54} = 10 \text{ kHz}$$

The bandwidth at the high-frequency end of the AM spectrum is centered around a carrier frequency of 1600 kHz and is

$$B = \frac{1600 \text{ kHz}}{54} = 29,630 \text{ Hz}$$

The −3-dB bandwidth at the low-frequency end of the AM spectrum is exactly 10 kHz, which is the desired value. However, the bandwidth at the high-frequency end is almost 30 kHz, which is three times the desired range. Consequently, when tuning for stations at the high end of the spectrum, three stations would be received simultaneously.

To achieve a bandwidth of 10 kHz at the high-frequency end of the spectrum, a Q of 160 is required (1600 kHz/10 kHz). With a Q of 160, the bandwidth at the low-frequency end is

$$B = \frac{540 \text{ kHz}}{160} = 3375 \text{ Hz}$$

which is obviously too selective because it would block approximately two-thirds of the information bandwidth.

Superheterodyne Receiver

The nonuniform selectivity of the TRF led to the development of the *superheterodyne receiver* near the end of World War I. Although the quality of the superheterodyne receiver has improved greatly since its original design, its basic configuration has not changed much and it is still used today for a wide variety of radio communications services. The superheterodyne receiver has remained in use because its gain, selectivity, and sensitivity characteristics are superior to those of other receiver configurations.

Heterodyne means to mix two frequencies together in a nonlinear device or to translate one frequency to another using nonlinear mixing. A block diagram of a noncoherent superheterodyne receiver is shown in Figure 4-5. Essentially, there are five sections to a superheterodyne receiver: the RF section, the mixer/converter section, the IF section, the audio detector section, and the audio amplifier section.

RF section. The RF section generally consists of a preselector and an amplifier stage. They can be separate circuits or a single combined circuit. The preselector is a broad-tuned bandpass filter with an adjustable center frequency that is tuned to the desired carrier frequency. The primary purpose of the preselector is to provide enough initial bandlimiting to prevent a specific unwanted radio frequency called the *image frequency* from entering the receiver (image frequency is explained later in this chapter). The preselector also reduces the noise bandwidth of the receiver and provides the initial step toward reducing the overall receiver bandwidth to the minimum bandwidth required to pass the

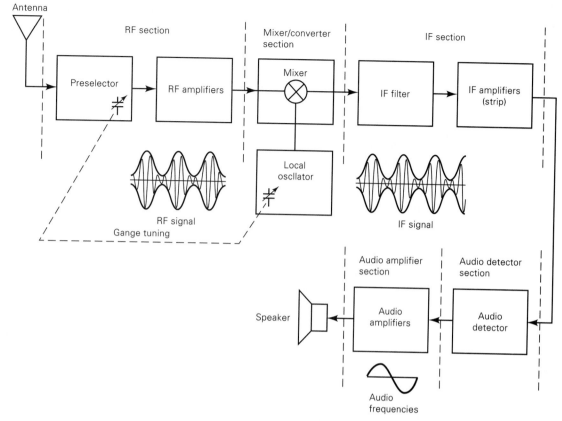

Figure 4-5 AM superheterodyne receiver block diagram.

information signals. The RF amplifier determines the sensitivity of the receiver (that is, sets the signal threshold). Also, because the RF amplifier is the first active device encountered by a received signal, it is the primary contributor of noise and therefore a predominant factor in determining the noise figure for the receiver. A receiver can have one or more RF amplifiers or it may not have any, depending on the desired sensitivity. Several advantages of including RF amplifiers in a receiver are as follows:

1. Greater gain, thus better sensitivity
2. Improved image-frequency rejection
3. Better signal-to-noise ratio
4. Better selectivity

Mixer/converter section. The mixer/converter section includes a radio-frequency oscillator stage (commonly called a *local oscillator*) and a mixer/converter stage (commonly called the *first detector*). The local oscillator can be any of the oscillator circuits discussed in Chapter 2, depending on the stability and accuracy desired. The mixer stage is a nonlinear device and its purpose is to convert radio frequencies to intermediate frequencies (RF-to-IF frequency translation). Heterodyning takes place in the mixer stage, and radio frequencies are down-converted to intermediate frequencies. Although the carrier and sideband frequencies are translated from RF to IF, the shape of the envelope remains the same, and therefore the original information contained in the envelope remains unchanged. It is important to note that, although the carrier and upper and lower

AM Receivers

side frequencies change frequency, the bandwidth is unchanged by the heterodyning process. The most common intermediate frequency used in AM broadcast-band receivers is 455 kHz.

IF section. The IF section consists of a series of IF amplifiers and bandpass filters and is often called the *IF strip*. Most of the receiver gain and selectivity is achieved in the IF section. The IF center frequency and bandwidth are constant for all stations and are chosen so that their frequency is less than any of the RF signals to be received. The IF is always lower in frequency than the RF because it is easier and less expensive to construct high-gain, stable amplifiers for the low-frequency signals. Also, low-frequency IF amplifiers are less likely to oscillate than their RF counterparts. Therefore, it is not uncommon to see a receiver with five or six IF amplifiers and a single RF amplifier or possibly no RF amplification.

Detector section. The purpose of the detector section is to convert the IF signals back to the original source information. The detector is generally called an *audio detector* or the *second detector* in a broadcast-band receiver because the information signals are audio frequencies. The detector can be as simple as a single diode or as complex as a phase-locked loop or balanced demodulator.

Audio section. The audio section comprises several cascaded audio amplifiers and one or more speaker. The number of amplifiers used depends on the audio signal power desired.

Receiver operation. During the demodulation process in a superheterodyne receiver, the received signals undergo two or more frequency translations: first, the RF is converted to IF; then the IF is converted to the source information. The terms RF and IF are system dependent and are often misleading because they do not necessarily indicate a specific range of frequencies. For example, RF for the commercial AM broadcast band are frequencies between 535 and 1605 kHz, and IF signals are frequencies between 450 and 460 kHz. In commercial broadcast-band FM receivers, intermediate frequencies as high as 10.7 MHz are used, which are considerably higher than AM broadcast-band RF signals. Intermediate frequencies simply refer to frequencies that are used within a transmitter or receiver that fall somewhere between the radio frequencies and the original source information frequencies.

Frequency conversion. *Frequency conversion* in the mixer/converter stage is identical to frequency conversion in the modulator stage of a transmitter except that in the receiver the frequencies are down-converted rather than up-converted. In the mixer/converter, RF signals are combined with the local oscillator frequency in a nonlinear device. The output of the mixer contains an infinite number of harmonic and cross-product frequencies, which include the sum and difference frequencies between the desired RF carrier and local oscillator frequencies. The IF filters are tuned to the difference frequencies. The local oscillator is designed such that its frequency of oscillation is always above or below the desired RF carrier by an amount equal to the IF center frequency. Therefore, the difference between the RF and the local oscillator frequency is always equal to the IF. The adjustment for the center frequency of the preselector and the adjustment for the local oscillator frequency are *gang tuned*. Gang tuning means that the two adjustments are mechanically tied together so that a single adjustment will change the center frequency of the preselector and, at the same time, change the local oscillator frequency. When the local oscillator frequency is tuned above the RF, it is called *high-side injection* or *high-beat injection*. When the local oscillator is tuned below the RF, it is called *low-side injection* or *low-beat injection*. In AM broadcast-band receivers, high-side injection is always used (the reason for this is explained later in this chapter). Mathematically, the local oscillator frequency is:

For high-side injection: $f_{lo} = f_{rf} + f_{if}$ (4-6a)

For low-side injection: $f_{lo} = f_{rf} - f_{if}$ (4-6b)

where f_{lo} = local oscillator frequency (hertz)
f_{rf} = radio frequency (hertz)
f_{if} = intermediate frequency (hertz)

EXAMPLE 4-4

For an AM superheterodyne receiver that uses high-side injection and has a local oscillator frequency of 1355 kHz, determine the IF carrier, upper side frequency, and lower side frequency for an RF wave that is made up of a carrier and upper and lower side frequencies of 900, 905, and 895 kHz, respectively.

Solution Refer to Figure 4-6. Because high-side injection is used, the intermediate frequencies are the difference between the radio frequencies and the local oscillator frequency. Rearranging Equation 4-6a yields

$$f_{if} = f_{lo} - f_{rf}$$

$$= 1355\ \text{kHz} - 900\ \text{kHz} = 455\ \text{kHz}$$

The upper and lower intermediate frequencies are

$$f_{if(usf)} = f_{lo} - f_{rf(lsf)}$$

$$= 1355\ \text{kHz} - 895\ \text{kHz} = 460\ \text{kHz}$$

$$f_{if(lsf)} = f_{lo} - f_{rf(usf)}$$

$$= 1355\ \text{kHz} - 905\ \text{kHz} = 450\ \text{kHz}$$

Note that the side frequencies undergo a sideband reversal during the heterodyning process (that is, the RF upper side frequency is translated to an IF lower side frequency, and the RF lower side frequency is translated to an IF upper side frequency). This is commonly called *sideband inversion*. Sideband inversion is not detrimental to conventional double-sideband AM because exactly the same information is contained in both sidebands.

Local oscillator tracking. *Tracking* is the ability of the local oscillator in a receiver to oscillate either above or below the selected radio frequency carrier by an amount equal to the intermediate frequency throughout the entire radio-frequency band. With high-side injection, the local oscillator should track above the incoming RF carrier by a fixed frequency equal to $f_{rf} + f_{if}$, and with low-side injection, the local oscillator should track below the RF carrier by a fixed frequency equal to $f_{rf} - f_{if}$.

Figure 4-7a shows the schematic diagram of the preselector and local oscillator tuned circuit in a broadcast-band AM receiver. The broken lines connecting the two tuning capacitors indicate that they are *ganged* together (connected to a single tuning control). This is called *gang tuning* a receiver. The tuned circuit in the preselector is tunable from a center frequency of 540 to 1600 kHz (a ratio of 2.96 to 1), and the local oscillator is tunable from 995 to 2055 kHz (a ratio of 2.06 to 1). Because the resonant frequency of a

AM Receivers

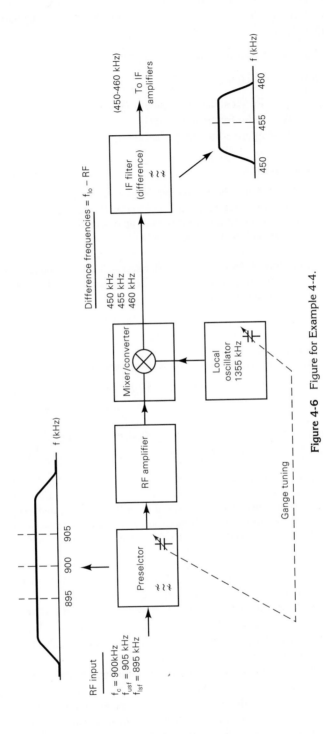

Figure 4-6 Figure for Example 4-4.

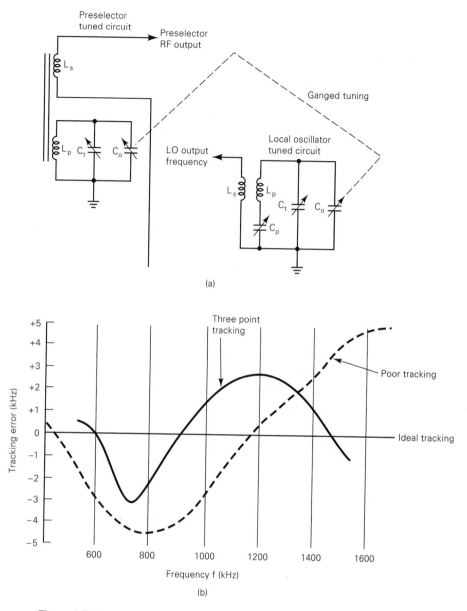

Figure 4-7 Receiver tracking: (a) preselector and local oscillator schematic; (b) tracking curve.

tuned circuit is inversely proportional to the square root of the capacitance, the capacitance in the preselector must change by a factor of 8.8, whereas, at the same time, the capacitance in the local oscillator must change by a factor of only 4.26. The local oscillator should oscillate 455 kHz above the preselector center frequency over the entire AM frequency band, and there should be a single tuning control. Fabricating such a circuit is difficult if not impossible. Therefore, perfect tracking over the entire AM band is unlikely to occur. The difference between the actual local oscillator frequency and the desired frequency is called *tracking error*. Typically, the tracking error is not uniform over the entire RF spectrum. A maximum tracking error of ±3 kHz is about the best that can be expected from a domestic AM broadcast band receiver with a 455-kHz intermediate frequency.

AM Receivers

Figure 4-7b shows a typical tracking curve. A tracking error of +3 kHz corresponds to an IF center frequency of 458 kHz, and a tracking error of −3 kHz corresponds to an IF center frequency of 452 kHz.

The tracking error is reduced by a technique called *three-point tracking*. The preselector and local oscillator each have a trimmer capacitor (C_t) in parallel with the primary tuning capacitor (C_o) that compensates for minor tracking errors at the high end of the AM spectrum. The local oscillator has an additional padder capacitor (C_p) in series with the tuning coil that compensates for minor tracking errors at the low end of the AM spectrum. With three-point tracking, the tracking error is adjusted to 0 Hz at approximately 600, 950, and 1500 kHz.

With low-side injection, the local oscillator would have to be tunable from 85 to 1145 kHz (a ratio of 13.5 to 1). Consequently, the capacitance must change by a factor of 182. Standard variable capacitors seldom tune over more than a 10 to 1 range. This is why low-side injection is impractical for commercial AM broadcast-band receivers. With high-side injection, the local oscillator must be tunable from 995 to 2055 kHz, which corresponds to a capacitance ratio of only 4.63 to 1.

Ganged capacitors are relatively large, expensive, and inaccurate, and they are somewhat difficult to compensate. Consequently, they are being replaced by solid-state electronically tuned circuits. Electronically tuned circuits are smaller, less expensive, more accurate, relatively immune to environmental changes, more easily compensated, and more easily adapted to digital remote control and push-button tuning than their mechanical counterparts. As with the crystal oscillator modules explained in Chapter 2, electronically tuned circuits use solid-state variable-capacitance diodes (varactor diodes). Figure 4-8 shows a schematic diagram for an electronically tuned preselector and local

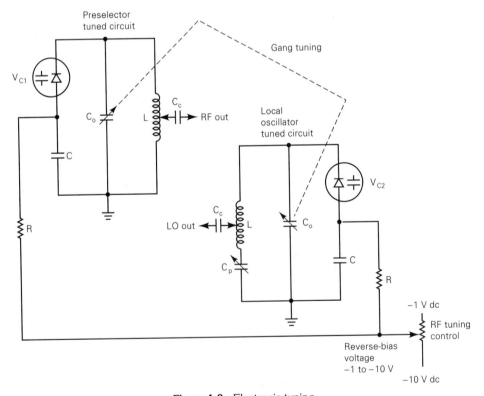

Figure 4-8 Electronic tuning.

Chap. 4   Amplitude Modulation Reception

oscillator. The -1- to -10-V reverse-biased voltage comes from a single tuning control. By changing the position of the wiper arm on a precision variable resistor, the dc reverse bias for the two tuning diodes (V_{C1} and V_{C2}) is changed. The diode capacitance and, consequently, the resonant frequency of the tuned circuit vary with the reverse bias. Three-point compensation with electronic tuning is accomplished the same as with mechanical tuning.

In a superheterodyne receiver, most of the receiver's selectivity is accomplished in the IF stage. For maximum noise reduction, the bandwidth of the IF filters is equal to the minimum bandwidth required to pass the information signal, which with double-sideband transmission is equal to two times the highest modulating signal frequency. For a maximum modulating signal frequency of 5 kHz, the minimum IF bandwidth with perfect tracking is 10 kHz. For a 455-kHz IF center frequency, a 450- to 460-kHz passband is necessary. In reality, however, some RF carriers are tracked as much as ± 3 kHz above or below 455 kHz. Therefore, the RF bandwidth must be expanded to allow the IF signals from the off-track stations to pass through the IF filters.

EXAMPLE 4-5

For the tracking curve shown in Figure 4-9a, a 455-kHz IF center frequency, and a maximum modulating signal frequency of 5 kHz, determine the minimum IF bandwidth.

Solution The maximum intermediate frequency occurs for the RF carrier with the most positive tracking error (1400 kHz) and a 5-kHz modulating signal.

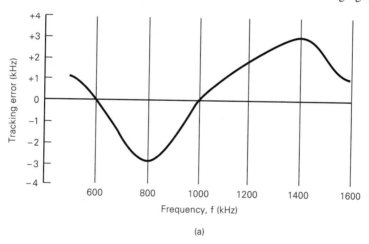

(a)

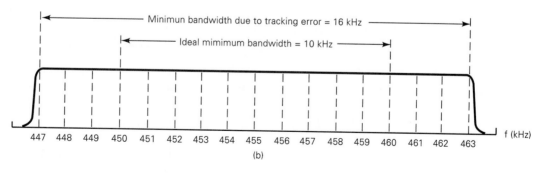

(b)

Figure 4-9 Tracking error for Example 4-5: (a) tracking curve; (b) bandpass characteristics.

AM Receivers

$$f_{\text{if(maximum)}} = f_{\text{if}} + \text{tracking error} + f_{m(\text{max})}$$

$$= 455 \text{ kHz} + 3 \text{ kHz} + 5 \text{ kHz} = 463 \text{ kHz}$$

The minimum intermediate frequency occurs for the RF carrier with the most negative tracking error (800 kHz) and a 5-kHz modulating signal.

$$f_{\text{if(minimum)}} = f_{\text{if}} + \text{tracking error} - f_{m(\text{max})}$$

$$= 455 \text{ kHz} + (-3 \text{ kHz}) - 5 \text{ kHz} = 447 \text{ kHz}$$

The minimum IF bandwidth necessary to pass the two sidebands is the difference between the maximum and minimum intermediate frequencies or

$$B_{\text{min}} = 463 \text{ kHz} - 447 \text{ kHz} = 16 \text{ kHz}$$

Figure 4-9b shows the IF bandpass characteristics for Example 4-5.

Image frequency. An *image frequency* is any frequency other than the selected radio-frequency carrier that, if allowed to enter a receiver and mix with the local oscillator, will produce a cross-product frequency that is equal to the intermediate frequency. An image frequency is equivalent to a second radio frequency that will produce an IF that will interfere with the IF from the desired radio frequency. Once an image frequency has been mixed down to IF, it cannot be filtered out or suppressed. If the selected RF carrier and its image frequency enter a receiver at the same time, they both mix with the local oscillator frequency and produce difference frequencies that are equal to the IF. Consequently, two different stations are received and demodulated simultaneously, producing two sets of information frequencies. For a radio frequency to produce a cross product equal to the IF, it must be displaced from the local oscillator frequency by a value equal to the IF. With high-side injection, the selected RF is below the local oscillator by an amount equal to the IF. Therefore, the image frequency is the radio frequency that is located in the IF frequency above the local oscillator. Mathematically, for high-side injection, the image frequency (f_{im}) is

$$f_{\text{im}} = f_{\text{lo}} + f_{\text{if}} \tag{4-7a}$$

and, since the desired RF equals the local oscillator frequency minus the IF,

$$f_{\text{im}} = f_{\text{rf}} + 2f_{\text{if}} \tag{4-7b}$$

Figure 4-10 shows the relative frequency spectrum for the RF, IF, local oscillator, and image frequencies for a superheterodyne receiver using high-side injection. Here we see that the higher the IF, the farther away in the frequency spectrum the image frequency is from the desired RF. Therefore, for better *image-frequency rejection*, a high intermediate frequency is preferred. However, the higher the IF, the more difficult it is to build stable amplifiers with high gain. Therefore, there is a trade-off when selecting the IF for a radio receiver between image-frequency rejection and IF gain and stability.

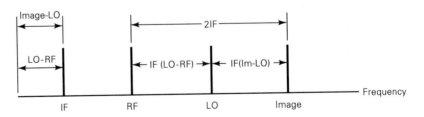

Figure 4-10 Image frequency.

Image-frequency rejection ratio. The *image frequency rejection ratio (IFRR)* is a numerical measure of the ability of a preselector to reject the image frequency. For a single-tuned preselector, the ratio of its gain at the desired RF to the gain at the image frequency is the IFRR. Mathematically, IFRR is

$$\text{IFRR} = \sqrt{(1 + Q^2\rho^2)} \qquad (4\text{-}8a)$$

where

$$\rho = \frac{f_{im}}{f_{rf}} - \frac{f_{rf}}{f_{im}}$$

$$\text{IFRR}_{dB} = 20 \log \text{IFRR} \qquad (4\text{-}8b)$$

If there is more than one tuned circuit in the front end of a receiver (perhaps a preselector filter and a separately tuned RF amplifier), the total IFRR is simply the product of the two ratios.

EXAMPLE 4-6

For an AM broadcast-band superheterodyne receiver with IF, RF, and local oscillator frequencies of 455, 600, and 1055 kHz, respectively, refer to Figure 4-11 and determine:

(a) Image frequency.
(b) IFRR for a preselector Q of 100.

Solution (a) From Equation 4-7a, $f_{im} = 1055 \text{ kHz} + 455 \text{ kHz} = 1510 \text{ kHz}$

or from Equation 4-7b $\qquad f_{im} = 600 \text{ kHz} + 2(455 \text{ kHz}) = 1510 \text{ kHz}$

(b) From Equation 4-8a and 4-8b,

$$\rho = \frac{1510 \text{ kHz}}{600 \text{ kHz}} - \frac{600 \text{ kHz}}{1510 \text{ kHz}}$$

$$= 2.51 - 0.397 = 2.113$$

$$\text{IFRR} = \sqrt{1 + (100^2)(2.113^2)}$$

$$= 211.3 \quad \text{or} \quad 46.5 \text{ dB}$$

Once an image frequency has been down-converted to IF, it cannot be removed. Therefore, to reject the image frequency, it has to be blocked prior to the mixer/converter stage. Image-frequency rejection is the primary purpose for the RF preselector. If the bandwidth of the preselector is sufficiently narrow, the image frequency is prevented from entering the receiver. Figure 4-12 illustrates how proper RF and IF filtering can prevent an image frequency from interfering with the desired radio frequency.

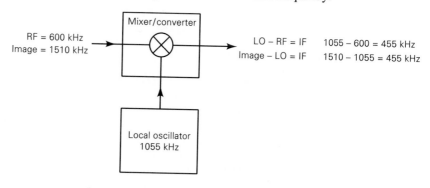

Figure 4-11 Frequency conversion for Example 4-6.

AM Receivers

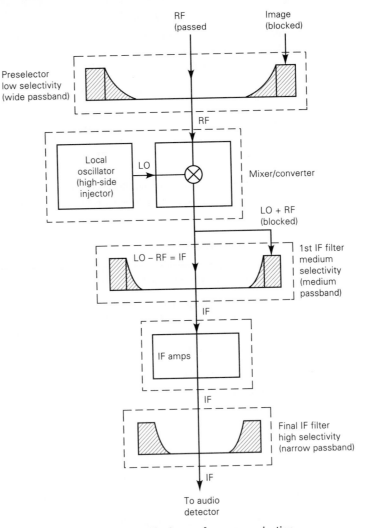

Figure 4-12 Image-frequency rejection.

The ratio of the RF to the IF is also an important consideration for image-frequency rejection. The closer the RF is to the IF, the closer the RF is to the image frequency.

EXAMPLE 4-7

For a citizens-band receiver using high-side injection with an RF carrier of 27 MHz and an IF center frequency of 455 kHz, determine:

(a) Local oscillator frequency.
(b) Image frequency.
(c) IFRR for a preselector Q of 100.
(d) Preselector Q required to achieve the same IFRR as that achieved for an RF carrier of 600 kHz in Example 4-6.

Solution (a) From Equation 4-6a, $f_{lo} = 27$ MHz $+ 455$ kHz $= 27.455$ MHz

(b) From Equation 4-7a, $\qquad f_{im} = 27.455$ MHz $+ 455$ kHz $= 27.91$ MHz

(c) From Equation 4-8a and 4-8b, $\qquad$ IFRR $= 6.7$ or 16.5 dB

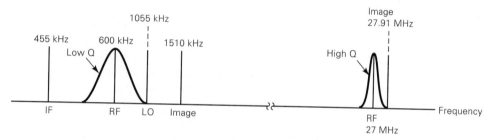

Figure 4-13 Frequency spectrum for Example 4-7.

(d) Rearranging Equation 4-8a,

$$Q = \frac{\sqrt{(\mathrm{IFRR}^2 - 1)}}{\rho^2} = 3187$$

From Examples 4-6 and 4-7, it can be seen that the higher the RF carrier, the more difficult it is to prevent the image frequency from entering the receiver. For the same IFRR, the higher RF carriers require a much higher-quality preselector filter. This is illustrated in Figure 4-13.

Double spotting.   *Double spotting* occurs when a receiver picks up the same station at two nearby points on the receiver tuning dial. One point is the desired location and the other point is called the *spurious point*. Double spotting is caused by poor front-end selectivity or inadequate image-frequency rejection.

Double spotting is harmful because weak stations can be overshadowed by the reception of a nearby strong station at the spurious location in the frequency spectrum. Double spotting may be used to determine the intermediate frequency of an unknown receiver because the spurious point on the dial is precisely two times the IF center frequency below the correct receive frequency.

AM RECEIVER CIRCUITS

RF Amplifier Circuits

An RF amplifier is a high-gain, low-noise, tuned amplifier that, when used, is the first active stage encountered by the received signal. The primary purposes of an RF stage are selectivity, amplification, and sensitivity. Therefore, the following characteristics are desirable for RF amplifiers:

1. Low thermal noise
2. Low noise figure
3. Moderate to high gain
4. Low intermodulation and harmonic distortion (that is, linear operation)
5. Moderate selectivity
6. High image-frequency rejection ratio

Two of the most important parameters for a receiver are amplification and noise figure, which both depend on the RF stage. An AM demodulator (or detector as it is sometimes called) detects amplitude variations in the modulated wave and converts them to amplitude changes in its output. Consequently, amplitude variations that were caused by noise are converted to erroneous fluctuations in the demodulator output, and the quality of the demodulated signal is degraded. The more gain that a signal experiences as it passes through a receiver, the more pronounced are the amplitude variations at the demodulator input, and the less noticeable are the variations caused by noise. The narrower the bandwidth is, the less noise propagated through the receiver and, consequently, the less noise demodulated by the detector. From Equation 1-28 ($V_N = \sqrt{4RKTB}$), noise voltage is directly proportional to the square root of the temperature, bandwidth, and equivalent noise resistance. Therefore, if these three parameters are minimized, the thermal noise is reduced. The temperature of an RF stage can be reduced by artificially cooling the front end of the receiver with air fans or even liquid helium in the more expensive receivers. The bandwidth is reduced by using tuned amplifiers and filters, and the equivalent noise resistance is reduced by using specially constructed solid-state components for the active devices. Noise figure is essentially a measure of the noise added by an amplifier. Therefore, the noise figure is improved (reduced) by reducing the amplifier's internal noise.

Intermodulation and harmonic distortion are both forms of nonlinear distortion that increase the magnitude of the noise figure by adding correlated noise to the total noise spectrum. The more linear an amplifier's operation is, the less nonlinear distortion produced, and the better the receiver's noise figure. The image-frequency reduction by the RF amplifier combines with the image frequency reduction of the preselector to reduce the receiver input bandwidth sufficiently to help prevent the image frequency from entering the mixer/converter stage. Consequently, moderate selectivity is all that is required from the RF stage.

Figure 4-14 shows several commonly used RF amplifier circuits. Keep in mind that RF is a relative term and simply means that the frequency is high enough to be efficiently radiated by an antenna and propagated through free space as an electromagnetic wave. RF for the AM broadcast band is between 535 and 1605 kHz, whereas RF for microwave radio is in excess of 1 GHz (1000 MHz). A common intermediate frequency used for FM broadcast-band receivers is 10.7 MHz, which is considerably higher than the radio frequencies associated with the AM broadcast band. RF is simply the radiated or received signal, and IF is an intermediate signal within a transmitter or receiver. Therefore, many of the considerations for RF amplifiers also apply to IF amplifiers, such as neutralization, filtering, and coupling.

Figure 4-14a shows a schematic diagram for a bipolar transistor RF amplifier. C_a, C_b, C_c, and L_1 form the coupling circuit from the antenna. Q_1 is class A biased to reduce nonlinear distortion. The collector circuit is transformer coupled to the mixer/converter through T_1, which is double tuned for more selectivity. C_x and C_y are RF bypass capacitors. Their symbols indicate that they are specially constructed *feedthrough* capacitors. Feedthrough capacitors offer less inductance, which prevents a portion of the signal from radiating from their leads. C_n is a *neutralization* capacitor. A portion of the collector signal is fed back to the base circuit to offset (or neutralize) the signal fed back through the transistor collector-to-base lead capacitance to prevent oscillations from occurring. C_f, in conjunction with C_n, form an ac voltage divider for the feedback signal. This neutralization configuration is called *off-ground* neutralization.

Figure 4-14b shows an RF amplifier using dual-gate field-effect transistors. This configuration uses DEMOS (depletion-enhancement metal-oxide semiconductor) FETs. The FETs feature high input impedance and low noise. A FET is a square-law device that generates only second-order harmonic and intermodulation distortion components, therefore producing less nonlinear distortion than a bipolar transistor. Q_1 is again biased class A for linear operation. T_1 is single tuned to the desired RF carrier frequency to enhance the receiver's selectivity and to improve the IFRR. L_5 is a radio frequency choke and, in conjunction with C_5, decouples RF signals from the dc power supply.

Figure 4-14c shows the schematic diagram for a special RF amplifier configuration

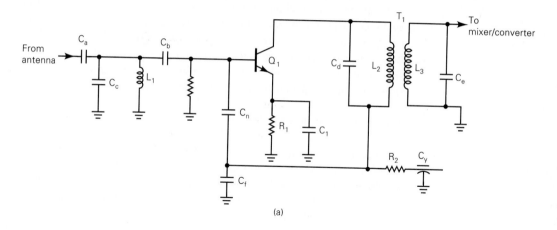

(a)

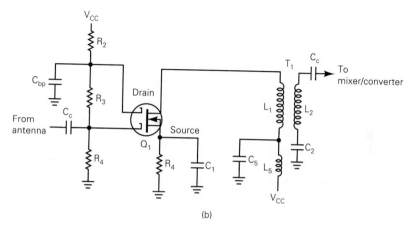

(b)

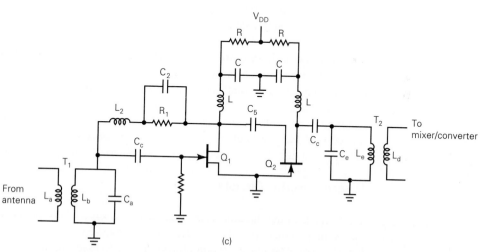

(c)

Figure 4-14 RF amplifier configurations: (a) bipolar transistor RF amplifier; (b) DEMOS-FET RF amplifier; (c) cascoded RF amplifier.

AM Receiver Circuits

called a *cascoded* amplifier. A cascoded amplifier offers higher gain and less noise than conventional cascaded amplifiers. The active devices can be either bipolar transistors or FETs. Q_1 is a common-source amplifier whose output is impedance coupled to the source of Q_2. Because of the low input impedance of Q_2, Q_1 does not need to be neutralized; however, neutralization reduces the noise figure even further. Therefore, L_2, R_1, and C_2 provide the feedback path for neutralization. Q_2 is a common-gate amplifier and because of its low input impedance requires no neutralization.

Low-noise Amplifiers

High-performance microwave receivers require a *low-noise amplifier* (*LNA*) as the input stage of the RF section to optimize their noise figure. Equation 1-37 showed that the first amplifier in a receiver is the most important in determining the receiver's noise figure. The first stage should have low noise and high gain. Unfortunately, this is difficult to achieve with a single amplifier stage; therefore, LNAs generally include two stages of amplification along with impedance-matching networks to enhance their performance. The first stage has moderate gain and minimum noise, and the second stage has high gain and moderate noise.

Low-noise RF amplifiers are biased class A and usually utilize silicon bipolar or field-effect transistors up to approximately 2 GHz and gallium arsenide FETs above this frequency. A special type of gallium arsenide FET most often used is the MESFET (MEsa Semiconductor FET). A MESFET is a FET with a metal–semiconductor junction at the gate of the device, called a Schottky barrier.

Integrated-circuit RF amplifiers. The NE/SA5200 is a wideband, unconditionally stable, low-power, dual-gain linear integrated-circuit RF amplifier manufactured by Signetics Corporation. The NE/SA5200 will operate from dc to approximately 1200 MHz and has a low noise figure. The NE/SA5200 has several inherent advantages over comparable discrete implementations; it does not need any external biasing components, it occupies little space on a printed circuit board, and the high level of integration improves its reliability over discrete counterparts. The NE/SA5200 is also equipped with a power-down mode that helps reduce power consumption in applications where the amplifiers can be disabled.

The block diagram for the SA5200 is shown in Figure 4-15a, and a simplified schematic diagram is shown in Figure 4-15b. Note that the two wideband amplifiers are biased from the same bias generator. Each amplifier stage has a noise figure of about 3.6 dB and a gain of approximately 11 dB. Several stages of NE/SA5200 can be cascaded and used as an IF strip, and the enable pin can be used to improve the dynamic range of the receiver. For extremely high input levels, the amplifiers in the NE/SA5200 can be disabled. When disabled, the input signal is attenuated 13 dB, preventing receiver overload.

Mixer/Converter Circuits

The purpose of the mixer/converter stage is to down-convert the incoming radio frequencies to intermediate frequencies. This is accomplished by mixing the RF signals with the local oscillator frequency in a nonlinear device. In essence, this is heterodyning. A mixer is a nonlinear amplifier similar to a modulator, except that the output is tuned to the difference between the RF and local oscillator frequencies. Figure 4-16 shows a block diagram for a mixer/converter stage. The output of a balanced mixer is the product of the RF and local oscillator frequencies and is expressed mathematically as

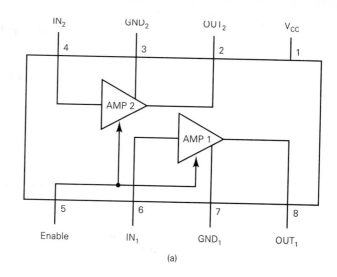

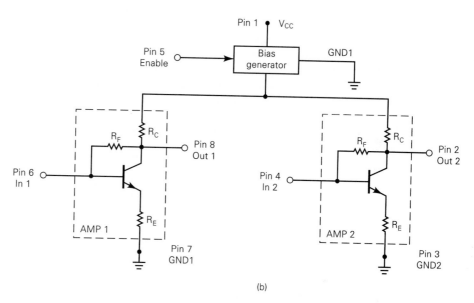

Figure 4-15 NE/SA5200 RF dual-gain stage: (a) block diagram; (b) simplified schematic diagram.

$$V_{out} = (\sin 2\pi f_{rf} t)(\sin 2\pi f_{lo} t)$$

where f_{rf} = incoming radio frequency (hertz)
f_{lo} = local oscillator frequency (hertz)

Therefore, using the trigonometric identity for the product of two sines, the output of a mixer is

$$V_{out} = \frac{1}{2} \cos [2\pi(f_{rf} - f_{lo})t] - \frac{1}{2} \cos [2\pi(f_{rf} + f_{if})t]$$

The absolute value of the difference frequency ($|f_{rf} - f_{lo}|$) is the intermediate frequency.

Although any nonlinear device can be used for a mixer, a transistor or FET is generally preferred over a simple diode because it is also capable of amplification. However,

AM Receiver Circuits

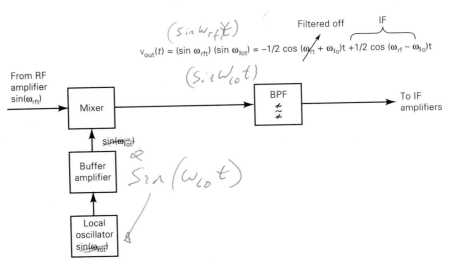

$$V_{out}(t) = (\sin \omega_{rf}t)(\sin \omega_{lo}t) = -1/2 \cos(\omega_{rf} + \omega_{lo})t + 1/2 \cos(\omega_{rf} - \omega_{lo})t$$

Figure 4-16 Mixer/converter block diagram.

because the actual output signal from a mixer is a cross-product frequency, there is a net loss to the signal. This loss is called *conversion loss* (or sometimes *conversion gain*) because a frequency conversion has occurred and, at the same time, the IF output signal is lower in amplitude than the RF input signal. The conversion loss is generally 6 dB (which corresponds to a conversion gain of −6 dB). The conversion gain is the difference between the level of the IF output with a RF input signal to the level of the IF output with an IF input signal.

Figure 4-17 shows the schematic diagrams for several common mixer/converter circuits. Figure 4-17a shows what is probably the simplest mixer circuit available (other than a single diode mixer) and is used exclusively for inexpensive AM broadcast-band receivers. Radio-frequency signals from the antenna are filtered by the preselector tuned circuit (L_1 and C_1) and then transformer coupled to the base of Q_1. The active device for the mixer (Q_1) also provides amplification for the local oscillator. This configuration is commonly called a *self-excited* mixer because the mixer excites itself by feeding energy back to the local oscillator tank circuit (C_2 and L_2) to sustain oscillations. When power is initially applied, Q_1 amplifies both the incoming RF signals and any noise present and supplies the oscillator tank circuit with enough energy to begin oscillator action. The local oscillator frequency is the resonant frequency of the tank circuit. A portion of the resonant tank circuit energy is coupled through L_2 and L_5 to the emitter of Q_1. This signal drives Q_1 into its nonlinear operating region and, consequently, produces sum and difference frequencies at its collector. The difference frequency is the IF. The output tank circuit (L_3 and C_3) is tuned to the IF band. Therefore, the IF signal is transformer coupled to the input of the first IF amplifier. The process is regenerative as long as there is an incoming RF signal. The tuning capacitors in the RF and local oscillator tank circuits are ganged into a single tuning control. C_p and C_t are for three-point tracking. This configuration has poor selectivity and poor image frequency rejection because there is no amplifier tuned to the RF signal frequency and, consequently, the only RF selectivity is in the preselector. In addition, there is essentially no RF gain, and the transistor nonlinearities produce harmonic and intermodulation components that may fall within the IF passband.

The mixer/converter circuit shown in Figure 4-17b is a *separately excited* mixer. Its operation is essentially the same as the self-excited mixer except that the local oscillator and the mixer have their own gain devices. The mixer itself is a FET, which has nonlinear characteristics that are better suited for IF conversion than those of a bipolar transistor.

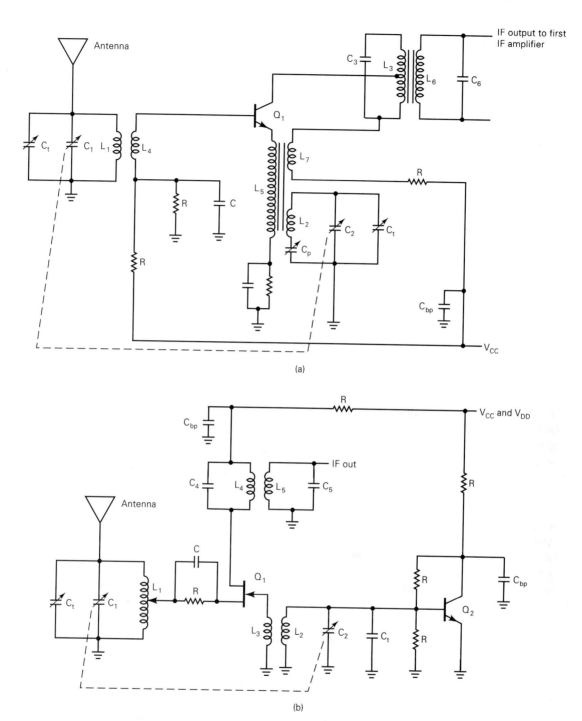

Figure 4-17 Mixer/converter circuits: (a) self-excited mixer; (b) separately excited mixer; *(Continued on next page.)*

Feedback is from L_2 to L_3 of the transformer in the source of Q_1. This circuit is commonly used for high-frequency (HF) and very high frequency (VHF) receivers.

The mixer converter circuit shown in Figure 4-17c is a *single-diode mixer*. The concept is quite simple: the RF and local oscillator signals are coupled into the diode, which is

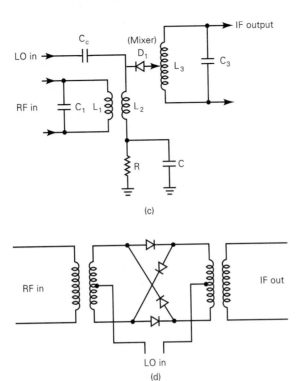

Figure 4-17 *(continued)* (c) diode mixer; (d) balanced diode mixer.

a nonlinear device. Therefore, nonlinear mixing occurs and the sum and difference frequencies are produced. The output tank circuit (C_3 and L_3) is tuned to the difference (IF) frequency. A single-diode mixer is inefficient because it has a net loss. However, a diode mixer is commonly used for the audio detector in AM receivers and to produce the audio subcarrier in television receivers.

Figure 4-17d shows the schematic diagram for a *balanced diode mixer*. Balanced mixers are one of the most important circuits used in communications systems today. Balanced mixers are also called *balanced modulators*, *product modulators*, and *product detectors*. The phase detectors used in phase-locked loops and explained in Chapter 2 are balanced modulators. Balanced mixers are used extensively in both transmitters and receivers for AM, FM, and many of the digital modulation schemes, such as PSK and QAM. Balanced mixers have two inherent advantages over other types of mixers: noise reduction and carrier suppression.

Integrated-circuit mixer/oscillator. Figure 4-18 shows the block diagram for the Signetics NE/SA602A *double-balanced mixer and oscillator*. The NE/SA602A is a low-power VHF monolithic double-balanced mixer with input amplifier, on-board oscillator, and voltage regulator. It is intended to be used for high-performance, low-power communications systems; it is particularly well suited for *cellular radio* applications. The mixer is a *Gilbert cell* multiplier configuration, which typically provides 18 dB of gain at 45 MHz. A Gilbert cell is a differential amplifier that drives a balanced switching cell. The differential input stage provides gain and determines the noise figure and signal-handling performance of the system. The oscillator will operate up to 200 MHz and can be configured as a crystal or tuned LC tank circuit oscillator or a buffer amplifier for an external oscillator. The noise figure for the NE/SA602A at 45 MHz is typically less than 5 dB. The gain, third-order intercept performance, and low-power and noise characteristics make the

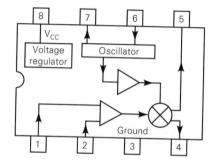

Figure 4-18 NE/SA602A double-balanced mixer and oscillator block diagram.

NE/SA602A a superior choice for high-performance, battery-operated equipment. The input, RF mixer output, and oscillator ports can support a variety of input configurations. The RF inputs (pins 1 and 2) are biased internally and they are symmetrical. Figure 4-19 shows three typical input configurations: single-ended tuned input, balanced input, and single-ended untuned.

IF Amplifier Circuits

Intermediate frequency (IF) amplifiers are relatively high gain tuned amplifiers that are very similar to RF amplifiers, except that IF amplifiers operate over a relatively narrow, fixed frequency band. Consequently, it is easy to design and build IF amplifiers that are stable, do not radiate, and are easily neutralized. Because IF amplifiers operate over a fixed frequency band, successive amplifiers can be inductively coupled with *double-tuned* circuits (with double-tuned circuits, both the primary and secondary sides of the transformer are tuned tank circuits). Therefore, it is easier to achieve an optimum (low) shape factor and good selectivity. Most of a receiver's gain and selectivity is achieved in the IF amplifier section. An IF stage generally has between two and five IF amplifiers. Figure 4-20 shows a schematic diagram for a three-stage IF section. T_1 and T_2 are double-tuned transformers; and L_1, L_2, and L_3 are tapped to reduce the effects of loading. The base of Q_3 is fed from the tapped capacitor pair, C_9 and C_{10}, for the same reason. C_1 and C_6 are neutralization capacitors.

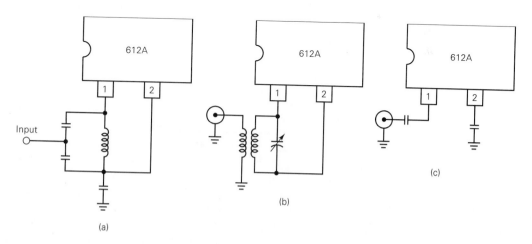

Figure 4-19 NE/SA61219 Typical input configurations: (a) single-ended tuned input; (b) balanced input; (c) single-ended untuned input.

AM Receiver Circuits

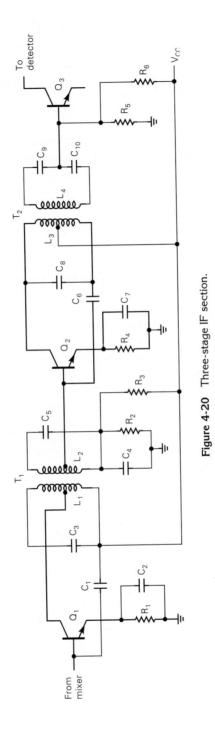

Figure 4-20 Three-stage IF section.

Inductive coupling.   *Inductive* or *transformer coupling* is the most common technique used for coupling IF amplifiers. With inductive coupling, voltage that is applied to the primary windings of a transformer is transferred to the secondary windings. The proportion of the primary voltage that is coupled across to the secondary depends on the number of turns in both the primary and secondary windings (the turns ratio), the amount of *magnetic flux* in the primary winding, the *coefficient of coupling*, and the speed at which the flux is changing (angular velocity). Mathematically, the magnitude of the voltage induced in the secondary windings is

$$E_s = \omega M I_p \qquad \qquad (4\text{-}9)$$

where E_s = voltage magnitude induced in the secondary winding (volts)
ω = angular velocity of the primary voltage wave (radians per second)
M = mutual inductance (henrys)
I_p = primary current (Ampere)

The ability of a coil to induce a voltage within its own windings is called *self-inductance* or simply *inductance* (*L*). When one coil induces a voltage through *magnetic induction* into another coil, the two coils are said to be *coupled* together. The ability of one coil to induce a voltage in another coil is called *mutual inductance* (*M*). Mutual inductance in a transformer is caused by the *magnetic lines of force* (*flux*) that are produced in the primary windings and cut through the secondary windings and is directly proportional to the coefficient of coupling. Coefficient of coupling is the ratio of the secondary flux to the primary flux and is expressed mathematically as

$$k = \frac{\phi_s}{\phi_p} \qquad \qquad (4\text{-}10)$$

where k = coefficient of coupling (unitless)
ϕ_p = primary flux (webers)
ϕ_s = secondary flux (webers)

If all the flux produced in the primary windings cuts through the secondary windings, the coefficient of coupling is 1. If none of the primary flux cuts through the secondary windings, the coefficient of coupling is 0. A coefficient of coupling of 1 is nearly impossible to attain unless the two coils are wound around a common high-permeability iron core. Typically, the coefficient of coupling for standard IF transformers is much less than 1. The transfer of flux from the primary to the secondary windings is called *flux linkage* and is directly proportional to the coefficient of coupling. The mutual inductance of a transformer is directly proportional to the coefficient of coupling and the square root of the product of the primary and secondary inductances. Mathematically, mutual inductance is

$$M = k\left(\sqrt{L_s L_p}\right) \qquad \qquad (4\text{-}11)$$

where M = mutual inductance (henrys)
L_s = inductance of the secondary winding (henrys)
L_p = inductance of the primary winding (henrys)
k = coefficient of coupling (unitless)

Transformer-coupled amplifiers are divided into two general categories: single and double tuned.

AM Receiver Circuits

Single-tuned transformers. Figure 4-21a shows a schematic diagram for a *single-tuned inductively coupled amplifier*. This configuration is called *untuned primary-tuned secondary*. The primary side of T_1 is simply the inductance of the primary winding, whereas a capacitor is in parallel with the secondary winding, creating a tuned secondary. The transformer windings are not tapped because the loading effect of the FET is insignificant. Figure 4-21b shows the response curve for an untuned primary-tuned secondary transformer. E_s increases with frequency until the resonant frequency (f_o) of the secondary is reached; then E_s begins to decrease with further increases in frequency. The peaking of the response curve at the resonant frequency is caused by the reflected impedance. The impedance of the secondary is reflected back into the primary due to the mutual inductance between the two windings. For frequencies below resonance, the increase in ωM is greater than the decrease in I_p; therefore, E_s increases. For frequencies above resonance, the increase in ωM is less than the decrease in I_p; therefore, E_s decreases.

Figure 4-21c shows the effect of coupling on the response curve of an untuned primary-tuned secondary transformer. With *loose coupling* (low coefficient of coupling), the secondary voltage is relatively low and the bandwidth is narrow. As the degree of coupling increases (coefficient of coupling increases), the secondary induced voltage increases and the bandwidth widens. Therefore, for a high degree of selectivity, loose coupling is desired; however, signal amplitude is sacrificed. For high gain and a broad bandwidth, *tight coupling* is necessary. Another single-tuned amplifier configuration is the *tuned primary-untuned secondary*, which is shown in Figure 4-21d.

Double-tuned transformers. Figure 4-22a shows the schematic diagram for a *double-tuned* inductively coupled amplifier. This configuration is called *tuned primary-tuned secondary* because both the primary and secondary windings of transformer T_1 are tuned tank circuits. Figure 4-22b shows the effect of coupling on the response curve for a double-tuned inductively coupled transformer. The response curve closely resembles that

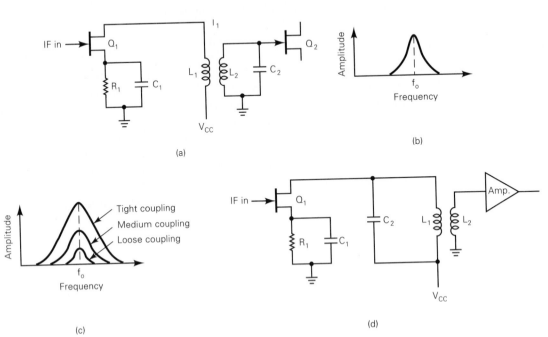

Figure 4-21 Single-tuned transformer: (a) schematic diagram; (b) response curve; (c) effects of coupling; (d) tuned primary, untuned secondary.

Chap. 4   Amplitude Modulation Reception

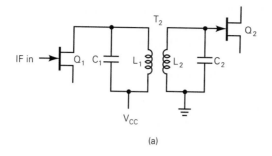

(a)

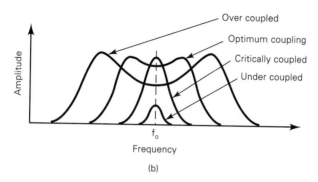

(b)

Figure 4-22 Double-tuned transformer: (a) schematic diagram; (b) response curve.

of a single-tuned circuit for coupling values below critical *coupling* (k_c). Critical coupling is the point where the reflected resistance is equal to the primary resistance and the Q of the primary tank circuit is halved and the bandwidth doubled. If the coefficient of coupling is increased beyond the critical point, the response at the resonant frequency decreases and two new peaks occur on either side of the resonant frequency. This *double peaking* is caused by the reactive element of the reflected impedance being significant enough to change the resonant frequency of the primary tuned circuit. If the coefficient of coupling is increased further, the dip at resonance becomes more pronounced and the two peaks are spread even farther away from the resonant frequency. Increasing coupling beyond the critical value broadens the bandwidth but, at the same time, produces a ripple in the response curve. An ideal response curve has a rectangular shape (a flat top with vertical skirts). From Figure 4-22b it can be seen that a coefficient of coupling approximately 50% greater than the critical value yields a good compromise between flat response and steep skirts. This value of coupling is called *optimum coupling* (k_{opt}) and is expressed mathematically as

$$k_{opt} = 1.5k_c \tag{4-12}$$

where

$$k_{opt} = \text{optimum coupling}$$

$$k_c = \text{critical coupling} = \frac{1}{\sqrt{Q_pQ_s}}$$

where Q_p and Q_s are uncoupled values.

The bandwidth of a double-tuned amplifier is

$$BW_{dt} = kf_o \tag{4-13}$$

AM Receiver Circuits

Bandwidth reduction. When several tuned amplifiers are cascaded together, the total response is the product of the amplifiers' individual responses. Figure 4-23a shows a response curve for a tuned amplifier. The gain at f_1 and f_2 is 0.707 of the gain at f_o. If two identical tuned amplifiers are cascaded, the gain at f_1 and f_2 will be reduced to 0.5 (0.707 $\times$ 0.707), and if three identical tuned amplifiers are cascaded, the gain at f_1 and f_2 is reduced to 0.353. Consequently, as additional tuned amplifiers are added, the overall shape of the response curve narrows and the bandwidth is reduced. This bandwidth reduction is shown in Figures 4-23b and c. Mathematically, the overall bandwidth of n single-tuned stages is given as

$$BW_n = BW_1\left[\sqrt{2^{1/n} - 1}\right]$$ (4-14)

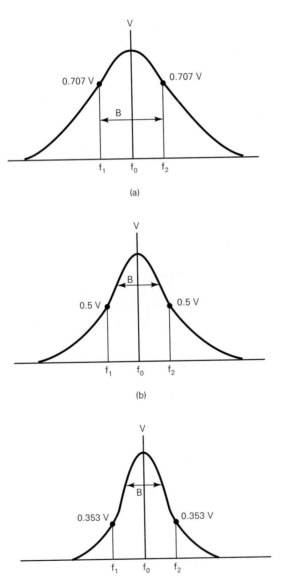

(a)

(b)

(c)

Figure 4-23 Bandwidth reduction: (a) single-tuned stage; (b) two cascaded stages; (c) three cascaded stages.

Chap. 4 **Amplitude Modulation Reception**

where BW_n = bandwidth of n single-tuned stages (hertz)
BW_1 = bandwidth of one single-tuned stage (hertz)
n = number of stages (any positive integer)

The bandwidth for n double-tuned stages is

$$BW_{ndt} = BW_{1dt}[2^{1/n} - 1]^{1/4}$$

(4-15)

where BW_{ndt} = overall bandwidth of n double-tuned amplifiers (hertz)
BW_{1dt} = bandwidth of a single double-tuned amplifier (hertz)
n = number of double-tuned stages (any positive integer)

EXAMPLE 4-8

Determine the overall bandwidth for (a) two single-tuned amplifiers each with a bandwidth of 10 kHz, (b) three single-tuned amplifiers each with a bandwidth of 10 kHz, (c) four single-tuned amplifiers each with a bandwidth of 10 kHz, and (d) a double-tuned amplifier with optimum coupling, a critical coupling of 0.02, and a resonant frequency of 1 MHz; (e) repeat parts a, b, and c for the double-tuned amplifier of part d.

Solution   (a) From Equation 4-14,

$$BW_2 = 10 \text{ kHz}\left[\sqrt{2^{1/2} - 1}\right]$$

$$= 6436 \text{ Hz}$$

(b) Again from Equation 4-14,

$$BW_3 = 10 \text{ kHz}\left[\sqrt{2^{1/3} - 1}\right]$$

$$= 5098 \text{ Hz}$$

(c) Again from Equation 4-14,

$$BW_4 = 10 \text{ kHz}\left[\sqrt{2^{1/4} - 1}\right]$$

$$= 4350 \text{ Hz}$$

(d) From Equation 4-12, $K_{opt} = 1.5(0.02) = 0.03$

From Equation 4-13, $BW_{dt} = 0.03(1 \text{ MHz}) = 30 \text{ kHz}$

(e) From Equation 4-15,

n	BW (Hz)
2	24,067
3	21,420
4	19,786

IF transformers come as specially designed tuned circuits in groundable metal packages called *IF cans*. Figure 4-24 shows the physical and schematic diagrams for a typical IF can. The primary winding comes with a shunt 125-pF capacitor. The vertical arrow shown between the primary and secondary windings indicates that the ferrite core is tunable with a nonmetallic screwdriver or tuning tool. Adjusting the ferrite core changes the mutual inductance, which controls the magnitude of the voltage induced into the

AM Receiver Circuits

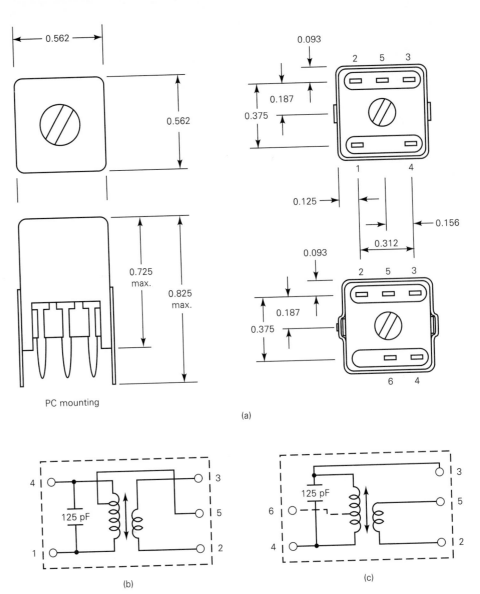

Figure 4-24 IF can: (a) physical diagram; (b) schematic diagram untapped coil; and (c) schematic diagram tapped coil.

secondary windings. The tap in the primary winding can be used to increase the Q of the collector circuit of the driving transistor. If the tap is not used, the equivalent circuit is shown in Figure 4-24b; the effective $Q = R_L/X_L$ and the bandwidth $B = f/Q$, where f is the resonant frequency. If ac ground is connected to the tap, the equivalent circuit is shown in Figure 4-24c. With the tap at ac ground, the effective Q increases and the overall response is more selective.

Integrated-circuit IF amplifiers. In recent years, integrated-circuit IF amplifiers have seen universal acceptance in mobile radio systems such as two-way radio. Integrated circuits offer the obvious advantages of small size and low power consumption. One of the most popular IC intermediate frequency amplifiers is the CA3028A. The CA3028A is a

Chap. 4 **Amplitude Modulation Reception**

differential cascoded amplifier designed for use in communications and industrial equipment as an IF or RF amplifier at frequencies from dc to 120 MHz. The CA3028A features a controlled input offset voltage, offset current, and input bias current. It uses a balanced differential amplifier configuration with a controlled constant-current source and can be used for both single- and dual-ended operation. The CA3028A has balanced-AGC (automatic gain control) capabilities and a wide operating-current range.

Figure 4-25a shows the schematic diagram for the CA3028A, and Figure 4-25b

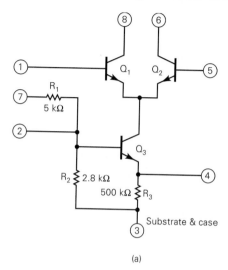

(a)

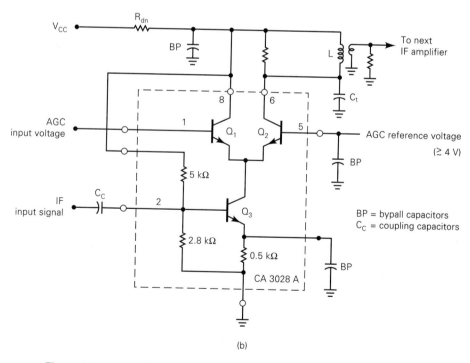

(b)

Figure 4-25 CA3028A linear integrated-circuit differential/cascoded amplifier: (a) schematic diagram; (b) cascoded amplifier configuration.

shows the CA3028A used as a cascoded IF amplifier. The IF input is applied to pin 2 and the output is taken from pin 6. When the AGC voltage on pin 1 is equal to the AGC reference voltage on pin 5, the emitter currents flowing in Q_1 and Q_2 are equal and the amplifier has high gain. If the AGC voltage on pin 1 increases, Q_2 current decreases and the stage gain decreases.

AM Detector Circuits

The function of an AM detector is to demodulate the AM signal and recover or reproduce the original source information. The recovered signal should contain the same frequencies as the original information signal and have the same relative amplitude characteristics. The AM detector is sometimes called the *second detector*, with the mixer/converter being the first detector because it precedes the AM detector.

Peak detector. Figure 4-26a shows a schematic diagram for a simple noncoherent AM demodulator, which is commonly called a *peak detector*. Because a diode is a nonlinear device, nonlinear mixing occurs in D_1 when two or more signals are applied to its input. Therefore, the output contains the original input frequencies, their harmonics, and

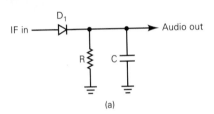

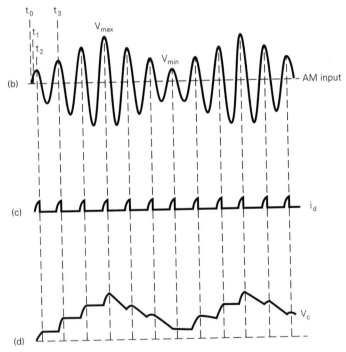

Figure 4-26 Peak detector: (a) schematic diagram; (b) AM input waveform; (c) diode current waveform; (d) output voltage waveform.

Chap. 4   Amplitude Modulation Reception

their cross products. If a 300-kHz carrier is amplitude modulated by a 2-kHz sine wave, the modulated wave is made up of a lower side frequency, carrier, and upper side frequency of 298, 300, and 302 kHz, respectively. If the resultant signal is the input to the AM detector shown in Figure 4-26a, the output will comprise the three input frequencies, the harmonics of all three frequencies, and the cross products of all possible combinations of the three frequencies and their harmonics. Mathematically, the output is

$$V_{\text{out}} = \text{input frequencies} + \text{harmonics} + \text{sums and differences}$$

Because the RC network is a lowpass filter, only the difference frequencies are passed on to the audio section. Therefore, the output is simply

$$V_{\text{out}} = 300 - 298 \text{ kHz} = 2 \text{ kHz}$$

$$= 302 - 300 \text{ kHz} = 2 \text{ kHz}$$

$$= 302 - 298 \text{ kHz} = 4 \text{ kHz}$$

Because of the relative amplitude characteristics of the upper and lower side frequencies and the carrier, the difference between the carrier frequency and either the upper or lower side frequency is the predominant output signal. Consequently, for practical purposes, the original modulating signal (2 kHz) is the only component that is contained in the output of the peak detector.

In the preceding analysis, the diode detector was analyzed as a simple mixer, which it is. Essentially, the difference between an AM modulator and an AM demodulator is that the output of a modulator is tuned to the sum frequencies (up-converter), whereas the output of a demodulator is tuned to the difference frequencies (down-converter). The demodulator circuit shown in Figure 4-26a is commonly called a *diode detector* because the nonlinear device is a diode, or a *peak detector* because it detects the peaks of the input envelope, or a *shape* or *envelope detector* because it detects the shape of the input envelope. Essentially, the carrier signal *captures* the diode and forces it to turn on and off (rectify) synchronously (both frequency and phase). Thus the side frequencies mix with the carrier, and the original baseband signals are recovered.

Figures 4-26b, c, and d show a detector input voltage waveform, the corresponding diode current waveform, and the detector output voltage waveform. At time t_0 the diode is reverse biased and off ($i_d = 0$ A), the capacitor is completely discharged ($V_C = 0$ V), and thus the output is 0 V. The diode remains off until the input voltage exceeds the barrier potential of D_1 (approximately 0.3 V). When V_{in} reaches 0.3 V (t_1), the diode turns on and diode current begins to flow, charging the capacitor. The capacitor voltage remains 0.3 V below the input voltage until V_{in} reaches its peak value. When the input voltage begins to decrease, the diode turns off and i_d goes to 0 A (t_2). The capacitor begins to discharge through the resistor, but the RC time constant is made sufficiently long so that the capacitor cannot discharge as rapidly as V_{in} is decreasing. The diode remains off until the next input cycle, when V_{in} goes 0.3 V more positive than V_C (t_3). At this time the diode turns on, current flows, and the capacitor begins to charge again. It is relatively easy for the capacitor to charge to the new value because the RC charging time constant is $R_d C$, where R_d is the *on* resistance of the diode, which is quite small. This sequence repeats itself on each successive positive peak of V_{in}, and the capacitor voltage follows the positive peaks of V_{in} (hence the name peak detector). The output waveform resembles the shape of the input envelope (hence the name shape detector). The output waveform has a high-frequency ripple that is equal to the carrier frequency. This is due to the diode turning on during the positive peaks of the envelope. The ripple is easily removed by the audio amplifiers because the carrier frequency is much higher than the highest modulating signal frequency. The circuit shown in Figure 4-26 responds only to the positive peaks of V_{in} and is therefore called a *positive peak detector*. By simply turning the diode around, the circuit becomes a negative peak detector. The output voltage reaches its peak positive amplitude at the same time that the input envelope reaches its maximum positive value (V_{max}), and the output voltage goes to its minimum peak amplitude at the same time that the input voltage goes to its minimum value (V_{min}). For 100% modulation, V_{out} swings from 0 V to a value equal to $V_{max} - 0.3$ V.

Figure 4-27 shows the input and output waveforms for a peak detector with various percentages of modulation. With no modulation, a peak detector is simply a filtered half-wave rectifier and the output voltage is approximately equal to the peak input voltage minus the 0.3 V. As the percent modulation changes, the variations in the output voltage increase and decrease proportionately; the output waveform follows the shape of the AM envelope. However, regardless of whether modulation is present or not, the average value of the output voltage is approximately equal to the peak value of the unmodulated carrier.

Detector distortion. When successive positive peaks of the detector input waveform are increasing, it is important that the capacitor hold its charge between peaks (that is, a relatively long RC time constant is necessary). However, when the positive peaks are decreasing in amplitude, it is important that the capacitor discharge between successive peaks to a value less than the next peak (a short RC time constant is necessary). Obviously, a trade-off between a long and a short time constant is in order. If the RC time constant is too short, the output waveform resembles a half-wave rectified signal. This is sometimes called *rectifier distortion* and is shown in Figure 4-28b. If the RC time constant is too long, the slope of the output waveform cannot follow the trailing slope of the envelope. This type of distortion is called *diagonal clipping* and is shown in Figure 4-28c.

The RC network following the diode in a peak detector is a low-pass filter. The slope of the envelope depends on both the modulating signal frequency and the modulation coefficient (m). Therefore, the maximum slope (fastest rate of change) occurs when the envelope is crossing its zero axis in the negative direction. The highest modulating signal frequency that can be demodulated by a peak detector without attenuation is given as

$$f_{m(max)} = \frac{\sqrt{(1/m^2) - 1}}{2\pi RC} \qquad (4\text{-}16a)$$

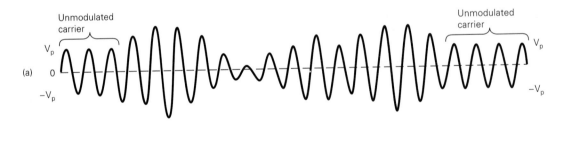

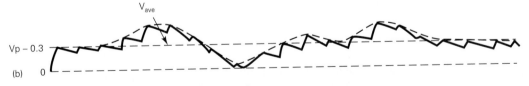

Figure 4-27 Positive peak detector: (a) input waveform; (b) output waveform.

Chap. 4 **Amplitude Modulation Reception**

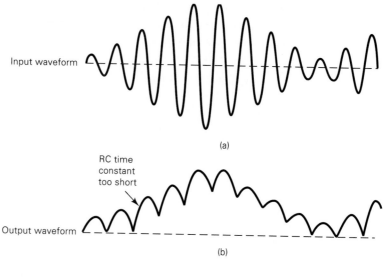

Input waveform

(a)

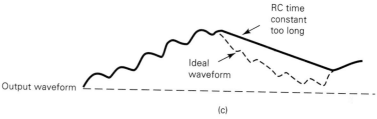

RC time
constant
too short

Output waveform

(b)

RC time
constant
too long

Ideal
waveform

Output waveform

(c)

Figure 4-28 Detector distortion: (a) input envelope; (b) rectifier distortion; (c) diagonal clipping.

where $f_{m(max)}$ = maximum modulating signal frequency (hertz)
m = modulation coefficient (unitless)
RC = time constant (seconds)

For 100% modulation, the numerator in Equation 4-16a goes to zero, which essentially means that all modulating signal frequencies are attenuated as they are demodulated. Typically, the modulating signal amplitude in a transmitter is limited or compressed such that approximately 90% modulation is the maximum that can be achieved. For 70.7% modulation, Equation 4-16a reduces to

$$f_{m(max)} = \frac{1}{2\pi RC} \qquad (4\text{-}16b)$$

Equation 4-16b is commonly used when designing peak detectors to determine an approximate maximum modulating signal.

Automatic Gain Control Circuits

An *automatic gain control* (*AGC*) circuit compensates for minor variations in the received RF signal level. The AGC circuit automatically increases the receiver gain for weak RF input levels and automatically decreases the receiver gain when a strong RF signal is received. Weak signals can be buried in receiver noise and, consequently, be impossible to

AM Receiver Circuits

detect. An excessively strong signal can overdrive the RF and/or IF amplifiers and produce excessive nonlinear distortion and even saturation. There are several types of AGC, which include direct or simple AGC, delayed AGC, and forward AGC.

Simple AGC. Figure 4-29 shows a block diagram for an AM superheterodyne receiver with simple AGC. The automatic gain control circuit monitors the received signal level and sends a signal back to the RF and IF amplifiers to adjust their gain automatically. AGC is a form of degenerative or negative feedback. The purpose of AGC is to allow a receiver to detect and demodulate, equally well, signals that are transmitted from different stations whose output power and distance from the receiver vary. For example, an AM radio in a vehicle does not receive the same signal level from all the transmitting stations in the area or, for that matter, from a single station when the automobile is moving. The AGC circuit produces a voltage that adjusts the receiver gain and keeps the IF carrier power at the input to the AM detector at a relatively constant level. The AGC circuit is not a form of *automatic volume control* (*AVC*); AGC is independent of modulation and totally unaffected by normal changes in the modulating signal amplitude.

Figure 4-30 shows a schematic diagram for a simple AGC circuit. As you can see, an AGC circuit is essentially a peak detector. In fact, very often the AGC correction voltage is taken from the output of the audio detector. In Figure 4-27, it was shown that the dc voltage at the output of a peak detector is equal to the peak unmodulated carrier amplitude minus the barrier potential of the diode and is totally independent of the depth of modulation. If the carrier amplitude increases, the AGC voltage increases, and if the carrier amplitude decreases, the AGC voltage decreases. The circuit shown in Figure 4-30 is a negative peak detector and produces a negative voltage at its output. The greater the amplitude of the input carrier is, the more negative the output voltage. The negative voltage from the AGC detector is fed back to the IF stage where it controls the bias voltage on the base of Q_1. When the carrier amplitude increases, the voltage on the base of Q_1 becomes less positive, causing the emitter current to decrease. As a result, r'_e increases and the amplifier gain (r_c/r'_e) decreases, which in turn causes the carrier amplitude to decrease. When the carrier amplitude decreases, the AGC voltage becomes less negative,

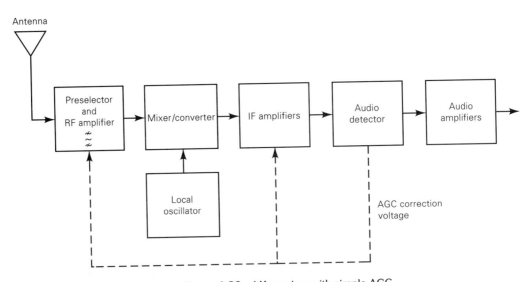

Figure 4-29 AM receiver with simple AGC.

Chap. 4   Amplitude Modulation Reception

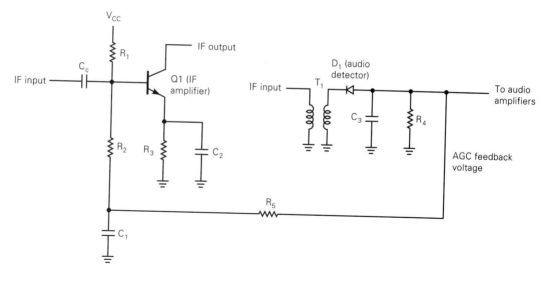

Figure 4-30 Simple AGC circuit.

the emitter current increases, r'_e decreases, and the amplifier gain increases. Capacitor C_1 is an audio bypass capacitor that prevents changes in the AGC voltage due to modulation from affecting the gain of Q_1.

Delayed AGC. Simple AGC is used in most inexpensive broadcast-band radio receivers. However, with simple AGC, the AGC bias begins to increase as soon as the received signal level exceeds the thermal noise of the receiver. Consequently, the receiver becomes less sensitive (this is sometimes called *automatic desensing*). *Delayed AGC* prevents the AGC feedback voltage from reaching the RF or IF amplifiers until the RF level exceeds a predetermined magnitude. Once the carrier signal has exceeded the threshold level, the delayed AGC voltage is proportional to the carrier signal strength. Figure 4-31a shows the response characteristics for both simple and delayed AGC. It can be seen that with delayed AGC the receiver gain is unaffected until the AGC threshold level is exceeded, whereas with simple AGC the receiver gain is immediately affected. Delayed AGC is used with more sophisticated communications receivers. Figure 4-31b shows IF gain-versus-RF input signal level for both simple and delayed AGC.

Forward AGC. An inherent problem with both simple and delayed AGC is the fact that they are both forms of *post-AGC* (after-the-fact compensation). With post-AGC, the circuit that monitors the carrier level and provides the AGC correction voltage is located after the IF amplifiers; therefore, the simple fact that the AGC voltage changed indicates that it may be too late (the carrier level has already changed and propagated through the receiver). Therefore, neither simple or delayed AGC can accurately compensate for rapid changes in the carrier amplitude. *Forward AGC* is similar to conventional AGC except that the receive signal is monitored closer to the front end of the receiver and the correction voltage is fed forward to the IF amplifiers. Consequently, when a change in signal level is detected, the change can be compensated for in succeeding stages. Figure 4-32 shows an AM superheterodyne receiver with forward AGC. For a more sophisticated method of accomplishing AGC, refer to Chapter 7 under the heading "Two-way FM Receivers."

AM Receiver Circuits

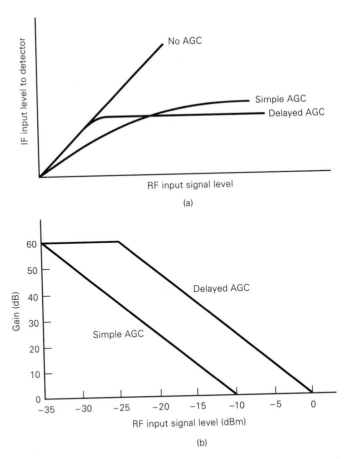

Figure 4-31 Automatic gain control (AGC): (a) response characteristics; (b) IF gain versus RF input signal level.

Squelch Circuits

The purpose of a *squelch circuit* is to *quiet* a receiver in the absence of a received signal. If an AM receiver is tuned to a location in the RF spectrum where there is no RF signal, the AGC circuit adjusts the receiver for maximum gain. Consequently, the receiver amplifies and demodulates its own internal noise. This is the familiar crackling and sputtering heard on the speaker in the absence of a received carrier. In domestic AM systems, each station is continuously transmitting a carrier regardless of whether there is any modulation or not. Therefore, the only time the idle receiver noise is heard is when tuning between stations. However, in two-way radio systems, the carrier in the transmitter is generally turned off except when a modulating signal is present. Therefore, during idle transmission times, a receiver is simply amplifying and demodulating noise. A squelch circuit keeps the audio section of the receiver turned off or *muted* in the absence of a received signal (the receiver is squelched). A disadvantage of a squelch circuit is weak RF signals will not produce an audio output.

Figure 4-33 shows a schematic diagram for a typical squelch circuit. This squelch circuit uses the AGC voltage to monitor the received RF signal level. The greater the AGC voltage, the stronger the RF signal. When the AGC voltage drops below a preset level, the squelch circuit is activated and disables the audio section of the receiver. In Figure 4-33

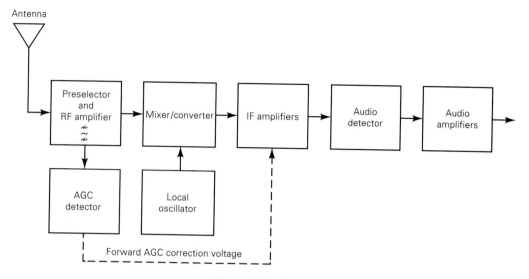

Figure 4-32 Forward AGC.

it can be seen that the squelch detector uses a resistive voltage divider to monitor the AGC voltage. When the RF signal drops below the squelch threshold, Q_1 turns on and shuts off the audio amplifiers. When the RF signal level increases above the squelch threshold, the AGC voltage becomes more negative, turning off Q_1, and enabling the audio amplifiers. The squelch threshold level can be adjusted with R_1. A more sophisticated method of squelching a receiver is described in Chapter 7 under the heading "Two-way FM Receivers."

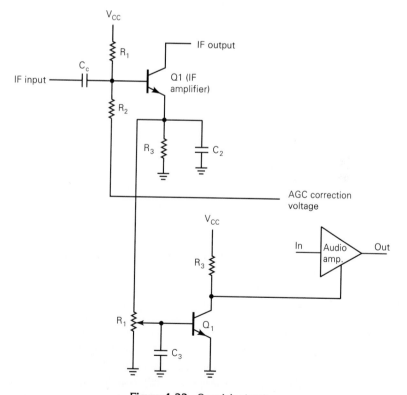

Figure 4-33 Squelch circuit.

AM Receiver Circuits

Linear Integrated-circuit AM Receivers

Linear integrated circuits are now available from several manufacturers that perform all receiver functions except RF and IF filtering and volume control on a single chip. Figure 4-34 shows the schematic diagram of an AM receiver that uses the National Semiconductor Corporation LM1820 linear integrated-circuit AM radio chip. The LM1820 has onboard RF amplifier, mixer, local oscillator, and IF amplifier stages. However, RF and IF selectivity is accomplished by adjusting tuning coils in externally connected tuned circuits or cans. Also, a LIC audio amplifier, such as the LM386, and a speaker are necessary to complete a functional receiver.

LIC AM radios are not widely used because the physical size reduction made possible by reducing the component count through integration is offset by the size of the external components necessary for providing bandlimiting and channel selection. Alternatives to *LC* tank circuits and IF cans, such as ceramic filters, may be integrable in the near future. Also, new receiver configurations (other than TRF or superheterodyne) may be possible in the future using phase-locked-loop technology. Phase-locked-loop receivers would need only two external components: a volume control and a station tuning control.

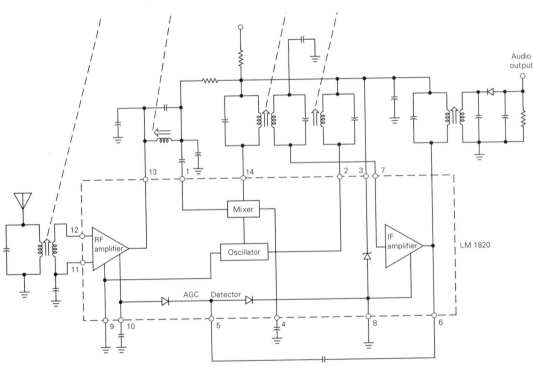

Figure 4-34 LM1820 linear integrated-circuit AM radio system.

DOUBLE-CONVERSION AM RECEIVERS

For good image-frequency rejection, a relatively high intermediate frequency is desired. However, for high-gain selective amplifiers that are stable and easily neutralized, a low intermediate frequency is necessary. The solution is to use two intermediate frequencies. The *first IF* is a relatively high frequency for good image-frequency rejection, and the *second IF* is a relatively low frequency for easy amplification. Figure 4-35 shows a block dia-

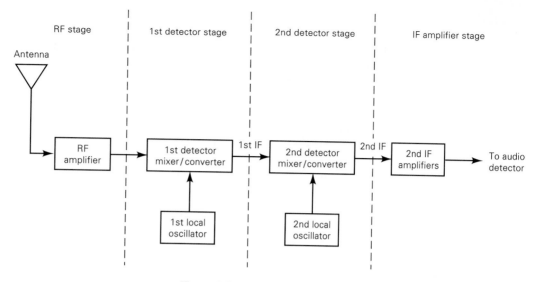

Figure 4-35 Double-conversion AM receiver.

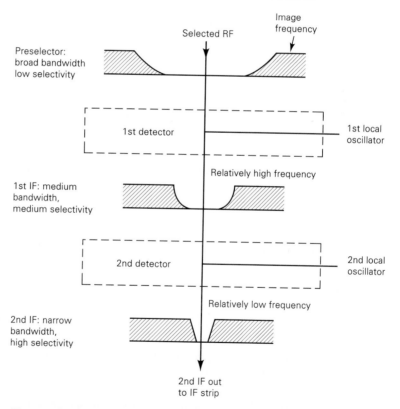

Figure 4-36 Filtering requirements for the double-conversion AM receiver shown in Figure 4-35.

gram for a *double-conversion* AM receiver. The first IF is 10.625 MHz, which pushes the image frequency 21.25 MHz away from the desired RF. The first IF is immediately downconverted to 455 kHz and fed to a series of high-gain IF amplifiers. Figure 4-36 illustrates the filtering requirements for a double-conversion AM receiver.

Double-conversion AM Receivers

NET RECEIVER GAIN

Thus far we have discussed RF gain, conversion gain, and IF gain. However, probably the most important gain is *net receiver gain*. The net receiver gain is simply the ratio of the demodulated signal level at the output of the receiver (audio) to the RF signal level at the input to the receiver, or the difference between the audio signal level in dBm and the RF signal level in dBm.

In essence, net receiver gain is the dB sum of all the gains in the receiver minus the dB sum of all the losses. Receiver losses typically include preselector loss, mixer loss (that is, conversion gain), and detector losses. Gains include RF gain, IF gain, and audio-amplifier gain. Mathematically, net receiver gain is

$$G_{dB} = \text{gains}_{dB} - \text{losses}_{dB}$$

where   gains = RF amplifier gain + IF amplifier gain + audio-amplifier gain
         losses = preselector loss + mixer loss + detector loss

EXAMPLE 4-9

For an AM receiver with a −60-dBm RF input signal level and the following gains and losses, determine the net receiver gain and the audio signal level.

    Gains: RF amplifier = 33 dB, IF amplifier = 27 dB, audio amplifier = 25 dB
    Losses: preselector loss = 3 dB, mixer loss = 6 dB, detector loss = 8 dB

Solution   The sum of the gains is $33 + 27 + 25 = 85$ dB

The sum of the losses is $3 + 6 + 8 = 17$ dB

thus   net receiver gain $G = 85 - 17 = 68$ dB

and the audio signal level is   −60 dBm + 68 dB = 8 dBm

Net receiver gain should not be confused with overall *system gain*. Net receiver gain includes only components within the receiver beginning at the input to the preselector. System gain includes all the gains and losses incurred by a signal as it propagates from the transmitter output stage to the output of the detector in the receiver and includes antenna gain and transmission line and propagation losses. System gain is discussed in detail in Chapter 17.

QUESTIONS

4-1. What is meant by the *front end* of a receiver?

4-2. What are the primary functions of the front end of a receiver?

4-3. Define *selectivity* and *shape factor*. What is the relationship between receiver noise and selectivity?

4-4. Describe *bandwidth improvement*. What is the relationship between bandwidth improvement and receiver noise?

4-5. Define *sensitivity*.

4-6. What is the relationship among receiver noise, bandwidth, and temperature?

4-7. Define *fidelity*.

4-8. List and describe the three types of distortion that reduce the fidelity of a receiver.

4-9. Define *insertion loss*.

Chap. 4   Amplitude Modulation Reception

4-10. Define *noise temperature* and *equivalent noise temperature*.

4-11. Describe the difference between a *coherent* and a *noncoherent* radio receiver.

4-12. Draw the block diagram for a TRF radio receiver and briefly describe its operation.

4-13. What are the four predominant disadvantages of a TRF receiver?

4-14. Draw the block diagram for an AM superheterodyne receiver and describe its operation and the primary functions of each stage.

4-15. Define *heterodyning*.

4-16. What is meant by the terms *high-* and *low-side injection*.

4-17. Define *local oscillator tracking* and *tracking error*.

4-18. Describe *three-point tracking*.

4-19. What is meant by *gang* tuning?

4-20. Define *image frequency*.

4-21. Define *image-frequency rejection ratio*.

4-22. List six characteristics that are desirable in an RF amplifier.

4-23. What advantage do FET RF amplifiers have over BJT RF amplifiers?

4-24. Define *neutralization*. Describe the neutralization process.

4-25. What is a *cascoded* amplifier?

4-26. Define *conversion gain*.

4-27. What is the advantage of a relatively high frequency intermediate frequency; a relatively low frequency intermediate frequency?

4-28. Define the following terms: *inductive coupling*; *self-inductance*; *mutual inductance*; *coefficient of coupling*; *critical coupling*; *optimum coupling*.

4-29. Describe *loose* coupling; *tight* coupling.

4-30. Describe the operation of a *peak detector*.

4-31. Describe *rectifier distortion* and what causes it.

4-32. Describe *diagonal clipping* and what causes it.

4-33. Describe the following terms: *simple AGC*; *delayed AGC*; *forward AGC*.

4-34. What is the purpose of a *squelch* circuit?

4-35. Explain the operation of a double-conversion superheterodyne receiver.

PROBLEMS

4-1. Determine the shape factor and percent selectivity for the gain-versus-frequency curve shown.

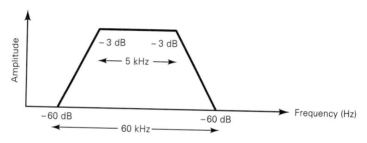

4-2. Determine the improvement in the noise figure for a receiver with an RF bandwidth equal to 40 kHz and IF bandwidth of 16 kHz.

4-3. Determine the equivalent noise temperature for an amplifier with a noise figure of 6 dB and an environmental temperature $T = 27°C$.

4-4. For an AM commercial broadcast-band receiver with an input filter Q-factor of 85, determine the bandwidth at the low and high ends of the RF spectrum.

4-5. For an AM superheterodyne receiver using high-side injection with a local oscillator frequency of 1200 kHz, determine the IF carrier and upper and lower side frequencies for an RF envelope that is made up of a carrier and upper and lower side frequencies of 600, 604, and 596 kHz, respectively.

4-6. For a receiver with a ± 2.5-kHz tracking error, a 455-kHz IF, and a maximum modulating signal frequency $f_m = 6$ kHz, determine the minimum IF bandwidth.

4-7. For a receiver with IF, RF, and local oscillator frequencies of 455, 900, and 1355 kHz, respectively, determine:
 (a) Image frequency.
 (b) IFRR for a preselector Q of 80.

4-8. For a citizens-band receiver using high-side injection with an RF carrier of 27.04 MHz and a 10.645 MHz first IF, determine:
 (a) Local oscillator frequency.
 (b) Image frequency.

4-9. For a three-stage double-tuned RF amplifier with an RF carrier equal to 800 kHz and a coefficient of coupling $k_{opt} = 0.025$, determine:
 (a) Bandwidth for each individual stage.
 (b) Overall bandwidth for the three stages.

4-10. Determine the maximum modulating signal frequency for a peak detector with the following parameters: $C = 1000$ pF, $R = 10$ kΩ, and $m = 0.5$. Repeat the problem for $m = 0.707$.

Chap. 4   **Amplitude Modulation Reception**

5

Single-sideband Communications Systems

INTRODUCTION

Conventional AM double-sideband systems, such as those discussed in Chapters 3 and 4, have several inherent and outstanding disadvantages. First, in conventional AM systems, at least two-thirds of the transmitted power is in the carrier. However, there is no information in the carrier; the sidebands contain the information. Also, the information contained in the upper sideband is identical to the information contained in the lower sideband. Therefore, transmitting both sidebands is redundant. Consequently, conventional AM is both power and bandwidth inefficient, which are two of the most important considerations when designing an electronic communications system.

The purpose of this chapter is to introduce the reader to several single-sideband AM systems and explain the advantages and disadvantages of choosing them over conventional double-sideband AM.

SINGLE-SIDEBAND SYSTEMS

Single-sideband was mathematically recognized and understood as early as 1914; however, not until 1923 was the first patent granted and a successful communications link established between England and the United States. There are many different types of *sideband* communications systems. Some of them conserve bandwidth, some conserve power, and some conserve both. Figure 5-1 compares the frequency spectra and relative power distributions for conventional AM and several of the more common single-sideband (SSB) systems.

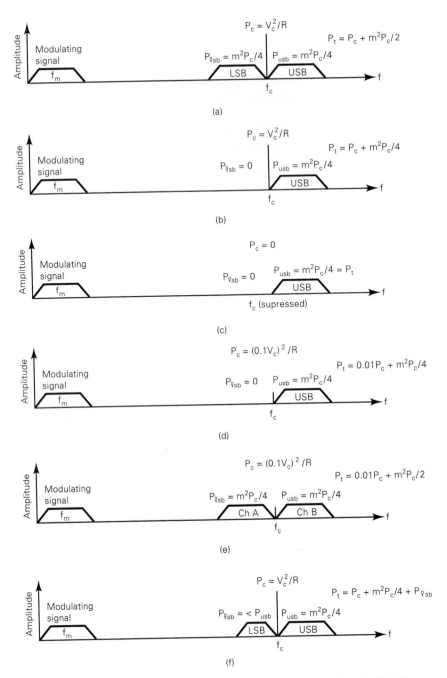

Figure 5-1 Single-sideband systems: (a) conventional DSBFC AM; (b) full-carrier single sideband; (c) suppressed-carrier single sideband; (d) reduced-carrier single sideband; (e) independent sideband; (f) vestigal sideband.

AM Single-sideband Full Carrier

AM *single-sideband full carrier* (*SSBFC*) is a form of amplitude modulation in which the carrier is transmitted at full power, but only one of the sidebands is transmitted. Therefore, SSBFC transmissions require only half as much bandwidth as conventional double-sideband AM. The frequency spectrum and relative power distribution for SSBFC are shown in Figure 5-1b. Note that with 100% modulation the carrier power (P_c) constitutes four-fifths (80%) of the total transmitted power (P_t), and only one-fifth (20%) of the total power is in the sideband. For conventional double-sideband AM with 100% modulation, two-thirds (67%) of the total transmitted power is in the carrier and one-third (33%) is in the sidebands. Therefore, although SSBFC requires less total power, it actually utilizes a smaller percentage of that power for the information-carrying portion of the signal.

Figure 5-2 shows the waveform for a 100% modulated SSBFC wave with a single-frequency modulating signal. The 100% modulated single-sideband full-carrier envelope looks identical to a 50% modulated double-sideband full-carrier envelope. Recall from Chapter 3 that the maximum positive and negative peaks of an AM DSBFC wave occur when the carrier and both sidebands reach their respective peaks at the same time, and the peak change in the envelope is equal to the sum of the amplitudes of the upper and lower side frequencies. With single-sideband transmission, there is only one sideband (either the upper or lower) to add to the carrier. Therefore, the peak change in the envelope is only half of what it is with double-sideband transmission. Consequently, with single-sideband full-carrier transmission, the demodulated signals have only half the amplitude of a double-sideband demodulated wave. Thus, a trade-off is made. SSBFC requires less bandwidth than DSBFC but also produces a demodulated signal with a lower amplitude. However, when the bandwidth is halved, the total noise power is also halved (that is, reduced by 3 dB). And if one sideband is removed, the power in the information portion of the wave is also halved. Consequently, the signal-to-noise ratios for single and double sideband are the same.

With SSBFC, the repetition rate of the envelope is equal to the frequency of the modulating signal, and the depth of modulation is proportional to the amplitude of the modulating signal. Therefore, as with double-sideband transmission, the information is contained in the envelope of the full carrier modulated signal.

AM Single-sideband Suppressed Carrier

AM *single-sideband suppressed carrier* (*SSBSC*) is a form of amplitude modulation in which the carrier is totally suppressed and one of the sidebands removed. Therefore, SSBSC requires half as much bandwidth as conventional double-sideband AM and considerably less transmitted power. The frequency spectrum and relative power distribution for SSBSC with upper sideband transmission are shown in Figure 5-1c. It can be seen that the sideband power makes up 100% of the total transmitted power. Figure 5-3 shows a SSBSC waveform for a single-frequency modulating signal. As you can see, the waveform is not an envelope; it is simply a sine wave at a single frequency equal to the carrier

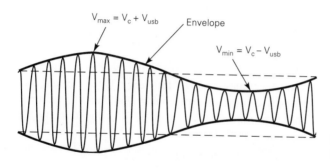

Figure 5-2   SSBFC waveform, 100% modulation.

Single-sideband Systems

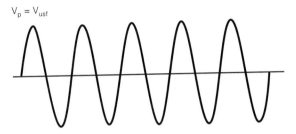

$V_p = V_{usf}$

Figure 5-3 SSBSC waveform.

frequency plus the modulating-signal frequency or the carrier frequency minus the modulating-signal frequency, depending on which sideband is transmitted.

AM Single-sideband Reduced Carrier

AM *single-sideband reduced carrier* (*SSBRC*) is a form of amplitude modulation in which one sideband is totally removed and the carrier voltage is reduced to approximately 10% of its unmodulated amplitude. Consequently, as much as 96% of the total power transmitted is in the unsuppressed sideband. To produce a reduced carrier component, the carrier is totally suppressed during modulation and then reinserted at a reduced amplitude. Therefore, SSBRC is sometimes called single-sideband *reinserted* carrier. The reinserted carrier is often called a pilot carrier and is reinserted for demodulation purposes, which is explained later in this chapter. The frequency spectrum and relative power distribution for SSBRC are shown in Figure 5-1d. The figure shows that the sideband power constitutes almost 100% of the transmitted power. Figure 5-4a shows the transmitted waveform for a single-frequency modulating signal when the carrier and sideband amplitudes are equal, and Figure 5-4b shows the waveform when the carrier amplitude is less than the sideband amplitude. As with double-sideband full-carrier AM, the repetition rate of the envelope is equal to the frequency of the modulating signal. To demodulate a reduced carrier waveform with a conventional peak detector, the carrier must be separated, amplified, and then reinserted at a higher level in the receiver. Therefore, suppressed-carrier transmission is sometimes called *exalted* carrier because the carrier is elevated in the receiver prior to demodulation. With exalted-carrier detection, the amplification of the carrier in the receiver must be sufficient to raise the level of the carrier to a value greater than that of the sideband signal. SSBRC requires half as much bandwidth as conventional AM and, because the carrier is transmitted at a reduced level, also conserves considerable power.

AM Independent Sideboard

AM *independent sideband* (*ISB*) is a form of amplitude modulation in which a single carrier frequency is independently modulated by two different modulating signals. In essence, ISB is a form of double-sideband transmission in which the transmitter consists of two independent single-sideband suppressed-carrier modulators. One modulator produces only the upper sideband and the other produces only the lower sideband. The single-sideband output signals from the two modulators are combined to form a double-sideband signal in which the two sidebands are totally independent of each other except that they are symmetrical about a common carrier frequency. One sideband is positioned above the carrier in the frequency spectrum and one below. For demodulation purposes, the carrier is generally reinserted at a reduced level as with SSBRC transmission. Figure 5-1e shows the frequency spectrum and power distribution for ISB, and Figure 5-5 shows the transmitted waveform for two independent single-frequency information signals (f_{m1} and f_{m2}). The

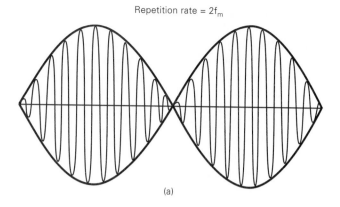

Repetition rate = $2f_m$

(a)

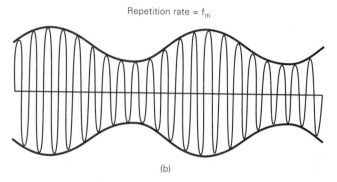

Repetition rate = f_m

(b)

Figure 5-4 SSBRC waveform: (a) carrier level equal to the sideband level; (b) carrier level less than the sideband level.

two information signals are equal in frequency; therefore, the waveform is identical to a double-sideband suppressed-carrier waveform except with a repetition rate equal to twice the modulating signal frequency. ISB conserves both transmit power and bandwidth as two information sources are transmitted within the same frequency spectrum as would be required by a single source using conventional double-sideband transmission. ISB is one technique that is used in the United States for stereo AM transmission. One channel (the left) is transmitted in the lower sideband, and the other channel (the right) is transmitted in the upper sideband.

$f_{m1} = f_{m2}$
Repetition rate = $2f_{m1} = 2f_{m2}$

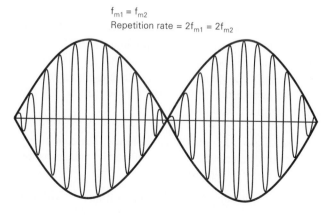

Figure 5-5 ISB waveform.

Single-sideband Systems

AM Vestigial Sideband

AM *vestigial sideband* (*VSB*) is a form of amplitude modulation in which the carrier and one complete sideband are transmitted, but only part of the second sideband is transmitted. The carrier is transmitted at full power. In VSB, the lower modulating-signal frequencies are transmitted double sideband and the higher modulating-signal frequencies are transmitted single sideband. Consequently, the lower frequencies can appreciate the benefit of 100% modulation, whereas the higher frequencies cannot achieve more than the effect of 50% modulation. Consequently, the low-frequency modulating signals are emphasized and produce larger-amplitude signals in the demodulator than the high frequencies. The frequency spectrum and relative power distribution for VSB are shown in Figure 5-1f. Probably the most widely known VSB system is the picture portion of a commercial television broadcasting signal.

Comparison of Single-sideband AM to Double-sideband AM

From the preceding discussion and Figure 5-1, it can be seen that bandwidth conservation and power efficiency are obvious advantages of single-sideband transmission over conventional double-sideband transmission. Single sideband requires only half as much bandwidth and considerably less total transmitted power. The total power transmitted that is necessary to produce a given signal-to-noise ratio at the output of the receiver is a convenient and useful means of comparing the power requirement and relative performance of single sideband to conventional AM systems. The received signal-to-noise ratio determines the degree of intelligibility of a received signal.

Peak envelope power (*PEP*) is the rms power developed at the crest of the modulation envelope (that is, when the modulating signal frequency components are at their maximum amplitudes). A conventional AM wave with 100% modulation contains 1 unit of carrier power and 0.25 unit of power in each sideband for a total transmitted peak power of 1.5 units. A single-sideband transmitter rated at 0.5 units of power will produce the same S/N ratio at the output of a receiver as 1.5 units of carrier plus sideband power from a double-sideband full-carrier signal. In other words, the same performance is achieved with SSBSC using only one-third as much transmitted power and half the bandwidth. Table 5-1 compares conventional AM to single-sideband suppressed carrier for a single-frequency modulating signal. The voltage vectors for the power requirements stated are also shown. It can be seen that it requires 0.5 unit of voltage per sideband and 1 unit for the carrier with conventional AM for a total of 2 PEV (peak envelope volts) and only 0.707 PEV for single sideband. The RF envelopes are also shown, which correspond to the voltage and power relationships previously outlined. The demodulated signal at the output from a conventional AM receiver is proportional to the quadratic sum of the voltages from the upper and lower sideband signals, which equals 1 PEV unit. For single-sideband reception, the demodulated signal is $0.707 \times 1 = 0.707$ PEV. If the noise voltage for conventional AM is arbitrarily chosen as 0.1 V/kHz, the noise voltage for single-sideband signal with half the bandwidth is 0.707 V/kHz. Consequently, the S/N performance for SSBSC is equal to that of conventional AM.

Advantages of single-sideband transmission. There are four predominant advantages of single-sideband suppressed- or reduced-carrier transmission over conventional double-sideband full-carrier transmission.

Bandwidth conservation. Single-sideband transmission requires half as much

TABLE 5-1 CONVENTIONAL AM VERSUS SINGLE-SIDEBAND

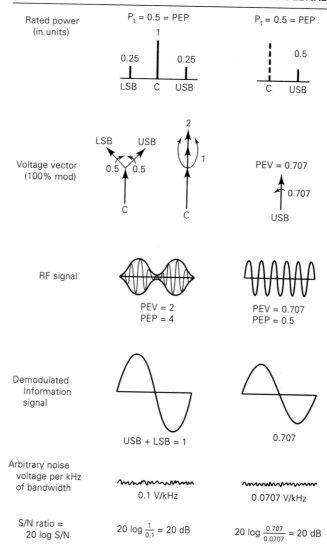

Rated power (in units)	$P_t = 0.5 = PEP$	$P_t = 0.5 = PEP$
Voltage vector (100% mod)		$PEV = 0.707$
RF signal	$PEV = 2$ $PEP = 4$	$PEV = 0.707$ $PEP = 0.5$
Demodulated Information signal	$USB + LSB = 1$	0.707
Arbitrary noise voltage per kHz of bandwidth	0.1 V/kHz	0.0707 V/kHz
S/N ratio = 20 log S/N	$20 \log \frac{1}{0.1} = 20$ dB	$20 \log \frac{0.707}{0.0707} = 20$ dB

bandwidth as conventional AM double-sideband transmission. This advantage is especially important today with an already overcrowded radio-frequency spectrum.

Power conservation.   With single-sideband transmission, only one sideband is transmitted and normally either a suppressed or reduced carrier. As a result, much less total transmitted power is necessary to produce essentially the same quality signal as is achieved with double-sideband full-carrier transmission. Consequently, smaller, more reliable transmitters can be used with single sideband.

Selective fading.   With double-sideband transmission, the two sidebands and carrier may propagate through the transmission media by different paths and, therefore, experience different transmission impairments. This condition is called *selective fading*. One type of selective fading is called *sideband fading*. With sideband fading, one sideband is significantly attenuated. This loss results in a reduced signal amplitude at the output of the receiver demodulator and, consequently, a 3-dB reduced signal-to-noise ratio. This loss

Single-sideband Systems

causes some distortion but is not entirely detrimental to the signal because the two sidebands contain the same information.

The most common and most serious form of selective fading is *carrier-amplitude fading*. Reduction of the carrier level of a 100% modulated wave will make the carrier voltage less than the vector sum of the two sidebands. Consequently, the envelope resembles an overmodulated envelope, causing severe distortion to the demodulated signal.

A third cause of selective fading is carrier or sideband phase shift. When the relative positions of the carrier and sideband vectors of the received signal change, a decided change in the shape of the envelope will occur, causing a severely distorted demodulated signal.

When only one sideband and either a reduced or totally suppressed carrier are transmitted, carrier phase shift and carrier fading cannot occur, and sideband fading only changes the amplitude and frequency response of the demodulated signal. These changes do not generally produce enough distortion to cause loss of intelligibility in the received signal. With single-sideband transmission, it is not necessary to maintain a specific amplitude or phase relationship between the carrier and sideband signals.

Noise reduction. Because single-sideband system utilizes half as much bandwidth as conventional AM, the thermal noise power is reduced to half that of double-sideband system. Taking into consideration both the bandwidth reduction and the immunity to selective fading, SSB systems enjoy approximately a 12-dB S/N ratio advantage over conventional AM (that is, a conventional AM system must transmit a 12-dB more powerful signal to achieve the same performance as a comparable single-sideband system).

Disadvantages of single-sideband transmission. There are two major disadvantages of single-sideband reduced- or suppressed-carrier transmission as compared to conventional double-sideband full-carrier transmission.

Complex receivers. Single-sideband systems require more complex and expensive receivers than conventional AM transmission. This is because most single-sideband transmissions include either a reduced or suppressed carrier; thus envelope detection cannot be used unless the carrier is regenerated at an exalted level. Single-sideband receivers require a carrier recovery and synchronization circuit, such as a PLL frequency synthesizer, which adds to their cost, complexity, and size.

Tuning difficulties. Single-sideband receivers require more complex and precise tuning than conventional AM receivers. This is undesirable for the average user. This disadvantage can be overcome by using more accurate, complex, and expensive tuning circuits.

MATHEMATICAL ANALYSIS OF SUPPRESSED-CARRIER AM

An AM modulator is a *product modulator*; the output signal is the product of the modulating signal and the carrier. In essence, the carrier is multiplied by the modulating signal. Equation 3-8c was given as

$$v_{am}(t) = [1 + m \sin(2\pi f_m t)][E_c \sin(2\pi f_c t)]$$

where $[1 + m \sin(2\pi f_m t)] = $ constant + modulating signal
$[\sin(2\pi f_c t)] = $ unmodulated carrier

If the constant component is removed from the modulating signal, then

$$v_{am}(t) = [m \sin (2\pi f_m t)][E_c \sin (2\pi f_c t)]$$

Multiplying yields $v_{am}(t) = -\dfrac{mE_c}{2} \cos [2\pi(f_c + f_m)t] + \dfrac{mE_c}{2} \cos [2\pi(f_c - f_m)t]$

where $-\dfrac{mE_c}{2} \cos [2\pi(f_c + f_m)t] =$ upper side frequency component

$\dfrac{mE_c}{2} \cos [2\pi(f_c - f_m)t] =$ lower side frequency component

From the preceding mathematical operation, it can be seen that, if the constant component is removed prior to performing the multiplication, the carrier component is removed from the modulated wave and the output signal is simply two cosine waves, one at the sum frequency ($f_c + f_m = f_{usf}$) and the other at the difference frequency ($f_c - f_m = f_{lsf}$). The carrier has been suppressed in the modulator. To convert to single sideband, simply remove either the sum or the difference frequency.

SINGLE-SIDEBAND GENERATION

In the preceding sections it was shown that with most single-sideband systems the carrier is either totally suppressed or reduced to only a fraction of its original value, and one sideband is removed. To remove the carrier from the modulated wave or to reduce its amplitude using conventional notch filters is extremely difficult if not impossible, because the filters simply do not have sufficient Q-factors to remove the carrier without also removing a portion of the sideband. However, it was also shown that removing the constant component suppressed the carrier in the modulator itself. Consequently, modulator circuits that inherently remove the carrier during the modulation process have been developed. Such circuits are called *double-sideband suppressed carrier (DSBSC) modulators*. It will be shown later in this chapter how one of the sidebands can be removed once the carrier has been suppressed.

A circuit that produces a double-sideband suppressed-carrier signal is a *balanced modulator*. The balanced modulator has rapidly become one of the most useful and widely used circuits in electronic communications. In addition to suppressed-carrier AM systems, balanced modulators are widely used in frequency and phase modulation systems as well as in digital modulation systems, such as phase-shift keying and quadrature amplitude modulation.

Balanced Ring Modulator

Figure 5-6a shows the schematic diagram for a *balanced ring modulator* constructed with diodes and transformers. Semiconductor diodes are ideally suited for use in balanced modulator circuits because they are stable, require no external power source, have a long life, and require virtually no maintenance. The balanced ring modulator is sometimes called a *balanced lattice modulator* or simply *balanced modulator*. A balanced modulator has two inputs: a single-frequency carrier and the modulating signal, which may be a single frequency or a complex waveform. For the balanced modulator to operate properly, the amplitude of the carrier must be sufficiently greater than the amplitude of the modulating signal (approximately six to seven times greater). This ensures that the carrier and not the modulating signal controls the *on* or *off* condition of the four diode switches (D_1 to D_4).

Single-sideband Generation

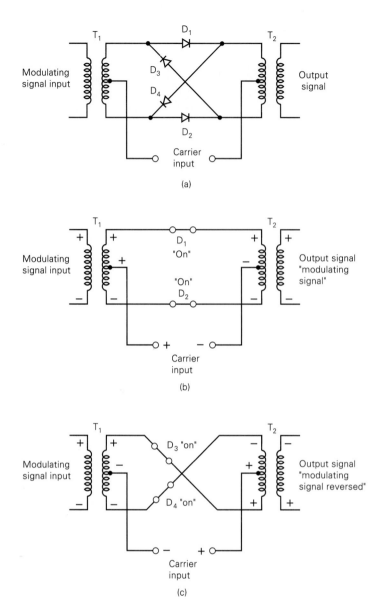

Figure 5-6 Balanced ring modulator: (a) schematic diagram; (b) D_1 and D_2 biased *on*; (c) D_3 and D_4 biased *on*.

Circuit operation. Essentially, diodes D_1 to D_4 are electronic switches that control whether the modulating signal is passed from input transformer T_1 to output transformer T_2 as is or with a 180° phase shift. With the carrier polarity as shown in Figure 5-6b, diode switches D_1 and D_2 are forward biased and *on*, while diode switches D_3 and D_4 are reverse biased and *off*. Consequently, the modulating signal is transferred across the closed switches to T_2 without a phase reversal. When the polarity of the carrier reverses, as shown in Figure 5-6c, diode switches D_1 and D_2 are reverse biased and *off*, while diode switches D_3 and D_4 are forward biased and *on*. Consequently, the modulating signal undergoes a 180° phase reversal before reaching T_2. Carrier current flows from its source to the center taps of T_1 and T_2, where it splits and goes in opposite directions through the upper and lower halves of the transformers. Thus, their magnetic fields cancel in the secondary windings of the transformer and the carrier is suppressed. If the diodes are not perfectly matched or if the transformers are not exactly center tapped, the circuit is out of balance and the carrier is not totally suppressed. It is virtually impossible to achieve perfect balance; thus, a small carrier component is always present in the output signal. This is commonly called *carrier leak*. The amount of carrier suppression is typically between 40 and 60 dB.

Figure 5-7 shows the input and output waveforms associated with a balanced modulator for a single-frequency modulating signal. It can be seen that D_1 and D_2 conduct only during the positive half-cycles of the carrier input signal, and D_3 and D_4 conduct only during the negative half-cycles. The output from a balanced modulator consists of a series of RF pulses whose repetition rate is determined by the RF carrier switching frequency, and amplitude is controlled by the level of the modulating signal. Consequently, the output waveform takes the shape of the modulating signal, except with alternating positive and negative polarities that correspond to the polarity of the carrier signal.

FET Push–Pull Balanced Modulator

Figure 5-8 shows a schematic diagram for a balanced modulator that uses FETs rather than diodes for the nonlinear devices. A FET is a nonlinear device that exhibits square-law properties and produces only second-order cross-product frequencies. Like the diode bal-

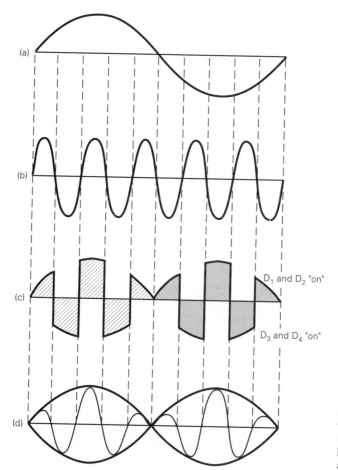

Figure 5-7 Balanced modulator waveforms: (a) modulating signal; (b) carrier signal; (c) output waveform before filtering; (d) output waveform after filtering.

Single-sideband Generation

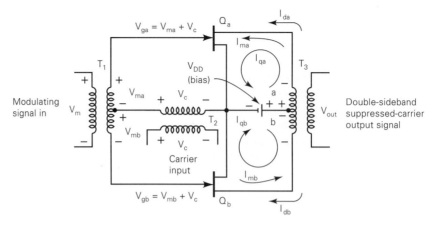

Figure 5-8 FET balanced modulator. For the polarities shown:

$$I_{ta} = I_{qa} + I_{da} + I_{ma}$$

$$I_{tb} = -I_{qb} - I_{db} + I_{mb}$$

$$I_t = I_{ma} + I_{mb} = 2I_m$$

V_{out} is proportional to the modulating current (I_{ma} and I_{mb}).

anced modulator, a FET modulator is a product modulator and produces only the side-bands at its output and suppresses the carrier. The FET balanced modulator is similar to a standard push–pull amplifier except that the modulator circuit has two inputs (the carrier and the modulating signal).

Circuit operation. The carrier is fed into the circuit in such a way that it is applied simultaneously and in phase to the gates of both FET amplifiers (Q_1 and Q_2). The carrier produces currents in both the top and bottom halves of output transformer T_3 that are equal in magnitude but 180° out of phase. Therefore, they cancel and no carrier component appears in the output waveform. The modulating signal is applied to the circuit in such a way that it is applied simultaneously to the gates of the two FETs 180° out of phase. The modulating signal causes an increase in the drain current in one FET and a decrease in the drain current in the other FET.

Figure 5-9 shows the phasor diagram for the currents produced in the output transformer of a FET balanced modulator. Figure 5-9a shows that the quiescent dc drain currents from Q_a and Q_b (I_{qa} and I_{qb}) pass through their respective halves of the primary winding of T_3 180° out of phase with each other. Figure 5-9a also shows that an increase in drain current due to the carrier signal (I_{da} and I_{db}) adds to the quiescent current in both halves of the transformer windings, producing currents (I_{qa} and I_{qb}) that are equal and simply the sum of the quiescent and carrier currents. I_{qa} and I_{qb} are equal but travel in opposite directions; consequently, they cancel each other. Figure 5-9b shows the phasor sum of the quiescent and carrier currents when the carrier currents travel in the opposite direction to the quiescent currents. The total currents in both halves of the windings are still equal in magnitude, but now they are equal to the difference between the quiescent and carrier currents. Figure 5-9c shows the phasor diagram when a current component is added due to a modulating signal. The modulating signal currents (I_{ma} and I_{mb}) produce in their respective halves of the output transformer currents that are in phase with each other. However,

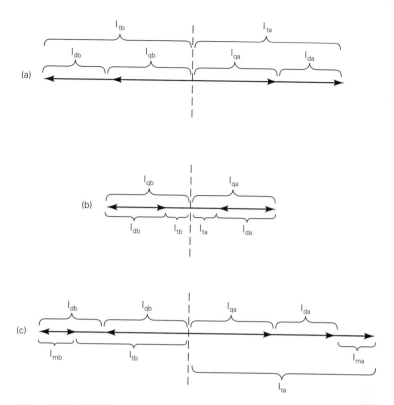

Figure 5-9 FET balanced modulator phasor diagrams: (a) in-phase sum of dc and carrier currents; (b) out-of-phase sum of dc and carrier currents; (c) sum of dc, carrier, and modulating signal currents.

it can be seen that in one-half of the windings the total current is equal to the difference between the dc and carrier currents and the modulating signal current, and in the other half of the winding, the total current is equal to the sum of the dc, carrier, and modulating signal currents. Thus, the dc and carrier currents cancel in the secondary windings, while the difference components add. The continuously changing carrier and modulating signal currents produce the cross-product frequencies.

The carrier and modulating signal polarities shown in Figure 5-8 produce an output current that is proportional to the carrier and modulating signal voltages. The carrier signal (V_c) produces a current in both FETs $(I_{da}$ and $I_{db})$ that is in the same direction as the quiescent currents $(I_{qa}$ and $I_{qb})$. The modulating signal $(V_{ma}$ and $V_{mb})$ produces a current in Q_a (I_{ma}) that is in the same direction as I_{da} and I_{qa} and a current in Q_b (I_{mb}) that is in the opposite direction as I_{db} and I_{qb}. Therefore, the total current through the a side of T_3 is $I_{ta} = I_{da} + I_{qa} + I_{ma}$ and the total current through the b side of T_3 is $I_{tb} = -I_{db} -I_{qb} + I_{mb}$. Thus, the net current through the primary winding of T_3 is $I_{ta} + I_{tb} = I_{ma} + I_{mb}$. For a modulating signal with the opposite polarity, the drain current in Q_b will increase and in Q_a it will decrease. Ignoring the quiescent dc current $(I_{qa}$ and $I_{qb})$, the drain current in one FET is the sum of the carrier and modulating signal currents $(I_d + I_m)$, and the drain current in the other FET is the difference $(I_d - I_m)$.

T_1 is an audio transformer while T_2 and T_3 are radio-frequency transformers. Therefore, any audio component that appears at the drain circuits of Q_1 and Q_2 is not passed on to the output. To achieve total carrier suppression, Q_a and Q_b must be perfectly matched and T_1 and T_3 must be exactly center tapped. As with the diode balanced modulators, the FET balanced modulator typically adds between 40 and 60 dB of attenuation to the carrier.

Balanced Bridge Modulator

Figure 5-10a shows the schematic diagram for a *balanced bridge modulator*. The operation of the bridge modulator, like the balanced ring modulator, is completely dependent on the switching action of diodes D_1 through D_4 under the influence of the carrier and modulating signal voltages. Again, the carrier voltage controls the *on* or *off* condition of the diodes and therefore must be appreciably larger than the modulating signal voltage.

   Circuit operation. For the carrier polarities shown in Figure 5-10b, all four diodes are reverse biased and *off*. Consequently, the audio signal voltage is transferred directly to the load resistor (R_L). Figure 5-10c shows the equivalent circuit for a carrier with the opposite polarity. All four diodes are forward biased and *on*, and the load resistor is bypassed (that is, *shorted out*). As the carrier voltage changes from positive to negative,

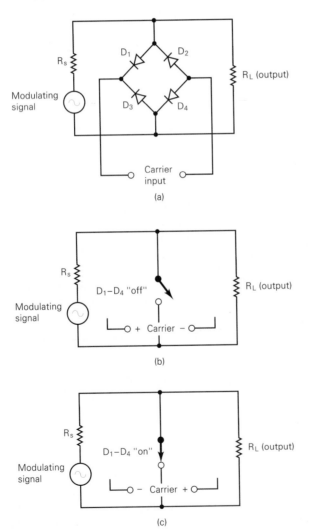

Figure 5-10 Balanced bridge modulator: (a) schematic diagram; (b) diodes biased *off*; (c) diodes biased *on*; *(Continued on next page.)*

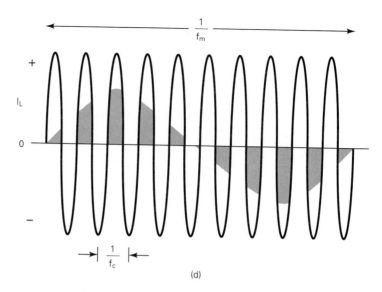

Figure 5-10 *(continued)* (d) output waveform.

and vice versa, the output waveform contains a series of pulses that is comprised mainly of the upper and lower sideband frequencies. The output waveform is shown in Figure 5-10d. The series of pulses is shown as the shaded area in the figure.

Linear Integrated-circuit Balanced Modulators

Linear integrated-circuit (LIC) balanced modulators are available, such as the LM1496, that can provide carrier suppression of 50 dB at 10 MHz and up to 65 dB at 500 kHz. The LM1496 balanced modulator integrated circuit is a *double-balanced modulator–demodulator* that produces an output signal that is proportional to the product of its input signals. Integrated circuits are ideally suited for applications that require balanced operation.

Circuit operation. Figure 5-11 shows a simplified schematic diagram for a differential amplifier, which is the fundamental circuit of an LIC balanced modulator because of its excellent *common-mode rejection ratio* (typically 85 dB or more). When a carrier signal is applied to the base of Q_1, the emitter currents in both transistors will vary by the same amount. Because the emitter current for both Q_1 and Q_2 comes from a common constant-current source (Q_4), any increase in Q_1's emitter current results in a corresponding decrease in Q_2's emitter current, and vice versa. Similarly, when a carrier signal is applied to the base of Q_2, the emitter currents of Q_1 and Q_2 vary by the same magnitude, except in opposite directions. Consequently, if the same carrier signal is fed simultaneously to the bases of Q_1 and Q_2, the respective increases and decreases are equal and thus cancel. Therefore, the collector currents and output voltage remain unchanged. If a modulating signal is applied to the base of Q_3, it causes a corresponding increase or decrease (depending on its polarity) in the collector currents of Q_1 and Q_2. However, the carrier and modulating signal frequencies mix in the transistors and produce cross-product frequencies in the output. Therefore, the carrier and modulating signal frequencies are canceled in the balanced transistors, while the sum and difference frequencies appear in the output.

Figure 5-12 shows the schematic diagram for a typical AM DSBSC modulator using the LM1496 integrated circuit. The LM1496 is a balanced modulator/demodulator for

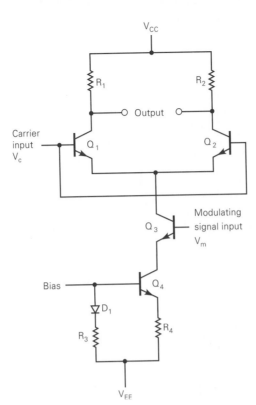

Figure 5-11 Differential amplifier schematic.

which the output is the product of its two inputs. The LM1496 offers excellent carrier suppression (65 dB at 0.5 MHz), adjustable gain, balanced inputs and outputs, and a high common-mode rejection ratio (85 dB). When used as a product detector, the LM1496 has a sensitivity of 3.0 μV and a dynamic range of 90 dB when operating at an intermediate frequency of 9 MHz.

The carrier signal is applied to pin 10, which, in conjunction with pin 8, provides an input to a quad cross-coupled differential output amplifier. This configuration is used to ensure that full-wave multiplication of the carrier and modulating signal occurs. The modulating signal is applied to pin 1, which, in conjunction with pin 4, provides a differential input to the current driving transistors for the output difference amplifier. The 50-kΩ potentiometer, in conjunction with V_{EE} (-8 V dc), is used to balance the bias currents for the difference amplifiers and null the carrier. Pins 6 and 12 are single-ended outputs that contain carrier and sideband components. When one of the outputs is inverted and added to the other, the carrier is suppressed and a double-sideband suppressed-carrier wave is produced. Such a process is accomplished in the op-amp subtractor. The subtractor inverts the signal at the inverting ($-$) input and adds it to the signal at the noninverting ($+$) input. Thus, a double-sideband suppressed-carrier wave appears at the output of the op-amp. The 6.8-kΩ resistor connected to pin 5 is a bias resistor for the internal constant-current supply.

The XR-2206 linear integrated-circuit AM DSBFC modulator described in Chapter 3 and shown in Figure 3-20a can also be used to produce a double-sideband suppressed carrier wave by simply setting the dc bias to $V^+/2$ and limiting the modulating signal amplitude to ± 4 Vp. As the modulating signal passes through its zero crossings, the phase of the carrier undergoes a 180° phase reversal. This property also makes the XR-2206 ideally suited as a phase-shift modulator. The dynamic range of amplitude modulation for the XR-2206 is approximately 55 dB.

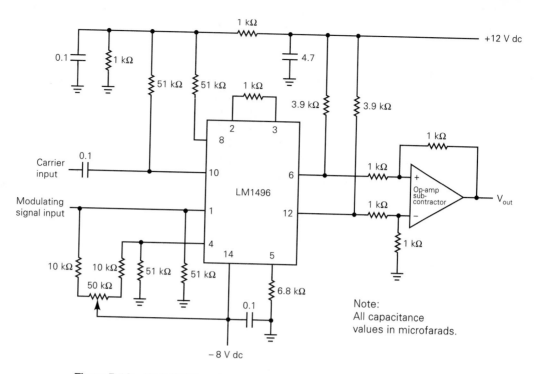

Figure 5-12 AM DSBSC modulator using the LM1496 linear integrated circuit: (a) schematic diagram; (b) 1496 specification sheets. (Copyright Motorola, Inc., 1982. Used by permission.)

SINGLE-SIDEBAND TRANSMITTERS

The transmitters used for single-sideband suppressed- and reduced-carrier transmission are identical except that the reinserted carrier transmitters have an additional circuit that adds a low-amplitude carrier to the single-sideband waveform after suppressed carrier modulation has been performed and one of the sidebands has been removed. The reinserted carrier is called a *pilot carrier*. The circuit where the carrier is reinserted is called a *linear summer* if it is a resistive network and a *hybrid coil* if the SSB waveform and pilot carrier are inductively combined in a transformer bridge circuit. Three transmitter configurations are commonly used for single-sideband generation: the filter method, the phase-shift method, and the so-called *third method*.

Single-sideband Transmitter: Filter Method

Figure 5-13 shows a block diagram for a SSB transmitter that uses balanced modulators to suppress the unwanted carrier, and filters to suppress the unwanted sideband. The modulating signal is an audio spectrum that extends from 0 to 5 kHz. The modulating signal mixes with a low-frequency (LF) 100-kHz carrier in balanced modulator 1 to produce a double-sideband frequency spectrum centered around the suppressed 100-kHz IF carrier. Bandpass filter 1 (BPF 1) is tuned to a 5-kHz bandwidth centered around 102.5 kHz, which is the center of the upper sideband frequency spectrum. The pilot or reduced-amplitude carrier is added to the single-sideband waveform in the carrier reinsertion stage, which is simply a linear summer. The summer is a simple adder circuit that combines the 100-kHz pilot carrier with the 100- to 105-kHz upper sideband frequency spectrum. Thus,

Single-sideband Transmitters

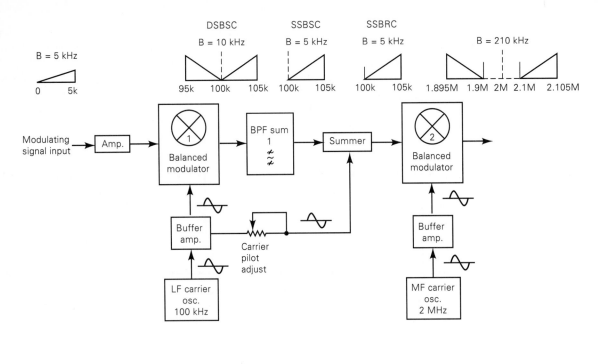

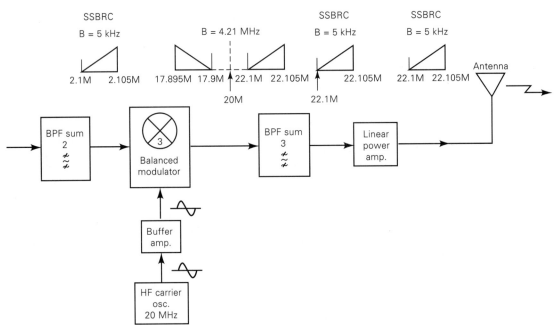

Figure 5-13 Single-sideband transmitter: filter method.

the output of the summer is a SSBRC waveform. (If suppressed-carrier transmission is desired, the carrier pilot and summer circuit can be omitted.)

The low-frequency IF is converted to the final operating frequency band through a series of frequency translations. First, the SSBRC waveform is mixed in balanced modulator 2 with a 2-MHz medium-frequency (MF) carrier. The output is a double-sideband suppressed-carrier signal in which the upper and lower sidebands each contain the original SSBRC frequency spectrum. The upper and lower sidebands are separated by a 200-kHz frequency band that is void of information. The center frequency of BPF 2 is 2.1025 MHz with a 5-kHz bandwidth. Therefore, the output of BPF 2 is once again a single-sideband reduced-carrier waveform. Its frequency spectrum comprises a reduced 2.1-MHz second IF carrier and a 5-kHz-wide upper sideband. The output of BPF 2 is mixed with a 20-MHz high-frequency (HF) carrier in balanced modulator 3. The output is a double-sideband suppressed carrier signal in which the upper and lower sidebands again each contain the original SSBRC frequency spectrum. The sidebands are separated by a 4.2-MHz frequency band that is void of information. BPF 3 is centered on 22.1025 MHz with a 5-kHz bandwidth. Therefore, the output of BPF 3 is once again a single-sideband waveform with a reduced 22.1-MHz RF carrier and a 5-kHz-wide upper sideband. The output waveform is amplified in the linear power amplifier and then transmitted.

In the transmitter just described, the original modulating signal frequency spectrum was up-converted in three modulation steps to a final carrier frequency of 22.1 MHz and a single upper sideband that extended from the carrier to 22.105 MHz. After each up-conversion (frequency translation), the desired sideband is separated from the double-sideband spectrum with a BPF. The same final output spectrum can be produced with a single heterodyning process: one balanced modulator, one bandpass filter, and a single HF carrier supply. Figure 5-14a shows the block diagram and output frequency spectrum for a single-conversion transmitter. The output of the balanced modulator is a double-sideband frequency spectrum centered around a suppressed-carrier frequency of 22.1 MHz. To separate the 5-kHz-wide upper sideband from the composite frequency spectrum, a multiple-pole BPF with an extremely high Q is required. A BPF that meets this criterion is in itself difficult to construct, but suppose that this were a multiple-channel transmitter and the carrier frequency were tunable; then the BPF must also be tunable. Constructing a tunable BPF in the megahertz frequency range with a passband of only 5 kHz is beyond economic and engineering feasibility. The only BPF in the transmitter shown in Figure 5-13 that has to separate sidebands that are immediately adjacent to each other is BPF 1. To construct a 5-kHz-wide, steep-skirted BPF at 100 kHz is a relatively simple task, as only a moderate Q is required. The sidebands separated by BPF 2 are 200 kHz apart; thus a low-Q filter with gradual roll-off characteristics can be used with no danger of passing any portion of the undesired sideband. BPF 3 separates sidebands that are 4.2 MHz apart. If multiple channels are used and the HF carrier is tunable, a single broadband filter can be used for BPF 3 with no danger of any portion of the undesired sideband leaking through the filter. For single-channel operation, the single conversion transmitter is the simplest design, but for multiple-channel operation, the three-conversion system is more practical. Figures 5-14b and c show the output spectrum and filtering requirements for both methods.

Single-sideband filters.   It should be evident that filters are an essential part of any electronic communications system and especially single-sideband systems. Transmitters as well as receivers have requirements for highly selective networks for limiting both the signal and noise frequency spectrums. Conventional *LC* filters do not have a high enough Q for most single-sideband transmitters. Therefore, filters used for single-sideband generation are usually constructed from either *crystal* or *ceramic* materials, *mechanical* filters, or *surface acoustic wave (SAW)* filters.

Crystal filters.   The *crystal lattice filter* is commonly used in single-sideband systems. The schematic diagram for a typical crystal lattice bandpass filter is shown in Figure 5-15a. The lattice comprises two sets of matched crystal pairs (X_1 and X_2, X_3 and X_4)

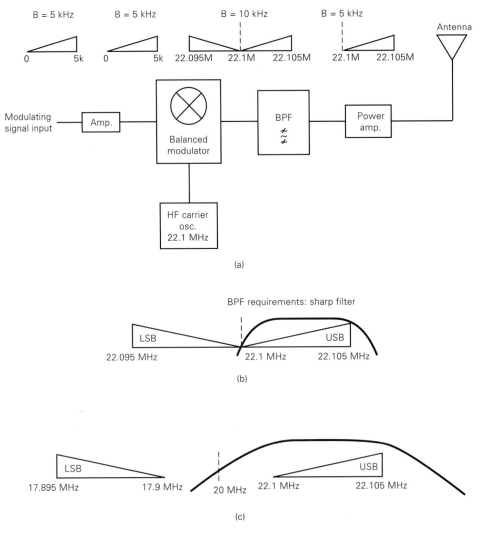

Figure 5-14 Single conversion SSBSC transmitter, filter method: (a) block diagram; (b) output spectrum and filtering requirements for a single-conversion transmitter; (c) output spectrum and filtering requirements for a three-conversion transmitter.

connected between tuned input and output transformers T_1 and T_2. Crystals X_1 and X_2 are series connected, while X_3 and X_4 are connected in parallel. Each pair of crystals is matched in frequency within 10 to 20 Hz. X_1 and X_2 are cut to operate at the filter lower cutoff frequency, and X_3 and X_4 are cut to operate at the upper cutoff frequency. The input and output transformers are tuned to the center of the desired passband, which tends to spread the difference between the series and parallel resonant frequencies. C_1 and C_2 are used to correct for any overspreading of frequency difference under matched crystal conditions.

The operation of the crystal filter is similar to the operation of a bridge circuit. When the reactances of the bridge arms are equal and have the same sign (either inductive or capacitive), the signals propagating through the two possible paths of the bridge cancel each other out. At the frequency where the reactances have equal magnitudes and opposite signs (one inductive and the other capacitive), the signal is propagated through the network with maximum amplitude.

Figure 5-15b shows a typical characteristic curve for a crystal lattice bandpass filter. Crystal filters are available with a Q as high as 100,000. The filter shown in Figure 5-15a is a single-element filter. However, for a crystal filter to adequately pass a specific band of frequencies and reject all others, at least two elements are necessary. Typical insertion losses for crystal filters are between 1.5 and 3 dB.

Ceramic filter.   *Ceramic filters* are made from lead zinconate-titanate, which exhibits the piezoelectric effect. Therefore, they operate quite similar to crystal filters except that ceramic filters do not have as high a Q-factor. Typical Q values for ceramic filters go up to about 2000. Ceramic filters are less expensive, smaller, and more rugged than their crystal lattice counterparts. However, ceramic filters have more loss. The insertion loss for ceramic filters is typically between 2 and 4 dB.

Ceramic filters typically come in one-element, three-terminal packages; two-element, eight-terminal packages; and four-element fourteen-terminal packages. Ceramic filters feature small size, low profile, symmetrical selectivity characteristics, low spurious response, and excellent immunity to variations in environmental conditions with minimum variation in operating characteristics. However, certain precautions must be taken with ceramic filters, which include the following:

1. *Impedance matching and load conditions:* Ceramic filters differ from coils in that

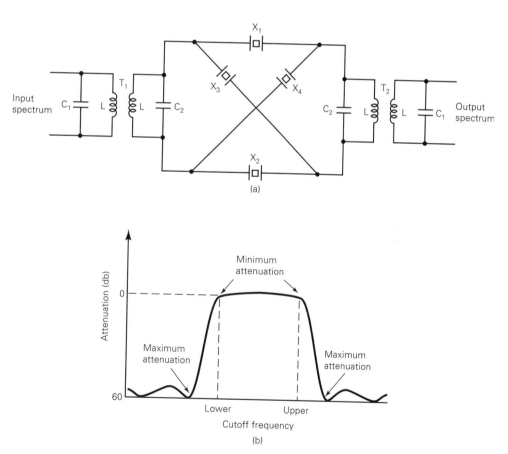

Figure 5-15   Crystal lattice filter: (a) schematic diagram; (b) characteristic curve.

their impedance cannot readily be changed. When using ceramic filters, it is very important that impedances be properly matched.

2. *Spurious signals:* In practically all cases where ceramic filters are used, spurious signals are generated. To suppress these responses, impedance matching with IF transformers is the simplest and most effective way.

3. *Matching coils:* When difficulties arise in spurious response suppression or for improvement in selectivity or impedance matching in IF stages, use of an impedance matching coil is advised.

4. *Error in wiring input and output connections:* Care must be taken when connecting the input and output terminals of a ceramic filter. Any error will cause waveform distortion and possibly frequency deviation of the signal.

5. *Use of two ceramic filters in cascade:* For best performance, a coil should be used between two ceramic filter units. When cost is a factor and a direct connection is necessary, a suitable capacitor or resistor can be used.

Mechanical filters. A *mechanical filter* is a *mechanically resonant transducer*. It receives electrical energy, converts it to mechanical vibrations and then converts the vibrations back to electrical energy at its output. Essentially, four elements comprise a mechanical filter: an input transducer that converts the input electrical energy to mechanical vibrations, a series of mechanical resonant metal disks that vibrate at the desired resonant frequency, a coupling rod that couples the metal disks together, and an output transducer that converts the mechanical vibrations back to electrical energy. Figure 5-16 shows the electrical equivalent circuit for a mechanical filter. The series resonant circuits (LC combinations) represent the metal disks, coupling capacitor C_1 represents the coupling rod, and R represents the matching mechanical loads. The resonant frequency of the filter is determined by the series LC disks, and C_1 determines the bandwidth. Mechanical filters are more rugged than either ceramic or crystal filters and have comparable frequency-response characteristics. However, mechanical filters are larger and heavier and therefore are impractical for mobile communications equipment.

Surface acoustic wave filters. *Surface acoustic wave (SAW) filters* were first developed in the 1960s but did not become commercially available until the 1970s. SAW filters use acoustic energy rather than electromechanical energy to provide excellent performance for precise bandpass filtering. In essence, SAW filters trap or guide acoustical waves along a surface. They can operate at center frequencies up to several gigahertz and bandwidths up to 50 MHz with more accuracy and reliability than their predecessor, the mechanical filter, and they do it at a lower cost. SAW filters have extremely steep roll-off characteristics and typically attenuate frequencies outside their passband between 30 and 50 dB more than the signals within their passband. SAW filters are used in both single- and multiple-conversion superheterodyne receivers for both RF and IF filters and in single-sideband systems for a multitude of filtering applications.

A SAW filter consists of transducers patterned from a thin aluminum film deposited

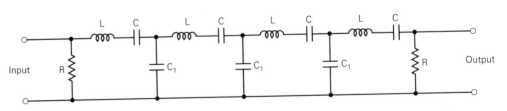

Figure 5-16 Mechanical filter equivalent circuit.

on the surface of a semiconductor crystal material that exhibits the piezoelectric effect. This results in a physical deformation (rippling) on the surface of the substrate. These ripples vary at the frequency of the applied signal, but travel along the surface of the material at the speed of sound. With SAW filters, an oscillating electrical signal is applied across a small piece of semiconductor crystal that is part of a larger, flat surface, as shown in Figure 5-17a. The piezoelectric effect causes the crystal material to vibrate. These vibrations are in the form of acoustic energy that travels across the surface of the substrate until it reaches a second crystal at the opposite end, where the acoustic energy is converted back to electrical energy.

To provide filter action, a precisely spaced row of metallic *fingers* is deposited on the flat surface of the substrate, as shown in Figure 5-17a. The finger centers are spaced at either a half- or quarter-wavelength of the desired center frequency. As the acoustic waves travel across the surface of the substrate, they reflect back and forth as they impinge on the fingers. Depending on the acoustical wavelength and the spacing between the fingers, some of the reflected energy cancels and attenuates the incident wave energy (this is called *destructive interference*), while some of the energy aids (*constructive interference*). The exact frequencies of acoustical energy that are canceled depend on the spacing between the fingers. The bandwidth of the filter is determined by the thickness and number of fingers.

The basic SAW filter is *bidirectional*. That is, half the power is radiated toward the output transducer while the other half is radiated toward the end of the crystal substrate and is lost. By reciprocity, half the power is lost at the output transducer. Consequently, SAW filters have a relatively high insertion loss. This shortcoming can be overcome to a certain degree by using a more complex structure called a *unidirectional transducer*, which launches the acoustic wave in only one direction.

SAW filters are inherently very rugged and reliable. Because their operating frequencies and bandpass responses are set by the photolithographic process, they do not require complicated tuning operations nor do they become detuned over a period of time. The semiconductor wafer processing techniques used in manufacturing SAW filters permit

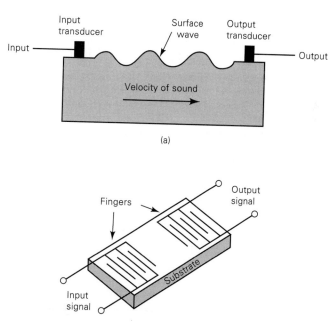

Figure 5-17 SAW filter: (a) surface wave; (b) metallic fingers.

Single-sideband Transmitters

large-volume production of economical and reproducible devices. Finally, their excellent performance capabilities are achieved with significantly reduced size and weight when compared to competing technologies.

The predominant disadvantage of SAW filters is their extremely high insertion loss, which is typically between 25 and 35 dB. For this reason, SAW filters cannot be used to filter low-level signals. SAW filters also exhibit a much longer delay time than their electronic counterparts (approximately 20,000 times as long). Consequently, SAW filters are sometimes used for *delay lines*.

Single-sideband Transmitter: Phase-shift Method

With the phase-shift method of single-sideband generation, the undesired sideband is canceled in the output of the modulator; therefore, sharp filtering is unnecessary. Figure 5-18 shows a block diagram for a SSB transmitter that uses the phase-shift method to remove the upper sideband. Essentially, there are two separate double-sideband modulators (balanced modulators 1 and 2). The modulating signal and carrier are applied directly to one of the modulators and then both are shifted 90° and applied to the second modulator. The outputs from the two balanced modulators are double-sideband suppressed-carrier signals with the proper phase such that, when they are combined in a linear summer, the upper sideband is canceled.

Phasor representation.   The phasors shown in Figure 5-18 illustrate how the upper sideband is canceled by rotating both the carrier and the modulating signal 90° prior to modulation. The output phase from balanced modulator 1 shows the relative position and direction of rotation of the upper (ω_{usf}) and lower (ω_{lsf}) side frequencies to the suppressed carrier (ω_c). The phasors at the output of balanced modulator 2 are essentially the same

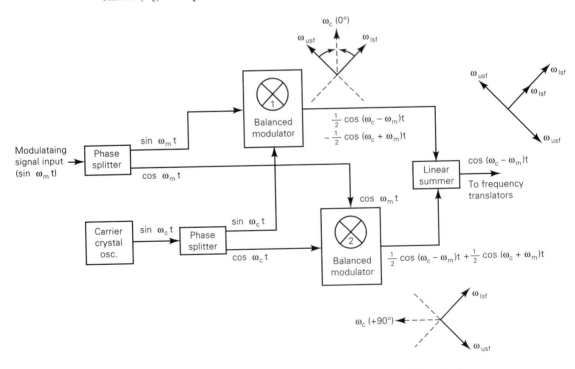

Figure 5-18   SSB transmitter: phase-shift method.

except that the phase of the carrier and the modulating signal are each rotated 90°. The output of the summer shows the sum of the phasors from the two balanced modulators. The two phasors for the lower sideband are in phase and additive, whereas the phasors for the upper sideband are 180° out of phase and thus cancel. Consequently, only the lower sideband appears at the output of the summer.

Mathematical analysis. In Figure 5-18 the input modulating signal ($\sin \omega_m t$) is fed directly to balanced modulator 1 and shifted 90° ($\cos \omega_m t$) and fed to balanced modulator 2. The low-frequency carrier ($\sin \omega_c t$) is also fed directly to balanced modulator 1 and shifted 90° ($\cos \omega_c t$) and fed to balanced modulator 2. The balanced modulators are product modulators and their outputs are expressed mathematically as

$$\text{output from balanced modulator } 1 = (\sin \omega_m t)(\sin \omega_c t)$$

$$= \frac{1}{2} \cos (\omega_c - \omega_m)t - \frac{1}{2} \cos (\omega_c + \omega_m)t$$

$$\text{output from balanced modulator } 2 = (\cos \omega_m t)(\cos \omega_c t)$$

$$= \frac{1}{2} \cos (\omega_c - \omega_m)t + \frac{1}{2} \cos (\omega_c + \omega_m)t$$

and the output from the linear summer is

$$\frac{1}{2} \cos (\omega_c - \omega_m)t - \frac{1}{2} \cos (\omega_c + \omega_m)t$$

$$+ \frac{1}{2} \cos (\omega_c - \omega_m)t + \frac{1}{2} \cos (\omega_c + \omega_m)t$$

$$\overline{\cos (\omega_c - \omega_m)t \qquad\qquad \text{canceled}}$$

lower sideband
(difference signal)

Single-sideband Transmitter: The Third Method

The *third method* of single-sideband generation, developed by D. K. Weaver in the 1950s, is similar to the phase-shift method in that it uses phase shifting and summing to cancel the undesired sideband. However, it has an advantage in that the information signal is initially modulated onto an audio *subcarrier*, thus eliminating the need for a *wideband* phase shifter (a phase shifter that has to shift a band of frequencies by the same amount, which is difficult to built in practice). The block diagram for a third-method SSB modulator is shown in Figure 5-19. Note that the inputs and outputs of the two phase shifters are single frequencies ($f_o, f_o + 90°, f_c$, and $f_c + 90°$). The input audio signals mix with the audio subcarrier in balanced modulators 1 and 2, which are supplied with quadrature (90° out of phase) subcarrier signals (f_o and $f_o + 90°$). The output from balanced modulator 2 contains the upper and lower sidebands ($f_o \pm f_m$), while the output from balanced modulator number 1 contains the upper and lower sidebands, each shifted in phase 90° ($f_o \pm f_m + 90°$). The upper sidebands are removed by their respective low-pass filters, which have an upper cutoff frequency equal to that of the suppressed audio subcarrier. The output from LPF 1 ($f_o - f_m + 90°$) is mixed with the RF carrier (f_c) in balanced modulator 3, and the output

Single-sideband Transmitters

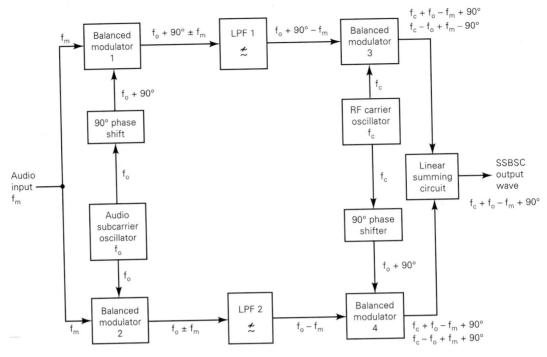

Figure 5-19 Single-sideband suppressed-carrier modulator: the "third method."

from LPF 2 $(f_o - f_m)$ is mixed with a 90° phase-shifted RF carrier $(f_c + 90°)$ in balanced modulator 4. The RF carriers are suppressed in balanced modulators 3 and 4. Therefore, the sum and difference output signals from balanced modulator 3, $(f_c + f_o - f_m + 90°)$ + $(f_c - f_o + f_m - 90°)$, are combined in the linear summer with the sum and difference output signals from balanced modulator 4 $(f_c + f_o - f_m + 90°)$ + $(f_c - f_o + f_m + 90°)$. The output from the summer is

$$(f_c + f_o - f_m + 90°) + (f_c - f_o + f_m - 90°)$$
$$+(f_c + f_o - f_m + 90°) + (f_c - f_o + f_m + 90°)$$

$$\overline{(f_c + f_o - f_m + 90°) \qquad \text{canceled}}$$

The final RF output frequency is $f_c + f_o - f_m$, which is essentially the lower sideband of RF carrier $f_c + f_o$. The 90° offset phase is an absolute phase shift that all frequencies undergo and is therefore insignificant. If the RF upper sideband is desired, simply interchange the carrier inputs to balanced modulators 3 and 4, in which case the final RF carrier is $f_c - f_o$.

Independent Sideband Transmitter

Figure 5-20 shows a block diagram for an *independent sideband (ISB)* transmitter with three stages of modulation. The transmitter uses the filter method to produce two independent single-sideband channels (channel A and channel B). The two channels are combined; then a pilot carrier is reinserted. The composite ISB reduced-carrier waveform is up-converted to RF with two additional stages of frequency translation. There are two 5-kHz-wide information signals that originate from two independent sources. The channel A

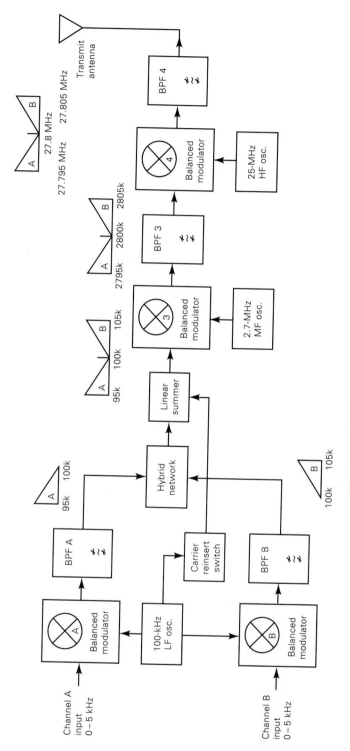

Figure 5-20 ISB transmitter: block diagram.

information signals modulate a 100-kHz LF carrier in balanced modulator A. The output from balanced modulator A passes through BPF A, which is tuned to the lower sideband (95 to 100 kHz). The channel B information signals modulate the same 100-kHz LF carrier in balanced modulator B. The output from balanced modulator B passes through BPF B, which is tuned to the upper sideband (100 to 105 kHz). The two single-sideband frequency spectrums are combined in a hybrid network to form a composite ISB suppressed-carrier spectrum (95 to 105 kHz). The LF carrier (100 kHz) is reinserted in the linear summer to form an ISB reduced-carrier waveform. The ISB spectrum is mixed with a 2.7-MHz MF carrier in balanced modulator 3. The output from balanced modulator 3 passes through BPF 3 to produce an ISB reduced-carrier spectrum that extends from 2.795 to 2.805 MHz with a reduced 2.8-MHz pilot carrier. Balanced modulator 4, BPF 4, and the HF carrier translate the MF spectrum to an RF band that extends from 27.795 to 27.8 MHz (channel A) and 27.8 to 27.805 MHz (channel B) with a 27.8-MHz reduced-amplitude carrier.

SINGLE-SIDEBAND RECEIVERS

Single-sideband BFO Receiver

Figure 5-21 shows the block diagram for a simple noncoherent single-sideband *BFO receiver*. The selected radio-frequency spectrum is amplified and then mixed down to intermediate frequencies for further amplification and band reduction. The output from the IF amplifier stage is heterodyned (beat) with the output from a *beat frequency oscillator* (*BFO*). The BFO frequency is equal to the IF carrier frequency; thus, the difference between the IF and the BFO frequencies is the information signal. Demodulation is accomplished through several stages of mixing and filtering. The receiver is noncoherent because the RF local oscillator and BFO signals are not synchronized to each other or to the oscillators in the transmitter. Consequently, any difference between the transmit and receive local oscillator frequencies produces a frequency offset error in the demodulated information signal. For example, if the receive local oscillator is 100 Hz above its designated frequency and the BFO is 50 Hz above its designated frequency, the restored information is offset 150 Hz from its original frequency spectrum. Fifty hertz or more offset is distinguishable by a normal listener as a tonal variation.

The RF mixer and second detector shown in Figure 5-21 are product detectors. Like

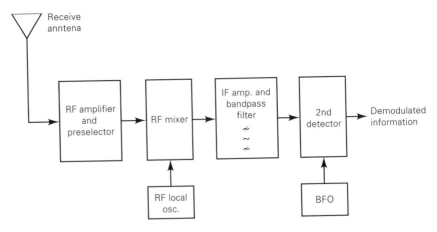

Figure 5-21 Noncoherent BFO SSB receiver.

the balanced modulators in the transmitter, their outputs are the product of their inputs. A product modulator and product detector are essentially the same circuit. The only difference is that the input to a product modulator is tuned to a low-frequency modulating signal and the output is tuned to a high-frequency carrier, whereas with a product detector, the input is tuned to a high-frequency modulated carrier and the output is tuned to a low-frequency information signal. With both the modulator and detector, the single-frequency carrier is the switching signal. In a receiver, the input signal, which is a suppressed or reduced RF carrier and one sideband, is mixed with the RF local oscillator frequency to produce an intermediate frequency. The output from the second product detector is the sum and difference frequencies between the IF and the beat frequency. The difference frequency band is the original input information.

EXAMPLE 5-1

For the BFO receiver shown in Figure 5-21, a received frequency band of 30 to 30.005 MHz, an RF local oscillator frequency of 20 MHz, an IF frequency band of 10 to 10.005 MHz, and a BFO frequency of 10 MHz, determine:

(a) Demodulated first IF frequency band and demodulated information frequency band.
(b) Demodulated information frequency band if the RF local oscillator frequency drifts down 0.001%.

Solution (a) The IF output from the RF mixer is the difference between the received signal frequency and the RF local oscillator frequency or

$$f_{IF} = (30 \text{ to } 30.005 \text{ MHz}) - 20 \text{ MHz}$$

$$= 10 \quad \text{to} \quad 10.005 \text{ MHz}$$

The demodulated information signal spectrum is the difference between the intermediate frequency band and the BFO frequency or

$$f_m = (10 \text{ to } 10.005 \text{ MHz}) - 10 \text{ MHz} = 0 \quad \text{to} \quad 5 \text{ kHz}$$

(b) A 0.001% drift would cause a decrease in the RF local oscillator frequency of

$$\Delta f = (0.00001)(20 \text{ MHz}) = 200 \text{ Hz}$$

Thus the RF local oscillator frequency would drift down to 19.9998 Hz, and the output from the RF mixer is

$$f_{IF} = (30 \text{ to } 30.005 \text{ MHz}) - 19.9998 \text{ MHz}$$

$$= 10.0002 \quad \text{to} \quad 10.0052 \text{ MHz}$$

The demodulated information signal spectrum is the difference between the intermediate frequency band and the BFO or

$$f_m = (10.0002 \text{ to } 10.0052 \text{ MHz}) - 10 \text{ MHz}$$

$$= 200 \quad \text{to} \quad 5200 \text{ Hz}$$

The 0.001% drift in the RF local oscillator frequency caused a corresponding 200-Hz error in the demodulated information signal spectrum.

Coherent Single-sideband BFO Receiver

Figure 5-22 shows a block diagram for a coherent single-sideband BFO receiver. This receiver is identical to the BFO receiver shown in Figure 5-21 except that the LO and BFO frequencies are synchronized to the carrier oscillators in the transmitter. The carrier *recov-*

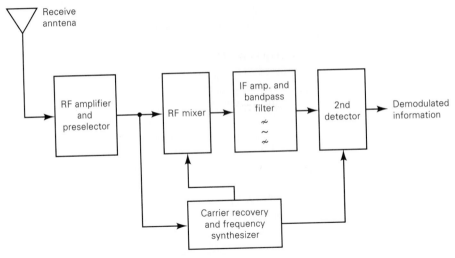

Figure 5-22 Coherent SSB BFO receiver.

ery circuit is a narrowband PLL that tracks the pilot carrier in the composite SSBRC receive signal and uses the recovered carrier to regenerate coherent local oscillator frequencies in the synthesizer. The synthesizer circuit produces a coherent RF local oscillator and BFO frequency. The carrier recovery circuit tracks the received pilot carrier. Therefore, minor changes in the carrier frequency in the transmitter are compensated for in the receiver, and the frequency offset error is eliminated. If the coherent receiver shown in Figure 5-22 had been used in Example 5-1, the RF local oscillator would not have been allowed to drift independently.

Single-sideband Envelope Detection Receiver

Figure 5-23 shows the block diagram for a single-sideband receiver that uses synchronous carriers and envelope detection to demodulate the received signals. The reduced carrier pilot is detected, separated from the demodulated spectrum, and regenerated in the carrier

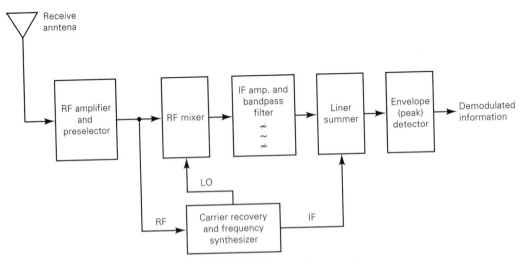

Figure 5-23 Single-sideband envelope detection receiver.

recovery circuit. The regenerated pilot is divided and used as the stable frequency source for a frequency synthesizer, which supplies the receiver with frequency coherent local oscillators. The receive RF is mixed down to IF in the first detector. A regenerated IF carrier is added to the IF spectrum in the last linear summer, which produces a SSB full-carrier envelope. The envelope is demodulated in a conventional peak diode detector to produce the original information signal spectrum. This type of receiver is often called an exalted carrier receiver.

Multichannel Pilot Carrier Single-sideband Receiver

Figure 5-24 shows a block diagram for a multichannel pilot carrier SSB receiver that uses a PLL carrier recovery circuit and a frequency synthesizer to produce coherent local and beat frequency oscillator frequencies. The RF input range extends from 4 to 30 MHz, and the VCO natural frequency is coarsely adjusted with an external channel selector switch over a frequency range of 6 to 32 MHz. The VCO frequency tracks above the incoming RF by 2 MHz, which is the first IF. A 1.8-MHz beat frequency sets the second IF to 200 kHz.

The VCO frequency is coarsely set with the channel selector switch and then mixed with the incoming RF signal in the first detector to produce a first IF difference frequency of 2 MHz. The first IF mixes with the 1.8-MHz beat frequency to produce a 200-kHz second IF. The PLL locks onto the 200-kHz pilot and produces a dc correction voltage that fine-tunes the VCO. The second IF is beat down to audio in the third detector, which is passed on to the audio preamplifier for further processing. The AGC detector produces an AGC voltage that is proportional to the amplitude of the 200-kHz pilot. The AGC voltage is fed back to the RF and/or IF amplifiers to adjust their gains proportionate to the received pilot level and to the squelch circuit to turn the audio preamplifier off in the absence of a received pilot. The PLL compares the 200-kHz pilot to a stable crystal-controlled reference. Consequently, although the receiver carrier supply is not directly synchronized to the transmit oscillators, the first and second IFs are, thus compensating for any frequency offset in the demodulated audio spectrum.

AMPLITUDE COMPANDORING SINGLE SIDEBAND

Amplitude compandoring single-sideband (ACSSB) systems provide narrowband voice communications for land-mobile services with nearly the quality achieved with FM systems and do it using less than one-third the bandwidth. With ACSSB, the audio signals are compressed before modulation by amplifying the higher-magnitude signals less than the lower-magnitude signals. After demodulation in the receiver, the audio signals are expanded by amplifying the higher-magnitude signals more than the lower-magnitude signals. A device that performs compression and expansion is called a *compandor* (*compres*-sor-exp*ander*).

Companding an information signal increases the dynamic range of a system by reducing the dynamic range of the information signals prior to transmission and then expanding them after demodulation. For example, when companding is used, information signals with an 80-dB dynamic range can be propagated through a communications system with only a 50-dB dynamic range. Companding slightly decreases the signal-to-noise ratios for the high-amplitude signals, while considerably increasing the signal-to-noise ratios of the low-amplitude signals.

ACSSB systems require that a pilot carrier signal be transmitted at a reduced ampli-

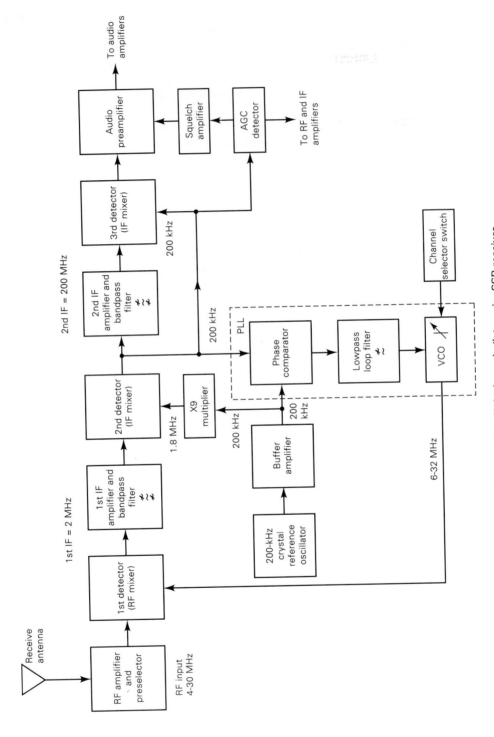

Figure 5-24 Multichannel pilot carrier SSB receiver.

tude along with the information signals. The pilot is used to synchronize the oscillators in the receiver and provides a signal for the AGC which monitors and adjusts the gain of the receiver and silences the receiver when no pilot is received.

SINGLE-SIDEBAND MEASUREMENTS

Single-sideband transmitters are rated in peak envelope power (PEP) and peak envelope voltage (PEV), rather than simply rms power and voltage. For a single-frequency modulating signal, the modulated output signal with single-sideband suppressed-carrier transmission is not an envelope, but rather a continuous, single-frequency signal. A single frequency is not representative of a typical information signal. Therefore, for test purposes, a *two-frequency* test signal is used for the modulating signal for which the two tones have equal amplitudes. Figure 5-25a shows the waveform produced in a SSBSC modulator with a two-tone modulating signal. The waveform is the vector sum of the two equal-amplitude side frequencies and is similar to a conventional AM waveform except that the repetition rate is equal to the difference between the two modulating signal frequencies. Figure 5-25b shows the envelope for a two-tone test signal when a low-amplitude pilot carrier is added. The envelope has basically the same shape except with the addition of a low-amplitude sine-wave ripple at the carrier frequency.

The envelope out of two-tone SSB is an important consideration because it is from this envelope that the output power for a SSB transmitter is determined. The PEP for a SSBSC transmitter is analogous to the total output power from a conventional double-sideband full-carrier transmitter. The rated PEP is the output power measured at the peak of the envelope when the input is a two-tone test signal and the two tones are equal in

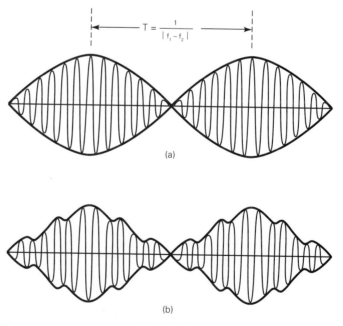

Figure 5-25 Two-tone SSB test signal: (a) without reinserted carrier; (b) with reinserted carrier.

magnitude. With such an output signal, the actual power dissipated in the load is equal to half the PEP. Therefore, the voltage developed across the load is

$$e_{total} = \sqrt{E_1^2 + E_2^2}$$

where E_1 and E_2 are the rms voltages of the two test tones. Therefore,

$$PEP = \frac{\sqrt{E_1^2 + E_2^2}}{R}$$

and since $E_1 = E_2$
$$PEP = \frac{(2E)^2}{R} = \frac{4E^2}{R} \tag{5-1}$$

However, the average power dissipated in the load is equal to the sum of the powers of the two tones:

$$P_{ave} = \frac{E_1^2}{R} + \frac{E_2^2}{R} = \frac{2E^2}{R} \tag{5-2}$$

which simplifies to
$$P_{ave} = \frac{PEP}{2} \tag{5-3}$$

Two equal-amplitude test tones are used for the test signal for the following reasons:

1. One tone produces a continuous single-frequency output that does not produce intermodulation.
2. A single-frequency output signal is not analogous to a normal information signal.
3. More than two tones makes analysis impractical.
4. Two tones of equal amplitude place a more demanding requirement on the transmitter than is likely to occur during normal operation.

EXAMPLE 5-2

For a two-tone test signal of 1.5 and 3 kHz and a carrier frequency of 100 kHz, determine for a single-sideband suppressed carrier transmission:

(a) Output frequency spectrum if only the upper sideband is transmitted.
(b) For $e_1 = e_2 = 5$ V and a load resistance of 50 Ω, the PEP and average output power.

Solution (a) The output frequency spectrum contains the two upper side frequencies:

$$f_{usf1} = 100 \text{ kHz} + 1.5 \text{ kHz} = 101.5 \text{ kHz}$$

$$f_{usf2} = 100 \text{ kHz} + 3 \text{ kHz} = 103 \text{ kHz}$$

(b) Substituting into Equation 5-1 yields

$$PEP = \frac{4(0.707 \times 5)^2}{50} = 1 \text{ W}$$

Substituting into Equation 5-3 yields

$$P_{ave} = \frac{1}{2} = 0.5 \text{ W}$$

QUESTIONS

5-1. Describe AM SSBFC. Compare SSBFC to conventional AM.

5-2. Describe AM SSBSC. Compare SSBSC to conventional AM.

5-3. Describe AM SSBRC. Compare SSBRC to conventional AM.

5-4. What is a *pilot carrier*?

5-5. What is an *exalted* carrier?

5-6. Describe AM ISB. Compare ISB to conventional AM.

5-7. Describe AM VSB. Compare VSB to conventional AM.

5-8. Define *peak envelope power*.

5-9. Describe the operation of a balanced ring modulator.

5-10. What is a product modulator?

5-11. Describe the operation of a FET push-pull balanced modulator.

5-12. Describe the operation of a balanced bridge modulator.

5-13. What are the advantages of an LIC balanced modulator over a discrete circuit?

5-14. Describe the operation of a filter-type SSB transmitter.

5-15. Contrast crystal, ceramic, and mechanical filters.

5-16. Describe the operation of a phase-shift-type SSB transmitter.

5-17. Describe the operation of the "third type" of SSB transmitter.

5-18. Describe the operation of an independent sideband transmitter.

5-19. What is the difference between a product modulator and a product detector?

5-20. What is the difference between a coherent and a noncoherent receiver?

5-21. Describe the operation of a multichannel pilot carrier SSBRC receiver.

5-22. Why is a two-tone test signal used for making PEP measurements?

PROBLEMS

5-1. For the balanced ring modulator shown in Figure 5-6a, a carrier input frequency $f_c = 400$ kHz and a modulating signal frequency range $f_m = 0$ to 4 kHz; determine:
   (a) Output frequency spectrum.
   (b) Output frequency for a single frequency input $f_m = 2.8$ kHz.

5-2. For the LIC balanced modulator shown in Figure 5-12, a carrier input frequency of 200 kHz, and a modulating signal frequency range $f_m = 0$ to 3 kHz, determine:
   (a) Output frequency spectrum.
   (b) Output frequency for a single frequency input $f_m = 1.2$ kHz.

5-3. For the SSB transmitter shown in Figure 5-13, a low-frequency carrier of 100 kHz, a medium-frequency carrier of 4 MHz, a high-frequency carrier of 30 MHz, and a modulating signal frequency range of 0 to 4 kHz:
   (a) Sketch the frequency spectrums for the following points: balanced modulator 1 out, BPF 1 out, summer out, balanced modulator 2 out, BPF 2 out, balanced modulator 3 out, and BPF 3 out.
   (b) For a single-frequency input $f_m = 1.5$ kHz, determine the translated frequency for the following points: BPF 1 out, BPF 2 out, and BPF 3 out.

5-4. Repeat Problem 5-3, except change the low-frequency carrier to 500 kHz. Which transmitter has the more stringent filtering requirements?

5-5. For the SSB transmitter shown in Figure 5-14a, a modulating input frequency range of 0 to 3 kHz, and a high-frequency carrier of 28 MHz:

(a) Sketch the output frequency spectrum.

(b) For a single-frequency modulating signal input of 2.2 kHz, determine the output frequency.

5-6. Repeat Problem 5-5, except change the audio input frequency range to 300 to 5000 Hz.

5-7. For the SSB transmitter shown in Figure 5-18, an input carrier frequency of 500 kHz, and a modulating signal frequency range of 0 to 4 kHz.

(a) Sketch the frequency spectrum at the output of the linear summer.

(b) For a single modulating signal frequency of 3 kHz, determine the output frequency.

5-8. Repeat Problem 5-7, except change the input carrier frequency to 400 kHz and the modulating signal frequency range to 300 to 5000 Hz.

5-9. For the ISB transmitter shown in Figure 5-20, channel A input frequency range of 0 to 4 kHz, channel B input frequency range of 0 to 4 kHz, a low-frequency carrier of 200 kHz, a medium-frequency carrier of 4 MHz, and a high-frequency carrier of 32 MHz:

(a) Sketch the frequency spectrums for the following points: balanced modulator A out, BPF A out, balanced modulator B out, BPF B out, hybrid network out, linear summer out, balanced modulator 3 out, BPF 3 out, balanced modulator 4 out, and BPF 4 out.

(b) For an A-channel input frequency of 2.5 kHz and a B-channel input frequency of 3 kHz, determine the frequency components at the following points: BPF A out, BPF B out, BPF 3 out, and BPF 4 out.

5-10. Repeat Problem 5-9, except change the channel A input frequency range to 0 to 10 kHz and the channel B input frequency range to 0 to 6 kHz.

5-11. For the SSB receiver shown in Figure 5-21, an RF input frequency of 35.602 MHz, a RF local oscillator frequency of 25 MHz, and a 2-kHz modulating signal frequency, determine the IF and BFO frequencies.

5-12. For the multichannel pilot carrier SSB receiver shown in Figure 5-24, a crystal oscillator frequency of 300 kHz, a first IF frequency of 3.3 MHz, an RF input frequency of 23.303 MHz, and modulating signal frequency of 3 kHz; determine the following: VCO output frequency, multiplication factor, and second IF frequency.

5-13. For a two-tone test signal of 2 and 3 kHz and a carrier frequency of 200 kHz:

(a) Determine the output frequency spectrum.

(b) For $e_1 = e_2 = 12$ Vp and a load resistance $R_L = 50$ Ω, determine the PEP and average power.

6

Angle Modulation Transmission

Three properties of an analog signal can be varied: amplitude, frequency, and phase. Chapters 3, 4, and 5 dealt with amplitude modulation. This chapter and Chapter 7 deal with *frequency modulation* (*FM*) and *phase modulation* (*PM*). Frequency and phase modulation are both forms of *angle modulation*. Unfortunately, both forms of angle modulation are often referred to as simply FM when, actually, there is a distinct (although subtle) difference between the two. There are several distinct advantages to using angle modulation over amplitude modulation, such as noise reduction, improved system fidelity, and more efficient use of power. However, FM and PM have several important drawbacks, which include requiring an extended bandwidth and more complex circuits in both the transmitter and receiver.

Angle modulation was first introduced in 1931 as an alternative to amplitude modulation. It was suggested that an angle-modulated wave was less susceptible to noise than AM and, consequently, could improve the performance of radio communications. Major E. H. Armstrong (who also developed the superheterodyne receiver) developed the first successful FM radio system in 1936, and in July 1939 the first regularly scheduled broadcasting of FM signals began in Alpine, New Jersey. Today, angle modulation is used extensively for commercial radio broadcasting, television sound transmission, two-way mobile radio, cellular radio, and microwave and satellite communications systems.

The purposes of this chapter are to introduce the reader to the basic concepts of frequency and phase modulation and how they relate to each other, to show some of the common circuits used to produce angle-modulated waves, and to compare the performance of angle modulation to amplitude modulation.

ANGLE MODULATION

Angle modulation results whenever the phase angle (θ) of a sinusoidal wave is varied with respect to time. An angle-modulated wave is expressed mathematically as

$$m(t) = V_c \cos [\omega_c t + \theta(t)] \qquad (6\text{-}1)$$

where $m(t)$ = angle-modulated wave
        V_c = peak carrier amplitude (volts)
        ω_c = carrier radian frequency (that is, angular velocity, $2\pi f_c$)
        $\theta(t)$ = instantaneous phase deviation (radians)

With angle modulation, it is necessary that $\theta(t)$ be a prescribed function of the modulating signal. Therefore, if $v_m(t)$ is the modulating signal, the angle modulation is expressed mathematically as

$$\theta(t) = F[v_m(t)] \qquad (6\text{-}2)$$

where $v_m(t) = V_m \sin (\omega_m t)$
        ω_m = angular velocity of the modulating signal (radians/second)
        f_m = modulating signal frequency (hertz)
        V_m = peak amplitude of the modulating sign (volts)

In essence, the difference between frequency and phase modulation lies in which property of the carrier (the frequency or the phase) is directly varied by the modulating signal and which property is indirectly varied. Whenever the frequency of a carrier is varied, the phase is also varied, and vice versa. Therefore, FM and PM must both occur whenever either form of angle modulation is performed. If the frequency of the carrier is varied directly in accordance with the modulating signal, FM results. If the phase of the carrier is varied directly in accordance with the modulating signal, PM results. Therefore, direct FM is indirect PM and direct PM is indirect FM. Frequency and phase modulation can be defined as follows:

Direct frequency modulation (FM): varying the frequency of a constant-amplitude carrier directly proportional to the amplitude of the modulating signal at a rate equal to the frequency of the modulating signal.

Direct phase modulation (PM): varying the phase of a constant-amplitude carrier directly proportional to the amplitude of the modulating signal at a rate equal to the frequency of the modulating signal.

Figure 6-1 shows the waveform for a sinusoidal carrier for which angle modulation is occurring. The frequency and phase of the carrier are changing proportional to the amplitude of the modulating signal (v_m). The change in frequency (Δf) is called *frequency deviation*, and the change in phase ($\Delta\theta$) is called phase deviation. Frequency deviation is the relative displacement of the carrier frequency in hertz, and phase deviation is the relative angular displacement (in radians) of the carrier in respect to a reference phase. The magnitude of the frequency and phase deviation is proportional to the amplitude of the modulating signal (v_m), and the rate at which the deviation occurs is equal to the frequency of the modulating signal (f_m). Whenever the period (T) of a sinusoidal carrier changes, its frequency also changes, and if the changes are continuous, the wave is no longer a single frequency. It will be shown that the resultant waveform comprises the original carrier frequency (sometimes called the *carrier rest frequency*) and an infinite number of pairs of

Chap. 6   Angle Modulation Transmission

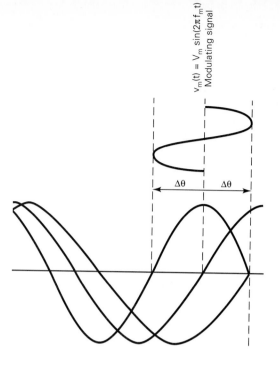

$$v_m(t) = V_m \sin(2\pi f_m t)$$
Modulating signal

Figure 6-1 Frequency changing with time.

side frequencies displaced on either side of the carrier by an integral multiple of the modulating signal frequency.

Figure 6-2 shows a sinusoidal carrier in which the frequency (f) is changed (*deviated*) over a period of time. The fat portion of the waveform corresponds to the peak-to-peak change in the period of the carrier (ΔT). The minimum period (T_{min}) corresponds to the maximum frequency (f_{max}), and the maximum period (T_{max}) corresponds to the minimum frequency (f_{min}). The peak-to-peak frequency deviation is determined by simply measuring the difference between the maximum and minimum frequencies ($\Delta f_{p\text{-}p} = 1/T_{min} - 1/T_{max}$).

Mathematical Analysis

The difference between FM and PM is more easily understood by defining the following four terms with reference to Equation 6-1: instantaneous phase deviation, instantaneous phase, instantaneous frequency deviation, and instantaneous frequency.

Instantaneous phase deviation. The *instantaneous phase deviation* is the instantaneous change in the phase of the carrier at a given instant of time and indicates how much the phase of the carrier is changing with respect to its reference phase. Instantaneous phase deviation is expressed mathematically as

$$\text{instantaneous phase deviation} = \theta(t) \quad \text{radians} \tag{6-3}$$

Instantaneous phase. The *instantaneous phase* is the precise phase of the carrier at a given instant of time and is expressed mathematically as

$$\text{instantaneous phase} = \omega_c t + \theta(t) \tag{6-4}$$

Angle Modulation

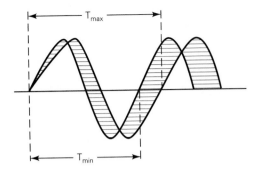

Figure 6-2 Phase changing with frequency.

where $\quad \omega_c t$ = carrier reference phase

$$= \left(2\pi \frac{\text{radians}}{\text{cycle}}\right)\left(f_c \frac{\text{cycles}}{\text{second}}\right)(t \text{ seconds}) = 2\pi f_c t \quad \text{radians}$$

f_c = carrier frequency (hertz)

$\theta(t)$ = instantaneous phase deviation (radians)

Instantaneous frequency deviation. The *instantaneous frequency deviation* is the instantaneous change in the frequency of the carrier and is defined as the first time derivative of the instantaneous phase deviation. Therefore, the instantaneous phase deviation is the first integral of the instantaneous frequency deviation. In terms of Equation 6-3, the instantaneous frequency deviation is expressed mathematically as

$$\text{instantaneous frequency deviation} = \theta'(t) \text{ rad/sec} \qquad (6\text{-}5a)$$

or

$$= \frac{\theta'(t) \text{ rad/sec}}{2\pi \text{ rad/cycle}} = \frac{\text{cycles}}{\text{second}} = \text{hertz}$$

The prime (') is used to denote the first derivative with respect to time.

Instantaneous frequency. The *instantaneous frequency* is the precise frequency of the carrier at a given instant of time and is defined as the first time derivative of the instantaneous phase. In terms of Equation 6-4, the instantaneous frequency is expressed mathematically as

$$\omega_i(t) = \text{instantaneous frequency} = \frac{d}{dt}[\omega_c t + \theta(t)] \qquad (6\text{-}6a)$$

$$= \omega_c + \theta'(t) \quad \text{rad/sec} \qquad (6\text{-}6b)$$

Substituting $2\pi f_c$ for ω_c gives

instantaneous frequency = $f_i(t)$

$$= \left(2\pi \frac{\text{rad}}{\text{cycle}}\right)\left(f_c \frac{\text{cycles}}{\text{sec}}\right) + \theta'(t) = 2\pi f_c + \theta'(t) \quad \text{rad/sec}$$

or

$$= \frac{2\pi f_c + \theta(t) \text{ rad/sec}}{2\pi \text{ rad/cycle}} = f_c + \frac{\theta'(t)}{2\pi} \frac{\text{cycles}}{\text{second}} = f_c + \frac{\theta'(t)}{2\pi} \quad \text{hertz} \qquad (6\text{-}6c)$$

Phase modulation can then be defined as angle modulation in which the instantaneous phase deviation, $\theta(t)$, is proportional to the modulating signal voltage. Similarly, frequency modulation is angle modulation in which the instantaneous frequency deviation, $\theta'(t)$, is proportional to the modulating signal voltage.

For a modulating signal $v_m(t)$, the phase and frequency modulation are

$$\text{phase modulation} = \theta(t) = Kv_m(t) \quad \text{rad} \tag{6-7}$$

$$\text{frequency modulation} = \theta'(t) = K_1 v_m(t) \quad \text{rad/sec} \tag{6-8}$$

where K and K_1 are constants that are the deviation sensitivities of the phase and frequency modulators, respectively. The deviation sensitivities are the output-versus-input transfer functions for the modulators. The deviation sensitivity for a phase modulator is

$$K = \frac{\text{radians}}{\text{volt}}$$

and for a frequency modulator $\quad K_1 = \dfrac{\text{radians/second}}{\text{volt}} \quad \text{or} \quad \dfrac{\text{radians}}{\text{volt-second}}$

Phase modulation is the first integral of the frequency modulation. Therefore, from Equations 6-7 and 6-8

$$\text{phase modulation} = \theta(t) = \int \theta'(t) \, dt$$

$$= \int K_1 v_m(t) \, dt$$

$$= K_1 \int v_m(t) \, dt \tag{6-9}$$

Therefore, substituting a modulating signal $v_m(t) = V_m \cos(\omega_m t)$ into Equation 6-1 yields

For phase modulation
$$v(t) = V_c \cos[\omega_c t + \theta(t)]$$
$$= V_c \cos[\omega_c t + KV_m \cos(\omega_m t)]$$

For frequency modulation
$$v(t) = V_c \cos[\omega_c t + \int \theta'(t)]$$
$$= V_c \cos[\omega_c t + \int K_1 v_m(t) \, dt]$$
$$= V_c \cos[\omega_c t + K_1 \int V_m \cos(\omega_m t)]$$
$$= V_c \cos\left[\omega_c t + \frac{K_1 V_m}{\omega_m} \sin(\omega_m t)\right]$$

The preceding mathematical relationships are summarized in Table 6-1. Also, the expressions for the FM and PM waves that result when the modulating signal is a single-frequency sinusoidal wave are shown.

FM and PM Waveforms

Figure 6-3 illustrates both frequency and phase modulation of a sinusoidal carrier by a single frequency-modulating signal. It can be seen that the FM and PM waveforms are identical except for their time relationship (phase). Thus, it is impossible to distinguish an FM

TABLE 6-1 EQUATIONS FOR PHASE- AND FREQUENCY-MODULATED CARRIERS

Type of Modulation	Modulating Signal	Angle-modulated Wave, $m(t)$
(a) Phase	$v_m(t)$	$V_c \cos[\omega_c t + Kv_m(t)]$
(b) Frequency	$v_m(t)$	$V_c \cos[\omega_c t + K_1 \int v_m(t) \, dt]$
(c) Phase	$V_m \cos(\omega_m t)$	$V_c \cos[\omega_c t + KV_m \cos(\omega_m t)]$
(d) Frequency	$-V_m \sin(\omega_m t)$	$V_c \cos\left[\omega_c t + \dfrac{K_1 V_m}{\omega_m} \cos(\omega_m t)\right]$
(e) Frequency	$V_m \cos(\omega_m t)$	$V_c \cos\left[\omega_c t + \dfrac{K_1 V_m}{\omega_m} \sin(\omega_m t)\right]$

Angle Modulation

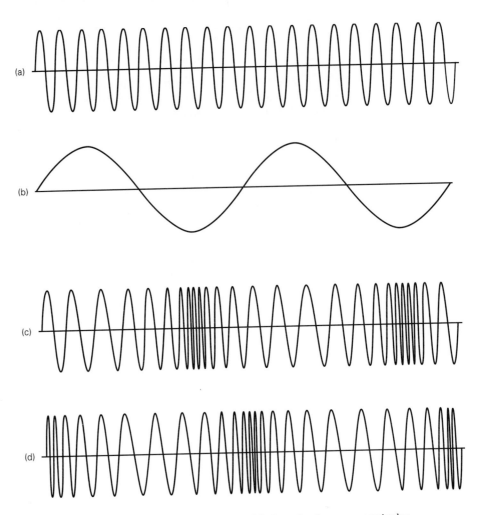

Figure 6-3 Phase and frequency modulation of a sine-wave carrier by a sine-wave signal: (a) unmodulated carrier; (b) modulating signal; (c) frequency-modulated wave; (d) phase-modulated wave.

waveform from a PM waveform without knowing the characteristics of the modulating signal. With FM, the maximum frequency deviation (change in the carrier frequency) occurs during the maximum positive and negative peaks of the modulating signal (that is, the frequency deviation is proportional to the amplitude of the modulating signal). With PM, the maximum frequency deviation occurs during the zero crossings of the modulating signal (that is, the frequency deviation is proportional to the slope or first derivative of the modulating signal). For both frequency and phase modulation, the rate at which the frequency changes occur is equal to the modulating-signal frequency.

Similarly, it is not apparent from Equation 6-1 whether an FM or PM wave is represented. It could be either. However, knowledge of the modulating signal will permit correct identification. If $\theta(t) = K v_m(t)$, it is phase modulation, and if $\theta'(t) = K_1 v_m(t)$, it is frequency modulation. In other words, if the instantaneous frequency is directly proportional to the amplitude of the modulating signal, it is frequency modulation, and if the instantaneous phase is directly proportional to the amplitude of the modulating frequency, it is phase modulation.

Chap. 6   Angle Modulation Transmission

Phase Deviation, Modulation Index, and Frequency Deviation

Comparing expressions (c), (d), and (e) for the angle-modulated carrier in Table 6-1 shows that the expression for a carrier that is being phase or frequency modulated by a single frequency-modulating signal can be written in a general form by modifying Equation 6-1 as follows:

$$m(t) = V_c \cos [\omega_c t + m \cos (\omega_m t)] \qquad (6\text{-}10)$$

where $m \cos (\omega_m t) =$ instantaneous phase deviation, $\theta(t)$.

When the modulating signal is a single-frequency sinusoid, it is evident from Equation 6-10 that the phase angle of the carrier varies from its unmodulated value in a simple sinusoidal fashion.

In Equation 6-10, m represents the *peak phase deviation* in radians for a phase-modulated carrier. Peak phase deviation is called the *modulation index* (or sometimes *index of modulation*). One primary difference between frequency and phase modulation is the way in which the modulation index is defined. For PM, the modulation index is proportional to the amplitude of the modulating signal, independent of its frequency. The modulation index for a phase-modulated carrier is expressed mathematically as

$$m = KV_m \quad \text{radians} \qquad (6\text{-}11)$$

where $V_m =$ peak modulating signal voltage (volts)
$KV_m =$ peak phase deviation (radians)

For a frequency-modulated carrier, the modulation index is directly proportional to the amplitude of the modulating signal and inversely proportional to its frequency and is expressed mathematically as

$$m = \frac{K_1 V_m}{\omega_m} \qquad (6\text{-}12a)$$

$$= \frac{K_1 V_m}{2\pi f_m} \quad \text{unitless ratio for FM} \qquad (6\text{-}12b)$$

where $K_1 V_m =$ frequency deviation (radians/second)
$\dfrac{K_1 V_m}{2\pi} =$ frequency deviation (hertz)

From Equation 6-12b, it can be seen that with FM the modulation index is a unitless ratio and is used only to describe the depth of modulation achieved for a given amplitude and frequency modulating signal. *Frequency deviation* is the change in frequency that occurs in the carrier when it is acted on by a modulating signal. Frequency deviation is typically given as a peak frequency shift in hertz (Δf). The peak-to-peak frequency deviation is sometimes called *carrier swing*.

For an FM modulator, the deviation sensitivity is often given in hertz per volt. Therefore, the frequency deviation is simply the product of the deviation sensitivity and the modulating signal voltage. Also, with FM it is common to express the modulation index as simply the ratio of the peak frequency deviation divided by the modulating signal frequency, or rearranging Equation 6-12b gives

$$m = \frac{\Delta f}{f_m} \frac{\text{Hz}}{\text{Hz}} \quad \text{(unitless ratio)} \qquad (6\text{-}13)$$

where $\Delta f = \dfrac{K_1 V_m}{2\pi}$ (hertz)

EXAMPLE 6-1

(a) Determine the peak frequency deviation (Δf) and modulation index (m) for an FM modulator with a deviation sensitivity $K_1 = 5$ kHz/V and a modulating signal $v_m(t) = 2 \cos(2\pi 2000t)$.

(b) Determine the peak phase deviation (m) for a PM modulator with a deviation sensitivity $K = 2.5$ rad/V and a modulating signal $v_m(t) = 2 \cos(2\pi 2000t)$.

Solution (a) The peak frequency deviation is simply the product of the deviation sensitivity and the peak amplitude of the modulating signal, or

$$\Delta f = \frac{5 \text{ kHz}}{V} \times 2 \text{ V} = 10 \text{ kHz}$$

The modulation index is determined by substituting into Equation 6-13.

$$m = \frac{10 \text{ kHz}}{2 \text{ kHz}} = 5$$

(b) The peak phase shift for a phase-modulated wave is the modulation index and is found by substituting into Equation 6-11.

$$m = \frac{2.5 \text{ rad}}{V} \times 2 \text{ V} = 5 \text{ rad}$$

In Example 6-1, the modulation index for the frequency-modulated carrier was equal to the modulation index of the phase-modulated carrier (5). If the amplitude of the modulating signal is changed, the modulation index for both the frequency- and phase-modulated waves will change proportionally. However, if the frequency of the modulating signal changes, the modulation index for the frequency-modulated wave will change inversely proportional, while the modulation index of the phase-modulated wave is unaffected. Therefore, under identical conditions, FM and PM are indistinguishable for a single-frequency modulating signal; however, when the frequency of the modulating signal changes, the PM modulation index remains constant, whereas the FM modulation index increases as the modulating signal frequency decreases, and vice versa.

Percent modulation.   The percent modulation for an angle-modulated wave is determined in a different manner than it was with an amplitude-modulated wave. With angle modulation, percent modulation is simply the ratio of the frequency deviation actually produced to the maximum frequency deviation allowed by law stated in percent form. Mathematically, percent modulation is

$$\% \text{ modulation} = \frac{\Delta f \text{ (actual)}}{\Delta f \text{ (maximum)}} \times 100 \tag{6-14}$$

For example, in the United States the Federal Communications Commission (FCC) limits the frequency deviation for commercial FM broadcast-band transmitters to ± 75 kHz. If a given modulating signal produces ± 50-kHz frequency deviation, the percent modulation is

$$\% \text{ modulation} = \frac{50 \text{ kHz}}{75 \text{ kHz}} \times 100 = 67\%$$

Phase and Frequency Modulators and Demodulators

A *phase modulator* is a circuit in which the carrier is varied in such a way that its instantaneous phase is proportional to the modulating signal. The unmodulated carrier is a single-frequency sinusoid and is commonly called the *rest* frequency. A *frequency modulator* (often called a *frequency deviator*) is a circuit in which the carrier is varied in such a way that its instantaneous phase is proportional to the integral of the modulating signal. Therefore, with a frequency modulator, if the modulating signal $v(t)$ is differentiated prior to being applied to the modulator, the instantaneous phase deviation is proportional to the integral of $v(t)$ or, in other words, proportional to $v(t)$ because $\int v'(t) = v(t)$. Similarly, an FM modulator that is preceded by a differentiator produces an output wave in which the phase deviation is proportional to the modulating signal and is, therefore, equivalent to a phase modulator. Several other interesting equivalences are possible. For example, a frequency demodulator followed by an integrator is equivalent to a phase demodulator. Four commonly used equivalences are listed next and illustrated in Figure 6-4.

1. PM modulator = differentiator followed by an FM modulator
2. PM demodulator = FM demodulator followed by an integrator
3. FM modulator = integrator followed by a PM modulator
4. FM demodulator = PM demodulator followed by a differentiator

Frequency Analysis of Angle-modulated Waves

With angle modulation, the frequency components of the modulated wave are much more complexly related to the frequency components of the modulating signal than with amplitude modulation. In a frequency or phase modulator, a single-frequency modulating signal

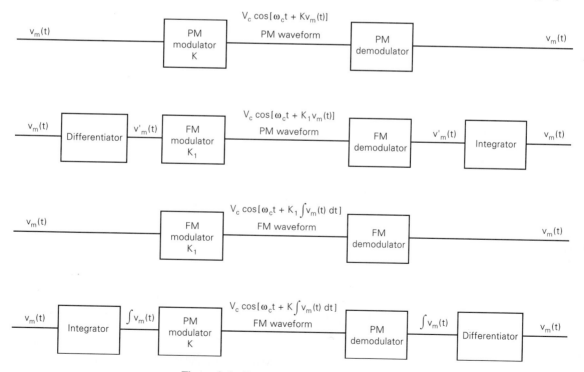

Figure 6-4 Frequency and phase modulation and demodulation.

produces an infinite number of pairs of side frequencies and thus has an infinite band-width. Each side frequency is displaced from the carrier by an integral multiple of the modulating signal frequency. However, generally most of the side frequencies are negligibly small in amplitude and can be ignored.

Modulation by a single-frequency sinusoid. Frequency analysis of an angle-modulated wave by a single-frequency sinusoid produces a peak phase deviation of m radians, where m is the modulation index. Again, from Equation 6-10 and for a modulating frequency equal to ω_m, $m(t)$ is written as

$$m(t) = V_c \cos \left[\omega_c t + m \cos (\omega_m t) \right]$$

From Equation 6-10, the individual frequency components that make up the modulated wave are not obvious. However, *Bessel function identities* are available that may be applied directly. One such identity is

$$\cos (\alpha + m \cos \beta) = \sum_{n = -\infty}^{\infty} J_n(m) \cos \left(\alpha + n\beta + \frac{n\pi}{2} \right) \qquad (6\text{-}15)$$

$J_n(m)$ is the Bessel function of the first kind of nth order with argument m. If Identify 6-15 is applied to Equation 6-11, $m(t)$ may be rewritten as

$$m(t) = V_c \sum_{n = -\infty}^{\infty} J_n(m) \cos \left(\omega_c t + n\omega_m t + \frac{n\pi}{2} \right) \qquad (6\text{-}16)$$

Expanding Equation 6-16 for the first four terms yields

$$m(t) = V_c \{ J_0(m) \cos \omega_c t + J_1(m) \cos \left[(\omega_c + \omega_m)t + \frac{\pi}{2} \right]$$

$$-J_1(m) \cos \left[(\omega_c - \omega_m)t - \frac{\pi}{2} \right] - J_2(m) \cos \left[(\omega c + 2\omega_m)t \right] \qquad (6\text{-}17)$$

$$+J_2(m) \cos \left[(\omega_c - 2\omega_m)t \right] + \cdots$$

Equations 6-16 and 6-17 show that with angle modulation a single-frequency modulating signal produces an infinite number of sets of side frequencies, each displaced from the carrier by an integral multiple of the modulating signal frequency. A sideband set includes an upper and a lower side frequency ($f_c \pm f_m, f_c \pm 2f_m, f_c \pm nf_m$, and so on). Successive sets of sidebands are called first-order sidebands, second-order sidebands, and so on, and their magnitudes are determined by the coefficients $J_1(m)$, $J_2(m)$, and so on, respectively. Table 6-2 shows the Bessel functions of the first kind for several values of modulation index. We see that a modulation index of 0 (no modulation) produces zero side frequencies, and the larger the modulation index, the more sets of side frequencies produced. The values shown for J_n are relative to the amplitude of the unmodulated carrier. For example, $J_2 = 0.35$ indicates that the amplitude of the second set of side frequencies is equal to 35% of the unmodulated carrier amplitude ($0.35\ V_c$). It can be seen that the amplitude of the higher-order side frequencies rapidly becomes insignificant as the modulation index decreases below unity. For larger values of m, the value of $J_n(m)$ starts to decrease rapidly as soon as $n = m$. As the modulation index increases from zero, the magnitude of the carrier $J_0(m)$ decreases. When m is equal to approximately 2.4, $J_0(m) = 0$

TABLE 6-2 BESSEL FUNCTIONS OF THE FIRST KIND, $J_n(m)$

m	J_0	J_1	J_2	J_3	J_4	J_5	J_6	J_7	J_8	J_9	J_{10}	J_{11}	J_{12}	J_{13}	J_{14}
0.00	1.00	—	—	—	—	—	—	—	—	—	—	—	—	—	—
0.25	0.98	0.12	—	—	—	—	—	—	—	—	—	—	—	—	—
0.5	0.94	0.24	0.03	—	—	—	—	—	—	—	—	—	—	—	—
1.0	0.77	0.44	0.11	0.02	—	—	—	—	—	—	—	—	—	—	—
1.5	0.51	0.56	0.23	0.06	0.01	—	—	—	—	—	—	—	—	—	—
2.0	0.22	0.58	0.35	0.13	0.03	—	—	—	—	—	—	—	—	—	—
2.4	0	0.52	0.43	0.20	0.06	0.02	—	—	—	—	—	—	—	—	—
2.5	−0.05	0.50	0.45	0.22	0.07	0.02	—	—	—	—	—	—	—	—	—
3.0	−0.26	0.34	0.49	0.31	0.13	0.04	0.01	—	—	—	—	—	—	—	—
4.0	−0.40	−0.07	0.36	0.43	0.28	0.13	0.05	0.02	—	—	—	—	—	—	—
5.0	−0.18	−0.33	0.05	0.36	0.39	0.26	0.13	0.05	0.02	—	—	—	—	—	—
6.0	0.15	−0.28	−0.24	0.11	0.36	0.36	0.25	0.13	0.06	0.02	—	—	—	—	—
7.0	0.30	0.00	−0.30	−0.17	0.16	0.35	0.34	0.23	0.13	0.06	0.02	—	—	—	—
8.0	0.17	0.23	−0.11	−0.29	−0.10	0.19	0.34	0.32	0.22	0.13	0.06	0.03	—	—	—
9.0	−0.09	0.25	0.14	−0.18	−0.27	−0.06	0.20	0.33	0.31	0.21	0.12	0.06	0.03	0.01	—
10.0	−0.25	0.05	0.25	0.06	−0.22	−0.23	−0.01	0.22	0.32	0.29	0.21	0.12	0.06	0.03	0.01

and the carrier component goes to zero (this is called the *first carrier null*). This property is often used to determine the modulation index or set the deviation sensitivity of an FM modulator. The carrier reappears as m increases beyond 2.4. When m reaches 5.4, the carrier component once again disappears (this is called the *second carrier null*). Further increases in the modulation index will produce additional carrier nulls at periodic intervals.

Figure 6-5 shows the curves for the relative amplitudes of the carrier and several sets of side frequencies for values of m up to 10. It can be seen that the amplitude of both the carrier and the side frequencies vary at a periodic rate that resembles a damped sine wave. The negative values for $J(m)$ simply indicate the relative phase of that side frequency set.

In Table 6-2, only the significant side frequencies are listed. A side frequency is not considered significant unless it has an amplitude equal to or greater than 1% of the unmodulated carrier amplitude ($J_n \geq 0.01$). From Table 6-2 it can be seen that as m increases the number of significant side frequencies increases. Consequently, the bandwidth of an angle-modulated wave is a function of the modulation index.

EXAMPLE 6-2

For an FM modulator with a modulation index $m = 1$, a modulating signal $v_m(t) = V_m \sin(2\pi 1000t)$, and an unmodulated carrier $v_c(t) = 10 \sin(2\pi 5 \times 10^5 t)$, determine:

(a) Number of sets of significant side frequencies.
(b) Their amplitudes.
(c) Draw the frequency spectrum showing their relative amplitudes.

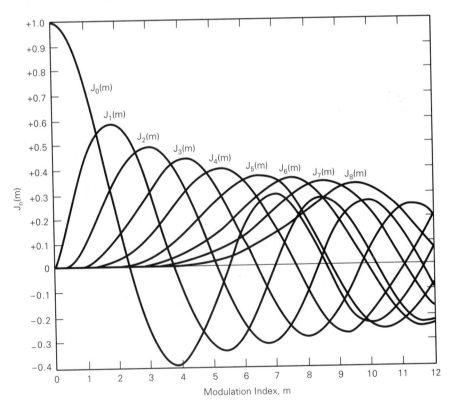

Figure 6-5 $J_n(m)$ versus m.

Solution (a) From Table 6-2, a modulation index of 1 yields a reduced carrier component and three sets of significant side frequencies.

(b) The relative amplitudes of the carrier and side frequencies are

$$J_0 = 0.77(10) = 7.7 \text{ V}$$

$$J_1 = 0.44(10) = 4.4 \text{ V}$$

$$J_2 = 0.11(10) = 1.1 \text{ V}$$

$$J_3 = 0.02(10) = 0.2 \text{ V}$$

(c) The frequency spectrum is shown in Figure 6-6.

If the FM modulator used in Example 6-2 were replaced with a PM modulator and the same carrier and modulating signal frequencies were used, a peak phase deviation of 1 rad would produce exactly the same frequency spectrum.

Bandwidth Requirements for Angle-modulated Waves

In 1922, J. R. Carson mathematically proved that for a given modulating signal frequency a frequency-modulated wave cannot be accommodated in a narrower bandwidth than an amplitude-modulated wave. From the preceding discussion and Example 6-2, it can be seen that the bandwidth of an angle-modulated wave is a function of the modulating signal frequency and the modulation index. With angle modulation, multiple sets of sidebands are produced and, consequently, the bandwidth can be significantly wider than that of an amplitude-modulated wave with the same modulating signal. The modulator output waveform in Example 6-2 requires 6 kHz of bandwidth to pass the carrier and all the significant side frequencies. A conventional double-sideband AM modulator would require only 2 kHz of bandwidth, and a single-sideband system, only 1 kHz.

Angle-modulated waveforms are generally classified as either *low-*, *medium-*, or *high-index*. For the low-index case, the peak phase deviation (modulation index) is less than 1 rad, and the high-index case occurs when the peak phase deviation is greater than 10 rad. Modulation indexes greater than 1 and less than 10 are classified as medium index. From Table 6-2 it can be seen that with low-index angle modulation most of the signal information is carried by the first set of sidebands, and the minimum bandwidth required is approximately equal to twice the highest modulating signal frequency. For this reason, low-index FM systems are sometimes called *narrowband FM*. For a high-index signal, a method of determining the bandwidth called the *quasi-stationary* approach may be used. With this approach, it is assumed that the modulating signal is changing very slowly. For example, for an FM modulator with a deviation sensitivity $K_1 = 2$ kHz/V and a 1-Vp modulating signal, the peak frequency deviation $\Delta f = 2000$ Hz. If the frequency of the modulating signal is very low, the bandwidth is determined by the peak-to-peak frequency

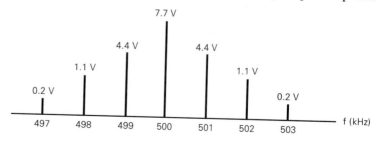

Figure 6-6 Frequency spectrum for Example 6-2.

Angle Modulation

deviation. Therefore, for large modulation indexes, the minimum bandwidth required to propagate a frequency-modulated wave is approximately equal to the peak-to-peak frequency deviation ($2\Delta f$).

Thus, for low-index modulation, the frequency spectrum resembles double-sideband AM and the minimum bandwidth is approximated by

$$B = 2f_m \text{ (hertz)} \tag{6-18}$$

and for high-index modulation, the minimum bandwidth is approximated by

$$B = 2\Delta f \text{ (hertz)} \tag{6-19}$$

The actual bandwidth required to pass all the significant sidebands for an angle-modulated wave is equal to two times the product of the highest modulating signal frequency and the number of significant sidebands determined from the table of Bessel functions. Mathematically, the rule for determining the minimum bandwidth for an angle-modulated wave using the Bessel table is

$$B = 2(n \times f_m) \text{ (hertz)} \tag{6-20}$$

where n = number of significant sidebands
$\quad\quad f_m$ = modulating signal frequency (hertz)

In an unpublished memorandum dated August 28, 1939, Carson established a general rule to estimate the bandwidth for all angle-modulated systems regardless of the modulation index. This is called *Carson's rule*. Simply stated, Carson's rule approximates the minimum bandwidth of an angle-modulated wave as twice the sum of the peak frequency deviation and the highest modulating signal frequency. Mathematically stated, Carson's rule is

$$B = 2[\Delta f + f_{m\,(\text{max})}] \text{ (hertz)} \tag{6-21}$$

where Δf = peak frequency deviation (hertz)
$\quad\quad f_{m\,(\text{max})}$ = highest modulating signal frequency (hertz)

Carson's rule is an approximation and gives transmission bandwidths that are slightly narrower than the bandwidths determined using the Bessel table and Equation 6-20. Carson's rule defines a bandwidth that includes approximately 98% of the total power in the modulated wave. The actual bandwidth necessary is a function of the modulating signal waveform and the quality of transmission desired.

EXAMPLE 6-3

For an FM modulator with a peak frequency deviation $\Delta f = 10$ kHz, a modulating signal frequency $f_m = 10$ kHz, $V_c = 10$V, and a 500-kHz carrier, determine:

(a) Actual minimum bandwidth from the Bessel function table.
(b) Approximate minimum bandwidth using Carson's rule.
(c) Plot the output frequency spectrum for the Bessel approximation.

Solution   (a) Substituting into Equation 6-13 yields

$$m = \frac{10 \text{ kHz}}{10 \text{ kHz}} = 1$$

From Table 6-2, a modulation index of 1 yields three sets of significant sidebands. Substituting into Equation 6-20, the bandwidth is

$$B = 2(3 \times 10 \text{ kHz}) = 60 \text{ kHz}$$

(b) Substituting into Equation 6-21, the minimum bandwidth is

$$B = 2(10 \text{ kHz} + 10 \text{ kHz}) = 40 \text{ kHz}$$

(c) The output frequency spectrum for the Bessel approximation is shown in Figure 6-7.

From Example 6-3, it can be seen that there is a significant difference in the minimum bandwidth determined from Carson's rule and the minimum bandwidth determined from the Bessel table. The bandwidth from Carson's rule is less than the actual minimum bandwidth required to pass all the significant sideband sets as defined by the Bessel table. Therefore, a system that was designed using Carson's rule would have a narrower bandwidth and thus poorer performance than a system designed using the Bessel table. For modulation indexes above 5, Carson's rule is a close approximation to the actual bandwidth required.

Deviation ratio.   For a given FM system, the minimum bandwidth is greatest when the maximum frequency deviation is obtained with the maximum modulating signal frequency (that is, the highest modulating frequency occurs with the maximum amplitude allowed). By definition, *deviation ratio (DR)* is the *worst-case* modulation index and is equal to the maximum peak frequency deviation divided by the maximum modulating signal frequency. The worst-case modulation index produces the widest output frequency spectrum. Mathematically, the deviation ratio is

$$DR = \frac{\Delta f_{(\text{max})}}{f_{m(\text{max})}} \tag{6-22}$$

where DR = deviation ratio (unitless)
$\Delta f_{(\text{max})}$ = maximum peak frequency deviation (hertz)
$f_{m(\text{max})}$ = maximum modulating signal frequency (hertz)

For example, for the sound portion of a commercial TV broadcast band station, the maximum frequency deviation set by the FCC is 50 kHz, and the maximum modulating signal frequency is 15 kHz. Therefore, the deviation ratio for a television broadcast station is

$$DR = \frac{50 \text{ kHz}}{15 \text{ kHz}} = 3.33$$

This does not mean that whenever a modulation index of 3.33 occurs the widest bandwidth

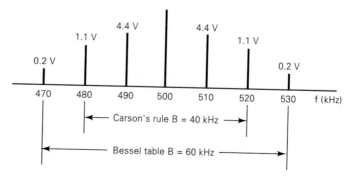

Figure 6-7   Frequency spectrum for Example 6-3.

also occurs at the same time. It means that whenever a modulation index of 3.33 occurs for a maximum modulating signal frequency the widest bandwidth occurs.

EXAMPLE 6-4

(a) Determine the deviation ratio and bandwidth for the worst-cast (widest-bandwidth) modulation index for an FM broadcast-band transmitter with a maximum frequency deviation of 75 kHz and a maximum modulating signal frequency of 15 kHz.

(b) Determine the deviation ratio and maximum bandwidth for an equal modulation index with only half the peak frequency deviation and modulating signal frequency.

Solution (a) The deviation ratio is found by substituting into Equation 6-22.

$$DR = \frac{75 \text{ kHz}}{15 \text{ kHz}} = 5$$

From Table 6-2, a modulation index of 5 produces eight significant sidebands. Substituting into Equation 6-20 yields

$$B = 2(8 \times 15,000) = 240 \text{ kHz}$$

(b) For a 37.5-kHz frequency deviation and a modulating signal frequency $f_m = 7.5$ kHz, the modulation index is

$$m = \frac{37.5 \text{ kHz}}{7.5 \text{ kHz}} = 5$$

and the bandwidth is $B = 2(8 \times 7500) = 120 \text{ kHz}$

From Example 6-4, it can be seen that, although the same modulation index (5) was achieved with two different modulating signal frequencies and amplitudes, two different bandwidths were produced. An infinite number of combinations of modulating signal frequency and frequency deviation will produce a modulation index of 5. However, the case produced from the maximum modulating signal frequency and maximum frequency deviation will always yield the widest bandwidth.

At first it may seem that a higher modulation index with a lower modulating signal frequency would generate a wider bandwidth because more sideband sets are produced; but remember that the sidebands would be closer together. For example, a 1-kHz modulating signal that produces 10 kHz of frequency deviation has a modulation index of $m = 10$ and produces 14 significant sets of sidebands. However, the sidebands are only displaced from each other by 1 kHz, and therefore the total bandwidth is only 28,000 Hz [2(14 × 1000)].

Phasor Representation of an Angle-modulated Wave

As with amplitude modulation, an angle-modulated wave can be shown in phasor form. The phasor diagram for a low-index-angle modulated wave with a single-frequency modulating signal is shown in Figure 6-8. For this special case ($m < 1$), only the first set of sideband pairs is considered, and the phasor diagram closely resembles that of an AM wave except for a phase reversal of one of the side frequencies. The resultant vector has an amplitude that is close to unity at all times and a peak phase deviation of m radians. It is important to note that if the side frequencies from the higher-order terms were included the vector would have no amplitude variations. The dashed line shown in Figure 6-8e is the locus of the resultant formed by the carrier and the first set of side frequencies.

Figure 6-9 shows the phasor diagram for a high-index-angle modulated wave with five sets of side frequencies (for simplicity only the vectors for the first two sets are

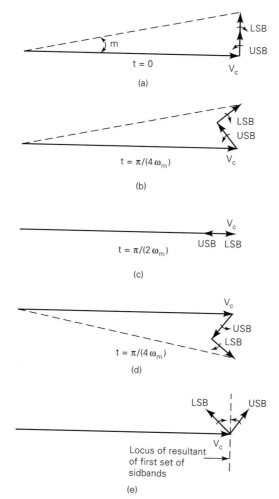

Figure 6-8 Angle modulation phasor representation, low modulation index.

shown). The resultant vector is the sum of the carrier component and the components of the significant side frequencies with their magnitudes adjusted according to the Bessel table. Each side frequency is shifted an additional 90° from the preceding side frequency. The locus of the resultant five-component approximation is curved and closely follows the signal locus. By definition, the locus is a segment of the circle with a radius equal to the amplitude of the unmodulated carrier. It should be noted that the resultant signal amplitude and, consequently, the signal power remain constant.

Average Power of an Angle-modulated Wave

One of the most important differences between angle modulation and amplitude modulation is the distribution of power in the modulated wave. Unlike AM, the total power in an angle-modulated wave is equal to the power of the unmodulated carrier (that is, the sidebands do not add power to the composite modulated signal). Therefore, with angle modulation, the power that was originally in the unmodulated carrier is redistributed among the carrier and its sidebands. The average power of an angle-modulated wave is independent of the modulating signal, the modulation index, and the frequency deviation. It is equal to

Angle Modulation

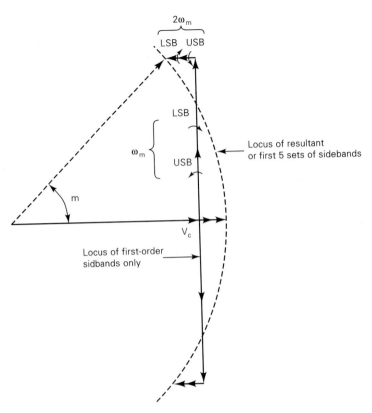

Figure 6-9 Angle modulation phasor representation, high modulation index.

the average power of the unmodulated carrier, regardless of the depth of modulation. Mathematically, the average power in the unmodulated carrier is

$$P_c = \frac{V_c^2}{2R} \quad \text{watts} \tag{6-23}$$

where P_c = carrier power (watts)
V_c = peak unmodulated carrier voltage (volts)
R = load resistance (ohms)

The total instantaneous power in an angle-modulated carrier is

$$p_t = \frac{m(t)^2}{R} \quad \text{watts} \tag{6-24a}$$

Substituting for $m(t)$ gives $$p_t = \frac{V_c^2}{R} \cos^2 [\omega_c t + \theta(t)] \tag{6-24b}$$

and expanding yields $$= \frac{V_c^2}{R} \left\{ \frac{1}{2} + \frac{1}{2} \cos [2\omega_c t + 2\theta(t)] \right\} \tag{6-24c}$$

In Equation 6-24c, the second term consists of an infinite number of sinusoidal side-frequency components about a frequency equal to twice the carrier frequency ($2\omega_c$). Consequently, the average value of the second term is zero, and the average power of the modulated wave reduces to

Chap. 6   **Angle Modulation Transmission**

$$P_t = \frac{V_c^2}{2R} \quad \text{watts} \tag{6-25}$$

Note that Equations 6-23 and 6-25 are identical, so the average power of the modulated carrier must be equal to the average power of the unmodulated carrier. The modulated carrier power is the sum of the powers of the carrier and the side-frequency components. Therefore, the total modulated wave power is

$$P_t = P_c + P_1 + P_2 + P_3 + P_n \tag{6-26}$$

$$P_t = \frac{V_c^2}{2R} + \frac{2(V_1)^2}{2R} + \frac{2(V_2)^2}{2R} + \frac{2(V_3)^2}{2R} + \frac{2(V_n)^2}{2R} \tag{6-27}$$

where P_t = total power
P_c = carrier power
P_1 = power in the first set of sidebands
P_2 = power in the second set of sidebands
P_3 = power in the third set of sidebands
P_n = power in the nth set of sidebands

EXAMPLE 6-5

(a) Determine the unmodulated carrier power for the FM modulator and conditions given in Example 6-2 (assume a load resistance $R_L = 50 \ \Omega$).
(b) Determine the total power in the angle-modulated wave.

Solution (a) Substituting into Equation 6-22 yields

$$P_c = \frac{10^2}{2(50)} = 1 \ \text{W}$$

(b) Substituting into Equation 6-26 gives us

$$P_c = \frac{7.7^2}{2(50)} + \frac{2(4.4)^2}{2(50)} + \frac{2(1.1)^2}{2(50)} + \frac{2(0.2)^2}{2(50)}$$

$$= 0.5929 + 0.3872 + 0.0242 + 0.0008$$

$$= 1.0051 \ \text{W}$$

The results of (a) and (b) are not exactly equal because the values given in the Bessel table have been rounded off. However, the results are close enough to illustrate that the power in the modulated wave and the unmodulated carrier are equal.

Noise and Angle Modulation

When thermal noise with a constant spectral density is added to an FM signal, it produces an unwanted deviation of the carrier frequency. The magnitude of this unwanted frequency deviation depends on the relative amplitude of the noise with respect to the carrier. When this unwanted carrier deviation is demodulated, it becomes noise if it has frequency components that fall within the information-frequency spectrum. The spectral shape of the demodulated noise depends on whether an FM or PM demodulator is used. The noise voltage at the output of a PM demodulator is constant with frequency, whereas the noise voltage at the output of an FM demodulator increases linearly with frequency. This is commonly called the FM *noise triangle* and is illustrated in Figure 6-10. It can be seen

Angle Modulation

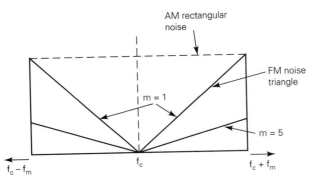

Figure 6-10 FM noise triangle.

that the demodulated noise voltage is inherently higher for the higher-modulating signal frequencies.

Phase modulation due to an interfering sinusoid. Figure 6-11 shows phase modulation caused by a single-frequency noise signal. The noise component V_n is separated in frequency from the signal component V_c by frequency f_n. This is shown in Figure 6-11b. Assuming that $V_c > V_n$, the peak phase deviation due to an interfering single-frequency sinusoid occurs when the signal and noise voltages are in quadrature and is approximated for small angles as

$$\Delta\theta \text{ (peak)} \simeq \frac{V_n}{V_c} \quad \text{rad} \tag{6-28}$$

Figure 6-11c shows the effect of *limiting* the amplitude of the composite FM signal on noise. (Limiting is commonly used in angle-modulation receivers and is explained in Chapter 7.) It can be seen that the single-frequency noise signal has been transposed into a noise sideband pair each with amplitude $V_n/2$. These sidebands are coherent; therefore, the peak phase deviation is still V_n/V_c radians. However, the unwanted amplitude variations have been removed, which reduces the total power, but does not reduce the interference in the demodulated signal due to the unwanted phase deviation.

Frequency modulation due to an interfering sinusoid. From Equation 6-6a, the instantaneous frequency deviation $\Delta f(t)$ is the first time derivative of the instantaneous phase deviation $\theta(t)$. When the carrier component is much larger than the interfering noise voltage, the instantaneous phase deviation is approximately

$$\theta(t) = \frac{V_n}{V_c} \sin(\omega_n t + \theta_n) \quad \text{radians} \tag{6-29}$$

and, taking the first derivative, we obtain

$$\Delta\omega(t) = \frac{V_n}{V_c} \omega_n \cos(\omega_n t + \theta_n) \quad \text{radians/second} \tag{6-30}$$

Therefore, the peak frequency deviation is

$$\Delta\omega_{\text{peak}} = \frac{V_n}{V_c} \omega_n \quad \text{radians/second} \tag{6-31}$$

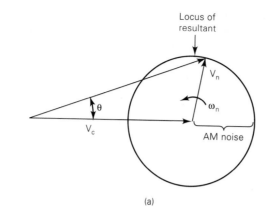

(a)

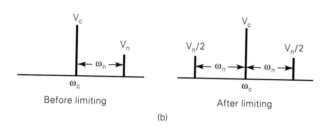

Before limiting   After limiting

(b)

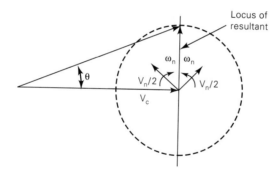

Figure 6-11 Interfering sinusoid of noise: (a) before limiting; (b) frequency spectrum; (c) after limiting.

$$\Delta f_{peak} = \frac{V_n}{V_c} f_n \quad \text{hertz} \tag{6-32}$$

Rearranging Equation 6-13, it can be seen that the peak frequency deviation (Δf) is a function of the modulating signal frequency and the modulation index. Therefore, for a noise modulating frequency f_n, the peak frequency deviation is

$$\Delta f_{peak} = m f_n \quad \text{hertz} \tag{6-33}$$

where m = modulation index ($m \ll 1$).

From Equation 6-33, it can be seen that the farther the noise frequency is displaced from the carrier frequency, the larger the frequency deviation. Therefore, noise frequencies

Angle Modulation

249

that produce components at the high end of the modulating signal frequency spectrum produce more frequency deviation for the same phase deviation than frequencies that fall at the low end. FM demodulators generate an output voltage that is proportional to the frequency deviation and of frequency equal to the difference between the carrier frequency and the interfering signal frequency. Therefore, high-frequency noise components produce more demodulated noise than do low-frequency components.

The signal-to-noise ratio at the output of an FM demodulator due to unwanted frequency deviation from an interfering sinusoid is the ratio of the peak frequency deviation due to the information signal to the peak frequency deviation due to the interfering signal.

$$\frac{S}{N} = \frac{\Delta f_{\text{due to signal}}}{\Delta f_{\text{due to noise}}} \tag{6-34}$$

EXAMPLE 6-6

For an angle-modulated carrier $V_c = 6 \cos (2\pi 110 \text{ MHz } t)$ with 75-kHz frequency deviation due to the information signal and a single-frequency interfering signal $V_n = 0.3 \cos (2\pi 109.985 \text{ MHz } t)$, determine:

(a) Frequency of the demodulated interference signal.
(b) Peak phase and frequency deviations due to the interfering signal.
(c) Voltage signal-to-noise ratio at the output of the demodulator.

Solution (a) The frequency of the noise interference is the difference between the carrier frequency and the frequency of the single-frequency interfering signal.

$$f_c - f_n = 110 \text{ MHz} - 109.985 \text{ MHz} = 15 \text{ kHz}$$

(b) Substituting into Equation 6-28 yields

$$\Delta\theta_{\text{peak}} = \frac{0.3}{6} = 0.05 \text{ rad}$$

Substituting into Equation 6-32 gives us

$$\Delta f_{\text{peak}} = \frac{0.3 \times 15 \text{ kHz}}{6} = 750 \text{ Hz}$$

(c) The voltage S/N ratio due to the interfering tone is the ratio of the carrier amplitude to the amplitude of the interfering signal, or

$$\frac{6}{0.3} = 20$$

The voltage S/N ratio after demodulation is found by substituting into Equation 6-34:

$$\frac{S}{N} = \frac{75 \text{ kHz}}{750 \text{ Hz}} = 100$$

Thus, there is a voltage signal-to-noise improvement of $100/20 = 5$ or $20 \log 5 = 14 \text{ dB}$.

Preemphasis and Deemphasis

The noise triangle shown in Figure 6-10 shows that, with FM, there is a nonuniform distribution of noise. Noise at the higher modulating signal frequencies is inherently greater in amplitude than noise at the lower frequencies. This includes both single-frequency inter-

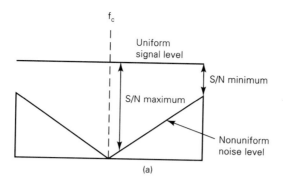

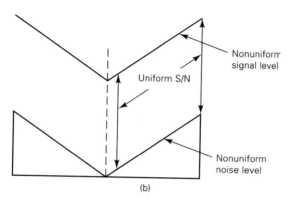

Figure 6-12 FM signal-to-noise:
(a) without preemphasis; (b) with
preemphasis.

ference and thermal noise. Therefore, for information signals with a uniform signal level, a nonuniform signal-to-noise ratio is produced, and the higher modulating signal frequencies have a lower signal-to-noise ratio than the lower frequencies. This is shown in Figure 6-12a. It can be seen that the S/N ratio is lower at the high-frequency ends of the triangle. To compensate for this, the high-frequency modulating signals are emphasized or boosted in amplitude in the transmitter prior to performing modulation. To compensate for this boost, the high-frequency signals are attenuated or deemphasized in the receiver after demodulation has been performed. Deemphasis is the reciprocal of preemphasis, and therefore a deemphasis network restores the original amplitude-versus-frequency characteristics to the information signals. In essence, the preemphasis network allows the high-frequency modulating signals to modulate the carrier at a higher level and thus cause more frequency deviation than their original amplitudes would have produced. The high-frequency signals are propagated through the system at an elevated level (increased frequency deviation), demodulated, and then restored to their original amplitude proportions. Figure 6-12b shows the effects of pre- and deemphasis on the signal-to-noise ratio. The figure shows that pre- and deemphasis produces a more uniform signal-to-noise ratio throughout the modulating signal frequency spectrum.

A preemphasis network is a high-pass filter (that is, a differentiator) and a deemphasis network is a low-pass filter (an integrator). Figure 6-13a shows the schematic diagrams for an active preemphasis network and a passive deemphasis network. Their corresponding frequency-response curves are shown in Figure 6-13b. A preemphasis network provides a constant increase in the amplitude of the modulating signal with an increase in

Angle Modulation

251

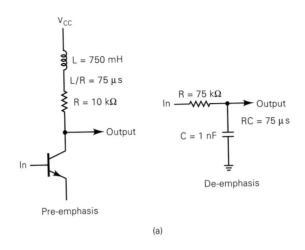

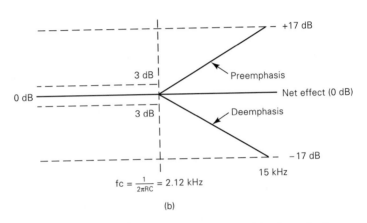

Figure 6-13 Preemphasis and deemphasis: (a) schematic diagrams; (b) attenuation curves.

frequency. With FM, approximately 12 dB of improvement in noise performance is achieved using pre- and deemphasis. The break frequency (the frequency where pre- and deemphasis begins) is determined by the RC or L/R time constant of the network. The break frequency occurs at the frequency where X_C or X_L equals R. Mathematically, the break frequency is

$$f_b = \frac{1}{2\pi RC} \qquad (6\text{-}35\text{a})$$

$$f_b = \frac{1}{2\pi L/R} \qquad (6\text{-}35\text{b})$$

The networks shown in Figure 6-13 are for the FM broadcast band, which uses a 75-μs time constant. Therefore, the break frequency is approximately

$$f_b = \frac{1}{2\pi 75\ \mu s} = 2.12\ \text{kHz}$$

The FM transmission of the audio portion of commercial television broadcasting uses a 50-μs time constant.

FREQUENCY MODULATION TRANSMISSION

Direct FM Modulators

Direct FM is angle modulation in which the frequency of the carrier is varied (deviated) directly by the modulating signal. With direct FM, the instantaneous frequency deviation is directly proportional to the amplitude of the modulating signal. Figure 6-14 shows a schematic diagram for a simple (although highly impractical) direct FM generator. The tank circuit (L and C_m) is the frequency-determining section for a standard LC oscillator. The capacitor microphone is a transducer that converts acoustical energy to mechanical energy, which is used to vary the distance between the plates of C_m and, consequently, change its capacitance. As C_m is varied, the resonant frequency is varied. Thus, the oscillator output frequency varies directly with the external sound source. This is direct FM because the oscillator frequency is changed directly by the modulating signal, and the magnitude of the frequency change is proportional to the amplitude of the modulating signal voltage.

Varactor diode modulators.   Figure 6-15 shows the schematic diagram for a more practical direct FM generator that uses a varactor diode to deviate the frequency of a crystal oscillator. R_1 and R_2 develop a dc voltage that reverse biases varactor diode VD_1 and determines the rest frequency of the oscillator. The external modulating signal voltage adds to and subtracts from the dc bias, which changes the capacitance of the diode and thus the frequency of oscillation. Positive alternations of the modulating signal increase the reverse bias on VD_1, which decreases its capacitance and increases the frequency of oscillation. Conversely, negative alternations of the modulating signal decrease the frequency of oscillation. Varactor diode FM modulators are extremely popular because they are simple to use and reliable and have the stability of a crystal oscillator. However, because a crystal is used, the peak frequency deviation is limited to relatively small values. Consequently, they are used primarily for low-index applications, such as two-way mobile radio.

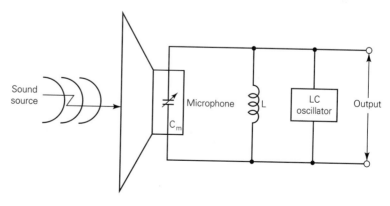

Figure 6-14 Simple direct FM modulator.

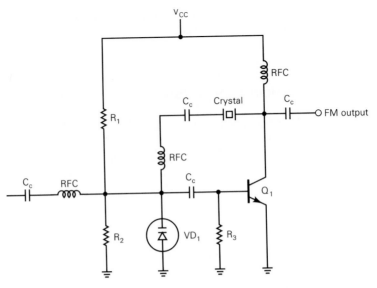

Figure 6-15 Varactor diode direct FM modulator.

Figure 6-16 shows a simplified schematic diagram for a voltage-controlled oscillator (VCO) FM generator. Again, a varactor diode is used to transform changes in the modulating signal amplitude to changes in frequency. The center frequency for the oscillator is determined as follows

$$f_c = \frac{1}{2\pi \sqrt{LC}} \quad \text{hertz} \tag{6-36}$$

where L = inductance of the primary winding of T_1 (henries)
           C = varactor diode capacitance (farads)

With a modulating signal applied, the frequency is

$$f = \frac{1}{2\pi \sqrt{L(C + \Delta C)}} \quad \text{hertz} \tag{6-37}$$

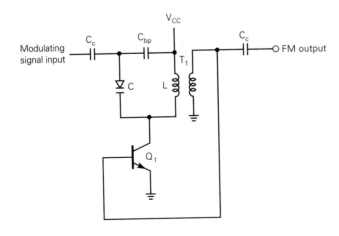

Figure 6-16 Varactor diode VCO FM modulator.

where f is the new frequency of oscillation and ΔC is the change in varactor diode capacitance due to the modulating signal. The change in frequency is

$$\Delta f = |f_c - f| \qquad (6\text{-}38)$$

FM reactance modulator. Figure 6-17a shows a schematic diagram for a reactance modulator using a JFET as the active device. This circuit configuration is called a reactance modulator because the JFET looks like a variable-reactance load to the LC tank circuit. The modulating signal varies the reactance of Q_1, which causes a corresponding change in the resonant frequency of the oscillator tank circuit.

Figure 6-17b shows the ac equivalent circuit. R_1, R_3, R_4, and R_C provide the dc bias for Q_1. R_E is bypassed by C_c and is, therefore, omitted from the ac equivalent circuit. Circuit operation is as follows. Assuming an ideal JFET (gate current $i_g = 0$),

$$v_g = i_g R$$

where

$$i_g = \frac{v}{R - jX_C}$$

Therefore,

$$v_g = \frac{v}{R - jX_C} \times R$$

and the JFET drain current is

$$i_d = g_m v_g = g_m \frac{v}{R - jX_C} \times R$$

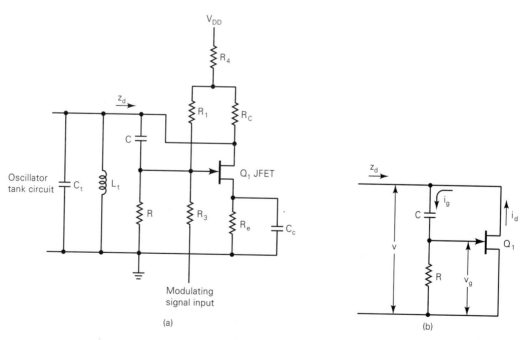

Figure 6-17 JFET reactance modulator: (a) schematic diagram; (b) ac equivalent circuit.

Frequency Modulation Transmission

where g_m is the transconductance of the JFET, and the impedance between the drain and ground is

$$z_d = \frac{v}{i_d}$$

Substituting and rearranging gives us

$$z_d = \frac{R - jX_C}{g_m R} = \frac{1}{g_m}\left[1 - \frac{jX_C}{R}\right]$$

Assuming that $R <<< X_C$, $\quad z_d = -j\dfrac{X_C}{g_m R} = \dfrac{-j}{2\pi f_m g_m RC}$

$g_m RC$ is equivalent to a variable capacitance and is inversely proportional to resistance (R), the angular velocity of the modulating signal $(2\pi f_m)$, and the transconductance (g_m) of Q_1, which varies with the gate-to-source voltage. When a modulating signal is applied to the bottom of R_3, the gate-to-source voltage is varied accordingly, causing a proportional change in g_m. As a result, the equivalent circuit impedance (z_d) is a function of the modulating signal. Therefore, the resonant frequency of the oscillator tank circuit is a function of the amplitude of the modulating signal, and the rate at which it changes is equal to f_m. Interchanging R and C causes the variable reactance to be inductive rather than capacitive, but does not affect the output FM waveform. The maximum frequency deviation obtained with a reactance modulator is approximately 5 kHz.

Linear integrated-circuit direct FM modulators. *Linear integrated-circuit voltage-controlled oscillators* and *function generators* can generate a direct FM output waveform that is relatively stable, accurate, and directly proportional to the input modulating signal. The primary disadvantage of using LIC VCOs and function generators for direct FM modulation is their low output power and the need for several additional external components for them to function, such as timing capacitors and resistors for frequency determination and power supply filters.

Figure 6-18 shows a simplified block diagram for a linear integrated-circuit monolithic function generator that can be used for direct FM generation. The VCO center frequency is determined by external resistor and capacitor (R and C). The input modulating signal deviates the VCO frequency, which produces an FM output waveform. The analog multiplier and sine shaper convert the square-wave VCO output signal to a sine wave, and the unity gain amplifier provides a buffered output. The modulator output frequency is

$$f_{\text{out}} = (f_c + \Delta f)N$$

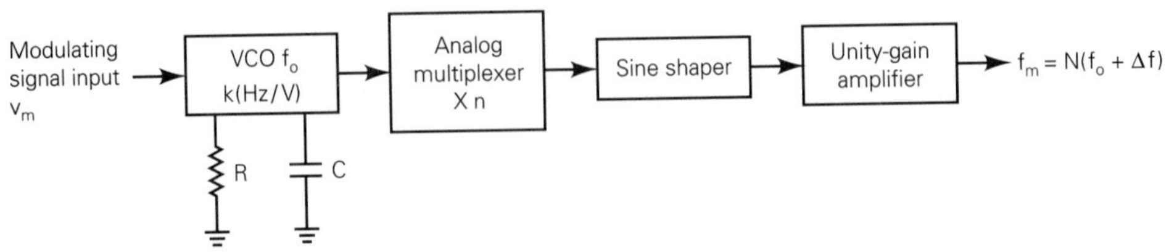

Figure 6-18 LIC direct FM generator: simplified schematic diagram.

Chap. 6   **Angle Modulation Transmission**

where the peak frequency deviation (Δf) is equal to the peak modulating signal amplitude times the deviation sensitivity of the VCO. Linear integrated-circuit function generators and voltage controlled oscillators can generally be used for either sweep frequency operation, frequency shift keying, or direct FM generation.

Figure 6-19a shows the schematic diagram for the Motorola MC1376 monolithic FM transmitter. The MC1376 is a complete FM modulator on a single 8-pin DIP integrated-circuit chip. The MC1376 can operate with carrier frequencies between 1.4 and 14 MHz and is intended to be used for producing direct FM waves for low-power applications such as cordless telephones. When the auxiliary transistor is connected to a 12-V supply voltage, output powers as high as 600 mW can be achieved. Figure 6-19b shows the output frequency-versus-input voltage curve for the internal VCO. As the figure shows, the curve is fairly linear between 2 and 4 V and can produce a peak frequency deviation of nearly 150 kHz.

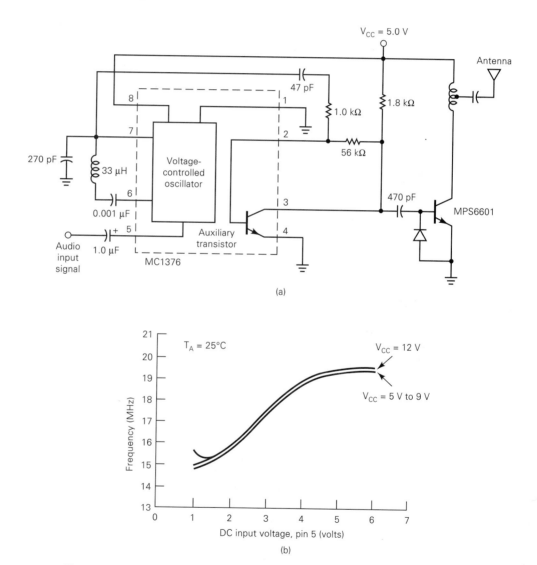

Figure 6-19   MC1376 FM transmitter LFC: (a) schematic diagram; (b) VCO output-versus-input frequency-response curve.

Frequency Modulation Transmission

Indirect FM Modulators

Indirect FM is angle modulation in which the frequency of the carrier is deviated indirectly by the modulating signal. Indirect FM is accomplished by directly changing the phase of the carrier and is therefore a form of direct phase modulation. The instantaneous phase of the carrier is directly proportional to the modulating signal.

Figure 6-20 shows a schematic diagram for an indirect FM modulator. The modulator comprises a varactor diode VD_1 in series with an inductive network (tunable coil L_1 and resistor R_1). The combined series–parallel network appears as a series resonant circuit to the output frequency from the crystal oscillator. A modulating signal is applied to VD_1, which changes its capacitance and, consequently, the phase angle of the impedance seen by the carrier varies, which results in a corresponding phase shift in the carrier. The phase shift is directly proportional to the amplitude of the modulating signal. An advantage of indirect FM is that a buffered crystal oscillator is used for the source of the carrier signal. Consequently, indirect FM transmitters are more frequency stable than their direct counterparts. A disadvantage is that the capacitance-versus-voltage characteristics of a varactor diode are nonlinear. In fact, they closely resemble a square-root function. Consequently, to minimize distortion in the modulated waveform, the amplitude of the modulating signal must be keep quite small, which limits the phase deviation to rather small values and its uses to low-index, narrowband applications.

Direct FM Transmitters

Direct FM transmitters produce an output waveform in which the frequency deviation is directly proportional to the modulating signal. Consequently, the carrier oscillator must be deviated directly. Therefore, for medium- and high-index FM systems, the oscillator cannot be a crystal because the frequency at which a crystal oscillates cannot be significantly varied. As a result, the stability of the oscillators in direct FM transmitters often cannot meet FCC specifications. To overcome this problem, *automatic frequency control (AFC)* is used. An AFC circuit compares the frequency of the noncrystal carrier oscillator to a crystal reference oscillator and then produces a correction voltage proportional to the difference between the two frequencies. The correction voltage is fed back to the carrier oscillator to automatically compensate for any drift that may have occurred.

Crosby direct FM transmitter. Figure 6-21 shows the block diagram for a commercial broadcast-band transmitter. This particular configuration is called a *Crosby direct FM transmitter* and includes an *AFC loop*. The frequency modulator can be either a reactance modulator or a voltage-controlled oscillator. The carrier rest frequency is the unmodulated output frequency from the master oscillator (f_c). For the transmitter shown in Figure 6-21, the center frequency of the master oscillator $f_c = 5.1$ MHz, which is multiplied by 18 in three steps ($3 \times 2 \times 3$) to produce a final transmit carrier frequency $f_t = 91.8$ MHz. At this time, three aspects of frequency conversion should be noted. First, when the frequency of a frequency-modulated carrier is multiplied, its frequency and phase deviations are multiplied as well. Second, the rate at which the carrier is deviated (that is, the modulating signal frequency, f_m) is unaffected by the multiplication process. Therefore, the modulation index is also multiplied. Third, when an angle-modulated carrier is heterodyned with another frequency in a nonlinear mixer, the carrier can be either up- or down-converted, depending on the passband of the output filter. However, the frequency deviation, phase deviation, and rate of change are unaffected by the heterodyning

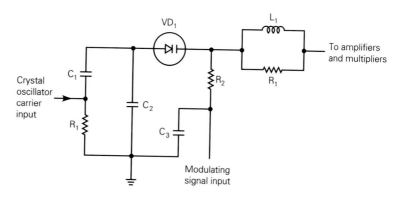

Figure 6-20 Indirect FM modulator schematic diagram.

process. Therefore, for the transmitter shown in Figure 6-21, the frequency and phase deviations at the output of the modulator are also multiplied by 18. To achieve the maximum frequency deviation allowed FM broadcast-band stations at the antenna (75 kHz), the deviation at the output of the modulator must be

$$\Delta f = \frac{75 \text{ kHz}}{18} = 4166.7 \text{ Hz}$$

and the modulation index must be

$$m = \frac{4166.7 \text{ Hz}}{f_m}$$

For the maximum modulating signal frequency allowed, $f_m = 15$ kHz,

$$m = \frac{4166.7 \text{ Hz}}{15,000 \text{ Hz}} = 0.2778$$

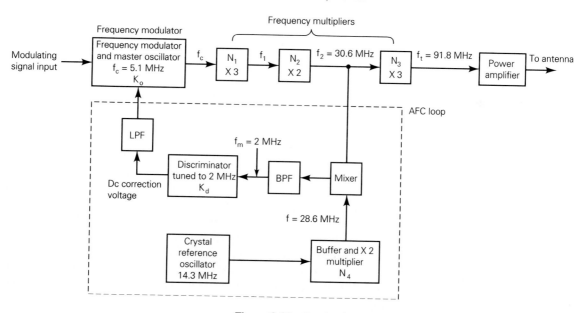

Figure 6-21 Crosby direct FM transmitter.

Frequency Modulation Transmission

259

Thus, the modulation index at the antenna is

$$m = 0.2778(18) = 5$$

which is the deviation ratio for commercial FM broadcast transmitters with a 15 kHz modulating signal.

AFC loop. The purpose of the *AFC loop* is to achieve near-crystal stability of the transmit carrier frequency without using a crystal in the carrier oscillator. With AFC, the carrier signal is mixed with the output signal from a crystal reference oscillator in a non-linear device, down-converted in frequency, and then fed back to the input of a *frequency discriminator*. A frequency discriminator is a frequency-selective device whose output voltage is proportional to the difference between the input frequency and its resonant frequency (discriminator operation is explained in Chapter 7). For the transmitter shown in Figure 6-21, the output from the doubler $f_2 = 30.6$ MHz, which is mixed with a crystal-controlled reference frequency $f_r = 28.6$ MHz to produce a difference frequency $f_d = 2$ MHz. The discriminator is a relatively high Q (narrowband) tuned circuit that reacts only to frequencies near its center frequency (2 MHz in this case). Therefore, the discriminator responds to long-term, low-frequency changes in the carrier center frequency due to master oscillator frequency drift and because of low pass filtering does not respond to the frequency deviation produced by the modulating signal. If the discriminator responded to the frequency deviation, the feedback loop would cancel the deviation and thus remove the modulation from the FM wave (this effect is called *wipe off*). The dc correction voltage is added to the modulating signal to automatically adjust the master oscillator's center frequency to compensate for the low-frequency drift.

EXAMPLE 6-7

Use the transmitter model shown in Figure 6-21 to answer the following questions. For a total frequency multiplication of 20 and a transmit carrier frequency $f_t = 88.8$ MHz, determine:

(a) Master oscillator center frequency.
(b) Frequency deviation at the output of the modulator for a frequency deviation of 75 kHz at the antenna.
(c) Deviation ratio at the output of the modulator for a maximum modulating signal frequency $f_m = 15$ kHz.
(d) Deviation ratio at the antenna.

Solution (a)
$$f_c = \frac{f_t}{N_1 N_2 N_3} = \frac{88.8 \text{ MHz}}{20} = 4.43 \text{ MHz}$$

(b)
$$\Delta f = \frac{\Delta f_t}{N_1 N_2 N_3} = \frac{75 \text{ kHz}}{20} = 3750 \text{ Hz}$$

(c)
$$\text{DR} = \frac{\Delta f_{(\text{maximum})}}{f_{m\,(\text{maximum})}} = \frac{3750 \text{ Hz}}{15 \text{ kHz}} = 0.25$$

(d)
$$\text{DR} = 0.25 \times 20 = 5$$

Automatic frequency control. Because the Crosby transmitter uses either a VCO, a reactance oscillator, or a linear integrated circuit oscillator to generate the carrier frequency, it is more susceptible to frequency drift due to temperature change, power supply fluctuations, and so on, than if it were a crystal oscillator. As stated in Chapter 2, the stability of an oscillator is often given in parts per million (ppm) per degree celsius. For example, for the transmitter shown in Figure 6-21, an oscillator stability of ±40 ppm could produce ±204 Hz (5.1 MHz × ±40 Hz/million) of frequency drift per degree celsius at the output of the master oscillator. This would correspond to a ± 3672-Hz drift at the antenna ($18 \times \pm 204$), which far exceeds the ± 2-kHz maximum set by the FCC for commercial FM broadcasting. Although an AFC circuit does not totally eliminate frequency drift, it can substantially reduce it. Assuming a rock-stable crystal reference oscillator and a perfectly tuned discriminator, the frequency drift at the output of the second multiplier without feedback (that is, open loop) is

$$\text{open-loop drift} = df_{ol} = N_1 N_2 df_c \tag{6-39}$$

where d denotes drift. The closed-loop drift is

$$\text{closed-loop drift} = df_{cl} = df_{ol} - N_1 N_2 k_d k_o df_{cl} \tag{6-40}$$

Therefore, $$df_{cl} + N_1 N_2 k_d k_o df_{cl} = df_{ol}$$

and $$df_{cl}(1 + N_1 N_2 k_d k_o) = df_{ol}$$

Thus $$df_{cl} = \frac{df_{ol}}{1 + N_1 N_2 k_d k_o} \tag{6-41}$$

where k_d = discriminator transfer function (volts/hertz)
    k_o = master oscillator transfer function (hertz/volt)

From Equation 6-42 it can be seen that the frequency drift at the output of the second multiplier and consequently at the input to the discriminator is reduced by a factor of $1 + N_1 N_2 k_d k_o$ when the AFC loop is closed. The carrier frequency drift is multiplied by the AFC loop gain and fed back to the master oscillator as a correction voltage. The total frequency error cannot be canceled because then there would not be any error voltage at the output of the discriminator to feed back to the master oscillator. In addition, Equations 6-39, 6-40, and 6-41 were derived assuming that the discriminator and crystal reference oscillator were perfectly stable, which of course they are not.

EXAMPLE 6-8

Use the transmitter block diagram and values given in Figure 6-21 to answer the following questions. Determine the reduction in frequency drift at the antenna for a transmitter without AFC compared to a transmitter with AFC. Use a VCO stability = $+200$ ppm, k_o = 10 kHz/V, and k_d = 2 V/kHz.

Solution With the feedback loop open, the master oscillator output frequency is

$$f_c = 5.1 \text{ MHz} + (200 \text{ ppm} \times 5.1 \text{ MHz}) = 5{,}101{,}020 \text{ Hz}$$

and the frequency at the output of the second multiplier is

$$f_2 = N_1 N_2 f_c = (5{,}101{,}020)(6) = 30{,}606{,}120 \text{ Hz}$$

Thus the frequency drift is

$$df_2 = 30{,}606{,}120 - 30{,}600{,}000 = 6120 \text{ Hz}$$

Therefore, the antenna transmit frequency is

$$f_t = 30{,}606{,}120(3) = 91.81836 \text{ MHz}$$

which is 18.36 kHz above the assigned frequency and well out of limits.

Frequency Modulation Transmission

With the feedback loop closed, the frequency drift at the output of the second multiplier is reduced by a factor of $1 + N_1 N_2 k_d k_o$ or

$$1 + \frac{(2)(3)(10 \text{ kHz})}{\text{V}} \frac{2 \text{ V}}{\text{kHz}} = 121$$

Therefore

$$df_2 = \frac{6120}{121} = 51 \text{ Hz}$$

Thus

$$f_2 = 30{,}600{,}051 \text{ Hz}$$

and the antenna transmit frequency is

$$f_t = 30{,}600{,}051 \times 3 = 91{,}800{,}153 \text{ Hz}$$

The frequency drift at the antenna has been reduced from 18,360 Hz to 153 Hz, which is now well within the ± 2-kHz FCC requirements.

The preceding discussion and Example 6-9 assumed a perfectly stable crystal reference oscillator and a perfectly tuned discriminator. In actuality, both the discriminator and the reference oscillator are subject to drift, and the worst-case situation is when they both drift in the same direction as the master oscillator. The drift characteristics for a typical discriminator are on the order of ± 100 ppm. Perhaps now it can be seen why the output frequency from the second multiplier was mixed down to a relatively low frequency prior to being fed to the discriminator. For a discriminator tuned to 2 MHz with a stability of ± 200 ppm, the maximum discriminator drift is

$$df_d = \pm 100 \text{ ppm} \times 2 \text{ MHz} = \pm 200 \text{ Hz}$$

If the 30.6-MHz signal were fed directly into the discriminator, the maximum drift would be

$$df_d = \pm 100 \text{ ppm} \times 30.6 \text{ MHz} = \pm 3060 \text{ Hz}$$

Frequency drift due to discriminator instability is multiplied by the AFC open-loop gain. Therefore, the change in the second multiplier output frequency due to discriminator drift is

$$df_2 = df_d N_1 N_2 k_d k_o \tag{6-42}$$

Similarly, the crystal reference oscillator can drift and also contribute to the total frequency drift at the output of the second multiplier. The drift due to crystal instability is multiplied by 2 before entering the nonlinear mixer; therefore,

$$df_2 = N_4 df_o N_1 N_2 k_d k_o \tag{6-43}$$

and the maximum open-loop frequency drift at the output of the second multiplier is

$$Df_{(\text{total})} = N_1 N_2 (df_c + k_o k_d f_d + k_o k_d N_4 df_o) \tag{6-44}$$

Phase-locked-loop direct FM transmitter. Figure 6-22 shows a *wideband* FM transmitter that uses a phase-locked loop to achieve crystal stability from a VCO master oscillator and, at the same time, generate a high-index, wideband FM output signal. The VCO output frequency is divided by N and fed back to the PLL phase comparator, where it is compared to a stable crystal reference frequency. The phase comparator generates a correction voltage that is proportional to the difference between the two frequencies. The correction voltage is added to the modulating signal and applied to the VCO input. The correction voltage adjusts the VCO center frequency to its proper value. Again, the low-pass filter prevents changes in the VCO output frequency due to the modulating signal from

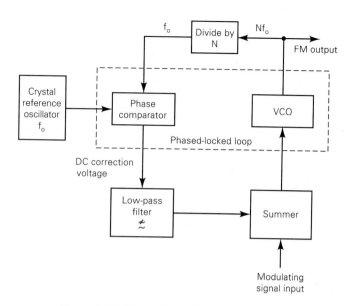

Figure 6-22 Phase-locked-loop FM transmitter.

being converted to a voltage, fed back to the VCO, and wiping out the modulation. The low-pass filter also prevents the loop from locking onto a side frequency.

PM from FM. As shown in Figure 6-4, an FM modulator preceded by a differentiator generates a PM waveform. If the transmitters shown in Figures 6-21 and 6-22 are preceded by a preemphasis network, which is a differentiator (high-pass filter), an interesting situation occurs. For a 75-μs time constant, the amplitude of frequencies above 2.12 kHz is emphasized through differentiation. Therefore, for modulating frequencies below 2.12 kHz, the output waveform is proportional to the modulating signal, and for frequencies above 2.12 kHz, the output waveform is proportional to the derivative of the input signal. In other words, frequency modulation occurs for frequencies below 2.12 kHz, and phase modulation occurs for frequencies above 2.12 kHz. Because the gain of a differentiator increases with frequency above the break frequency (2.12 kHz) and since the frequency deviation is proportional to the modulating signal amplitude, the frequency deviation also increases with frequencies above 2.12 kHz. From Equation 6-13, it can be seen that if Δf and f_m increase proportionately, the modulation index remains constant, which is a characteristic of phase modulation.

Indirect FM Transmitters

Indirect FM transmitters produce an output waveform in which the phase deviation is directly proportional to the modulating signal. Consequently, the carrier oscillator is not directly deviated. Therefore, the carrier oscillator can be a crystal because the oscillator itself is not the modulator. As a result, the stability of the oscillators with indirect FM transmitters can meet FCC specifications without using an AFC circuit.

Armstrong indirect FM transmitter. With indirect FM, the modulating signal directly deviates the phase of the carrier, which indirectly changes the frequency. Figure 6-23 shows the block diagram for a wideband *Armstrong indirect FM transmitter*. The

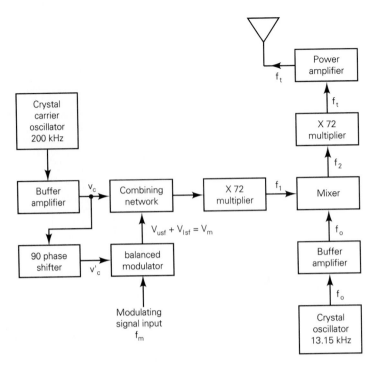

Figure 6-23 Armstrong indirect FM transmitter.

carrier source is a crystal. Therefore, the stability requirements for the carrier frequency set by the FCC can be achieved without using an AFC loop.

With an Armstrong transmitter, a relatively low frequency subcarrier (f_c) is phase shifted 90° (f_c') and fed to a balanced modulator, where it is mixed with the input modulating signal (f_m). The output from the balanced modulator is a double-sideband suppressed-carrier wave that is combined with the original carrier in a combining network to produce a low-index phase-modulated waveform. Figure 6-24a shows the phasor for the original carrier (V_c), and Figure 6-24b shows the phasors for the side-frequency components of the suppressed-carrier wave (V_{usf} and V_{lsf}). Because the suppressed-carrier voltage (V_c') is 90° out of phase with V_c, the upper and lower sidebands combine to produce a component (V_m) that is always in quadrature (at right angles) with V_c. Figures 6-24c through f show the progressive phasor addition of V_c, V_{usf}, and V_{lsf}. It can be seen that the output from the combining network is a signal whose phase is varied at a rate equal to f_m and whose magnitude is directly proportional to the magnitude of V_m. From Figure 6-24 it can be seen that the peak phase deviation (modulation index) can be calculated as follows:

$$\theta = m = \arctan \frac{V_m}{V_c} \qquad (6\text{-}45\text{a})$$

For very small angles, the tangent of the angle is approximately equal to the angle; therefore,

$$\theta = m = \frac{V_m}{V_c} \qquad (6\text{-}45\text{b})$$

EXAMPLE 6-9

For the Armstrong transmitter shown in Figure 6-23 and the phase-shifted carrier (V_c'), upper side frequency (V_{usf}), and lower side frequency (V_{lsf}) components shown in Figure 6-24, determine:

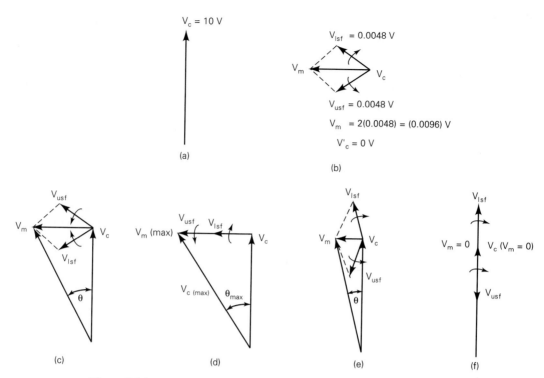

Figure 6-24 Phasor addition of V_c, V_{usf}, and V_{lsf}: (a) carrier phasor; (b) sideband phasors; (c)–(f) progressive phasor addition. Part (d) shows the peak phase shift.

(a) Peak carrier phase shift in both radians and degrees.
(b) Frequency deviation for a modulating signal frequency $f_m = 15$ kHz.

Solution (a) The peak amplitude of the modulating component is

$$V_m = V_{usf} + V_{lsf}$$

$$= 0.0048 + 0.0048 = 0.0096$$

Peak phase deviation is the modulation index and can be determined by substituting into Equation 6-46a.

$$\theta = m = \arctan \frac{0.0096}{10} = 0.055°$$

$$= 0.055° \times \frac{\pi \text{ rad}}{180°} = 0.00096 \text{ rad}$$

(b) Rearranging Equation 6-13 gives us

$$\Delta f = m f_m = (0.00096)(15 \text{ kHz}) = 14.4 \text{ Hz}$$

From the phasor diagrams shown in Figure 6-24, it can be seen that the carrier amplitude is varied, which produces unwanted amplitude modulation in the output waveform, and $V_{c(\max)}$ occurs when V_{usf} and V_{lsf} are in phase with each other and with V_c. The maximum phase deviation that can be produced with this type of modulator is approximately 1.67 milliradians. Therefore, from Equation 6-13 and a maximum modulating signal frequency $f_{m(\max)} = 15$ kHz, the maximum frequency deviation possible is

$$\Delta f_{\max} = (0.00167)(15,000) = 25 \text{ Hz}$$

Frequency Modulation Transmission

From the preceding discussion it is evident that the modulation index at the output of the combining network is insufficient to produce a wideband FM frequency spectrum and therefore must be multiplied considerably before being transmitted. For the transmitter shown in Figure 6-23, a 200-kHz phase-modulated subcarrier with a peak phase deviation $m = 0.00096$ rad, only produces a frequency deviation of 14.4 Hz at the output of the combining network. To achieve 75-kHz frequency deviation at the antenna, the frequency must be multiplied by approximately 5208. However, this would produce a transmit carrier frequency at the antenna of

$$f_t = 5208 \times 200 \text{ kHz} = 1041.6 \text{ MHz}$$

which is well beyond the frequency limits for the commercial FM broadcast band. It is apparent that multiplication by itself is inadequate. Therefore, a combination of multiplying and mixing is necessary to develop the desired transmit carrier frequency with 75-kHz frequency deviation. The waveform at the output of the combining network is multiplied by 72, producing the following signal:

$$f_1 = 72 \times 200 \text{ kHz} = 14.4 \text{ MHz}$$

$$m = 72 \times 0.00096 = 0.06912 \text{ rad}$$

$$\Delta f = 72 \times 14.4 \text{ Hz} = 1036.8 \text{ Hz}$$

The output from the first multiplier is mixed with a 13.15-MHz crystal-controlled frequency (f_o) to produce a difference signal (f_2) with the following characteristics:

$$f_2 = 14.4 \text{ MHz} - 13.15 \text{ MHz} = 1.25 \text{ MHz (down-converted)}$$

$$m = 0.6912 \text{ rad (unchanged)}$$

$$\Delta f = 1036.8 \text{ Hz (unchanged)}$$

Note that only the carrier frequency is affected by the heterodyning process. The output from the mixer is once again multiplied by 72 to produce a transmit signal with the following characteristics:

$$f_t = 1.25 \text{ MHz} \times 72 = 90 \text{ MHz}$$

$$m = 0.06912 \times 72 = 4.98 \text{ rad}$$

$$\Delta f = 1036.8 \times 72 = 74,650 \text{ Hz}$$

In the preceding example with the use of both the multiplying and heterodyning processes, the carrier was increased by a factor of 450; at the same time, the frequency deviation and modulation index were increased by a factor of 5184.

With the Armstrong transmitter, the phase of the carrier is directly modulated in the combining network through summation, producing indirect frequency modulation. The magnitude of the phase deviation is directly proportional to the amplitude of the modulating signal but independent of its frequency. Therefore, the modulation index remains constant for all modulating signal frequencies of a given amplitude. For example, for the transmitter shown in Figure 6-24, if the modulating signal amplitude is held constant while its frequency is decreased to 5 kHz, the modulation index remains at 5, while the frequency deviation is reduced to $\Delta f = 5 \times 5000 = 25,000$ Hz.

FM from PM.   As shown in Figure 6-4, a PM modulator preceded by an integrator produces an FM waveform. If the PM transmitter shown in Figure 6-23 is preceded by a low-pass filter (which is an integrator), FM results. The low-pass filter is simply a $1/f$ filter, which is commonly called a *predistorter* or *frequency correction network*.

FM versus PM

From a purely theoretical viewpoint, the difference between FM and PM is quite simple: the modulation index for FM is defined differently than for PM. With PM, the modulation index is directly proportional to the amplitude of the modulating signal and independent of its frequency. With FM, the modulation index is directly proportional to the amplitude of the modulating signal and inversely proportional to its frequency.

Considering FM as a form of phase modulation, the larger the frequency deviation is, the larger the phase deviation. Therefore, the latter depends, at least to a certain extent, on the amplitude of the modulating signal, just as with PM. With PM, the modulation index is proportional to the amplitude of the modulating signal voltage only, whereas with FM the modulation index is also inversely proportional to the modulating signal frequency. If FM transmissions are received on a PM receiver, the bass frequencies would have considerably more phase deviation than a PM modulator would have given them. Since the output voltage from a PM demodulator is proportional to the phase deviation, the signal appears excessively bass-boosted. Alternatively (and this is the more practical situation), PM demodulated by an FM receiver produces an information signal in which the higher-frequency modulating signals are boosted.

QUESTIONS

6-1. Define *angle modulation*.

6-2. Define *direct FM* and *indirect FM*.

6-3. Define *direct PM* and *indirect PM*.

6-4. Define *frequency deviation* and *phase deviation*.

6-5. Define *instantaneous phase*, *instantaneous phase deviation*, *instantaneous frequency*, and *instantaneous frequency deviation*.

6-6. Define *deviation sensitivity* for a frequency modulator and for a phase modulator.

6-7. Describe the relationship between the instantaneous carrier frequency and the modulating signal for FM.

6-8. Describe the relationship between the instantaneous carrier phase and the modulating signal for PM.

6-9. Describe the relationship between frequency deviation and the amplitude and frequency of the modulating signal.

6-10. Define *carrier swing*.

6-11. Define *modulation index* for FM and for PM.

6-12. Describe the relationship between modulation index and the modulating signal for FM; for PM.

6-13. Define percent modulation for angle-modulated signals.

6-14. Describe the difference between a direct frequency modulator and a direct phase modulator.

6-15. How can a frequency modulator be converted to a phase modulator; a phase modulator to a frequency modulator?

6-16. How many sets of sidebands are produced when a carrier is frequency modulated by a single input frequency?

6-17. What are the requirements for a side frequency to be considered significant?

6-18. Define a *low*, a *medium*, and a *high* modulation index.

6-19. Describe the significance of the *Bessel* table.

6-20. State *Carson's general rule* for determining the bandwidth for an angle-modulated wave.

6-21. Define deviation ratio.

6-22. Describe the relationship between the power in the unmodulated carrier and the power in the modulated wave for FM.

6-23. Describe the significance of the FM *noise triangle*.

6-24. What effect does *limiting* have on the composite FM waveform?

6-25. Define *preemphasis* and *deemphasis*.

6-26. Describe a preemphasis network; a deemphasis network.

6-27. Describe the basic operation of a varactor diode FM generator.

6-28. Describe the basic operation of a reactance FM modulator.

6-29. Describe the basic operation of a linear integrated-circuit FM modulator.

6-30. Draw the block diagram for a Crosby direct FM transmitter and describe its operation.

6-31. What is the purpose of an AFC loop? Why is one required for the Crosby transmitter?

6-32. Draw the block diagram for a phase-locked-loop FM transmitter and describe its operation.

6-33. Draw the block diagram for an Armstrong indirect FM transmitter and describe its operation.

6-34. Compare FM to PM.

PROBLEMS

6-1. If a frequency modulator produces 5 kHz of frequency deviation for a 10-V modulating signal, determine the deviation sensitivity. How much frequency deviation is produced for a 2-V modulating signal?

6-2. If a phase modulator produces 2 rad of phase deviation for a 5-V modulating signal, determine the deviation sensitivity. How much phase deviation would a 2-V modulating signal produce?

6-3. Determine (a) the peak frequency deviation, (b) the carrier swing, and (c) the modulation index for an FM modulator with deviation sensitivity $K_1 = 4$ kHz/V and a modulating signal $v_m(t) = 10 \sin (2\pi 2000t)$. What is the peak frequency deviation produced if the modulating signal were to double in amplitude?

6-4. Determine the peak phase deviation for a PM modulator with deviation sensitivity $K = 1.5$ rad/V and a modulating signal $v_m(t) = 2 \sin (2\pi 2000t)$. How much phase deviation is produced for a modulating signal with twice the amplitude?

6-5. Determine the percent modulation for a television broadcast station with a maximum frequency deviation $\Delta f = 50$ kHz when the modulating signal produces 40 kHz of frequency deviation at the antenna. How much deviation is required to reach 100% modulation of the carrier?

6-6. From the Bessel table, determine the number of sets of sidebands produced for the following modulation indexes: 0.5, 1.0, 2.0, 5.0, and 10.0.

6-7. For an FM modulator with modulation index $m = 2$, modulating signal $v_m(t) = V_m \sin (2\pi 2000t)$, and an unmodulated carrier $v_c(t) = 8 \sin (2\pi 800kt)$:

(a) Determine the number of sets of significant sidebands.

(b) Determine their amplitudes.

(c) Draw the frequency spectrum showing the relative amplitudes of the side frequencies.

(d) Determine the bandwidth.

(e) Determine the bandwidth if the amplitude of the modulating signal increases by a factor of 2.5.

6-8. For an FM transmitter with 60-kHz carrier swing, determine the frequency deviation. If the amplitude of the modulating signal decreases by a factor of 2, determine the new frequency deviation.

6-9. For a given input signal, an FM broadcast-band transmitter has a frequency deviation $\Delta f = 20$ kHz. Determine the frequency deviation if the amplitude of the modulating signal increases by a factor of 2.5.

6-10. An FM transmitter has a rest frequency $f_c = 96$ MHz and a deviation sensitivity $K_1 = 4$ kHz/V. Determine the frequency deviation for a modulating signal $v_m(t) = 8 \sin (2\pi 2000t)$. Determine the modulation index.

6-11. Determine the deviation ratio and worst-case bandwidth for an FM signal with a maximum frequency deviation $\Delta f = 25$ kHz and a maximum modulating signal $f_{m(max)} = 12.5$ kHz.

6-12. For an FM modulator with 40-kHz frequency deviation and a modulating signal frequency $f_m = 10$ kHz, determine the bandwidth using both the Bessel table and Carson's rule.

6-13. For an FM modulator with an unmodulated carrier amplitude $V_c = 20$ V, a modulation index $m = 1$, and a load resistance $R_L = 10$ Ω, determine the power in the modulated carrier and each side frequency, and sketch the power spectrum for the modulated wave.

6-14. For an angle-modulated carrier $v_c(t) = 2 \cos (2\pi 200$ MHz $t)$ with 50 kHz of frequency deviation due to the modulating signal and a single-frequency interfering signal $Vn(t) = 0.5 \cos (2\pi 200.01$ MHz $t)$, determine:

(a) Frequency of the demodulated interference signal.

(b) Peak phase and frequency deviation due to the interfering signal.

(c) Signal-to-noise ratio at the output of the demodulator.

6-15. Determine the total peak phase deviation produced by a 5-kHz band of random noise with a peak voltage $V_n = 0.08$ V and a carrier $v_c(t) = 1.5 \sin (2\pi 40$ MHz $t)$.

6-16. For a Crosby direct FM transmitter similar to the one shown in Figure 6-21 with the following parameters, determine:

(a) Frequency deviation at the output of the VCO and the power amplifier.

(b) Modulation index at the same two points.

(c) Bandwidth at the output of the power amplifier.

$N_1 = \times 3$

$N_2 = \times 3$

$N_3 = \times 3$

Crystal reference oscillator frequency $= 13$ MHz

Reference multiplier $= \times 3$

VCO deviation sensitivity $K_1 = 450$ Hz/V

Modulating signal $v_m(t) = 3 \sin (2\pi 5 \times 10^3 t)$

VCO rest frequency $f_c = 4.5$ MHz

Discriminator resonant frequency $f_d = 1.5$ MHz

6-17. For an Armstrong indirect FM transmitter similar to the one show in Figure 6-23 with the following parameters, determine:

(a) Modulation index at the output of the combining network and the power amplifier.

(b) Frequency deviation at the same two points.

(c) Transmit carrier frequency.

Crystal carrier oscillator $= 210$ kHz

Crystal reference oscillator $= 10.2$ MHz

Sideband voltage $V_m = 0.018$ V

Carrier input voltage to combiner $V_c = 5$ V

First multiplier $= \times 40$

Second multiplier $= \times 50$

Modulating signal frequency $f_m = 2$ kHz

7

Angle Modulation Receivers and Systems

INTRODUCTION

Receivers used for angle-modulated signals are very similar to those used for conventional AM or SSB reception, except for the method used to extract the audio information from the composite IF waveform. In FM receivers, the voltage at the output of the audio detector is directly proportional to the frequency deviation at its input. With PM receivers, the voltage at the output of the audio detector is directly proportional to the phase deviation at its input. Because frequency and phase modulation both occur with either angle modulation system, FM signals can be demodulated by PM receivers, and vice versa. Therefore, the circuits used to demodulate FM and PM signals are both described under the heading "FM Receivers."

With conventional AM, the modulating signal is impressed onto the carrier in the form of amplitude variations. However, noise introduced into the system also produces changes in the amplitude of the envelope. Therefore, the noise cannot be removed from the composite waveform without also removing a portion of the information signal. With angle modulation, the information is impressed onto the carrier in the form of frequency or phase variations. Therefore, with angle modulation receivers, amplitude variations caused by noise can be removed from the composite waveform simply by *limiting* (*clipping*) the peaks of the envelope prior to detection. With angle modulation, an improvement in the signal-to-noise ratio is achieved during the demodulation process; thus, system performance in the presence of noise can be improved by limiting. In essence, this is the major advantage of angle modulation over conventional AM.

The purposes of this chapter are to introduce the reader to the basic receiver configurations and circuits used for the reception and demodulation of FM and PM signals, and to describe how they function and how they differ from conventional AM or single-sideband receivers. In addition, several FM communications systems are described, including two-way FM and cellular radio.

FM RECEIVERS

Figure 7-1 shows a simplified block diagram for a double-conversion superheterodyne FM receiver. It is very similar to a conventional AM receiver. The RF, mixer, and IF stages are almost identical to those used in AM receivers, although FM receivers generally have more IF amplification. Also, due to the noise-suppression characteristics inherent in FM receivers, RF amplifiers are often not required. However, the audio detector stage in an FM receiver is quite different from those used in AM receivers. The envelope (peak) detector used in conventional AM receivers is replaced by a *limiter, frequency discriminator,* and *deemphasis network*. The limiter circuit and deemphasis network contribute to the improvement in the S/N ratio that is achieved in the audio demodulator stage. For FM broadcast-band receivers, the first IF is a relatively high frequency (generally, 10.7 MHz) for good image-frequency rejection, and the second IF is a relatively low frequency (very often 455 kHz) that allows the IF amplifiers to have a relatively high gain and still not be susceptible to breaking into oscillations.

FM Demodulators

FM demodulators are frequency-dependent circuits that produce an output voltage that is directly proportional to the instantaneous frequency at its input ($V_{out} = \Delta f K$, where K is in volts per hertz and is the transfer function for the demodulator, and Δf is the difference between the input frequency and the center frequency of the demodulator). Several circuits are used for demodulating FM signals. The most common are the *slope detector, Foster–Seeley discriminator, ratio detector, PLL demodulator,* and *quadrature detector*. The slope detector, Foster–Seeley discriminator, and ratio detector are all forms of *tuned circuit frequency discriminators*. Tuned circuit frequency discriminators convert FM to AM and then demodulate the AM envelope with conventional peak detectors. Also, most frequency discriminators require a 180° phase inverter, an adder circuit, and one or more frequency-dependent circuits.

Slope detector. Figure 7-2a shows the schematic diagram for a *single-ended slope detector*, which is the simplest form of tuned circuit frequency discriminator. The single-ended slope detector has the most nonlinear voltage-versus-frequency characteristics and is therefore seldom used. However, its circuit operation is basic to all tuned circuit frequency discriminators.

In Figure 7-2a, the tuned circuit (L_a and C_a) produces an output voltage that is proportional to the input frequency. The maximum output voltage occurs at the resonant frequency of the tank circuit (f_c), and its output decreases proportionately as the input frequency deviates above or below f_o. The circuit is designed so that the IF center frequency (f_c) falls in the center of the most linear portion of the voltage-versus-frequency curve, as shown in Figure 7-2b. When the intermediate frequency deviates above f_c, the output voltage increases; when the intermediate frequency deviates below f_c, the output voltage decreases. Therefore, the tuned circuit converts frequency variations to amplitude variations (FM-to-AM conversion). D_i, C_i, and R_i make up a simple peak detector that

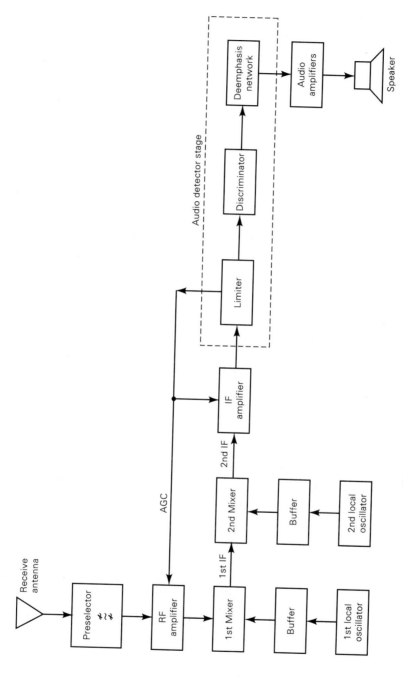

Figure 7-1 Double-conversion FM receiver block diagram.

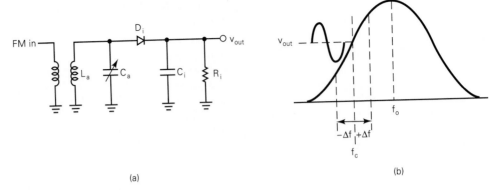

(a)   (b)

Figure 7-2 Slope detector: (a) schematic diagram; (b) voltage-versus-frequency curve.

converts the amplitude variations to an output voltage that varies at a rate equal to that of the input frequency changes and whose amplitude is proportional to the magnitude of the frequency changes.

    Balanced slope detector.   Figure 7-3a shows the schematic diagram for a *balanced slope detector*. A single-ended slope detector is a tuned circuit frequency discriminator, and a balanced slope detector is simply two single-ended slope detectors connected in parallel and fed 180° out of phase. The phase inversion is accomplished by center tapping the tuned secondary windings of transformer T_1. In Figure 7-3a, the tuned circuits (L_a, C_a, and L_b, C_b) perform the FM-to-AM conversion, and the balanced peak detectors

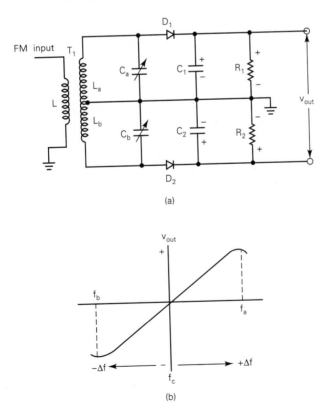

(a)

(b)

Figure 7-3 Balanced slope detector: (a) schematic diagram; (b) voltage-versus-frequency response curve.

               Chap. 7   **Angle Modulation Receivers and Systems**

(D_1, C_1, R_1 and D_2, C_2, R_2) remove the information from the AM envelope. The top tuned circuit (L_a and C_a) is tuned to a frequency (f_a) that is above the IF center frequency (f_o) by approximately $1.33 \times \Delta f$ (for the FM broadcast band this is approximately 1.33×75 kHz $= 100$ kHz). The lower tuned circuit (L_b and C_b) is tuned to a frequency (f_b) that is below the IF center frequency by an equal amount.

Circuit operation is quite simple. The output voltage from each tuned circuit is proportional to the input frequency, and each output is rectified by its respective peak detector. Therefore, the closer the input frequency is to the tank circuit resonant frequency, the greater the tank circuit output voltage. The IF center frequency falls exactly halfway between the resonant frequencies of the two tuned circuits. Therefore, at the IF center frequency, the output voltages from the two tuned circuits are equal in amplitude but opposite in polarity. Consequently, the rectified output voltage across R_1 and R_2, when added, produce a differential output voltage $V_{out} = 0$ V. When the IF deviates above resonance, the top tuned circuit produces a higher output voltage than the lower tank circuit and V_{out} goes positive. When the IF deviates below resonance, the output voltage from the lower tank circuit is larger than the output voltage from the upper tank circuit and V_{out} goes negative. The output-versus-frequency response curve is shown in Figure 7-3b.

Although the slope detector is probably the simplest FM detector, it has several inherent disadvantages, which include poor linearity, difficulty in tuning, and lack of provisions for limiting. Because limiting is not provided, a slope detector produces an output voltage that is proportional to amplitude, as well as frequency variations in the input signal, and, consequently, must be preceded by a separate limiter stage. A balanced slope detector is aligned by injecting a frequency equal to the IF center frequency and tuning C_a and C_b for 0 V at the output. Then frequencies equal to f_a and f_b are alternately injected while C_a and C_b are tuned for maximum and equal output voltages with opposite polarities.

Foster–Seeley discriminator.   A *Foster–Seeley discriminator* (sometimes called a *phase-shift discriminator*) is a tuned circuit frequency discriminator whose operation is very similar to that of the balanced slope detector. The schematic diagram for a Foster–Seeley discriminator is shown in Figure 7-4a. The capacitance value for C_c, C_1, and C_2 are chosen such that they are short circuits for the IF center frequency. Therefore, the right side of L_3 is at ac ground potential, and the IF signal (V_{in}) is fed directly (in phase) across L_3 (V_{L3}). The incoming IF is inverted 180° by transformer T_1 and divided equally between L_a and L_b. At the resonant frequency of the secondary tank circuit (the IF center frequency), the secondary current (I_s) is in phase with the total secondary voltage (V_s) and 180° out of phase with V_{L3}. Also, due to loose coupling, the primary of T_1 acts like an inductor, and the primary current I_p is 90° out of phase with V_{in}, and, since magnetic induction depends on primary current, the voltage induced in the secondary is 90° out of phase with V_{in} (V_{L3}). Therefore, V_{La} and V_{Lb} are 180° out of phase with each other and in quadrature, or 90°, out of phase with V_{L3}. The voltage across the top diode (V_{D1}) is the vector sum of V_{L3} and V_{La}, and the voltage across the bottom diode V_{D2} is the vector sum of V_{L3} and V_{Lb}. The corresponding vector diagrams are shown in Figure 7-4b. The figure shows that the voltages across D_1 and D_2 are equal. Therefore, at resonance, I_1 and I_2 are equal and C_1 and C_2 charge to equal magnitude voltages except with opposite polarities. Consequently, $V_{out} = V_{C1} - V_{C2} = 0$ V. When the IF goes above resonance ($X_L > X_C$), the secondary tank circuit impedance becomes inductive, and the secondary current lags the secondary voltage by some angle θ, which is proportional to the magnitude of the frequency deviation. The corresponding phasor diagram is shown in Figure 7-4c. The figure shows that the vector sum of the voltage across D_1 is greater than the vector sum of the voltages across D_2. Consequently, C_1 charges while C_2 discharges and V_{out} goes positive.

FM Receivers

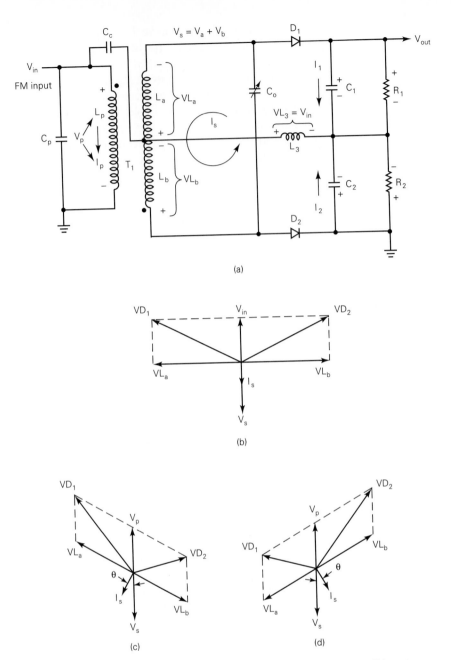

Figure 7-4 Foster–Seeley discriminator: (a) schematic diagram; (b) vector diagram, $f_{in} = f_o$; (c) vector diagram, $f_{in} > f_o$; (d) vector diagram, $f_{in} < f_o$.

When the IF goes below resonance ($X_L < X_C$), the secondary current leads the secondary voltage by some angle θ, which is again proportional to the magnitude of the change in frequency. The corresponding phasors are shown in Figure 7-4d. It can be seen that the vector sum of the voltages across D_1 is now less than the vector sum of the voltages across D_2. Consequently, C_1 discharges while C_2 charges and V_{out} goes negative. A Foster–Seeley discriminator is tuned by injecting a frequency equal to the IF center frequency and tuning C_0 for 0 V out.

Chap. 7   Angle Modulation Receivers and Systems

The preceding discussion and Figure 7-4 show that the output voltage from a Foster–Seeley discriminator is directly proportional to the magnitude and direction of the frequency deviation. Figure 7-5 shows a typical voltage-versus-frequency response curve for a Foster–Seeley discriminator. For obvious reasons, it is often called an *S-curve*. It can be seen that the output voltage-versus-frequency deviation curve is more linear than that of a slope detector, and because there is only one tank circuit, it is easier to tune. For distortionless demodulation, the frequency deviation should be restricted to the linear portion of the secondary tuned circuit frequency response curve. Like the slope detector, a Foster–Seeley discriminator responds to amplitude as well as frequency variations and therefore must be preceded by a separate limiter circuit.

Ratio detector. The *ratio detector* has one major advantage over the slope detector and Foster–Seeley discriminator for FM demodulation; a ratio detector is relatively immune to amplitude variations in its input signal. Figure 7-6a shows the schematic diagram for a ratio detector. Like the Foster–Seeley discriminator, a ratio detector has a single tuned circuit in the transformer secondary. Therefore, the operation of a ratio detector is very similar to that of the Foster–Seeley discriminator. In fact, the voltage vectors for D_1 and D_2 are identical to those of the Foster–Seeley discriminator circuit shown in Figure 7-4. However, with the ratio detector, one diode is reversed (D_2), and current (I_d) can flow around the outermost loop of the circuit. Therefore, after several cycles of the input signal, shunt capacitor C_s charges to approximately the peak voltage across the secondary winding of T_1. The reactance of C_s is low, and R_s simply provides a dc path for diode current. Therefore, the time constant for R_s and C_s is sufficiently long so that rapid changes in the amplitude of the input signal due to thermal noise or other interfering signals are shorted to ground and have no effect on the average voltage across C_s. Consequently, C_1 and C_2 charge and discharge proportional to frequency changes in the input signal and are relatively immune to amplitude variations. Also, the output voltage from a ratio detector is taken with respect to ground, and for the diode polarities shown in Figure 7-6a, the average output voltage is positive. At resonance, the output voltage is divided equally between C_1 and C_2 and redistributed as the input frequency is deviated above and below resonance. Therefore, changes in V_{out} are due to the changing ratio of the voltage across C_1 and C_2, while the total voltage is clamped by C_s.

Figure 7-6b shows the output frequency response curve for the ratio detector shown in Figure 7-6a. It can be seen that at resonance, V_{out} is not equal to 0 V but, rather, to one-half of the voltage across the secondary windings of T_1. Because a ratio detector is relatively immune to amplitude variations, it is often selected over a discriminator. However, a discriminator produces a more linear output voltage-versus-frequency response curve.

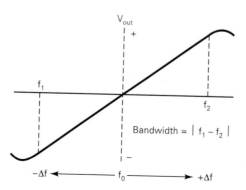

Figure 7-5 Discriminator voltage-versus-frequency response curve.

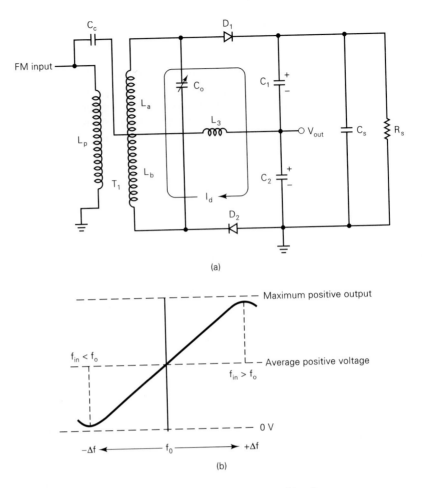

(a)

(b)

Figure 7-6 Ratio detector: (a) schematic diagram; (b) voltage-versus-frequency response curve.

Phase-locked-loop FM demodulator. Since the development of LSI linear integrated circuits, FM demodulation can be accomplished quite simply with a phase-locked loop (PLL). Although the operation of a PLL is quite involved, the operation of a *PLL FM demodulator* is probably the simplest and easiest to understand. A PLL frequency demodulator requires no tuned circuits and automatically compensates for changes in the carrier frequency due to instability in the transmit oscillator. Figure 7-7a shows the simplified block diagram for a PLL FM demodulator.

In Chapter 2, a detailed description of PLL operation was given. It was shown that after frequency lock had occurred the VCO would track frequency changes in the input signal by maintaining a phase error at the input of the phase comparator. Therefore, if the PLL input is a deviated FM signal and the VCO natural frequency is equal to the IF center frequency, the correction voltage produced at the output of the phase comparator and fed back to the input of the VCO is proportional to the frequency deviation and is thus the demodulated information signal. If the IF amplitude is sufficiently limited prior to reaching the PLL and the loop is properly compensated, the PLL loop gain is constant and equal to K_v. Therefore, the demodulated signal can be taken directly from the output of the internal buffer and is mathematically given as

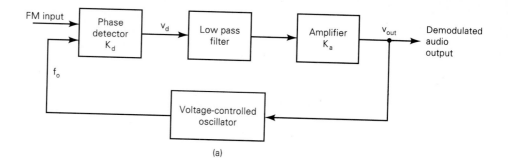

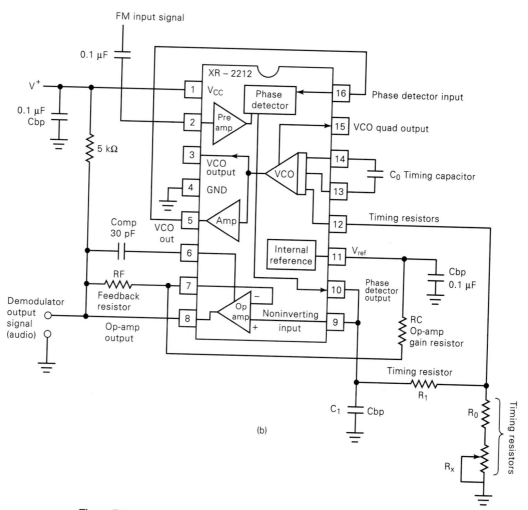

Figure 7-7 (a) Block diagram for a PLL FM demodulator; (b) PLL FM demodulator using the XR-2212 PLL.

$$V_{\text{out}} = \Delta f K_d K_a \tag{7-1}$$

Figure 7-7b shows a schematic diagram for an FM demodulator using the XR-2212. R_0 and C_0 are course adjustments for setting the VCO's free-running frequency. R_x is for

FM Receivers

fine tuning, and R_F and R_C set the internal op-amp voltage gain (K_a). The PLL closed-loop frequency response should be compensated to allow unattenuated demodulation of the entire information signal bandwidth. The PLL op-amp buffer provides voltage gain and current drive stability.

Quadrature FM demodulator. A *quadrature FM demodulator* (sometimes called a *coincidence detector*) extracts the original information signal from the composite IF waveform by multiplying two quadrature (90° out of phase) signals. A quadrature detector uses a 90° phase shifter, a single tuned circuit, and a product detector to demodulate FM signals. The 90° phase shifter produces a signal that is in quadrature with the received IF signals. The tuned circuit converts frequency variations to phase variations, and the product detector multiplies the received IF signals by the phase-shifted IF signal.

Figure 7-8 shows a simplified schematic diagram for an FM quadrature detector. C_i is a high-reactance capacitor that, when placed in series with tank circuit (R_o, L_o, and C_o), produces a 90° phase shift at the IF center frequency. The tank circuit is tuned to the IF center frequency and produces an additional phase shift (θ) that is proportional to the frequency deviation. The IF input signal (v_i) is multiplied by the quadrature signal (v_o) in the product detector and produces an output signal that is proportional to the frequency deviation. At the resonant frequency, the tank circuit impedance is resistive. However, frequency variations in the IF signal produce an additional positive or negative phase shift. Therefore, the product detector output voltage is proportional to the phase difference between the two input signals and is expressed mathematically as

$$v_{\text{out}} = v_i v_o$$

$$= [V_i \sin (\omega_i t + \theta)][V_o \cos (\omega_0 t)]$$

Substituting into the trigonometric identity for the product of a sine and a cosine wave of equal frequency gives us

$$v_{\text{out}} = \frac{V_i V_o}{2}[\sin (2\omega_i t + \theta) + \sin (\theta)]$$

The second harmonic ($2\omega_i$) is filtered out, leaving

$$v_{\text{out}} = \frac{V_i V_o}{2} \sin (\theta)$$

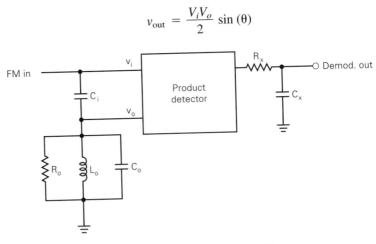

Figure 7-8 Quadrature FM demodulator.

Chap. 7   Angle Modulation Receivers and Systems

where $\theta = \tan^{-1} \rho Q$ $\rho = 2\pi f/f_o$ (fractional frequency deviation)

                                           Q = tank circuit quality factor

Amplitude Limiters and FM Thresholding

The vast majority of terrestrial FM radio communications systems use conventional non-coherent demodulation because most standard frequency discriminators use envelope detection to remove the intelligence from the FM waveform. Unfortunately, envelope detectors (including ratio detectors) will demodulate incidental amplitude variations as well as frequency variations. Transmission noise and interference add to the signal and produce unwanted amplitude variations. Also, frequency modulation is generally accompanied by small amounts of residual amplitude modulation. In the receiver, the unwanted AM and random noise interference are demodulated along with the signal and produce unwanted distortion in the recovered information signal. The noise is more prevalent at the peaks of the FM waveform and relatively insignificant during the zero crossings. A limiter is a circuit that produces a constant-amplitude output for all input signals above a pre-scribed minimum input level, which is often called the *threshold*, *quieting*, or *capture* level. Limiters are required in most FM receivers because many of the demodulators discussed earlier in this chapter demodulate amplitude as well frequency variations. With amplitude limiters, the signal to noise ratio at the output of the demodulator (postdetection) can be improved by as much as 20 dB or more over the input (predetection) signal to noise.

Essentially, an amplitude limiter is an additional IF amplifier that is overdriven. Limiting begins when the IF signal is sufficiently large that it drives the amplifier alternately into saturation and cutoff. Figure 7-9 shows the input and output waveforms for a typical limiter. In Figure 7-9b, it can be seen that for IF signals that are below threshold the AM noise is not reduced, and for IF signals above threshold there is a large reduction in the AM noise level. The purpose of the limiter is to remove all amplitude variations from the IF signal.

Figure 7-10a shows the limiter output when the noise is greater than the signal (that is, the noise has captured the limiter). The irregular widths of the serrations are caused by noise impulses saturating the limiter. Figure 7-10b shows the limiter output when the signal is sufficiently greater than the noise (the signal has captured the limiter). The peaks of the signal have the limiter so far into saturation that the weaker noise is totally eliminated. The improvement in the S/N ratio is called *FM thresholding*, *FM quieting*, or the *FM capture effect*. Three criteria must be satisfied before FM thresholding can occur:

1. The predetection signal-to-noise ratio must be 10 dB or greater.
2. The IF signal must be sufficiently amplified to overdrive the limiter.
3. The signal must have a modulation index equal to or greater than unity ($m \geq 1$).

Figure 7-11 shows typical FM thresholding curves for low ($m = 1$) and medium ($m = 4$) index signals. The output voltage from an FM detector is proportional to m^2. Therefore, doubling m increases the S/N ratio by a factor of 4 (6 dB). The quieting ratio for $m = 1$ is an input S/N = 13 dB, and for $m = 4$, 22 dB. For S/N ratios below threshold, the receiver is said to be captured by the noise, and for S/N ratios above threshold, the receiver is said to be captured by the signal. Figure 7-11 shows that IF signals at the input to the limiter with 13 dB or more S/N undergo 17 dB of S/N improvement. FM quieting begins with an input S/N ratio of 10 dB, but does not produce the full 17-dB improvement until the input signal-to-noise ratio reaches 13 dB.

FM Receivers

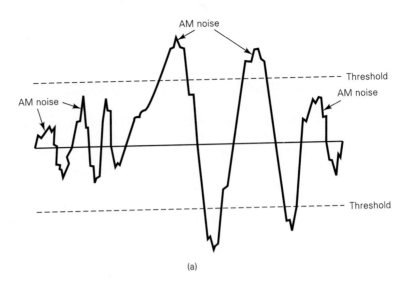

AM noise

Threshold

AM noise

AM noise

Threshold

(a)

AM noise removed

Threshold

AM noise

AM noise

Threshold

(b)

Figure 7-9 Amplitude limiter input and output waveforms: (a) input waveform; (b) output waveform.

Limiter circuits. Figure 7-12a shows a schematic diagram for a single-stage limiter circuit with a built-in output filter. This configuration is commonly called a *bandpass limiter/amplifier* (BPL). A BPL is essentially a class A biased tuned IF amplifier, and for limiting and FM quieting to occur, it requires an IF input signal sufficient enough to drive

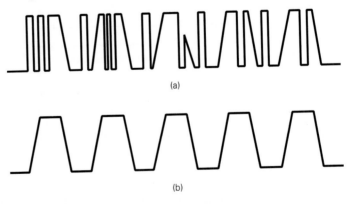

(a)

(b)

Figure 7-10 Limiter output: (a) captured by noise; (b) captured by signal.

Chap. 7   **Angle Modulation Receivers and Systems**

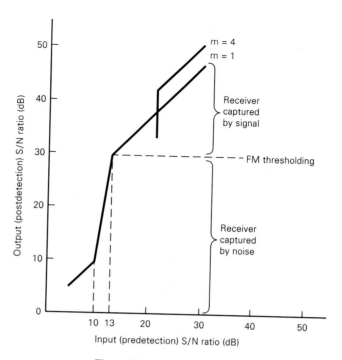

Figure 7-11 FM thresholding.

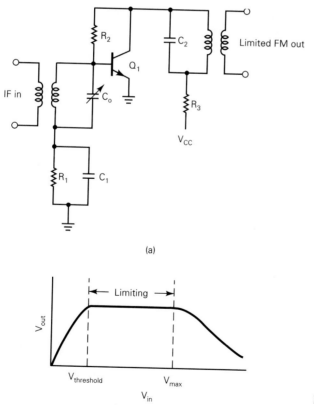

(a)

(b)

Figure 7-12 Single-stage tuned limiter: (a) schematic diagram; (b) limiter action.

FM Receivers

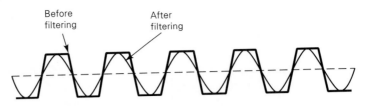

Figure 7-13 Filtered limiter output.

it into both saturation and cutoff. The output tank circuit is tuned to the IF center frequency. Filtering removes the harmonic and intermodulation distortion present in the rectangular pulses due to *hard limiting*. The effect of filtering is shown in Figure 7-13. If resistor R_2 were removed entirely, the amplifier would be biased for class C operation, which is also appropriate for this type of circuit, but requires more filtering. Figure 7-12b shows limiter action for the circuit shown in Figure 7-12a. For small signals (below the threshold voltage), no limiting occurs. When V_{in} reaches $V_{threshold}$, limiting begins, and for input amplitudes above V_{max}, there is actually a decrease in V_{out} with increases in V_{in}. This is because with high input drive levels the collector current pulses are sufficiently narrow that they actually develop less tank circuit power. The problem of overdriving the limiter can be rectified by incorporating AGC into the circuit.

When two limiter stages are used, it is called *double limiting*; three stages, *triple limiting*; and so on. Figure 7-14 shows a three-stage *cascaded limiter* without a built-in filter. This type of limiter circuit must be followed by either a ceramic or crystal filter to remove the nonlinear distortion. The limiter shown has three RC-coupled limiter stages that are dc series connected to reduce the current drain. Cascaded amplifiers combine several of the advantages of common-emitter and common-gate amplifiers. Cascading amplifiers also decrease the thresholding level and thus improve the quieting capabilities of the stage. The effects of double and triple limiting are shown in Figure 7-15. Because FM receivers have sufficient gain to saturate the limiters over a relatively large range of RF input signal levels, AGC is usually unnecessary. In fact, very often AGC actually degrades the performance of an FM receiver.

EXAMPLE 7-1

For an FM receiver with a bandwidth $B = 200$ kHz, a power noise figure NF $= 8$ dB, and an input noise temperature $T = 100$ K, determine the minimum receive carrier power necessary to achieve a postdetection signal-to-noise ratio of 37 dB. Use the receiver block diagram

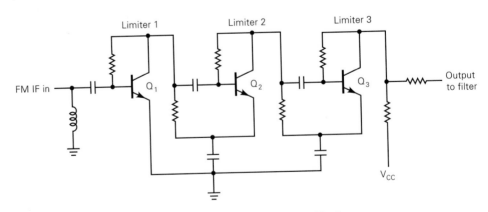

Figure 7-14 Three-stage cascaded limiter.

Chap. 7   **Angle Modulation Receivers and Systems**

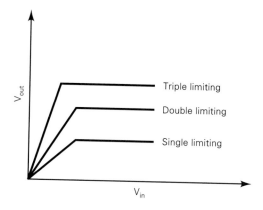

Figure 7-15 Limiter response curves.

shown in Figure 7-1 as the receiver model and the FM thresholding curve shown in Figure 7-11 for $m = 1$.

Solution From Figure 7-11, it can be seen that 17 dB of signal-to-noise improvement is evident in the detector, assuming the limiters are saturated and the input signal-to-noise is greater than 13 dB. Therefore, to achieve a postdetection signal-to-noise ratio of 37 dB, the predetection signal-to-noise ratio must be at least

$$37 \text{ dB} - 17 \text{ dB} = 20 \text{ dB}$$

Therefore, for an overall receiver noise figure equal to 8 dB, the S/N ratio at the input to the receiver must be at least

$$20 \text{ dB} + 8 \text{ dB} = 28 \text{ dB}$$

The receiver input noise power is

$$N_{\text{(dBm)}} = 10 \log \frac{KTB}{0.001} = 10 \log \frac{(1.38 \times 10^{-23})(100)(200,000)}{0.001}$$

$$N_{\text{(dBm)}} = -125.6 \text{ dBm}$$

Consequently, the minimum receiver signal power for a 28-dB S/N ratio is

$$S = -125.6 \text{ dBm} + 28 \text{ dB} = -97.6 \text{ dBm}$$

LINEAR INTEGRATED-CIRCUIT FM RECEIVERS

In recent years, several manufacturers of integrated circuits such as Signetics, RCA, and Motorola have developed reliable, low-power monolithic integrated circuits that perform virtually all the receiver functions for both AM and FM communications systems. These integrated circuits offer the advantages of being reliable, predictable, miniaturized, and easy to design with. The development of these integrated circuits is one of the primary reasons for the tremendous growth of both portable two-way FM and cellular radio communications systems that has occurred in the past few years.

Low-power Integrated-circuit FM IF System

The NE/SA614A is an improved monolithic low-power FM IF system manufactured by Signetics Corporation. The NE/SA614A is a high-gain, high-frequency device that offers low power consumption (3.3-mA typical current drain) and excellent input sensitivity

(1.5 μV across its input pins) at 455 kHz. The NE/SA614A has an onboard temperature-compensated *received signal strength indicator* (*RSSI*) with a logarithmic output and a dynamic range in excess of 90 dB. It has two audio outputs (one muted and one not). The NE/SA614A requires a low number of external components to function and meets cellular radio specifications. The NE/SA614A can be used for the following applications:

1. FM cellular radio
2. High-performance FM communications receivers
3. Intermediate frequency amplification and detection up to 25 MHz
4. RF signal strength meter
5. Spectrum analyzer applications
6. Instrumentation circuits
7. Data transceivers

The block diagram for the NE/SA614A is shown in Figure 7-16. As the figure shows, the NE/SA614A includes two limiting intermediate-frequency amplifiers, an FM quadrature detector, an audio muting circuit, a logarithmic received signal strength indicator (RSSI), and a voltage regulator. The NE/SA614A is an IF signal-processing system suitable for frequencies as high as 21.4 MHz.

IF amplifiers. Figure 7-17 shows the equivalent circuit for the NE/SA614A. The IF amplifier section consists of two log-limiting amplifier stages. The first consists of two differential amplifiers with 39 dB of gain and a small-signal ac bandwidth of 41 MHz when driven from a 50-Ω source. The output of the first limiter is a low-impedance emitter follower with 1 kΩ of equivalent series resistance. The second limiting stage consists of three differential amplifiers with a total gain of 62 dB and a small-signal ac bandwidth of 28 MHz. The outputs of the final differential amplifier are buffered to the internal quadrature detector. One output is available to drive an external quadrature capacitor and L/C quadrature tank. Both limiting stages are dc biased with feedback. The buffered outputs of

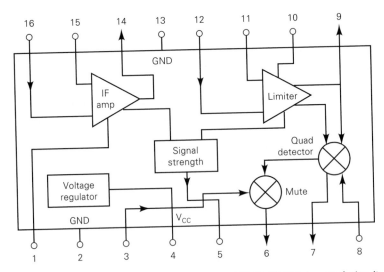

Figure 7-16 Block diagram for the Signetics NE/SA614A integrated-circuit low-power FM IF system.

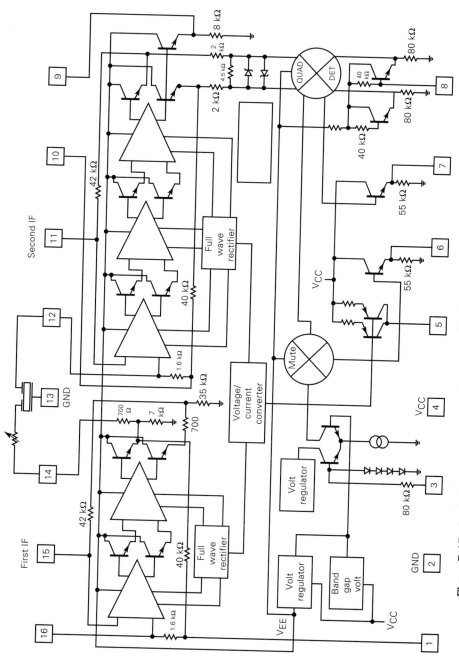

Figure 7-17 Equivalent circuit for the Signetics NE/SA614A integrated circuit low-power FM IF system.

287

the final differential amplifier in each stage are fed back to the input of that stage through a 42-kΩ resistor. Because of the very high gain, wide bandwidth, and high input impedance of the limiters, the limiter stage is potentially unstable at IF frequencies above 455 kHz. The stability can be improved by reducing the gain. This is accomplished by adding attenuators between amplifier stages. The IF amplifiers also feature low phase shift (typically only a few degrees over a wide range of input frequencies).

Quadrature detector. Figure 7-18 shows the block diagram for the equivalent circuit for the quadrature detector in the NE/SA614A. A quadrature detector is a multiplier cell similar to a mixer stage, but instead of mixing two different frequencies, it mixes two signals with the same frequencies but with different phases. A constant-amplitude (amplitude-limited) signal is applied to the lower part of the multiplier. The same signal is applied single-ended to an external capacitor connected to pin 9. There is a 90° phase shift across the plates of the capacitor. The phase-shifted signal is applied to the upper port of the multiplier at pin 8. A quadrature tank (a parallel *LC* network) permits frequency selective phase shifting at the IF signal. The quadrature detector produces an output signal whose amplitude is proportional to the magnitude of the frequency deviation of the input FM signal.

Audio outputs. The NE/SA614A has two audio outputs. Both are PNP current-to-voltage converters with 55-kΩ nominal internal loads. The unmuted output is always active to permit the use of signaling tones such as for cellular radio. The other output can be muted with 70-dB typical attenuation. The two outputs have an internal 180° phase difference and can therefore be applied to the differential inputs of an op-amp amplifier or comparator. Once the threshold of the reference frequency has been established, the two output amplitudes will shift in opposite directions as the input frequency shifts.

RSSI. The received signal-strength indicator demonstrates a monotonic logarithmic output over a range of 90 dB. The signal-strength output is derived from the summed stage currents in the limiting amplifiers. It is essentially independent of the IF frequency. Thus, unfiltered signals at the limiter input, such as spurious products or regenerated signals, will manifest themselves as an RSSI output. At low frequencies, the RSSI makes an excellent logarithmic ac voltmeter. The RSSI output is a current-to-voltage converter similar to the audio outputs.

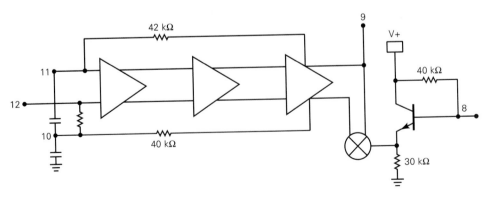

Figure 7-18 Quadrature detector block diagram.

Low-voltage High-performance Mixer FM IF System

The NE/SA616 is a low-voltage, high-performance monolithic FM IF system similar to the NE/SA614A except with the addition of a mixer/oscillator circuit. The NE/SA616 will operate at frequencies up to 150 MHz and with as little as 2.7 Vdc. The NE/SA616 features low power consumption, a mixer conversion power gain of 17 dB at 45 MHz, 102 dB of IF amplifier/limiter gain, and a 2 MHz IF amplifier/limiter small-signal ac bandwidth. The NE/SA614A can be used for the following applications:

1. Portable FM cellular radio
2. Cordless telephones
3. Wireless communications systems
4. RF signal-strength meter
5. Spectrum analyzer applications
6. Instrumentation circuits
7. Data transceivers
8. Log amps
9. Single-conversion VHF receivers

The block diagram for the NE/SA616 is shown in Figure 7-19. The NE/SA616 is similar to the NE/SA614A with the addition of a mixer and local oscillator stage. The input stage is a Gilbert cell mixer with an oscillator. Typical mixer characteristics include a noise figure of 6.2 dB, conversion gain of 17 dB, and input third-order intercept of −9 dBm. The oscillator will operate in excess of 200 MHz in an *LC* tank circuit configuration. The output impedance of the mixer is a 1.5-kΩ resistor, permitting direct connection to a 455-kHz ceramic filter. The IF amplifier has 43 dB of gain and a 5.5-MHz bandwidth. The IF limiter has 60 dB of gain and a 4.5-MHz bandwidth. The quadrature detector also uses a Gilbert cell. One port of the cell is internally driven by the IF signal, and the other output of the IF is ac-coupled to a tuned quadrature network, where it undergoes a 90° phase shift before being fed back to the other port of the Gilbert cell. The demodulator output of the quadrature detector drives an internal op-amp. The op-amp can

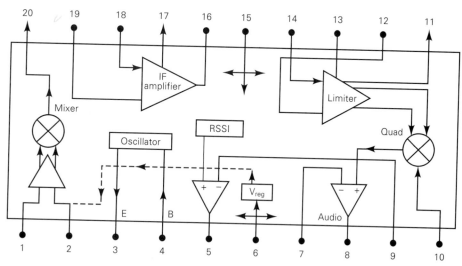

Figure 7-19 Block diagram for the Signetics NE/SA616 monolithic FM IF system.

Linear Integrated-circuit FM Receivers

be configured as a unity gain buffer, or for simultaneous gain, filtering, and second-order temperature compensation if needed.

Single-chip FM Radio System

The TDA7000 is a monolithic integrated-circuit FM radio system manufactured by Signetics Corporation for monophonic FM portable radios. In essence, the TDA7000 is a complete FM radio receiver on a single integrated-circuit chip. The TDA7000 features

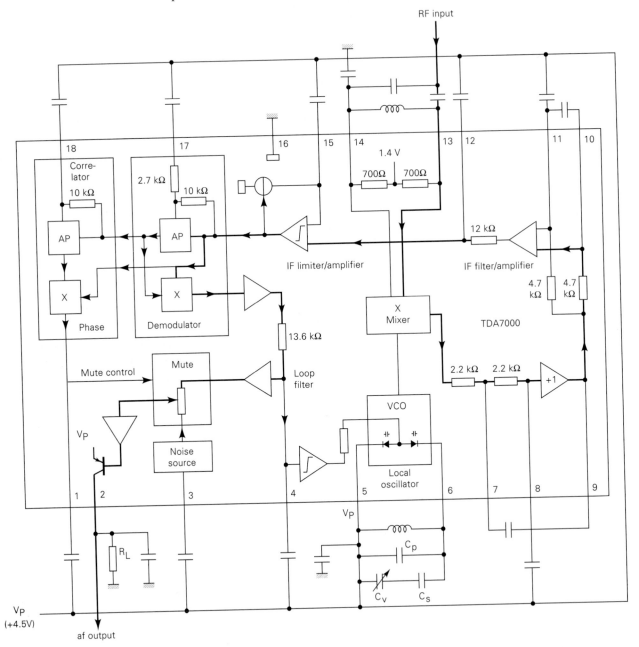

Figure 7-20 Block diagram for the Signetics TDA7000 integrated-circuit FM radio

small size, lack of IF coils, easy assembly, and low power consumption. External to the IC is only one tunable *LC* tank circuit for the local oscillator, a few inexpensive ceramic plate capacitors, and one resistor. Using the TDA7000, a complete FM radio can be made small enough to fit inside a calculator, cigarette lighter, key-ring fob, or even a slim watch. The TDA7000 can also be used in equipment such as cordless telephones, radio-controlled models, paging systems, or the sound channel of a television receiver.

The block diagram for the TDA7000 is shown in Figure 7-20. The TDA7000 includes the following functional blocks: RF input stage, mixer, local oscillator, IF amplitude/limiter, phase demodulator, mute detector, and mute switch. The IC has an internal FLL (frequency-locked-loop) system with an intermediate frequency of 70 MHz. The FLL is used to reduce the total harmonic distortion (THD) by compressing the IF frequency swing (deviation). This is accomplished by using the audio output from the FM demodulator to shift the local oscillator frequency in opposition to the IF deviation. The principle is to compress 75 kHz of frequency deviation down to approximately 15 kHz. This limits the total harmonic distortion to 0.7% with ± 22.5-kHz deviation and to 2.3% with ± 75-kHz deviation. The IF selectivity is obtained with active *RC* Sallen–Key filters. The only function that needs alignment is the resonant circuit for the oscillator.

FM STEREO BROADCASTING

Until 1961, all commercial FM broadcast-band transmissions were *monophonic*. That is, a single 50-Hz to 15-kHz audio channel made up the entire voice and music information frequency spectrum. This single audio channel modulated a high-frequency carrier and was transmitted through a 200-kHz-bandwidth FM communications channel. With *mono* transmission, each speaker assembly at the receiver reproduces exactly the same information. It is possible to separate the information frequencies with special speakers, such as *woofers* for low frequencies and *tweeters* for high frequencies. However, it is impossible to separate monophonic sound *spatially*. The entire information signal sounds as though it is coming from the same direction (that is, from a *point source*, there is no directivity to the sound). In 1961, the FCC authorized *stereophonic* transmission for the commercial FM broadcast band. With stereophonic transmission, the information signal is spatially divided into two 50-Hz to 15-kHz audio channels (a left and a right). Music that originated on the left side is reproduced only on the left speaker, and music that originated on the right side is reproduced only on the right speaker. Therefore, with stereophonic transmission, it is possible to reproduce music with a unique directivity and spatial dimension that before was possible only with live entertainment (that is, from an *extended* source). Also, with stereo transmission, it is possible to separate music or sound by *tonal quality*, such as percussion, strings, horns, and so on.

A primary concern of the FCC before authorizing stereophonic transmission was its compatibility with monophonic receivers. Stereo transmission was not to affect mono reception. Also, monophonic receivers must be able to receive stereo transmission as monaural without any perceptible degradation in program quality. In addition, stereophonic receivers were to receive stereo programming with nearly perfect separation (40 dB or more) between the left and right channels.

The original FM audio spectrum is shown in Figure 7-21a. The audio channel extended from 50 Hz to 15 kHz. In 1955, the FCC approved subcarrier transmission under the Subsidiary Communications Authorization (SCA). SCA is used to broadcast uninterrupted music to private subscribers, such as department stores, restaurants, and medical

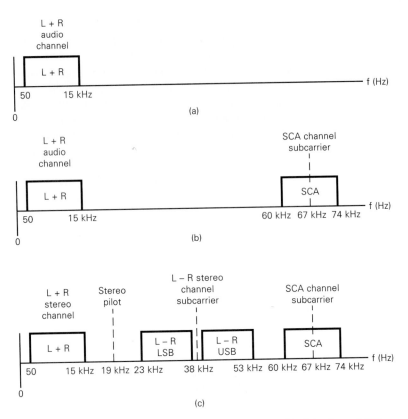

Figure 7-21 FM baseband spectrum: (a) prior to 1955; (b) prior to 1961; (c) since 1961.

offices equipped with special SCA receivers. This is the music we sometimes cordially refer to as "elevator music." Originally, the SCA subcarrier ranged from 25 to 75 kHz, but has since been standardized at 67 kHz. The subcarrier and its associated sidebands become part of the total signal that modulates the main carrier. At the receiver, the subcarrier is demodulated along with the primary channel, but cannot be heard because of its high frequency. The process of placing two or more independent channels next to each other in the frequency domain (stacking the channels), and then modulating a single high-frequency carrier with the combined signal is called *frequency division multiplexing* (*FDM*). With FM stereophonic broadcasting, three voice or music channels are frequency-division multiplexed onto a single FM carrier. Figure 7-21b shows the total baseband frequency spectrum for FM broadcasting prior to 1961 (the composite baseband comprises the total modulating signal spectrum). The primary audio channel remained at 50 Hz to 15 kHz, while an additional SCA channel is frequency translated to the 50- to 74-kHz passband. The SCA subcarrier may be AM single- or double-sideband transmission or FM with a maximum modulating signal frequency of 7 kHz. However, the SCA modulation of the main carrier is low-index narrowband FM and, consequently, is a much lower quality transmission than the primary FM channel. The total frequency deviation remained at 75 kHz with 90% (67.5 kHz) reserved for the primary channel and 10% (7.5 kHz) reserved for SCA.

Figure 7-21c shows the FM baseband frequency spectrum as it has been since 1961. It comprises the 50-Hz to 15-kHz stereo channel plus an additional stereo channel frequency division multiplexed into a composite baseband signal with a 19-kHz pilot. The three channels are (1) the left (L) plus the right (R) audio channels (the L + R stereo channel), (2) the left plus the inverted right audio channels (the L − R stereo channel), and (3) the SCA subcarrier and its associated sidebands. The L + R stereo channel occupies the 0- to 15-kHz passband (in essence, the unaltered L and R audio information combined). The L − R audio channel amplitude modulates a 38-kHz subcarrier and produces the L − R stereo channel, which is a double-sideband suppressed carrier signal that occupies the 23- to 53-kHz passband, used only for FM stereo transmission. SCA transmissions occupy the 60- to 74-kHz frequency spectrum. The information contained in the L + R and L − R stereo channels is identical except for their phase. With this scheme, mono receivers can demodulate the total baseband spectrum, but only the 50- to 15-kHz L + R audio channel is amplified and fed to all its speakers. Therefore, each speaker reproduces the total original sound spectrum. Stereophonic receivers must provide additional demodulation of the 23- to 53-kHz L − R stereo channel, separate the left and right audio channels, and then feeds them to their respective speakers. Again, the SCA subcarrier is demodulated by all FM receivers, although only those with special SCA equipment further demodulate the subcarrier to audio frequencies.

With stereo transmission, the maximum frequency deviation is still 75 kHz; 7.5 kHz (10%) is reserved for SCA transmission and another 7.5 kHz (10%) is reserved for a 19-kHz stereo pilot. This leaves 60 kHz of frequency deviation for the actual stereophonic transmission of the L + R and L − R stereo channels. However, the L + R and L − R stereo channels are not necessarily limited to 30-kHz frequency deviation each. A rather simple but unique technique is used to interleave the two channels such that at times either the L + R or the L − R stereo channel may deviate the main carrier 60 kHz by themselves. However, the total deviation will never exceed 60 kHz. This interleaving technique is explained later in this chapter.

FM Stereo Transmission

Figure 7-22 shows a simplified block diagram for a stereo FM transmitter. The L and R audio channels are combined in a matrix network to produce the L + R and L − R audio channels. The L − R audio channel modulates a 38-kHz subcarrier and produces a 23- to 53-kHz L − R stereo channel. Because there is a time delay introduced in the L − R signal path as it propagates through the balanced modulator, the L + R stereo channel must be artificially delayed somewhat to maintain phase integrity with the L − R stereo channel for demodulation purposes. Also for demodulation purposes, a 19-kHz pilot is transmitted rather than the 38-kHz subcarrier because it is considerably more difficult to recover the 38-kHz subcarrier in the receiver. The composite baseband signal is fed to the FM transmitter, where it modulates the main carrier.

L + R and L − R channel interleaving. Figure 7-23 shows the development of the composite stereo signal for equal-amplitude L and R audio channel signals. For illustration purposes, rectangular waveforms are shown. Table 7-1 is a tabular summary of the individual and total signal voltages for Figure 7-23. Note that the L − R audio channel does not appear in the composite waveform. The L − R audio channel modulates the 38-kHz subcarrier to form the L − R stereo sidebands, which are part of the composite spectrum.

For the FM modulator in this example, it is assumed that 10 V of baseband signal

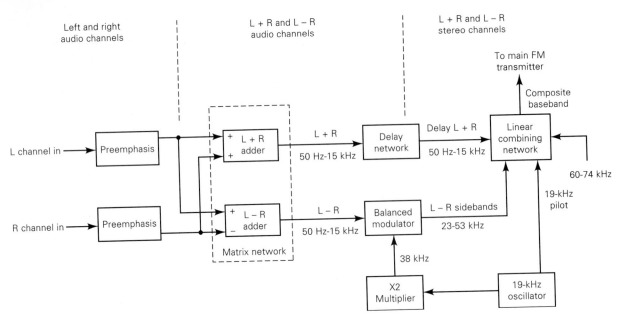

Figure 7-22 Stereo FM transmitter using frequency-division multiplexing.

will produce 75 kHz of frequency deviation of the main carrier, and the SCA and 19-kHz pilot polarities shown are for maximum frequency deviation. The L and R audio channels are each limited to a maximum value of 4 V; 1 V is for SCA, and 1 V is for the 19-kHz stereo pilot. Therefore, 8 V is left for the L + R and L − R stereo channels. Figure 7-19 shows the L, R, L + R, and L − R channels, the SCA and 19-kHz pilot, and the composite stereo waveform. It can be seen that the L + R and L − R stereo channels interleave and never produce more than 8 V of total amplitude and therefore never produce more than 60 kHz of frequency deviation. The total composite baseband never exceeds 10 V (75-kHz deviation).

Figure 7-24 shows the development of the composite stereo waveform for unequal values for the L and R signals. Again, it can be seen that the composite stereo waveform never exceeds 10 V or 75 kHz of frequency deviation. For the first set of waveforms, it appears that the sum of the L + R and L − R waveforms completely cancels. Actually, this is not true; it only appears that way because rectangular waveforms are used in this example.

FM Stereo Reception

FM stereo receivers are identical to standard FM receivers up to the output of the audio detector stage. The output of the discriminator is the total baseband spectrum that was shown in Figure 7-21c.

Figure 7-25 shows a simplified block diagram for an FM receiver that has both mono and stereo audio outputs. In the mono section of the signal processor, the L + R stereo channel, which contains all of the original information from both the L and R audio channels, is simply filtered, amplified, and then fed to both the L and R speakers. In the stereo section of the signal processor, the baseband signal is fed to a stereo demodulator where the L and R audio channels are separated and then fed to their respective speakers. The L + R and L − R stereo channels and the 19-kHz pilot are separated from the com-

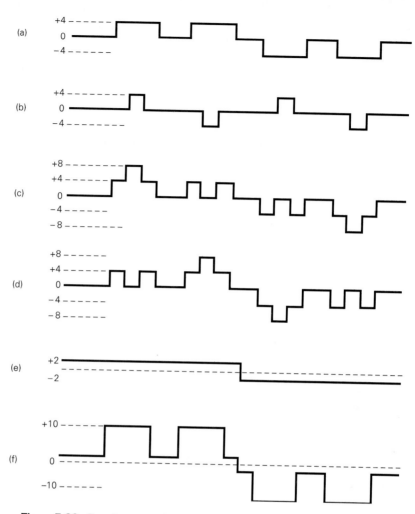

Figure 7-23 Development of the composite stereo signal for equal-amplitude L and R signals: (a) L audio signal; (b) R audio signal; (c) L + R stereo channel; (d) L − R stereo channel; (e) SCA + 19-kHz pilot; (f) composite baseband waveform.

posite baseband signal with filters. The 19-kHz pilot is filtered with a high-Q bandpass filter, multiplied by 2, amplified, and then fed to the L − R demodulator. The L + R stereo channel is filtered off by a low-pass filter with an upper cutoff frequency of 15 kHz. The L − R double-sideband signal is separated with a broadly tuned bandpass filter and then

TABLE 7-1 COMPOSITE FM VOLTAGES

L	R	L + R	L − R	SCA and pilot	Total
0	0	0	0	2	2
4	0	4	4	2	10
0	4	4	−4	2	2
4	4	8	0	2	10
4	−4	0	8	2	10
−4	4	0	−8	−2	−10
−4	−4	−8	0	−2	−10

FM Stereo Broadcasting

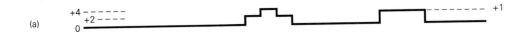

(a)

(b)

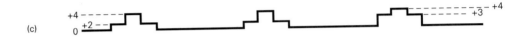

(c)

(d)

(e)

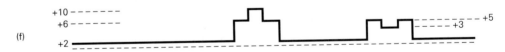

(f)

Figure 7-24 Development of the composite stereo signal for unequal amplitude L and R signals: (a) L audio signal; (b) R audio signal; (c) L + R stereo channel; (d) L − R stereo channel; (e) SCA + 19-kHz pilot; (f) composite baseband waveform.

mixed with the recovered 38-kHz carrier in a balanced modulator to produce the L − R audio information. The matrix network combines the L + R and L − R signals in such a way as to separate the L and R audio information signals, which are fed to their respective deemphasis networks and speakers.

Figure 7-26 shows the block diagram for a stereo matrix decoder. The L − R audio channel is added directly to the L + R audio channel. The output from the adder is

$$
\begin{array}{r}
L + R \\
+\ (L - R) \\
\hline
2L
\end{array}
$$

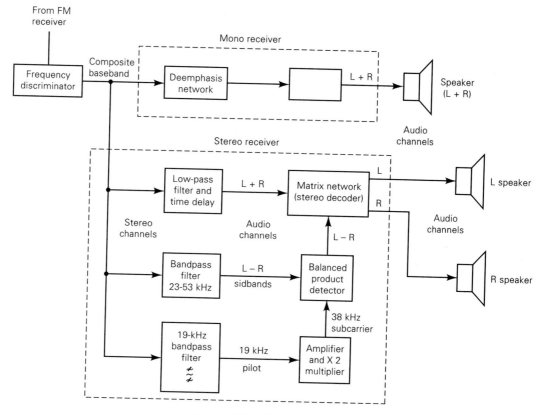

Figure 7-25 FM stereo and mono receiver.

The L − R audio channel is inverted and then added to the L + R audio channel. The output from the adder is

$$
\begin{array}{r}
L + R \\
- (L - R) \\
\hline
2R
\end{array}
$$

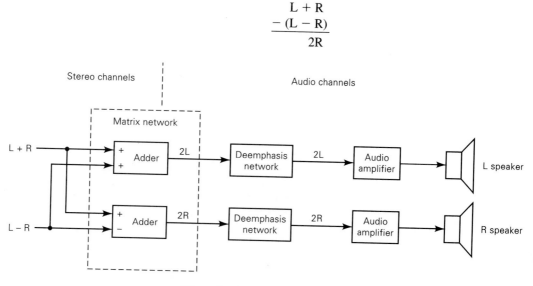

Figure 7-26 Stereo matrix network decoder.

FM Stereo Broadcasting

Large-scale integration stereo demodulator. Figure 7-27 shows the specification sheet for the XR-1310 stereo demodulator/decoder. The XR-1310 is a monolithic FM stereo demodulator that uses phase-locked loop techniques to derive the right and left audio channels from the composite stereo signal. The XR-1310 uses a phase-locked loop to lock onto the 19-kHz pilot and regenerate the 38-kHz subcarrier. The XR-1310 requires no external *LC* tank circuits for tuning, and alignment is accomplished with a single potentiometer. The XR-1310 features simple noncritical tuning, excellent channel separation, low distortion, and a wide dynamic range.

STEREO DEMODULATOR

GENERAL DESCRIPTION

The XR-1310 is a unique FM stereo demodulator which uses phase-locked techniques to derive the right and left audio channels from the composite signal. Using a phase-locked loop to regenerate the 38 kHz subcarrier, it requires no external L-C tanks for tuning. Alignment is accomplished with a single potentiometer.

FEATURES

Requires No Inductors
Low External Part Count
Simple, Noncritical Tuning by Single Potentiometer Adjustment
Internal Stereo/Monaural Switch with 100 mA Lamp Driving Capability
Wide Dynamic Range: 600 mV (RMS) Maximum Composite Input Signal
Wide Supply Voltage Range: 8 to 14 Volts
Excellent Channel Separation
Low Distortion
Excellent SCA Rejection

FUNCTIONAL BLOCK DIAGRAM   March 1982

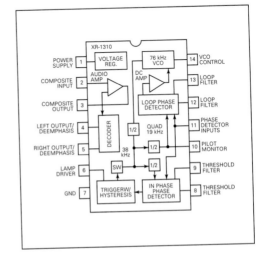

ORDERING INFORMATION

Part Number	Package	Operating Temperature
XR-1310CP	Plastic	−40°C to +85°C

APPLICATIONS

FM Stereo Demodulation

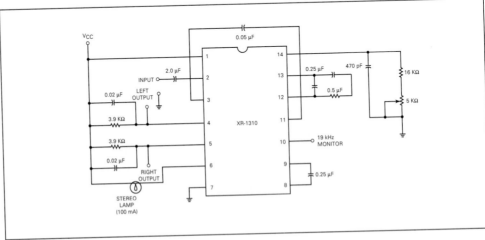

Figure 1. Typical Application

Figure 7-27 XR-1310 stereo demodulator. (Courtesy of EXAR Corporation.)
(Continued on next page.)

ELECTRICAL CHARACTERISTICS

Test conditions: Unless otherwise noted; V_{CC}* = +12Vdc, T_A = +25°C, 560mV(RMS)(2.8Vp-p) standard multiplex composite signal with L or R channel only modulated at 1.0 kHz and with 100 mV (RMS) (10 % pilot level), using circuit of Figure 1.

PARAMETERS	MIN.	TYP.	MAX.	UNIT
Maximum Standard Composite Input Signal (0.5 % THD)	2.8			V (p-p)
Maximum Monural Input Signal (1.0 % THD)	2.8			V (p-p)
Input Impedance		50		kΩ
Stereo Channel Separation (50 Hz – 15 kHz)	30	40		dB
Audio Output Voltage (desired channel)		485		mV (rms)
Monaural Channel Balance (pilot tone "off")			1.5	dB
Total Harmonic Distortion		0.3		%
Ultrasonic Frequency Rejection 19 kHz	50	34.4		dB
38 kHz		45		
Inherent SCA Rejection		80		dB
(f = 67 kHz; 9.0 kHz beat note measured with 1.0 kHz				
modulation "off")				
Stereo Switch Level				
(19 kHz input for lamp "on")	13		20	mV (rms)
Hysteresis		6		dB
Capture Range (permissable tuning error of internal oscillator,		±3.5		%
reference circuit values of Figure 1)				
Operating Supply Voltage (loads reduced to 2.7 kΩ for 8.0-volt operation	8.0		14	V (dc)
Current Drain (lamp "off")		13		mA (dc)

*Symbols conform to JEDEC Engineering Bulletin No. 1 when applicable.

ABSOLUTE MAXIMUM RATINGS

(TA = +25°C unless otherwise noted)

Power Supply Voltage	14 V
Lamp Current	75 mA
(nominal rating, 12 V lamp)	

Power Dissipation	625 mW
(package limitation)	
Derate above TA = +25°C	5.0 mW/°C
Operating Temperayure	−40 to +85°C
Range (Ambient)	
Storage Temperature Range	−65 to +150°C

Figure 7-27 *(continued)*

TWO-WAY FM RADIO COMMUNICATIONS

Two-way FM radio communication is used extensively for *public safety* mobile communications, such as police and fire departments and emergency medical services. Three primary frequency bands are allocated by the FCC for two-way FM radio communications: 132 to 174 MHz, 450 to 470 MHz, and 806 to 947 MHz. The maximum frequency deviation for two-way FM transmitters is typically 5 kHz, and the maximum modulating signal frequency is 3 kHz. These values give a deviation ratio of 1.67 and a maximum Bessel bandwidth of approximately 24 kHz. However, the allocated FCC channel spacing is 30 kHz. Two-way FM radio is half-duplex, which supports two-way communications but not simultaneously; only one party can transmit at a time. Transmissions are initiated by closing a *push-to-talk* (PTT) switch, which turns on the transmitter and shuts off the receiver. During idle conditions, the transmitter is shut off and the receiver is turned on to allow monitoring the radio channel for transmissions from other stations' transmitters.

Two-way FM Radio Communications

Two-way FM Radio Transmitter

The simplified block diagram for a *modular integrated circuit* two-way indirect FM radio transmitter is shown in Figure 7-28. Indirect FM is generally used because direct FM transmitters do not have the frequency stability necessary to meet FCC standards without using AFC loops. The transmitter shown is a four-channel unit that operates in the 150- to 174-MHz frequency band. The channel selector switch applies power to one of four crystal oscillator modules that operates at a frequency between 12.5 and 14.5 MHz, depending on the final transmit carrier frequency. The oscillator frequency is temperature compensated by the compensation module to ensure a stability of $\pm 0.0002\%$. The phase modulator uses a varactor diode that is modulated by the audio signal at the output of the audio limiter. The audio signal amplitude is limited to ensure that the transmitter is not overdeviated. The modulated IF carrier is amplified and then multiplied by 12 to produce the desired RF carrier frequency. The RF signal is further amplified and filtered prior to transmission. The *electronic push-to-talk* (PTT) is used rather than a simple mechanical switch to reduce the static noise associated with *contact bounce* in mechanical switches. Keying the PTT applies dc power to the selected transmit oscillator module and the RF power amplifiers.

Figure 7-29 shows the schematic diagram for a typical electronic PTT module. Keying the PTT switch grounds the base of Q_1, causing it to conduct and turn off Q_2. With Q_2 off, V_{CC} is applied to the transmitter and removed from the receiver. With the PTT switch released, Q_1 shuts off, removing V_{CC} from the transmitter, turning on Q_2, and applying V_{CC} to the receiver.

Transmitters equipped with VOX (*voice-operated transmitter*) are automatically keyed each time the operator speaks into the microphone, regardless of whether or not the PTT button is depressed. Transmitters equipped with VOX require an external microphone. The schematic diagram for a typical VOX module is shown in Figure 7-30. Audio signal power in the 400- to 600-Hz passband is filtered and amplified by Q_1, Q_2, and Q_3. The output from Q_3 is rectified and used to turn on Q_4, which places a ground on the PTT circuit, enabling the transmitter and disabling the receiver. With no audio input signal, Q_4 is off and the PTT pin is open, disabling the transmitter and enabling the receiver.

Two-way FM Radio Receiver

The block diagram for a typical two-way FM radio receiver is shown in Figure 7-31. The receiver shown is a four-channel integrated-circuit modular receiver with four separate crystal oscillator modules. Whenever the receiver is on, one of the four oscillator modules is activated, depending on the position of the channel selector switch. The oscillator frequency is temperature compensated and then multiplied by 9. The output from the multiplier is applied to the mixer, where it heterodynes with the incoming RF signal to produce a 20-MHz intermediate frequency. This receiver uses low-side injection, and the crystal oscillator frequency is determined as follows:

$$\text{crystal frequency} = \frac{\text{RF frequency} - 20\text{ MHz}}{9}$$

The IF signal is filtered, amplified, limited, and then applied to the frequency discriminator for demodulation. The discriminator output voltage is amplified and then

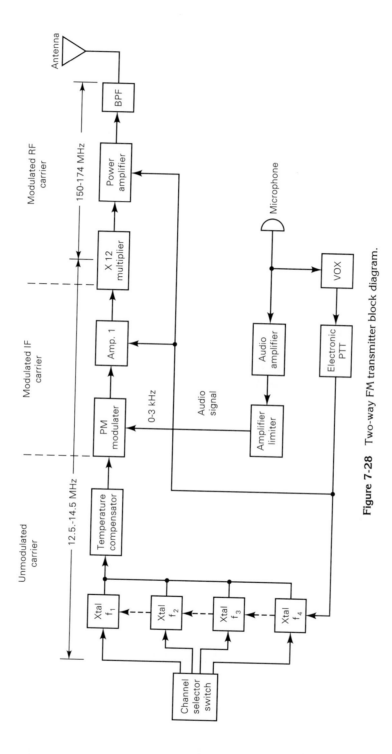

Figure 7-28 Two-way FM transmitter block diagram.

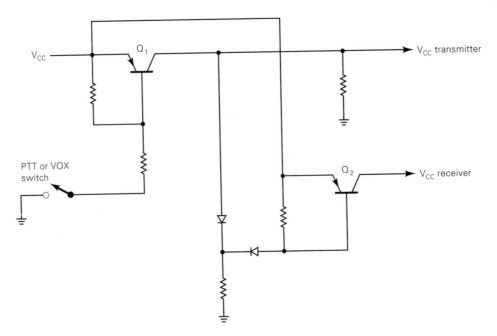

Figure 7-29 Electronic PTT schematic diagram.

applied to the speaker. A typical noise amplifier/squelch circuit is shown in Figure 7-32. The squelch circuit is keyed by out-of-band noise at the output of the audio amplifier. With no receive RF signal, AGC causes the gain of the IF amplifiers to increase to maximum, which increases the receiver noise in the 3- to 5-kHz band. Whenever excessive

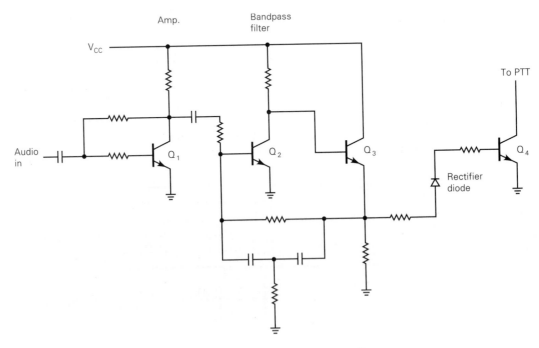

Figure 7-30 VOX schematic diagram.

　　Chap. 7　**Angle Modulation Receivers and Systems**

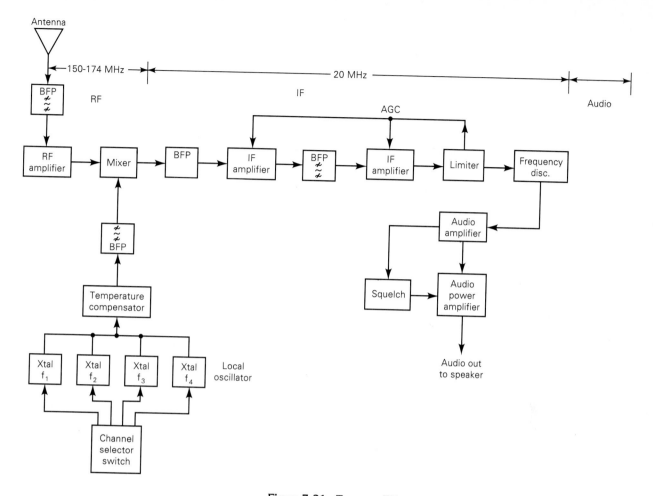

Figure 7-31 Two-way FM receiver block diagram.

noise is present, the audio amplifier is turned off and the receiver is quieted. The input bandpass filter passes the 3- to 5-kHz noise signal, which is amplified and rectified. The rectified output voltage determines the off/on condition of squelch switch Q_3. When Q_3 is on, V_{CC} is applied to the audio amplifier. When Q_3 is off, V_{CC} is removed from the audio amplifier, quieting the receiver. R_x is a squelch sensitivity adjustment.

MOBILE TELEPHONE SERVICE

Mobile telephone service is best described by explaining the differences between it and two-way radio. As previously stated, mobile radio is half-duplex and all transmissions (unless scrambled) can be heard by any listener tuned to that channel. Mobile radio is a *one-to-many* communications system. Mobile telephone uses full-duplex transmission and operates much the same as the *wireline* telephone service provided by local telephone companies. Mobile telephone permits two-way simultaneous transmission and, for privacy, each mobile unit is assigned a unique telephone number. Coded transmissions from the base station activate only the intended receiver.

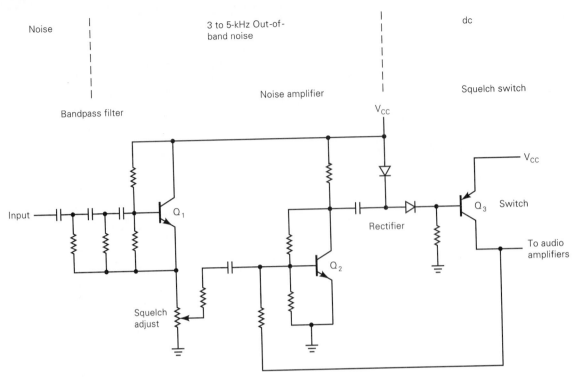

Figure 7-32 Squelch circuit.

Historical Perspective

Mobile radio was used as early as 1921 when the Detroit Police Department used a mobile radio system that operated at a frequency close to 2 MHz. In 1940, the FCC made available new frequencies for mobile radio in the 30- to 40-MHz frequency band. However, not until researchers developed frequency modulation techniques to improve reception in the presence of electrical noise and signal fading did mobile radio become useful. The first commercial mobile telephone system in the United States was established in 1946 in St. Louis, Missouri, when the FCC allocated six 60-kHz mobile telephone channels in the 150-MHz frequency range. In 1947, a public mobile telephone system was established along the highway between New York City and Boston that operated in the 35- to 40-MHz frequency range. In 1949, the FCC authorized six additional mobile channels to *radio common carriers*, which they defined as companies that do not provide public wireline telephone service but do interconnect to the public telephone network and provide equivalent *nonwireline* telephone service. The FCC later increased the number of channels from 6 to 11 by reducing the bandwidth to 30 kHz and spacing the new channels between the old ones. In 1950, the FCC added 12 new channels in the 450-MHz band.

Until 1964, mobile telephone systems operated only in the *manual mode*; a special mobile telephone operator handled every call to and from each *mobile unit*. In 1964, *automatic channel selection systems* were placed in service for mobile telephone systems. This eliminated the need for push-to-talk operation and allowed customers to *direct dial* their calls without the aid of an operator. *Automatic call completion* was extended to the 450-MHz band in 1969, and *improved mobile telephone systems* (IMTS) became the United States' standard mobile telephone service. Presently, there are more than 200,000 *mobile telephone service* (MTS) subscribers nationwide. MTS uses FM radio channels to establish communication links between mobile telephones and central *base station* transceivers, which are linked to the local telephone exchange via normal metallic telephone lines. Most MTS systems serve an area approximately 40 miles in diameter, and each channel operates similarly to a *party line*. Each channel may be assigned to several subscribers, but only one subscriber can use it at a time. If the preassigned channel is busy, the subscriber must wait until it is idle before either placing or receiving a call.

The growing demand for the overcrowded mobile telephone frequency spectrum prompted the FCC to issue Docket 18262, which inquired into a means for providing a higher frequency-spectrum efficiency. In 1971, AT&T submitted a proposal on the technical feasibility of providing efficient use of the mobile telephone frequency spectrum. AT&T's report, entitled *High Capacity Mobile Phone Service*, outlined the principles of cellular radio.

In April 1981, the FCC approved a licensing scheme for *cellular radio* markets. Each market services one *coverage area*, defined according to modified 1980 Census Bureau Standard Metropolitan Statistical Areas (SMSAs). In early 1982, the FCC approved a final plan for accepting cellular license applications beginning in June 1982 and a final round of applications by March 1983. The ensuing legal battles for cellular licenses between AT&T, MCI, GTE, and numerous other common carriers go well beyond the scope of this book.

CELLULAR RADIO

Cellular radio corrects many of the problems of traditional two-way mobile telephone service and creates a totally new environment for both mobile and traditional wireline telephone service. The key concepts of cellular radio were uncovered by researchers at Bell Telephone Laboratories in 1947. It was determined that, by subdividing a relatively large geographical area into smaller sections called *cells*, a concept of *frequency reuse* could be employed to dramatically increase the capacity of a mobile telephone channel. Frequency reuse is when the same set of frequencies (channels) can be allocated to more than one cell, provided the cells are a certain distance apart. In essence, cellular telephone systems allow a large number of users to share a limited number of *common usage* channels available in a region. In addition, integrated-circuit technology and *microprocessors* have recently enabled complex radio and logic circuits to be used in *electronic switching machines* to store programs that provide faster and more efficient call processing.

In 1974, the FCC allocated an additional 40 MHz of bandwidth for cellular radio service (825 to 845 MHz and 870 to 890 MHz). These frequency bands were previously allocated to UHF television channels 70 to 83. In 1975, AT&T was granted the first license to operate a developmental cellular radio service in Chicago, and AT&T subsequently formed the *Advanced Mobile Phone Service* (AMPS). The following year *American Radio Telephone Service* (ARTS) was granted authorization from the FCC to install a second developmental system in the Baltimore-Washington, D.C., area.

Frequency Allocation

In 1980, the FCC decided to license two common carriers per service area. The idea was to eliminate the possibility of a *monopoly* and provide the advantages that generally accompany a competitive environment. Subsequently, two frequency allocation systems emerged, each with its own group of channels, system A and system B, to share the

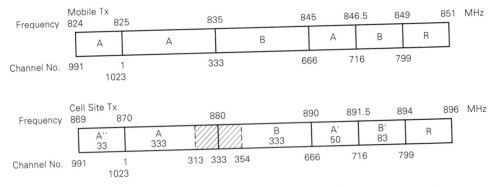

Figure 7-33 Advanced Mobile Phone Service (AMPS) frequency spectrum.

allocated frequency spectrum. System A is defined for the nonwireline companies, and system B is defined for the wireline companies.

Figures 7-33 and 7-34 show the frequency management systems for the Advanced Mobile Phone Service (AMPS) and Total Access Communications System (TACS), respectively. The AMPS cellular system uses a 20-MHz frequency band made up of 666 channels with 30-kHz channel spacing. For mobile units, channel 1 has a transmit frequency of 825.03 MHz and channel 666 has a transmit frequency of 844.98 MHz. The receivers for each channel operate 45 MHz above the transmitter; therefore, channel 1 receives at 870.03 MHz and channel 666, at 889.98 MHz. An additional 5-MHz frequency spectrum was subsequently added to the existing 20-MHz band, which increases the total number of channels available to 832. The TACS cellular standard uses a 15-MHz frequency band that comprises 600 channels with 25-kHz channel spacing. The transmit frequency for channel 1 is 890.0125 MHz, and 904.9875 MHz, for channel 600. Both the AMPS and TACS channel spectrums are divided into two basic groups. A set of channels is dedicated for control information exchange between mobile units and the cell site and is termed *control channels* (shaded areas on figure). The second group, termed voice or user channels, is made up from the remaining channels and is used for actual conversations. As with the AMPS system, TACS receivers operate 45 MHz above the transmit frequency. Therefore, for mobile units, channel 1 receives at 935.0125 MHz and channel 600, at 959.9875 MHz. Figures 7-33 and 7-34 show the additional frequency spectrum for the 166 additional channels for AMPS and 400 channels for TACS. It might be noted that the

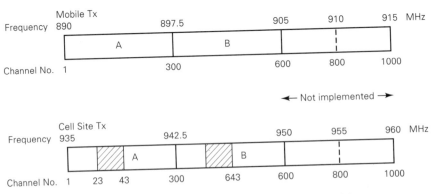

Figure 7-34 Total Access Communications System (TACS) frequency spectrum.

TACS additional frequency spectrum has not been implemented, and the dedicated control channels are for the 600-channel system. The shaded area outlines the set of dedicated control channels. Table 7-2 summarizes the frequency parameters for AMPS and TACS. The set of control channels may be split by the system operator into subsets of dedicated control channels, paging channels, and access channels.

There are several types of cellular telephones: mobiles, or car mount; portable, or pocket phone; and handheld, or transportable phone. There are three classes of cellular telephones (four for TACS). Which class a particular radio falls into is determined by the type of telephone it is and how much transmit power it is capable of producing. Mobiles (class 1) radiate the most power and then transportables (class 2); pocket phones (class 3; classes 3

TABLE 7-2 AMPS AND TACS FREQUENCY ALLOCATION

	AMPS	TACS
Channel spacing	30 kHz	25 kHz
Spectrum allocation	20MHz	15 MHz
Additional spectrum	5 MHz	10 MHz
Total number of channels	832	1000

System A Frequency Allocation					
AMPS			TACS		
Channel number	Mobile TX, MHz	Mobile RX, MHz	Channel number	Mobile TX, MHz	Mobile RX, MHz
1	825.030	870.030	1	890.0125	935.0125
313[a]	834.390	879.390	23[b]	890.5625	935.5625
333[b]	843.990	879.990	43[a]	891.5625	936.5625
667	845.010	890.010	300	897.5625	942.5625
716	846.480	891.480			
991	824.040	869.040			
1023	825.000	870.000			

System B Frequency Allocation					
AMPS			TACS		
334[c]	835.020	880.020	323[c]	897.0625	942.0625
354[d]	835.620	880.620	343[d]	897.5625	942.5625
666	844.980	890.000	600	904.9875	949.9875
717	846.510	891.000			
799	848.970	894.000			

[a] Last dedicated control channel for system A

[b] First dedicated control channel for system A

[c] Last dedicated control channel for system B

[d] First dedicated control channel for system B

Cellular Radio

and 4 for TACS) have the lowest power output capabilities. Tables 7-3a and b show the classes of cellular phones and their power levels for AMPS and TACS, respectively.

Basic Cellular Radio Concept

The basic cellular radio concept is quite simple. The FCC defined geographic cellular radio coverage areas based on modified 1980 census figures. With the cellular concept, each area is further divided into *hexagonal cells* that fit together to form a *honeycomb* pattern. The hexagon shape was chosen because it provides the most effective transmission by approximating a circular pattern, while eliminating gaps present between adjacent circles. A cell is defined by its physical size and, more importantly, by the size of its population and traffic patterns. The number of cells per system is not defined by the FCC and has been left to the provider to establish in accordance with anticipated traffic patterns. Each geographic mobile service area is allocated 666 cellular radio channels. Each transceiver within a covered area has a fixed subset of the 666 available radio channels based on expected traffic flow.

Figure 7-35 shows a simplified cellular telephone system that includes all the basic components necessary for cellular radio communications. There is an FM radio network covering a set of geographical areas (cells) inside of which mobile two-way radio units, like cellular telephones, can communicate. The radio network is defined by a set of radio-frequency transceivers located at the physical center of each cell. The locations of these radio-frequency transceivers are called the *base stations*. A base station serves as a central control for all users within that cell. Mobile units communicate directly with the base station, which serves as a high-power relay station. Mobile units transmit to the base station, and the base station rebroadcasts those transmissions at a higher power. The base station can improve the transmission quality, but it cannot increase the channel capacity within the fixed bandwidth of the network. Base stations are distributed over the area of system coverage and are managed and controlled by a computerized *cell-site controller* that handles all cell-site control and switching functions. The switch itself is called a *Mobile Tele-*

TABLE 7-3A POWER OF MOBILE PHONE FOR AMPS

	Power of Mobile Phone for AMPS					
	Mobile Station Power Class					
	I		II		III	
Power Level	dBW	mW	dBW	mW	dBW	mW
0	6	4000	2	1600	-2	630
1	2	1600	2	1600	-2	630
2	2	630	-2	630	-2	630
3	-6	250	-6	250	-6	250
4	-10	100	-10	100	-10	100
5	-14	40	-14	40	-14	40
6	-19	15	-18	15	-18	15
7	-22	6	-22	6	-22	6

Chap. 7   Angle Modulation Receivers and Systems

TABLE 7-3B POWER OF MOBILE PHONE FOR TACS

Power Level	Power of Mobile Phone for TACS							
	Mobile Station Power Class							
	I		II		III		IV	
	dBW	mW	dBW	mW	dBW	mW	dBW	mW
0	10	10000	6	4000	2	1600	-2	630
1	2	1600	2	1600	2	1600	-2	630
2	-2	6300	-2	630	-2	630	-2	630
3	-6	250	-6	250	-6	250	-6	250
4	-10	100	-10	100	-10	100	-10	100
5	-14	40	-14	40	-14	40	-14	40
6	-19	15	-18	15	-18	15	-18	15
7	-22	6	-22	6	-22	6	-22	6

phone Switching Office (MTSO). A base station is comprised of a low-power FM transceiver, power amplifiers, control unit, and other hardware depending on the system configuration. Cellular radio uses several moderately powered transceivers over a relatively wide service area, as opposed to MTS, which uses a single high-powered transceiver at a high elevation. The function of the base station is to interface between cellular mobile telephones and the MTSO. It communicates with the MTSO over dedicated data links,

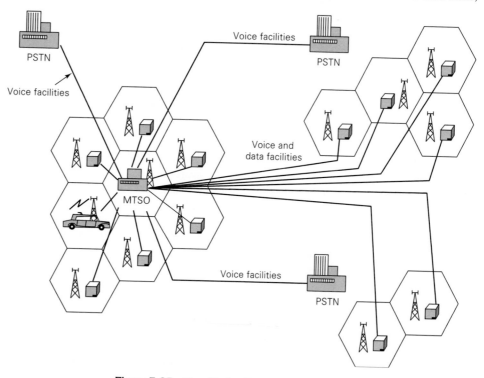

Figure 7-35 Simplified cellular telephone system.

Cellular Radio

both metallic and nonmetallic, and communicates with mobile units over the air waves using the control channel. The MTSO's function is to control call processing and call setup and release, which includes *signaling, supervision, switching*, and allocating RF channels. The MTSO also provides a centralized administration and maintenance point for the entire network, and it interfaces with the *Public Switched Telephone Network* (PSTN) over wireline voice facilities to honor services from conventional wireline telephones.

An MTSO is known by several different names depending on the manufacturer and system configuration. MTSO (*Mobile Telephone Switching Office*) was the name given by Bell Labs, EMX (*Electronic Mobile Xchange*) by Motorola, AEX by Ericcson, NEAX by NEC, and SMC (*Switching Mobile Center*) and MMC (*Master Mobile Center*) by Novatel.

Each geographical area or cell can generally accommodate up to 70 different user channels simultaneously. Within a cell, each channel can support only one mobile telephone user at a time. Channels are dynamically assigned and dedicated to a single user for the duration of the call, and any user may be assigned to any user channel. This is called frequency reuse and allows a cellular telephone system in a single area to handle considerably more than the 666 available channels. Thus, cellular radio makes more efficient use of the available frequency spectrum than does traditional MTS service.

As a car moves away from the transceiver in the center of a cell, the received signal strength begins to decrease. The maximum output power from a cellular transceiver is 35 dBm (3 W) and can be adjusted in 4-dB increments down to 7.8 dBm (0.7 W). The output power of the mobile units is controlled by the base station through the transmission of up/down commands, which depends on the signal strength it is currently receiving. When the signal strength drops below a predetermined threshold level, the electronic switching center locates the cell in the honeycomb that is receiving the strongest signal from the mobile unit and transfers the mobile unit to the transceiver in the new cell. The transfer includes converting the call to an available frequency within the new cell's allocated channel subset. This transfer is called a *handoff* and is completely transparent to the subscriber (the subscriber does not know that his or her facility has been switched). The transfer takes approximately 0.2 s, which is imperceptible to voice telephone users. However, this delay may be disruptive to data communications.

The six primary components of a cellular radio system are the electronic switching center, cell-site controller, radio transceivers, system interconnections, mobile telephone units, and a communications protocol.

Electronic switching center. The *electronic switching center* is a digital telephone exchange and is the heart of the cellular system. The switch performs two essential functions: (1) it controls the switching between the public telephone network and the cell sites for all wireline-to-mobile, mobile-to-wireline, and mobile-to-mobile calls; and (2) it processes data received from the cell-site controllers concerning mobile unit status, diagnostic data, and bill-compiling information. The electronic switch communicates with the cell-cite controllers with a data link using the X.25 protocol and a transmission rate of 9.6 kbps full-duplex.

Cell-site controller. Each cell contains one *cell-site controller* that operates under the direction of the switching center. The cell-site controller manages each of the radio channels at the site, supervises calls, turns the radio transmitter and receiver on and off, injects data onto the control and user channels, and performs diagnostic tests on the cell-site equipment.

Radio transceiver.   The radio *transceivers* used for cellular radio are narrowband FM with an audio-frequency band of 300 Hz to 3 kHz and ± 12-kHz frequency deviation for 100% modulation. This corresponds to a bandwidth of 30 kHz using Carson's rule. Each cell contains one radio transmitter and two radio receivers tuned to the same frequency. Whichever radio receiver detects the strongest signal is selected.

System interconnections.   Four-wire leased telephone lines are generally used to connect the switching centers to each of the cell sites. There is one dedicated four-wire trunk circuit for each of the cell's user channels. Also, there must be at least one four-wire circuit to connect the switch to the cell-site controller as a control channel.

Mobile and portable telephone units.   *Mobile* and *portable telephone units* are essentially the same thing. The only difference is that the portable units have a lower output power and a less efficient antenna. Each mobile telephone unit consists of a control unit, a radio transceiver, a logic unit, and a mobile antenna. The control unit houses all the user interfaces, including a handset. The transceiver uses a frequency synthesizer to tune into any designated cellular system channel. The logic unit interrupts subscriber actions and system commands and manages the transceiver and control units.

Communications protocol.   The last constituent of a cellular system is the *communication protocol* that governs the way a telephone call is established. Cellular protocols differ between countries. In the United States the Advanced Mobile Phone Service (AMPS) standard is used, while in Canada the AURORA 800 system is used. Each country in Europe has its own standard. Total Access Communication System (TACS) is used in the United Kingdom; NMT or Nordic system, in the Scandinavian countries; RC2000, in France; NETZ C-450, in Germany; and NTT is the Japanese standard for cellular telephone.

Call Processing

A telephone call over a cellular network requires using two full-duplex voice channels simultaneously, one called the *user channel* and one called the *control channel*. The base station transmits and receives on what is called the *forward control channel* and the *forward voice channel*, and the mobile unit transmits and receives on the *reverse* control and voice channels.

Completing a call within a cellular radio system is quite similar to the public switched telephone network. When a mobile unit is first turned on, it performs a series of start-up procedures and then samples the received signal strength on all prescribed user channels. The unit automatically tunes to the channel with the strongest receive signal strength and synchronizes to the control data transmitted by the cell-site controller. The mobile unit interprets the data and continues monitoring the control channel(s). The mobile unit automatically rescans periodically to ensure that it is using the best control channel.

Within a cellular system, calls can take place between a wireline party and a mobile telephone or between two mobile telephones.

Wireline-to-mobile calls.   The cellular system's switching center receives a call from a wireline party through a dedicated interconnect line from the public switched telephone network. The switch translates the received dialing digits and determines whether the mobile unit to which the call is destined is on or off hook (busy). If the mobile unit is available, the switch *pages* the mobile subscriber. Following a *page response* from the mobile unit, the switch assigns an idle channel and instructs the mobile unit to tune into that channel. The mobile unit sends a verification of channel tuning via the controller in the cell site and then sends an audible *call progress tone* to the subscriber's mobile telephone, causing it to ring. The switch terminates the call progress tones when it receives positive indication that the subscriber has answered the phone and the conversation between the two parties has begun.

Mobile-to-wireline calls. A mobile subscriber who desires to call a wireline party first enters the called number into the unit's memory using Touch-Tone buttons or a dial on the telephone unit. The subscriber then presses a *send key*, which transmits the called number as well as the mobile subscriber's identification number to the switch. If the identification number is valid, the switch routes the call over a leased wireline interconnection to the public telephone network, which completes the connection to the wireline party. Using the cell-site controller, the switch assigns the mobile unit a nonbusy user channel and instructs the mobile unit to tune into that channel. After the switch receives verification that the mobile unit is tuned to the assigned channel, the mobile subscriber receives an audible *call progress tone* from the switch. After the called party picks up the phone, the switch terminates the call progress tones and the conversation can begin.

Mobile-to-mobile calls. Calls between two mobile units are also possible in the cellular radio system. To originate a call to another mobile unit, the calling party enters the called number into the unit's memory via the touchpad on the telephone set and then presses the send key. The switch receives the caller's identification number and the called number and then determines if the called unit is free to receive a call. The switch sends a *page command* to all cell-site controllers, and the called party (who may be anywhere in the service area) receives a page. Following a positive page from the called party, the switch assigns each party an idle user channel and instructs each party to tune into their respective user channel. Then the called party's phone rings. When the system receives notice that the called party has answered the phone, the switch terminates the call progress tone, and the conversation may begin between the two mobile units.

If a mobile subscriber wishes to initiate a call and all user channels are busy, the switch sends a *directed retry command* instructing the subscriber to reattempt the call through a neighboring cell. If the system cannot allocate a user channel through the neighboring cell, the switch transmits an *intercept message* to the calling mobile unit over the control channel. Whenever the called party is *off hook*, the calling party receives a busy signal. Also, if the called number is invalid, the system either sends a reorder message via the control channel or provides an announcement that the call cannot be processed.

Handoff feature. One of the most important features of a cellular system is its ability to transfer calls that are already in progress from one cell site controller to another as mobile units move from cell to cell within the cellular network. This transfer process is called a *handoff.* Computers at the cell-site controller stations transfer calls from cell to cell with minimal disruption and no degradation in the quality of transmission. The *handoff decision algorithm* is based on variations in signal strength. When a call is in progress, the switching center monitors the received signal strength of each user channel. If the signal level on an occupied channel drops below some predetermined threshold level for more than a given time interval, the switch performs a handoff, provided there is a vacant channel. The handoff operation reroutes the call through a new cell site.

The handoff process takes approximately 200 ms. Handoff parameters allow for optimized transfer based on cell-site traffic load and the surrounding terrain. *Blocking* occurs when the signal level drops below a usable level and there are no usable channels available to switch to. To help avoid blocking or loss of a call during the handoff process, the system employs a load-balancing scheme that frees channels for handoff and sets handoff priorities. Programmers at the central switch site continually update the switching algorithm to amend the system to accommodate changing traffic loads.

Cellular Telephone Block Diagram

Figure 7-36 shows the block diagram for a typical cellular telephone transceiver. Note that a cellular radio transceiver is very similar to a standard FM transceiver except for the addition of several stages. Because a cellular telephone must be capable of full-duplex transmissions, it must be designed such that the transmitter and receiver can be on at the same time.

The receiver shown in Figure 7-36 is a double-conversion superheterodyne type. The receiver uses high-side injection and a local oscillator circuit consisting of a single-chip frequency synthesizer with an onboard prescaler and loop filter and a 915- to 937-MHz VCO (45 MHz above the incoming RF). The first IF (45 MHz) is sufficiently high to push the image frequency well beyond the passband of the preamplifier filter, and the second IF (455 kHz) is low enough to be sufficiently amplified to drive the FM detector. In the transmitter, the modulating signal is applied directly to the transmit VCO to produce a low-index, relatively low frequency direct FM carrier signal that is multiplied and then amplified significantly before reaching the transmit antenna.

Note that the microprocessor has control lines to every major functional block of the transceiver. The microprocessor determines the frequencies at the outputs of the frequency synthesizer. It also monitors the output from the FM detector to determine received signal quality and for detecting received setup information. The microprocessor also supplies setup information to the transmitter and controls the gain of the final power amplifier.

Integrated-circuit Cellular Radio

Although cellular radio has been with us for quite some time, the early equipment was complex, very expensive, and affordable only by the business sector. However, a rapid increase in the number of users, fueled by competition, has resulted in a steadily decreas-

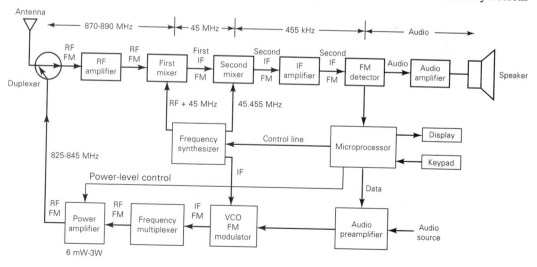

Figure 7-36 Block diagram for a typical cellular transceiver.

ing price for the consumer. To remain competitive, manufacturers of cellular radio equipment have continually strived to reduce costs. Probably the key to achieving reliable, yet affordable components, is the development of integrated-circuit chips for the major cellular functions.

Integrated-circuit manufacturers have developed integrated-circuit chips that incorporate most of the functions needed in a cellular telephone system. For example, Signetics Corporation has a set of integrated-circuit chips that perform all the cellular transceiver functions except the RF stages. Signetic's chip set is suitable for both the American (AMPS) and United Kingdom (TACS) standards. All the ICs in this set are connected together through a mini local area network the manufacturer calls an I²C (*inter-integrated circuit*) bus. The I²C uses bus topology, and *master/slave protocol* and requires only two wires to interconnect all the ICs in the transceiver. Figure 7-37 shows the set of chips manufactured by Signetics Corporation and how they integrate to form almost a complete cellular radio transceiver.

The hardware design is divided into two major sections: and RF section and a data/audio control section, also known as the baseband section. The RF section consists of a receive and transmit local oscillator, a receiver front-end circuit, a duplexer, and a transmit module. The duplexer is a directional device that directs the relatively high transmit signal power to the antenna and, at the same time, directs the relatively low receiver power to the receiver input. The baseband section consists of the logic and program control circuit, a Data PROCessing circuit, and an Audio PROCessing circuit, which includes an audio power amplifier.

System architecture.   The receiver section includes a UMA1014 low-power frequency synthesizer and a NE/SA605 single-chip second mixer/oscillator/IF amplifier/demodulator. A UMA1000 single-chip CMOS data processor performs all functions associated with control data, supervisory, and signaling tones. All voice alert and DTMF func-

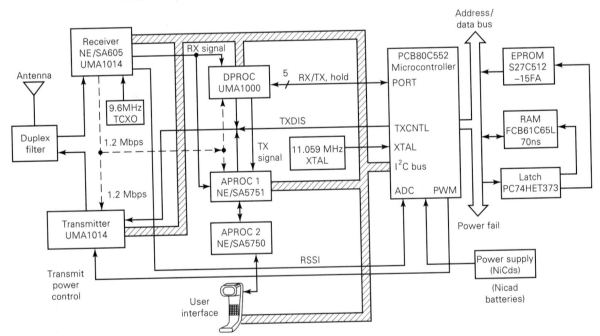

Figure 7-37 Transceiver using Signetics Corporation's cellular radio chip set.

tions are contained on two audio processor devices: one CMOS (NE/SA5751) and the other bipolar (NE/SA5750). Switched capacitor integrated filter technology is used in all baseband processing functions to achieve a minimum number of external passive components. The system's master controller is the 80C552, which is a derivative of the Intel 80C52 microcontroller with integrated analog-to-digital converter (ADC), pulse width modulator (PWM), I^2C bus, and UART interfaces. The DPROC (data processor for cellular radio), APROC (audio processor for cellular radio), and LOPSY (low-power frequency synthesizer for cellular radio) are interconnected via an I^2C bus, which is a two-wire serial bus that can transfer data at 100 kbps.

Receiver frontend. The receiver uses a double-conversion superheterodyne design with a 45-MHz first IF and a 455-kHz second IF. A single transistor is used for the RF amplifier and first mixer stages. The second mixer, second local oscillator, all IF amplifiers, limiter, demodulator, and RSSI (receiver signal-strength indicator) functions are provided by the NE/SA605 IC chip. The receiver's local oscillator circuit is an UMA1014 single-chip frequency synthesizer with on-board prescaler and loop filter and a 926-MHz VCO (45 MHz above the receive RF signal). An on-chip divide-by-8 circuit provides the 1.2-MHz clock for the data and audio processors from a 9.6-MHz VTCXO (voltage-controlled, temperature-compensated crystal oscillator).

Transmitter. The modulating signal is applied to the transmit VCO in which a loop bandwidth of approximately 200 Hz is used in the UMA1014 transmit synthesizer to achieve a flat modulation characteristic above 300 Hz, permitting sufficiently fast switching time. Hybrid power modules are used to amplify the transmit signal to the level required according to the class of the transceiver.

Audio processors. The NE5750 and NE5751 is an audio processor system that uses a pair of ICs to perform all the audio signal functions associated with the voice channel for the transmitter and receiver. The NE5750 provides a low-noise, adjustable-gain microphone preamplifier, a voice-operated transmit/receive switch (VOX), deviation limiting, a speaker amplifier, an earphone amplifier, pre- and deemphasis, signal path switching, and tone generation. The NE5751 contains the I^2C bus interface and the 300- to 3000-Hz transmit and receive audio bandpass filters, a transmit limiter, a pre- and deemphasis network, a digitally controlled volume control, and the power-up and power-down circuitry. Switched capacitor techniques have been used to fully integrate all filter functions.

Data processor. All functions associated with control data, supervision (SAT), and signaling (ST) are incorporated in the UMA1000 data processor chip. The UMA1000 is a special data processor chip for cellular radio. It handles all the signaling and supervision tasks associated with normal telephone operation. The UMA1000 relieves the microcontroller of many of the smaller tasks. In the receive path, a dedicated two-wire serial link (RX_{clock} and RX_{line}) passes the fully decoded data word to the microcontroller chip. Clocking is under the control of the microcontroller and is not time critical. When operating on a voice channel, the dotting sequence detector, which precedes a data burst, is used to blank the audio path directly to prevent data bursts from being heard by the user. In the transmit path, a 40-bit (36 data bits and 4 framing bits) transmit word is passed from the microcontroller to the data processor via another dedicated two-wire serial link (TX_{clock} and TX_{line}), where it is held in a buffer. The microcontroller can trigger immediate transmission of the data at the appropriate time. The base station returns the status of the channel access within the busy/idle data stream. If the busy/idle data stream does not revert to the busy condition during the specified window, transmission is aborted with the TX_{ctrl}

signal. This time-critical channel arbitration sequence is performed by the data processor IC independently of the microcontroller.

Microcontroller. The PCB80C552 is an 8-bit microcontroller chip that includes a microprocessor plus the additional circuitry necessary to control the operation of the telephone. The chip is sometimes called the *brain* of the cellular telephone. It has six 8-bit input/output ports, an 8-bit analog-to-digital converter (ADC), a timer circuit to measure elapsed time, a UART, and a watchdog timer. The UART (universal asynchronous transmitter/receiver) performs serial-to-parallel and parallel-to-serial conversion of data. The PCB80C552 can operate with an instruction cycle time of less than 1 μs, and about 50% of its operations execute in a single machine cycle. This device also supports extensive power-saving modes, giving approximately 75% power saving in the standby mode, and a micropower power-down mode for backup battery operation. A software package implementing the AMPS and TACS standards, written entirely in PL/M51, has been developed to introduce a fully functional cellular radio.

I^2C bus. The I^2C bus is essentially a local area network (LAN) for integrated circuits. Each device has its own 70-bit address and is connected to a two-wire serial bus. The interface protocol is self-checking and self-arbitrating, allowing a multimaster control of the bus. Data are transferred at a rate of 100 kbps as a message block consisting of device address, a read/write indication, and the data block.

QUESTIONS

7-1. Describe the basic differences between AM and FM receivers.

7-2. Draw the schematic diagram for a *single-ended slope detector* and describe its operation.

7-3. Draw the schematic diagram for a *double-ended slope detector* and describe its operation.

7-4. Draw the schematic diagram for a *Foster–Seeley discriminator* and describe its operation.

7-5. Draw the schematic diagram for a *ratio detector* and describe its operation.

7-6. Describe the operation of a PLL FM demodulator.

7-7. Draw the schematic diagram for a *quadrature FM demodulator* and describe its operation.

7-8. Compare the advantages and disadvantages of the FM demodulator circuits discussed in Questions 7-1 through 7-7.

7-9. What is the purpose of a *limiter* in an FM receiver?

7-10. Describe *FM thresholding*.

7-11. Describe the operation of an FM *stereo* transmitter; an FM stereo receiver.

7-12. Draw the block diagram for a two-way FM radio transmitter and explain its operation.

7-13. Draw the block diagram for a two-way FM radio receiver and explain its operation.

7-14. Describe the operation of an electronic *push-to-talk circuit*.

7-15. Describe the operation of a *VOX circuit*.

7-16. Briefly explain how a composite *FM stereo* signal is produced.

7-17. What is meant by the term *interleaving* of L and R signals in stereo transmission?

7-18. What is the purpose of the 19-kHz *pilot* in FM stereo broadcasting?

7-19. What is the difference between *mobile radio* and *mobile telephone*?

7-20. Describe the basic concept of *cellular telephone*.

7-21. What is the purpose of the *Mobile Telephone Switching Office* (MTSO)?

7-22. What is the purpose of the *electronic switching center* in a cellular radio system?

7-23. What is the purpose of the *cell-site controller* in a cellular radio system?

7-24. What is the purpose of the *radio transceivers* in a cellular radio system?

7-25. What is the purpose of the *communications protocol* in a cellular radio system?

7-26. Briefly describe how a *wireline-to-mobile* call is established in a cellular radio system.

7-27. Briefly describe how a *mobile-to-wireline* call is established in a cellular radio system.

7-28. Briefly describe how a *mobile-to-mobile* call is established in a cellular radio system.

7-29. Briefly describe the *handoff* feature in a cellular radio system.

PROBLEMS

7-1. Determine the minimum input S/N ratio required for a receiver with 15 dB of FM improvement, a noise figure NF = 4 dB, and a desired postdetection S/N = 33 dB.

7-2. For an FM receiver with a 100-kHz bandwidth, a noise figure NF = 6 dB, and an input noise temperature T = 200°C, determine the minimum receive carrier power to achieve a postdetection S/N = 40 dB. Use the receiver block diagram shown in Figure 7-1 as the receiver model and the FM thresholding curve show in Figure 7-12.

7-3. For an FM receiver tuned to 92.75 MHz using high-side injection and a first IF of 10.7 MHz, determine the image frequency and the local oscillator frequency.

7-4. For an FM receiver with an input frequency deviation Δf = 40 kHz and a transfer ratio K = 0.01 V/kHz, determine V_{out}.

7-5. For the balanced slope detector shown in Figure 7-3a, a center frequency f_c = 20.4 MHz, and a maximum input frequency deviation Δf = 50 kHz, determine the upper and lower cutoff frequencies for the tuned circuit.

7-6. For the Foster–Seeley discriminator shown in Figure 7-4, V_{C1} = 1.2 V and V_{C2} = 0.8 V, determine V_{out}.

7-7. For the ratio detector shown in Figure 7-6, V_{C1} = 1.2 V and V_{C2} = 0.8 V, determine V_{out}.

7-8. For an FM demodulator with an FM improvement factor of 23 dB and an input S/N = 26 dB, determine the postdetection S/N.

7-9. From Figure 7-12, determine the approximate FM improvement factor for an input S/N = 10.5 dB and m = 1.

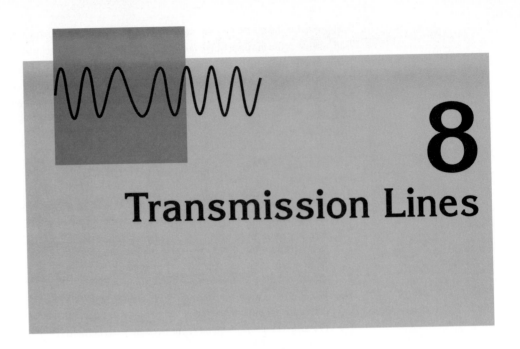

8

Transmission Lines

INTRODUCTION

A *transmission line* is a *metallic conductor system* that is used to transfer electrical energy from one point to another. More specifically, a transmission line is two or more conductors separated by an insulator, such as a pair of wires or a system of wire pairs. A transmission line can be as short as a few inches or it can span several thousand miles. Transmission lines can be used to propagate dc or low-frequency ac (such as 60-cycle electrical power and audio signals); they can also be used to propagate very high frequencies (such as intermediate and radio-frequency signals). When propagating low-frequency signals, transmission-line behavior is rather simple and quite predictable. However, when propagating high-frequency signals, the characteristics of transmission lines become more involved and their behavior is somewhat peculiar to a student of lumped constant circuits and systems.

TRANSVERSE ELECTROMAGNETIC WAVES

Propagation of electrical power along a transmission line occurs in the form of *transverse electromagnetic* (TEM) *waves*. A wave is an *oscillatory motion*. The vibration of a particle excites similar vibrations in nearby particles. A TEM wave propagates primarily in the nonconductor (dielectric) that separates the two conductors of a transmission line. Therefore, a wave travels or propagates itself through a medium. For a transverse wave, the direction of displacement is perpendicular to the direction of propagation. A surface wave of water is a longitudinal wave. A wave in which the displacement is in the direction of propagation is called a *longitudinal wave*. Sound waves are longitudinal. An electromagnetic (EM) wave is produced by the acceleration of an electric charge. In a conductor, current and voltage are always accompanied by an electric (E) and a magnetic (H) field in the adjoining region of space. Figure 8-1a shows the spatial relationships between the E and H fields of an electromagnetic wave. Figure 8-1b shows the cross-sectional views of the E and H fields that surround a parallel two-wire and a coaxial line. It can be seen that the E and H fields are perpendicular to each other (at 90° angles) at all points. This is referred to as *space quadrature*. Electromagnetic waves that travel along a transmission line from the source toward the load are called *incident waves*, and those that travel from the load back toward the source are called *reflected waves*.

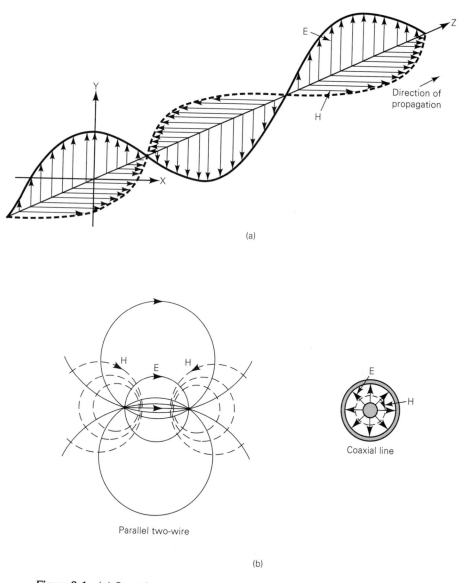

(a)

(b)

Figure 8-1 (a) Spatial and (b) cross-sectional views showing the relative displacement of the E and H fields on a transmission line.

Transverse Electromagnetic Waves

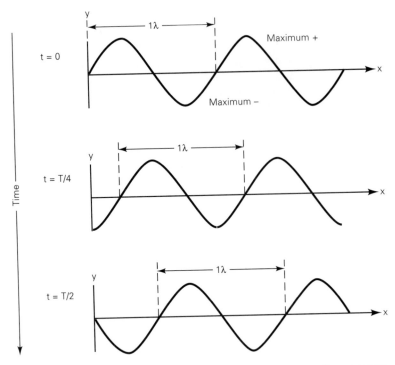

Figure 8-2 Displacement and velocity of a transverse wave as it propagates down a transmission line.

Characteristics of Electromagnetic Waves

Wave velocity.   Waves travel at various speeds, depending on the type of wave and the characteristics of the propagation medium. Sound waves travel at approximately 1100 ft/s in the normal atmosphere. Electromagnetic waves travel much faster. In free space (a vacuum), TEM waves travel at the speed of light, c = 186,283 statute mi/s or 299,793,000 m/s, rounded off to 186,000 mi/s and 3×10^8 m/s. However, in air (such as Earth's atmosphere), TEM waves travel slightly more slowly, and along a transmission line, electromagnetic waves travel considerably more slowly.

Frequency and wavelength.   The oscillations of an electromagnetic wave are periodic and repetitive. Therefore, they are characterized by a frequency. The rate at which the periodic wave repeats is its frequency. The distance of one cycle occurring in space is called the *wavelength* and is determined from the following fundamental equation:

$$\text{distance} = \text{velocity} \times \text{time} \tag{8-1}$$

If the time for one cycle is substituted into Equation 8-1, we get the length of one cycle, which is called the wavelength and whose symbol is the Greek lowercase letter lambda (λ)

$$\lambda = \text{velocity} \times \text{period}$$

$$= v \times T$$

And since $T = 1/f$,

$$\lambda = \frac{v}{f} \quad (8\text{-}2)$$

For free-space propagation, $v = c$; therefore, the length of one cycle is

$$\lambda = \frac{c}{f} = \frac{3 \times 10^8 \text{ m/s}}{f \text{ cycles/s}} = \frac{\text{meters}}{\text{cycle}} \qquad (8\text{-}3)$$

Figure 8-2 shows a graph of the displacement and velocity of a transverse wave as it propagates along a transmission line from a source to a load. The horizontal (x) axis is distance and the vertical (y) axis is displacement. One wavelength is the distance covered by one cycle of the wave. It can be seen that the wave moves to the right or propagates down the line with time. If a voltmeter is placed at any stationary point on the line, the voltage measured will fluctuate from 0 to maximum positive, back to zero, to maximum negative, back to zero again, and then the cycle repeats.

TYPES OF TRANSMISSION LINES

Transmission lines can be generally classified as *balanced* or *unbalanced*. With two-wire balanced lines, both conductors carry current; one conductor carries the signal and the other is the return. This type of transmission is called *differential* or *balanced* signal transmission. The signal propagating down the wire is measured as the potential difference between the two wires. Figure 8-3 shows a balanced transmission system. Both conductors in a balanced line carry signal current, and the currents are equal in magnitude with respect to electrical ground but travel in opposite directions. Currents that flow in opposite directions in a balanced wire pair are called *metallic circuit currents*. Currents that flow in the same directions are called *longitudinal currents*. A balanced wire pair has the advantage that most noise interference (sometimes called *common-mode voltage*) is induced equally in both wires, producing longitudinal currents that cancel in the load. Any pair of wires can operate in the balanced mode provided neither wire is at ground potential. This includes coaxial cable that has two center conductors and a shield. The shield is generally connected to ground to prevent static interference from penetrating the center conductors.

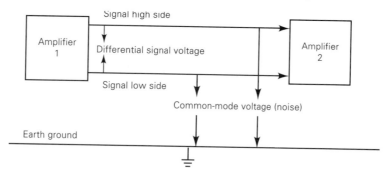

Figure 8-3 Differential or balanced transmission system.

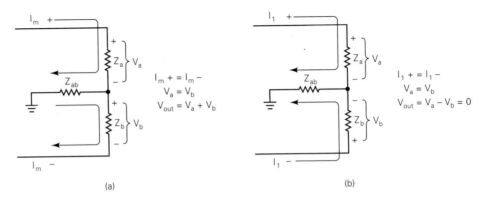

$$I_m + = I_m -$$
$$V_a = V_b$$
$$V_{out} = V_a + V_b$$

$$I_1 + = I_1 -$$
$$V_a = V_b$$
$$V_{out} = V_a - V_b = 0$$

(a)   (b)

Figure 8-4 Results of metallic and longitudinal currents on a balanced transmission line: (a) metallic currents due to signal voltages; (b) longitudinal currents due to noise voltages.

Figure 8-4 shows the result of metallic and logitudinal currents on a balanced transmission line. It can be seen that the longitudinal currents (often produced by static interference) cancel in the load.

With an unbalanced transmission line, one wire is at ground potential, while the other wire is at signal potential. This type of transmission is called *single-ended* or *unbalanced* signal transmission. With unbalanced signal transmission, the ground wire may also be the reference for other signal-carrying wires. If this is the case, the ground wire must go wherever any of the signal wires go. Sometimes this creates a problem because a length of wire has resistance, inductance, and capacitance and, therefore, a small potential difference may exist between any two points on the ground wire. Consequently, the ground wire is not a perfect reference point and is capable of having noise induced into it. A standard two-conductor coaxial cable is an unbalanced line. The second wire is the shield, which is generally connected to ground.

Figure 8-5 shows two unbalanced transmission systems. The potential difference on each signal wire is measured from that wire to ground. Balanced transmission lines can be

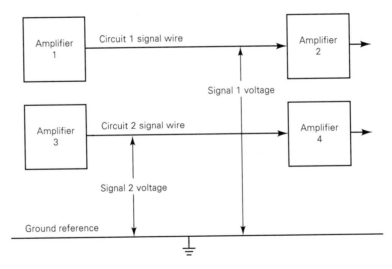

Figure 8-5 Single-ended or unbalanced transmission system.

connected to unbalanced lines, and vice versa, with special transformers called *baluns* which are discussed later in this chapter.

Parallel-conductor Transmission Lines

Open-wire transmission line. An *open-wire transmission line* is a *two-wire parallel conductor*; it is shown in Figure 8-6a. It consists simply of two parallel wires, closely spaced and separated by air. Nonconductive spacers are placed at periodic intervals for support and to keep the distance between the conductors constant. The distance between the two conductors is generally between 2 and 6 inches. The dielectric is simply the air between and around the two conductors in which the TEM wave propagates. The only real advantage of this type of transmission line is its simple construction. Because there is no shielding, radiation losses are high and it is susceptible to noise pickup. These are the primary disadvantages of an open-wire transmission line. Therefore, open-wire transmission lines are normally operated in the balanced mode.

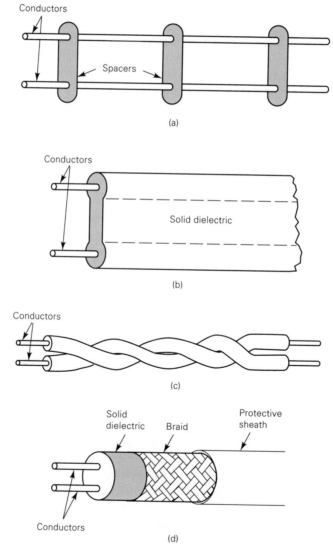

Figure 8-6 Transmission lines: (a) open wire; (b) twin lead; (c) twisted pair; (d) shielded pair.

Types of Transmission Lines

Twin lead. *Twin lead* is another form of two-wire parallel conductor transmission line and is shown in Figure 8-6b. Twin lead is often called *ribbon cable*. Twin lead is essentially the same as an open-wire transmission line except that the spacers between the two conductors are replaced with a continuous solid dielectric. This assures uniform spacing along the entire cable, which is a desirable characteristic for reasons that are explained later in the chapter. Typically, the distance between the two conductors is $\frac{5}{16}$ in. for television transmission cable. Common dielectric materials are Teflon and polyethylene.

Twisted-pair cable. A *twisted-pair cable* is formed by twisting together two insulated conductors. Pairs are often stranded in *units*, and the units are then cabled into *cores*. The cores are covered with various types of *sheaths*, depending on their intended use. Neighboring pairs are twisted with different *pitch* (twist length) to reduce interference between pairs due to mutual induction. The *primary constants* of twisted-pair cable are its electrical parameters (resistance, inductance, capacitance, and conductance), which are subject to variations with the physical environment such as temperature, moisture, and mechanical stress and depend on manufacturing deviations. A twisted-pair cable is shown in Figure 8-6c.

Shielded cable pair. To reduce radiation losses and interference, parallel two-wire transmission lines are often enclosed in a conductive metal *braid*. The braid is connected to ground and acts like a shield. The braid also prevents signals from radiating beyond its boundaries and keeps electromagnetic interference from reaching the signal conductors. A shielded parallel wire pair is shown in Figure 8-6d. It consists of two parallel wire conductors separated by a solid dielectric material. The entire structure is enclosed in a braided conductive tube and then covered with a protective plastic coating.

Concentric or Coaxial Transmission Lines

Parallel-conductor transmission lines are suitable for low-frequency applications. However, at high frequencies, their radiation and dielectric losses, as well as their susceptibility to external interference, are excessive. Therefore, *coaxial conductors* are used extensively for high-frequency applications to reduce losses and to isolate transmission paths. The basic coaxial cable consists of a center conductor surrounded by a *concentric* (uniform distance from the center) *outer conductor*. At relatively high operating frequencies, the coaxial outer conductor provides excellent shielding against external interference. However, at lower operating frequencies, the use of shielding is usually not cost effective. Also, a coaxial cable's outer conductor is generally grounded, which limits its use to unbalanced applications.

Essentially, there are two types of coaxial cables: *rigid air filled* or *solid flexible* lines. Figure 8-7a shows a rigid air coaxial line. It can be seen that the center conductor is surrounded coaxially by a tubular outer conductor and the insulating material is air. The outer conductor is physically isolated and separated from the center conductor by a spacer, which is generally made of Pyrex, polystyrene, or some other nonconductive material. Figure 8-7b shows a solid flexible coaxial cable. The outer conductor is braided, flexible, and coaxial to the center conductor. The insulating material is a solid nonconductive polyethylene material that provides both support and electrical isolation between the inner and outer conductors. The inner conductor is a flexible copper wire that can be either solid or hollow.

Rigid air-filled coaxial cables are relatively expensive to manufacture, and to minimize losses, the air insulator must be relatively free of moisture. Solid coaxial cables have

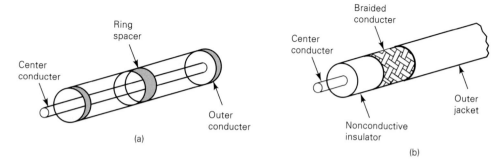

Figure 8-7 Concentric or coaxial transmission lines; (a) rigid air filled; (b) solid flexible line.

lower losses and are easier to construct and to install and maintain. Both types of coaxial cables are relatively immune to external radiation, radiate little themselves, and can operate at higher frequencies than can their parallel-wire counterparts. The basic disadvantages of coaxial transmission lines is that they are expensive and must be used in the unbalanced mode.

Baluns. A circuit device used to connect a balanced transmission line to an unbalanced load is called a *balun* (balanced to unbalanced). Or more commonly, an unbalanced transmission line, such as a coaxial cable, can be connected to a balanced load, such as an antenna, using a special transformer with an unbalanced primary and a center-tapped secondary winding. The outer conductor (*shield*) of an unbalanced coaxial transmission line is generally connected to ground. At relatively low frequencies, an ordinary transformer can be used to isolate the ground from the load, as shown in Figure 8-8a. The balun must

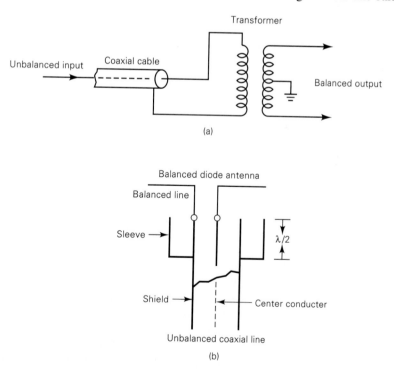

Figure 8-8 Baluns: (a) transformer balun; (b) bazooka balun.

Types of Transmission Lines

have an electrostatic shield connected to earth ground to minimize the effects of stray capacitances.

For relatively high frequencies, several different kinds of transmission-line baluns exist. The most common type is a *narrowband* balun, sometimes called a *choke*, *sleeve*, or *bazooka* balun, which is shown in Figure 8-8b. A quarter-wavelength sleeve is placed around and connected to the outer conductor of a coaxial cable. Consequently, the impedance seen looking back into the transmission line is formed by the sleeve and the outer conductor and is equal to infinity (that is, the outer conductor no longer has a zero impedance to ground). Thus, one wire of the balanced pair can be connected to the sleeve without short-circuiting the signal. The second conductor is connected to the inner conductor of the coaxial cable.

TRANSMISSION-LINE EQUIVALENT CIRCUIT

Uniformly Distributed Lines

The characteristics of a transmission line are determined by its electrical properties, such as wire conductivity and insulator dielectric constant, and its physical properties, such as wire diameter and conductor spacing. These properties, in turn, determine the primary electrical constants: series dc resistance (R), series inductance (L), shunt capacitance (C), and shunt conductance (G). Resistance and inductance occur along the line, whereas capacitance and conductance occur between the two conductors. The primary constants are uniformly distributed throughout the length of the line and are therefore commonly called *distributed parameters*. To simplify analysis, distributed parameters are commonly *lumped* together per a given unit length to form an artificial electrical model of the line. For example, series resistance is generally given in ohms per mile or kilometer.

Figure 8-9 shows the electrical equivalent circuit for a metallic two-wire transmission line showing the relative placement of the various lumped parameters. The conductance between the two wires is shown in reciprocal form and given as a shunt leakage resistance (R_s).

Transmission Characteristics

The transmission characteristics of a transmission line are called *secondary constants* and are determined from the four primary constants. The secondary constants are characteristic impedance and propagation constant.

Characteristic impedance. For maximum power transfer from the source to the load (that is, no reflected energy), a transmission line must be terminated in a purely resistive load equal to the *characteristic impedance* of the line. The characteristic impedance (Z_o) of a transmission line is a complex quantity that is expressed in ohms, is ideally independent of line length, and cannot be measured. Characteristic impedance (which is sometimes called *surge impedance*) is defined as the impedance seen looking into an infinitely long line or the impedance seen looking into a finite length of line that is terminated in a purely resistive load equal to the characteristic impedance of the line. A transmission line stores energy in its distributed inductance and capacitance. If the line is infinitely long, it can store energy indefinitely; energy from the source is entering the line and none is returned. Therefore, the line acts like a resistor that dissipates all the energy. An infinite

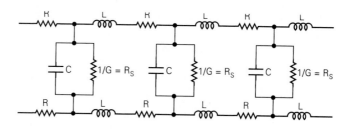

C = capacitance – two conductors separated
      by an insulator
R = resistance – opposition to current flow
L = self inductance
1/G = leakage resistance of dielectric
R_s = shunt leakage resistance

Figure 8-9 Two-wire parallel transmission line, electrical equivalent circuit.

line can be simulated if a finite line is terminated in a purely resistive load equal to Z_o; all the energy that enters the line from the source is dissipated in the load (this assumes a totally lossless line).

Figure 8-10 shows a single section of a transmission line terminated in a load Z_L that is equal to Z_o. The impedance seen looking into a line of n such sections is determined from the following expression:

$$Z_o^2 = Z_1 Z_2 + \frac{Z_L^2}{n} \tag{8-4}$$

where n is the number of sections. For an infinite number of sections Z_L^2/n approaches 0 if

$$\lim \frac{Z_L^2}{n} \bigg|_{n \to \infty} = 0$$

Then

$$Z_o = \sqrt{Z_1 Z_2}$$

where

$$Z_1 = R + j\omega L$$

$$Y_2 = \frac{1}{Z_2} = \frac{1}{R_s} + \frac{1}{1/j\omega C}$$

$$= G + j\omega C$$

$$Z_2 = \frac{1}{G + j\omega C}$$

Therefore,

$$Z_o = \sqrt{(R + j\omega L)\frac{1}{G + j\omega C}}$$

or

$$Z_o = \sqrt{\frac{R + j\omega L}{G + j\omega C}} \tag{8-5}$$

Transmission-line Equivalent Circuit

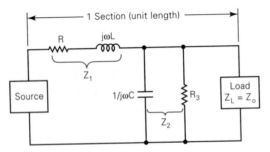

Figure 8-10 Equivalent circuit for a single section of transmission line terminated in a load equal to Z_o.

For extremely low frequencies, the resistances dominate and Equation 8-5 simplifies to

$$Z_o = \sqrt{\frac{R}{G}} \tag{8-6}$$

For extremely high frequencies, the inductance and capacitance dominate and Equation 8-5 simplifies to

$$Z_o = \sqrt{\frac{j\omega L}{j\omega C}} = \sqrt{\frac{L}{C}} \tag{8-7}$$

From Equation 8-7 it can be seen that for high frequencies the characteristic impedance of a transmission line approaches a constant, is independent of both frequency and length, and is determined solely by the distributed inductance and capacitance. It can also be seen that the phase angle is 0°. Therefore, Z_o looks purely resistive and all of the incident energy is absorbed by the line.

From a purely resistive approach, it can easily be seen that the impedance seen looking into a transmission line made up of an infinite number of sections approaches the characteristic impedance. This is shown in Figure 8-11. Again, for simplicity, only the series resistance R and the shunt resistance R_{sh} are considered. The impedance seen looking into the last section of the line is simply the sum of R and R_{sh}. Mathematically, Z_1 is

$$Z_1 = R + R_{sh} = 10 + 100 = 110$$

Adding a second section, Z_2, gives

$$Z_2 = R + \frac{R_{sh}Z_1}{R_{sh} + Z_1} = 10 + \frac{100 \times 110}{100 + 110} = 10 + 52.38 = 62.38$$

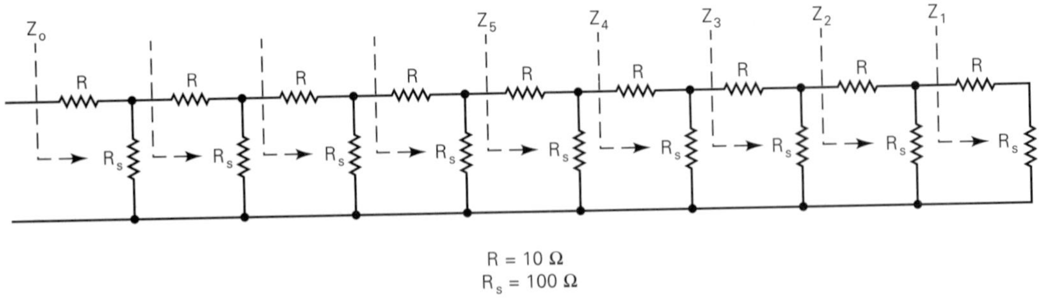

R = 10 Ω
R$_s$ = 100 Ω

Figure 8-11 Characteristic impedance of a transmission line of infinite sections or terminated in load equal to Z_o.

Chap. 8   Transmission Lines

and a third section, Z_3, is

$$Z_3 = R + \frac{R_{sh}Z_2}{R_{sh} + Z_2}$$

$$= 10 + \frac{100 \times 62.38}{100 + 62.38} = 10 + 38.42 = 48.32$$

A fourth section, Z_4, is

$$Z_4 = 10 + \frac{100 \times 48.32}{100 + 48.32} = 10 + 32.62 = 42.62$$

It can be seen that after each additional section the total impedance seen looking into the line decreases from its previous value. However, each time the magnitude of the decrease is less than the previous value. If the process shown above were continued, the impedance seen looking into the line will decrease asymptotically toward 37 Ω, which is the characteristic impedance of the line.

If the transmission line shown in Figure 8-11 were terminated in a load resistance Z_L = 37 Ω, the impedance seen looking into any number of sections would equal 37 Ω, the characteristic impedance. For a single section of line, Z_o is

$$Z_o = Z_1 = R + \frac{R_{sh} \times Z_L}{R_{sh} + Z_L} = 10 + \frac{100 \times 37}{100 + 37} = 10 + \frac{3700}{137} = 37 \ \Omega$$

Adding a second section, Z_2, is

$$Z_o = Z_2 = R + \frac{R_{sh} \times Z_1}{R_{sh} + Z_1} = 10 + \frac{100 \times 37}{100 + 37} = 10 + \frac{3700}{137} = 37 \ \Omega$$

Therefore, if this line were terminated into a load resistance Z_L = 37 Ω, Z_o = 37 Ω no matter how many sections are included.

The characteristic impedance of a transmission line can also be determined using Ohm's law. When a source is connected to an infinitely long line and a voltage is applied, a current flows. Even though the load is open, the circuit is complete through the distributed constants of the line. The characteristic impedance is simply the ratio of source voltage (E_o) to line current (I_o). Mathematically, Z_o is

$$Z_o = \frac{E_o}{I_o} \tag{8-8}$$

The characteristic impedance of a two-wire parallel transmission line with an air dielectric can be determined from its physical dimensions (see Figure 8-12a) and the formula

$$Z_o = 276 \log \frac{D}{r} \tag{8-9}$$

where D = distance between the centers of the two conductors
       r = radius of the conductor

and $D \gg r$.

Transmission-line Equivalent Circuit

329

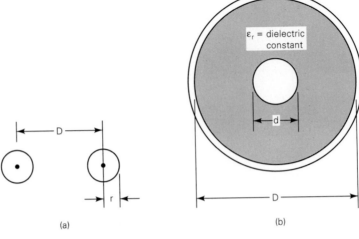

Figure 8-12 Physical dimensions of transmission lines: (a) two-wire parallel transmission line; (b) coaxial-cable transmission line.

EXAMPLE 8-1

Determine the characteristic impedance for an air dielectric two-wire parallel transmission line with a D/r ratio $= 12.22$.

Solution Substituting into Equation 8-9, we obtain

$$Z_o = 276 \log 12.22 = 300 \ \Omega$$

The characteristic impedance of a concentric coaxial cable can also be determined from its physical dimensions (see Figure 8-12b) and the formula

$$Z_o = \frac{138}{\sqrt{\epsilon_r}} \log \frac{D}{d} \tag{8-10}$$

where $D =$ inside diameter of the outer conductor
$d =$ outside diameter of the inner conductor
$\epsilon_r =$ dielectric constant (relative permativity) of the insulating material

EXAMPLE 8-2

Determine the characteristic impedance for a RG-59A coaxial cable with the following specifications: $L = 0.118 \ \mu H/ft$, $C = 21 \ pF/ft$, $d = 0.025$ in., $D = 0.15$, and $\epsilon = 2.23$.

Solution Substituting into Equation 8-8 yields

$$Z_o = \sqrt{\frac{L}{C}} = \sqrt{\frac{0.118 \times 10^{-6} H/ft}{21 \times 10^{-12} \ F/ft}} = 75 \ \Omega$$

Substituting into Equation 8-10 gives us

$$Z_o = \frac{138}{\sqrt{2.23}} \log \frac{0.15 \ in.}{0.25 \ in.} = 71.9 \ \Omega$$

Transmission lines can be summarized thus far as follows:

Chap. 8   Transmission Lines

1. The input impedance of an infinitely long line at radio frequencies is resistive and equal to Z_o.

2. Electromagnetic waves travel down the line without reflections; such a line is called *nonresonant*.

3. The ratio of voltage to current at any point along the line is equal to Z_o.

4. The incident voltage and current at any point along the line are in phase.

5. Line losses on a nonresonant line are minimum per unit length.

6. Any transmission line that is terminated in a purely resistive load equal to Z_o acts like an infinite line.

    a. $Z_i = Z_o$.

    b. There are no reflected waves.

    c. V and I are in phase.

    d. There is maximum transfer of power from source to load.

Propagation constant. *Propagation constant* (sometimes called *propagation coefficient*) is used to express the attenuation (signal loss) and the phase shift per unit length of a transmission line. As a wave propagates down a transmission line, its amplitude decreases with distance traveled. The propagation constant is used to determine the reduction in voltage or current with distance as a TEM wave propagates down a transmission line. For an infinitely long line, all the incident power is dissipated in the resistance of the wire as the wave propagates down the line. Therefore, with an infinitely long line or a line that looks infinitely long, such as a finite line terminated in a matched load ($Z_o = Z_L$), no energy is returned or reflected back toward the source. Mathematically, the propagation constant is

$$\gamma = \alpha + j\beta \tag{8-11a}$$

where γ = propagation constant

        α = attenuation coefficient (nepers per unit length)

        β = phase shift coefficient (radians per unit length)

The propagation constant is a complex quantity defined by

$$\gamma = \sqrt{(R + j\omega L)(G + j\omega C)} \tag{8-11b}$$

Since a phase shift of 2π rad occurs over a distance of one wavelength,

$$\beta = \frac{2\pi}{\lambda} \tag{8-12}$$

At intermediate and radio frequencies, $\omega L > R$ and $\omega C > G$; thus

$$\alpha = \frac{R}{2Z_o} + \frac{GZ_o}{2} \tag{8-13}$$

and

$$\beta = \omega \sqrt{LC} \tag{8-14}$$

The current and voltage distribution along a transmission line that is terminated in a load equal to its characteristic impedance (a matched line) are determined from the formulas

$$I = I_s e^{-l\gamma} \tag{8-15}$$

$$V = V_s e^{-l\gamma} \tag{8-16}$$

where I_s = current at the source end of the line

Transmission-line Equivalent Circuit

V_s = voltage at the source end of the line

γ = propagation constant

l = distance from the source at which the current or voltage is determined

For a matched load $Z_L = Z_o$, and for a given length of cable l, the loss in signal voltage or current is the real part of γl, and the phase shift is the imaginary part.

TRANSMISSION-LINE WAVE PROPAGATION

As stated previously, electromagnetic waves travel at the speed of light when propagating through a vacuum and nearly at the speed of light when propagating through air. However, in metallic transmission lines where the conductor is generally copper and the dielectric materials vary considerably with cable type, an electromagnetic wave travels much more slowly.

Velocity Factor

Velocity factor (sometimes called *velocity constant*) is defined simply as the ratio of the actual velocity of propagation through a given medium to the velocity of propagation through free space. Mathematically, the velocity factor is

$$V_f = \frac{V_p}{c} \tag{8-17}$$

where V_f = velocity factor

  V_p = actual velocity of propagation

  c = velocity of propagation through free space, $c = 3 \times 10^8$ m/s

and $$V_f \times c = V_p$$

The velocity at which an electromagnetic wave travels through a transmission line depends on the dielectric constant of the insulating material separating the two conductors. The velocity factor is closely approximated with the formula

$$V_f = \frac{1}{\sqrt{\epsilon_r}} \tag{8-18}$$

where ϵ_r is the dielectric constant of a given material the (permittivity of the material relative to the permittivity of a vacuum—the ratio ϵ/ϵ_o).

Dielectric constant is simply the *relative permittivity* of a material. The relative dielectric constant of air is 1.0006. However, the dielectric constant of materials commonly used in transmission lines range from 1.2 to 2.8, giving velocity factors from 0.6 to 0.9. The velocity factors for several common transmission-line configurations are given in Table 8-1, and the dielectric constants for several insulating materials are listed in Table 8-2.

Dielectric constant depends on the type of material used. Inductors store magnetic energy and capacitors store electric energy. It takes a finite amount of time for an inductor or a capacitor to take on or give up energy. Therefore, the velocity at which an electromagnetic wave propagates along a transmission line varies with the inductance and capacitance of the cable. It can be shown that time $T = \sqrt{LC}$. Therefore, inductance, capacitance, and velocity of propagation are mathematically related by the formula

TABLE 8-1 VELOCITY FACTORS

Material	Velocity Factor
Air	0.95–0.975
Rubber	0.56–0.65
Polyethylene	0.66
Teflon	0.70
Teflon foam	0.82
Teflon pins	0.81
Teflon spiral	0.81

$$\text{velocity} \times \text{time} = \text{distance}$$

Therefore,

$$V_p = \frac{\text{distance}}{\text{time}} = \frac{D}{T} \tag{8-19}$$

Substituting for time yields

$$V_p = \frac{D}{\sqrt{LC}} \tag{8-20}$$

If distance is normalized to 1 m, the velocity of propagation for a lossless line is

$$V_p = \frac{1 \text{ m}}{\sqrt{LC}} = \frac{1}{\sqrt{LC}} \quad \text{m/s} \tag{8-21}$$

EXAMPLE 8-3

For a given length of RG8A/U coaxial cable with a distributed capacitance $C = 96.6$ pF/m, a distributed inductance $L = 241.56$ nH/m, and a dielectric constant $\epsilon_r = 2.3$, determine the velocity of propagation and the velocity factor.

Solution From Equation 8-21,

$$V_p = \frac{1}{\sqrt{96.6 \times 10^{-12} \times 241.56 \times 10^{-9}}}$$

$$= 2.07 \times 10^8 \text{ m/s}$$

From Equation 8-17,

$$V_f = \frac{2.07 \times 10^8 \text{ m/s}}{3 \times 10^8 \text{ m/s}} = 0.69$$

TABLE 8-2 DIELECTRIC CONSTANTS

Material	Dielectric Constant
Vacuum	1.0
Air	1.0006
Teflon	2.0
Paper, paraffined	2.5
Rubber	3.0
Mica	5.0
Glass	7.5

Transmission-line Wave Propagation

From Equation 8-18,

$$V_f = \frac{1}{\sqrt{2.3}} = 0.66$$

Because wavelength is directly proportional to velocity and the velocity of propagation of a TEM wave varies with dielectric constant, the wavelength of a TEM wave also varies with dielectric constant. Therefore, for transmission media other than free space, Equation 8-3 can be rewritten as

$$\lambda = \frac{V_p}{f} = \frac{cV_f}{f} = \frac{c}{f\sqrt{\epsilon_r}} \tag{8-22}$$

Electrical Length of a Transmission Line

The length of a transmission line relative to the length of the wave propagating down it is an important consideration when analyzing transmission line behavior. At low frequencies (long wavelengths), the voltage along the line remains relatively constant. However, for high frequencies, several wavelengths of the signal may be present on the line at the same time. Therefore, the voltage along the line may vary appreciably. Consequently, the length of a transmission line is often given in wavelengths rather than in linear dimensions. Transmission-line phenomena apply to long lines. Generally, a transmission line is defined as long if its length exceeds one-sixteenth of a wavelength; otherwise, it is considered short. A given length of transmission line may appear short at one frequency and long at another frequency. For example, a 10-m length of transmission line at 1000 Hz is short ($\lambda = 300,000$ m; 10 m is only a small fraction of a wavelength). However, the same line at 6 GHz is long ($\lambda = 5$ cm; the line is 200 wavelengths long). It will be apparent later in this chapter, in Chapter 9, and in Appendix A that electrical length is used extensively for transmission line calculations and antenna design.

TRANSMISSION-LINE LOSSES

For analysis purposes, transmission lines are often considered totally lossless. In reality, however, there are several ways in which power is lost in a transmission line. They are conductor loss, radiation loss, dielectric heating loss, coupling loss, and corona.

Conductor Loss

Because current flows through a transmission line and the transmission line has a finite resistance, there is an inherent and unavoidable power loss. This is sometimes called *conductor* or *conductor heating loss* and is simply an I^2R loss. Because resistance is distributed throughout a transmission line, conductor loss is directly proportional to the square of the line length. Also, because power dissipation is directly proportional to the square of the current, conductor loss is inversely proportional to characteristic impedance. To reduce conductor loss, simply shorten the transmission line or use a larger-diameter wire (keep in mind that changing the wire diameter also changes the characteristic impedance and consequently the current).

Conductor loss depends somewhat on frequency. This is because of an action called the *skin effect*. When current flows through an isolated round wire, the magnetic flux asso-

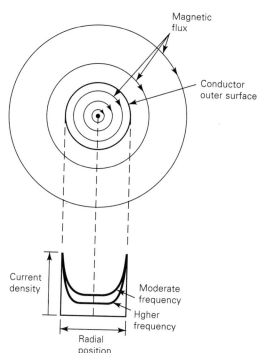

Figure 8-13 Isolated round conductor showing magnetic lines of flux, current distributions, and the skin effect.

ciated with it is in the form of concentric circles. This is shown in Figure 8-13. It can be seen that the flux density near the center of the conductor is greater than it is near the surface. Consequently, the lines of flux near the center of the conductor encircle the current and reduce the mobility of the encircled electrons. This is a form of self-inductance and causes the inductance near the center of the conductor to be greater than at the surface. Therefore, at radio frequencies, most of the current flows along the surface (outer skin) rather than near the center of the conductor. This is equivalent to reducing the cross-sectional area of the conductor and increasing the opposition to current flow (that is, resistance). The additional opposition has a 0° phase angle and is therefore a resistance and not a reactance. Therefore, the ac resistance of the conductor is proportional to the square root of the frequency. The ratio of the ac resistance to the dc resistance of a conductor is called the *resistance ratio*. Above approximately 100 MHz, the center of a conductor can be completely removed and have absolutely no effect on the total conductor loss or EM wave propagation.

Conductor loss in transmission lines varies from as low as a fraction of a decibel per 100 m for rigid air dielectric coaxial cable to as high as 200 dB per 100 m for a solid dielectric flexible line.

Radiation Loss

If the separation between conductors in a transmission line is an appreciable fraction of a wavelength, the electrostatic and electromagnetic fields that surround the conductor cause the line to act like an antenna and transfer energy to any nearby conductive material. The amount of energy radiated depends on the dielectric material, the conductor spacing, and the length of the line. *Radiation losses* are reduced by properly shielding the cable. Therefore, coaxial cables have less radiation loss than do two-wire parallel lines. Radiation loss is also directly proportional to frequency.

Transmission-line Losses

Dielectric Heating Loss

A difference of potential between the two conductors of a transmission line causes *dielectric heating*. Heat is a form of energy and must be taken from the energy propagating down the line. For air dielectric lines, the heating loss is negligible. However, for solid lines, dielectric heating loss increases with frequency.

Coupling Loss

Coupling loss occurs whenever a connection is made to or from a transmission line or when two separate pieces of transmission line are connected together. Mechanical connections are discontinuities (places where dissimilar materials meet). Discontinuities tend to heat up, radiate energy, and dissipate power.

Corona

Corona is a luminous discharge that occurs between the two conductors of a transmission line when the difference of potential between them exceeds the *breakdown* voltage of the dielectric insulator. Generally, once corona has occurred, the transmission line may be destroyed.

INCIDENT AND REFLECTED WAVES

An ordinary transmission line is bidirectional; power can propagate equally well in both directions. Voltage that propagates from the source toward the load is called *incident voltage*, and voltage that propagates from the load toward the source is called *reflected voltage*. Similarly, there are incident and reflected currents. Consequently, incident power propagates toward the load, and reflected power propagates toward the source. Incident voltage and current are always in phase for a resistive characteristic impedance. For an infinitely long line, all the incident power is stored by the line and there is no reflected power. Also, if the line is terminated in a purely resistive load equal to the characteristic impedance of the line, the load absorbs all the incident power (this assumes a lossless line). For a more practical definition, reflected power is the portion of the incident power that was not absorbed by the load. Therefore, the reflected power can never exceed the incident power.

Resonant and Nonresonant Lines

A line with no reflected power is called a *flat* or *nonresonant line*. On a flat line, the voltage and current are constant throughout its length, assuming no losses. When the load is either a short or an open circuit, all the incident power is reflected back toward the source. If the source were replaced with an open or a short and the line were lossless, energy present on the line would reflect back and forth (oscillate) between the load and source ends similar to the power in a tank circuit. This is called a *resonant line*. In a resonant line, the energy is alternately transferred between the magnetic and electric fields of the distributed inductance and capacitance. Figure 8-14 shows a source, transmission line, and load with their corresponding incident and reflected waves.

Chap. 8   Transmission Lines

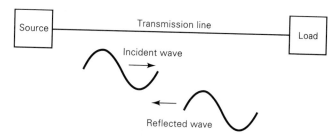

Figure 8-14 Source, load, transmission line, and their corresponding incident and reflected waves.

Reflection Coefficient

The reflection coefficient (sometimes called the *coefficient of reflection*) is a vector quantity that represents the ratio of reflected voltage to incident voltage or reflected current to incident current. Mathematically, the reflection coefficient is gamma, Γ, defined by

$$\Gamma = \frac{E_r}{E_i} \quad \text{or} \quad \frac{I_r}{I_i} \tag{8-23}$$

where Γ = reflection coefficient
          E_i = incident voltage
          E_r = reflected voltage
          I_i = incident current
          I_r = reflected current

From Equation 8-23 it can be seen that the maximum and worst-case value for Γ is 1 ($E_r = E_i$), and the minimum value and ideal condition occur when $\Gamma = 0$ ($E_r = 0$).

STANDING WAVES

When $Z_o = Z_L$, all the incident power is absorbed by the load. This is called a *matched line*. When $Z_o \neq Z_L$, some of the incident power is absorbed by the load and some is returned (reflected) to the source. This is called an *unmatched* or *mismatched line*. With a mismatched line, there are two electromagnetic waves, traveling in opposite directions, present on the line at the same time (these waves are in fact called *traveling waves*). The two traveling waves set up an interference pattern known as a *standing wave*. This is shown in Figure 8-15. As the incident and reflected waves pass each other, stationary patterns of voltage and current are produced on the line. These stationary waves are called standing waves because they appear to remain in a fixed position on the line, varying only in amplitude. The standing wave has minima (nodes) separated by a half-wavelength of the traveling waves and maxima (antinodes) also separated by a half wavelength.

Standing-wave Ratio

The *standing-wave ratio* (SWR) is defined as the ratio of the maximum voltage to the minimum voltage or the maximum current to the minimum current of a standing wave on a transmission line. SWR is often called the *voltage standing-wave ratio* (VSWR). Essentially,

Standing Waves

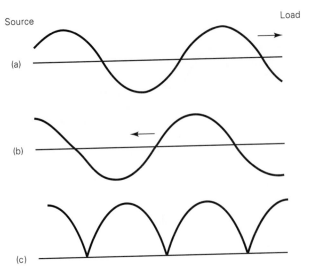

Source
Load

(a)

(b)

(c)

Figure 8-15 Developing a standing wave on a transmission line: (a) incident wave; (b) reflected wave; (c) standing wave.

SWR is a measure of the mismatch between the load impedance and the characteristic impedance of the transmission line. Mathematically, SWR is

$$SWR = \frac{V_{max}}{V_{min}} \qquad (8\text{-}24)$$

The voltage maxima (V_{max}) occur when the incident and reflected waves are in phase (that is, their maximum peaks pass the same point on the line with the same polarity), and the voltage minima (V_{min}) occur when the incident and reflected waves are 180° out of phase. Mathematically, V_{max} and V_{min} are

$$V_{max} = E_i + E_r \qquad (8\text{-}25)$$

$$V_{min} = E_i - E_r \qquad (8\text{-}26)$$

Therefore, Equation 8-24 can be rewritten as

$$SWR = \frac{V_{max}}{V_{min}} = \frac{E_i + E_r}{E_i - E_r} \qquad (8\text{-}27)$$

From Equation 8-27, it can be seen that when the incident and reflected waves are equal in amplitude (a total mismatch) SWR = infinity. This is the worst-case condition. Also, from Equation 8-27, it can be seen that when there is no reflected wave ($E_r = 0$) SWR = E_i/E_i or 1. This condition occurs when $Z_o = Z_L$ and is the ideal situation.

The standing-wave ratio can also be written in terms of Γ. Rearranging Equation 8-24 yields

$$\Gamma E_i = E_r$$

Substituting into Equation 8-27 gives us

$$SWR = \frac{E_i + E_i\Gamma}{E_i - E_i\Gamma}$$

Factoring out E_i yields

$$SWR = \frac{E_i(1 + \Gamma)}{E_i(1 - \Gamma)} = \frac{1 + \Gamma}{1 - \Gamma} \qquad (8\text{-}28)$$

Chap. 8   Transmission Lines

Cross-multiplying gives

$$SWR(1 - \Gamma) = 1 + \Gamma$$

$$SWR - SWR\,\Gamma = 1 + \Gamma$$

$$SWR = 1 + \Gamma + (SWR)\,\Gamma$$

$$SWR - 1 = \Gamma(1 + SWR) \qquad (8\text{-}29)$$

$$\Gamma = \frac{SWR - 1}{SWR + 1} \qquad (8\text{-}30)$$

EXAMPLE 8-4

For a transmission line with incident voltage $E_i = 5$ V and reflected voltage $E_r = 3$ V, determine:

(a) Reflection coefficient.
(b) SWR.

Solution (a) Substituting into Equation 8-23 yields

$$\Gamma = \frac{E_r}{E_i} = \frac{3}{5} = 0.6$$

(b) Substituting into Equation 8-27 gives us

$$SWR = \frac{E_i + E_r}{E_i - E_r} = \frac{5 + 3}{5 - 3} = \frac{8}{2} = 4$$

Substituting into Equation 8-30, we obtain

$$\Gamma = \frac{4 - 1}{4 + 1} = \frac{3}{5} = 0.6$$

When the load is purely resistive, SWR can also be expressed as a ratio of the characteristic impedance to the load impedance, or vice versa. Mathematically, SWR is

$$SWR = \frac{Z_o}{Z_L} \quad \text{or} \quad \frac{Z_L}{Z_o} \quad \text{(whichever gives an SWR greater than 1)} \qquad (8\text{-}31)$$

The numerator and denominator for Equation (8-31) are chosen such that the SWR is always a number greater than 1, to avoid confusion and comply with the convention established in Equation 8-27. From Equation 8-31 it can be seen that a load resistance $Z_L = 2Z_o$ gives the same SWR as a load resistance $Z_L = Z_o/2$; the degree of mismatch is the same.

The disadvantages of not having a matched (flat) transmission line can be summarized as follows:

1. One hundred percent of the source incident power does not reach the load.
2. The dielectric separating the two conductors can break down and cause corona as a result of the high voltage standing wave ratio.
3. Reflections and rereflections cause more power loss.
4. Reflections cause ghost images.
5. Mismatches cause noise interference.

Although it is highly unlikely that a transmission line will be terminated in a load that is either an open or a short circuit, these conditions are examined because they illustrate

Standing Waves

the worst possible conditions that could occur and produce standing waves that are representative of less severe conditions.

Standing Waves on an Open Line

When incident waves of voltage and current reach an open termination, none of the power is absorbed; it is all reflected back toward the source. The incident voltage wave is reflected in exactly the same manner as if it were to continue down an infinitely long line. However, the incident current is reflected 180° reversed from how it would have continued if the line were not open. As the incident and reflected waves pass, standing waves are produced on the line. Figure 8-16 shows the voltage and current standing waves on a transmission line that is terminated in an open circuit. It can be seen that the voltage standing wave has a maximum value at the open end and a minimum value one-quarter wavelength from the open. The current standing wave has a minimum value at the open end and a maximum value one-quarter wavelength from the open. It stands to reason that maximum voltage occurs across an open and there is minimum current.

The characteristics of a transmission line terminated in an open can be summarized as follows:

1. The voltage incident wave is reflected back just as if it were to continue (that is, no phase reversal).

2. The current incident wave is reflected back 180° from how it would have continued.

3. The sum of the incident and reflected current waveforms is minimum at the open.

4. The sum of the incident and reflected voltage waveforms is maximum at the open.

From Figure 8-16 it can also be seen that the voltage and current standing waves repeat every one-half wavelength. The impedance at the open end $Z = V_{max}/I_{min}$ and is maximum. The impedance one-quarter wavelength from the open $Z = V_{min}/I_{max}$ and is minimum. Therefore, one-quarter wavelength from the open an impedance inversion occurs and additional impedance inversions occur each quarter-wavelength.

Figure 8-16 shows the development of a voltage standing wave on a transmission line that is terminated in an open circuit. Figure 8-17 shows an incident wave propagating down a transmission line toward the load. The wave is traveling at approximately the speed of light; however, for illustration purposes, the wave has been frozen at eighth-wavelength intervals. In Figure 8-17a it can be seen that the incident wave has not reached the open. Figure 8-17b shows the wave one time unit later (for this example, the wave travels one-eighth wavelength per time unit). As you can see, the wave has moved one-quarter wavelength closer to the open. Figure 8-17c shows the wave just as it arrives at the open. Thus far there has been no reflected wave and, consequently, no standing wave. Figure 8-17d shows the incident and reflected waves one time unit after the incident wave has

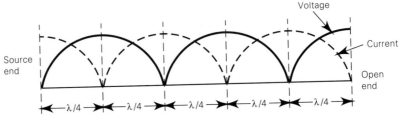

Figure 8-16 Voltage and current standing waves on a transmission line that is terminated in an open circuit.

Chap. 8   Transmission Lines

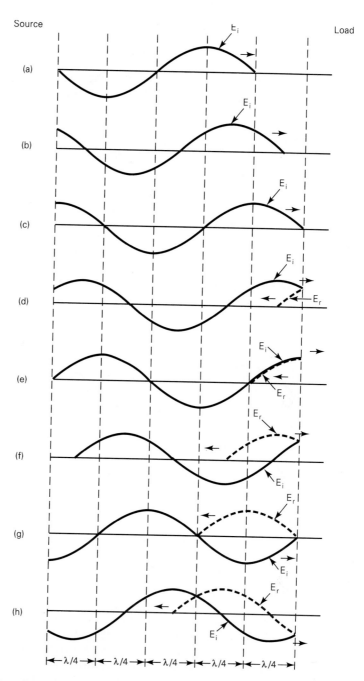

Figure 8-17 Incident and reflected waves on a transmission line terminated in an open circuit. *(Continued on next page.)*

reached the open; the reflected wave is propagating away from the open. Figures 8-17e, f, and g show the incident and reflected waves for the next three time units. In Figure 8-17e it can be seen that the incident and reflected waves are at their maximum positive values at the same time, thus producing a voltage maximum at the open. It can also be seen that one-quarter wavelength from the open the sum of the incident and reflected waves (the standing wave) is always equal to 0 V (a minimum). Figures 8-17h through m show propagation

Standing Waves

341

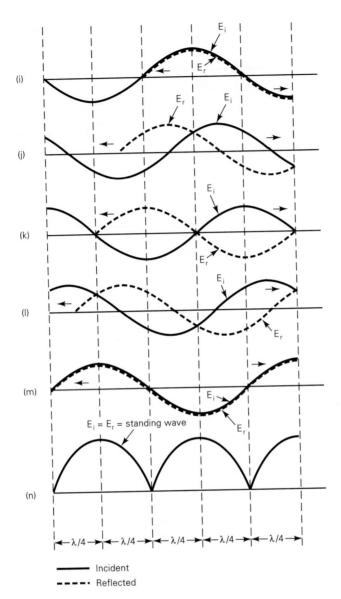

(i)

(j)

(k)

(l)

(m)

$E_i = E_r$ = standing wave

(n)

$\leftarrow \lambda/4 \rightarrow \leftarrow \lambda/4 \rightarrow \leftarrow \lambda/4 \rightarrow \leftarrow \lambda/4 \rightarrow \leftarrow \lambda/4 \rightarrow$

—— Incident
- - - - Reflected

Figure 8-17 *(Continued)*

of the incident and reflected waves until the reflected wave reaches the source, and Figure 8-17n shows the resulting standing wave. It can be seen that the standing wave remains stationary (the voltage nodes and antinodes remain at the same points). However, the amplitude of the antinodes varies from maximum positive to zero to maximum negative and then repeats. For an open load, all the incident voltage is reflected ($E_r = E_i$); therefore, $V_{max} = E_i + E_r$ or $2E_i$. A similar illustration can be shown for a current standing wave (however, remember that the current reflects back with a 180° phase inversion).

Standing Waves on a Shorted Line

As with an open line, none of the incident power is absorbed by the load when a transmission line is terminated in a short circuit. However, with a shorted line, the incident voltage and current waves are reflected back in the opposite manner. The voltage wave is reflected

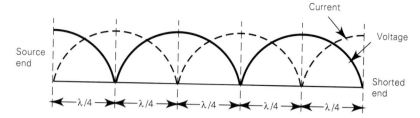

Figure 8-18 Voltage and current standing waves on a transmission line that is terminated in a short circuit.

180° reversed from how it would have continued down an infinitely long line, and the current wave is reflected in exactly the same manner as if there were no short.

Figure 8-18 shows the voltage and current standing waves on a transmission line that is terminated in a short circuit. It can be seen that the voltage standing wave has a minimum value at the shorted end and a maximum value one-quarter wavelength from the short. The current standing wave has a maximum value at the short and a minimum value one-quarter wavelength back. The voltage and current standing waves repeat every quarter-wavelength. Therefore, there is an impedance inversion every quarter-wavelength. The impedance at the short $Z = V_{min}/I_{max} = $ minimum, and one-quarter wavelength back $Z = V_{max}/I_{min} = $ maximum. Again, it stands to reason that a voltage minimum will occur across a short and there is maximum current.

The characteristics of a transmission line terminated in a short can be summarized as follows:

1. The voltage standing wave is reflected back 180° reversed from how it would have continued.

2. The current standing wave is reflected back the same as if it had continued.

3. The sum of the incident and reflected current waveforms is maximum at the short.

4. The sum of the incident and reflected voltage waveforms is zero at the short.

For a transmission line terminated in either a short or an open circuit, the reflection coefficient is 1 (the worst case) and the SWR is infinity (also the worst-case condition).

TRANSMISSION-LINE INPUT IMPEDANCE

In the preceding section it was shown that, when a transmission line is terminated in either a short or an open circuit, there is an *impedance inversion* every quarter-wavelength. For a lossless line, the impedance varies from infinity to zero. However, in a more practical situation where power losses occur, the amplitude of the reflected wave is always less than that of the incident wave except at the termination. Therefore, the impedance varies from some maximum to some minimum value, or vice versa, depending on whether the line is terminated in a short or an open. The input impedance for a lossless line seen looking into a transmission line that is terminated in a short or an open can be resistive, inductive, or capacitive, depending on the distance from the termination.

Transmission-line Input Impedance

Phasor Analysis of Input Impedance: Open Line

Phasor diagrams are generally used to analyze the input impedance of a transmission line because they are relatively simple and give a pictorial representation of the voltage and current phase relationships. Voltage and current phase relations refer to variations in time. Figures 8-16, 8-17, and 8-18 show standing waves of voltage and current plotted versus distance and are therefore not indicative of true phase relationships. The succeeding sections use phasor diagrams to analyze the input impedance of several transmission-line configurations.

Quarter-wavelength transmission line. Figure 8-19a shows the phasor diagram for the voltage and Figure 8-19b shows the phasor diagram for the current at the input to a quarter-wave section of a transmission line terminated in an open circuit. I_i and V_i are the in-phase incident current and voltage waveforms at the input (source) end of the line at a given instant in time, respectively. Any reflected voltage (E_r) present at the input of the line has traveled one-half wavelength (from the source to the open and back) and is, consequently, 180° behind the incident voltage. Therefore, the total voltage (E_t) at the input end is the sum of E_i and E_r. $E_t = E_i + E_r \underline{/-180°}$, and, assuming a small line loss, $E_t = E_i - E_r$. The reflected current is delayed 90° propagating from the source to the load and another 90° from the load back to the source. Also, the reflected current undergoes an 180° phase reversal at the open. The reflected current has effectively been delayed 360°. Therefore, when the reflected current reaches the source end, it is in phase with the incident current and the total current $I_t = I_i + I_r$. By examining Figures 8-19, it can be seen that E_t and I_t are in phase. Therefore, the input impedance seen looking into a transmission line one-quarter wavelength long that is terminated in an open circuit $Z_{in} = E_t \underline{/0°}/I_t \underline{/0°} = Z_{in} \underline{/0°}$. Z_{in} has a 0° phase angle, is resistive, and is minimum. Therefore, a quarter-wavelength transmission line terminated in an open circuit is equivalent to a series resonant LC circuit.

Figure 8-20 shows several voltage phasors for the incident and reflected waves on a transmission line that is terminated in an open circuit and how they produce a voltage standing wave.

Transmission line less than one-quarter wavelength long. Figure 8-21a shows the voltage phasor diagram and Figure 8-21b shows the current phasor diagram for a transmission line that is less than one-quarter wavelength long ($\lambda/4$) and terminated in an open circuit. Again, the incident current (I_i) and voltage (E_i) are in phase. The reflected

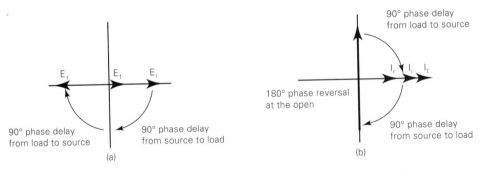

Figure 8-19 Voltage and current phase relationships for a quarter-wave line terminated in an open circuit: (a) voltage phase relationships; (b) current phase relationships.

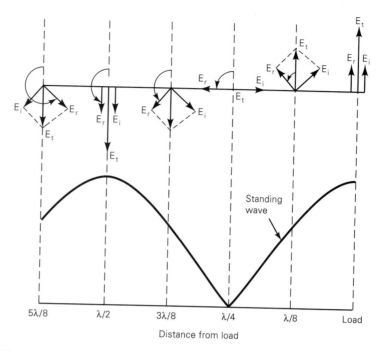

Figure 8-20 Vector addition of incident and reflected waves producing a standing wave.

voltage wave is delayed 45° traveling from the source to the load (a distance of one-eighth wavelength) and another 45° traveling from the load back to the source (an additional one-eighth wavelength). Therefore, when the reflected wave reaches the source end, it lags the incident wave by 90°. The total voltage at the source end is the vector sum of the incident and reflected waves. Thus, $E_t = \sqrt{E_i^2 + E_r^2} = E_t \underline{/-45°}$. The reflected current wave is delayed 45° traveling from the source to the load and another 45° from the load back to the source (a total distance of one-quarter wavelength). In addition, the reflected current wave has undergone a 180° phase reversal at the open prior to being reflected. The reflected current wave has been delayed a total of 270°. Therefore, the reflected wave effectively leads

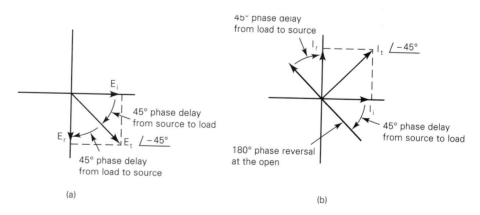

Figure 8-21 Voltage and current phase relationships for a transmission line less than one-quarter wavelength terminated in an open circuit: (a) voltage phase relationships; (b) current phase relationships.

Transmission-line Input Impedance

the incident wave by 90°. The total current at the source end is the vector sum of the present and reflected waves. Thus, $I_t = \sqrt{I_i^2 + I_r^2} = I_t \underline{/+45°}$. By examining Figure 8-21, it can be seen that E_t lags I_t by 90°. Therefore, $Z_{in} = E_t \underline{/-45°} /I_t \underline{/+45°} = Z_{in} \underline{/-90°}$. Z_{in} has a $-90°$ phase angle and is therefore capacitive. Any transmission line that is less than one-quarter wavelength and terminated in an open circuit is equivalent to a capacitor. The amount of capacitance depends on the exact electrical length of the line.

Transmission line more than one-quarter wavelength long. Figure 8-22a shows the voltage phasor diagram and Figure 8-22b shows the current phasor diagram for a transmission line that is more than one-quarter wavelength long and terminated in an open circuit. For this example, a three-eighths wavelength transmission line is used. The reflected voltage is delayed three-quarters wavelength or 270°. Therefore, the reflected voltage effectively leads the incident voltage by 90°. Consequently, the total voltage $E_t = \sqrt{E_i^2 + E_r^2} \underline{/+45°} = E_t \underline{/+45°}$. The reflected current wave has been delayed 270° and undergone an 180° phase reversal. Therefore, the reflected current effectively lags the incident current by 90°. Consequently, the total current $I_t = \sqrt{I_i^2 + I_r^2} \underline{/-45°} = I_t \underline{/-45°}$. Therefore, $Z_{in} = E_t \underline{/+45°} /I_t \underline{/-45°} = Z_{in} \underline{/+90°}$. Z_{in} has a $+90°$ phase angle and is therefore inductive. The magnitude of the input impedance equals the characteristic impedance at eighth wavelength points. A transmission line between one-quarter and one-half wavelength that is terminated in an open circuit is equivalent to an inductor. The amount of inductance depends on the exact electrical length of the line.

Open transmission line as a circuit element. From the preceding discussion and Figures 8-19 through 8-22, it is obvious that an open transmission line can behave like a resistor, an inductor, or a capacitor, depending on its electrical length. Because standing-wave patterns on an open line repeat every half-wavelength, the input impedance also repeats. Figure 8-23 shows the variations in input impedance for an open transmission line of various electrical lengths. It can be seen that an open line is resistive and maximum at the open and at each successive half-wavelength interval, and resistive and minimum one-quarter wavelength from the open and at each successive half-wavelength interval. For electrical lengths less than one-quarter wavelength, the input impedance is capacitive and decreases with length. For electrical lengths between one-quarter and one-half wavelength, the input impedance is inductive and increases with length. The capacitance and inductance patterns also repeat every half-wavelength.

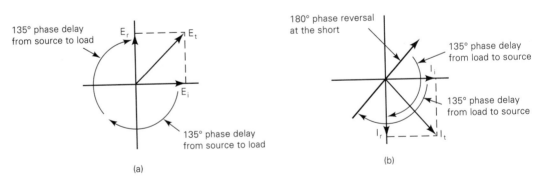

(a)

(b)

Figure 8-22 Voltage and current phase relationships for a transmission more than one-quarter wavelength terminated in an open circuit: (a) voltage phase relationships; (b) current phase relationships.

Chap. 8   Transmission Lines

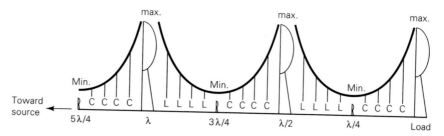

Figure 8-23 Input impedance variations for an open-circuited transmission line.

Phasor Analysis of Input Impedance: Shorted Line

The following explanations use phasor diagrams to analyze shorted transmission lines in the same manner as with open lines. The difference is that with shorted transmission lines the voltage waveform is reflected back with a 180° phase reversal and the current waveform is reflected back as if there were no short.

Quarter-wavelength transmission line. The voltage and current phasor diagrams for a quarter-wavelength transmission line terminated in a short circuit are identical to those shown in Figure 8-19, except reversed. The incident and reflected voltages are in phase; therefore, $E_t = E_i + E_r$ and maximum. The incident and reflected currents are 180° out of phase; therefore, $I_t = I_i - I_r$ and minimum. $Z_{in} = E_i \underline{/0°}/I_t \underline{/0°} = Z_{in} \underline{/0°}$ and maximum. Z_{in} has a 0° phase angle, is resistive, and is maximum. Therefore, a quarter-wavelength transmission line terminated in a short circuit is equivalent to a parallel *LC* circuit.

Transmission line less than one-quarter wavelength long. The voltage and current phasor diagrams for a transmission line less than one-quarter wavelength long and terminated in a short circuit are identical to those shown in Figure 8-21, except reversed. The voltage is reversed 180° at the short, and the current is reflected with the same phase as if it had continued. Therefore, the total voltage at the source end of the line leads the current by 90° and the line looks inductive.

Transmission line more than one-quarter wavelength long. The voltage and current phasor diagrams for a transmission line more than one-quarter wavelength long and terminated in a short circuit are identical to those shown in Figure 8-22, except reversed. The total voltage at the source end of the line lags the current by 90° and the line looks capacitive.

Shorted transmission line as a circuit element. From the preceding discussion it is obvious that a shorted transmission line can behave like a resistor, an inductor, or a capacitor, depending on its electrical length. On a shorted transmission line, standing waves repeat every half-wavelength; therefore, the input impedance also repeats. Figure 8-24 shows the variations in input impedance of a shorted transmission line for various electrical lengths. It can be seen that a shorted line is resistive and minimum at the short and at each successive half-wavelength interval, and resistive and maximum one-quater wavelength from the short and at each successive half-wavelength interval. For electrical lengths less than one-quarter wavelength, the input impedance is inductive and increases with length. For electrical lengths between one-quarter and one-half wavelength, the input impedance is capacitive and decreases with length. The inductance and capacitance patterns also repeat every half-wavelength.

Transmission-line Input Impedance

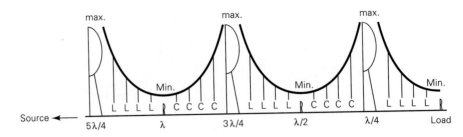

Figure 8-24 Input impedance variations for a short-circuited transmission line.

Transmission-line input impedance summary. Figure 8-25 summarizes the transmission-line configurations described in the preceding sections, their input impedance characteristics, and their equivalent *LC* circuits. It can be seen that both shorted and open sections of transmission lines can behave as resistors, inductors, or capacitors, depending on their electrical length.

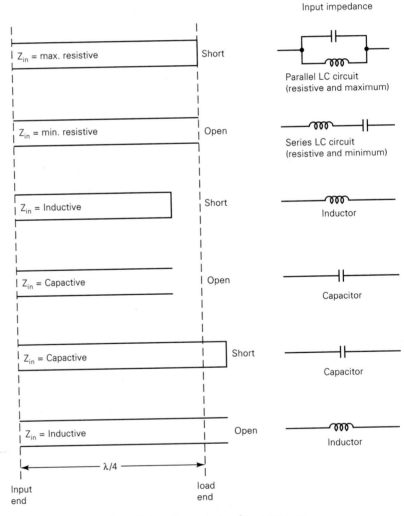

Figure 8-25 Transmission-line summary.

Transmission-line Impedance Matching

Power is transferred most efficiently to a load when there are no reflected waves, that is, when the load is purely resistive and equal to Z_o. Whenever the characteristic impedance of a transmission line and its load are not matched (equal), standing waves are present on the line and maximum power is not transferred to the load. Standing waves cause power loss, dielectric breakdown, noise, radiation, and *ghost signals*. Therefore, whenever possible a transmission line should be matched to its load. Two common transmission-line techniques are used to match a transmission line to a load having an impedance that is not equal to Z_o. They are quarter-wavelength transformer matching and stub matching.

Quarter-wavelength transformer matching. *Quarter-wavelength transformers* are used to match transmission lines to purely resistive loads whose resistance is not equal to the characteristic impedance of the line. Keep in mind that a quarter-wavelength transformer is not actually a transformer, but rather a quarter-wavelength section of transmission line that acts like a transformer. The input impedance to a transmission line varies from some maximum value to some minimum value, or vice versa, every quarter-wavelength. Therefore, a transmission line one-quarter wavelength long acts like a *step-up* or *step-down transformer*, depending on whether Z_L is greater than or less than Z_o. A quarter-wavelength transformer is not a broadband impedance-matching device; it is a quarter-wavelength at only a single frequency. The impedance transformations for a quarter-wavelength transmission line are as follows:

1. $R_L = Z_o$: The quarter-wavelength line acts like a transformer with a 1:1 turns ratio.
2. $R_L > Z_o$: The quarter-wavelength line acts like a step-down transformer.
3. $R_L < Z_o$: The quarter-wavelength line acts like a step-up transformer.

Like a transformer, a quarter-wavelength transformer is placed between a transmission line and its load. A quarter-wavelength transformer is simply a length of transmission line one-quarter wavelength long. Figure 8-26 shows how a quarter-wavelength transformer is used to match a transmission line to a purely resistive load. The characteristic impedance of the quarter-wavelength section is determined mathematically from the formula

$$Z'_o = \sqrt{Z_o Z_L} \qquad (8-32)$$

where $\quad Z'_o$ = characteristic impedance of a quarter-wavelength transformer
$\qquad Z_o$ = characteristic impedance of the transmission line that is being matched
$\qquad Z_L$ = load impedance

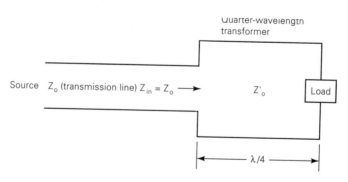

Figure 8-26 Quarter-wavelength transformer.

Transmission-line Input Impedance

349

EXAMPLE 8-5

Determine the physical length and characteristic impedance for a quarter-wavelength transformer that is used to match a section of RG-8A/U ($Z_o = 50 \, \Omega$) to a 150-Ω resistive load. The frequency of operation is 150 MHz and the velocity factor $V_f = 1$.

Solution The physical length of the transformer depends on the wavelength of the signal. Substituting into Equation 8-2 yields

$$\lambda = \frac{c}{f} = \frac{3 \times 10^8 \text{ m/s}}{150 \text{ MHz}} = 2 \text{ m}$$

$$\frac{\lambda}{4} = \frac{2 \text{ m}}{4} = 0.5 \text{ m}$$

The characteristic impedance of the 0.5-m transformer is determined from Equation 8-32.

$$Z'_o = \sqrt{Z_o Z_L} = \sqrt{(50)(150)} = 86.6 \, \Omega$$

Stub matching. When a load is purely inductive or purely capacitive, it absorbs no energy. The reflection coefficient is 1 and the SWR is infinity. When the load is a complex impedance (which is usually the case), it is necessary to remove the reactive component to match the transmission line to the load. Transmission-line *stubs* are commonly used for this purpose. A transmission-line stub is simply a piece of additional transmission line that is placed across the primary line as close to the load as possible. The susceptance of the stub is used to tune out the susceptance of the load. With *stub matching*, either a shorted or an open stub can be used. However, shorted stubs are preferred because open stubs have a tendency to radiate, especially at the higher frequencies.

Figure 8-27 shows how a shorted stub is used to cancel the susceptance of the load and match the load resistance to the characteristic impedance of the transmission line. It has been shown how a shorted section of transmission line can look resistive, inductive, or

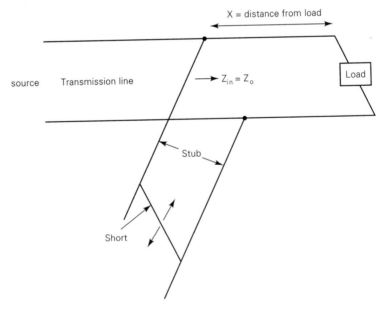

Figure 8-27 Shorted stub impedance matching.

capacitive, depending on its electrical length. A transmission line that is one-half wavelength or shorter can be used to tune out the reactive component of a load.

The process of matching a load to a transmission line with a shorted stub is as follows:

1. Locate a point as close to the load as possible where the conductive component of the input admittance is equal to the characteristic admittance of the transmission line.

$$Y_{in} = G - jB, \quad \text{where } G = \frac{1}{Z_o}$$

2. Attach the shorted stub to the point on the transmission line identified in step 1.

3. Depending on whether the reactive component at the point indentified in step 1, due to the load is inductive or capacitive, the stub length is adjusted accordingly.

$$Y_{in} = G_o - jB + jB_{stub}$$

$$= G_o$$

if

$$B = B_{stub}$$

For a more complete explanation of stub matching using the Smith chart, refer to Appendix A.

TIME-DOMAIN REFLECTOMETRY

Metallic cables, like all components within an electronic communications system, can develop problems that inhibit their ability to perform as expected. Cable problems often create unique situations because cables often extend over large distances, sometimes as far as several thousand feet, or longer. Cable problems are often attributed to chemical erosion at cross-connect points and mechanical failure. When a problem occurs in a cable, it can be extremely time consuming and, consequently, quite expensive, to determine the type and exact location of the problem.

A technique that can be used to locate an impairment in a metallic cable is called *time-domain reflectometry (TDR)*. With TDR, transmission-line impairments can be pinpointed within several feet at distances of 10 miles. TDR makes use of the well-established theory that transmission line impairments, such as shorts and opens, cause a portion of the incident signal to return to the source. How much of the transmitted signal returns depends on the type and magnitude of the impairment. The point in the line where the impairment is located represents a discontinuity to the signal. This discontinuity causes a portion of the transmitted signal to be reflected rather than continuing down the cable. If no energy is returned (that is, the transmission line and load are perfectly matched), the line is either infinitely long or it is terminated in a resistive load with an impedance equal to the characteristic impedance of the line. TDR operates in a fashion similar to *radar*. A short duration pulse with a fast rise time is propagated down a cable; then the time for a portion of that signal to return to the source is measured. This return signal is sometimes called an *echo*. Knowing the velocity of propagation on the cable, the exact distance between the impairment and the source can be determined using the following mathematical relationships:

$$d = \frac{v \times t}{2} \tag{8-33}$$

Time-domain Reflectometry

where d = distance to the discontinuity (meters)
       v = velocity (meters/second)
       v = $k \times c$ (meters/second)
       k = velocity factory (v/c)
       c = 3×10^8 meters/second
       t = elapsed time (seconds)

The elapsed time is measured from the leading edge of the transmitted pulse to the reception of the reflected signal as shown in Figure 8-28a. It is important that the transmitted pulse be as narrow as possible. Otherwise, when the impairment is located close to the source, the reflected signal could return while the pulse is still being transmitted (Figure 8-28b), making it difficult to detect it. For signals traveling at the speed of light (c), the velocity of propagation is 3×10^8 m/s or approximately 1 ns/ft. Consequently, a pulse width of several microseconds would limit the usefulness of TDR only to cable impairments that occurred several thousand feet or farther away. Producing an extremely narrow pulse was one of the limiting factors in the development of TDR for locating cable faults on short cables.

EXAMPLE 8-6

A pulse is transmitted down a cable that has a velocity of propagation of $0.8c$. The reflected signal is received 1 μs later. How far down the cable is the impairment?

Solution   Substituting into Equation 8-33,

$$d = \frac{(0.8\ c) \times 1\ \mu s}{2}$$

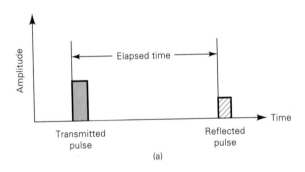

(a)

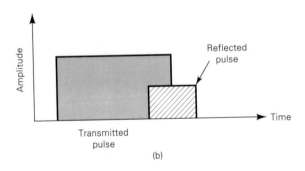

(b)

Figure 8-28   Time-domain reflectometry: (a) elapsed time; (b) transmitted pulse too long.

$$= \frac{0.8 \times (3 \times 10^8 \text{ m/s}) \times 1 \times 10^{-6} \text{ s}}{2}$$

$$= 120 \text{ m}$$

EXAMPLE 8-7

Using TDR, a transmission-line impairment is located 3000 m from the source. For a velocity of propagation of $0.9c$, determine the time elapsed from the beginning of the pulse to the reception of the echo.

Solution   Rearranging Equation 8-33 gives

$$t = \frac{2d}{v} = \frac{2d}{k \times c}$$

$$t = \frac{2 (3000 \text{ m})}{0.9 (3 \times 10^8 \text{ m/s})}$$

$$= 22.22 \text{ μs}$$

QUESTIONS

8-1. Define *transmission line.*

8-2. Describe a transverse electromagnetic wave.

8-3. Define *wave velocity.*

8-4. Define *frequency* and *wavelength* for a transverse electromagnetic wave.

8-5. Describe balanced and unbalanced transmission lines.

8-6. Describe an open-wire transmission line.

8-7. Describe a twin-lead transmission line.

8-8. Describe a twisted-pair transmission line.

8-9. Describe a shielded-cable transmission line.

8-10. Describe a concentric transmission line.

8-11. Describe the electrical and physical properties of a transmission line.

8-12. List and describe the four primary constants of a transmission line.

8-13. Define *characteristic impedance* for a transmission line.

8-14. What properties of a transmission line determine its characteristic impedance?

8-15. Define *propagation constant* for a transmission line.

8-16. Define *velocity factor* for a transmission line.

8-17. What properties of a transmission line determine its velocity factor?

8-18. What properties of a transmission line determine its dielectric constant?

8-19. Define *electrical length* for a transmission line.

8-20. List and describe five types of transmission-line losses.

8-21. Describe an incident wave; a reflected wave.

8-22. Describe a resonant transmission line; a nonresonant transmission line.

8-23. Define *reflection coefficient.*

8-24. Describe standing waves; standing-wave ratio.

8-25. Describe the standing waves present on an open transmission line.

8-26. Describe the standing waves present on a shorted transmission line.

Questions

8-27. Define *input impedance* for a transmission line.

8-28. Describe the behavior of a transmission line that is terminated in a short circuit that is greater than one-quarter wavelength long; less than one-quarter wavelength.

8-29. Describe the behavior of a transmission line that is terminated in an open circuit that is greater than one-quarter wavelength long; less than one-quarter wavelength long.

8-30. Describe the behavior of an open transmission line as a circuit element.

8-31. Describe the behavior of a shorted transmission line as a circuit element.

8-32. Describe the input impedance characteristics of a quarter-wavelength transmission line.

8-33. Describe the input impedance characteristics of a transmission line that is less than one-quarter wavelength long; greater than one-quarter wavelength long.

8-34. Describe quarter-wavelength transformer matching.

8-35. Describe how stub matching is accomplished.

8-36. Describe time-domain reflectometry.

PROBLEMS

8-1. Determine the wavelengths for electromagnetic waves in free space with the following frequencies: 1 kHz, 100 kHz, 1 MHz, and 1 GHz.

8-2. Determine the frequencies for electromagnetic waves in free space with the following wavelengths: 1 cm, 1 m, 10 m, 100 m, and 1000 m.

8-3. Determine the characteristic impedance for an air-dielectric transmission line with D/r ratio of 8.8.

8-4. Determine the characteristic impedance for a concentric transmission line with D/d ratio of 4.

8-5. Determine the characteristic impedance for a coaxial cable with inductance $L = 0.2$ μH/ft and capacitance $C = 16$ pF/ft.

8-6. For a given length of coaxial cable with distributed capacitance $C = 48.3$ pF/m and distributed inductance $L = 241.56$ nH/m, determine the velocity factor and velocity of propagation.

8-7. Determine the reflection coefficient for a transmission line with incident voltage $E_i = 0.2$ V and reflected voltage $E_r = 0.01$ V.

8-8. Determine the standing-wave ratio for the transmission line described in Problem 8-7.

8-9. Determine the SWR for a transmission line with maximum voltage standing-wave amplitude $V_{max} = 6$ V and minimum voltage standing-wave amplitude $V_{min} = 0.5$.

8-10. Determine the SWR for a 50-Ω transmission line that is terminated in a load resistance $Z_L = 75$ Ω.

8-11. Determine the SWR for a 75-Ω transmission line that is terminated in a load resistance $Z_L = 50$ Ω.

8-12. Determine the characteristic impedance for a quarter-wavelength transformer that is used to match a section of 75-Ω transmission line to a 100-Ω resistive load.

8-13. Using TDR, a pulse is transmitted down a cable with a velocity of propagation of 0.7c. The reflected signal is received 1.2 μs later. How far down the cable is the impairment?

8-14. Using TDR, a transmission-line impairment is located 2500 m from the source. For a velocity of propagation of 0.95c, determine the elapsed time from the beginning of the pulse to the reception of the echo.

8-15. Using TDR, a transmission-line impairment is located 200 m from the source. If the elapsed time from the beginning of the pulse to the reception of the echo is 833 ns, determine the velocity factor.

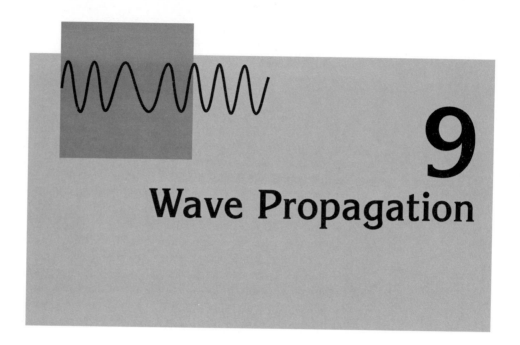

9

Wave Propagation

INTRODUCTION

In Chapter 8 we explained how a metallic wire is used as a transmission medium to propagate electromagnetic waves from one point to another. However, very often in communications systems it is impractical or impossible to interconnect two pieces of equipment with a physical facility such as a wire, for example, across large spans of water, rugged mountains, or desert terrain, or to and from satellite transponders parked 22,000 mi above the earth. Also, when the transmitters and receivers are mobile, as with two-way radio or mobile telephone, metallic facilities are impossible. Therefore, free space or Earth's atmosphere is often used as a transmission medium. Free-space propagation of electromagnetic waves is often called *radio-frequency* (RF) *propagation* or simply *radio propagation*.

To propagate TEM waves through Earth's atmosphere, it is necessary that energy be radiated from the source; then the energy must be *captured* at the receive end. Radiating and capturing energy are antenna functions and are explained in Chapter 10, and the properties of electromagnetic waves were explained in Chapter 8.

RAYS AND WAVEFRONTS

Electromagnetic waves are invisible. Therefore, they must be analyzed by indirect methods using schematic diagrams. The concepts of *rays* and *wavefronts* are aids to illustrating the effects of electromagnetic wave propagation through free space. A ray is a line drawn along the direction of propagation of an electromagnetic wave. Rays are used to show the relative direction of electromagnetic wave propagation. However, a ray does not necessarily represent the propagation of a single electromagnetic wave. Several rays are shown in

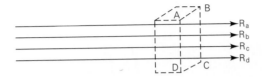

Figure 9-1 Plane wave.

Figure 9-1 (R_a, R_b, R_c, and so on). A wavefront shows a surface of constant phase of a wave. A wavefront is formed when points of equal phase on rays propagated from the same source are joined together. Figure 9-1 shows a wavefront with a surface that is perpendicular to the direction of propagation (rectangle *ABCD*). When a surface is plane, its wavefront is perpendicular to the direction of propagation. The closer to the source, the more complicated the wavefront becomes.

Most wavefronts are more complicated than a simple plane wave. Figure 9-2 shows a point source, several rays propagating from it, and the corresponding wavefront. A *point source* is a single location from which rays propagate equally in all directions (an *isotropic source*). The wavefront generated from a point source is simply a sphere with radius R and its center located at the point of origin of the waves. In free space and a sufficient distance from the source, the rays within a small area of a spherical wavefront are nearly parallel. Therefore, the farther from a source, the more wave propagation appears as a plane wavefront.

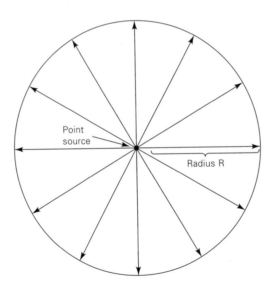

Figure 9-2 Wavefront from a point source.

ELECTROMAGNETIC RADIATION

Power Density and Field Intensity

Electromagnetic waves represent the flow of energy in the direction of propagation. The rate at which energy passes through a given surface area in free space is called *power density*. Therefore, power density is energy per unit time per unit of area and is usually given in watts per square meter. *Field intensity* is the intensity of the electric and magnetic fields of an electromagnetic wave propagating in free space. Electric field intensity is usually given in volts per meter and magnetic field intensity in ampere-turns per meter. Mathematically, power density is

$$\mathscr{P} = \mathscr{E}H \text{ watts per meter squared} \tag{9-1}$$

where $\mathscr{P}$ = power density (W/m^2)
 $\mathscr{E}$ = rms electric field intensity (V/m)
 H = rms magnetic field intensity (At/m)

Characteristic Impedance of Free Space

The electric and magnetic field intensities of an electromagnetic wave in free space are related through the characteristic impedance (resistance) of free space. The characteristic impedance of a lossless transmission medium is equal to the square root of the ratio of its magnetic permeability to its electric permittivity. Mathematically, the characteristic impedance of free space (Z_s) is

$$Z_s = \sqrt{\frac{\mu_0}{\epsilon_0}} \tag{9-2}$$

where Z_s = characteristic impedance of free space (ohms)
 μ_0 = magnetic permeability of free space (1.26×10^{-6} H/m)
 ϵ_0 = electric permittivity of free space (8.85×10^{-12} F/m)

Substituting into Equation 9-2, we have

$$Z_s = \sqrt{\frac{1.26 \times 10^{-6}}{8.85 \times 10^{-12}}} = 377 \ \Omega$$

Therefore, using Ohm's law, we obtain

$$\mathscr{P} = \frac{\mathscr{E}^2}{377} = 377H^2 \quad \text{W/m}^2 \tag{9-3}$$

$$H = \frac{\mathscr{E}}{377} \quad \text{At/m} \tag{9-4}$$

SPHERICAL WAVEFRONT AND THE INVERSE SQUARE LAW

Spherical Wavefront

Figure 9-3 shows a point source that radiates power at a constant rate uniformly in all directions. Such a source is called an *isotropic radiator*. A true isotropic radiator does not exist. However, it is closely approximated by an *omnidirectional antenna*. An isotropic radiator produces a spherical wavefront with radius R. All points distance R from the source lie on the surface of the sphere and have equal power densities. For example, in Figure 9-3 points A and B are an equal distance from the source. Therefore, the power densities at points A and B are equal. At any instant of time, the total power radiated, P_r watts, is uniformly distributed over the total surface of the sphere (this assumes a lossless transmission medium). Therefore, the power density at any point on the sphere is the total

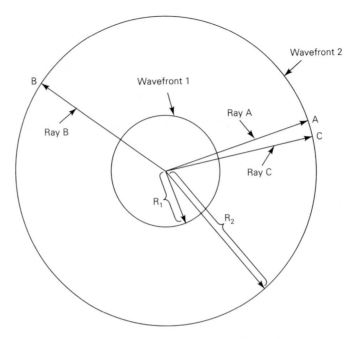

Figure 9-3 Spherical wavefront from an isotropic source.

radiated power divided by the total area of the sphere. Mathematically, the power density at any point on the surface of a spherical wavefront is

$$\mathscr{P}_a = \frac{P_r}{4\pi R^2} \tag{9-5}$$

where P_r = total power radiated (watts)
R = radius of the sphere (which is equal to the distance from any point on the surface of the sphere to the source)
$4\pi R^2$ = area of the sphere

and for a distance R_a meters from the source, the power density is

$$\mathscr{P}_a = \frac{P_r}{4\pi R_a^2}$$

Equating Equations 9-3 and 9-5 gives

$$\frac{P_r}{4\pi R^2} = \frac{\mathscr{E}^2}{377}$$

Therefore,

$$\mathscr{E}^2 = \frac{377 P_r}{4\pi R^2} \quad \text{and} \quad \mathscr{E} = \frac{\sqrt{30 P_r}}{R} \tag{9-6}$$

Inverse Square Law

From Equation 9-5 it can be seen that the farther the wavefront moves from the source, the smaller the power density (R_a and R_c move farther apart). The total power distributed over the surface of the sphere remains the same. However, because the area of the sphere increases in direct proportion to the distance from the source squared (that is, the radius of the sphere squared), the power density is inversely proportional to the square of the distance from the source. This relationship is called the *inverse square law*. Therefore, the power density at any point on the surface of the outer sphere is

$$\mathscr{P}_2 = \frac{P_r}{4\pi R_2^2}$$

and the power density at any point on the inner sphere is

$$\mathscr{P}_1 = \frac{P_r}{4\pi R_1^2}$$

Therefore,

$$\frac{\mathscr{P}_2}{\mathscr{P}_1} = \frac{P_r/4\pi R_2^2}{P_r/4\pi R_1^2} = \frac{R_1^2}{R_2^2} = \left(\frac{R_1}{R_2}\right)^2 \tag{9-7}$$

From Equation 9-7 it can be seen that as the distance from the source doubles the power density decreases by a factor of 2^2, or 4. When deriving the inverse square law of radiation (Equation 9-7), it was assumed that the source radiates isotropically, although it is not necessary. However, it is necessary that the velocity of propagation in all directions be uniform. Such a propagation medium is called an *isotropic medium*.

EXAMPLE 9-1

For an isotropic antenna radiating 100 W of power, determine:

(a) Power density 1000 m from the source.
(b) Power density 2000 m from the source.

Solution (a) Substituting into Equation 9-5 yields

$$\mathscr{P}_1 = \frac{100}{4\pi 1000^2} = 7.96\ \mu\text{W/m}^2$$

(b) Again, substituting into Equation 9-5 gives

$$\mathscr{P}_2 = \frac{100}{4\pi 2000^2} = 1.99\ \mu\text{W/m}^2$$

or, substituting into Equation 9-7, we have

$$\frac{\mathscr{P}_2}{\mathscr{P}_1} = \frac{1000^2}{2000^2} = 0.25$$

or

$$\mathscr{P}_2 = 7.96\ \mu\text{W/m}^2\ 0.25 = 1.99\ \mu\text{W/m}^2$$

WAVE ATTENUATION AND ABSORPTION

Attenuation

The inverse square law for radiation mathematically describes the reduction in power density with distance from the source. As a wavefront moves away from the source, the continuous electromagnetic field that is radiated from that source spreads out. That is, the waves move farther away from each other and, consequently, the number of waves per unit area decreases. None of the radiated power is lost or dissipated because the wavefront

Wave Attenuation and Absorption

is moving away from the source; the wave simply spreads out or disperses over a larger area, decreasing the power density. The reduction in power density with distance is equivalent to a power loss and is commonly called *wave attenuation*. Because the attenuation is due to the spherical spreading of the wave, it is sometimes called the *space attenuation* of the wave. Wave attenuation is generally expressed in terms of the common logarithm of the power density ratio (dB loss). Mathematically, wave attenuation (γ_a) is

$$\gamma_a = 10 \log \frac{\mathscr{P}_1}{\mathscr{P}_2} \tag{9-8}$$

The reduction in power density due to the inverse-square law presumes free-space propagation (a vacuum or nearly a vacuum) and is called wave attenuation. The reduction in power density due to nonfree-space propagation is called *absorption*.

Absorption

The earth's atmosphere is not a vacuum. Rather, it is made up of atoms and molecules of various substances, such as gases, liquids, and solids. Some of these materials are capable of absorbing electromagnetic waves. As an electromagnetic wave propagates through the earth's atmosphere, energy is transferred from the wave to the atoms and molecules of the atmosphere. Wave absorption by the atmosphere is analogous to an I^2R power loss. Once absorbed, the energy is lost forever and causes an attenuation in the voltage and magnetic field intensities and a corresponding reduction in power density.

Absorption of radio frequencies in a normal atmosphere depends on frequency and is relatively insignificant below approximately 10 GHz. Figure 9-4 shows atmospheric absorption in decibels per kilometer due to oxygen and water vapor for radio frequencies above 10 GHz. It can be seen that certain frequencies are affected more or less by absorption, creating peaks and valleys in the curves. Wave attenuation due to absorption does not depend on distance from the radiating source, but rather the total distance that the wave propagates through the atmosphere. In other words, for a *homogeneous medium* (one with uniform properties throughout), the absorption experienced during the first mile of propa-

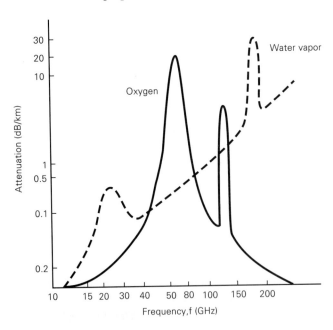

Figure 9-4 Atmospheric absorption of electromagnetic waves.

gation is the same as for the last mile. Also, abnormal atmospheric conditions such as heavy rain or dense fog absorb more energy than a normal atmosphere. Atmospheric absorption (η) for a wave propagating from R_1 to R_2 is $\gamma(R_2 - R_1)$, where γ is the absorption coefficient. Therefore, wave attenuation depends on the ratio R_2/R_1, and wave absorption depends on the distance between R_1 and R_2. In a more practical situation (that is, an *inhomogeneous medium*), the absorption coefficient varies considerably with location, thus creating a difficult problem for radio systems engineers.

OPTICAL PROPERTIES OF RADIO WAVES

In Earth's atmosphere, ray–wavefront propagation may be altered from free-space behavior by *optical* effects such as *refraction, reflection, diffraction,* and *interference.* Using rather unscientific terminology, refraction can be thought of as *bending*, reflection as *bouncing*, diffraction as *scattering*, and interference as *colliding*. Refraction, reflection, diffraction, and interference are called optical properties because they were first observed in the science of optics, which is the behavior of light waves. Because light waves are high-frequency electromagnetic waves, it stands to reason that optical properties will also apply to radio-wave propagation. Although optical principles can be analyzed completely by application of Maxwell's equations, this is necessarily complex. For most applications, *geometric ray tracing* can be substituted for analysis by Maxwell's equations.

Refraction

Electromagnetic *refraction* is the change in direction of a ray as it passes obliquely from one medium to another with different velocities of propagation. The velocity at which an electromagnetic wave propagates is inversely proportional to the density of the medium in which it is propagating. Therefore, refraction occurs whenever a radio wave passes from one medium into another medium of different density. Figure 9-5 shows refraction of a wavefront at a *plane* boundary between two media with different densities. For this example,

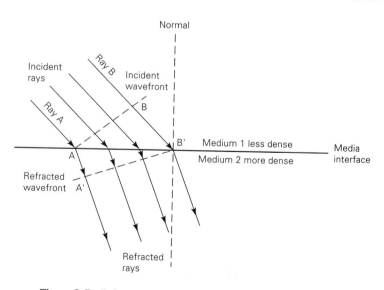

Figure 9-5 Refraction at a plane boundary between two media.

Optical Properties of Radio Waves

medium 1 is less dense than medium 2 ($v_1 > v_2$). It can be seen that ray A enters the more dense medium before ray B. Therefore, ray B propagates more rapidly than ray A and travels distance B–B' during the same time that ray A travels distance A–A'. Therefore, wavefront (A'B') is *tilted* or bent in a downward direction. Since a ray is defined as being perpendicular to the wavefront at all points, the rays in Figure 9-5 have changed direction at the interface of the two media. Whenever a ray passes from a less dense to a more dense medium, it is effectively bent toward the *normal*. (The normal is simply an imaginary line drawn perpendicular to the interface at the point of incidence.) Conversely, whenever a ray passes from a more dense to a less dense medium, it is effectively bent away from the normal. The *angle of incidence* is the angle formed between the incident wave and the normal, and the *angle of refraction* is the angle formed between the refracted wave and the normal.

The amount of bending or refraction that occurs at the interface of two materials of different densities is quite predictable and depends on the *refractive index* (also called the *index of refraction)* of the two materials. The refractive index is simply the ratio of the velocity of propagation of a light ray in free space to the velocity of propagation of a light ray in a given material. Mathematically, the refractive index is

$$n = \frac{c}{v} \qquad (9\text{-}9)$$

where n = refractive index (unitless)
   c = speed of light in free space (3×10^8 m/s)
   v = speed of light in a given material (m/s)

The refractive index is also a function of frequency. However, the variation in most applications is insignificant and is therefore omitted from this discussion. How an electromagnetic wave reacts when it meets the interface of two transmissive materials that have different indexes of refraction can be explained with *Snell's law*. Snell's law simply states that

$$n_1 \sin \theta_1 = n_2 \sin \theta_2 \qquad (9\text{-}10)$$

and

$$\frac{\sin \theta_1}{\sin \theta_2} = \frac{n_2}{n_1}$$

where n_1 = refractive index of material 1
   n_2 = refractive index of material 2
   θ_1 = angle of incidence (degrees)
   θ_2 = angle of refraction (degrees)

and since the refractive index of a material is equal to the square root of its dielectric constant,

$$\frac{\sin \theta_1}{\sin \theta_2} = \sqrt{\frac{\epsilon_{r2}}{\epsilon_{r1}}} \qquad (9\text{-}11)$$

where ϵ_{r1} = dielectric constant of medium 1
   ϵ_{r2} = dielectric constant of medium 2

Refraction also occurs when a wavefront propagates in a medium that has a *density gradient* that is perpendicular to the direction of propagation (that is, parallel to the wavefront). Figure 9-6 shows wavefront refraction in a transmission medium that has a gradual variation in its refractive index. The medium is more dense near the bottom and less dense at the top. Therefore, rays traveling near the top travel faster than rays near the bottom

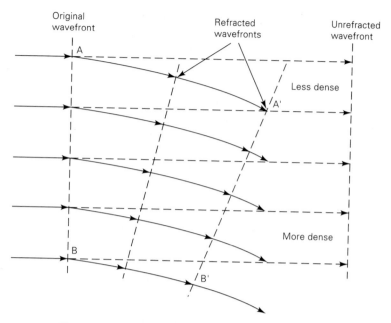

Figure 9-6 Wavefront refraction in a gradient medium.

and, consequently, the wavefront tilts downward. The tilting occurs in a gradual fashion as the wave progresses, as shown.

Reflection

Reflect means to cast or turn back, and *reflection* is the act of reflecting. Electromagnetic reflection occurs when an incident wave strikes a boundary of two media and some or all of the incident power does not enter the second material. The waves that do not penetrate the second medium are reflected. Figure 9-7 shows electromagnetic wave reflection at a plane boundary between two media. Because all the reflected waves remain in medium 1, the velocities of the reflected and incident waves are equal. Consequently, the *angle of reflection* equals the *angle of incidence* ($\theta_i = \theta_r$). However, the reflected voltage field intensity is less than the incident voltage field intensity. The ratio of the reflected to the incident voltage intensities is called the *reflection coefficient*, Γ (sometimes called the *coefficient of reflection*). For a perfect conductor, $\Gamma = 1$. Γ is used to indicate both the

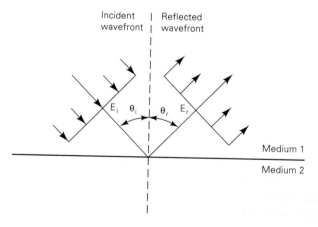

Figure 9-7 Electromagnetic reflection at a plane boundary of two media.

Optical Properties of Radio Waves

relative amplitude of the incident and reflected fields and also the phase shift that occurs at the point of reflection. Mathematically, the reflection coefficient is

$$\Gamma = \frac{E_r e^{j\theta_r}}{E_i e^{j\theta_i}} = \frac{E_r}{E_i} \, e^{j(\theta_r - \theta_i)} \tag{9-12}$$

where Γ = reflection coefficient (unitless)
E_i = incident voltage intensity (volts)
E_r = reflected voltage intensity (volts)
θ_i = incident phase (degrees)
θ_r = reflected phase (degrees)

The ratio of the reflected and incident power densities is Γ. The portion of the total incident power that is not reflected is called the *power transmission coefficient* (T) (or simply the *transmission coefficient*). For a perfect conductor, $T = 0$. The *law of conservation of energy* states that for a perfect reflective surface the total reflected power must equal the total incident power. Therefore,

$$T + |\Gamma|^2 = 1 \tag{9-13}$$

For imperfect conductors, both $|\Gamma|^2$ and T are functions of the angle of incidence, the electric field polarization, and the dielectric constants of the two materials. If medium 2 is not a perfect conductor, some of the incident waves penetrate it and are absorbed. The absorbed waves set up currents in the resistance of the material and the energy is converted to heat. The fraction of power that penetrates medium 2 is called the *absorption coefficient* (or sometimes the *coefficient of absorption*).

When the reflecting surface is not plane (that is, it is curved), the curvature of the reflected wave is different from that of the incident wave. When the wavefront of the incident wave is curved and the reflective surface is plane, the curvature of the reflected wavefront is the same as that of the incident wavefront.

Reflection also occurs when the reflective surface is *irregular* or *rough*. However, such a surface may destroy the shape of the wavefront. When an incident wavefront strikes an irregular surface, it is randomly scattered in many directions. Such a condition is called *diffuse reflection*, whereas reflection from a perfectly smooth surface is called *specular* (mirrorlike) *reflection*. Surfaces that fall between smooth and irregular are called *semirough surfaces*. Semirough surfaces cause a combination of diffuse and specular reflection. A semirough surface will not totally destroy the shape of the reflected wavefront. However, there is a reduction in the total power. The *Rayleigh criterion* states that a semirough surface will reflect as if it were a smooth surface whenever the cosine of the angle of incidence is greater than $\lambda/8d$, where d is the depth of the surface irregularity and λ is the wavelength of the incident wave. Reflection from a semirough surface is shown in Figure 9-8. Mathematically, Rayleigh's criterion is

$$\cos \theta_i > \frac{\lambda}{8d} \tag{9-14}$$

Diffraction

Diffraction is defined as the modulation or redistribution of energy within a wavefront when it passes near the edge of an *opaque* object. Diffraction is the phenomenon that allows light or radio waves to propagate (*peek*) around corners. The previous discussions

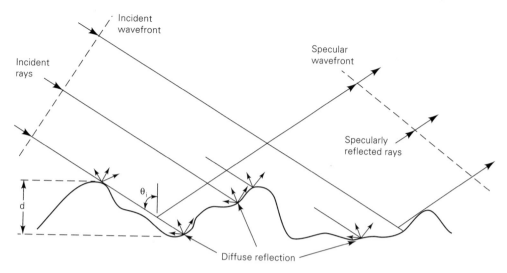

Figure 9-8 Reflection from a semirough surface.

of refraction and reflection assumed that the dimensions of the refracting and reflecting surfaces were large with respect to a wavelength of the signal. However, when a wavefront passes near an obstacle or discontinuity with dimensions comparable in size to a wavelength, simple geometric analysis cannot be used to explain the results and *Huygens's principle* (which is deduced from Maxwell's equations) is necessary.

Huygens's principle states that every point on a given spherical wavefront can be considered as a secondary point source of electromagnetic waves from which other secondary waves (wavelets) are radiated outward. Huygens's principle is illustrated in Figure 9-9. Normal wave propagation considering an infinite plane is shown in Figure 9-9a. Each secondary point source (p_1, p_2, and so on) radiates energy outward in all directions. However, the wavefront continues in its original direction rather than spreading out, because cancellation of the secondary wavelets occurs in all directions except straight forward. Therefore, the wavefront remains plane.

When a finite plane wavefront is considered, as in Figure 9-9b, cancellation in random directions is incomplete. Consequently, the wavefront spreads out or *scatters*. This scattering effect is called *diffraction*. Figure 9-9c shows diffraction around the edge of an obstacle. It can be seen that wavelet cancellation occurs only partially. Diffraction occurs around the edge of the obstacle, which allows secondary waves to "sneak" around the corner of the obstacle into what is called the *shadow zone*. This phenomenon can be observed when a door is opened into a dark room. Light rays diffract around the edge of the door and illuminate the area behind the door.

Interference

Interfere means to come into opposition, and *interference* is the act of interfering. Radiowave interference occurs when two or more electromagnetic waves combine in such a way that system performance is degraded. Refraction, reflection, and diffraction are categorized as geometric optics, which means that their behavior is analyzed primarily in terms of rays and wavefronts. Interference, on the other hand, is subject to the principle of *linear superposition* of electromagnetic waves and occurs whenever two or more waves simultaneously occupy the same point in space. The principle of linear superposition

Optical Properties of Radio Waves

365

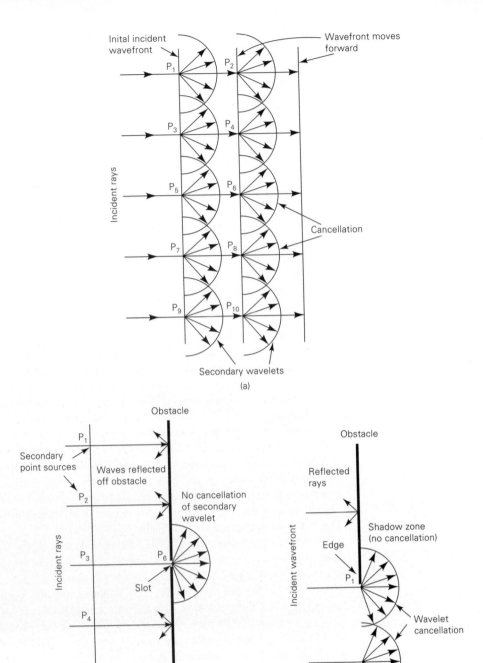

Figure 9-9 Electromagnetic wave diffraction: (a) Huygens's principle for a plane wavefront; (b) finite wavefront through a slot; (c) around an edge.

states that the total voltage intensity at a given point in space is the sum of the individual wave vectors. Certain types of propagation media have nonlinear properties; however, in an ordinary medium (such as air or Earth's atmosphere), linear superposition holds true.

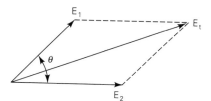

Figure 9-10 Linear addition of two vectors with differing phase angles.

Figure 9-10 shows the linear addition of two instantaneous voltage vectors whose phase angles differ by angle θ. It can be seen that the total voltage is not simply the sum of the two vector magnitudes, but rather the phasor addition of the two. With free-space propagation, a phase difference may exist simply because the electromagnetic polarizations of two waves differ. Depending on the phase angles of the two vectors, either addition or subtraction can occur. (This implies simply that the result may be more or less than either vector because the two electromagnetic waves can reinforce or cancel.)

Figure 9-11 shows interference between two electromagnetic waves in free space. It can be seen that at point *X* the two waves occupy the same area of space. However, wave *B* has traveled a different path than wave *A*, and therefore their relative phase angles may be different. If the difference in distance traveled is an odd integral multiple of one-half wavelength, reinforcement takes place. If the difference is an even integral multiple of one-half wavelength, total cancellation occurs. More likely the difference in distance falls somewhere between the two and partial cancellation occurs. For frequencies below VHF, the relatively large wavelengths prevent interference from being a significant problem. However, with UHF and above, wave interference can be severe.

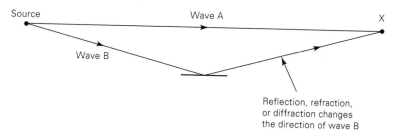

Figure 9-11 Electromagnetic wave interference.

PROPAGATION OF WAVES

In radio communications systems, waves can be propagated in several ways, depending on the type of system and the environment. Also, as previously explained, electromagnetic waves travel in straight lines except when the earth and its atmosphere alter their path. There are three ways of propagating electromagnetic waves: ground-wave, space-wave (which includes both direct and ground-reflected waves), and sky-wave propagation.

Figure 9-12 shows the normal modes of propagation between two radio antennas. Each of these modes exists in every radio system; however, some are negligible in certain frequency ranges or over a particular type of terrain. At frequencies below 1.5 MHz, ground waves provide the best coverage. This is because ground losses increase rapidly with frequency. Sky waves are used for high-frequency applications, and space waves are used for very high frequencies and above.

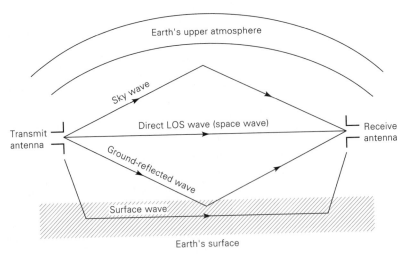

Figure 9-12 Normal modes of wave propagation.

Ground-wave Propagation

A *ground wave* is an electromagnetic wave that travels along the surface of the earth. Therefore, ground waves are sometimes called *surface waves*. Ground waves must be vertically polarized. This is because the electric field in a horizontally polarized wave would be parallel to the earth's surface, and such waves would be short-circuited by the conductivity of the ground. With ground waves, the changing electric field induces voltages in Earth's surface, which cause currents to flow that are very similar to those in a transmission line. Earth's surface also has resistance and dielectric losses. Therefore, ground waves are attenuated as they propagate. Ground waves propagate best over a surface that is a good conductor, such as salt water, and poorly over dry desert areas. Ground-wave losses increase rapidly with frequency. Therefore, ground-wave propagation is generally limited to frequencies below 2 MHz.

Figure 9-13 shows ground-wave propagation. Earth's atmosphere has a *gradient density* (that is, the density decreases gradually with distance from the earth's surface), which causes the wavefront to tilt progressively forward. Therefore, the ground wave propagates around the earth, remaining close to its surface, and if enough power is transmitted, the wavefront could propagate beyond the horizon or even around the entire circumference of Earth. However, care must be taken when selecting the frequency and the terrain over which the ground wave will propagate to ensure that the wavefront does not tilt excessively and simply turn over, lie flat on the ground, and cease to propagate.

Ground-wave propagation is commonly used for ship-to-ship and ship-to-shore communications, for radio navigation, and for maritime mobile communications. Ground waves are used at frequencies as low as 15 kHz.

The disadvantages of ground-wave propagation are as follows:

1. Ground waves require a relatively high transmission power.
2. Ground waves are limited to very low, low, and medium frequencies (VLF, LF, and MF) requiring large antennas (the reason for this is explained in Chapter 11).
3. Ground losses vary considerably with surface material.

The advantages of ground-wave propagation are as follows:

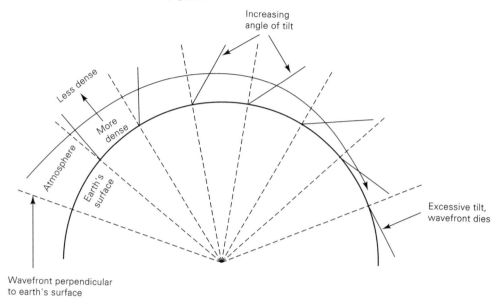

Figure 9-13 Ground-wave propagation.

1. Given enough transmit power, ground waves can be used to communicate between any two locations in the world.

2. Ground waves are relatively unaffected by changing atmospheric conditions.

Space-wave Propagation

Space-wave propagation includes radiated energy that travels in the lower few miles of Earth's atmosphere. Space waves include both direct and ground-reflected waves (see Figure 9-14). *Direct waves* travel essentially in a straight line between the transmit and receive antennas. Space-wave propagation with direct waves is commonly called *line-of-sight* (LOS) *transmission*. Therefore, space-wave propagation is limited by the curvature of the earth. Ground-reflected waves are waves reflected by Earth's surface as they propagate between the transmit and receive antennas.

Figure 9-14 shows space-wave propagation between two antennas. It can be seen that the field intensity at the receive antenna depends on the distance between the two

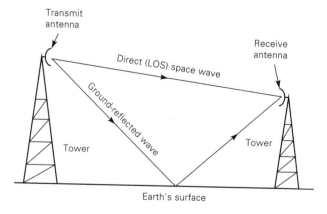

Figure 9-14 Space-wave propagation.

Propagation of Waves

antennas (attenuation and absorption) and whether the direct and ground-reflected waves are in phase (interference).

The curvature of the earth presents a horizon to space-wave propagation commonly called the *radio horizon*. Due to atmospheric refraction, the radio horizon extends beyond the *optical horizon* for the common *standard atmosphere*. The radio horizon is approximately four-thirds that of the optical horizon. Refraction is caused by the troposphere, due to changes in its density, temperature, water-vapor content, and relative conductivity. The radio horizon can be lengthened simply by elevating the transmit or receive antennas (or both) above Earth's surface with towers or by placing them on top of mountains or high buildings.

Figure 9-15 shows the effect of antenna height on the radio horizon. The line-of-sight radio horizon for a single antenna is given as

$$d = \sqrt{2h} \qquad (9\text{-}15)$$

where d = distance to radio horizon (miles)
$\quad\quad\quad h$ = antenna height above sea level (feet)

Therefore, for a transmit and receive antenna, the distance between the two antennas is

$$d = d_t + d_r$$

or

$$d = \sqrt{2h_t} + \sqrt{2h_r} \qquad (9\text{-}16)$$

where d = total distance (miles)
$\quad\quad\quad d_t$ = radio horizon for transmit antenna (miles)
$\quad\quad\quad d_r$ = radio horizon for receive antenna (miles)
$\quad\quad\quad h_t$ = transmit antenna height (feet)
$\quad\quad\quad h_r$ = receive antenna height (feet)

or

$$d = \sqrt{2}\sqrt{h_t} + \sqrt{2}\sqrt{h_r} \qquad (9\text{-}17)$$

where d_t, and d_r are distance in kilometers and h_t and h_r are height in meters.

From Equations 9-16 and 9-17, it can be seen that the space-wave propagation distance can be extended simply by increasing either the transmit or receive antenna height, or both.

Because the conditions in the earth's lower atmosphere are subject to change, the degree of refraction can vary with time. A special condition called *duct propagation*

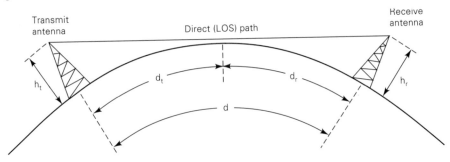

Figure 9-15 Space waves and radio horizon.

Chap. 9   **Wave Propagation**

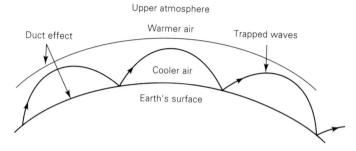

Figure 9-16 Duct propagation.

occurs when the density of the lower atmosphere is such that electromagnetic waves are trapped between it and Earth's surface. The layers of the atmosphere act like a duct, and an electromagnetic wave can propagate for great distances around the curvature of Earth within this duct. Duct propagation is shown in Figure 9-16.

Sky-wave Propagation

Electromagnetic waves that are directed above the horizon level are called *sky waves*. Typically, sky waves are radiated in a direction that produces a relatively large angle with reference to Earth. Sky waves are radiated toward the sky, where they are either reflected or refracted back to Earth by the ionosphere. The ionosphere is the region of space located approximately 50 to 400 km (30 to 250 mi) above Earth's surface. The ionosphere is the upper portion of Earth's atmosphere. Therefore, it absorbs large quantities of the sun's radiant energy, which ionizes the air molecules, creating free electrons. When a radio wave passes through the ionosphere, the electric field of the wave exerts a force on the free electrons, causing them to vibrate. The vibrating electrons decrease current, which is equivalent to reducing the dielectric constant. Reducing the dielectric constant increases the velocity of propagation and causes electromagnetic waves to bend away from the regions of high electron density toward regions of low electron density (that is, increasing refraction). As the wave moves farther from Earth, ionization increases. However, there are fewer air molecules to ionize. Therefore, in the upper atmosphere, there is a higher percentage of ionized molecules than in the lower atmosphere. The higher the ion density is, the more refraction. Also, due to the ionosphere's nonuniform composition and its temperature and density variations, it is *stratified*. Essentially, three layers comprise the ionosphere (the D, E, and F layers), which are shown in Figure 9-17. It can be seen that all three layers of the ionosphere vary in location and in *ionization density* with the time of day. They also fluctuate in a cyclic pattern throughout the year and according to the 11-year *sunspot cycle*. The ionosphere is most dense during times of maximum sunlight (during the daylight hours and in the summer).

D layer. The *D layer* is the lowest layer of the ionosphere and is located between 30 and 60 mi (50 to 100 km) above Earth's surface. Because it is the layer farthest from the sun, there is very little ionization in this layer. Therefore, the D layer has very little effect on the direction of propagation of radio waves. However, the ions in the D layer can absorb appreciable amounts of electromagnetic energy. The amount of ionization in the D layer depends on the altitude of the sun above the horizon. Therefore, it disappears at night. The D layer reflects VLF and LF waves and absorbs MF and HF waves. (See Table 1-1 for VLF, LF, MF, and HF frequency regions.)

Propagation of Waves

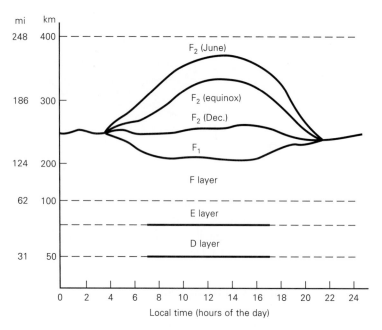

Figure 9-17 Ionospheric layers.

E layer. The *E layer* is located between 60 and 85 mi (100 to 140 km) above Earth's surface. The E layer is sometimes called the *Kennelly–Heaviside layer* after the two scientists who discovered it. The E layer has its maximum density at approximately 70 mi at noon, when the sun is at its highest point. Like the D layer, the E layer almost totally disappears at night. The E layer aids MF surface-wave propagation and reflects HF waves somewhat during the daytime. The upper portion of the E layer is sometimes considered separately and is called the sporadic E layer because it seems to come and go rather unpredictably. The sporadic E layer is caused by *solar flares* and *sunspot activity*. The sporadic E layer is a thin layer with a very high ionization density. When it appears, there generally is an unexpected improvement in long-distance radio transmission.

F layer. The *F layer* is actually made up of two layers, the F_1 and F_2 layers. During the daytime, the F_1 layer is located between 85 and 155 mi (140 to 250 km) above Earth's surface, and the F_2 layer is located 85 to 185 mi (140 to 300 km) above Earth's surface during the winter and 155 to 220 mi (250 to 350 km) in the summer. During the night, the F_1 layer combines with the F_2 layer to form a single layer. The F_1 layer absorbs and attenuates some HF waves, although most of the waves pass through to the F_2 layer, where they are refracted back to Earth.

PROPAGATION TERMS AND DEFINITIONS

Critical Frequency and Critical Angle

Frequencies above the UHF range are virtually unaffected by the ionosphere because of their extremely short wavelengths. At these frequencies, the distances between ions are appreciably large and, consequently, the electromagnetic waves pass through them with

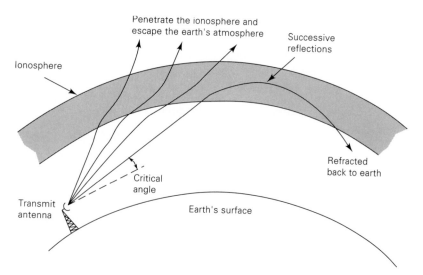

Figure 9-18 Critical angle.

little noticeable effect. Therefore, it stands to reason that there must be an upper frequency limit for sky-wave propagation. *Critical frequency* (f_c) is defined as the highest frequency that can be propagated directly upward and still be returned to Earth by the ionosphere. The critical frequency depends on the ionization density and therefore varies with the time of day and the season. If the vertical angle of radiation is decreased, frequencies at or above the critical frequency can still be refracted back to Earth's surface because they will travel a longer distance in the ionosphere and thus have a longer time to be refracted. Therefore, critical frequency is used only as a point of reference for comparison purposes. However, every frequency has a maximum vertical angle at which it can be propagated and still be refracted back by the ionosphere. This angle is called the *critical angle*. The critical angle θ_c is shown in Figure 9-18.

Virtual Height

Virtual height is the height above the earth's surface from which a refracted wave appears to have been reflected. Figure 9-19 shows a wave that has been radiated from Earth's surface toward the ionosphere. The radiated wave is refracted back to Earth and follows path

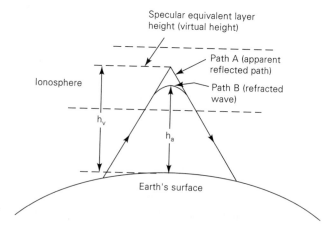

Figure 9-19 Virtual and actual height.

Propagation Terms and Definitions

373

B. The actual maximum height that the wave reached is height h_a. However, path *A* shows the projected path that a reflected wave could have taken and still been returned to Earth at the same location. The maximum height that this hypothetical reflected wave would have reached is the virtual height (h_v).

Maximum Usable Frequency

The *maximum usable frequency* (MUF) is the highest frequency that can be used for sky-wave propagation between two specific points on Earth's surface. It stands to reason, then, that there are as many values possible for MUF as there are points on Earth and frequencies—an infinite number. MUF, like the critical frequency, is a limiting frequency for sky-wave propagation. However, the maximum usable frequency is for a specific angle of incidence (the angle between the incident wave and the normal). Mathematically, MUF is

$$\text{MUF} = \frac{\text{critical frequency}}{\cos\theta} \tag{9-18a}$$

$$= \text{critical frequency} \times \sec\theta \tag{9-18b}$$

where θ is the angle of incidence.

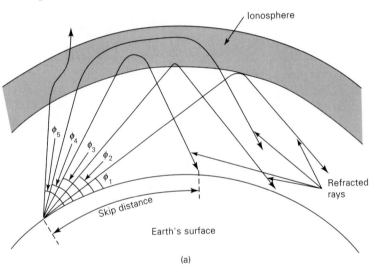

(a)

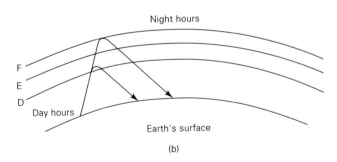

(b)

Figure 9-20 (a) Skip distance; (b) daytime versus nighttime propagation.

Equation 9-18 is called the *secant law*. The secant law assumes a flat Earth and a flat reflecting layer, which, of course, can never exist. Therefore, MUF is used only for making preliminary calculations.

Skip Distance

The *skip distance* (d_s) is the minimum distance from a transmit antenna that a sky wave of given frequency (which must be less than the MUF) will be returned to Earth. Figure 9-20a shows several rays with different elevation angles being radiated from the same point on Earth. It can be seen that the point where the wave is returned to Earth moves closer to the transmitter as the elevation angle (ϕ) is increased. Eventually, however, the angle of elevation is sufficiently high that the wave penetrates through the ionosphere and totally escapes Earth's atmosphere.

Figure 9-20b shows the effect on the skip distance of the disappearance of the D and E layers during nighttime. Effectively, the *ceiling* formed by the ionosphere is raised, allowing sky waves to travel higher before being refracted back to Earth. This effect explains how faraway radio stations are sometimes heard during the night that cannot be heard during daylight hours.

QUESTIONS

9-1. Describe an electromagnetic ray; a wavefront.

9-2. Describe power density; voltage intensity.

9-3. Describe a spherical wavefront.

9-4. Explain the inverse square law.

9-5. Describe wave attenuation.

9-6. Describe wave absorption.

9-7. Describe refraction. Explain Snell's law for refraction.

9-8. Describe reflection. Explain Snell's law for reflection.

9-9. Describe diffraction. Explain Huygens's principle.

9-10. Describe the composition of a good reflector.

9-11. Describe the atmospheric conditions that cause electromagnetic refraction.

9-12. Define electromagnetic wave interference.

9-13. Describe ground-wave propagation. List its advantages and disadvantages.

9-14. Describe space-wave propagation.

9-15. Explain why the radio horizon is at a greater distance than the optical horizon.

9-16. Describe the various layers of the ionosphere.

9-17. Describe sky-wave propagation.

9-18. Explain why ionospheric conditions vary with time of day, month of year, and so on.

9-19. Define *critical frequency*; *critical angle*.

9-20. Describe virtual height.

9-21. Define *maximum usable frequency*.

9-22. Define *skip distance* and give the reasons that it varies.

PROBLEMS

9-1. Determine the power density for a radiated power of 1000 W at distance of 20 km from an isotropic antenna.

9-2. Determine the power density for Problem 9-1 for a point that is 30 km from the antenna.

9-3. Describe the effects on power density if the distance from a transmit antenna is tripled.

9-4. Determine the radio horizon for a transmit antenna that is 100 ft high and a receiving antenna that is 50 ft high; 100 m and 50 m.

9-5. Determine the maximum usable frequency for a critical frequency of 10 MHz and an angle of incidence of 45°.

9-6. Determine the electric field intensity for the same point in Problem 9-1.

9-7. Determine the electric field intensity for the same point in Problem 9-2.

9-8. For a radiated power $P_r = 10$ kW, determine the voltage intensity at a distance 20 km from the source.

9-9. Determine the change in power density when the distance from the source increases by a factor of 4.

9-10. If the distance from the source is reduced to one-half its value, what effect does this have on the power density?

9-11. The power density at a point from a source is 0.001 μW and the power density at another point is 0.00001 μW; determine the attenuation in decibels.

9-12. For a dielectric ratio $\sqrt{\epsilon_{r2}/\epsilon_{r1}} = 0.8$ and an angle of incidence $\theta_i = 26°$, determine the angle of refraction, θ_r.

9-13. Determine the distance to the radio horizon for an antenna located 40 ft above sea level.

9-14. Determine the distance to the radio horizon for an antenna that is 40 ft above the top of a 4000-ft mountain peak.

9-15. Determine the maximum distance between identical antennas equally distant above sea level for Problem 9-13.

Chap. 9   Wave Propagation

Antennas
and Waveguides

INTRODUCTION

In essence, an *antenna* is a metallic conductor system capable of radiating and receiving electromagnetic waves, and a *waveguide* is a conducting metallic tube through which high-frequency electromagnetic energy is propagated, usually between an antenna and a transmitter, a receiver, or both. An antenna is used to interface a transmitter to free space or free space to a receiver. A waveguide, like a transmission line, is simply used to efficiently interconnect an antenna with the transceiver. An antenna couples energy from the output of a transmitter to Earth's atmosphere or from Earth's atmosphere to a receiver. An antenna is a *passive reciprocal device*; passive in that it cannot actually amplify a signal, at least not in the true sense of the word (however, as you will see later in this chapter, an antenna can have *gain*), and reciprocal in that the transmit and receive characteristics of an antenna are identical, except where *feed currents* to the antenna element are tapered to modify the transmit pattern.

BASIC ANTENNA OPERATION

Basic antenna operation is best understood by looking at the voltage standing-wave patterns on a transmission line, which are shown in Figure 10-1a. The transmission line is terminated in an open circuit, which represents an abrupt discontinuity to the incident voltage wave in the form of a phase reversal. The phase reversal results in some of the incident voltage to be radiated, not reflected back toward the source. The radiated energy propagates away from the antenna in the form of transverse electromagnetic waves. The *radiation efficiency* of an open transmission line is extremely low. Radiation efficiency is the

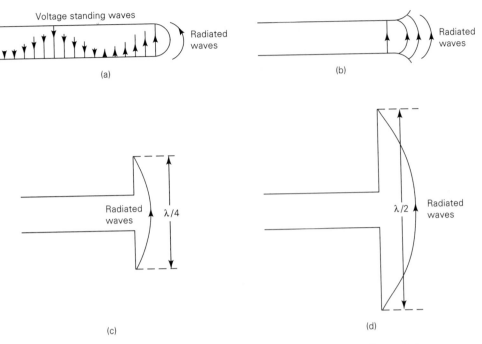

Figure 10-1 Radiation from a transmission line: (a) transmission-line radiation; (b) spreading conductors; (c) Marconi antenna; (d) Hertz antenna.

ratio of radiated to reflected energy. To radiate more energy, simply spread the conductors farther apart. Such an antenna is called a *dipole* (meaning two poles) and is shown in Figure 10-1b.

In Figure 10-1c, the conductors are spread out in a straight line to a total length of one-quarter wavelength. Such an antenna is called a basic *quarter-wave dipole* or a *vertical monopole* (sometimes called a Marconi antenna). A half-wave dipole is called a *Hertz antenna* and is shown in Figure 10-1d.

TERMS AND DEFINITIONS

Radiation Pattern

A *radiation pattern* is a *polar* diagram or graph representing field strengths or power densities at various angular positions relative to an antenna. If the radiation pattern is plotted in terms of electric field strength ($\mathscr{E}$) or power density ($\mathscr{P}$), it is called an *absolute* radiation pattern. If it plots field strength or power density with respect to the value at a reference point, it is called a *relative* radiation pattern. Figure 10-2a shows an absolute radiation pattern for an unspecified antenna. The pattern is plotted on *polar* coordinate paper with the heavy solid line representing points of equal power density (10 μW/m^2). The circular gradients indicate distance in 2-km steps. It can be seen that maximum radiation is in a direction 90° from the reference. The power density 10 km from the antenna in a 90° direction is 10 μW/m^2. In a 45° direction, the point of equal power density is 5 km from the antenna; at 180°, only 4 km; and in a $-90°$ direction, there is essentially no radiation.

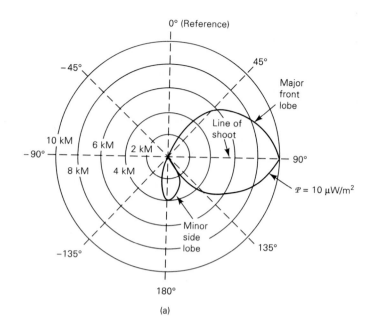

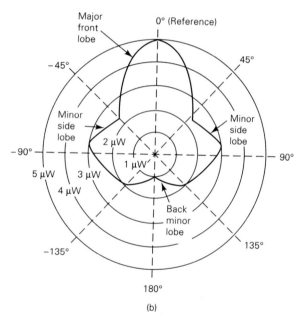

Figure 10-2 Radiation patterns: (a) absolute radiation pattern; (b) relative radiation pattern; *(Continued on page 381.)*

In Figure 10-2a the primary beam is in a 90° direction and is called the *major lobe*. There can be more than one major lobe. There is also a *secondary* beam or *minor* lobe in a −180° direction. Normally, minor lobes represent undesired radiation or reception. Because the major lobe propagates and receives the most energy, that lobe is called the *front* lobe (the front of the antenna). Lobes adjacent to the front lobe are called *side* lobes (the 180° minor lobe is a side lobe), and lobes in a direction exactly opposite the front lobe

Terms and Definitions

379

are called *back* lobes (there is no back lobe shown on this pattern). The ratio of the front lobe power to the back lobe power is simply called the *front-to-back-ratio*, and the ratio of the front lobe to a side lobe is called the *front-to-side ratio*. The line bisecting the major lobe or pointing from the center of the antenna in the direction of maximum radiation is called the *line of shoot*.

Figure 10-2b shows a relative radiation pattern for an unspecified antenna. The heavy solid line represents points of equal distance from the antenna (10 km), and the circular gradients indicate power density in 1-μW/m^2 divisions. It can be seen that maximum radiation (5 μW/m^2) is in the direction of the reference (0°), and the antenna radiates the least power (1 μW/m^2) in a direction 180° from the reference. Consequently, the front-to-back ratio is 5:1 = 5. Generally, relative field strength and power density are plotted in decibels (dB), where dB = 20 log ($\mathscr{E}/\mathscr{E}_{max}$) or 10 log ($\mathscr{P}/\mathscr{P}_{max}$). Figure 10-2c shows a relative radiation pattern for power density in decibels. In a direction $\pm45°$ from the reference, the power density is -3 dB (half-power) relative to the power density in the direction of maximum radiation (0°). Figure 10-2d shows a relative radiation pattern for power density for an omnidirectional antenna. An omnidirectional antenna radiates energy equally in all directions; therefore, the radiation pattern is simply a circle (actually, a sphere). Also, with an omnidirectional antenna, there are no front, back, or side lobes because radiation is equal in all directions.

The radiation patterns shown in Figure 10-2 are two dimensional. However, radiation from an actual antenna is three dimensional. Therefore, radiation patterns are taken in both the horizontal (from the top) and the vertical (from the side) planes. For the omnidirectional antenna shown in Figure 10-2d, the radiation patterns in the horizontal and vertical planes are circular and equal because the actual radiation pattern for an isotropic radiator is a sphere.

Near and Far Fields

The radiation field that is close to an antenna is not the same as the radiation field that is at a great distance. The term *near field* refers to the field pattern that is close to the antenna, and the term *far field* refers to the field pattern that is at great distance. During one half of a cycle, power is radiated from an antenna where some of the power is stored temporarily in the near field. During the second half of the cycle, power in the near field is returned to the antenna. This action is similar to the way in which an inductor stores and releases energy. Therefore, the near field is sometimes called the *induction field*. Power that reaches the far field continues to radiate outward and is never returned to the antenna. Therefore, the far field is sometimes called the *radiation field*. Radiated power is usually the more important of the two; therefore, antenna radiation patterns are generally given for the far field. The near field is defined as the area within a distance D^2/λ from the antenna, where λ is the wavelength and D the antenna diameter in the same units.

Radiation Resistance and Antenna Efficiency

All the power supplied to an antenna is not radiated. Some of it is converted to heat and dissipated. *Radiation resistance* is somewhat "unreal" in that it cannot be measured directly. Radiation resistance is an ac antenna resistance and is equal to the ratio of the power radiated by the antenna to the square of the current at its feed point. Mathematically, radiation resistance is

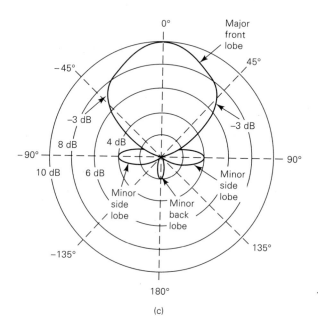

(c)

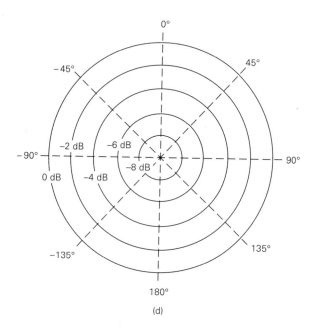

180°

(d)

Figure 10-2 *(continued)* (c) relative radiation pattern in decibels; (d) relative radiation pattern for an omnidirectional antenna.

$$R_r = \frac{P}{i^2} \tag{10-1}$$

where R_r = radiation resistance (ohms)
           P = power radiated by the antenna (watts)
           i = antenna current at the feedpoint (Ampere)

Radiation resistance is the resistance that, if it replaced the antenna, would dissipate exactly the same amount of power that the antenna radiates.

Antenna efficiency is the ratio of the power radiated by an antenna to the sum of the

Terms and Definitions

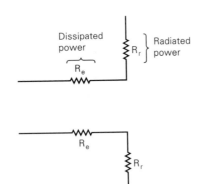

Figure 10-3 Simplified equivalent circuit of an antenna.

power radiated and the power dissipated or the ratio of the power radiated by the antenna to the total input power. Mathematically, antenna efficiency is

$$\eta = \frac{P_r}{P_r + P_d} \times 100 \qquad (10\text{-}2)$$

where η = antenna efficiency (%)
  P_r = power radiated by antenna (watts)
  P_d = power dissipated in antenna (watts)

Figure 10-3 shows a simplified electrical equivalent circuit for an antenna. Some of the input power is dissipated in the effective resistances (ground resistance, corona, imperfect dielectrics, eddy currents, and so on) and the remainder is radiated. The total antenna power is the sum of the dissipated and radiated powers. Therefore, in terms of resistance and current, antenna efficiency is

$$\eta = \frac{i^2 R_r}{i^2 (R_r + R_e)} = \frac{R_r}{R_r + R_e} \qquad (10\text{-}3)$$

where η = antenna efficiency
  i = antenna current (Ampere)
  R_r = radiation resistance (ohms)
  R_e = effective antenna resistance (ohms)

DIRECTIVE GAIN AND POWER GAIN

The terms *directive gain* and *power gain* are often misunderstood and, consequently, misused. *Directive gain* is the ratio of the power density radiated in a particular direction to the power density radiated to the same point by a reference antenna, assuming both antennas are radiating the same amount of power. The relative power density radiation pattern for an antenna is actually a directive gain pattern if the power density reference is taken from a standard reference antenna, which is generally an isotropic antenna. The maximum directive gain is called *directivity*. Mathematically, directive gain is

$$\mathscr{D} = \frac{\mathscr{P}}{\mathscr{P}_{\text{ref}}} \qquad (10\text{-}4)$$

where $\mathscr{D}$ = directive gain (unitless)
          $\mathscr{P}$ = power density at some point with a given antenna (W/m^2)
          $\mathscr{P}_{\text{ref}}$ = power density at the same point with a reference antenna (W/m^2)

Power gain is the same as directive gain except that the total power fed to the antenna is used (that is, antenna efficiency is taken into account). It is assumed that the given antenna and the reference antenna have the same input power and that the reference antenna is lossless ($\eta = 100\%$). Mathematically, power gain (A_p) is

$$A_p = \mathscr{D}\eta \tag{10-5}$$

If an antenna is lossless, it radiates 100% of the input power and the power gain is equal to the directive gain. The power gain for an antenna is also given in decibels relative to some reference antenna. Therefore, power gain is

$$A_p = 10 \log \frac{\mathscr{P}\eta}{\mathscr{P}_{\text{ref}}} \tag{10-6}$$

For an isotropic reference, the directivity of a half-wave dipole is approximately 1.64 (2.15 dB). It is usual to state the power gain in decibels if referred to a $\lambda/2$ dipole. However, if referenced to an isotropic radiator, the decibel figure is stated as dBi, or dB/isotropic radiator, and is 2.15 dB greater than if a half-wave dipole were used for the reference. It is important to note that the power radiated from an antenna can never exceed the input power. Therefore, the antenna does not actually amplify the input power. An antenna simply concentrates its radiated power in a particular direction. Therefore, points that are located in areas where the radiated power is concentrated realize an apparent gain relative to the power density at the same points had an isotropic antenna been used. If gain is realized in one direction, a corresponding reduction in power density (a loss) must be realized in another direction. The direction in which an antenna is "pointing" is always the direction of maximum radiation. Because an antenna is a reciprocal device, its radiation pattern is also its reception pattern. For maximum *captured* power, a receive antenna must be pointing in the direction from which reception is desired. Therefore, receive antennas have directivity and power gain just like transmit antennas.

Effective Isotropic Radiated Power

Effective isotropic radiated power (EIRP) is defined as an equivalent transmit power and is expressed mathematically as

$$\text{EIRP} = P_r A_t \quad \text{watts} \tag{10-7a}$$

where P_r = total radiated power (watts)
          A_t = transmit antenna directive gain (unitless)

or

$$\text{EIRP (dBm)} = 10 \log \frac{P_r}{0.001} + 10 \log A_t \tag{10-7b}$$

Equation 10-7a can be rewritten using antenna input power and power gain as

$$\text{EIRP} = P_{\text{in}} A_p \tag{10-7c}$$

EIRP or simply ERP (effective radiated power) is the equivalent power that an isotropic antenna would have to radiate to achieve the same power density in the chosen

Directive Gain and Power Gain

direction at a given point as another antenna. For instance, if a given antenna has a power gain of 10, the power density is 10 times greater than it would have been had the antenna been an isotropic radiator. An isotropic antenna would have to radiate 10 times as much power to achieve the same power density. Therefore, the given antenna effectively radiates 10 times as much power as an isotropic antenna with the same input power and efficiency.

To determine the power density at a given point, Equation 9-5 is expanded to include the transmit antenna gain and rewritten as

$$\mathscr{P} = \frac{P_r A_t}{4\pi R^2} \qquad (10\text{-}8a)$$

EXAMPLE 10-1

If a transmit antenna has a directive gain $A_t = 10$ and radiated power $P_r = 100$ W, determine:

(a) EIRP.
(b) Power density at a point 10 km away.
(c) Power density had an isotropic antenna been used with the same input power and efficiency.

Solution (a) Substituting into Equation 10-7a yields

$$\text{EIRP} = P_r A_t = 100 \text{ W} \times 10 = 1000 \text{ W}$$

(b) Substituting into Equation 10-8a gives us

$$\mathscr{P} = \frac{P_r A_t}{4\pi R^2} = \frac{\text{EIRP}}{4\pi R^2} = \frac{1000 \text{ W}}{4\pi(10{,}000 \text{ m})^2} = 0.796 \ \mu\text{W/m}^2$$

(c) Substituting into Equation 9-5, we obtain

$$\mathscr{P} = \frac{P_r}{4\pi R^2} = \frac{100}{4\pi(10{,}000 \text{ m})^2} = 0.0796 \ \mu\text{W/m}^2$$

It can be seen that the power density at a point 10,000 m from the transmit antenna is 10 times greater with the first antenna than with the isotropic radiator. To achieve the same power density, the isotrope would have to radiate 1000 W. Therefore, the first antenna effectively radiates 1000 W.

Antennas are reciprocal devices; thus, an antenna has the same power gain and directivity when it is used to receive electromagnetic waves as it has for transmitting electromagnetic waves. Consequently, the power received or captured by an antenna is the product of the power density in the space immediately surrounding the antenna and the antenna directive gain. Therefore, Equation 10-8a can be expanded to

$$C = \frac{P_r A_t A_r}{4\pi R^2} \qquad (10\text{-}8b)$$

where C = power density (W/m^2)
$\quad A_t$ = transmit antenna gain
$\quad A_r$ = receive antenna gain
$\quad R$ = distance between antennas (meters)

In Example 10-1, if an antenna that was identical to the transmit antenna were used to receive the signal, the captured power would be

$$C = \mathscr{P} A_r$$

$$= (0.0796 \ \mu\text{W/m}^2)(10) = 7.96 \text{W/m}^2$$

The captured power is not all useful; some of it is dissipated in the receive antenna. The actual useful received power is the product of the received power density, the receive antenna's direct gain, and the receive antenna's efficiency or the receive power density times the receive antenna's power gain.

ANTENNA POLARIZATION

The *polarization* of an antenna refers simply to the orientation of the electric field radiated from it. An antenna may be *linearly* (generally, either horizontally or vertically polarized, assuming that the antenna elements lie in a horizontal or vertical plane), *elliptically*, or *circularly polarized*. If an antenna radiates a vertically polarized electromagnetic wave, the antenna is defined as vertically polarized; if an antenna radiates a horizontally polarized electromagnetic wave, the antenna is said to be horizontally polarized; if the radiated electric field rotates in an elliptical pattern, it is elliptically polarized; and if the electric field rotates in a circular pattern, it is circularly polarized.

ANTENNA BEAMWIDTH

Antenna *beamwidth* is simply the angular separation between the two half-power (-3 dB) points on the major lobe of an antenna's plane radiation pattern, usually taken in one of the "principal" planes. The beamwidth for the antenna whose radiation pattern is shown in Figure 10-4 is the angle formed between points *A, X,* and *B* (angle θ). Points *A* and *B* are the half-power points (the power density at these points is one-half of what it is an equal distance from the antenna in the direction of maximum radiation). Antenna beamwidth is sometimes called -3-dB beamwidth or half-power beamwidth.

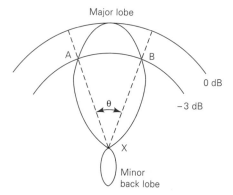

Figure 10-4 Antenna beamwidth.

ANTENNA BANDWIDTH

Antenna *bandwidth* is vaguely defined as the frequency range over which antenna operation is "satisfactory." This is normally taken between the half-power points, but it sometimes refers to variations in the antenna's input impedance.

Antenna Bandwidth

ANTENNA INPUT IMPEDANCE

Radiation from an antenna is a direct result of the flow of RF current. The current flows to the antenna through a transmission line, which is connected to a small gap between the conductors that make up the antenna. The point on the antenna where the transmission line is connected is called the antenna input terminal or simply the *feedpoint*. The feedpoint presents an ac load to the transmission line called the *antenna input impedance*. If the transmitter's output impedance and the antenna's input impedance are equal to the characteristic impedance of the transmission line, there will be no standing waves on the line, and maximum power is transferred to the antenna and radiated.

Antenna input impedance is simply the ratio of the antenna's input voltage to input current. Mathematically, input impedance is

$$Z_{in} = \frac{E_i}{I_i} \qquad (10\text{-}9)$$

where Z_{in} = antenna input impedance (ohms)
E_i = antenna input voltage (volts)
I_i = antenna input current (Ampere)

Antenna input impedance is generally complex. However, if the feedpoint is at a current maximum and there is no reactive component, the input impedance is equal to the sum of the radiation resistance and the effective resistance.

BASIC ANTENNAS

Elementary Doublet

The simplest type of antenna is the *elementary doublet*. The elementary doublet is an electrically short dipole and is often referred to simply as a *short dipole*. "Electrically short" means short compared to one-half wavelength (generally, any dipole that is less than one-tenth wavelength long is considered electrically short). In reality, an elementary doublet cannot be achieved. However, the concept of a short dipole is useful in understanding more practical antennas.

An elementary doublet is a short dipole that has uniform current throughout its length. However, the current is assumed to vary sinusoidally in time and at any instant is

$$i(t) = I \sin(2\pi f t + \theta)$$

where $i(t)$ = instantaneous current
I = peak amplitude of the RF current (Amperes)
f = frequency (hertz)
t = instantaneous time (seconds)
θ = phase angle (radians)

With the aid of Maxwell's equations, it can be shown that the far (radiation) field is

$$\mathscr{E} = \frac{60\pi I l \sin\phi}{\lambda R} \qquad (10\text{-}10)$$

where $\mathscr{E}$ = electric field intensity (volts/meter)
I = dipole current (amperes rms)

l = length of the dipole (meters)
R = distance from the dipole (meters)
λ = wavelength (meters)
ϕ = angle between the axis of the antenna and the direction of radiation

Plotting Equation 10-10 gives the relative electric field intensity pattern for an elementary dipole, which is shown in Figure 10-5. It can be seen that radiation is maximum at right angles to the dipole and falls off to zero at the ends.

The relative power density pattern can be derived from Equation 10-10 by substituting $\mathscr{P} = \mathscr{E}^2/120\pi$. Mathematically, we have

$$\mathscr{P} = \frac{30\pi I^2 l^2 \sin^2 \phi}{\lambda^2 R^2} \tag{10-11}$$

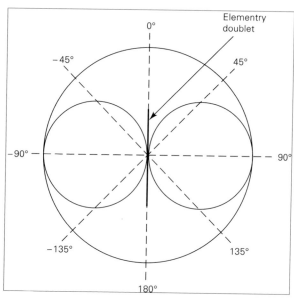

Figure 10-5 Relative radiation pattern for an elementary doublet in a plane perpendicular to the dipole axis.

HALF-WAVE DIPOLE

The linear half-wave dipole is one of the most widely used antennas at frequencies above 2 MHz. At frequencies below 2 MHz, the physical length of a half-wavelength antenna is prohibitive. The half-wave dipole is generally referred to as a *Hertz antenna*.

A Hertz antenna is a *resonant* antenna. That is, it is a multiple of quarter-wavelengths long and open circuited at the far end. Standing waves of voltage and current exist along a resonant antenna. Figure 10-6 shows the idealized voltage and current distributions along a half-wave dipole. Each pole of the antenna looks like an open quarter-wavelength section of transmission line. Therefore, there is a voltage maximum and current minimum at the ends and a voltage minimum and current maximum in the middle. Consequently, assuming that the feedpoint is in the center of the antenna, the input impedance is E_{min}/I_{max} and a minimum value. The impedance at the ends of the antenna is E_{max}/I_{min} and a maximum value. Figure 10-7 shows the impedance curve for a center-fed half-wave dipole. The impedance varies from a maximum value at the ends of approximately 2500 Ω to a minimum value at the feedpoint of approximately 73 Ω (of which between 68 and 70 Ω is the radiation resistance).

Half-wave Dipole

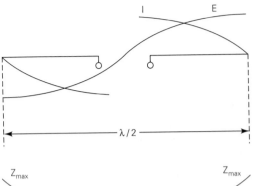

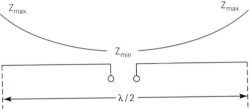

Figure 10-6 Idealized voltage and current distributions along a half-wave dipole.

Figure 10-7 Impedance curve for a center-fed half-wave dipole.

A wire radiator such as a half-wave dipole can be thought of as an infinite number of elementary doublets placed end to end. Therefore, the radiation pattern can be obtained by integrating Equation 10-10 over the length of the antenna. The free-space radiation pattern for a half-wave dipole depends on whether the antenna is placed horizontally or vertically with respect to Earth's surface. Figure 10-8a shows the vertical (from the side) radiation pattern for a vertically mounted half-wave dipole. Note that two major lobes radiate in

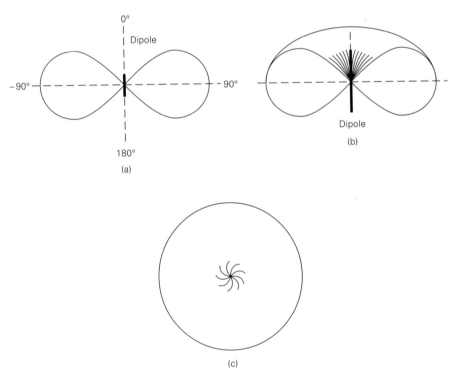

Figure 10-8 Half-wave dipole radiation patterns: (a) vertical view of a vertically mounted dipole; (b) cross-sectional view; (c) horizontal view.

Chap. 10   Antennas and Waveguides

opposite directions that are at right angles to the antenna. Also note that the lobes are not circles. Circular lobes are obtained only for the ideal case when the current is constant throughout the antenna's length, and this is unachievable in a practical antenna. Figure 10-8b shows the cross-sectional view. Note that the radiation pattern has a figure-eight pattern and resembles the shape of a doughnut. Maximum radiation is in a plane parallel to Earth's surface. The higher the angle of elevation is, the less the radiation, and for 90° there is no radiation. Figure 10-8c shows the horizontal (from the top) radiation pattern for a vertically mounted half-wave dipole. The pattern is circular because radiation is uniform in all directions perpendicular to the antenna.

Ground effects on a half-wave dipole. The radiation patterns shown in Figure 10-8 are for free-space conditions. In Earth's atmosphere, wave propagation is affected by antenna orientation, atmospheric absorption, and ground effects such as reflection. The effect of ground reflection for an ungrounded half-wave dipole is shown in Figure 10-9. The antenna is mounted an appreciable number of wavelengths (height h) above the surface of Earth. The field strength at any given point in space is the sum of the direct and ground-reflected waves. The ground-reflected wave appears to be radiating from an image antenna distance h below Earth's surface. This apparent antenna is a mirror image of the actual antenna. The ground-reflected wave is inverted 180° and travels a distance $2h \sin \theta$ farther than the direct wave to reach the same point in space (point P). The resulting radiation pattern is a summation of the radiations from the actual antenna and the mirror antenna. Note that this is the classical ray-tracing technique used in optics.

Figure 10-10 shows the vertical radiation patterns for a horizontally mounted half-wave dipole one-quarter and one-half wavelength above the ground. For an antenna mounted one-quarter wavelength above the ground, the lower lobe is completely gone and the field strength directly upward is doubled. Figure 10-10a shows in the dotted line the free space pattern and in the solid line the vertical distribution in a plane through the antenna, and Figure 10-10b shows the vertical distribution in a plane at right angles to the antenna. Figure 10-10c shows the vertical radiation pattern for a horizontal dipole one-half wavelength above the ground. The figure shows that the pattern is now broken into two lobes, and the direction of maximum radiation (end view) is now at 30° to the horizontal instead

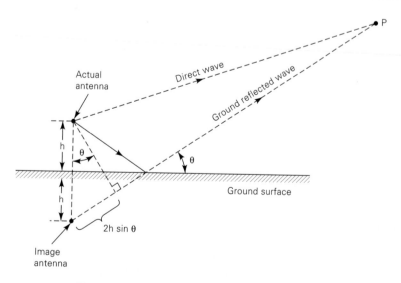

Figure 10-9 Ground effects on a half-wave dipole.

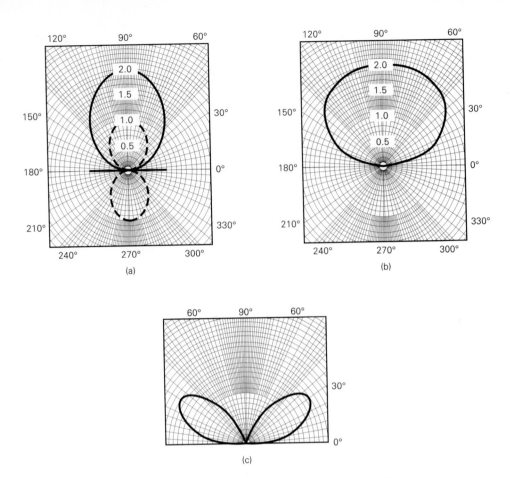

(a)

(b)

(c)

Figure 10-10 Vertical radiation pattern for a half-wave dipole.

of directly upward. There is no component along the ground for horizontal polarization because of the phase shift of the reflected component. Ground-reflected waves have similar effects on all antennas. The best way to eliminate or reduce the effect of ground reflected waves is to mount the antenna far enough above Earth's surface to obtain free-space conditions. However, in many applications, this is impossible. Ground reflections are sometimes desirable to get the desired elevation angle for the major lobe maximum response.

The height of an ungrounded antenna above Earth's surface also affects the antenna's radiation resistance. This is due to the reflected waves cutting through or intercepting the antenna and altering its current. Depending on the phase of the ground-reflected wave, the antenna current can increase or decrease, causing a corresponding increase or decrease in the input impedance.

GROUNDED ANTENNA

A *monopole* (single pole) antenna one-quarter wavelength long mounted vertically with the lower end either connected directly to ground or grounded through the antenna coupling network is called a *Marconi antenna*. The characteristics of a Marconi antenna are

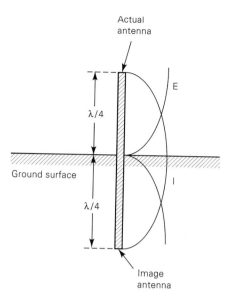

Figure 10-11 Voltage and current standing waves for a quarter-wave grounded antenna.

similar to those of the Hertz antenna because of the ground-reflected waves. Figure 10-11 shows the voltage and current standing waves for a quarter-wave grounded antenna. It can be seen that if the Marconi antenna is mounted directly on Earth's surface the actual antenna and its *image* combine and produce exactly the same standing-wave patterns as those of the half-wave ungrounded (Hertz) antenna. Current maxima occur at the grounded ends, which causes high current flow through ground. To reduce power losses, the ground should be a good conductor, such as rich, loamy soil. If the ground is a poor conductor, such as sandy or rocky terrain, an artificial *ground plane* system made of heavy copper wires spread out radially below the antenna may be required. Another way of artificially improving the conductivity of the ground area below the antenna is with a *counterpoise*. A counterpoise is a wire structure placed below the antenna and erected above the ground. The counterpoise should be insulated from earth ground. A counterpoise is a form of capacitive ground system; capacitance is formed between the counterpoise and Earth's surface.

Figure 10-12 shows the radiation pattern for a quarter-wave grounded (Marconi) antenna. It can be seen that the lower half of each lobe is canceled by the ground-reflected waves. This is generally of no consequence because radiation in the horizontal direction is increased, thus increasing radiation along Earth's surface (ground waves) and improving area coverage. It can also be seen that increasing the antenna length improves horizontal radiation at the expense of sky-wave propagation. This is also shown in Figure 10-12. Optimum horizontal radiation occurs for an antenna that is approximately five-eighths wavelength long. For a one-wavelength antenna, there is no ground-wave propagation.

A Marconi antenna has the obvious advantage over a Hertz antenna of being only

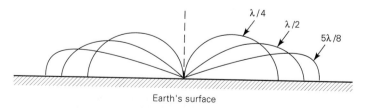

Figure 10-12 Grounded antenna radiation patterns.

Grounded Antenna

half as long. The disadvantage of a Marconi antenna is that it must be located close to the ground.

ANTENNA LOADING

Thus far we have considered antenna length in terms of wavelengths rather than physical dimensions. By the way, how long is a quarter-wavelength antenna? For a transmit frequency of 1 GHz, one-quarter wavelength is 0.075 m (2.95 in.). However, for a transmit frequency of 1 MHz, one-quarter wavelength is 75 m, and at 100 kHz, one-quarter wavelength is 750 m. It is obvious that the physical dimensions for low-frequency antennas are not practical, especially for mobile radio applications. However, it is possible to increase the electrical length of an antenna by a technique called *loading*. When an antenna is loaded, its physical length remains unchanged although its effective electrical length is increased. Several techniques are used for loading antennas.

Loading coils.   Figure 10-13a shows how a coil (inductor) added in series with a dipole antenna effectively increases the antenna's electrical length. Such a coil is appropriately called a *loading coil*. The loading coil effectively cancels out the capacitance component of the antenna input impedance. Thus the antenna looks like a resonant circuit, is resistive, and can now absorb 100% of the incident power. Figure 10-13b shows the current standing-wave patterns on an antenna with a loading coil. The loading coil is generally placed at the bottom of the antenna, allowing the antenna to be easily tuned to resonance. A loading coil effectively increases the radiation resistance of the antenna by approximately 5 Ω. Note also that the current standing wave has a maximum value at the coil, increasing power losses, creating a situation of possible corona, and effectively reducing the radiation efficiency of the antenna.

Top loading.   Loading coils have several shortcomings that can be avoided by using a technique called antenna *top loading*. With top loading, a metallic array that resembles a spoked wheel is placed on top of the antenna. The wheel increases the shunt

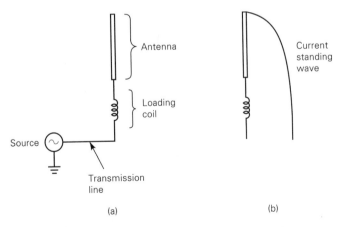

Figure 10-13 Loading coil: (a) antenna with loading coil; (b) current standing wave with loading coil.

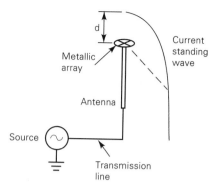

Figure 10-14 Antenna top loading.

capacitance to ground, reducing the overall antenna capacitance. Antenna top loading is shown in Figure 10-14. Notice that the current standing-wave pattern is pulled up along the antenna as though the antenna length had been increased distance d, placing the current maximum at the base. Top loading results in a considerable increase in the radiation resistance and radiation efficiency. It also reduces the voltage of the standing wave at the antenna base. Unfortunately, top loading is awkward for mobile applications.

The current loop of the standing wave can be raised even further (improving the radiation efficiency even more) if a *flat top* is added to the antenna. If a vertical antenna is folded over on top to form an L or T, as shown in Figure 10-15, the current loop will occur nearer the top of the radiator. If the flat-top and vertical portions are each one-quarter wavelength long, the current maximum will occur at the top of the vertical radiator.

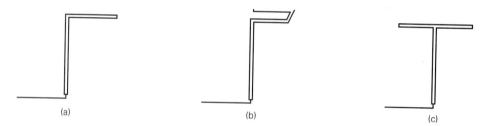

| (a) | (b) | (c) |

Figure 10-15 Flat-top antenna loading.

ANTENNA ARRAYS

An antenna *array* is formed when two or more antenna elements are combined to form a single antenna. An antenna element is an individual radiator such as a half- or quarter-wave dipole. The elements are physically placed in such a way that their radiation fields interact with each other, producing a total radiation pattern that is the vector sum of the individual fields. The purpose of an array is to increase the directivity of an antenna system and concentrate the radiated power within a smaller geographic area.

In essence, there are two types of antenna elements: *driven* and *parasitic* (non-driven). Driven elements are directly connected to the transmission line and receive power from or are driven by the source. Parasitic elements are not connected to the transmission line; they receive energy only through mutual induction with a driven element or another parasitic element. A parasitic element that is longer than the driven element from which it receives energy is called a *reflector*. A reflector effectively reduces the signal strength in its direction and increases it in the opposite direction. Therefore, it acts like a concave

Antenna Arrays

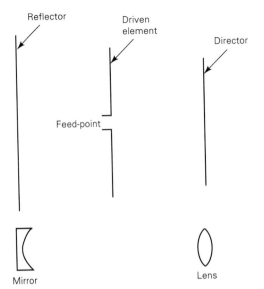

Figure 10-16   Antenna array.

mirror. This action occurs because the wave passing through the parasitic element induces a voltage that is reversed 180° with respect to the wave that induced it. The induced voltage produces an in-phase current and the element radiates (it actually reradiates the energy it just received). The reradiated energy sets up a field that cancels in one direction and reinforces in the other. A parasitic element that is shorter than its associated driven element is called a *director*. A director increases field strength in its direction and reduces it in the opposite direction. Therefore, it acts like a convergent convex lens. This is shown in Figure 10-16.

Radiation directivity can be increased in either the horizontal or vertical plane, depending on the placement of the elements and whether they are driven. If not driven, the pattern depends on whether the elements are directors or reflectors. If driven, the pattern depends on the relative phase of the feeds.

BROADSIDE ARRAY

A *broadside array* is one of the simplest types of antenna arrays. It is made by simply placing several resonant dipoles of equal size (both length and diameter) in parallel with each other and in a straight line (collinear). All elements are fed in phase from the same source. As the name implies, a broadside array radiates at right angles to the plane of the array and radiates very little in the direction of the plane. Figure 10-17a shows a broadside array that is comprised of four driven half-wave elements separated by one-half wavelength. Therefore, the signal that is radiated from element 2 has traveled one-half wavelength farther than the signal radiated from element 1 (that is, they are radiated 180° out of phase). Crisscrossing the transmission line produces an additional 180° phase shift. Therefore, the currents in all the elements are in phase, and the radiated signals are in phase and additive in a plane at right angles to the plane of the array. Although the horizontal radiation pattern for each element by itself is omnidirectional, when combined their fields produce a highly directive bidirectional radiation pattern (10-17b). Directivity can be increased even further by increasing the length of the array by adding more elements.

Chap. 10   **Antennas and Waveguides**

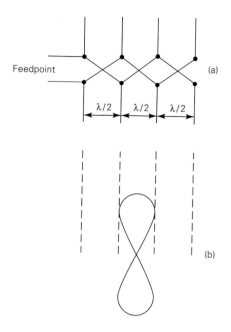

Feedpoint

$\lambda/2$ $\lambda/2$ $\lambda/2$

(a)

(b)

Figure 10-17 Broadside antenna: (a) broadside array; (b) radiation pattern.

End-Fire Array

An *end-fire array* is essentially the same element configuration as the broadside array except that the transmission line is not crisscrossed between elements. As a result, the fields are additive in line with the plane of the array. Figure 10-18 shows an end-fire array and its resulting radiation pattern.

Nonresonant Array: The Rhombic Antenna

The *rhombic antenna* is a nonresonant antenna that is capable of operating satisfactorily over a relatively wide bandwidth, making it ideally suited for HF transmission (range 3 to 30 MHz). The rhombic antenna is made up of four nonresonant elements each several wavelengths long. The entire array is terminated in a resistor if unidirectional operation is desired. The most widely used arrangement for the rhombic antenna resembles a transmission line that has been pinched out in the middle: it is shown in Figure 10-19. The antenna is mounted horizontally and placed one-half wavelength or more above the ground. The exact height depends on the precise radiation pattern desired. Each set of elements acts like a transmission line terminated in its characteristic impedance; thus, waves are radiated only in the forward direction. The terminating resistor absorbs approximately one-third of the total antenna input power. Therefore, a rhombic antenna has a maximum efficiency of 67%. Gains of over 40 (16 dB) have been achieved with rhombic antennas.

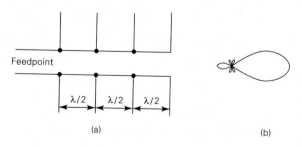

Feedpoint

$\lambda/2$ $\lambda/2$ $\lambda/2$

(a)

(b)

Figure 10-18 End-fire antenna: (a) end-fire array; (b) radiation pattern.

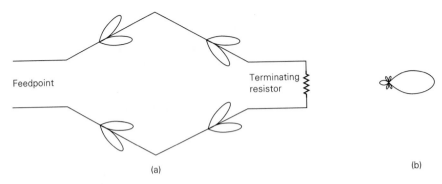

Figure 10-19 Rhombic antenna: (a) rhombic array; (b) radiation pattern.

SPECIAL-PURPOSE ANTENNAS

Folded Dipole

A two-wire *folded dipole* and its associated voltage standing-wave pattern are shown in Figure 10-20a. The folded dipole is essentially a single antenna made up of two elements. One element is fed directly, while the other is conductively coupled at the ends. Each element is one-half wavelength long. However, because current can flow around corners, there is a full wavelength of current on the antenna. Therefore, for the same input power, the input current will be one-half that of the basic half-wave dipole and the input impedance is four times higher (4 × 72 = 288). The input impedance of a folded dipole is equal to the half-wave impedance (72 Ω) times the number of folded wires squared. For example, if there are three dipoles, as shown in Figure 10-20b, the input impedance is $3^2 \times 72 = 648$ Ω. Another advantage of a folded dipole over a basic half-wave dipole is wider bandwidth. The bandwidth can be increased even further by making the dipole elements larger in diameter (such an antenna is appropriately called a *fat dipole*). However, fat dipoles have slightly different current distributions and input impedance characteristics than thin ones.

Yagi–Uda antenna. A widely used antenna that commonly uses a folded dipole as the driven element is the *Yagi–Uda antenna*, named after two Japanese scientists who invented it and described its operation. (The Yagi–Uda is generally called simply Yagi.) A

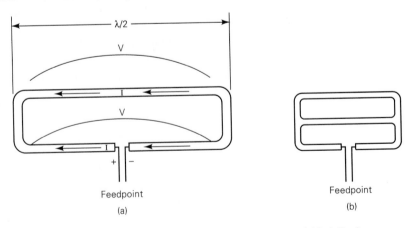

Figure 10-20 (a) Folded dipole; (b) three-element folded dipole.

Chap. 10   Antennas and Waveguides

Yagi antenna is a linear array consisting of a dipole and two or more parasitic elements: one reflector and one or more directors. A simple three-element Yagi is shown in Figure 10-21a. The driven element is a half-wavelength folded dipole. (This element is referred to as the driven element because it is connected to the transmission line. However, it is generally used for receiving only.) The reflector is a straight aluminum rod approximately 5% longer than the dipole, and the director is cut approximately 5% shorter than the driven element. The spacing between elements is generally between 0.1 and 0.2 wavelength. Figure 10-21b shows the radiation pattern for a Yagi antenna. The typical directivity for a Yagi is between 7 and 9 dB. The bandwidth of the Yagi can be increased by using more than one folded dipole, each cut to a slightly different length. Therefore, the Yagi antenna is commonly used for VHF television reception because of its wide bandwidth (the VHF TV band extends from 54 to 216 MHz).

Log-periodic Antenna

A class of frequency-independent antennas called *log periodics* evolved from the initial work of V. H. Rumsey, J. D. Dyson, R. H. DuHamel, and D. E. Isbell at the University of Illinois in 1957. The primary advantages of log-periodic antennas is the independence of their radiation resistance and radiation pattern to frequency. Log-periodic antennas have bandwidth ratios of 10:1 or greater. The bandwidth ratio is the ratio of the highest to the lowest frequency over which an antenna will satisfactorily operate. The bandwidth ratio is often used rather than simply stating the percentage of the bandwidth to the center frequency. Log periodics are not simply a type of antenna but rather a class of antenna, because there are many different types, some that are quite unusual. Log-periodic antennas can be unidirectional or bidirectional and have a low-to-moderate directive gain. High gains may also be achieved by using them as an element in a more complicated array.

The physical structure of a log-periodic antenna is repetitive, which results in repetitive behavior in its electrical characteristics. In other words, the design of a log-periodic antenna consists of a basic geometric pattern that repeats, except with a different size pattern. A basic log-periodic dipole array is probably the closest that a log-periodic comes to a conventional antenna, it is shown in Figure 10-22. It consists of several dipoles of different length and spacing that are fed from a single source at the small end. The transmission

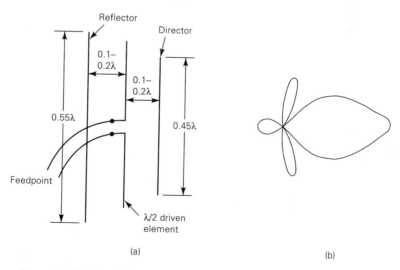

(a)

(b)

Figure 10-21 Yagi–Uda antenna: (a) three-element Yagi; (b) radiation pat-

Special-purpose Antennas

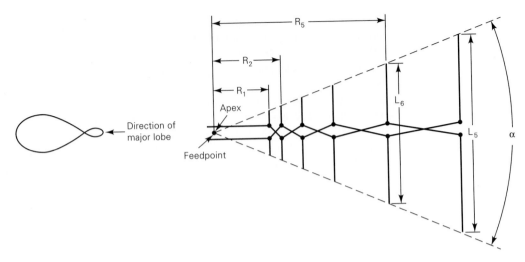

Figure 10-22 Log-periodic antenna.

line is crisscrossed between the feedpoints of adjacent pairs of dipoles. The radiation pattern for a basic log-periodic antenna has maximum radiation outward from the small end. The lengths of the dipoles and their spacing are related in such a way that adjacent elements have a constant ratio to each other. Dipole lengths and spacings are related by the formula

$$\frac{R_2}{R_1} = \frac{R_3}{R_2} = \frac{R_4}{R_3} = \frac{1}{\tau} = \frac{L_2}{L_1} = \frac{L_3}{L_2} = \frac{L_4}{L_3} \qquad (10\text{-}12)$$

or

$$\frac{1}{\tau} = \frac{R_n}{R_{n-1}} = \frac{L_n}{L_{n+1}}$$

where R = dipole spacing (inches)
L = dipole length (inches)
τ = design ratio (number < 1)

The ends of the dipoles lie along a straight line, and the angle where they meet is designated α. For a typical design, $\tau = 0.7$ and $\alpha = 30°$. With the preceding structural stipulations, the antenna input impedance varies repetitively when plotted as a function of frequency, and when plotted against the log of the frequency, varies periodically (hence the name "log periodic"). A typical plot of the input impedance is shown in Figure 10-23.

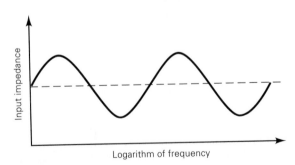

Figure 10-23 Log-periodic input impedance versus frequency.

Chap. 10   **Antennas and Waveguides**

Although the input impedance varies periodically, the variations are not necessarily sinusoidal. Also, the radiation pattern, directivity, power gain, and beamwidth undergo a similar variation with frequency.

The magnitude of a log-frequency period depends on the design ratio and, if two successive maxima occur at frequencies f_1 and f_2, they are related by the formula

$$\log f_2 - \log f_1 = \log \frac{f_2}{f_1} = \log \frac{1}{\tau} \qquad (10\text{-}13)$$

Therefore, the measured properties of a log-periodic antenna at frequency f will have identical properties at frequency τf, $\tau^2 f$, $\tau^3 f$, and so on. Log-periodic antennas, like rhombic antennas, are used mainly for HF and VHF communications. However, log-periodic antennas do not have a terminating resistor and are therefore more efficient. Very often, TV antennas advertised as "high-gain" or "high-performance" antennas are log-periodic antennas.

Loop Antenna

The most fundamental *loop antenna* is simply a single-turn coil of wire that is significantly shorter than one wavelength and carries RF current. Such a loop is shown in Figure 10-24. If the radius (r) is small compared to a wavelength, current is essentially in phase throughout the loop. A loop can be thought of as many elemental dipoles connected together. Dipoles are straight; therefore, the loop is actually a polygon rather than circular. However, a circle can be approximated if the dipoles are assumed to be sufficiently short. The loop is surrounded by a magnetic field that is at right angles to the wire, and the directional pattern is independent of its exact shape. Generally, loops are circular; however, any shape will work. The radiation pattern for a loop antenna is essentially the same as that of a short horizontal dipole.

The radiation resistance for a small loop is

$$R_r = \frac{31{,}200 A^2}{\lambda^4} \qquad (10\text{-}14)$$

where A is the area of the loop. For very low frequency applications, loops are often made with more than one turn of wire. The radiation resistance of a multiturn loop is simply the radiation resistance for a single-turn loop times the number of turns squared. The polarization of a loop antenna, like an elementel dipole, is linear. However, a vertical loop is vertically polarized and a horizontal loop is horizontally polarized.

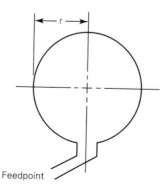

Feedpoint

Figure 10-24 Loop antenna.

Special-purpose Antennas

Small vertically polarized loops are very often used as direction-finding antennas. The direction of the received signal can be found by orienting the loop until a null or zero value is found. This is the direction of the received signal. Loops have an advantage over most other types of antennas in direction finding in that loops are generally much smaller and therefore more easily adapted to mobile communications applications.

Phased Array Antennas

A *phased array antenna* is a group of antennas or a group of antenna arrays that, when connected together, function as a single antenna whose beamwidth and direction (that is, radiation pattern) can be changed electronically without having to physically move any of the individual antennas or antenna elements within the array. The primary advantage of phased array antennas is that they eliminate the need for mechanically rotating antenna elements. In essence, a phased array is an antenna whose radiation pattern can be electronically adjusted or changed. The primary application of phased arrays is in radar when radiation patterns must be capable of being rapidly changed to follow a moving object. However, governmental agencies that transmit extremely high power signals to select remote locations all over the world, such as Voice of America, also use adjustable phased antenna arrays to direct their transmissions.

The basic principle of phased arrays is based on interference among electromagnetic waves in free space. When electromagnetic energies from different sources occupy the same space at the same time, they combine, sometimes constructively (aiding each other) and sometimes destructively (opposing each other).

There are two basic kinds of phased antenna arrays. In the first type, a single relatively high power output device supplies transmit power to a large number of antennas through a set of power splitters and phase shifters. How much of the total transmit power goes to each antenna and the phase of the signal are determined by an intricate combination of adjustable attenuators and time delays. The amount of loss in the attenuators and the phase shift introduced in the time delays is controlled by a computer. The time delays pass the RF signal without distorting it, other than to provide a specific amount of time delay (phase shift). The second kind of phased antenna arrays uses approximately as many low-power variable output devices as there are radiating elements, and the phase relationship among the output signals is controlled with phase shifters. In both types of phased arrays, the radiation pattern is selected by changing the phase delay introduced by each phase shifter. Figure 10-25 shows a phased antenna array that uses several identical antenna elements, each with its own adjustable phase delay.

Helical Antenna

A *helical antenna* is a broadband VHF or UHF antenna that is ideally suited for applications for which radiating circular rather than horizontal or vertical polarized electromagnetic waves are required. A helical antenna can be used as a single-element antenna or stacked horizontally or vertically in an array to modify its radiation pattern by increasing the gain and decreasing the beamwidth of the primary lobe.

A basic end-fire helical antenna is shown in Figure 10-26. The driven element of the antenna consists of a loosely wound rigid helix with an axis length approximately equal to the product of the number of turns and the distance between turns (pitch). A helical antenna is mounted on a ground plane made up of either solid metal or a metal screen that

Variable phase waveforms

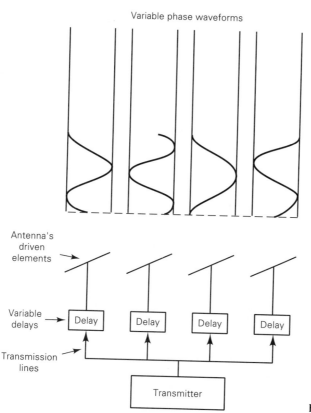

Figure 10-25 Phased array antenna.

resembles chicken wire. With a helical antenna, there are two modes of propagation: *normal* and *axial*. In the normal mode, electromagnetic radiation is in a direction at right angles to the axis of the helix. In the axial mode, radiation is in the axial direction and produces a broadband, relatively directional pattern. If the circumference of the helix is approximately equal to one wavelength, traveling waves propagate around the turns of the helix and radiate a circularly polarized wave. With the dimensions shown in Figure 10-26, frequencies within ±20% of the center frequency produce a directivity of almost 25 and a beamwidth of 90° between nulls.

The gain of a helical antenna depends on several factors, including the diameter of the helix, the number of turns in the helix, the pitch or spacing between turns, and the frequency of operation. Mathematically, the power gain of a helical antenna is

$$A_{p(\text{dB})} = 10 \log \left[15(\pi D/\lambda)^2 \, \frac{(NS)}{\lambda} \right] \qquad (10\text{-}15)$$

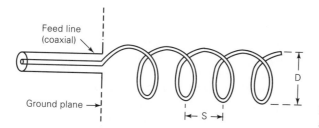

Figure 10-26 End-fire helical antenna.

Special-purpose Antennas

where $A_{p(dB)}$ = antenna power gain (unitless)
　　　　　　D = helix diameter (meters)
　　　　　　N = number of turns (any positive integer)
　　　　　　S = pitch (meters)
　　　　　　λ = wavelength (meters/cycle)

Typically, a helical antenna will have between a minimum of 3 or 4 and a maximum of 20 turns and power gains between 15 and 20 dB. The 3-dB beamwidth of a helical antenna can be determined with the following mathematical expression:

$$\theta = \frac{52}{(\pi D/\lambda)(\sqrt{NS/\lambda})} \tag{10-16}$$

where θ = beamwidth (degrees)
　　　　　　D = helix diameter (meters)
　　　　　　N = number of turns (any positive integer)
　　　　　　S = pitch (meters)
　　　　　　λ = wavelength (meters/cycle)

From Equations 10-15 and 10-16, it can be seen that, for a given helix diameter and pitch, the power gain increases proportional to the number of turns and the beamwidth decreases. Helical antennas provide bandwidths anywhere between ±20% of the center frequency up to as much as a 2:1 span between the maximum and minimum operating frequencies.

UHF AND MICROWAVE ANTENNAS

Antennas used for UHF (0.3 to 3 GHz) and microwave (1 to 100 GHz) must be highly directive. An antenna has an apparent gain because it concentrates the radiated power in a narrow beam rather than sending it uniformly in all directions, and the beamwidth decreases with increases in antenna gain. The relationship among antenna area, gain, and beamwidth are shown in Figure 10-27. Microwave antennas ordinarily have half-power beamwidths on the order of one 1° or less. A narrow beamwidth minimizes the effects of interference from outside sources and adjacent antennas. However, for line-of-site transmission, such as used with microwave radio, a narrow beamwidth imposes several limitations, such as mechanical stability and fading, that can lead to problems in antenna lineup.

All the electromagnetic energy emitted by a microwave antenna is not radiated in the direction of the *main lobe* (beam); some of it is concentrated in *minor lobes* called *sidelobes*, which can be sources of interference into or from other microwave signal paths. Figure 10-28 shows the relationship between the main beam and the sidelobes for a typical microwave antenna, such as a parabolic reflector.

Three important characteristics of microwave antennas are the front-to-back ratio, side-to-side coupling, and back-to-back coupling. The *front-to-back ratio* of an antenna is defined as the ratio of its maximum gain in the forward direction to its maximum gain in its backward direction. The front-to-back ratio of an antenna in an actual installation may be 20 dB or more below its isolated or free-space value because of foreground reflections from objects in or near the main transmission lobe. The front-to-back ratio of a microwave

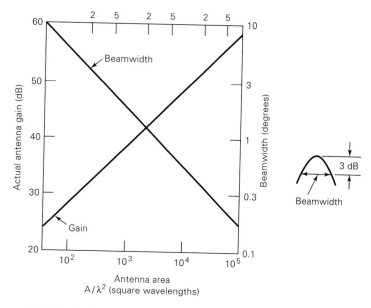

Figure 10-27 Antenna power gain and beamwidth relationship.

antenna is critical in radio system design because the transmit and receive antennas at repeater stations are often located opposite each other on the same structure (microwave radio systems and repeaters are discussed in more detail in Chapter 17). *Side-to-side* and *back-to-back coupling* express in decibels the coupling loss between antennas carrying transmitter output signals and nearby antennas carrying receiver input signals. Typically, transmitter output powers are 60 dB or higher in signal level than receiver input levels; accordingly, the coupling losses must be high to prevent a transmit signal from one antenna interfering with a receive signal of another antenna.

Highly directional (high gain) antennas are used with *point-to-point* microwave systems. By focusing the radio energy into a narrow beam that can be directed toward the receiving antenna, the transmitting antenna can increase the effective radiated power by several orders of magnitude over that of a nondirectional antenna. The receiving antenna, in a manner analogous to that of a telescope, can also increase the effective received power by a similar amount. The most common type of antenna used for microwave transmission and reception is the parabolic reflector.

Parabolic Reflector Antenna

Parabolic reflector antennas provide extremely high gain and directivity and are very popular for microwave radio and satellite communications links. A parabolic antenna is comprised of two main parts: a *parabolic reflector* and the active element called the *feed mechanism*. In essence, the feed mechanism houses the primary antenna (usually a dipole or a dipole array), which radiates electromagnetic waves toward the reflector. The reflector is a passive device that simply reflects the energy radiated by the feed mechanism into a concentrated, highly directional emission in which the individual waves are all in phase with each other (an in-phase wavefront).

UHF and Microwave Antennas

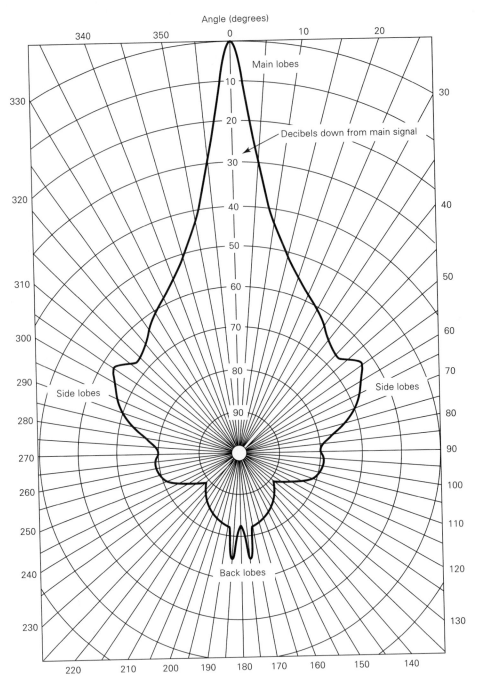

Figure 10-28 Main beam and sidelobes for a typical microwave antenna.

Parabolic reflectors. The parabolic reflector is probably the most basic component of a parabolic antenna. Parabolic reflectors resemble the shape of a plate or dish; therefore, they are sometimes called *parabolic dish* antennas or simply *dish* antennas. To understand how a parabolic reflector works, it is necessary to first understand the geometry of a *parabola*. A parabola is a plane curve that is expressed mathematically as $y = ax^2$ and defined as the locus of a point that moves so that its distance from another point (called the *focus*) added to its distance from a straight line (called the *directrix*) is of constant length. Figure 10-29 shows the geometry of a parabola whose focus is at point F and whose axis is line XY.

For the parabola shown in Figure 10-29, the following relationships exist:

$$FA + AA' = FB + BB' = FC + CC' = k \quad \text{(a constant length)}$$

and FX = focal length of the parabola (meters)
      k = a constant for a given parabola (meters)
   WZ = directrix length (meters)

The ratio of the focal length to the diameter of the mouth of the parabola (FX/WZ) is called the *aperture ratio* or simply *aperture* of the parabola; the same term is used to describe camera lenses. A parabolic reflector is obtained when the parabola is revolved around the XY axis. The resulting curved surface dish is called a paraboloid. The reflector behind the bulb of a flashlight or the headlamp of an automobile has a paraboloid shape to concentrate the light in a particular direction.

A parabolic antenna consists of a paraboloid reflector illuminated with microwave energy radiated by a feed system located at the focus point. If electromagnetic energy is radiating toward the parabolic reflector from the focus, all radiated waves will travel the same distance by the time they reach the directrix, regardless of which point on the parabola they are reflected from. Thus, all waves radiated toward the parabola from the focus will be in phase when they reach the directrix (line WZ). Consequently, radiation is concentrated along the XY axis, and cancellation take place in all other directions. A paraboloid reflector used to receive electromagnetic energy exhibits exactly the same behavior. Thus, a parabolic antenna exhibits the *principle of reciprocity* and works equally well as a receive antenna for waves arriving from the XY direction (normal to the directrix). Rays received from all other directions are cancelled at that point.

It is not necessary that the dish have a solid metal surface to efficiently reflect or receive the signals. The surface can be a mesh and still reflect or receive almost as much energy as a solid surface, provided the width of the openings is less than 0.1 wavelength. Using a mesh rather than a solid conductor considerably reduces the weight of the reflector.

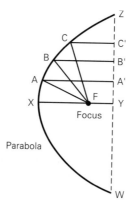

Figure 10-29 Geometry of a parabola.

Mesh reflectors are also easier to adjust, are affected less by wind, and in general provide a much more stable structure.

Parabolic antenna beamwidth. The three-dimensional radiation from a parabolic reflector has a main lobe that resembles the shape of a fat cigar in direction *XY*. The approximate −3-dB beamwidth for a parabolic antenna in degrees is given as

$$\theta = \frac{70\lambda}{D} \tag{10-17a}$$

or

$$= \frac{70c}{fD} \tag{10-17b}$$

where θ = beamwidth between half power points (degrees)
   λ = wavelength (meters)
   $c = 3 \times 10^8$ meters/second
   D = antenna mouth diameter (meters)
   f = frequency (hertz)

and

$$\phi_0 = 2\theta \tag{10-18}$$

where ϕ_0 = beamwidth between nulls in the radiation pattern (degrees)

Equations 10-17a and b and 10-18 are accurate when used for antennas with large apertures (that is, narrow beamwidths).

Parabolic antenna efficiency (η). In a parabolic reflector, reflectance from the surface of the dish is not perfect. Therefore, a small portion of the signal radiated from the feed mechanism is absorbed at the dish surface. In addition, energy near the edge of the dish does not reflect but rather is diffracted around the edge of the dish. This is called *spillover* or *leakage*. Due to dimensional imperfections, only about 50% to 75% of the energy emitted from the feed mechanism is actually reflected by the paraboloid. Also, in a real antenna the feed mechanism is not a point source; it occupies a finite area in front of the reflector and actually obscures a small area in the center of the dish and causes a shadow area in front of the antenna that is incapable of either gathering or focusing energy. These imperfections contribute to a typical efficiency for a parabolic antenna of only about 55% ($\eta = 0.55$). That is, only 55% of the energy radiated by the feed mechanism actually propagates forward in a concentrated beam.

Parabolic antenna power gain. For a transmit parabolic antenna, the power gain is approximated as

$$A_p = \eta \left(\frac{\pi D}{\lambda} \right)^2 \tag{10-19a}$$

where A_p = power gain with respect to an isotropic antenna
   D = mouth diameter of parabolic reflector (meters)
   η = antenna efficiency (antenna radiated power relative to the power radiated by the feed mechanism)
   λ = wavelength (meters/cycle)

and, for a typical antenna efficiency of 55% ($\eta = 0.55$), Equation 10-19a reduces to

$$A_p = \frac{5.4D^2f^2}{c^2}$$ (10-19b)

where c = velocity of propagation (3×10^8 meters/second).

In decibel form $A_{p(dB)} = 20 \log f \text{(MHz)} + 20 \log D \text{(meters)} - 42.2$ (10-19c)

where $A_{p(dB)}$ = power gain with respect to an isotropic antenna
   D = mouth diameter of parabolic reflector (meters)
   f = frequency (MHz)
   42.2 = constant (dB)

(For an antenna efficiency of 100%, add 2.66 dB to the value computed with Equation 10-19c.)

From Equations 10-19a, b, and c, it can be seen that the power gain of a parabolic antenna is inversely proportional to the wavelength squared. Consequently, the area (size) of the dish is an important factor when designing parabolic antennas. Very often, the area of the reflector itself is given in square wavelengths (sometimes called the *electrical* or *effective* area of the reflector). The larger the area is, the larger the ratio of the area to a wavelength, and the higher the power gain.

For a receive parabolic antenna, the surface of the reflector is again not completely illuminated, effectively reducing the area of the antenna. In a receiving parabolic antenna, the effective area is called the *capture area* and is always less than the actual mouth area. The capture area can be calculated by comparing the power received to the power density of the signal being received. Capture area is expressed mathematically as

$$A_c = kA$$ (10-20)

where A_c = capture area (square meters)
   A = actual area (square meters)
   k = a constant that is dependent on the type of antenna used and configuration
      (approximately 0.65 for a paraboloid fed by a half-wave dipole)

Therefore, the power gain for a receive parabolic antenna is

$$A_p = \frac{4\pi A_c}{\lambda^2} = \frac{4\pi kA}{\lambda^2}$$ (10-21a)

Substituting the area of the mouth of a paraboloid into Equation 10-21a, the power gain of a parabolic receive antenna can be closely approximated as

$$A_p = 6.4\left(\frac{D}{\lambda}\right)^2$$ (10-21b)

where D = dish diameter (meters)
   λ = wavelength (meters/cycle)

In decibel form, $$A_{p(dB)} = 10 \log\left[6.4\left(\frac{D}{\lambda}\right)^2\right]$$ (10-21c)

UHF and Microwave Antennas

The transmit power gain calculated using Equation 10-19c and the receive antenna power gain calculated using Equation 10-21c will yield approximately the same results for a given antenna, thus proving the reciprocity of parabolic antennas.

The radiation pattern shown in Figure 10-28 is typical for both transmit and receive parabolic antenna. The power gain within the main lobe is approximately 75 dB more than in the backward direction and almost 65 dB more than the maximum sidelobe gain.

EXAMPLE 10-2

Determine (a) beamwidth, (b) transmit power gain, (c) receive power gain, and (d) effective isotropic radiated power (EIRP) for an 2-m diameter parabolic reflector with 10 W of power radiated by the feed mechanism operating at 6 GHz with an efficiency of 55%.

Solution (a) The beamwidth is found by substituting into Equation 10-17b.

$$\theta = \frac{70(3 \times 10^8)}{(6 \times 10^9)(2)} = 1.75°$$

(b) The transmit power gain is found by substituting into Equation 10-19c.

$$A_{p(dB)} = 20 \log 6000 + 20 \log 2 - 42.2$$

$$= 39 \text{ dB}$$

(c) The receive power gain is found by substituting into Equation 10-21c.

$$\lambda = \frac{c \text{ (m/s)}}{\text{frequency (Hz)}} = \frac{3 \times 10^8}{6 \times 10^9} = 0.05 \text{ m/cycle}$$

$$A_{p(dB)} = 10 \log \left[6.4 \left(\frac{2}{0.05} \right)^2 \right] = 40.1 \text{ dB}$$

(d) The EIRP is the product of the radiated power times the transmit antenna gain or, in decibels,

$$\text{EIRP} = A_{p(dB)} + P_{\text{radiated(dBm)}}$$

$$= 40.1 + 10 \log \frac{10}{0.001}$$

$$= 40.1 \text{ dB} + 40 \text{ dBm}$$

$$= 80.1 \text{ dBm}$$

Feed mechanisms. The feed mechanism in a parabolic antenna actually radiates the electromagnetic energy and, is therefore often called the *primary antenna*. The feed mechanism is of primary importance because its function is to radiate the energy toward the reflector. An ideal feed mechanism should direct all the energy toward the parabolic reflector and have no shadow effect. In practice, this is impossible to accomplish, although if care is taken when designing the feed mechanism, most of the energy can be radiated in the proper direction, and the shadow effect can be minimized. There are three primary types of feed mechanisms for parabolic antennas: center feed, horn feed, and Cassegrain feed.

Center-feed. Figure 10-30 shows a diagram for a center-fed paraboloid reflector with an additional *spherical reflector*. The primary antenna is placed at the focus. Energy radiated toward the reflector is reflected outward in a concentrated beam. However,

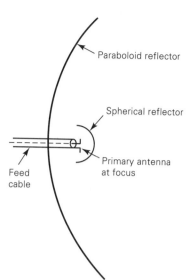

Figure 10-30 Parabolic antenna with a center feed.

energy not reflected by the paraboloid spreads in all directions and has the tendency of disrupting the overall radiation pattern. The spherical reflector redirects such emissions back toward the parabolic reflector, where they are rereflected in the proper direction. Although the additional spherical reflector helps to concentrate more energy in the desired direction, it also has a tendency to block some of the initial reflections. Consequently, the good it accomplishes is somewhat offset by its own shadow effect, and its overall performance is only marginally better than without the additional spherical reflector.

Horn-feed. Figure 10-31a shows a diagram for a parabolic reflector using a horn feed. With a horn-feed mechanism, the primary antenna is a small horn antenna rather than a simple dipole or dipole array. The horn is simply a flared piece of waveguide material that is placed at the focus and radiates a somewhat directional pattern toward the parabolic reflector. When a propagating electromagnetic field reaches the mouth of the horn, it continues to propagate in the same general direction, except that, in accordance with Huygens' principle, it spreads laterally, and the wavefront eventually becomes spherical. The horn structure can have several different shapes, as shown in Figure 10-31b: sectoral (flaring only in one direction), pyramidal, or conical. As with the center feed, a horn feed presents somewhat of an obstruction to waves reflected from the parabolic dish.

Cassegrain feed. The Cassegrain feed is named after an eighteenth-century astronomer and evolved directly from astronomical optical telescopes. Figure 10-32 shows the basic geometry of a Cassegrain-feed mechanism. The primary radiating source is located in or just behind a small opening at the vertex of the paraboloid, rather than at the focus. The primary antenna is aimed at a small secondary reflector (*Cassegrain subreflector*) located between the vertex and the focus.

The rays emitted from the primary antenna are reflected from the Cassegrain subreflector and then illuminate the main parabolic reflector just as if they had originated at the focus. The rays are collimated by the parabolic reflector in the same way as with the center- and horn-feed mechanisms. The subreflector must have a hyperboloidal curvature to reflect the rays from the primary antenna in such a way as to function as a *virtual source* at

UHF and Microwave Antennas

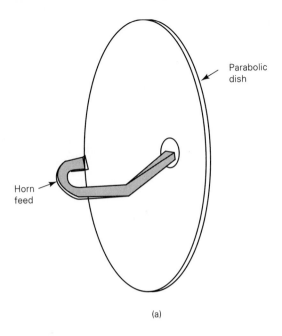

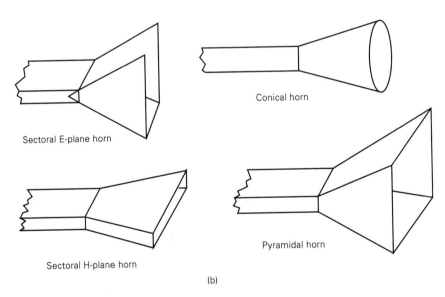

Sectoral E-plane horn

Conical horn

Sectoral H-plane horn

Pyramidal horn

(b)

Figure 10-31 Parabolic antenna with a horn feed: (a) horn feed; (b) wave-guide horn types.

the paraboloidal focus. The Cassegrain feed is commonly used for receiving extremely weak signals or when extremely long transmission lines or waveguide runs are required and it is necessary to place low-noise preamplifiers as close to the antenna as possible. With the Cassegrain feed, preamplifiers can be placed just before the feed mechanism and not be an obstruction to the reflected waves.

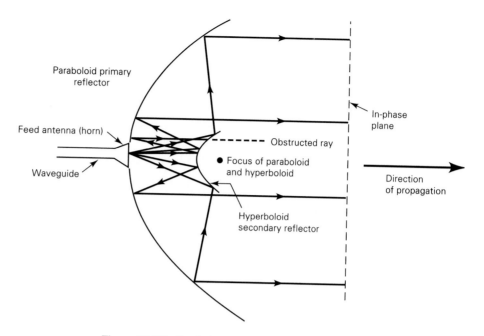

Figure 10-32 Parabolic antenna with a Cassegrain feed.

WAVEGUIDES

Parallel-wire transmission lines, including coaxial cables, cannot effectively propagate electromagnetic energy above approximately 1 GHz, and at frequencies above approximately 15 GHz, they are useless for distances greater than a few inches. This is because of the attenuation caused by skin effect and radiation losses. In addition, parallel-wire transmission lines cannot be used to propagate signals with high powers because the high voltages associated with them cause the dielectric separating the two conductors to break down. Consequently, parallel-wire transmission lines are impractical for many UHF and microwave applications. There are several alternatives, including optical fiber cables and waveguides. Optical fibers are discussed in detail in Chapter 20.

In its simplest form, a *waveguide* is a hollow conductive tube, usually rectangular in cross section, but sometimes circular or eliptical. The dimensions of the cross section are selected such that electromagnetic waves can propagate within the interior of the guide (hence the name waveguide). A waveguide does not conduct current in the true sense, but rather serves as a boundary that confines electromagnetic energy. The walls of the waveguide are conductors and therefore reflect electromagnetic energy from their surface. If the wall of the waveguide is a good conductor and very thin, little current flows in the interior walls and, consequently, very little power is dissipated. In a waveguide, conduction of energy does not occur in the walls of the waveguide, but rather through the dielectric within the waveguide, which is usually dehydrated air or inert gas. In essence, a waveguide is analogous to a metallic wire conductor with its interior removed. Electromagnetic energy propagates down a waveguide by reflecting back and forth in a zigzag pattern.

When discussing waveguide behavior, it is necessary to speak in terms of electromagnetic field concepts (that is, electric and magnetic fields), rather than currents and voltages as for transmission lines. The cross-sectional area of a waveguide must be on the

Waveguides

411

same order as the wavelength of the signal it is propagating. Therefore, waveguides are generally restricted to frequencies above 1 GHz.

Rectangular Waveguide

Rectangular waveguides are the most common form of waveguide. To understand how rectangular waveguides work, it is necessary to understand the basic behavior of waves reflecting from a conducting surface.

Electromagnetic energy is propagated through free space as transverse-electromagnetic (TEM) waves with a magnetic field, an electric field, and a direction of propagation that are mutually perpendicular. For an electromagnetic wave to exist in a waveguide, it must satisfy Maxwell's equations throughout the guide. Maxwell's equations are necessarily complex and beyond the intent of this book. However, a limiting factor of Maxwell's equations is that a TEM wave cannot have a tangential component of the electric field at the walls of the waveguide. The wave cannot travel straight down a waveguide without reflecting off the sides, because the electric field would have to exist next to a conductive wall. If that happened, the electric field would be short-circuited by the walls themselves. To successfully propagate a TEM wave through a waveguide, the wave must propagate down the guide in a zigzag manner, with the electric field maximum in the center of the guide and zero at the surface of the walls.

In transmission lines, wave velocity is independent of frequency, and for air or vacuum dielectrics, the velocity is equal to the velocity in free space. However, in waveguides the velocity varies with frequency. In addition, it is necessary to distinguish between two different kinds of velocity: *phase velocity* and *group velocity*. Group velocity is the velocity at which a wave propagates, and the phase velocity is the velocity at which the wave changes phase.

Phase velocity and group velocity. Phase velocity is the apparent velocity of a particular phase of the wave (for example, the crest or maximum electric intensity point). Phase velocity is the velocity with which a wave changes phase in a direction parallel to a conducting surface, such as the walls of a waveguide. Phase velocity is determined by measuring the wavelength of a particular frequency wave and then substituting it into the following formula:

$$v_{ph} = f\lambda \tag{10-22}$$

where v_{ph} = phase velocity (meters/second)
f = frequency (hertz)
λ = wavelength (meters/cycle)

Group velocity is the velocity of a group of waves (that is, a pulse). Group velocity is the velocity at which information signals of any kind are propagated. It is also the velocity at which energy is propagated. Group velocity can be measured by determining the time it takes for a pulse to propagate a given length of waveguide. Group and phase velocities have the same value in free space and in parallel wire transmission lines. However, if these two velocities are measured at the same frequency in a waveguide, it will be found that, in general, the two velocities are not the same. At some frequencies they will be nearly equal and at other frequencies they can be considerably different.

The phase velocity is always equal to or greater than the group velocity, and their product is equal to the square of the free-space propagation velocity. Thus,

$$v_g v_{ph} = c^2 \qquad (10\text{-}23)$$

where v_{ph} = phase velocity (meters/second)
           v_g = group velocity (meters/second)
           $c = 3 \times 10^8$ meters/second

Phase velocity may exceed the velocity of light. A basic principle of physics states that no form of energy can travel at a greater velocity than light (electromagnetic waves) in free space. This principle is not violated because it is group velocity, not phase velocity, that represents the velocity of propagation of energy.

Since the phase velocity in a waveguide is greater than its velocity in free space, the wavelength for a given frequency will be greater in the waveguide than in free space. The relationship among free-space wavelength, guide wavelength, and the free-space velocity of electromagnetic waves is

$$\lambda_g = \lambda_o \frac{v_{ph}}{c} \qquad (10\text{-}24)$$

where λ_g = guide wavelength (meters/cycle)
           λ_o = free-space wavelength (meters/cycle)
           v_{ph} = phase velocity (meters/second)
           c = free-space velocity of light (3×10^8 meters/second)

Cutoff frequency and cutoff wavelength. Unlike transmission lines that have a maximum frequency of operation, waveguides have a minimum frequency of operation called the *cutoff frequency*. The cutoff frequency is an absolute limiting frequency; frequencies below the cutoff frequency will not be propagated by the waveguide. Conversely, waveguides have a minimum wavelength that they can propagate called the *cutoff wavelength*. The cutoff wavelength is defined as the smallest free-space wavelength that is just unable to propagate in the waveguide. In other words, only frequencies with wavelengths less than the cutoff wavelength can propagate down the waveguide. The cutoff wavelength and frequency are determined by the cross-sectional dimensions of the waveguide.

The mathematical relationship between the guide wavelength at a particular frequency and the cutoff frequency is

$$\lambda_g = \frac{c}{\sqrt{f^2 - f_c^2}} \qquad (10\text{-}25)$$

where λ_g = guide wavelength (meters/cycle)
           f = frequency of operation (hertz)
           f_c = cutoff frequency (hertz)
           c = free-space propagation velocity (3×10^8 meters/second)

Equation 10-25 can be rewritten in terms of the free-space wavelength as

$$\lambda_g = \frac{\lambda_o}{\sqrt{1 - (f_c/f)^2}} \qquad (10\text{-}26)$$

where λ_g = guide wavelength (meters/cycle)
           λ_o = free-space wavelength (meters/cycle)
           f = frequency of operation (hertz)
           f_c = cutoff frequency (hertz)

Waveguides

Combining Equations 10-24 and 10-25 and rearranging gives

$$v_{ph} = \frac{c(\lambda_g)}{\lambda_o} = \frac{c}{\sqrt{1 - (f_c/f)^2}} \qquad (10\text{-}27)$$

It is evident from Equation 10-27 that if f becomes less than f_c the phase velocity becomes imaginary, which means that the wave is not propagated. Also, it can be seen that, as the frequency of operation approaches the cutoff frequency, the phase velocity and the guide wavelength become infinite, and the group velocity goes to zero.

Figure 10-33 shows a cross-sectional view of a piece of rectangular waveguide with dimensions a and b (a is normally designated the wider of the two dimensions). Dimension a determines the cutoff frequency of the waveguide according to the following mathematical relationship:

$$f_c = \frac{c}{2a} \qquad (10\text{-}28)$$

where f_c = cutoff frequency (hertz)
          a = cross-sectional length (meters)

or, in terms of wavelength $\lambda_c = 2a$ (10-29)

where λ_c = cutoff wavelength (meters/cycle)
          a = cross-sectional length (meters)

Equations 10-28 and 10-29 indicate that cutoff occurs at the frequency for which the largest transverse dimension of the guide is exactly one-half of the free-space wavelength.

Figure 10-34 shows the top view of a section of rectangular waveguide and illustrates how electromagnetic waves propagate down the guide. For frequencies above the cutoff frequency (Figures 10-34a, b, and c), the waves propagate down the guide by reflecting back and forth between the wall at various angles. Figure 10-34d shows what happens to the electromagnetic wave at the cutoff frequency.

EXAMPLE 10-3

For a rectangular waveguide with a wall separation of 3 cm and a desired frequency of operation of 6 GHz, determine (a) cutoff frequency, (b) cutoff wavelength, (c) group velocity, and (d) phase velocity.

Solution   (a) The cutoff frequency is determined by substituting into Equation 10-28.

$$f_c = \frac{3 \times 10^8 \text{ m/s}}{2(0.03 \text{ m})} = 5 \text{ GHz}$$

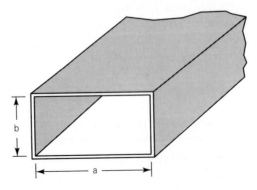

Figure 10-33 Cross-sectional view of a rectangular waveguide.

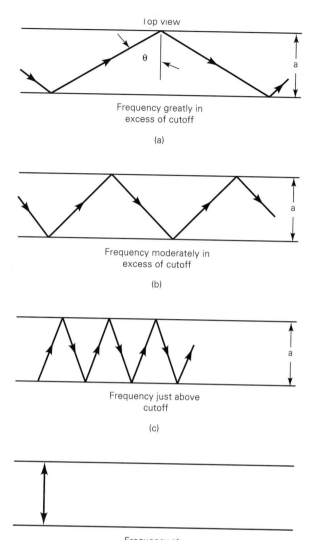

Top view

θ

a

Frequency greatly in
excess of cutoff

(a)

a

Frequency moderately in
excess of cutoff

(b)

a

Frequency just above
cutoff

(c)

Frequency at
cutoff

(d)

Figure 10-34 Electromagnetic
wave propagation in a rectangular
waveguide.

(b) The cutoff wavelength is determined by substituting into Equation 10-29.

$$\lambda_c = 2 \ (3 \ cm) = 6 \ cm$$

(c) The phase velocity is found using Equation 10-27.

$$v_{ph} = \frac{3 \times 10^8}{\sqrt{1 - (5 \ GHz/6 \ GHz)^2}} = 5.43 \times 10^8 \ m/s$$

(d) The group velocity is found by rearranging Equation 10-23.

$$v_g = \frac{c^2}{v_{ph}} = \frac{(3 \times 10^8)^2}{5.43 \times 10^8} = 1.66 \times 10^8 \ m/s$$

Modes of propagation.   Electromagnetic waves travel down a waveguide in different configurations called propagation *modes*. In 1955, the Institute of Radio Engineers (IRE) published a set of standards. These standards designated the modes for rectangular

Waveguides

waveguides as $TE_{m,n}$ for transverse-electric waves and $TM_{m,n}$ for transverse-magnetic waves. TE means that the electric field lines are everywhere transverse (that is, perpendicular to the guide walls), and TM means that the magnetic field lines are everywhere transverse. In both cases, m and n are integers designating the number of half-wavelengths of intensity (electric or magnetic) that exist between each pair of walls. m is measured along the X axis of the waveguide (the same axis the dimension a is measured on), and n is measured along the Y axis (the same as dimension b).

Figure 10-35 shows the electromagnetic field pattern for a $TE_{1,0}$ mode wave. The $TE_{1,0}$ mode is sometimes called the dominant mode because it is the most "*natural*" mode. A waveguide acts like a high-pass filter in that it passes only those frequencies above the minimum or cutoff frequency. At frequencies above the cutoff frequency, higher-order TE modes of propagation, with more complicated field configurations, are possible. However, it is undesirable to operate a waveguide at a frequency at which these higher modes can propagate. The next higher mode possible occurs when the free-space wavelength is equal to length a (that is, at twice the cutoff frequency). Consequently, a rectangular waveguide is normally operated within the frequency range between f_c and $2f_c$. Allowing higher modes to propagate is undesirable because they do not couple well to the load and thus cause reflections to occur and standing waves to be created. The $TE_{1,0}$ mode is also desired because it allows for the smallest possible size waveguide for a given frequency of operation.

In Figure 10-35a, the electric (E) field vectors are parallel to each other and perpendicular to the wide face of the guide. Their amplitude is greatest midway between the nar-

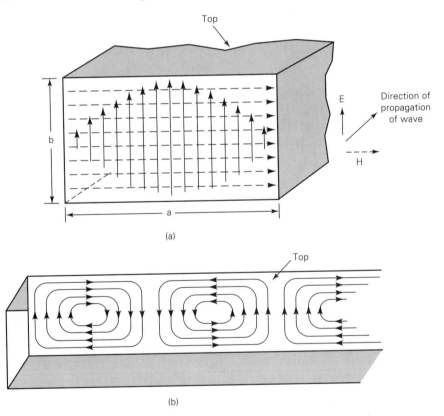

Figure 10-35 Electric and magnetic field vectors in a rectangular waveguide: (a) end view; (b) magnetic field configuration in a longitudinal section.

row walls and decreases to zero at the walls, in a cosinusoidal fashion. The magnetic (H) field vectors (shown by dashed lines) are also parallel to each other and perpendicular to the electric vectors. The magnetic intensity is constant in the vertical direction across the guide section. The wave is propagating in the longitudinal direction of the guide, perpendicular to the E and H vectors. Figure 10-35b shows the magnetic field configuration in a longitudinal section of waveguide for the $TE_{1,0}$ propagation mode.

Characteristic impedance.   Waveguides have a characteristic impedance that is analogous to the characteristic impedance of parallel wire transmission lines and closely related to the characteristic impedance of free space. The characteristic impedance of a waveguide has the same significance as the characteristic impedance of a transmission line, with respect to load matching, signal reflections, and standing waves. The characteristic impedance of a waveguide is expressed mathematically as

$$Z_o = \frac{377}{\sqrt{1 - (f_c/f)^2}} = 377\frac{\lambda_g}{\lambda_0} \qquad (10\text{-}30)$$

where Z_o = characteristic impedance (ohms)
   f_c = cutoff frequency (hertz)
   f = frequency of operation (hertz)

Z_o is generally greater than 377 Ω. In fact, at the cutoff frequency, Z_o becomes infinite, and at a frequency equal to twice the cutoff frequency ($2f_c$), Z_o = 435 Ω. Two waveguides with the same length a dimension but different length b dimensions will have the same value of cutoff frequency and the same value of characteristic impedance. However, if these two waveguides are connected together end to end and an electromagnetic wave is propagated down them, a discontinuity will occur at the junction point, and reflections will occur even though their impedances are matched.

Impedance matching.   Reactive stubs are used in waveguides for impedance transforming and impedance matching just as they are in parallel-wire transmission lines. Short-circuited waveguide stubs are used with waveguides in the same manner that they are used in transmission lines.

Figure 10-36 shows how inductive and capacitive irises are installed in rectangular waveguide to behave like shunt susceptances. The irises consist of thin metallic plates placed perpendicular to the walls of the waveguide and joined to them at the edges, with

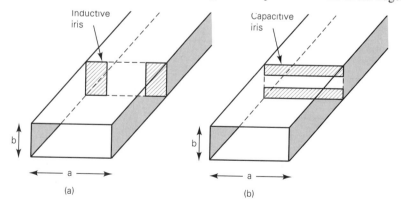

Figure 10-36 Waveguide impedance matching: (a) inductive iris; (b) capacitive iris.

Waveguides

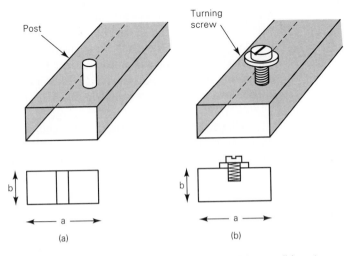

Figure 10-37 Waveguide impedance matching: (a) post; (b) tuning screw.

an opening between them. When the opening is parallel to the narrow walls, the susceptance is inductive; when it is parallel to the wide walls, it is capacitive. The magnitude of the susceptance is proportional to the size of the opening.

A post placed across the narrowest dimension of the waveguide, as shown in Figure 10-37a, acts like an inductive shunt susceptance whose value depends on its diameter and its position in the transverse plane. Tuning screws, shown in Figure 10-37b, project partway across the narrow guide dimension, act like a capacitance, and may be adjusted.

Transmission line-to-waveguide coupling. Figure 10-38 shows several ways in which a waveguide and transmission line can be joined together. The couplers shown can be used as wave launchers at the input end of a waveguide or as wave receptors at the load end of the guide. The dimensions labeled $\lambda_o/4$ and $\lambda_g/4$ are approximate. In practice, they are experimentally adjusted for best results.

TABLE 10-1 RECTANGULAR WAVEGUIDE DIMENSIONS AND ELECTRICAL CHARACTERISTICS

Useful Frequency Range, GHz	Outside Dimensions mm	Theoretical Average Attenuation, dB/m	Theoretical Average (CW) Power Rating, kW
1.12–1.70	169 × 86.6	0.0052	14,600
1.70–2.60	113 × 58.7	0.0097	6,400
2.60–3.95	76.2 × 38.1	0.019	2,700
3.95–5.85	50.8 × 25.4	0.036	1,700
5.85–8.20	38.1 × 19.1	0.058	635
8.20–12.40	25.4 × 12.7	0.110	245
12.40–18.00	17.8 × 9.9	0.176	140
18.0–26.5	12.7 × 6.4	0.37	51
26.5–40.0	9.1 × 5.6	0.58	27
40.0–60.0	6.8 × 4.4	0.95	13
60.0–90.0	5.1 × 3.6	1.50	5.1
90.0–140	4.0 (diam.)	2.60	2.2
140–220	4.0 (diam.)	5.20	0.9
220–325	4.0 (diam.)	8.80	0.4

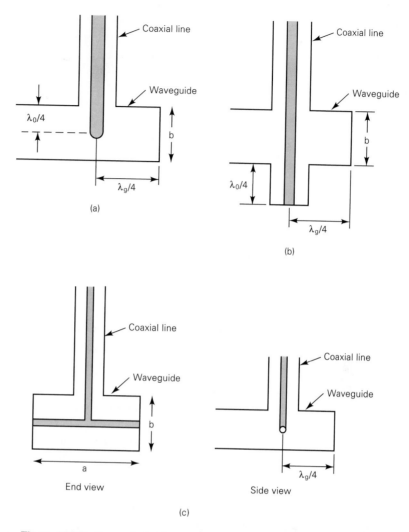

Figure 10-38 Transmission line-to-waveguide coupling: (a) quarter-wave probe coupler; (b) straight-through coupler; (c) cross-bar coupler.

Table 10-1 lists the frequency range, dimensions, and electrical characteristics for several common types of rectangular waveguide.

OTHER TYPES OF WAVEGUIDES

Circular Waveguide

Rectangular waveguide is by far the most common; however, circular waveguide is used in radar and microwave applications when it is necessary or advantageous to propagate both vertical and horizontally polarized waves in the same waveguide. Figure 10-39 shows two pieces of circular waveguide joined together by a rotation joint.

The behavior of electromagnetic waves in circular waveguide is the same as it is in rectangular waveguide. However, because of the different geometry, some of the calculations are performed in a slightly different manner.

Other Types of Waveguides

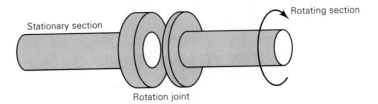

Figure 10-39 Circular waveguide with rotational joint.

The cutoff wavelength for circular waveguide is given as

$$\lambda_0 = \frac{2\pi r}{kr} \qquad (10\text{-}31)$$

where λ_0 = cutoff wavelength (meters/cycle)
r = internal radius of the waveguide (meters)
kr = solution of a Bessel function equation

Because the propagation mode with the largest cutoff wavelength is the one with the smallest value for kr (1.84), the $TE_{1,1}$ mode is dominant for circular waveguides. The cutoff wavelength for this mode reduces to

$$\lambda_0 = 1.7d \qquad (10\text{-}32)$$

where d = waveguide diameter (meters).

Circular waveguide is easier to manufacture than rectangular waveguide and easier to join together. However, circular waveguide has a much larger area than a corresponding rectangular waveguide used to carry the same signal. Another disadvantage of circular waveguide is that the plane of polarization may rotate while the wave is propagating down it (that is, a horizontally polarized wave may become vertically polarized, and vice versa).

Ridged Waveguide

Figure 10-40 shows two types of ridged waveguide. Ridged waveguide is more expensive to manufacture than standard rectangular waveguide; however, it also allows operation at lower frequencies for a given size. Consequently, smaller overall waveguide dimensions are possible using ridged waveguide. A ridged waveguide has more loss per unit length than rectangular waveguide. This characteristic combined with its increased cost limits its usefulness to specialized applications.

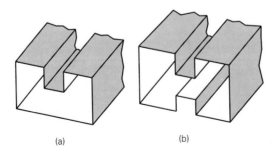

(a)   (b)

Figure 10-40 Ridged waveguide:
(a) single ridge; (b) double ridge.

Chap. 10   Antennas and Waveguides

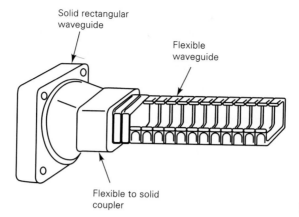

Solid rectangular
waveguide

Flexible
waveguide

Flexible to solid
coupler

Figure 10-41 Flexible waveguide.

Flexible Waveguide

Figure 10-41 shows a length of flexible rectangular waveguide. Flexible waveguide consists of spiral-wound ribbons of brass or copper. The outside is covered with a soft dielectric coating (often rubber) to keep the waveguide air- and watertight. Short pieces of flexible waveguide are used in microwave systems when several transmitters and receivers are interconnected to a complex combining or separating unit. Flexible waveguide is also used extensively in microwave test equipment.

QUESTIONS

10-1. Define *antenna*.

10-2. Describe basic antenna operation using standing waves.

10-3. Describe a relative radiation pattern; an absolute radiation pattern.

10-4. Define *front-to-back ratio*.

10-5. Describe an omnidirectional antenna.

10-6. Define *near field* and *far field*.

10-7. Define *radiation resistance* and *antenna efficiency*.

10-8. Define and contrast *directive gain* and *power gain*.

10-9. What is the directivity for an isotropic antenna?

10-10. Define *effective isotropic radiated power*.

10-11. Define *antenna polarization*.

10-12. Define *antenna beamwidth*.

10-13. Define *antenna bandwidth*.

10-14. Define *antenna input impedance*. What factors contribute to an antenna's input impedance?

10-15. Describe the operation of an elementary doublet.

10-16. Describe the operation of a half-wave dipole.

10-17. Describe the effects of ground on a half-wave dipole.

10-18. Describe the operation of a grounded antenna.

10-19. What is meant by *antenna loading?*

10-20. Describe an antenna loading coil.

10-21. Describe antenna top loading.

10-22. Describe an antenna array.

10-23. What is meant by *driven element; parasitic element?*

10-24. Describe the radiation pattern for a broadside array; an end-fire array.

10-25. Define *nonresonant antenna.*

10-26. Describe the operation of the rhombic antenna.

10-27. Describe a folded dipole antenna.

10-28. Describe a Yagi–Uda antenna.

10-29. Describe a log-periodic antenna.

10-30. Describe the operation of a loop antenna.

10-31. Describe briefly how a *phased array antenna* works and what it is primarily used for.

10-32. Describe briefly how a *helical* antenna works.

10-33. Define the following terms: *main lobe, sidelobes, side-to-side coupling,* and *back-to-back coupling.*

10-34. What are the two main parts of a *parabolic antenna?*

10-35. Describe briefly how a *parabolic reflector* works.

10-36. What is the purpose of the *feed mechanism* in a parabolic reflector antenna?

10-37. What is meant by the *capture area* of a parabolic antenna?

10-38. Describe how a *center-feed* mechanism works with a parabolic reflector.

10-39. Describe how a *horn-feed* mechanism works with a parabolic reflector.

10-40. Describe how a *Cassegrain feed* works with a parabolic reflector.

10-41. In its simplest form, what is a *waveguide?*

10-42. Describe *phase velocity; group velocity.*

10-43. Describe the *cutoff frequency* for a waveguide; *cutoff wavelength.*

10-44. What is meant by the TE mode of propagation, TM mode of propagation.

10-45. When is it advantageous to use circular waveguide?

PROBLEMS

10-1. For an antenna with input power $P_i = 100$ W, rms current $I = 2$ A, and effective resistance $R_e = 2\ \Omega$, determine:
   (a) Antenna's radiation resistance.
   (b) Antenna's efficiency.
   (c) Power radiated from the antenna, P_r.

10-2. Determine the directivity in decibels for an antenna that produces power density $\mathcal{P} = 2\ \mu W/m^2$ at a point when a reference antenna produces $0.5\ \mu W/m^2$ at the same point.

10-3. Determine the power gain in decibels for an antenna with directive gain $\mathcal{D} = 40$ and efficiency $\eta = 65\%$.

10-4. Determine the effective isotropic radiated power for an antenna with power gain $A_p = 43$ dB and radiated power $P_r = 200$ W.

10-5. Determine the effective isotropic radiated power for an antenna with directivity $\mathcal{D} = 33$ dB, efficiency $\eta = 82\%$, and input power $P_i = 100$ W.

10-6. Determine the power density at a point 20 km from an antenna that is radiating 1000 W and has power gain $A_p = 23$ dB.

10-7. Determine the power density at a point 30 km from an antenna that has input power $P_{in} = 40$ W, efficiency $\eta = 75\%$, and directivity $\mathcal{D} = 16$ dB.

10-8. Determine the power captured by a receiving antenna for the following parameters: power radiated $P_r = 50$ W; transmit antenna directive gain $A_t = 30$ dB; distance between transmit and receive antennas $d = 20$ km; receive antenna directive gain $A_r = 26$ dB.

10-9. Determine the directivity (in decibels) for an antenna that produces a power density at a point that is 40 times greater than the power density at the same point when the reference antenna is used.

10-10. Determine the effective radiated power for an antenna with directivity $\mathcal{D} = 400$, efficiency $\eta = 0.60$, and input power $P_{in} = 50$ W.

10-11. Determine the efficiency for an antenna with radiation resistance $R_r = 18.8$ Ω, effective resistance $R_e = 0.4$ Ω, and directive gain $\mathcal{D} = 200$.

10-12. Determine the power gain A_p for Problem 10-11.

10-13. Determine the efficiency for an antenna with radiated power $P_r = 44$ W, dissipated power $P_d = 0.8$ W, and directive gain $\mathcal{D} = 400$.

10-14. Determine power gain A_p for Problem 10-13.

10-15. Determine the power gain and beamwidth for an end-fire helical antenna with the following parameters: helix diameter = 0.1 m, number of turns = 10, pitch = 0.05 m, and frequency of operation = 500 MHz.

10-16. Determine the beamwidth and transmit and receive power gains of a parabolic antenna with the following parameters: dish diameter = 2.5 m, a frequency of operation of 4 GHz, and an efficiency of 55%.

10-17. For a rectangular waveguide with a wall separation of 2.5 cm and a desired frequency of operation of 7 GHz, determine (a) cutoff frequency, (b) cutoff wavelength, (c) group velocity, and (d) phase velocity.

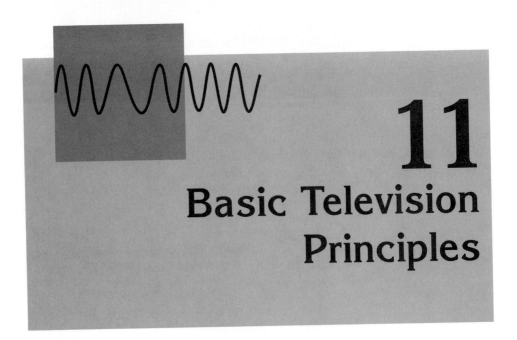

11

Basic Television Principles

INTRODUCTION

The word *television* comes from the Greek word *tele* (meaning distant) and the Latin word *vision* (meaning sight). Therefore, television simply means to see from a distance. In its simplest form, television is the process of converting *images* (either stationary or in motion) to electrical signals and then transmitting those signals to a distant receiver, where they are converted back to images that can be perceived with the human eye. Thus, television is a system in which images are transmitted from a central location and then received at distant receivers, where they are reproduced in their original form.

HISTORY OF TELEVISION

The idea of transmitting images or pictures was first experimented with in the 1880s when Paul Nipkow, a German scientist, conducted experiments using revolving *disks* placed between a powerful light source and the subject. A spiral row of holes was punched in the disk, which permitted light to scan the subject from top to bottom. After one complete revolution of the disk, the entire subject had been scanned. Light reflected from the subject was directed to a light-sensitive cell, producing current that was proportional in intensity to the reflected light. The fluctuating current operated a neon lamp, which gave off light in exact proportion to that reflected from the subject. A second disk exactly like the one in the transmitter was used in the receiver, and the two disks revolved in exact synchronization. The second disk was placed between the neon lamp and the eye of the observer, who thus saw a reproduction of the subject.

History of Television

The images reproduced with Nipkow's contraption were barely recognizable, although his scanning and synchronization principles are still used today.

In 1925, C. Francis Jenkins in the United States and John L. Baird in England, using scanning disks connected to vacuum-tube amplifiers and photoelectric cells, were able to reproduce images that were recognizable, although still of poor quality. Scientists worked for several years trying to develop effective mechanical scanning disks that, with improved mirrors and lenses and a more intense light source, would improve the quality of the reproduced image. However, in 1933, Radio Corporation of America (RCA) announced a television system, developed by Vladimir K. Zworykin, that used an electronic scanning technique. Zworykin's system required no mechanical moving parts and is essentially the system used today.

In 1941, commercial broadcasting of *monochrome* (black and white) television signals began in the United States. In 1945, the FCC assigned 13 VHF television channels: 6 low-band channels, 1 to 6 (44 to 88 MHz), and 7 high-band channels, 7 to 13 (174 to 216 MHz). However, in 1948 it was found that channel 1 (44 to 50 MHz) caused interference problems; consequently, this channel was reassigned to mobile radio services. In 1952, UHF channels 14 to 83 (470 to 890 MHz) were assigned by the FCC to provide even more television stations. In 1974, the FCC reassigned to cellular telephone frequency bands at 825 to 845 MHz and 870 to 890 MHz, thus eliminating UHF channels 73 to 83 (however, existing licenses are renewable). Table 11-1 shows a complete list of the FCC channel and frequency assignments used in the United States. In 1947, R. B. Dome of General Electric Corporation proposed the method of *intercarrier* sound transmission for television broadcasting that is used today. In 1949, experiments began with color transmission, and in 1953 the FCC adopted the *National Television Systems Committee* (NTSC) system for color television broadcasting, which is also still used today.

MONOCHROME TELEVISION TRANSMISSION

Television Transmitter

Monochrome television broadcasting involves the transmission of two separate signals: an *aural* (sound) and a *video* (picture) signal. Every television transmitter broadcasts two totally separate signals for the picture and sound information. Aural transmission uses frequency modulation and video transmission uses amplitude modulation. Figure 11-1 shows a simplified block diagram for a monochrome television transmitter. It shows two totally separate transmitters (an FM transmitter for the sound information and an AM transmitter for the picture information) whose outputs are combined in a *diplexer bridge* and fed to a single antenna. A diplexer bridge is a network that is used to combine the outputs from two transmitters operating at different frequencies that use the same antenna system. The video information is limited to frequencies below 4 MHz and can originate from either a *camera* (for live transmissions), a video tape or cassette recorder, or a video disk recorder. The video switcher is used to select the desired video information source for broadcasting. The audio information is limited to frequencies below 15 kHz and can originate from either a microphone (again, only for live transmissions), from sound tracks on tape or disk recorders, or from a separate audio cassette or disk recorder. The audio mixer/switcher is used to select the appropriate audio source for broadcasting. Figure 11-1 also shows

TABLE 11-1 FCC CHANNEL AND FREQUENCY ASSIGNMENTS

Channel number	Frequency band (MHz)	Channel number	Frequency band (MHz)	Channel number	Frequency band (MHz)
1[a]	44–50	29	560–566	57	728–734
2	54–60	30	566–572	58	734–740
3	60–66	31	572–578	59	740–746
4	66–72	32	578–584	60	746–752
5	76–82	33	584–590	61	752–758
6	82–88	34	590–596	62	758–764
7	174–180	35	596–602	63	764–770
8	180–186	36	602–608	64	770–776
9	186–192	37	608–614	65	776–782
10	192–198	38	614–620	66	782–788
11	198–204	39	620–626	67	788–794
12	204–210	40	626–632	68	794–800
13	210–216	41	632–638	69	800–806
14	470–476	42	638–644	70	806–812
15	476–482	43	644–650	71	812–818
16	482–488	44	650–656	72	818–824
17	488–494	45	656–662	73[a]	824–830
18	494–500	46	662–668	74[a]	830–836
19	500–506	47	668–674	75[a]	836–842
20	506–512	48	674–680	76[a]	842–848
21	512–518	49	680–686	77[a]	848–854
22	518–524	50	686–692	78[a]	854–860
23	524–530	51	692–698	79[a]	860–866
24	530–536	52	698–704	80[a]	866–872
25	536–542	53	704–710	81[a]	872–878
26	542–548	54	710–716	82[a]	878–884
27	548–554	55	716–722	83[a]	884–890
28	554–560	56	722–728		

[a]No longer assigned to television broadcasting.

horizontal and *vertical* synchronizing signals, which are combined with the picture information prior to modulation. These signals are used in the receivers to synchronize the horizontal and vertical *scanning rates* (synchronization is discussed in detail later in the chapter).

Television Broadcast Standards

Figure 11-2 shows the frequency spectrum for a standard television broadcast channel. Its total bandwidth is 6 Mhz. The picture carrier is spaced 1.25 MHz above the lower limit for the channel, and the sound carrier is spaced 0.25 MHz below the upper limit. Therefore, the picture and sound carriers are always 4.5 MHz apart. The color *subcarrier* is located 3.58 MHz above the picture carrier. Commercial television broadcasting uses AM vestigial sideband transmission for the picture information. The lower sideband is 0.75 MHz wide and the upper sideband, 4 MHz. Therefore, the low video frequencies (rough outline of the image) are emphasized relative to the high video frequencies (fine detail of the image). The FM sound carrier has a bandwidth of approximately 75 kHz ($\pm$25-kHz deviation for 100% modulation). Both amplitude and phase modulation are used to encode the color information onto the 3.58-MHz color subcarrier. The bandwidth and composition of the color spectrum are discussed later in the chapter. Also discussed is frequency interlacing, used to permit adding the color information without increasing the total bandwidth above 6 MHz.

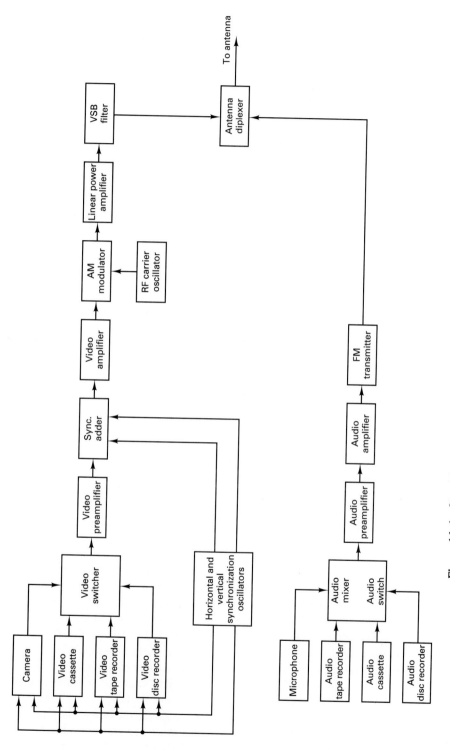

Figure 11-1 Simplified block diagram for a monochrome television transmitter.

427

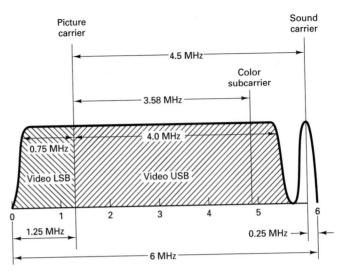

Figure 11-2 Standard television broadcast channel.

THE COMPOSITE VIDEO SIGNAL

Composite means made up of disparate or different parts. The composite video signal includes three separate parts: (1) the luminance signal, (2) the synchronization pulses, and (3) the blanking pulses. These three signals are combined in such a way as to form the composite or total video signal.

The Luminance Signal

The *luminance signal* is the picture information or video signal. This signal originates in the camera and varies in amplitude proportional to the intensity (brightness) of the image. With *negative* transmission, the lower amplitudes correspond to the whitest parts of the image and the higher amplitudes correspond to the darkest. With *positive* transmission, the lower amplitudes correspond to the darkest parts and the higher amplitudes to the whitest. Negative transmission is the FCC standard for modulation of the final picture carrier. However, at intermediate points in both the transmitter and receiver, both negative or positive transmission signals occur and can be observed with a standard oscilloscope.

Figure 11-3 shows a simplified diagram of a black-and-white television camera tube. Light is reflected from an image through an optical lens onto the surface of a photosensitive material. The surface can be made from either a photomissive or photoconductive material and is divided into smaller discrete segments called *picture elements*. A *photomissive* material emits photoelectrons proportional to the intensity of the light striking its surface. A *photoconductive* material has a resistance that is inversely proportional to the intensity of the light striking it. When excited by an electron beam, a picture element outputs a signal that is proportional to the intensity of the light striking it. Therefore, if elements are individually scanned (excited) in sequence, the amplitude of the output signal will vary in accordance with the intensity of the image being scanned.

Figure 11-4 shows the amplitude changes in the luminance signal for a single horizontal scan at the output of a black-and-white camera for an image with varying light intensities. Figure 11-4a shows the amplitude changes for a negative transmission signal as the image changes from black, to a gray scale, to pure white. Figure 11-4b shows the

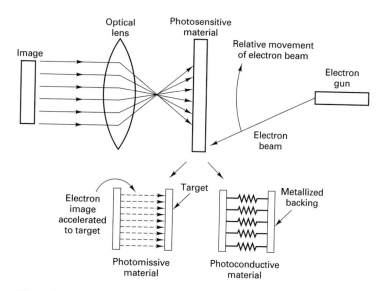

Figure 11-3 Simplified diagram of a black-and-white television camera.

same signal except for positive transmission. A positive transmission luminance signal for an alternating black/white checkerboard pattern is shown in Figure 11-4c. It can be seen that the amplitude of the luminance signal simply alternates from minimum to maximum with the brightness of the image.

Scanning

To produce the complete image, the entire surface of the camera tube must be *scanned*. Figure 11-5 shows a simple scanning sequence. The scanning is done in essentially the same manner in which a page from a book is read. This is called *sequential* horizontal scanning. The entire image is scanned in a sequential series of horizontal lines, one under

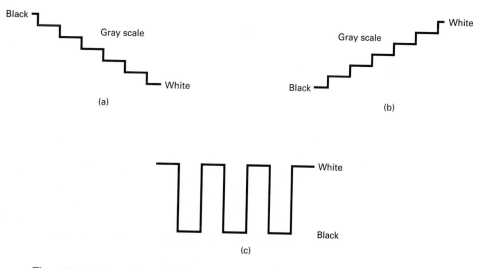

Figure 11-4 Luminance signal for a black-and-white camera: (a) gray scale, negative tansmission; (b) gray scale, positive transmission; (c) checkerboard pattern, positive transmission.

The Composite Video Signal

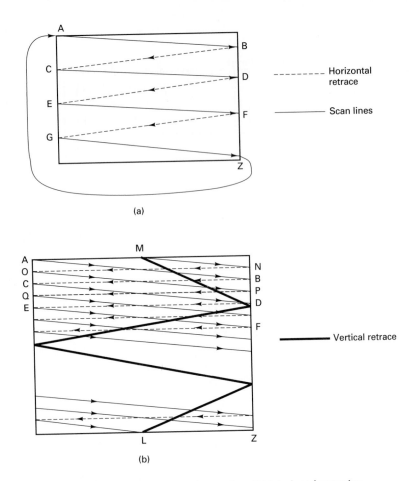

Figure 11-5 (a) Sequential scanning; (b) interlaced scanning.

the other. When the electron beam strikes the back of a picture element, a signal is produced whose amplitude is proportional to the light intensity striking the front of the element. Figure 11-5a shows the scanning beam beginning the active portion of the scan from the upper left corner and moving diagonally to the far right (line *A–B*). This is called the *active* portion of the scan line because this is the time in which the image is converted to electrical signals. Once the beam has reached the far-right side of the photosensitive surface, it immediately returns or retraces to the left side (point *C*). The return time is called horizontal *retrace* or *flyback time*. When the electron scanning beam has reached the bottom-right portion of the photosensitive surface (point *Z*), the beam is returned to top left (point *A*) and the sequence repeats. The return time is called the *vertical retrace time*. While the beam is retracing from the left to the right side of the image and from the bottom to the top, it is *shut off* or *blanked*. Consequently, no video signal is produced during either the horizontal or the vertical retrace times. The active and blanked portions of a single horizontal scan constitute one complete horizontal scan line. The number of horizontal scan lines depends on the detail desired and several other factors that are discussed later in the chapter.

In the United States, a total of 525 horizontal scan lines constitutes one *picture frame*, which is divided into two *fields* of 262.5 horizontal lines each. This scanning technique is called *interlaced scanning* and is shown in Figure 11-5b. Horizontal scanning produces the left-to-right movement of the electron beam, and vertical scanning produces the downward movement. The vertical scanning rate is 60 Hz. Therefore, 30 frames per second are produced. Because the human eye can barely perceive a 30-Hz flicker, the frame is divided into two fields. 262.5 horizontal scan lines beginning at the top left (point A) and ending at the bottom middle (point L) constitute one picture field (the odd field). The second field (the even field) comprises the remaining 262.5 horizontal scan lines interlaced between the scan lines of the first field. The second field begins at the top middle (point M) and ends at the bottom right (point Z). Between fields, the electron beam retraces from the bottom of the picture back to the top in a zigzag pattern. This is called the *vertical retrace time*. Each field is vertically scanned at a 60-Hz rate. Therefore, although the entire picture changes every $\frac{1}{30}$ s, only half of the picture changes every $\frac{1}{60}$ s. This scanning technique allows 525 lines to be scanned at a 30-Hz rate without producing a noticeable flicker in the picture. To scan 525 horizontal scan lines in $\frac{1}{30}$ s, a 15,750-Hz scanning frequency is required ($30 \times 525 = 15,750$).

Horizontal lines are called *raster lines*, and 525 horizontal lines constitute a *raster*. The raster is the luminance that you see when there is no picture (that is, when you are tuned to an unassigned channel). A raster simply means that there is horizontal and vertical scanning and brightness, but not necessarily a picture or an image on the screen.

Scanning waveforms. The scanning beam for the camera and the CRT in the receiver must move both horizontally and vertically at *uniform* rates. This is called *linear scanning*. Linear scanning is necessary to ensure that picture elements are not "squashed" together or "bunched" to one side or to the top or bottom of the screen. *Magnetic deflection* is generally used to move the electron beam. Magnetic deflection produces essentially the same results as *electrostatic deflection*, which is commonly used in the CRT circuitry of an oscilloscope. Electrostatic deflection cannot be used with large CRTs such as those found in most television sets. With magnetic deflection, a linear rise in current through the deflection coils produces a linear change in magnetic flux. The force from the magnetic flux pulls the scanning beam from left to right and from top to bottom of the screen in a continuous, uniform motion. Figure 11-6a shows an ideal horizontal scanning current waveform. The positive slope of the sawtooth moves the beam from left to right in a smooth, constant motion. The negative slope of the waveform produces a magnetic field with the opposite polarity; thus it pulls the beam back to the left side. This is the retrace time. The rate at which the beam moves is proportional to the slope. During retrace it is desirable that the beam move as rapidly as possible. Therefore, the slope during retrace is much steeper than during the active portion of the scan line. In Figure 11-6a, it can be seen that the most negative current corresponds to the left side of the screen and the most positive current to the right side. When zero current is flowing, the beam is in the center of the screen. Each scan line takes the time of one cycle of the sawtooth wave. Therefore, the frequency of the sawtooth is equal to the horizontal scan rate, 15,750 Hz, and the time for each scan line is 63.5 μs.

Figure 11-6b shows the vertical scanning waveform. Like the horizontal waveform, it is a sawtooth wave, which ensures a uniform movement of the beam in the vertical direction. The bottom of the waveform corresponds to the top of the screen, the top of the waveform corresponds to the bottom of the screen, and zero current again corresponds to the center of the screen. Each vertical cycle corresponds to one complete vertical scan plus the vertical retrace time. Therefore, the vertical sawtooth frequency is 60 Hz.

Figure 11-6c shows the interlaced scanning patterns for the odd and even fields of a standard U.S. television broadcast system. The vertical scanning waveforms are also shown superimposed over the horizontal scanning waveforms.

The Composite Video Signal

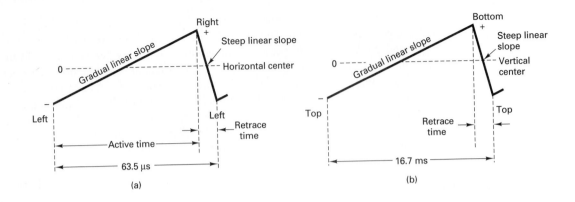

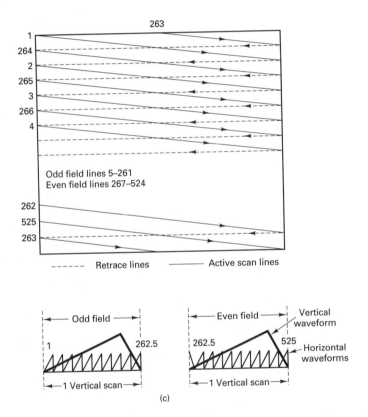

Figure 11-6 (a) Horizontal scanning waveform; (b) vertical scanning waveform; (c) interlaced scanning waveform.

Synchronizing pulses. To reproduce the original image, the horizontal and vertical scanning rates at the receiver must be equal to those at the camera. Also, the scan lines at the camera and the receiver must begin and end in exact time synchronization. Therefore, a horizontal synchronizing pulse of 15,750 Hz and a vertical synchronizing pulse of 60 Hz are added to the luminance signal at the transmitter. The synchronizing (sync) pulses are stripped off at the receiver and used to synchronize its scanning circuits.

Figure 11-7 shows the horizontal and vertical sync pulses for both the odd and even fields. A new field is scanned once every $\frac{1}{60}$ s. Therefore, the time between vertical sync

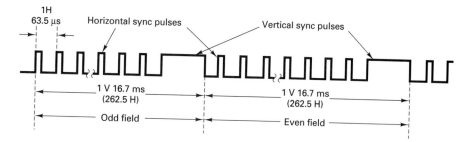

Figure 11-7 Horizontal and vertical sync pulses.

pulses is 1*V*, where $V = \frac{1}{60}$ s or 16.7 ms. The time between horizontal sync pulses is 1*H*, where *H* = 63.5 μs. 1*V* is sufficient time to scan 262.5 horizontal lines (1/15,750 = 63.5 μs and 16.7 ms/63.5 μs = 262.5). Each horizontal sync pulse produces one sawtooth horizontal scanning waveform, and each vertical sync pulse produces one sawtooth vertical scanning waveform.

Blanking Pulses

Blanking pulses are video signals that are added to the luminance and synchronizing pulses with the proper amplitude to ensure that the receiver is blacked out during the vertical and horizontal retrace times. The image is not scanned by the camera during retrace, and therefore no luminance information is transmitted for those times. Blanking pulses are essentially video signals with amplitudes that do not produce any luminance (brightness) on the CRT. Horizontal and vertical sync pulses occur during their respective blanking times.

THE COMPOSITE SIGNAL

The *composite video signal* includes luminance (brightness) signals, horizontal and vertical sync pulses, and blanking pulses. Figure 11-8 shows the composite video signal for a single horizontal scan line, 1*H* (63.5 μs). (Notice that a positive transmission signal is shown.) The figure shows that the active portion of the scan line occurs during the positive slope of the scanning waveform and horizontal retrace occurs during the negative slope (the blanking time). The brightness range for standard television broadcasting is 160 IEEE units peak to peak. 160 IEEE units is generally normalized to 1 Vp−p. The exact value of 1 IEEE unit is unimportant; however, the relative value of a video signal in IEEE units determines its brightness. For example, maximum brightness (pure white) is 120 IEEE units, and no brightness is produced for signals below the reference black level (7.5 IEEE units). The *reference* black level is also called the *pedestal* or *black setup* level. The blanking level is 0 IEEE units, which is below the black level or, in other words, *blacker than black*. Sync pulses are negative-going pulses that occupy 25% of the total IEEE range. A sync pulse has a maximum level of 0 IEEE units and a minimum level of −40 IEEE units. Therefore, the entire sync pulse is below black and thus produces no brightness. The brightness range occupies 75% of the total IEEE scale and extends from 0 to 120 IEEE units, with 120 units corresponding to 100% AM modulation of the RF carrier. However, to ensure that overmodulation does not occur, the FCC has established the maximum brightness (pure white) level to be 87.5% or 100 IEEE units (0.875 × 160 = 140 units, −40 + 140 = 100 units).

The Composite Signal

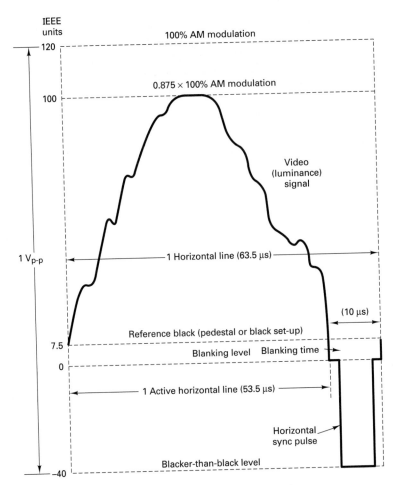

Figure 11-8 Composite video signal.

Figure 11-9 shows the composite video signal for the even field, which equals $1V$ or 16.7 ms and is sufficient time for 262.5 horizontal scan lines (262.5H). However, the vertical blanking pulse width is between 0.05 and 0.08V or 833 to 1333 μs. Therefore, the vertical blanking pulse occupies the time of 13 to 21 horizontal scan lines, which leaves 241.5 to 249.5 active horizontal scan lines. The figure also shows that most of the active scan lines occur during the positive slope of the vertical scanning waveform, and the vertical retrace occurs during the vertical blanking pulse.

Horizontal blanking time. Figure 11-10 shows the blanking time for a single horizontal scan line. The total blanking time is approximately 0.16H or 9.5 to 11.5 μs. Therefore, the active (visible) time for a horizontal line is approximately 0.84H or 52 to 54 μs. Figure 11-10 shows that the sync pulse does not occupy the entire blanking time. The width of the actual sync pulse is approximately 0.08H or 4.25 to 5.25 μs. The time between the beginning of the blanking time and the leading edge of the sync pulse is called the *front porch* and is approximately 0.02H with a minimum time of 1.27 μs. The time between the trailing edge of the sync pulse and the end of the blanking time is called the *back porch* and is approximately 0.06H with a minimum time of 3.81 μs.

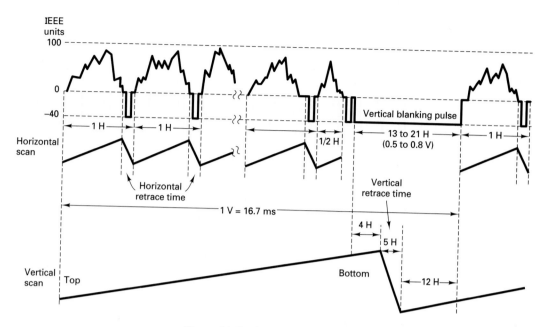

Figure 11-9 Composite video for the even field.

Vertical blanking time. Figure 11-11 shows the first 10*H* of a vertical blanking pulse for a negative transmission waveform. The figure shows that the entire blanking pulse is below the level for reference black (below 7.5 IEEE units). Each vertical blanking pulse begins with six *equalizing* pulses, a vertical sync pulse, and six more equalizing pulses. The equalizing pulses ensure a smooth, synchronized transition between the odd and even fields. Equalizing pulses are explained in more detail in a later section of the chapter. The equalizing pulse rate is 31.5 kHz, which is twice the horizontal

Period	Time
Horizontal line	1 H 63.5 μs
Horizontal blanking	0.16 H 9.5–11.5 μs
Sync pulse	0.08 H 4.75 ± 0.5 μs
Front porch	0.02 H 1.27 μs minimum
Back porch	0.06 H 3.81 μs minimum

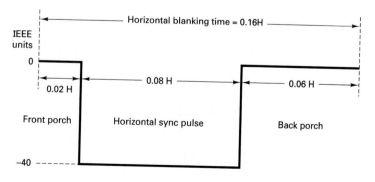

Figure 11-10 Horizontal blanking time.

The Composite Signal

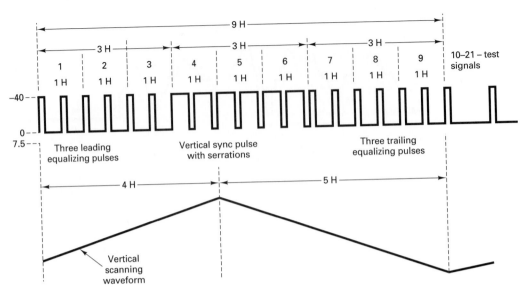

Figure 11-11 Vertical blanking pulse.

scanning rate. Therefore, each equalizing pulse takes $\frac{1}{2}H$, and the 12 pulses occupy a total time of $6H$. The actual vertical sync pulse occupies the time of $3H$. The *serrations* in the vertical sync pulse ensure that the receiver maintains horizontal synchronization during the vertical retrace time. Vertical serrations are explained in more detail in a later section.

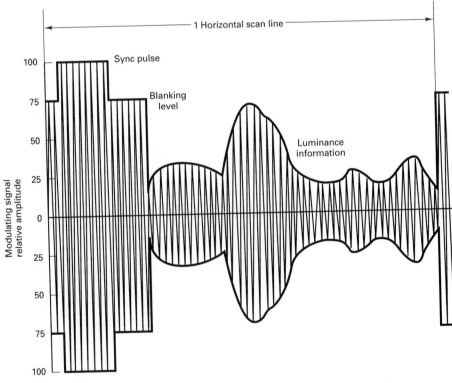

Figure 11-12 RF envelope for negative transmission video signal.

A total time of nine horizontal scan lines (9H) is required to transmit the equalizing and vertical sync pulses. From the figure it can be seen that the first 4H occur at the end of a vertical scan (at the bottom of the CRT). The following 5H occur during the retrace time, and all 9H occur during the vertical blanking time and are therefore not visible. The exact vertical blanking time is determined by the transmitting station; however, it is generally 21H. Horizontal lines 10 through 21 of each field are often used to send studio test signals and automatic color and brightness signals.

RF Transmission of the Composite Video

The transmitted AM picture carrier is shown in Figure 11-12 for negative polarity modulation, which is the FCC standard. Negative transmission of the RF carrier simply means that changes toward white in the picture decrease the amplitude of the AM picture carrier. The advantage of negative transmission is that noise pulses in the RF signal increase the carrier toward black, which makes the noise less annoying to the viewer than changes toward white. Also, with negative transmission, brighter images occur more often than darker ones; thus negative transmission uses less power than positive transmission. Notice that the AM envelope shown in Figure 11-12 has the shape of the composite video signal, and the luminance, blanking, and sync signals can be easily identified. Also, note that during the tips of the horizontal sync pulse there is no AM modulation, and the luminance signal never exceeds 87.5% AM modulation.

MONOCHROME TELEVISION RECEPTION

Figure 11-13 shows a block diagram for a monochrome television receiver. The receiver can be separated into five primary sections: RF, IF, video, horizontal and vertical deflection, and sound.

RF Section

A block diagram of an RF section is shown in Figure 11-14. The RF section includes the UHF and VHF antennas, the antenna coupling circuits, the preselectors, an RF amplifier, and a mixer/converter. A Yagi–Uda antenna is used for the VHF channels, and a simple loop antenna is used for the UHF channels.

The purposes of the RF or front-end section are to provide channel selection (that is, tuning), to provide image-frequency rejection, to isolate the local oscillator from the antenna (thus, preventing the local oscillator signal from radiating), to convert RF signals to IF signals, to provide amplification, and to provide antenna coupling. VHF signals are captured by the antenna, coupled to the receiver input, bandlimited by the preselector, and then amplified by the RF amplifier and fed to the mixer/converter. The mixer/converter beats the local oscillator frequency with the RF to produce the difference frequency, which is the IF. Channel selection is accomplished by changing the bandpass characteristics of the preselector and RF amplifier by switching capacitors or inductors in their tuned circuits and, at the same time, changing the local oscillator frequency. The preselector and local oscillator tuning circuits are ganged together. Commercial television receivers use high-side injection (the local oscillator is tuned to the IF above the desired RF channel frequency).

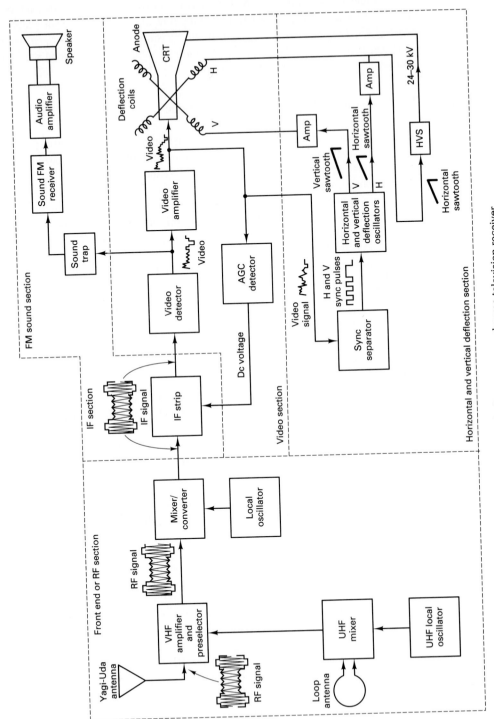

Figure 11-13 Block diagram monochrome television receiver.

438

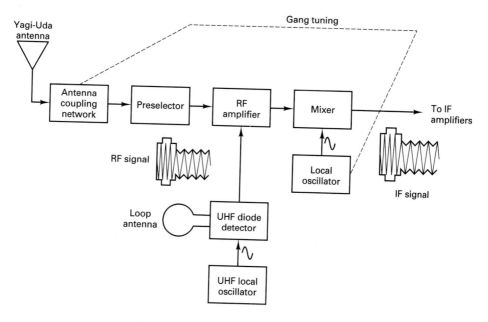

Figure 11-14 Block diagram RF section.

UHF signals are captured by the loop antenna and then immediately mixed down to IF. The UHF mixer is generally a simple diode mixer. When receiving UHF signals, the RF amplifier is simply an additional IF amplifier.

IF Section

The IF section of a television receiver provides most of the receiver's selectivity and gain. The block diagram for a three-stage IF amplifier is shown in Figure 11-15a. The IF section is generally several cascaded high-gain tuned amplifiers. In modern receivers, the IF section processes both the picture and sound IF signals. Such receivers are called *intercarrier receivers*. The standard IFs used in commercial television receivers are 45.75 MHz for the picture and 41.25 MHz for the sound. The IF carriers are separated by 4.5 MHz just as the RF carriers are. IF amplifiers use tuned bandpass filters that bandlimit the signal and prevent adjacent channel interference. A typical IF response curve is shown in Figure 11-15b. Special narrowband bandstop filters called *wavetraps* are used to trap or block the adjacent channel picture and sound carrier frequencies (39.75 and 47.25 MHz, respectively). Wavetraps are also used to attenuate the sound and picture carriers of the selected channel and limit the IF passband to approximately 3 MHz. 3 MHz is used rather than 4 MHz to minimize interference from the color signal, which is explained later in this chapter.

The Video Section

The video section includes a video detector and a series of video amplifiers. A simplified block diagram for a video section is shown in Figure 11-16a. The detector down-converts the picture IF signals to video frequencies and the first sound IF to a second sound IF. The second sound IF is fed to the FM receiver, where the aural information is removed and fed to the audio amplifiers. The video detector is generally a single-diode peak detector. The IF input signal provides the ac voltage necessary to drive the diode into conduction as a

Monochrome Television Reception

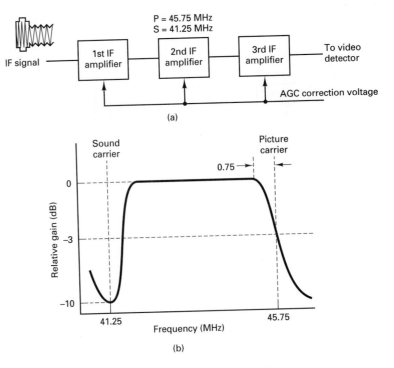

(a)

(b)

Figure 11-15 IF amplifiers: (a) block diagram; (b) frequency-response curve.

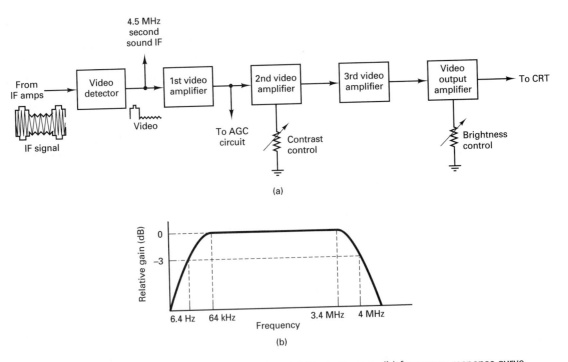

(a)

(b)

Figure 11-16 Video amplifiers: (a) block diagram; (b) frequency-response curve.

half-wave peak rectifier. The output from the video detector is the composite video signal, which is fed to the video amplifiers. The video amplifiers provide the gain necessary for the luminance signal to drive the CRT. Video amplifiers are generally direct coupled to provide dc restoration of the picture brightness. The contrast and brightness controls are located in the video section, and the AGC takeoff point is generally at the output of the first video amplifier. The brightness control simply allows the viewer to vary the dc bias voltage of the video signal. The contrast control adjusts the gain of the video amplifiers. The picture and sound IF signals mix in the diode detector, which is a nonlinear device, and produce a difference signal of 4.5 MHz, which is the second sound IF. A typical frequency-response curve for a video amplifier section is shown in Figure 11-6b.

Horizontal and Vertical Deflection Circuits

A simplified block diagram showing the vertical and horizontal deflection circuits is shown in Figure 11-17. The deflection section includes a sync separator, horizontal and vertical deflection oscillators, and a high-voltage stage. The horizontal and vertical synchronizing pulses are removed from the composite video signal by the sync separator circuit. The horizontal and vertical sync pulses are then further separated with filters and fed to their respective deflection circuits. The deflection circuits convert the sync pulses to sawtooth scanning signals and provide the dc high voltage required for the anode of the CRT.

Sync separator. Figure 11-18a shows the schematic diagram for a single-transistor sync separator, which is a simple clipper circuit. Q_1 is a class C amplifier and the R_1C_1 coupling circuit provides signal bias. The positive portion of the composite video signal (the sync pulses) forward biases Q_1, causing base current to flow, which charges C_1 to the polarity shown. Between sync pulses, C_1 discharges slightly through R_1. The long R_1C_1 time constant keeps C_1 charged to approximately 90% of the peak positive value. Therefore, once C_1 has charged, the luminance signal drives Q_1 further into cutoff. Thus Q_1 conducts only during the more positive sync pulses. Consequently, the sync pulses are the only portion of the composite video signal that appears at the collector of Q_1. The base–emitter circuit of Q_1 is effectively a diode rectifier. The rectifier operation is shown in Figure 11-18b. Once removed, the horizontal and vertical sync pulses are separated with filters. A high-pass filter (differentiator) detects the 15,750-Hz horizontal sync pulses, and a low-pass filter (integrator) detects the 60-Hz vertical sync pulses.

Vertical deflection oscillator. The output from the integrator is a 60-Hz waveform, which is fed to the vertical deflection oscillator. The deflection oscillator produces a 60-Hz linear sawtooth deflection voltage, which produces the vertical scan on the CRT. Figure 11-19 shows a schematic diagram for a transistorized *blocking oscillator*, which is a circuit often used to produce the sawtooth scanning waveform. A blocking oscillator is simply a *triggered oscillator* that produces a sawtooth output waveform that is synchronized to the incoming vertical sync pulse rate. The frequency control sets the *threshold* or *trigger level* for the oscillator. However, the frequency of oscillation is determined by the recovered vertical sync pulses. The output from the vertical oscillator is fed to a vertical output amplifier, which produces the sawtooth current wave required to drive the vertical deflection coils.

The integrator (low-pass filter) passes only the vertical sync pulses and produces a 60-Hz trigger pulse for the blocking oscillator. Figure 11-20a shows the operation of the

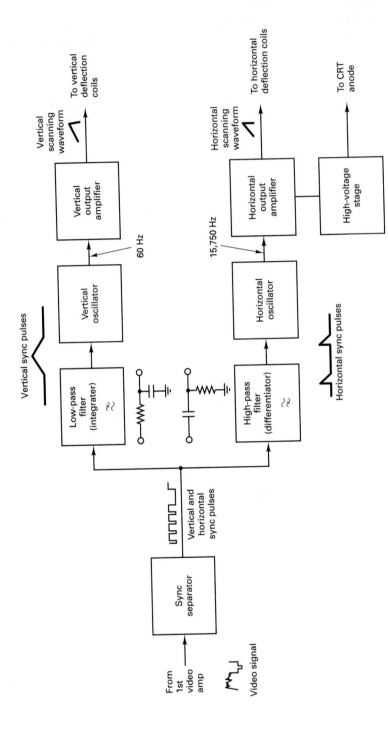

Figure 11-17 Horizontal and vertical deflection section.

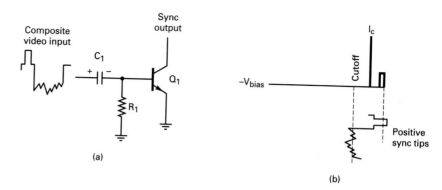

Figure 11-18 Sync separator circuit: (a) schematic diagram; (b) bias operation.

integrator without equalizing pulses. It can be seen that the sync pulses from the integrator without equalizing pulses. It can be seen that the sync pulses from the odd and even fields begin charging the capacitor from different initial levels; thus the two trigger pulses reach the threshold level for the oscillator at different times (Δt). This causes the vertical oscillator to change frequency and lose synchronization between fields, which is noticeable to the viewer as a slight vertical roll. To prevent loss of synchronization, 12 equalizing pulses are transmitted during the vertical blanking interval, 6 immediately before and 6 immediately after the vertical sync pulse. Figure 11-20b shows the waveform produced across the capacitor when the equalizing pulses are included. The sync pulses from each field begin charging the capacitor at precisely the same time with exactly the same initial voltage; thus the capacitor voltage reaches the threshold level at the same time for each field, preventing the vertical sync oscillator from losing synchronization during the transition between the odd and even fields.

Horizontal deflection oscillator. Deflection oscillators, such as the one shown in Figure 11-19, are highly susceptible to noise. Noise pulses can be mistaken for synchronizing pulses and trigger the oscillator at the wrong time, thus changing the horizontal scanning rate. To improve noise immunity, automatic frequency control (AFC) circuits are often used for the horizontal deflection oscillator in television receivers. Figure 11-21 shows the schematic diagram for a push–pull sync discriminator commonly used for

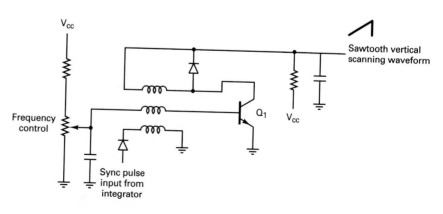

Figure 11-19 Vertical blocking oscillator.

Monochrome Television Reception

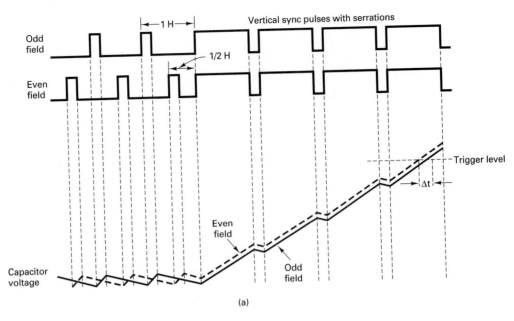

(a)

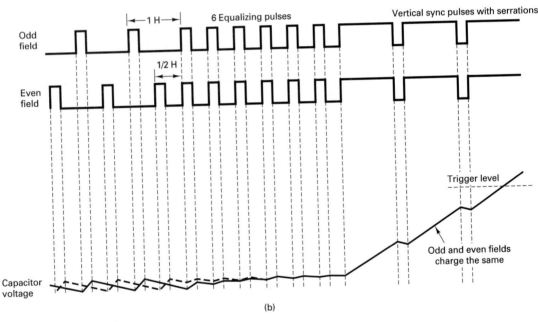

(b)

Figure 11-20 Integrator operation: (a) without equalizing pulses; (b) with equalizing pulses.

horizontal AFC. A phase splitter generates two 180° out-of-phase sync pulses, which are required for push–pull operation. The dual-diode sync discriminator produces a horizontal sawtooth waveform across output capacitor C_o. Consequently, the sawtooth frequency is synchronized to the recovered horizontal sync pulses. The output from the AFC circuit is fed to the horizontal deflection circuit, where it provides horizontal scanning current for the CRT. The output of the horizontal deflection amplifier is also fed to the receiver high-voltage section, where the anode voltage for the CRT is produced.

Chap. 11   **Basic Television Principles**

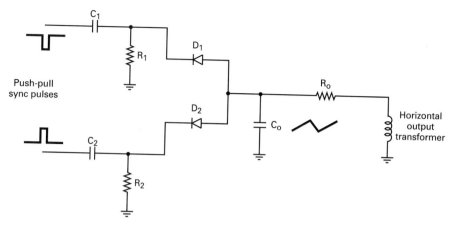

Figure 11-21 Push–pull horizontal AFC circuit.

COLOR TELEVISION TRANSMISSION AND RECEPTION

Color Television Transmitter

In essence, a color television transmitter is identical to the black-and-white transmitter shown in Figure 11-1, except that a color camera is used to produce the video signal. With color broadcasting, all the colors are produced by mixing different amounts of the three *primary colors*: red, blue, and green not to be confused with the three primary pigments—cyan, magenta, and yellow. A color camera is actually three cameras in one, each with separate video output signals. When an image is scanned, separate camera tubes are used for each of the primary colors. The red camera produces the R video signal, the green camera produces the G video signal, and the blue camera produces the B video signal. The R, G, and B video signals are combined in an encoder to produce the composite color signal, which when combined with the luminance signal, amplitude modulates the RF carrier.

Color Camera

Figure 11-22 shows a configuration of mirrors that can be used to split an image into the three primary colors. The *chromatic* mirrors reflect light of all colors. The *dichroic* mirrors are coated to reflect light of only one frequency (color) and allow all other frequencies (colors) to pass through. Light reflected from the image passes through a single camera lens, is reflected by the two achromatic mirrors, and passes through the relay lens. Dichroic mirrors *A* and *B* are mounted on opposing 45° angles. Mirror *A* reflects red light, while blue and green light pass straight through to mirror *B*. Mirror *B* reflects blue light and allows green light to pass through. Consequently, the image is separated into red, green, and blue light frequencies. Once separated, the three color-frequency signals modulate their respective camera tubes and produce the *R*, *G*, and *B* video signals.

Color Encoding

Figure 11-23 shows a simplified block diagram for a color television transmitter. The R, G, and B video signals are combined in specific proportions in the *color matrix* to produce the brightness (luminance) or *Y* video signal and the *I* and *Q* chrominance (color) video

Color Television Transmission and Reception

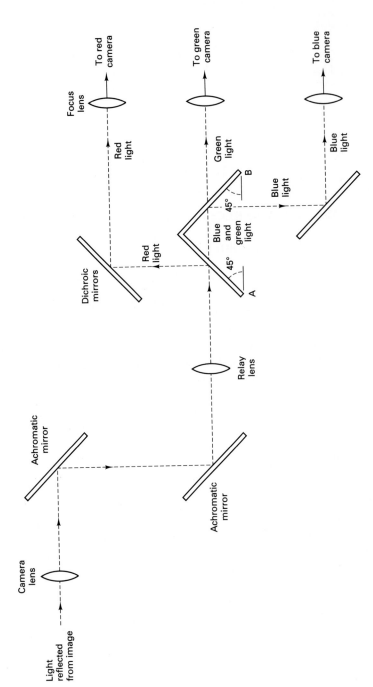

Figure 11-22 Mirror configuration used in color television camera to separate R, G, and B video signals.

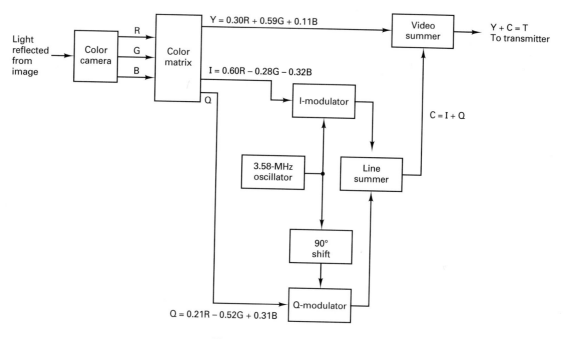

Figure 11-23 Color television transmitter.

signals. The luminance signal corresponds to a monochrome video signal. The I and Q color signals amplitude modulate a 3.58-MHz color subcarrier to produce the total color signal, C. The I signal modulates the subcarrier directly in the I balanced modulator, while the Q signal modulates a quadrature (90° out of phase) subcarrier in the Q balanced modulator. The I and Q modulated signals are linearly combined to produce a quadrature amplitude modulation (QAM) signal, C, which is a combination of both phase and amplitude modulation. The C signal is combined with the Y signal to produce the total composite video signal (T).

Luminance signal. The Y or *luminance signal* is formed by combining 30% of the R video signal, 59% of the G video signal, and 11% of the B video signal. Mathematically, Y is expressed as

$$Y = 0.30R + 0.59G + 0.11B \qquad (11\text{-}1)$$

The percentages shown in Equation 11-1 correspond to the relative brightness of the three primary colors. Consequently, a scene reproduced in black and white by the Y signal has exactly the same brightness as the original image. Figure 11-24 shows how the Y signal voltage is formed from several values of R, G, and B.

The Y signal has a maximum relative amplitude of unity or 1, which is 100% white. For maximum values of R, G, and B (1V each), the value for brightness is determined from Equation 12-1 as follows:

$$Y = 0.30(1) + 0.59(1) + 0.11(1) = 1.00$$

The voltage values for Y shown in Figure 11-24 are the relative luminance values for each color. If only the Y signal is used to reproduce the pattern in a receiver, it would appear on the CRT as seven monochrome bars shaded from white on the left to gray in the middle and black at the right. The Y signal is transmitted with a bandwidth of 0 to 4 MHz.

Color Television Transmission and Reception

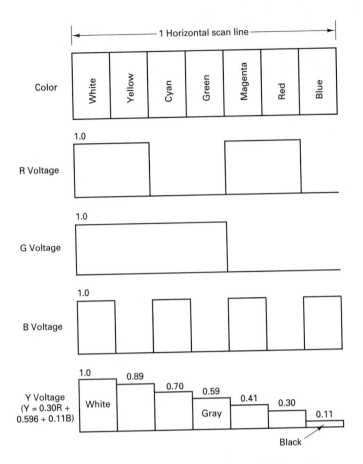

Figure 11-24 Relative luminance values for a color bar pattern.

However, most receivers bandlimit the *Y* signal to 3.2 MHz to minimize interference with the 3.58-MHz color signal. The *I* signal is transmitted with a bandwidth of 1.5 MHz, while the *Q* signal is transmitted with a bandwidth of 0.5 MHz. However, most receivers limit both the *I* and *Q* signals to 0.5-MHz bandwidth.

    Chrominance signal.   The *chrominance* or *C signal* is a combination of the *I* and *Q* color signals. The *I* or in-phase color signal is produced by combining 60% of the R video signal, 28% of the inverted G video signal, and 32% of the inverted B video signal. Mathematically, *I* is expressed as

$$I = 0.60R - 0.28G - 0.32B \qquad (11\text{-}2)$$

    The *Q* or in quadrature color signal is produced by combining 21% of the R video signal, 52% of the inverted G video signal, and 31% of the B video signal. Mathematically, *Q* is expressed as

$$Q = 0.21R - 0.52G + 0.31B \qquad (11\text{-}3)$$

    The *I* and *Q* signals are combined to produce the *C* signal, and since the *I* and *Q* signals are in quadrature, the *C* signal is the phasor sum of the two (that is, the magnitude of $C = \sqrt{I^2 + Q^2}$ and the phase is the $\tan^{-1} Q/I$). The amplitudes of the *I* and *Q* signals are in turn proportional to the R, G, and B video signals. Figure 11-25 shows the *color wheel*

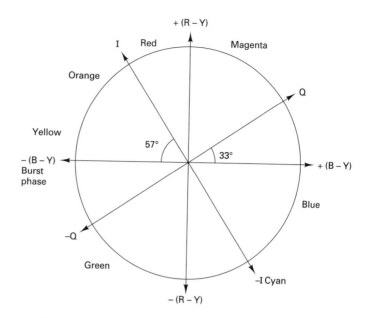

Figure 11-25 Standard television broadcasting color wheel.

for television broadcasting. The R–Y and B–Y signals are used in most color television receivers for demodulating the R, G, and B video signals and are explained later in the chapter. In the receiver, the C signal reproduces colors proportionate to the amplitudes of the I and Q signals. The hue (color tone) is determined by the phase of the C signal, and the depth or saturation is proportional to the magnitude of the C signal. The outside of the circle corresponds to a relative value of 1.0.

   Color burst.   The phase of the 3.58-MHz color subcarrier is the reference phase for color demodulation. Therefore, the color subcarrier must be transmitted together with the composite video so that a receiver can reconstruct the subcarrier with the proper frequency and reference phase and thus determine the phase (color) of the received signal. Eight to ten cycles of the 3.58-MHz subcarrier are inserted on the back porch of each horizontal blanking pulse. This is referred to as the *color burst*. In the receiver, the burst is removed and used to synchronize a local 3.58-MHz color oscillator. The color burst is shown in Figure 11-26.

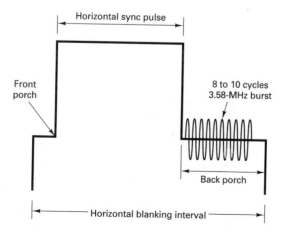

Figure 11-26 Horizontal blanking interval and 3.58-MHz burst.

Color Television Transmission and Reception

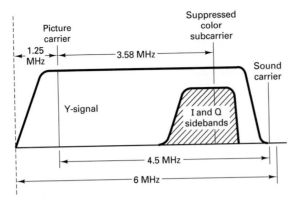

Figure 11-27 Composite RF frequency spectrum for color television broadcasting.

Figure 11-27 shows the composite RF frequency spectrum for color television broadcasting.

Scanning frequencies for color transmission. The frequency of the color subcarrier is determined by harmonic relations among the color subcarrier and the horizontal and vertical scanning rates. The exact value for the color subcarrier is 3.579545 MHz. The sound subcarrier (4.5 MHz) is the 286th harmonic of the horizontal line frequency. Therefore, the horizontal line rate (f_H) for color transmission is not exactly 15.750 kHz. Mathematically f_H is

$$f_H = \frac{4.5 \text{ MHz}}{286} = 15{,}734.26 \text{ Hz}$$

The exact value of the vertical scan rate (f_V) is

$$f_V = \frac{15{,}734.26}{262.5} = 59.94 \text{ Hz}$$

The color subcarrier frequency (C) is chosen as the 455th harmonic of one half of the horizontal scan rate. Therefore,

$$C = \frac{15{,}734.26}{2} \times 455 = 3.579545 \text{ MHz}$$

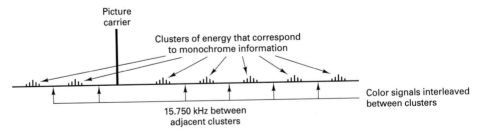

Figure 11-28 Frequency interleaving of color and luminance signals.

Chap. 11   **Basic Television Principles**

Frequency interlacing. The *Y* portion of the video signal produces clusters of energy at 15.73426-kHz intervals throughout the 4-MHz video bandwidth. By producing color signals around a 3.579545-MHz color subcarrier, the color energy is *clustered* within the void intervals between the black-and-white information. This is called frequency *interlacing* or sometimes frequency *interleaving* and is a form of *multiplexing* (that is, the color and black-and-white information is frequency division multiplexed into the total video spectrum). Figure 11-28 shows the spectrum for frequency interlacing.

Color Television Receivers

A color television receiver is essentially the same as a black-and-white receiver except for the picture tube and the addition of the color decoding circuits. Figure 11-29 shows the simplified block diagram for the color circuits in a color television receiver.

The composite video signal is fed to the *chroma* bandpass amplifier, which is tuned to the 3.58-MHz subcarrier and has a bandpass of 0.5 MHz. Therefore, only the *C* signal is amplified and passed on to the B–*Y* and R–*Y* demodulators. The 3.58-MHz color burst is separated from the horizontal blanking pulse by keying on the burst separator only during the horizontal flyback time. A synchronous 3.58-MHz color subcarrier is reproduced in the color AFC circuit, which consists of a 3.58-MHz color oscillator and a color AFPC (automatic frequency and phase control) circuit. The *color killer* shuts off the chroma amplifier during monochrome reception (no colors are better than wrong colors). The *C* signal is demodulated in the B–*Y* and R–*Y* demodulators by mixing it with the

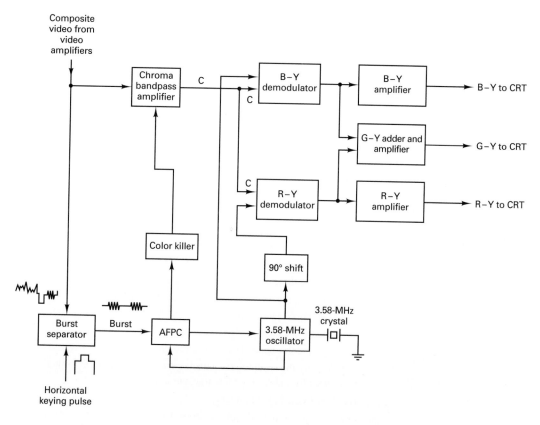

Figure 11-29 Color demodulator circuits.

Color Television Transmission and Reception

phase-coherent 3.58-MHz subcarrier. The B–Y and R–Y signals produce the R and B video signal by combining them with the Y signal in the following manner:

$$B - Y + Y = B$$

$$R - Y + Y = R$$

The G video signal is produced by combining the B–Y and R–Y signals in the proper proportions.

QUESTIONS

11-1. Briefly describe the meaning of the word *television*.

11-2. Describe a diplexer bridge.

11-3. What components make up the composite video signal?

11-4. Describe negative transmission; positive transmission. Which is the FCC standard for RF transmission?

11-5. Name and briefly describe two types of photosensitive materials used in television cameras.

11-6. Describe sequential horizontal scanning.

11-7. What is meant by the *active portion* of a horizontal line?

11-8. What is meant by *retrace time*?

11-9. Describe interlaced scanning.

11-10. Define *raster lines* and *raster*.

11-11. Why should the horizontal and vertical scanning waveforms be linear?

11-12. Why are horizontal and vertical synchronizing pulses included with the composite video signal?

11-13. Describe a blanking pulse.

11-14. Describe the IEEE scale used for television broadcasting.

11-15. What is meant by *reference black*; *black setup*; *pedestal*; *blacker than black*?

11-16. Describe the horizontal blanking time.

11-17. Describe the vertical blanking time.

11-18. Why are equalizing pulses transmitted during the vertical blanking interval?

11-19. Why are there serrations in the vertical sync pulses?

11-20. Draw the block diagram for a monochrome television receiver and describe its basic operation and the primary purpose of each section.

11-21. Describe the operation of a vertical blocking oscillator.

11-22. Describe the operation of a horizontal AFC circuit.

11-23. What are the three primary colors for television broadcasting?

11-24. Describe the basic operation of a color television camera.

11-25. Describe the Y or luminance signal.

11-26. Describe the I, Q, and C signals.

11-27. What is the color burst? How is it transmitted? What is its purpose?

11-28. Explain why the scanning frequencies used with color television are slightly different from those used for black-and-white transmission.

11-29. Describe frequency interlacing.

11-30. Draw the block diagram of the color decoding circuits in a color television receiver and briefly describe the decoding operation.

PROBLEMS

11.1. Draw the RF spectrum for channel 8. Include the picture, sound, and color frequencies and their respective bandwidths.

11.2. How long is $0.8H$? $0.8V$?

11.3. How many horizontal scan lines occur during $0.12V$?

11.4. If the back porch of the horizontal blanking pulse is $0.06H$, what is the maximum number of cycles of the color burst signal that can be transmitted?

11.5. If the vertical blanking interval is $20H$ long, how long is this?

11.6. What is the color subcarrier frequency for channel 55; channel 66?

11.7. What is the sound subcarrier frequency for channel 55; channel 66?

11.8. Determine the value for Y for the following R, G, and B signals: R = $0.8V$, G = $0.6V$, and B = $0.2V$.

11.9. Determine I and Q for the R, G, and B values given in Problem 11-8.

11.10. Determine the C signal for Problem 11-9.

12

Digital Communications

INTRODUCTION

During the past several years, the *electronic communications* industry has undergone some remarkable technological changes. Traditional electronic communications systems that use conventional analog modulation techniques, such as *amplitude modulation* (AM), *frequency modulation* (FM), and *phase modulation* (PM), are gradually being replaced with more modern *digital communications systems*. Digital communications systems offer several outstanding advantages over traditional analog systems: ease of processing, ease of multiplexing, and noise immunity.

In essence, electronic communications is the transmission, reception, and processing of *information* with the use of electronic circuits. Information is defined as knowledge or intelligence communicated or received. Figure 12-1 shows a simplified block diagram of an electronic communications system, which comprise three primary sections: a *source*, a *destination*, and a *transmission medium*. Information is propagated through a communications system in the form of symbols which can be *analog* (proportional), such as the human voice, video picture information, or music, or *digital* (discrete), such as binary-coded numbers, alpha/numeric codes, graphic symbols, microprocessor op-codes, or database information. However, very often the source information is unsuitable for transmission in its original form and must be converted to a more suitable form prior to transmission. For example, with digital communications systems, analog information is converted to digital form prior to transmission, and with analog communications systems, digital data are converted to analog signals prior to transmission.

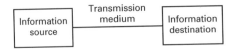

Figure 12-1 Simplified block diagram for an electronic communications system.

454

DIGITAL COMMUNICATIONS

The term *digital communications* covers a broad area of communications techniques, including *digital transmission* and *digital radio*. Digital transmission is the transmittal of digital pulses between two or more points in a communications system. Digital radio is the transmittal of digitally modulated analog carriers between two or more points in a communications system. Digital transmission systems require a physical facility between the transmitter and receiver, such as a metallic wire pair, a coaxial cable, or an optical fiber cable. In digital radio systems, the transmission medium is free space or Earth's atmosphere.

Figure 12-2 shows simplified block diagrams of both a digital transmission system and a digital radio system. In a digital transmission system, the original source information may be in digital or analog form. If it is in analog form, it must be converted to digital pulses prior to transmission and converted back to analog form at the receive end. In a digital radio system, the modulating input signal and the demodulated output signal are digital pulses. The digital pulses could originate from a digital transmission system, from a digital source such as a mainframe computer, or from the binary encoding of an analog signal.

SHANNON LIMIT FOR INFORMATION CAPACITY

The *information capacity* of a communications system represents the number of independent symbols that can be carried through the system in a given unit of time. The most basic symbol is the *binary digit* (bit). Therefore, it is often convenient to express the information capacity of a system in *bits per second* (bps). In 1928, R. Hartley of Bell Telephone Laboratories developed a useful relationship among bandwidth, transmission time, and information capacity. Simply stated, *Hartley's law* is

$$I \propto B \times T \qquad (12\text{-}1)$$

where I = information capacity (bps)
   B = bandwidth (Hz)
   T = transmission time (s)

From Equation 12-1 it can be seen that the information capacity is a linear function of bandwidth and transmission time and is directly proportional to both. If either the bandwidth or the transmission time is changed, a directly proportional change in information capacity will occur.

In 1948, C. E. Shannon (also of Bell Telephone Laboratories) published a paper in the *Bell System Technical Journal* relating the information capacity of a communications channel to bandwidth and signal-to-noise ratio. Mathematically stated, the *Shannon limit for information capacity* is

$$I = B \log_2 \left(1 + \frac{S}{N} \right) \qquad (12\text{-}2a)$$

or

$$I = 3.32 \, B \log_{10} \left(1 + \frac{S}{N} \right) \qquad (12\text{-}2b)$$

where I = information capacity (bps)
   B = bandwidth (Hz)
   $\dfrac{S}{N}$ = signal-to-noise power ratio (unitless)

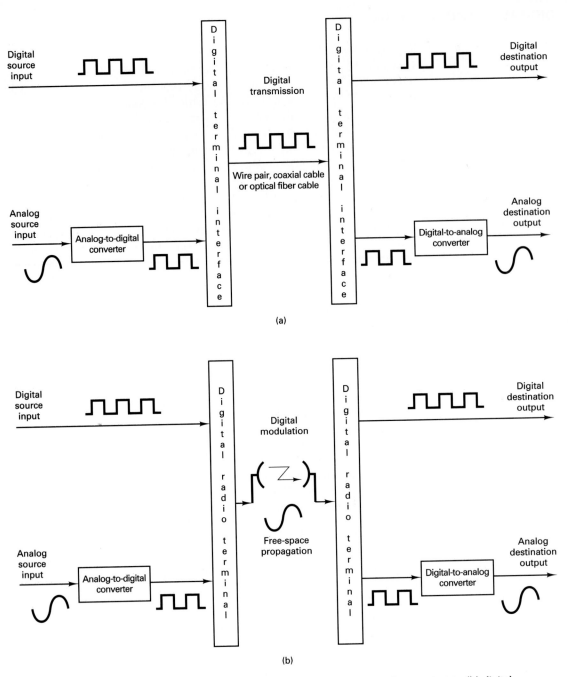

Figure 12-2 Digital communications systems: (a) digital transmission; (b) digital radio.

For a standard voice band communications channel with a signal-to-noise power ratio of 1000 (30 dB) and a bandwidth of 2.7 kHz, the Shannon limit for information capacity is

$$I = 2700 \log_2 (1 + 1000)$$

$$= 26.9 \text{ kbps}$$

Chap. 12 **Digital Communications**

Shannon's formula is often misunderstood. The results of the preceding example indicate that 26.9 kbps can be transferred through a 2.7-kHz channel. This may be true, but it cannot be done with a binary system. To achieve an information transmission rate of 26.9 kbps through a 2.7-kHz channel, each symbol transmitted must contain more than one bit of information. Therefore, to achieve the Shannon limit for information capacity, digital transmission systems that have more than two output conditions (symbols) must be used. Several such systems are described in the following chapters. These systems include both analog and digital modulation techniques and the transmission of both digital and analog signals.

DIGITAL RADIO

The property that distinguishes a digital radio system from a conventional AM, FM, or PM radio system is that in a digital radio system the modulating and demodulated signals are digital pulses rather than analog waveforms. Digital radio uses analog carriers just as conventional systems do. Essentially, there are three digital modulation techniques that are commonly used in digital radio systems: *frequency shift keying* (FSK), *phase shift keying* (PSK), and *quadrature amplitude modulation* (QAM).

FREQUENCY SHIFT KEYING

Frequency shift keying (FSK) is a relatively simple, low-performance form of digital modulation. Binary FSK is a form of constant-amplitude angle modulation similar to conventional frequency modulation except that the modulating signal is a binary pulse stream that varies between two discrete voltage levels rather than a continuously changing analog waveform. The general expression for a binary FSK signal is

$$v(t) = V_c \cos\left[\left(\omega_c + \frac{v_m(t)\,\Delta\omega}{2}\right)t\right] \qquad (12\text{-}3)$$

where $v(t)$ = binary FSK waveform
  V_c = peak unmodulated carrier amplitude
  ω_c = radian carrier frequency
  $v_m(t)$ = binary digital modulating signal
  $\Delta\omega$ = change in radian output frequency

From Equation 12-3 it can be seen that with binary FSK the carrier amplitude V_c remains constant with modulation. However, the output carrier radian frequency (ω_c) shifts by an amount equal to $\pm\Delta\omega/2$. The frequency shift ($\Delta\omega/2$) is proportional to the amplitude and polarity of the binary input signal. For example, a binary one could be +1 volt and a binary zero −1 volt producing frequency shifts of $+\Delta\omega/2$ and $-\Delta\omega/2$, respectively. In addition, the rate at which the carrier frequency shifts is equal to the rate of change of the binary input signal $v_m(t)$ (that is, the input bit rate). Thus the output carrier frequency deviates (shifts) between $\omega_c + \Delta\omega/2$ and $\omega_c - \Delta\omega/2$ at a rate equal to f_m.

Frequency Shift Keying

457

FSK Transmitter

With binary FSK, the center or carrier frequency is shifted (deviated) by the binary input data. Consequently, the output of a binary FSK modulator is a step function in the time domain. As the binary input signal changes from a logic 0 to a logic 1, and vice versa, the FSK output shifts between two frequencies: a *mark* or *logic 1 frequency* and a *space* or *logic 0 frequency*. With binary FSK, there is a change in the output frequency each time the logic condition of the binary input signal changes. Consequently, the output rate of change is equal to the input rate of change. In digital modulation, the rate of change at the input to the modulator is called the *bit rate* and has the units of bits per second (bps). The rate of change at the output of the modulator is called *baud* or *baud rate* and is equal to the reciprocal of the time of one output signaling element. In essence, baud is the line speed in symbols per second. In binary FSK, the input and output rates of change are equal; therefore, the bit rate and baud rate are equal. A simple binary FSK transmitter is shown in Figure 12-3.

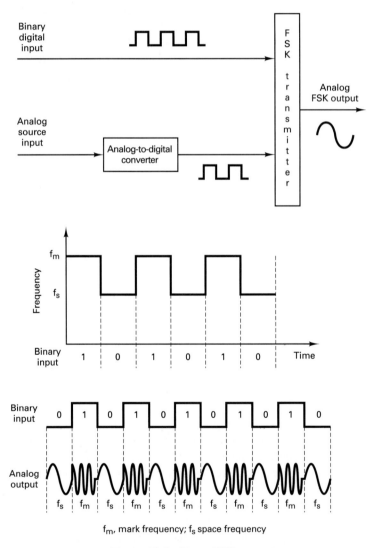

f_m, mark frequency; f_s space frequency

Figure 12-3   Binary FSK transmitter.

Bandwidth Considerations of FSK

As with all electronic communications systems, bandwidth is one of the primary considerations when designing a binary FSK transmitter. FSK is similar to conventional frequency modulation and so can be described in a similar manner.

Figure 12-4 shows a binary FSK modulator. FSK modulators are very similar to conventional FM modulators and are very often *voltage-controlled oscillators* (VCOs). The fastest input rate of change occurs when the binary input is a series of alternating 1's and 0's: namely, a square wave. Consequently, if only the *fundamental frequency* of the input is considered, the *highest modulating frequency* is equal to one-half of the input bit rate.

The rest frequency of the VCO is chosen such that it falls halfway between the mark and space frequencies. A logic 1 condition at the input shifts the VCO from its rest frequency to the mark frequency, and a logic 0 condition at the input shifts the VCO from its rest frequency to the space frequency. Consequently, as the input binary signal changes from a logic 1 to a logic 0, and vice versa, the VCO output frequency *shifts* or *deviates* back and forth between the mark and space frequencies. Because binary FSK is a form of frequency modulation, the formula for *modulation index* used in FM is also valid for binary FSK. Modulation index is given as

$$MI = \frac{\Delta f}{f_a} \tag{12-4}$$

where   MI = modulation index (unitless)
Δf = frequency deviation (Hz)
f_a = modulating frequency (Hz)

The worst-case modulation index is the modulation index that yields the widest output bandwidth, called the *deviation ratio*. The worst-case or widest bandwidth occurs when both the frequency deviation and the modulating frequencies are at their maximum values.

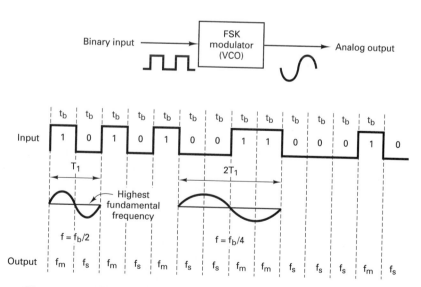

Figure 12-4 FSK modulator, t_b, Time of one bit = $1/f_b$; f_m, mark frequency; f_s, space frequency; T1, period of shortest cycle; 1/T1, fundamental frequency of binary square wave; f_b, input bit rate (bps).

Frequency Shift Keying

459

In a binary FSK modulator, Δf is the peak frequency deviation of the carrier and is equal to the difference between the rest frequency and either the mark or space frequency (or half the difference between the mark and space frequencies). The peak frequency deviation depends on the amplitude of the modulating signal. In a binary digital signal, all logic 1's have the same voltage and all logic 0's have the same voltage; consequently, the frequency deviation is constant and always at its maximum value. f_a is equal to the fundamental frequency of the binary input which under the worst-case condition (alternating 1's and 0's) is equal to one-half of the bit rate (f_b). Consequently, for binary FSK,

$$ \text{MI} = \frac{\left|\dfrac{f_m - f_s}{2}\right|}{\dfrac{f_b}{2}} = \frac{|f_m - f_s|}{f_b} \tag{12-5} $$

where
$$ \frac{|f_m - f_s|}{2} = \text{peak frequency deviation} $$

$$ f_b = \text{input bit rate} $$
$$ \frac{f_b}{2} = \text{fundamental frequency of the binary input signal} $$

With conventional narrowband FM, the bandwidth is a function of the modulation index. Consequently, in binary FSK the modulation index is generally kept below 1. 0, thus producing a relatively narrow band FM output spectrum. The minimum bandwidth required to propagate a signal is called the *minimum Nyquist bandwidth* (f_N). When modulation is used and a double-sided output spectrum is generated, the minimum bandwidth is called the *minimum double-sided Nyquist bandwidth* or the *minimum IF bandwidth*.

EXAMPLE 12-1

For a binary FSK modulator with space, rest, and mark frequencies of 60, 70, and 80 MHz, respectively and an input bit rate of 20 Mbps, determine the output baud and the minimum required bandwidth.

Solution   Substituting into Equation 12-5, we have
$$ \text{MI} = \frac{|f_m - f_s|}{f_b} = \frac{|80\ \text{MHz} - 60\ \text{MHz}|}{20\ \text{Mbps}} $$

$$ = \frac{20\ \text{MHz}}{20\ \text{Mbps}} = 1.0 $$

From the Bessel chart (Table 12-1), a modulation index of 1.0 yields three sets of significant side frequencies. Each side frequency is separated from the center frequency or an adjacent side frequency by a value equal to the modulating frequency, which in this example is 10

TABLE 12-1   BESSEL FUNCTION CHART

MI	J_0	J_1	J_2	J_3	J_4
0.0	1.00				
0.25	0.98	0.12			
0.5	0.94	0.24	0.03		
1.0	0.77	0.44	0.11	0.02	
1.5	0.51	0.56	0.23	0.06	0.01
2.0	0.22	0.58	0.35	0.13	0.03

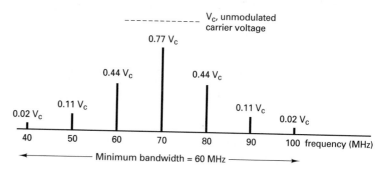

Figure 12-5 FSK output spectrum for Example 12-1.

MHz ($f_b/2$). The output spectrum for this modulator is shown in Figure 12-5 which shows that the minimum double-sided Nyquist bandwidth is 60 MHz. The baud rate is 20 megabaud, the same as the bit rate.

Because binary FSK is a form of narrowband frequency modulation, the minimum bandwidth is dependent on the modulation index. For a modulation index between 0.5 and 1, either two or three sets of significant side frequencies are generated. Thus the minimum bandwidth is two to three times the input bit rate.

FSK Receiver

The most common circuit used for demodulating binary FSK signals is the *phase-locked loop* (PLL), which is shown in block diagram form in Figure 12-6. A PLL-FSK demodulator works very much like a PLL-FM demodulator. As the input to the PLL shifts between

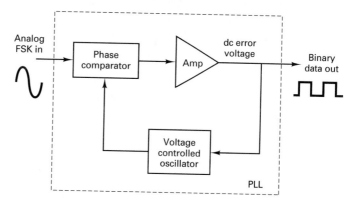

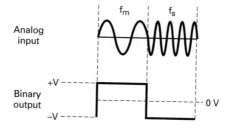

Figure 12-6 PLL-FSK demodulator.

Frequency Shift Keying

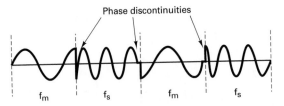

Figure 12-7 Noncontinuous FSK waveform.

the mark and space frequencies, the *dc error voltage* at the output of the phase comparator follows the frequency shift. Because there are only two input frequencies (mark and space), there are also only two output error voltages. One represents a logic 1 and the other a logic 0. Therefore, the output is a two-level (binary) representation of the FSK input. Generally, the natural frequency of the PLL is made equal to the center frequency of the FSK modulator. As a result, the changes in the dc error voltage follow the changes in the analog input frequency and are symmetrical around 0 V.

Binary FSK has a poorer error performance than PSK or QAM and, consequently, is seldom used for high-performance digital radio systems. Its use is restricted to low-performance, low-cost, asynchronous data modems that are used for data communications over analog, voice band telephone lines (see Chapter 13).

Minimum Shift-Keying FSK

Minimum shift-keying FSK (MSK) is a form of *continuous-phase* frequency shift keying (CPFSK). Essentially, MSK is binary FSK except that the mark and space frequencies are *synchronized* with the input binary bit rate. Synchronous simply means that there is a precise time relationship between the two; it does not mean they are equal. With MSK, the mark and space frequencies are selected such that they are separated from the center frequency by an exact odd multiple of one-half of the bit rate [f_m and $f_s = n(f_b/2)$, where n = any odd integer]. This ensures that there is a smooth phase transition in the analog output signal when it changes from a mark to a space frequency, or vice versa. Figure 12-7 shows a *noncontinuous* FSK waveform. It can be seen that when the input changes from a logic 1 to a logic 0, and vice versa, there is an abrupt phase discontinuity in the analog output signal. When this occurs, the demodulator has trouble following the frequency shift; consequently, an error may occur.

Figure 12-8 shows a continuous phase MSK waveform. Notice that when the output

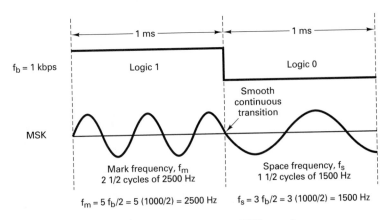

Figure 12-8 Continuous-phase MSK waveform.

Chap. 12 **Digital Communications**

frequency changes, it is a smooth, continuous transition. Consequently, there are no phase discontinuities. MSK has a better bit-error performance than conventional binary FSK for a given signal-to-noise ratio. The disadvantage of MSK is that it requires synchronizing circuits and is therefore more expensive to implement.

PHASE SHIFT KEYING

Phase shift keying (PSK) is another form of angle-modulated, constant-amplitude digital modulation. PSK is similar to conventional phase modulation except that with PSK the input signal is a binary digital signal and a limited number of output phases are possible.

BINARY PHASE SHIFT KEYING

With *binary phase shift keying* (BPSK), two output phases are possible for a single carrier frequency ("binary" meaning "2"). One output phase represents a logic 1 and the other a logic 0. As the input digital signal changes state, the phase of the output carrier shifts between two angles that are 180° out of phase. Other names for BPSK are *phase reversal keying* (PRK) and *biphase modulation.* BPSK is a form of suppressed carrier, square-wave modulation of a continuous wave (CW) signal.

BPSK Transmitter

Figure 12-9 shows a simplified block diagram of a BPSK modulator. The balanced modulator acts like a phase reversing switch. Depending on the logic condition of the digital input, the carrier is transferred to the output either in phase or 180° out of phase with the reference carrier oscillator.

Figure 12-10a shows the schematic diagram of a balanced ring modulator. The balanced modulator has two inputs: a carrier that is in phase with the reference oscillator and the binary digital data. For the balanced modulator to operate properly, the digital input voltage must be much greater than the peak carrier voltage. This ensures that the digital input controls the on/off state of diodes D1–D4. If the binary input is a logic 1 (positive voltage), diodes D1 and D2 are forward biased and "on," while diodes D3 and D4 are reverse biased and "off" (Figure 12-10b). With the polarities shown, the carrier voltage is developed across transformer T2 in phase with the carrier voltage across T1. Consequently, the output signal is in phase with the reference oscillator.

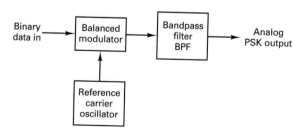

Figure 12-9   BPSK modulator.

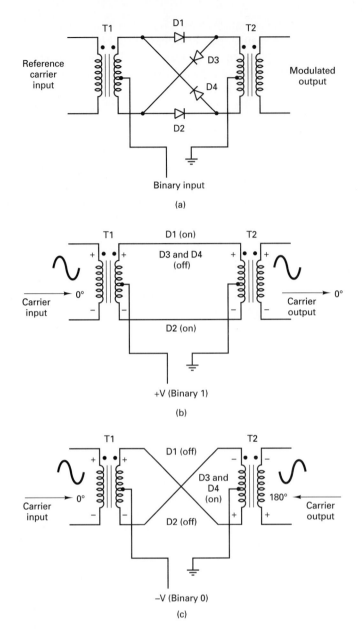

Figure 12-10 (a) Balanced ring modulator; (b) logic 1 input; (c) logic 0 input.

If the binary input is a logic 0 (negative voltage), diodes D1 and D2 are reverse biased and "off," while diodes D3 and D4 are forward biased and "on" (Figure 12-10c). As a result, the carrier voltage is developed across transformer T2 180° out of phase with the carrier voltage across T1. Consequently, the output signal is 180° out of phase with the reference oscillator. Figure 12-11 shows the truth table, phasor diagram, and constellation diagram for a BPSK modulator. A *constellation diagram,* which is sometimes called a *signal state-space diagram,* is similar to a phasor diagram except that the entire phasor is not drawn. In a constellation diagram, only the relative positions of the peaks of the phasors are shown.

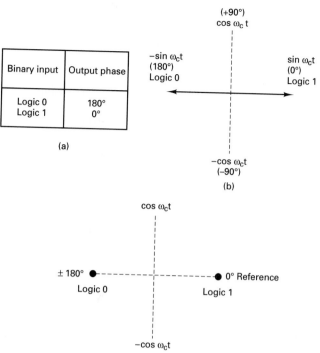

Binary input	Output phase
Logic 0	180°
Logic 1	0°

(a)

(+90°)
cos ω$_c$ t

−sin ω$_c$t
(180°)
Logic 0

sin ω$_c$t
(0°)
Logic 1

−cos ω$_c$t
(−90°)

(b)

cos ω$_c$t

± 180° ●- - - - - - - - - - - - - - ● 0° Reference

Logic 0   Logic 1

−cos ω$_c$t

(c)

Figure 12-11 BPSK modulator:
(a) truth table; (b) phasor diagram;
(c) constellation diagram.

Bandwidth Considerations of BPSK

A balanced modulator is a *product modulator;* the output signal is the product of the two input signals. In a BPSK modulator, the carrier input signal is multiplied by the binary data. If $+1$ V is assigned to a logic 1 and -1 V is assigned to a logic 0, the input carrier (sin ω$_c$t) is multiplied by either a $+$ or -1. Consequently, the output signal is either $+1$ sin ω$_c$t or -1 sin ω$_c$t; the first represents a signal that is *in phase* with the reference oscillator, the latter a signal that is 180° out of phase with the reference oscillator. Each time the input logic condition changes, the output phase changes. Consequently, for BPSK, the output rate of change (baud) is equal to the input rate of change (bps), and the widest output bandwidth occurs when the input binary data are an alternating 1/0 sequence. The fundamental frequency (f_a) of an alternating 1/0 bit sequence is equal to one-half of the bit rate ($f_b/2$). Mathematically, the output phase of a BPSK modulator is

$$\text{output} = \underbrace{(\sin \omega_a t)}_{\substack{\text{fundamental frequency} \\ \text{of the binary} \\ \text{modulating signal}}} \times \underbrace{(\sin \omega_c t)}_{\substack{\text{unmodulated} \\ \text{carrier}}} \tag{12-6}$$

or

$$\tfrac{1}{2}\cos (\omega_c - \omega_a)t - \tfrac{1}{2}\cos (\omega_c + \omega_a)t$$

Consequently, the minimum double-sided Nyquist bandwidth (f_N) is

$$\begin{array}{cc} \omega_c + \omega_a & \omega_c + \omega_a \\ -(\omega_c - \omega_a) & \text{or} \quad -\omega_c + \omega_a \\ \hline & \overline{2\omega_a} \end{array}$$

Binary Phase Shift Keying

and because $f_a = f_b/2$,

$$f_N = 2\left(\frac{f_b}{2}\right) = f_b$$

Figure 12-12 shows the output phase versus time relationship for a BPSK waveform. The output spectrum from a BPSK modulator is simply a double-sideband suppressed carrier signal where the upper and lower side frequencies are separated from the carrier frequency by a value equal to one-half of the bit rate. Consequently, the minimum bandwidth (f_N) required to pass the worst-case BPSK output signal is equal to the input bit rate.

EXAMPLE 12-2

For a BPSK modulator with a carrier frequency of 70 MHz and an input bit rate of 10 Mbps, determine the maximum and minimum upper and lower side frequencies, draw the output spectrum, determine the minimum Nyquist bandwidth, and calculate the baud.

Solution   Substituting into Equation 12-6 yields

output $= (\sin \omega_a t)\,(\sin \omega_c t)$

$\qquad = [\sin 2\pi(5\text{ MHz})t]\,|\,[\sin 2\pi(70\text{ MHz})t]$

$\qquad\qquad = \tfrac{1}{2}\cos 2\pi(70\text{ MHz} - 5\text{ MHz})t - \tfrac{1}{2}\cos 2\pi(70\text{ MHz} + 5\text{ MHz})t$

$\qquad\qquad\qquad \underbrace{\hphantom{==========}}_{\text{lower side frequency}}\ \underbrace{\hphantom{==========}}_{\text{upper side frequency}}$

Minimum lower side frequency (LSF):

$$\text{LSF} = 70\text{ MHz} - 5\text{ MHz} = 65\text{ MHz}$$

Maximum upper side frequency (USF):

$$\text{USF} = 70\text{ MHz} + 5\text{ MHz} = 75\text{ MHz}$$

Therefore, the output spectrum for the worst-case binary input conditions is as follows:

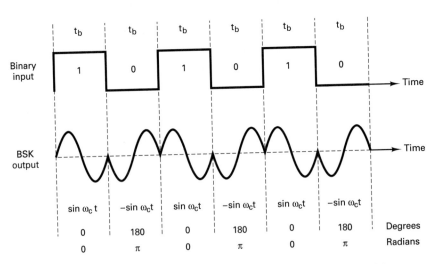

Figure 12-12   Output phase versus time relationship for a BPSK modulator.

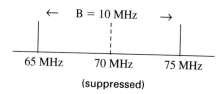

$$\leftarrow \quad B = 10 \text{ MHz} \quad \rightarrow$$

| 65 MHz | 70 MHz | 75 MHz |

(suppressed)

The minimum Nyquist bandwidth (f_N) is

$$f_N = 75 \text{ MHz} - 65 \text{ MHz} = 10 \text{ MHz}$$

and the baud $= f_b$ or 10 megabaud.

BPSK Receiver

Figure 12-13 shows the block diagram of a BPSK receiver. The input signal may be $+\sin \omega_c t$ or $-\sin \omega_c t$. The coherent carrier recovery circuit detects and regenerates a carrier signal that is both frequency and phase coherent with the original transmit carrier. The balanced modulator is a product detector; the output is the product of the two inputs (the BPSK signal and the recovered carrier). The low-pass filter (LPF) separates the recovered binary data from the complex demodulated signal. Mathematically, the demodulation process is as follows.

For a BPSK input signal of $+\sin \omega_c t$ (logic 1), the output of the balanced modulator is

$$\text{output} = (\sin \omega_c t)(\sin \omega_c t) = \sin^2 \omega_c t \qquad (12\text{-}7)$$

(filtered out)

or
$$\sin^2 \omega_c t = \tfrac{1}{2}(1 - \cos 2\omega_c t) = \tfrac{1}{2} - \tfrac{1}{2}\cos 2\omega_c t \nearrow$$

leaving
$$\text{output} = +\tfrac{1}{2}\text{V} = \text{logic 1}$$

It can be seen that the output of the balanced modulator contains a positive voltage ($+\tfrac{1}{2}$ V) and a cosine wave at twice the carrier frequency ($2\omega_c$). The LPF has a cutoff frequency much lower than $2\omega_c$ and thus blocks the second harmonic of the carrier and passes only the positive constant component. A positive voltage represents a demodulated logic 1.

For a BPSK input signal of $-\sin \omega_c t$ (logic 0), the output of the balanced modulator is

$$\text{output} = (-\sin \omega_c t)(\sin \omega_c t) = -\sin^2 \omega_c t \qquad (12\text{-}8)$$

(filtered out)

or
$$-\sin^2 \omega_c t = -\tfrac{1}{2}(1 - \cos 2\omega_c t) = -\tfrac{1}{2} + \tfrac{1}{2}\cos 2\omega_c t \nearrow$$

leaving
$$\text{output} = -\tfrac{1}{2}\text{V} = \text{logic 0}$$

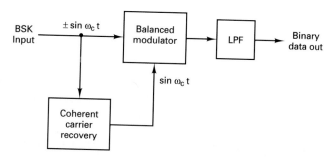

Figure 12-13 BPSK receiver.

Binary Phase Shift Keying

The output of the balanced modulator contains a negative voltage ($-\frac{1}{2}$ V) and a cosine wave at twice the carrier frequency ($2\omega_c$). Again, the LPF blocks the second harmonic of the carrier and passes only the negative constant component. A negative voltage represents a demodulated logic 0.

M-ary Encoding

M-ary is a term derived from the word "binary." *M* is simply a digit that represents the number of conditions possible. The two digital modulation techniques discussed thus far (binary FSK and BPSK) are binary systems; there are only two possible output conditions. One represents a logic 1 and the other a logic 0; thus they are *M*-ary systems where $M = 2$. With digital modulation, very often it is advantageous to encode at a level higher than binary. For example, a PSK system with four possible output phases is an *M*-ary system where $M = 4$. If there were eight possible output phases, $M = 8$, and so on. Mathematically,

$$N = \log_2 M \qquad (12\text{-}9)$$

where N = number of bits
M = number of output conditions possible with N bits

For example, if 2 bits were allowed to enter a modulator before the output were allowed to change,

$$2 = \log_2 M \quad \text{and} \quad 2^2 = M \quad \text{thus } M = 4$$

An $M = 4$ indicates that with 2 bits, four different output conditions are possible. For $N = 3$, $M = 2^3$ or 8, and so on.

QUATERNARY PHASE SHIFT KEYING

Quaternary phase shift keying (QPSK), or *quadrature PSK* as it is sometimes called, is another form of angle-modulated, constant-amplitude digital modulation. QPSK is an *M*-ary encoding technique where $M = 4$ (hence the name "quaternary," meaning "4"). With QPSK four output phases are possible for a single carrier frequency. Because there are four different output phases, there must be four different input conditions. Because the digital input to a QPSK modulator is a binary (base 2) signal, to produce four different input conditions it takes more than a single input bit. With 2 bits, there are four possible conditions: 00, 01, 10, and 11. Therefore, with QPSK, the binary input data are combined into groups of 2 bits called *dibits*. Each dibit code generates one of the four possible output phases. Therefore, for each 2-bit dibit clocked into the modulator, a single output change occurs. Therefore, the rate of change at the output (baud rate) is one-half of the input bit rate.

QPSK Transmitter

A block diagram of a QPSK modulator is shown in Figure 12-14. Two bits (a dibit) are clocked into the bit splitter. After both bits have been serially inputted, they are simultaneously parallel outputted. One bit is directed to the I channel and the other to the Q channel. The I bit modulates a carrier that is in phase with the reference oscillator (hence the name "I" for "in phase" channel), and the Q bit modulates a carrier that is 90° out of phase or in quadrature with the reference carrier (hence the name "Q" for "quadrature" channel).

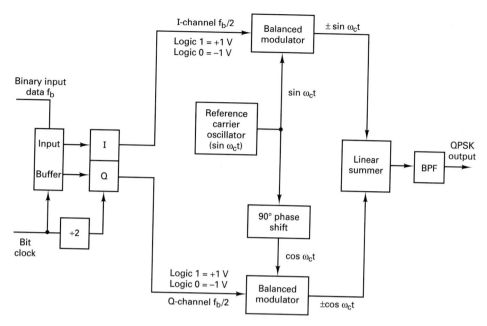

Figure 12-14 QPSK modulator.

It can be seen that once a dibit has been split into the I and Q channels, the operation is the same as in a BPSK modulator. Essentially, a QPSK modulator is two BPSK modulators combined in parallel. Again, for a logic 1 = +1 V and a logic 0 = −1 V, two phases are possible at the output of the I balanced modulator (+sin $\omega_c t$ and −sin $\omega_c t$), and two phases are possible at the output of the Q balanced modulator (+cos $\omega_c t$ and −cos $\omega_c t$). When the linear summer combines the two quadrature (90° out of phase) signals, there are four possible resultant phasors given by these expressions: +sin $\omega_c t$ + cos $\omega_c t$, +sin $\omega_c t$ − cos $\omega_c t$, −sin $\omega_c t$ + cos $\omega_c t$, and −sin $\omega_c t$ − cos $\omega_c t$.

EXAMPLE 12-3

For the QPSK modulator shown in Figure 12-14, construct the truth table, phasor diagram, and constellation diagram.

Solution   For a binary data input of Q = 0 and I = 0, the two inputs to the I balanced modulator are −1 and sin $\omega_c t$, and the two inputs to the Q balanced modulator are −1 and cos $\omega_c t$. Consequently, the outputs are

$$\text{I balanced modulator} = (-1)(\sin \omega_c t) = -1 \sin \omega_c t$$

$$\text{Q balanced modulator} = (-1)(\cos \omega_c t) = -1 \cos \omega_c t$$

and the output of the linear summer is

$$-1 \cos \omega_c t - 1 \sin \omega_c t = 1.414 \sin (\omega_c t - 135°)$$

For the remaining dibit codes (01, 10, and 11), the procedure is the same. The results are shown in Figure 12-15.

In Figure 12-15b it can be seen that with QPSK each of the four possible output phasors has exactly the same amplitude. Therefore, the binary information must be encoded entirely in the phase of the output signal. This constant amplitude characteristic is the most important characteristic of PSK that distinguishes it from QAM, which is explained later

Quaternary Phase Shift Keying

469

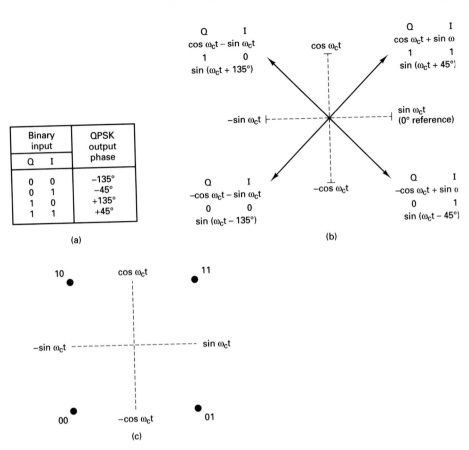

Binary input		QPSK output phase
Q	I	
0	0	−135°
0	1	−45°
1	0	+135°
1	1	+45°

(a)

(b)

(c)

Figure 12-15 QPSK modulator: (a) truth table; (b) phasor diagram; (c) constellation diagram.

in this chapter. Also, from Figure 12-15b it can be seen that the angular separation between any two adjacent phasors in QPSK is 90°. Therefore, a QPSK signal can undergo almost a +45° or −45° shift in phase during transmission and still retain the correct encoded information when demodulated at the receiver. Figure 12-16 shows the output phase versus time relationship for a QPSK modulator.

Bandwidth Considerations of QPSK

With QPSK, since the input data are divided into two channels, the bit rate in either the I or the Q channel is equal to one-half of the input data rate ($f_b/2$). (Essentially, the bit splitter stretches the I and Q bits to twice their input bit length.) Consequently, the highest fundamental frequency present at the data input to the I or the Q balanced modulator is equal

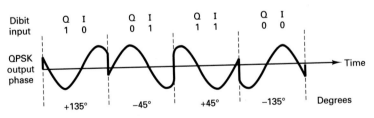

Figure 12-16 Output phase versus time relationship for a QPSK modulator.

Chap. 12   **Digital Communications**

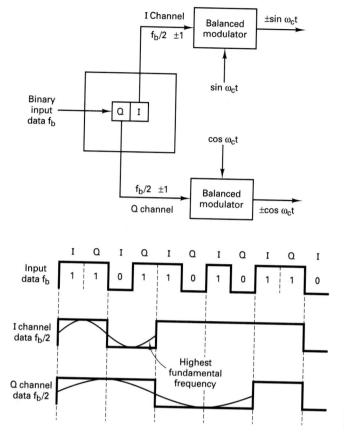

Figure 12-17 Bandwidth considerations of a QPSK modulator.

to one-fourth of the input data rate (one-half of $f_b/2 = f_b/4$). As a result, the output of the I and Q balanced modulators requires a minimum double-sided Nyquist bandwidth equal to one-half of the incoming bit rate (f_N = twice $f_b/4 = f_b/2$). Thus with QPSK, a bandwidth compression is realized (the minimum bandwidth is less than the incoming bit rate). Also, since the QPSK output signal does not change phase until 2 bits (a dibit) have been clocked into the bit splitter, the fastest output rate of change (baud) is also equal to one-half of the input bit rate. As with BPSK, the minimum bandwidth and the baud are equal. This relationship is shown in Figure 12-17.

In Figure 12-17 it can be seen that the worst-case input condition to the I or Q balanced modulator is an alternating 1/0 pattern, which occurs when the binary input data has a 1100 repetitive pattern. One cycle of the fastest binary transition (a 1/0 sequence) in the I or Q channel takes the same time as 4 input data bits. Consequently, the highest fundamental frequency at the input and fastest rate of change at the output of the balanced modulators is equal to one-fourth of the binary input bit rate.

The output of the balanced modulators can be expressed mathematically as

$$\text{output} = (\sin \omega_a t)(\sin \omega_c t)$$

where

$$\omega_a t = 2\pi \frac{f_b}{4} t \quad \text{and} \quad \omega_c t = 2\pi f_c t$$

$$\underbrace{\hphantom{\omega_a t = 2\pi \frac{f_b}{4} t}}_{\substack{\text{modulating} \\ \text{phase}}} \qquad \underbrace{\hphantom{\omega_c t = 2\pi f_c t}}_{\substack{\text{unmodulated} \\ \text{carrier phase}}}$$

Quaternary Phase Shift Keying

Thus

$$\text{output} = \left(\sin 2\pi \frac{f_b}{4} t \right) (\sin 2\pi f_c t)$$

$$\frac{1}{2} \cos 2\pi \left(f_c - \frac{f_b}{4} \right) t - \frac{1}{2} \cos 2\pi \left(f_c + \frac{f_b}{4} \right) t$$

The output frequency spectrum extends from $f_c + f_b/4$ to $f_c - f_b/4$ and the minimum bandwidth (f_N) is

$$\left(f_c + \frac{f_b}{4} \right) - \left(f_c - \frac{f_b}{4} \right) = \frac{2f_b}{4} = \frac{f_b}{2}$$

EXAMPLE 12-4

For a QPSK modulator with an input data rate (f_b) equal to 10 Mbps and a carrier frequency of 70 MHz, determine the minimum double-sided Nyquist bandwidth (f_N) and the baud. Also, compare the results with those achieved with the BPSK modulator in Example 12-2. Use the QPSK block diagram shown in Figure 12-14 as the modulator model.

Solution The bit rate in both the I and Q channels is equal to one-half of the transmission bit rate or

$$f_{bQ} = f_{bI} = \frac{f_b}{2} = \frac{10 \text{ Mbps}}{2} = 5 \text{ Mbps}$$

The highest fundamental frequency presented to either balanced modulator is

$$f_a = \frac{f_{bQ}}{2} \quad \text{or} \quad \frac{f_{bI}}{2} = \frac{5 \text{ Mbps}}{2} = 2.5 \text{ MHz}$$

The output wave from each balanced modulator is

$$(\sin 2\pi f_a t)(\sin 2\pi f_c t)$$

$$\tfrac{1}{2} \cos 2\pi (f_c - f_a)t - \tfrac{1}{2} \cos 2\pi (f_c + f_a)t$$

$$\tfrac{1}{2} \cos 2\pi [(70 - 2.5) \text{ MHz}]t - \tfrac{1}{2} \cos 2\pi [(70 + 2.4) \text{ MHz}]t$$

$$\tfrac{1}{2} \cos 2\pi (67.5 \text{ MHz})t - \tfrac{1}{2} \cos 2\pi (72.5 \text{ MHz})t$$

The minimum Nyquist bandwidth is

$$f_N = (72.5 - 67.5) \text{ MHz} = 5 \text{ MHz}$$

The symbol rate equals the bandwidth; thus

$$\text{symbol rate} = 5 \text{ megabaud}$$

The output spectrum is as follows:

$$f_N = 5 \text{ MHz}$$

It can be seen that for the same input bit rate the minimum bandwidth required to pass the output of the QPSK modulator is equal to one-half of that required for the BPSK modulator in Example 12-2. Also, the baud rate for the QPSK modulator is one-half that of the BPSK modulator.

QPSK Receiver

The block diagram of a QPSK receiver is shown in Figure 12-18. The power splitter directs the input QPSK signal to the I and Q product detectors and the carrier recovery circuit. The carrier recovery circuit reproduces the original transmit carrier oscillator signal. The recovered carrier must be frequency and phase coherent with the transmit reference carrier. The QPSK signal is demodulated in the I and Q product detectors, which generate the original I and Q data bits. The outputs of the product detectors are fed to the bit combining circuit, where they are converted from parallel I and Q data channels to a single binary output data stream.

The incoming QPSK signal may be any one of the four possible output phases shown in Figure 12-15. To illustrate the demodulation process, let the incoming QPSK signal be $-\sin \omega_c t + \cos \omega_c t$. Mathematically, the demodulation process is as follows.

The receive QPSK signal ($-\sin \omega_c t + \cos \omega_c t$) is one of the inputs to the I product detector. The other input is the recovered carrier ($\sin \omega_c t$). The output of the I product detector is

$$I = \underbrace{(-\sin \omega_c t + \cos \omega_c t)}_{\text{QPSK input signal}} \underbrace{(\sin \omega_c t)}_{\text{carrier}}$$

$$= (-\sin \omega_c t)(\sin \omega_c t) + (\cos \omega_c t)(\sin \omega_c t)$$

$$= -\sin^2 \omega_c t + (\cos \omega_c t)(\sin \omega_c t)$$

$$= -\tfrac{1}{2}(1 - \cos 2\omega_c t) + \tfrac{1}{2}\sin (\omega_c + \omega_c)t + \tfrac{1}{2}\sin (\omega_c - \omega_c)t$$

$$\text{(filtered out)} \quad \text{(equals 0)}$$

$$I = -\tfrac{1}{2} + \tfrac{1}{2}\cos 2\omega_c t + \tfrac{1}{2}\sin 2\omega_c t + \tfrac{1}{2}\sin 0$$

$$= -\tfrac{1}{2}\,\text{V (logic 0)}$$

Again, the receive QPSK signal ($-\sin \omega_c t + \cos \omega_c t$) is one of the inputs to the Q product detector. The other input is the recovered carrier shifted $90°$ in phase ($\cos \omega_c t$). The output of the Q product detector is

$$Q = \underbrace{(-\sin \omega_c t + \cos \omega_c t)}_{\text{QPSK input signal}} \underbrace{(\cos \omega_c t)}_{\text{carrier}}$$

$$= \cos^2 \omega_c t - (\sin \omega_c t)(\cos \omega_c t)$$

$$= \tfrac{1}{2}(1 + \cos 2\omega_c t) - \tfrac{1}{2}\sin (\omega_c + \omega_c)t - \tfrac{1}{2}\sin (\omega_c - \omega_c)t$$

$$\text{(filtered out)} \quad \text{(equals 0)}$$

$$Q = \tfrac{1}{2} + \tfrac{1}{2}\cos 2\omega_c t - \tfrac{1}{2}\sin 2\omega_c t - \tfrac{1}{2}\sin 0$$

$$= \tfrac{1}{2}\,\text{V (logic 1)}$$

Quaternary Phase Shift Keying

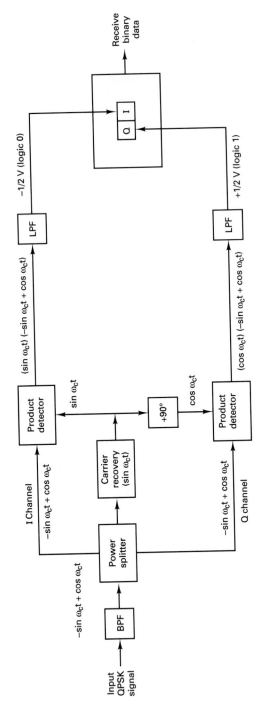

Figure 12-18 QPSK receiver.

The demodulated I and Q bits (0 and 1, respectively) correspond to the constellation diagram and truth table for the QPSK modulator shown in Figure 12-15.

Offset QPSK

Offset QPSK (OQPSK) is a modified form of QPSK where the bit waveforms on the I and Q channels are offset or shifted in phase from each other by one-half of a bit time.

Figure 12-19 shows a simplified block diagram, the bit sequence alignment, and the constellation diagram for a OQPSK modulator. Because changes in the I channel occur at

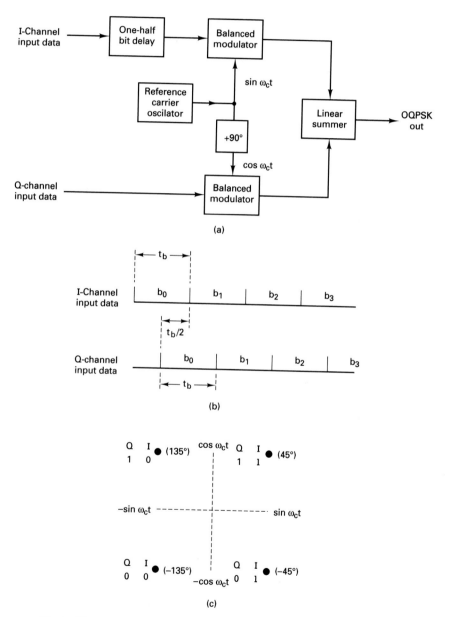

Figure 12-19 Offset keyed (OQPSK): (a) block diagram; (b) bit alignment; (c) constellation diagram.

the midpoints of the Q-channel bits, and vice versa, there is never more than a single bit change in the dibit code, and therefore there is never more than a 90° shift in the output phase. In conventional QPSK, a change in the input dibit from 00 to 11 or 01 to 10 causes a corresponding 180° shift in the output phase. Therefore, an advantage of OQPSK is the limited phase shift that must be imparted during modulation. A disadvantage of OQPSK is that changes in the output phase occur at twice the data rate in either the I or Q channels. Consequently, with OQPSK the baud and minimum bandwidth are twice that of conventional QPSK for a given transmission bit rate. OQPSK is sometimes called OKQPSK (*offset-keyed PSK*).

EIGHT-PHASE PSK

Eight-phase PSK (8-PSK) is an *M*-ary encoding technique where $M = 8$. With an 8-PSK modulator, there are eight possible output phases. To encode eight different phases, the incoming bits are considered in groups of 3 bits, called *tribits* ($2^3 = 8$).

8-PSK Transmitter

A block diagram of an 8-PSK modulator is shown in Figure 12-20. The incoming serial bit stream enters the bit splitter, where it is converted to a parallel, three-channel output (the I or in-phase channel, the Q or in-quadrature channel, and the C or control channel). Consequently, the bit rate in each of the three channels is $f_b/3$. The bits in the I and C channels enter the I-channel 2-to-4 level converter, and the bits in the Q and $\overline{C}$ channels enter the Q-channel 2-to-4-level converter. Essentially, the 2-to-4-level converters are parallel-input *digital-to-analog converters* (DACs). With 2 input bits, four output voltages are possible. The algorithm for the DACs is quite simple. The I or Q bit determines the polarity of the output analog signal (logic $1 = +V$ and logic $0 = -V$) while the C or $\overline{C}$ bit determines the magnitude (logic $1 = 1.307$ V and logic $0 = 0.541$ V). Consequently, with two magnitudes and two polarities, four different output conditions are possible.

Figure 12-21 shows the truth table and corresponding output conditions for the 2-to-4-level converters. Because the C and $\overline{C}$ bits can never be the same logic state, the outputs

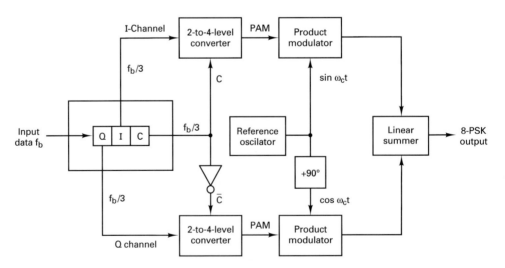

Figure 12-20 8-PSK modulator.

I	C	Output
0	0	−0.541 V
0	1	−1.307 V
1	0	+0.541 V
1	1	+1.307 V

Q	$\bar{C}$	Output
0	1	−1.307 V
0	0	−0.541 V
1	1	+1.307 V
1	0	+0.541 V

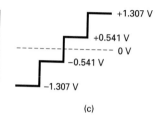

(a) (b) (c)

Figure 12-21 I- and Q-channel 2-to-4-level converters: (a) I-channel truth table; (b) Q-channel truth table; (c) PAM levels.

from the I and Q 2-to-4-level converters can never have the same magnitude, although they can have the same polarity. The output of a 2-to-4-level converter is an *M*-ary, *pulse-amplitude-modulated* (PAM) signal where $M = 4$.

EXAMPLE 12-5

For a tribit input of Q = 0, I = 0, and C = 0 (000), determine the output phase for the 8-PSK modulator shown in Figure 12-20.

Solution The inputs to the I-channel 2-to-4-level converter are I = 0 and C = 0. From Figure 12-21 the output is −0.541 V. The inputs to the Q-channel 2-to-4-level converter are Q = 0 and $\bar{C}$ = 1. Again from Figure 12-21, the output is −1.307 V.

Thus the two inputs to the I-channel product modulators are −0.541 and $\sin \omega_c t$. The output is

$$I = (-0.541)(\sin \omega_c t) = -0.541 \sin \omega_c t$$

The two inputs to the Q-channel product modulator are −1.307 V and $\cos \omega_c t$. The output is

$$Q = (-1.307)(\cos \omega_c t) = -1.307 \cos \omega_c t$$

The outputs of the I- and Q-channel product modulators are combined in the linear summer and produce a modulated output of

$$\text{summer output} = -0.541 \sin \omega_c t - 1.307 \cos \omega_c t$$

$$= 1.41 \sin (\omega_c t - 112.5°)$$

For the remaining tribit codes (001, 010, 011, 100, 101, 110, and 111), the procedure is the same. The results are shown in Figure 12-22.

From Figure 12-22 it can be seen that the angular separation between any two adjacent phasors is 45°, half what it is with QPSK. Therefore, an 8-PSK signal can undergo almost a ±22.5° phase shift during transmission and still retain its integrity. Also, each phasor is of equal magnitude; the tribit condition (actual information) is again contained only in the phase of the signal. The PAM levels of 1.307 and 0.541 are relative values. Any levels may be used as long as their ratio is 0.541/1.307 and their arc tangent is equal to 22.5°. For example, if their values were doubled to 2.614 and 1.082, the resulting phase angles would not change, although the magnitude of the phasor would increase proportionally.

It should also be noted that the tribit code between any two adjacent phases changes by only one bit. This type of code is called the *Gray code* or, sometimes, the *maximum distance code*. This code is used to reduce the number of transmission errors. If a signal were to undergo a phase shift during transmission, it would most likely be shifted to an adjacent phasor. Using the Gray code results in only a single bit being received in error.

Figure 12-23 shows the output phase-versus-time relationship of an 8-PSK modulator.

Eight-Phase PSK

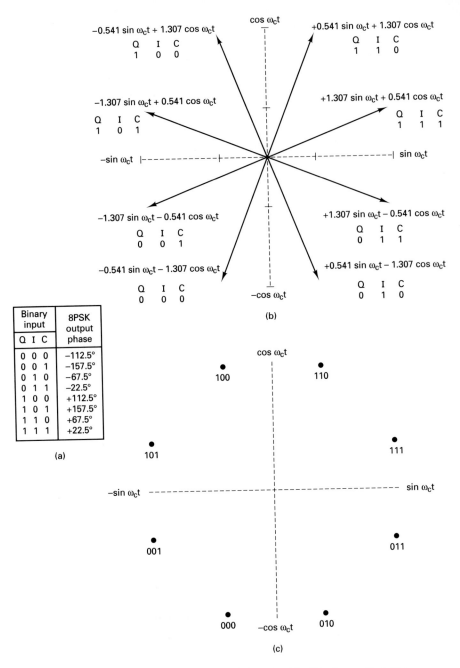

| Binary input | 8PSK output |
Q I C	phase
0 0 0	−112.5°
0 0 1	−157.5°
0 1 0	−67.5°
0 1 1	−22.5°
1 0 0	+112.5°
1 0 1	+157.5°
1 1 0	+67.5°
1 1 1	+22.5°

(a)

Figure 12-22 8-PSK modulator: (a) truth table; (b) phasor diagram; (c) constellation diagram.

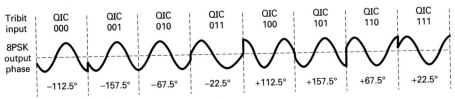

Figure 12-23   Output phase versus time relationship for an 8-PSK modulator.

Bandwidth Considerations of 8-PSK

With 8-PSK, since the data are divided into three channels, the bit rate in the I, Q, or C channel is equal to one-third of the binary input data rate ($f_b/3$). (The bit splitter stretches the I, Q, and C bits to three times their input bit length.) Because the I, Q, and C bits are outputted simultaneously and in parallel, the 2-to-4-level converters also see a change in their inputs (and consequently their outputs) at a rate equal to $f_b/3$.

Figure 12-24 shows the bit timing relationship between the binary input data; the I-, Q-, and C-channel data; and the I and Q PAM signals. It can be seen that the highest fundamental frequency in the I, Q or C channel is equal to one-sixth of the bit rate of the binary input (one cycle in the I, Q, or C channel takes the same amount of time as six input bits). Also, the highest fundamental frequency in either PAM signal is equal to one-sixth of the binary input bit rate.

With an 8-PSK modulator, there is one change in phase at the output for every 3 data input bits. Consequently, the baud for 8-PSK equals $f_b/3$, the same as the minimum

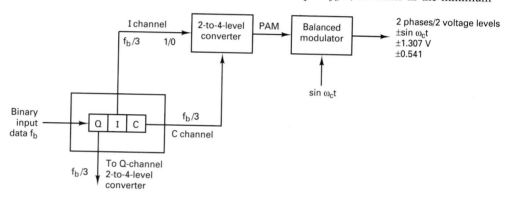

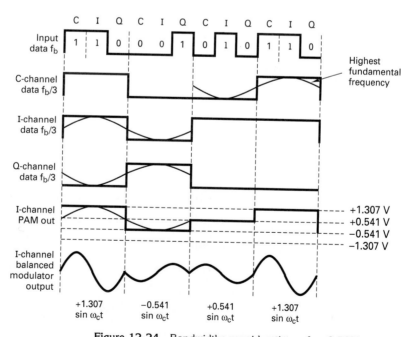

Figure 12-24 Bandwidths considerations of an 8-PSK modulator.

Eight-Phase PSK

bandwidth. Again, the balanced modulators are product modulators; their outputs are the product of the carrier and the PAM signal. Mathematically, the output of the balanced modulators is

$$\theta = (X \sin \omega_a t)(\sin \omega_c t) \tag{12-10}$$

where

$$\omega_a t = 2\pi \frac{f_b}{6} t \quad \text{and} \quad \omega_c t = 2\pi f_c t$$

modulating signal   carrier

and

$$X = \pm 1.307 \quad \text{or} \quad \pm 0.541$$

Thus

$$\theta = \left(X \sin 2\pi \frac{f_b}{6} t \right)(\sin 2\pi f_c t)$$

$$= \frac{X}{2} \cos 2\pi \left(f_c - \frac{f_b}{6} \right) t - \frac{X}{2} \cos 2\pi \left(f_c + \frac{f_b}{6} \right) t$$

The output frequency spectrum extends from $f_c + f_b/6$ to $f_c - f_b/6$ and the minimum bandwidth (f_N) is

$$\left(f_c + \frac{f_b}{6} \right) - \left(f_c - \frac{f_b}{6} \right) = \frac{2f_b}{6} = \frac{f_b}{3}$$

EXAMPLE 12-6

For an 8-PSK modulator with an input data rate (f_b) equal to 10 Mbps and a carrier frequency of 70 MHz, determine the minimum double-sided Nyquist bandwidth (f_N) and the baud. Also, compare the results with those achieved with the BPSK and QPSK modulators in Examples 12-2 and 12-4. Use the 8-PSK block diagram shown in Figure 12-20 as the modulator model.

Solution The bit rate in the I, Q, and C channels is equal to one-third of the input bit rate, or

$$f_{bC} = f_{bQ} = f_{bI} = \frac{10 \text{ Mbps}}{3} = 3.33 \text{ Mbps}$$

Therefore, the fastest rate of change and highest fundamental frequency presented to either balanced modulator is

$$f_a = \frac{f_{bC}}{2} \quad \text{or} \quad \frac{f_{bQ}}{2} \quad \text{or} \quad \frac{f_{bI}}{2} = \frac{3.33 \text{ Mbps}}{2} = 1.667 \text{ Mbps}$$

The output wave from the balance modulators is

$$(\sin 2\pi f_a t)(\sin 2\pi f_c t)$$

$$\tfrac{1}{2} \cos 2\pi (f_c - f_a)t - \tfrac{1}{2} \cos 2\pi (f_c + f_a)t$$

$$\tfrac{1}{2} \cos 2\pi [(70 - 1.667) \text{ MHz}]t - \tfrac{1}{2} \cos 2\pi [(70 + 1.667) \text{ MHz}]t$$

$$\tfrac{1}{2} \cos 2\pi (68.333 \text{ MHz})t - \tfrac{1}{2} \cos 2\pi (71.667 \text{ MHz})t$$

The minimum Nyquist bandwidth is

$$f_N = (71.667 - 68.333) \text{ MHz} = 3.333 \text{ MHz}$$

Again, the baud equals the bandwidth; thus

$$\text{baud} = 3.333 \text{ megabaud}$$

The output spectrum is as follows:

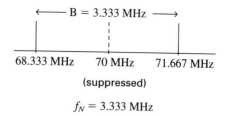

$$f_N = 3.333 \text{ MHz}$$

It can be seen that for the same input bit rate the minimum bandwidth required to pass the output of an 8-PSK modulator is equal to one-third that of the BPSK modulator in Example 12-2 and 50% less than that required for the QPSK modulator in Example 12-4. Also, in each case the baud has been reduced by the same proportions.

8-PSK Receiver

Figure 12-25 shows a block diagram of an 8-PSK receiver. The power splitter directs the input 8-PSK signal to the I and Q product detectors and the carrier recovery circuit. The carrier recovery circuit reproduces the original reference oscillator signal. The incoming 8-PSK signal is mixed with the recovered carrier in the I product detector and with a quadrature carrier in the Q product detector. The outputs of the product detectors are 4-level PAM signals that are fed to the 4-to-2-level *analog-to-digital converters* (ADCs). The outputs from the I-channel 4-to-2-level converter and the I and C bits, while the outputs from the Q-channel 4-to-2-level converter are the Q and $\overline{C}$ bits. The parallel-to-serial logic circuit converts the I/C and Q/$\overline{C}$ bit pairs to serial I, Q, and C output data streams.

SIXTEEN-PHASE PSK

Sixteen-phase PSK (16-PSK) is an *M*-ary encoding technique where $M = 16$; there are 16 different output phases possible. A 16-PSK modulator acts on the incoming data in groups of 4 bits ($2^4 = 16$), called *quadbits*. The output phase does not change until 4 bits have been inputted into the modulator. Therefore, the output rate of change (baud) and the minimum bandwidth are equal to one-fourth of the incoming bit rate ($f_b/4$). The truth table and constellation diagram for a 16-PSK transmitter are shown in Figure 12-26.

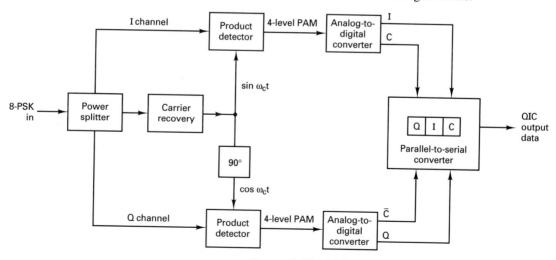

Figure 12-25 8-PSK receiver.

Bit code	Phase	Bit code	Phase
0000	11.25°	1000	191.25°
0001	33.75°	1001	213.75°
0010	56.25°	1010	236.25°
0011	78.75°	1011	258.75°
0100	101.25°	1100	281.25°
0101	123.75°	1101	303.75°
0110	146.25°	1110	326.25°
0111	168.75°	1111	348.75°

(a)

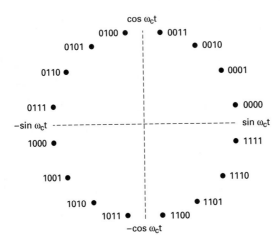

Figure 12-26 16-PSK: (a) truth table; (b) constellation diagram.

With 16-PSK, the angular separation between adjacent output phases is only 22.5°. Therefore, a 16-PSK signal can undergo almost a ±11.25° phase shift during transmission and still retain its integrity. Because of this, 16-PSK is highly susceptible to phase impairments introduced in the transmission medium and is therefore seldom used.

QUADRATURE AMPLITUDE MODULATION

Quadrature amplitude modulation (QAM) is a form of digital modulation where the digital information is contained in both the amplitude and phase of the transmitted carrier.

EIGHT QAM

Eight QAM (8-QAM) is an *M*-ary encoding technique where $M = 8$. Unlike 8-PSK, the output signal from an 8-QAM modulator is not a constant-amplitude signal.

8-QAM Transmitter

Figure 12-27 shows the block diagram of an 8-QAM transmitter. As you can see, the only difference between the 8-QAM transmitter and the 8-PSK transmitter shown in Figure 12-19 is the omission of the inverter between the C channel and the Q product modulator.

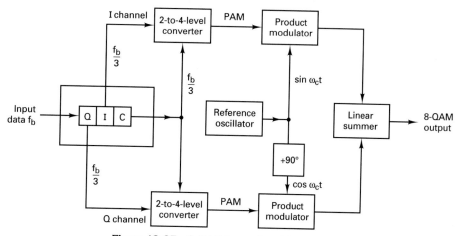

Figure 12-27   8-QAM transmitter block diagram.

As with 8-PSK, the incoming data are divided into groups of three bits (tribits): the I, Q, and C bit streams, each with a bit rate equal to one-third of the incoming data rate. Again, the I and Q bits determine the polarity of the PAM signal at the output of the 2-to-4-level converters, and the C channel determines the magnitude. Because the C bit is fed uninverted to both the I- and Q-channel 2-to-4-level converters, the magnitudes of the I and Q PAM signals are always equal. Their polarities depend on the logic condition of the I and Q bits and therefore may be different. Figure 12-28 shows the truth table for the I- and Q-channel 2-to-4-level converters; they are the same.

EXAMPLE 12-7

For a tribit input of Q = 0, I = 0, and C = 0 (000), determine the output amplitude and phase for the 8-QAM modulator shown in Figure 12-27.

Solution   The inputs to the I-channel 2-to-4-level converter are I = 0 and C = 0. From Figure 12-28 the output is -0.541 V. The inputs to the Q-channel 2-to-4-level converter are Q = 0 and C = 0. Again from Figure 12-28, the output is -0.541 V.

   Thus the two inputs to the I-channel product modulator are -0.541 and $\sin|\omega_c t$. The output is

$$I = (-0.541)(\sin \omega_c t) = -0.541 \sin \omega_c t$$

The two inputs to the Q-channel product modulator are -0.541 and $\cos \omega_c t$. The output is

$$Q = (-0.541)(\cos \omega_c t) = -0.541 \cos \omega_c t$$

The outputs from the I- and Q-channel product modulators are combined in the linear summer and produce a modulated output of

$$\text{summer output} = -0.541 \sin \omega_c t - 0.541 \cos \omega_c t$$

$$= 0.765 \sin (\omega_c t - 135°)$$

I/Q	C	Output
0	0	−0.541
0	1	−1.307 V
1	0	+0.541
1	1	+1.307 V

Figure 12-28   Truth table for the I-
and Q-channel 2-to-4-level converters.

Eight QAM

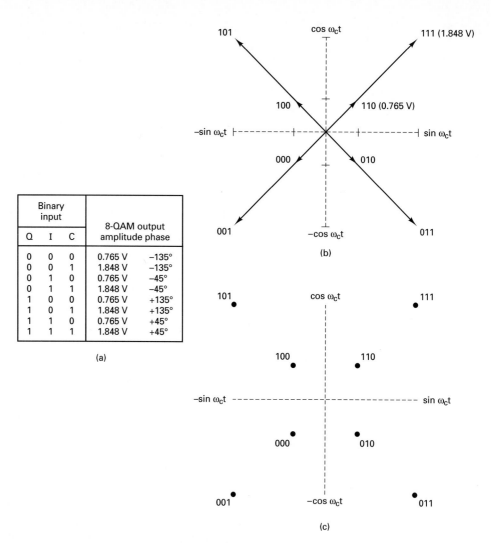

Binary input			8-QAM output	
Q	I	C	amplitude	phase
0	0	0	0.765 V	−135°
0	0	1	1.848 V	−135°
0	1	0	0.765 V	−45°
0	1	1	1.848 V	−45°
1	0	0	0.765 V	+135°
1	0	1	1.848 V	+135°
1	1	0	0.765 V	+45°
1	1	1	1.848 V	+45°

(a)

Figure 12-29 8-QAM modulator: (a) truth table; (b) phasor diagram; (c) constellation diagram.

For the remaining tribit codes (001, 010, 011, 100, 101, 110, and 111), the procedure is the same. The results are shown in Figure 12-29.

Figure 12-30 shows the output phase versus time relationship for an 8-QAM modulator. Note that there are two output amplitudes and only four phases are possible.

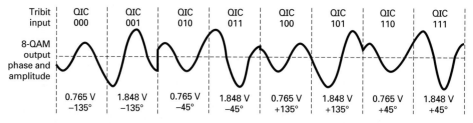

Figure 12-30 Output phase and amplitude versus time relationship for 8-QAM.

Chap. 12 Digital Communications

Bandwidth Considerations of 8-QAM

In 8-QAM, the bit rate in the I and Q channels is one-third of the input binary rate, the same as in 8-PSK. As a result, the highest fundamental modulating frequency and fastest output rate of change in 8-QAM are the same as with 8-PSK. Therefore, the minimum bandwidth required for 8-QAM is $f_b/3$, the same as in 8-PSK.

8-QAM Receiver

An 8-QAM receiver is almost identical to the 8-PSK receiver shown in Figure 12-25. The differences are the PAM levels at the output of the product detectors and the binary signals at the output of the analog-to-digital converters. Because there are two transmit amplitudes possible with 8-QAM that are different from those achievable with 8-PSK, the four demodulated PAM levels in 8-QAM are different from those in 8-PSK. Therefore, the conversion factor for the analog-to-digital converters must also be different. Also, with 8-QAM the binary output signals from the I-channel analog-to-digital converter are the I and C bits, and the binary output signals from the Q-channel analog-to-digital converter are the Q and C bits.

SIXTEEN QAM

Like 16-PSK, 16-QAM is an *M*-ary system where $M = 16$. The input data are acted on in groups of four ($2^4 = 16$). As with 8-QAM, both the phase and amplitude of the transmit carrier are varied.

16-QAM Transmitter

The block diagram for a 16-QAM transmitter is shown in Figure 12-31. The input binary data are divided into four channels: The I, I', Q, and Q'. The bit rate in each channel is equal to one-fourth of the input bit rate ($f_b/4$). Four bits are serially clocked into the bit splitter; then they are outputted simultaneously and in parallel with the I, I', Q, and Q' channels. The I and Q bits determine the polarity at the output of the 2-to-4-level convert-

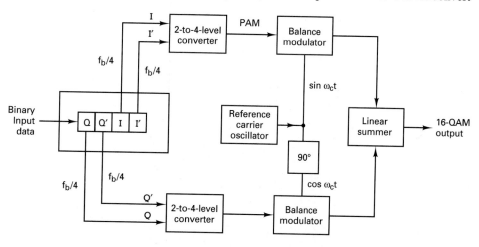

Figure 12-31   16-QAM transmitter block diagram.

I	I′	Output		Q	Q′	Output
0	0	−0.22 V		0	0	−0.22 V
0	1	−0.821 V		0	1	−0.821 V
1	0	+0.22 V		1	0	+0.22 V
1	1	+0.821 V		1	1	+0.821 V

Figure 12-32 Truth tables for the I- and Q-channel 2-to-4 level converters; (a) I channel; (b) Q channel.

ers (a logic 1 = positive and a logic 0 = negative). The I′ and Q′ bits determine the magnitude (a logic 1 = 0.821 V and a logic 0 = 0.22 V). Consequently, the 2-to-4-level converters generate a 4-level PAM signal. Two polarities and two magnitudes are possible at the output of each 2-to-4-level converter. They are ±0.22 V and ±0.821 V. The PAM signals modulate the in phase and quadrature carriers in the product modulators. Four outputs are possible for each product modulator. For the I product modulator they are $+0.821 \sin \omega_c t$, $-0.821 \sin \omega_c t$, $+0.22 \sin \omega_c t$, and $-0.22 \sin \omega_c t$. For the Q product modulator they are $+0.821 \cos \omega_c t$, $+0.22 \cos \omega_c t$, $-0.821 \cos \omega_c t$, and $-0.22 \cos \omega_c t$. The linear summer combines the outputs from the I- and Q-channel product modulators and produces the 16 output conditions necessary for 16-QAM. Figure 12-32 shows the truth table for the I- and Q-channel 2-to-4-level converters.

EXAMPLE 12-8

For a quadbit input of I = 0, I′ = 0, Q = 0, and Q′ = 0 (0000), determine the output amplitude and phase for the 16-QAM modulator shown in Figure 12-31.

Solution   The inputs to the I-channel 2-to-4-level converter are I = 0 and I′ = 0. From Figure 12-32 the output is −0.22 V. The inputs to the Q-channel 2-to-4-level converter are Q = 0 and Q′ = 0. Again from Figure 12-32, the output is −0.22 V.

Thus the two inputs to the I-channel product modulator are −0.22 V and $\sin \omega_c t$. The output is

$$I = (-0.22)(\sin \omega_c t) = -0.22 \sin \omega_c t$$

The two inputs to the Q-channel product modulator are −0.22 V and $\cos \omega_c t$. The output is

$$Q = (-0.22)(\cos \omega_c t) = -0.22 \cos \omega_c t$$

The outputs from the I- and Q-channel product modulators are combined in the linear summer and produce a modulated output of

$$\text{summer output} = -0.22 \sin \omega_c t - 0.22 \cos \omega_c t$$

$$= 0.311 \sin (\omega_c t - 135°)$$

For the remaining quadbit codes the procedure is the same. The results are shown in Figure 12-33.

Bandwidth Considerations of 16-QAM

With 16-QAM, since the input data are divided into four channels, the bit rate in the I, I′, Q, or Q′ channel is equal to one-fourth of the binary input data rate ($f_b/4$). (The bit splitter stretches the I, I′, Q, and Q′ bits to four times their input bit length.) Also, because the I, I′, Q, and Q′ bits are outputted simultaneously and in parallel, the 2-to-4-level converters see a change in their inputs and outputs at a rate equal to one-fourth of the input data rate.

Figure 12-34 shows the bit timing relationship between the binary input data; the I, I′, Q, and Q′ channel data; and the I PAM signal. It can be seen that the highest fundamental frequency in the I, I′, Q, or Q′ channel is equal to one-eighth of the bit rate of the binary input data (one cycle in the I, I′, Q, or Q′ channel takes the same amount of time as 8 input

Binary input				16-QAM output	
Q	Q′	I	I′		
0	0	0	0	0.311 V	−135°
0	0	0	1	0.850 V	−165°
0	0	1	0	0.311 V	−45°
0	0	1	1	0.850 V	−15°
0	1	0	0	0.850 V	−105°
0	1	0	1	1.161 V	−135°
0	1	1	0	0.850 V	−75°
0	1	1	1	1.161 V	−45°
1	0	0	0	0.311 V	135°
1	0	0	1	0.850 V	175°
1	0	1	0	0.311 V	45°
1	0	1	1	0.850 V	15°
1	1	0	0	0.850 V	105°
1	1	0	1	1.161 V	135°
1	1	1	0	0.850 V	75°
1	1	1	1	1.161 V	45°

(a)

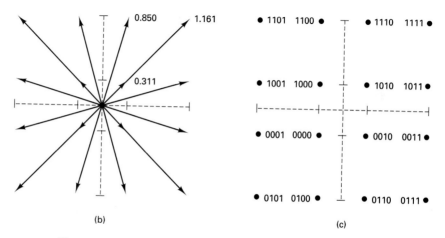

(b)   (c)

Figure 12-33   16-QAM modulator: (a) truth table; (b) phasor diagram; (c) constellation diagram.

bits). Also, the highest fundamental frequency of either PAM signal is equal to one-eighth of the binary input bit rate.

With a 16-QAM modulator, there is one change in the output signal (either its phase, amplitude, or both) for every 4 input data bits. Consequently, the baud equals $f_b/4$, the same as the minimum bandwidth.

Again, the balanced modulators are product modulators and their outputs can be represented mathematically as

$$\text{output} = (X \sin \omega_a t)(\sin \omega_c t) \tag{12-11}$$

where

$$\underbrace{\omega_a t = 2\pi \frac{f_b}{8} t}_{\substack{\text{modulating signal} \\ \text{phase}}} \quad \text{and} \quad \underbrace{\omega_c t = 2\pi f_c t}_{\text{carrier phase}}$$

and

$$X = \pm 0.22 \quad \text{or} \quad \pm 0.821$$

Thus
$$\text{output} = \left(X \sin 2\pi \frac{f_b}{8} t \right) (\sin 2\pi f_c t)$$

$$= \frac{X}{2} \cos 2\pi \left(f_c - \frac{f_b}{8} \right) t - \frac{X}{2} \cos 2\pi \left(f_c + \frac{f_b}{8} \right) t$$

The output frequency spectrum extends from $f_c + f_b/8$ to $f_c - f_b/8$ and the minimum bandwidth (f_N) is

$$\left(f_c + \frac{f_b}{8} \right) - \left(f_c - \frac{f_b}{8} \right) = \frac{2f_b}{8} = \frac{f_b}{4}$$

EXAMPLE 12-9

For a 16-QAM modulator with an input data rate (f_b) equal to 10 Mbps and a carrier frequency of 70 MHz, determine the minimum double-sided Nyquist frequency (f_N) and the baud. Also, compare the results with those achieved with the BPSK, QPSK, and 8-PSK

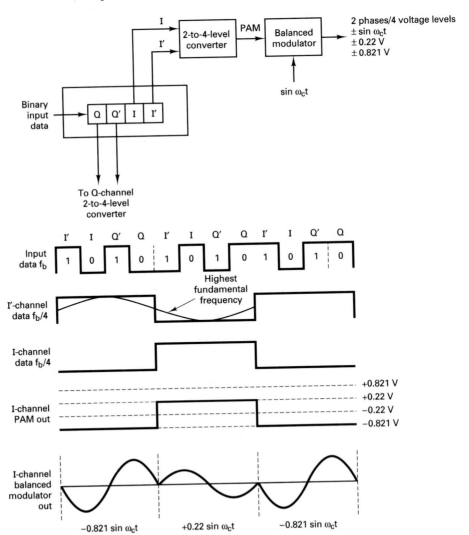

Figure 12-34   Bandwidth considerations of a 16-QAM modulator.

modulators in Examples 12-2, 12-4, and 12-6. Use the 16-QAM block diagram shown in Figure 12-27 as the modulator model.

Solution The bit rate in the I, I′, Q, and Q′ channels is equal to one-fourth of the input bit rate or

$$f_{bI} = f_{bI'} = f_{bQ} = f_{bQ'} = \frac{f_b}{4} = \frac{10 \text{ Mbps}}{4} = 2.5 \text{ Mbps}$$

Therefore, the fastest rate of change and highest fundamental frequency presented to either balanced modulator is

$$f_a = \frac{f_{bI}}{2} \quad \text{or} \quad \frac{f_{bI'}}{2} \quad \text{or} \quad \frac{f_{bQ}}{2} \quad \text{or} \quad \frac{f_{bQ'}}{2} = \frac{2.5 \text{ Mbps}}{2} = 1.25 \text{ MHz}$$

The output wave from the balanced modulator is

$$(\sin 2\pi f_a t)(\sin 2\pi f_c t)$$

$$\tfrac{1}{2}\cos 2\pi (f_c - f_a)t - \tfrac{1}{2}\cos 2\pi (f_c + f_a)t$$

$$\tfrac{1}{2}\cos 2\pi [(70 - 1.25) \text{ MHz}]t - \tfrac{1}{2}\cos 2\pi [(70 + 1.25) \text{ MHz}]t$$

$$\tfrac{1}{2}\cos 2\pi (68.75 \text{ MHz})t - \tfrac{1}{2}\cos 2\pi (71.25 \text{ MHz})t$$

The minimum Nyquist bandwidth is

$$f_N = (71.25 - 68.75) \text{ MHz} = 2.5 \text{ MHz}$$

The symbol rate equals the bandwidth; thus

$$\text{symbol rate} = 2.5 \text{ megabaud}$$

The output spectrum is as follows:

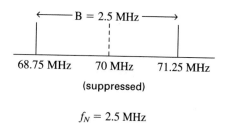

$$f_N = 2.5 \text{ MHz}$$

For the same input bit rate, the minimum bandwidth required to pass the output of a 16-QAM modulator is equal to one-fourth that of the BPSK modulator, one-half that of QPSK, and 25% less than with 8-PSK. For each modulation technique, the baud is also reduced by the same proportions.

BANDWIDTH EFFICIENCY

Bandwidth efficiency (or *information density* as it is sometimes called) is often used to compare the performance of one digital modulation technique to another. In essence, it is the ratio of the transmission bit rate to the minimum bandwidth required for a particular modulation scheme. Bandwidth efficiency is generally normalized to a 1-Hz bandwidth and thus indicates the number of bits that can be propagated through a medium for each hertz of bandwidth. Mathematically, bandwidth efficiency is

$$\text{BW efficiency} = \frac{\text{transmission rate (bps)}}{\text{minimum bandwidth (Hz)}} \tag{12-12}$$

$$= \frac{\text{bits/second}}{\text{hertz}} = \frac{\text{bits/second}}{\text{cycles/second}} = \frac{\text{bits}}{\text{cycle}}$$

EXAMPLE 12-10

Determine the bandwidth efficiencies for the following modulation schemes: BPSK, QPSK, 8-PSK, and 16-QAM.

Solution Recall from Examples 12-2, 12-4, 12-6, and 12-9 the minimum bandwidths required to propagate a 10-Mbps transmission rate with the following modulation schemes:

Modulation scheme	Minimum bandwidth (MHz)
BPSK	10
QPSK	5
8-PSK	3.33
16-QAM	2.5

Substituting into Equation 12-12, the bandwidth efficiencies are determined as follows:

$$\text{BPSK:} \quad \text{BW efficiency} = \frac{10\ \text{Mbps}}{10\ \text{MHz}} = \frac{1\ \text{bps}}{\text{Hz}} = \frac{1\ \text{bit}}{\text{cycle}}$$

$$\text{QPSK:} \quad \text{BW efficiency} = \frac{10\ \text{Mbps}}{5\ \text{MHz}} = \frac{2\ \text{bps}}{\text{Hz}} = \frac{2\ \text{bits}}{\text{cycle}}$$

$$\text{8-PSK:} \quad \text{BW efficiency} = \frac{10\ \text{Mbps}}{3.33\ \text{MHz}} = \frac{3\ \text{bps}}{\text{Hz}} = \frac{3\ \text{bits}}{\text{cycle}}$$

$$\text{16-QAM:} \quad \text{BW efficiency} = \frac{10\ \text{Mbps}}{2.5\ \text{MHz}} = \frac{4\ \text{bps}}{\text{Hz}} = \frac{4\ \text{bits}}{\text{cycle}}$$

The results indicate that BPSK is the least efficient and 16-QAM is the most efficient. 16-QAM requires one-fourth as much bandwidth as BPSK for the same input bit rate.

PSK AND QAM SUMMARY

The various forms of FSK, PSK, and QAM are summarized in Table 12-2.

TABLE 12-2 DIGITAL MODULATION SUMMARY

Modulation	Encoding	Bandwidth (Hz)	Baud	Bandwidth efficiency (bps/Hz)
FSK	Single bit	$\geq f_b$	f_b	≤ 1
BPSK	Single bit	f_b	f_b	1
QPSK	Dibit	$f_b/2$	$f_b/2$	2
8-PSK	Tribit	$f_b/3$	$f_b/3$	3
8-QAM	Tribit	$f_b/3$	$f_b/3$	3
16-PSK	Quadbit	$f_b/4$	$f_b/4$	4
16-QAM	Quadbit	$f_b/4$	$f_b/4$	4

Chap. 12   **Digital Communications**

CARRIER RECOVERY

Carrier recovery is the process of extracting a phase-coherent reference carrier from a received signal. This is sometimes called *phase referencing.*

In the phase modulation techniques described thus far, the binary data were encoded as a precise phase of the transmitted carrier. (This is referred to as *absolute phase encoding.*) Depending on the encoding method, the angular separation between adjacent phasors varied between 30 and 180°. To correctly demodulate the data, a phase-coherent carrier was recovered and compared with the received carrier in a product detector. To determine the absolute phase of the received carrier, it is necessary to produce a carrier at the receiver that is phase coherent with the transmit reference oscillator. This is the function of the carrier recovery circuit.

With PSK and QAM, the carrier is suppressed in the balanced modulators and is therefore not transmitted. Consequently, at the receiver the carrier cannot simply be tracked with a standard phase-locked loop. With suppressed carrier systems, such as PSK and QAM, sophisticated methods of carrier recovery are required such as a *squaring loop,* a *Costas loop,* or a *remodulator.*

Squaring Loop

A common method of achieving carrier recovery for BPSK is the *squaring loop.* Figure 12-35 shows the block diagram of a squaring loop. The received BPSK waveform is filtered and then squared. The filtering reduces the spectral width of the received noise. The squaring circuit removes the modulation and generates the second harmonic of the carrier frequency. This harmonic is phase tracked by the PLL. The VCO output frequency from the PLL is then divided by 2 and used as the phase reference for the product detectors.

With BPSK, only two output phases are possible: $+\sin \omega_c t$ and $-\sin \omega_c t$. Mathematically, the operation of the squaring circuit can be described as follows. For a receive signal of $+\sin \omega_c t$ the output of the squaring circuit is

$$\text{output} = (+\sin \omega_c t)(+\sin \omega_c t) = +\sin^2 \omega_c t$$

(filtered out)

$$= \tfrac{1}{2}(1 - \cos 2\omega_c t) = \tfrac{1}{2} - \tfrac{1}{2}\cos 2\omega_c t$$

For a received signal of $-\sin \omega_c t$ the output of the squaring circuit is

$$\text{output} = (-\sin \omega_c t)(-\sin \omega_c t) = +\sin^2 \omega_c t$$

(filtered out)

$$= \tfrac{1}{2}(1 - \cos 2\omega_c t) = \tfrac{1}{2} - \tfrac{1}{2}\cos 2\omega_c t$$

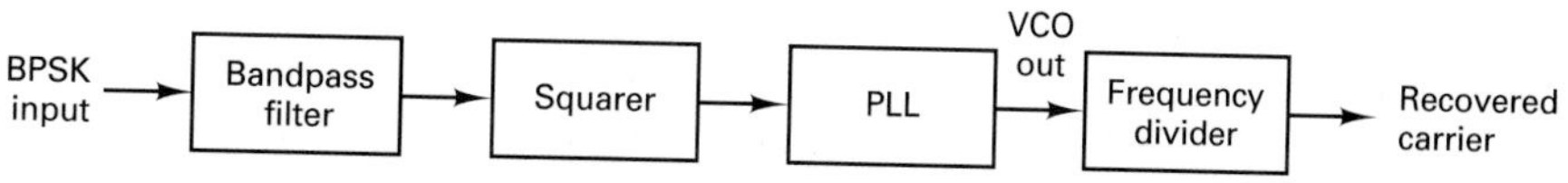

Figure 12-35 Squaring loop carrier recovery circuit for a BPSK receiver.

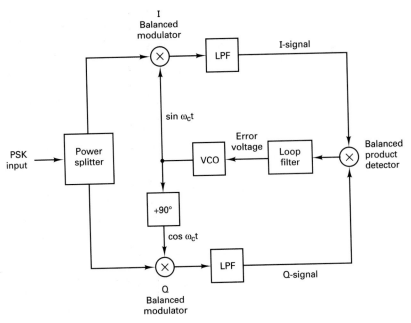

Figure 12-36 Costas loop carrier recovery circuit.

It can be seen that in both cases the output from the squaring circuit contained a constant voltage $(+\frac{1}{2} V)$ and a signal at twice the carrier frequency $(\cos 2\omega_c t)$. The constant voltage is removed by filtering, leaving only $\cos 2\omega_c t$.

Costas Loop

A second method of carrier recovery is the Costas, or quadrature, loop shown in Figure 12-36. The Costas loop produces the same results as a squaring circuit followed by an ordinary PLL in place of the BPF. This recovery scheme uses two parallel tracking loops (I and Q) simultaneously to derive the product of the I and Q components of the signal that drives the VCO. The in-phase (I) loop uses the VCO as in a PLL, and the quadrature (Q) loop uses a 90° shifted VCO signal. Once the frequency of the VCO is equal to the suppressed carrier frequency, the product of the I and Q signals will produce an error voltage proportional to any phase error in the VCO. The error voltage controls the phase and thus the frequency of the VCO.

Remodulator

A third method of achieving recovery of a phase and frequency coherent carrier is the remodulator, shown in Figure 12-37. The remodulator produces a loop error voltage that is proportional to twice the phase error between the incoming signal and the VCO signal. The remodulator has a faster acquisition time than either the squaring or the Costas loops.

Carrier recovery circuits for higher-than-binary encoding techniques are similar to BPSK except that circuits which raise the receive signal to the fourth, eighth, and higher powers are used.

Chap. 12   **Digital Communications**

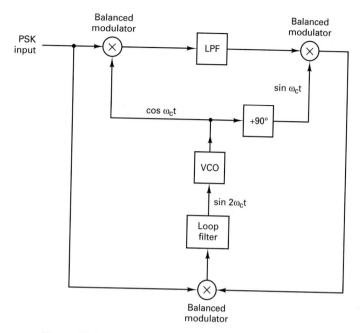

Figure 12-37 Remodulator loop carrier recovery circuit.

DIFFERENTIAL PHASE SHIFT KEYING

Differential phase shift keying (DPSK) is an alternative form of digital modulation where the binary input information is contained in the difference between two successive signaling elements rather than the absolute phase. With DPSK it is not necessary to recover a phase-coherent carrier. Instead, a received signaling element is delayed by one signaling element time slot and then compared to the next received signaling element. The difference in the phase of the two signaling elements determines the logic condition of the data.

DIFFERENTIAL BPSK

DBPSK Transmitter

Figure 12-38a shows a simplified block diagram of a *differential binary phase shift keying* (DBPSK) transmitter. An incoming information bit is XNORed with the preceding bit prior to entering the BPSK modulator (balanced modulator). For the first data bit, there is no preceding bit with which to compare it. Therefore, an initial reference bit is assumed. Figure 12-38b shows the relationship between the input data, the XNOR output data, and the phase at the output of the balanced modulator. If the initial reference bit is assumed a logic 1, the output from the XNOR circuit is simply the complement of that shown.

In Figure 12-38b the first data bit is XNORed with the reference bit. If they are the same, the XNOR output is a logic 1; if they are different, the XNOR output is a logic 0. The balanced modulator operates the same as a conventional BPSK modulator; a logic 1 produces $+\sin \omega_c t$ at the output and a logic 0 produces $-\sin \omega_c t$ at the output.

Differential BPSK

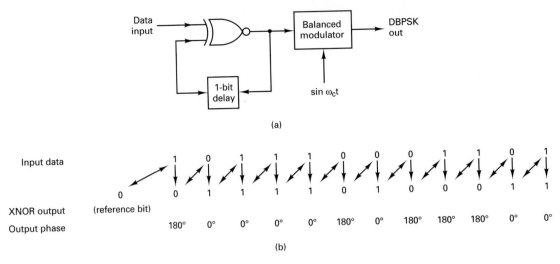

(a)

Input data

XNOR output

Output phase

(b)

Figure 12-38 DBPSK demodulator: (a) block diagram; (b) timing diagram.

DBPSK Receiver

Figure 12-39 shows the block diagram and timing sequence for a DBPSK receiver. The received signal is delayed by one bit time, then compared with the next signaling element in the balanced modulator. If they are the same, a logic 1 (+ voltage) is generated. If they are different, a logic 0 (− voltage) is generated. If the reference phase is incorrectly assumed, only the first demodulated bit is in error. Differential encoding can be implemented with higher-than-binary digital modulation schemes, although the differential algorithms are much more complicated than for DBPSK.

The primary advantage of DPSK is the simplicity with which it can be implemented. With DPSK, no carrier recovery circuit is needed. A disadvantage of DPSK is that it

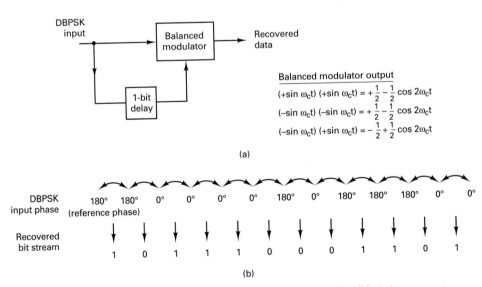

Figure 12-39 DBPSK demodulator: (a) block diagram; (b) timing sequence.

requires between 1 and 3 dB more signal-to-noise ratio to achieve the same bit error rate as that of absolute PSK.

CLOCK RECOVERY

As with any digital system, digital radio requires precise timing or clock synchronization between the transmit and the receive circuitry. Because of this, it is necessary to regenerate clocks at the receiver that are synchronous with those at the transmitter.

Figure 12-40a shows a simple circuit that is commonly used to recover clocking information from the received data. The recovered data are delayed by one-half a bit time and then compared with the original data in an XOR circuit. The frequency of the clock that is recovered with this method is equal to the received data rate (f_b). Figure 12-40b shows the relationship between the data and the recovered clock timing. From Figure 12-40b it can be seen that as long as the receive data contains a substantial number of transitions (1/0 sequences), the recovered clock is maintained. If the receive data were to undergo an extended period of successive 1's or 0's, the recovered clock would be lost. To prevent this from occurring, the data are scrambled at the transmit end and descrambled at the receive end. Scrambling introduces transitions (pulses) into the binary signal using a prescribed algorithm, and the descrambler uses the same algorithm to remove the transitions.

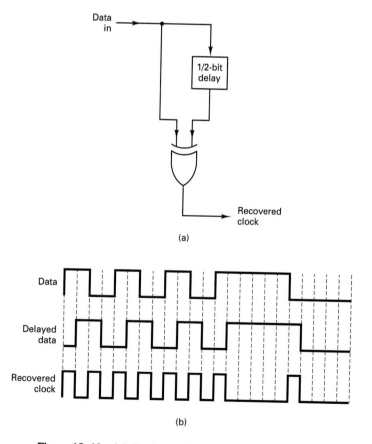

Figure 12-40 (a) Clock recovery circuit; (b) timing diagram.

Clock Recovery

PROBABILITY OF ERROR AND BIT ERROR RATE

Probability of error P(e) and *bit error rate* (BER) are often used interchangeably, although in practice they do have slightly different meanings. *P(e)* is a theoretical (mathematical) expectation of the bit error rate for a given system. BER is an empirical (historical) record of a system's actual bit error performance. For example, if a system has a *P(e)* of 10^{-5}, this means that mathematically, you can expect one bit error in every 100,000 bits transmitted ($1/10^5 = 1/100{,}000$). If a system has a BER of 10^{-5}, this means that in the past there was one bit error for every 100,000 bits transmitted. A bit error rate is measured, then compared to the expected probability of error to evaluate a system's performance.

Probability of error is a function of the *carrier-to-noise power ratio* (or more specifically, the average *energy per bit-to-noise power density ratio*) and the number of possible encoding conditions used (*M*-ary). Carrier-to-noise power ratio is the ratio of the average carrier power (the combined power of the carrier and its associated sidebands) to the *thermal noise power*. Carrier power can be stated in watts or dBm, where

$$C \text{ (dBm)} = 10 \log \frac{C \text{ watts}}{0.001} \tag{12-13}$$

Thermal noise power is expressed mathematically as

$$N = KTB \quad \text{(watts)} \tag{12-14a}$$

where N = thermal noise power (W)
         K = Boltzmann's proportionality constant (1.38×10^{-23} J/K)
         T = temperature (kelvin: 0 kelvin = -273 degrees Celsius, room temperature = 290 K)
         B = bandwidth (Hz)

Stated in dBm,

$$N \text{ (dBm)} = 10 \log \frac{KTB}{0.001} \tag{12-14b}$$

Mathematically, the carrier-to-noise power ratio is

$$\frac{C}{N} = \frac{C}{KTB} \quad \textit{(unitless ratio)} \tag{12-15a}$$

where C = carrier power (W)
         N = noise power (W)

Stated in dB,

$$\frac{C}{N} \text{ (dB)} = 10 \log \frac{C}{N} \tag{12-15b}$$

$$= C \text{ (dBm)} - N \text{ (dBm)}$$

Energy per bit is simply the energy of a single bit of information. Mathematically, energy per bit is

$$E_b = CT_b \quad \text{(J/bit)} \tag{12-16a}$$

where E_b = energy of a single bit (J/bit)
         T_b = time of a single bit (s)

$$C = \text{carrier power (W)}$$

Stated in dBJ,

$$E_b \text{ (dBJ)} = 10 \log E_b \qquad \text{(12-16b)}$$

and because $T_b = 1/f_b$, where f_b is the bit rate in bits per second, E_b can be rewritten as

$$E_b = \frac{C}{f_b} \quad \text{(J/bit)} \qquad \text{(12-16c)}$$

Stated in dBJ,

$$E_b \text{ (dBJ)} = 10 \log \frac{C}{f_b} \qquad \text{(12-16d)}$$

$$= 10 \log C - 10 \log f_b \qquad \text{(12-16e)}$$

Noise power density is the thermal noise power normalized to a 1-Hz bandwidth (i.e., the noise power present in a 1-Hz bandwidth). Mathematically, noise power density is

$$N_0 = \frac{N}{B} \quad \text{(W/Hz)} \qquad \text{(12-17a)}$$

where N_0 = noise power density (W/Hz)
   N = thermal noise power (W)
   B = bandwidth (Hz)

Stated in dBm,

$$N_0 \text{ (dBm)} = 10 \log \frac{N}{0.001} - 10 \log B \qquad \text{(12-17b)}$$

$$= N \text{ (dBm)} - 10 \log B \qquad \text{(12-17c)}$$

Combining Equations 12-14a and 12-17a yields

$$N_0 = \frac{KTB}{B} = KT \quad \text{(W/Hz)} \qquad \text{(12-17d)}$$

Stated in dBm,

$$N_0 \text{ (dBm)} = 10 \log \frac{K}{0.001} + 10 \log T \qquad \text{(12-17e)}$$

Energy per bit-to-noise power density ratio is used to compare two or more digital modulation systems that use different transmission rates (bit rates), modulation schemes (FSK, PSK, QAM), or encoding techniques (*M*-ary). The energy per bit-to-noise power density ratio is simply the ratio of the energy of a single bit to the noise power present in 1 Hz of bandwidth. Thus E_b/N_0 normalizes all multiphase modulation schemes to a common noise bandwidth allowing for a simpler and more accurate comparison of their error performance. Mathematically, E_b/N_0 is

$$\frac{E_b}{N_0} = \frac{C/f_b}{N/B} = \frac{CB}{Nf_b} \qquad \text{(12-18a)}$$

where E_b/N_0 is the energy per bit-to-noise power density ratio. Rearranging Equation 12-18a yields the following expression:

Probability of Error and Bit Error Rate

$$\frac{E_b}{N_0} = \frac{C}{N} \times \frac{B}{f_b} \qquad (12\text{-}18b)$$

where
$\dfrac{E_b}{N_0}$ = energy per bit-to-noise power density ratio

$\dfrac{C}{N}$ = carrier-to-noise power ratio

$\dfrac{B}{f_b}$ = noise bandwidth-to-bit rate ratio

Stated in dB,

$$\frac{E_b}{N_0}\ (\text{dB}) = 10 \log \frac{C}{N} + 10 \log \frac{B}{f_b} \qquad (12\text{-}18c)$$

or

$$= 10 \log E_b - 10 \log N_0 \qquad (12\text{-}18d)$$

From Equation 12-18b it can be seen that the E_b/N_0 ratio is simply the product of the carrier-to-noise power ratio and the noise bandwidth-to-bit rate ratio. Also, from Equation 12-18b, it can be seen that when the bandwidth equals the bit rate, $E_b/N_0 = C/N$.

In general, the minimum carrier-to-noise power ratio required for QAM systems is less than that required for comparable PSK systems. Also, the higher the level of encoding used (the higher the value of M), the higher the minimum carrier-to-noise power ratio. In Chapter 19, several examples are shown for determining the minimum carrier-to-noise power and energy per bit-to-noise power density ratios for a given M-ary system and desired $P(e)$.

EXAMPLE 12-11

For a QPSK system and the given parameters, determine (a) the carrier power in dBm, (b) the noise power in dBm, (c) the noise power density in dBm, (d) the energy per bit in dBJ, (e) the carrier-to-noise power ratio in dB, and (f) the E_b/N_0 ratio.

$$C = 10^{-12}\ \text{W} \qquad f_b = 60\ \text{kbps}$$
$$N = 1.2 \times 10^{-14}\ \text{W} \qquad B = 120\ \text{kHz}$$

Solution (a) The carrier power in dBm is determined by substituting into Equation 12-13.

$$C = 10 \log \frac{10^{-12}}{0.001} = -90\ \text{dBm}$$

(b) The noise power in dBm is determined by substituting into Equation 12-14b.

$$N = 10 \log \frac{1.2 \times 10^{-14}}{0.001} = -109.2\ \text{dBm}$$

(c) The noise power density is determined by substituting into Equation 12-17c.

$$N_0 = -109.2\ \text{dBm} - 10 \log 120\ \text{kHz} = -160\ \text{dBm}$$

(d) The energy per bit is determined by substituting into Equation 12-16d.

$$E_b = 10 \log \frac{10^{-12}}{60\ \text{kbps}} = -167.8\ \text{dBJ}$$

(e) The carrier-to-noise power ratio is determined by substituting into Equation 12-15b.

$$\frac{C}{N} = 10 \log \frac{10^{-12}}{1.2 \times 10^{-14}} = 19.2 \text{ dB}$$

(f) The energy per bit-to-noise density ratio is determined by substituting into Equation 12-18c.

498

$$\frac{E_b}{N_0} = 19.2 + 10 \log \frac{120 \text{ kHz}}{60 \text{ kbps}} = 22.2 \text{ dB}$$

PSK Error Performance

The bit error performance for the various multiphase digital modulation systems is directly related to the distance between points on a signal state-space diagram. For example, on the signal state-space diagram for BPSK shown in Figure 12-41a, it can be seen that the two signal points (logic 1 and logic 0) have maximum separation (d) for a given power level (D). In essence, one BPSK signal state is the exact negative of the other. As the figure shows, a noise vector (V_N), when combined with the signal vector (V_S), effectively shifts the phase of the signaling element (V_{SE}) alpha degrees. If the phase shift exceeds $\pm 90°$, the signal element is shifted beyond the threshold points into the error region. For BPSK, it would require a noise vector of sufficient amplitude and phase to produce more than a $\pm 90°$ phase shift in the signaling element to produce an error. For PSK systems, the general formula for the threshold points is

$$\text{TP} = \pm \frac{\pi}{M} \qquad (12\text{-}19)$$

where M is the number of signal states.

The phase relationship between signaling elements for BPSK (i.e., 180° out of phase) is the optimum signaling format, referred to as *antipodal signaling,* and occurs only when two binary signal levels are allowed and when one signal is the exact negative of the other. Because no other bit-by-bit signaling scheme is any better, antipodal performance is often used as a reference for comparison.

The error performance of the other multiphase PSK systems can be compared to that of BPSK simply by determining the relative decrease in error distance between points on a single state-space diagram. For PSK, the general formula for the maximum distance between signaling points is given by

$$\sin \theta = \sin \frac{360°}{2M} = \frac{d/2}{D} \qquad (12\text{-}20)$$

where d = error distance
  M = number of phases
  D = peak signal amplitude

Rearranging Equation 12-20 and solving for d yields

$$d = \left(2 \sin \frac{180°}{M} \right) \times D \qquad (12\text{-}21)$$

Figure 12-41b shows the signal state-space diagram for QPSK. From Figure 12-41b and Equation 12-2 it can be seen that QPSK can tolerate only a $\pm 45°$ phase shift. From Equation 12-19, the maximum phase shift for 8-PSK and 16-PSK is $\pm 22.5°$ and $\pm 11.25°$, respectively. Consequently, the higher levels of modulation (i.e., the greater the value of M) require a greater energy per bit-to-noise power density ratio to reduce the effect of

Probability of Error and Bit Error Rate

499

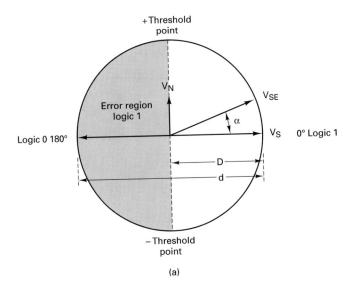

(a)

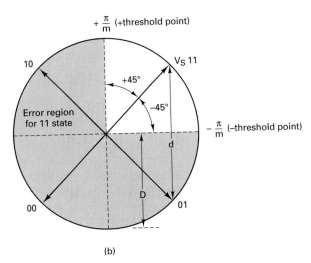

(b)

Figure 12-41 PSK error region: (a) BPSK; (b) QPSK.

noise interference. Hence the higher the level of modulation, the smaller the angular separation between signal points and the smaller the error distance.

The general expression for the bit-error probability of an *M*-phase PSK system is

$$P(e) = \frac{1}{\log_2 M} \, \text{erf} \, (z) \tag{12-22}$$

where   erf = error function

$$z = \sin \frac{\pi}{M} \left(\sqrt{\log_2 M} \right) \left(\sqrt{E_b/N_0} \right)$$

By substituting into Equation 12-22 it can be shown that QPSK provides the same error performance as BPSK. This is because the 3-dB reduction in error distance for QPSK is offset by the 3-dB decrease in its bandwidth (in addition to the error distance, the relative widths of the noise bandwidths must also be considered). Thus both systems provide optimum performance. Figure 12-41 shows the error performance for 2-, 4-, 8-, 16-, and 32-PSK systems as a function of E_b/N_0.

EXAMPLE 12-12

Determine the minimum bandwidth required to achieve a $P(e)$ of 10^{-7} for an 8-PSK system operating at 10 Mbps with a carrier-to-noise power ratio of 11.7 dB.

Solution From Figure 12-42, the minimum E_b/N_0 ratio to achieve a $P(e)$ of 10^{-7} for an 8-PSK system is 14.7 dB. The minimum bandwidth is found by rearranging Equation 12-18b.

$$\frac{B}{F_b} = \frac{E_b}{N_0} - \frac{C}{N}$$

$$= 14.7 \text{ dB} - 11.7 \text{ dB} = 3 \text{ dB}$$

$$\frac{B}{F_b} = \text{antilog } 3 = 2$$

$$B = 2 \times 10 \text{ Mbps} = 20 \text{ MHz}$$

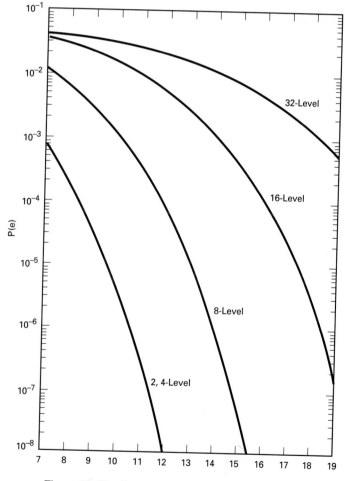

Figure 12-42 Error rates of PSK modulation systems.

Probability of Error and Bit Error Rate

QAM Error Performance

For a large numbers of signal points (i.e., *M*-ary systems greater than 4), QAM outperforms PSK. This is because the distance between signaling points in a PSK system is smaller than the distance between points in a comparable QAM system. The general expression for the distance between adjacent signaling points for a QAM system with *L* levels on each axis is

$$d = \frac{\sqrt{2}}{L-1} \times D \qquad (12\text{-}23)$$

where d = error distance
   L = number of levels on each axis
   D = peak signal amplitude

In comparing Equation 12-21 to Equation 12-23, it can be seen that QAM systems have an advantage over PSK systems with the same peak signal power level.
The general expression for the bit error probability of an *L*-level QAM system is

$$P(e) = \frac{1}{\log_2 L} \left(\frac{L-1}{L} \right) \mathrm{erfc}(z) \qquad (12\text{-}24)$$

where $\mathrm{erfc}(z)$ = complementary error function

$$z = \frac{\sqrt{\log_2 L}}{L-1} \sqrt{\frac{E_b}{N_0}}$$

Figure 12-43 shows the error performance for 4-, 16-, 32-, and 64-QAM systems as a function of E_b/N_0.
Table 12-3 lists the minimum carrier-to-noise power ratios and energy per bit-to-noise power density ratios required for a probability of error of 10^{-6} for several PSK and QAM modulation schemes.

EXAMPLE 12-13

Which system requires the highest E_b/N_0 ratio for a probability of error of 10^{-6}, a four-level QAM system or an 8-PSK system?

Solution From Figure 12-43, the minimum E_b/N_0 ratio required for a four-level QAM system is 10.6 dB. From Figure 12-42, the minimum E_b/N_0 ratio required for an 8-PSK system is 14 dB. Therefore, to achieve a $P(e)$ of 10^{-6}, a four-level QAM system would require 3.4-dB less E_b/N_0 ratio.

FSK Error Performance

The error probability for FSK systems is evaluated in a somewhat different manner than PSK and QAM. There are essentially only two types of FSK systems: noncoherent (asynchronous) and coherent (synchronous). With noncoherent FSK, the transmitter and receiver are not frequency or phase synchronized. With coherent FSK, local receiver reference signals are in frequency and phase lock with the transmitted signals. The probability of error for noncoherent FSK is

$$P(e) = \frac{1}{2} \exp \left(-\frac{E_b}{2N_0} \right) \qquad (12\text{-}25)$$

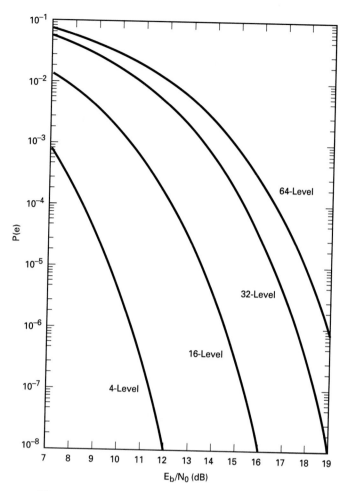

Figure 12-43 Error rates of QAM modulation systems.

The probability of error for coherent FSK is

$$P(e) = \text{erfc} \sqrt{\frac{E_b}{N_0}} \qquad (12\text{-}26)$$

TABLE 12-3 PERFORMANCE
COMPARISON OF VARIOUS DIGITAL
MODULATION SCHEMES (BER = 10^{-6})

Modulation technique	C/N ratio (dB)	E_b/N_0 ratio (dB)
BPSK	10.6	10.6
QPSK	13.6	10.6
4-QAM	13.6	10.6
8-QAM	17.6	10.6
8-PSK	18.5	14
16-PSK	24.3	18.3
16-QAM	20.5	14.5
32-QAM	24.4	17.4
64-QAM	26.6	18.8

Probability of Error and Bit Error Rate

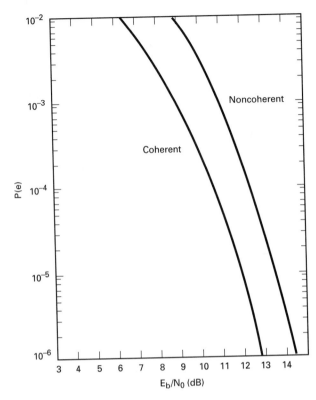

Figure 12-44 Error rates for FSK modulation systems.

Figure 12-44 shows probability of error curves for both coherent and noncoherent FSK for several values of E_b/N_0. From Equations 12-25 and 12-26 it can be determined that the probability of error for noncoherent FSK is greater than that of coherent FSK for equal energy per bit-to-noise power density ratios.

APPLICATIONS FOR DIGITAL MODULATION

A digitally modulated transceiver (*trans*mitter-re*ceiver*) that uses FSK, PSK, or QAM has many applications. They are used in digitally modulated microwave radio and satellite systems (Chapter 19) with carrier frequencies from tens of megahertz to several gigahertz, and they are also used for voice band data modems (Chapter 13) with carrier frequencies between 300 and 3000 Hz.

QUESTIONS

12-1. Explain *digital transmission* and *digital radio*.

12-2. Define *information capacity*.

12-3. What are the three most predominant modulation schemes used in digital radio systems?

12-4. Explain the relationship between bits per second and baud for an FSK system.

12-5. Define the following terms for FSK modulation: frequency deviation, modulation index, and deviation ratio.

Chap. 12   **Digital Communications**

12-6. Explain the relationship between (a) the minimum bandwidth required for an FSK system and the bit rate, and (b) the mark and space frequencies.

12-7. What is the difference between standard FSK and MSK? What is the advantage of MSK?

12-8. Define *PSK*.

12-9. Explain the relationship between bits per second and baud for a BPSK system.

12-10. What is a constellation diagram, and how is it used with PSK?

12-11. Explain the relationship between the minimum bandwidth required for a BPSK system and the bit rate.

12-12. Explain *M-ary*.

12-13. Explain the relationship between bits per second and baud for a QPSK system.

12-14. Explain the significance of the I and Q channels in a QPSK modulator.

12-15. Define *dibit*.

12-16. Explain the relationship between the minimum bandwidth required for a QPSK system and the bit rate.

12-17. What is a coherent demodulator?

12-18. What advantage does OQPSK have over conventional QPSK? What is a disadvantage of OQPSK?

12-19. Explain the relationship between bits per second and baud for an 8-PSK system.

12-20. Define *tribit*.

12-21. Explain the relationship between the minimum bandwidth required for an 8-PSK system and the bit rate.

12-22. Explain the relationship between bits per second and baud for a 16-PSK system.

12-23. Define *quadbit*.

12-24. Define *QAM*.

12-25. Explain the relationship between the minimum bandwidth required for a 16-QAM system and the bit rate.

12-26. What is the difference between PSK and QAM?

12-27. Define *bandwidth efficiency*.

12-28. Define *carrier recovery*.

12-29. Explain the differences between absolute PSK and differential PSK.

12-30. What is the purpose of a clock recovery circuit? When is it used?

12-31. What is the difference between probability of error and bit error rate?

PROBLEMS

12-1. For an FSK modulator with space, rest, and mark frequencies of 40, 50, and 60 MHz, respectively, and an input bit rate of 10 Mbps, determine the output baud and minimum bandwidth. Sketch the output spectrum.

12-2. Determine the minimum bandwidth and baud for a BPSK modulator with a carrier frequency of 40 MHz and an input bit rate of 500 kbps. Sketch the output spectrum.

12-3. For the QPSK modulator shown in Figure 12-14, change the $+90°$ phase-shift network to $-90°$ and sketch the new constellation diagram.

12-4. For the QPSK demodulator shown in Figure 12-18, determine the I and Q bits for an input signal of $\sin \omega_c t - \cos \omega_c t$.

12-5. For an 8-PSK modulator with an input data rate (f_b) equal to 20 Mbps and a carrier frequency of 100 MHz, determine the minimum double-sided Nyquist bandwidth (f_N) and the baud. Sketch the output spectrum.

12-6. For the 8-PSK modulator shown in Figure 12-20, change the reference oscillator to $\cos \omega_c t$ and sketch the new constellation diagram.

12-7. For a 16-QAM modulator with an input bit rate (f_b) equal to 20 Mbps and a carrier frequency of 100 MHz, determine the minimum double-sided Nyquist bandwidth (f_N) and the baud. Sketch the output spectrum.

12-8. For the 16-QAM modulator shown in Figure 12-31, change the reference oscillator to $\cos \omega_c t$ and determine the output expressions for the following I, I', Q, and Q' input conditions: 0000, 1111, 1010, and 0101.

12-9. Determine the bandwidth efficiency for the following modulators.
  (a) QPSK, $f_b = 10$ Mbps
  (b) 8-PSK, $f_b = 21$ Mbps
  (c) 16-QAM, $f_b = 20$ Mbps

12-10. For the DBPSK modulator shown in Figure 12-38, determine the output phase sequence for the following input bit sequence: 00110011010101 (assume that the reference bit = 1).

12-11. For a QPSK system and the given parameters, determine (a) the carrier power in dBm, (b) the noise power in dBm, (c) the noise power density in dBm, (d) the energy per bit in dBj, (e) the carrier-to-noise power ratio, and (f) the E_b/N_0 ratio.

$$C = 10^{-13} \text{ W} \qquad f_b = 30 \text{ kbps}$$
$$N = 0.06 \times 10^{-15} \text{ W} \qquad B = 60 \text{ kH}$$

12-12. Determine the minimum bandwidth required to achieve a $P(e)$ of 10^{-6} for an 8-PSK system operating at 20 Mbps with a carrier-to-noise power ratio of 11 dB.

13
Data
Communications

INTRODUCTION

Data communications is the process of transferring digital *information* (usually in binary form) between two or more points. Information is defined as knowledge or intelligence. Information that has been processed and organized is called *data*. Data can be any alphabetical, numeric, or symbolic information, including binary-coded alpha/numeric symbols, microprocessor op-codes, control codes, user addresses, program data, or data base information. At both the source and the destination, data are in digital form. However, during transmission, data may be in digital or analog form.

A data communications network can be as simple as two personal computers connected together through the public telephone network, or it can comprise a complex network of one or more mainframe computers and hundreds of remote terminals. Data communications networks are used to connect automatic teller machines (ATMs) to bank computers or they can be used to interface computer terminals (CTs) or keyboard displays (KDs) directly to application programs in mainframe computers. Data communications networks are used for airline and hotel reservation systems and for mass media and news networks such as the Associated Press (AP) or United Press International (UPI). The list of applications for data communications networks goes on almost indefinitely.

HISTORY OF DATA COMMUNICATIONS

It is highly likely that data communications began long before recorded time in the form of smoke signals or tom-tom drums, although it is improbable that these signals were binary coded. If we limit the scope of data communications to methods that use electrical

507

signals to transmit binary-coded information, then data communications began in 1837 with the invention of the *telegraph* and the development of the *Morse code* by Samuel F. B. Morse. With telegraph, dots and dashes (analogous to binary 1's and 0's) are transmitted across a wire using electromechanical induction. Various combinations of these dots and dashes were used to represent binary codes for letters, numbers, and punctuation. Actually, the first telegraph was invented in England by Sir Charles Wheatstone and Sir William Cooke, but their contraption required six different wires for a single telegraph line. In 1840, Morse secured an American patent for the telegraph and in 1844 the first telegraph line was established between Baltimore and Washington, D.C. In 1849, the first slow-speed telegraph printer was invented, but it was not until 1860 that high-speed (15 bps) printers were available. In 1850, the Western Union Telegraph Company was formed in Rochester, New York, for the purpose of carrying coded messages from one person to another.

In 1874, Emile Baudot invented a telegraph *multiplexer,* which allowed signals from up to six different telegraph machines to be transmitted simultaneously over a single wire. The telephone was invented in 1876 by Alexander Graham Bell and, consequently, very little new evolved in telegraph until 1899, when Marconi succeeded in sending radio telegraph messages. Telegraph was the only means of sending information across large spans of water until 1920, when the first commercial radio stations were installed.

Bell Laboratories developed the first special-purpose computer in 1940 using electromechanical relays. The first general-purpose computer was an automatic sequence-controlled calculator developed jointly by Harvard University and International Business Machines Corporation (IBM). The UNIVAC computer, built in 1951 by Remington Rand Corporation (now Sperry Rand), was the first mass-produced electronic computer. Since 1951, the number of mainframe computers, small business computers, personal computers, and computer terminals has increased exponentially, creating a situation where more and more people have the need to exchange digital information with each other. Consequently, the need for data communications has also increased exponentially.

Until 1968, the AT&T operating tariff allowed only equipment furnished by AT&T to be connected to AT&T lines. In 1968, a landmark Supreme Court decision, the Carterfone decision, allowed non-Bell companies to interconnect to the vast AT&T communications network. This decision started the *interconnect industry,* which has led to competitive data communications offerings by a large number of independent companies.

STANDARDS ORGANIZATIONS FOR DATA COMMUNICATIONS

During the past decade, the data communications industry has grown at an astronomical rate. Consequently, the need to provide communications between dissimilar computer systems has also increased. Thus, to ensure an orderly transfer of information between two or more data communications systems using different equipment with different needs, a consortium of organizations, manufacturers, and users meet on a regular basis to establish guidelines and standards. It is the intent that all data communications users comply with these standards. Several of the organizations are described below.

International Standards Organization (ISO):   The ISO is the international organization for standardization. The ISO creates the sets of rules and standards for graphics, document exchange, and related technologies. The ISO is responsible for endorsing and coordinating the work of the other standards organizations.

Consultative Committee for International Telephony and Telegraphy (CCITT):
The membership of the CCITT consists of government authorities and representatives
from many countries. The CCITT is now the standards organization for the United
Nations and develops the recommended sets of rules and standards for telephone and telegraph communications. The CCITT has developed three sets of specifications: the V series
for modem interfacing, the X series for data communications, and the I and Q series for
Integrated Services Digital Network (ISDN).

American National Standards Institute (ANSI): ANSI is the official standards agency
for the United States and is the U.S. voting representative for ISO.

Institute of Electrical and Electronics Engineers (IEEE): The IEEE is a U.S. professional organization of electronics, computer, and communications engineers.

Electronic Industries Association (EIA): The EIA is a U.S. organization that establishes and recommends industrial standards. The EIA is responsible for developing the RS
(recommended standard) series of standards for data and telecommunications.

Standards Council of Canada (SCC): The SCC is the official standards agency for
Canada with similar responsibilities to those of ANSI.

DATA COMMUNICATIONS CIRCUITS

Figure 13-1 shows a simplified block diagram of a data communications network. As the
figure shows, there is a source of digital information (primary station), a transmission
medium (facility), and a destination (*secondary* station). The *primary* (or host) location is
very often a mainframe computer with its own set of local terminals and peripheral equipment. For simplicity, there is only one secondary (or remote) station shown on the figure.
The secondary stations are the users of the network. How many secondary stations there are
and how they are interconnected to each other and the host station vary considerably
depending on the system and its applications. There are many different types of transmission media, including free-space radio transmission (terrestrial and satellite microwave),
metallic cable facilities (both digital and analog systems), and optical fiber cables (light
wave propagation).

Data terminal equipment (DTE) is a general term that describes the interface equipment used at the stations to adapt the digital signals from the computers and terminals to a

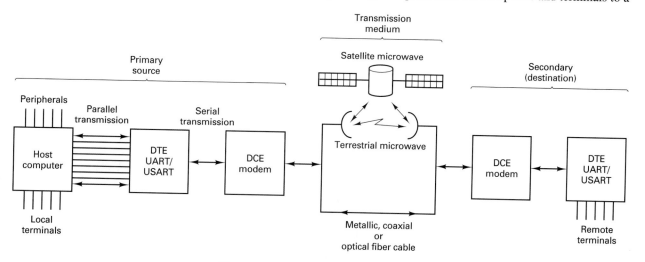

Figure 13-1 Simplified block diagram of a data communications network.

Data Communications Circuits

form more suitable for transmission. Essentially, any piece of equipment between the mainframe computer and the modem or the station equipment and its modem is classified as data terminal equipment. *Data communications equipment* (DCE) is a general term that describes the equipment that converts digital signals to analog signals and interfaces the data terminal equipment to the analog transmission medium. In essence, a DCE is a *modem* (*mod*ulator/*dem*odulator). A modem converts binary digital signals to analog signals such as FSK, PSK, and QAM, and vice versa.

Serial and Parallel Data Transmission

Binary information can be transmitted either in parallel or serially. Figure 13-2a shows how the binary code 0110 is transmitted from location A to location B in parallel. As the figure shows, each bit position (A_0 to A_3) has its own transmission line. Consequently, all 4 bits can be transmitted simultaneously during the time of a single clock pulse (*T*). This type of transmission is called *parallel-by-bit* or *serial-by-character*.

Figure 13-2b shows how the same binary code is transmitted serially. As the figure shows, there is a single transmission line, and thus only one bit can be transmitted at a time. Consequently, it requires four clock pulses (*4T*) to transmit the entire word. This type of transmission is often called *serial-by-bit.*

Obviously, the principal trade-off between parallel and serial transmission is speed versus simplicity. Data transmission can be accomplished much more quickly using parallel transmission. However, parallel transmission requires more lines between the source and destination. As a general rule, parallel transmission is used for short distance communications, and within a computer, and serial transmission is used for long-distance communications.

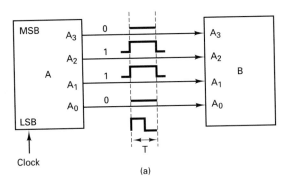

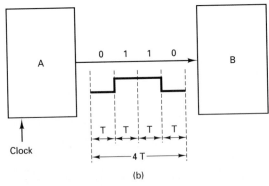

Figure 13-2 Data transmission: (a) parallel; (b) serial.

Chap. 13   Data Communications

Data Communications Circuit Configurations and Topologies

Configurations. Data communications circuits can be generally categorized as either two-point or multipoint. A *two-point* configuration involves only two locations or stations, whereas a *multipoint* configuration involves three or more stations. A two-point circuit can involve the transfer of information between a mainframe computer and a remote computer terminal, two mainframe computers, or two remote computer terminals. A multipoint circuit is generally used to interconnect a single mainframe computer (*host*) to many remote computer terminals, although any combination of three or more computers or computer terminals constitutes a multipoint circuit.

Topologies. The topology or architecture of a data communications circuit identifies how the various locations within the network are interconnected. The most common topologies used are the *point to point*, the *star*, the *bus* or *multidrop*, the *ring* or *loop*, and the *mesh*. These are all multipoint configurations except the point to point. Figure 13-3 shows the various circuit configurations and topologies used for data communications networks.

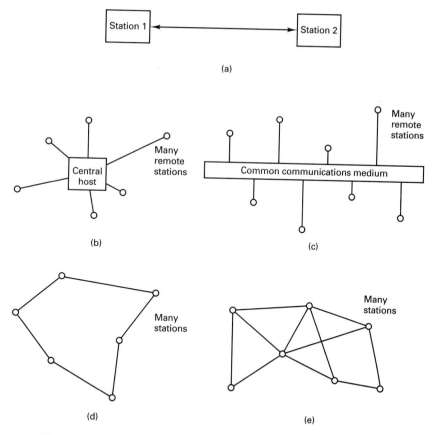

Figure 13-3 Data network topologies: (a) point to point; (b) star; (c) bus or multidrop; (d) ring or loop; (e) mesh.

Data Communications Circuits

Transmission Modes

Essentially, there are four modes of transmission for data communications circuits: *simplex, half duplex, full duplex,* and *full/full duplex.*

Simplex.   With simplex operation, data transmission is unidirectional; information can be sent only in one direction. Simplex lines are also called *receive-only, transmit-only,* or *one-way-only* lines. Commercial television and radio systems are examples of simplex transmission.

Half duplex (HDX).   In the half-duplex mode, data transmission is possible in both directions, but not at the same time. Half-duplex lines are also called two-way alternate or either way lines. Citizens band (CB) radio is an example of half-duplex transmission.

Full duplex (FDX).   In the full-duplex mode, transmissions are possible in both directions simultaneously, but they must be between the same two stations. Full-duplex lines are also called two-way-simultaneous, *duplex,* or both way lines. A standard telephone system is an example of full-duplex transmission.

Full/full duplex (F/FDX).   In the F/FDX mode, transmission is possible in both directions at the same time but not between the same two stations (i.e., one station is transmitting to a second station and receiving from a third station at the same time). F/FDX is possible only on multipoint circuits. The U.S. postal system is an example of full/full duplex transmission.

Two-Wire versus Four-Wire Operation

Two-wire, as the name implies, involves a transmission medium that either uses two wires (a signal and a reference lead) or a configuration that is equivalent to having only two wires. With two-wire operation, simplex, half-, or full-duplex transmission is possible. For full-duplex operation, the signals propagating in opposite directions must occupy different bandwidths; otherwise, they will mix linearly and interfere with each other.

Four-wire, as the name implies, involves a transmission medium that uses four wires (two are used for signals that are propagating in opposite directions and two are used for reference leads) or a configuration that is equivalent to having four wires. With four-wire operation, the signals propagating in opposite directions are physically separated and therefore can occupy the same bandwidths without interfering with each other. Four-wire operation provides more isolation and is preferred over two-wire, although four-wire requires twice as many wires and, consequently, twice the cost.

A transmitter and its associated receiver are equivalent to a two-wire circuit. A transmitter and a receiver for both directions of propagation is equivalent to a four-wire circuit. With full-duplex transmission over a two-wire line, the available bandwidth must be divided in half, thus reducing the information capacity in either direction to one-half of the half-duplex value. Consequently, full-duplex operation over two-wire lines requires twice as much time to transfer the same amount of information.

DATA COMMUNICATIONS CODES

Data communications codes are prescribed bit sequences used for encoding characters and symbols. Consequently, data communications codes are often called *character sets, character codes, symbol codes,* or *character languages.* In essence, there are only three types of characters used in data communications codes: *data link control characters,* which are used to facilitate the orderly flow of data from a source to a destination; *graphic control characters,* which involve the syntax or presentation of the data at the receive terminal; and *alpha/numeric* characters, which are used to represent the various symbols used for letters, numbers, and punctuation in the English language.

The first data communications code that saw widespread usage was the Morse code. The Morse code used three unequal-length symbols (dot, dash, and space) to encode alpha/numeric characters, punctuation marks, and an interrogation word.

The Morse code is inadequate for use in modern digital computer equipment because all characters do not have the same number of symbols or take the same length of time to send, and each Morse code operator transmits code at a different rate. Also, with Morse code, there is an insufficient selection of graphic and data link control characters to facilitate the transmission and presentation of the data typically used in contemporary computer applications.

The three most common character sets presently used for character encoding are the Baudot code, the American Standard Code for Information Interchange (ASCII), and the Extended Binary-Coded Decimal Interchange Code (EBCDIC).

Baudot Code

The *Baudot code* (sometimes called the *Telex code*) was the first fixed-length character code. The Baudot code was developed by a French postal engineer, Thomas Murray, in 1875 and named after Emile Baudot, an early pioneer in telegraph printing. The Baudot code is a 5-bit character code that is used primarily for low-speed teletype equipment such as the TWX/Telex system. With a 5-bit code there are only 2^5 or 32 combinations possible, which is insufficient to represent the 26 letters of the alphabet, the 10 digits, and the various punctuation marks and control characters. Therefore, the Baudot code uses *figure* shift and *letter* shift characters to expand its capabilities to 58 characters. The latest version of the Baudot code is recommended by the CCITT as the International Alphabet No. 2. The Baudot code is still used by Western Union Company for the TWX and Telex teletype systems. The AP and UPI news services for years used the Baudot code for sending news information around the world. The most recent version of the Baudot code is shown in Table 13-1.

ASCII Code

In 1963, in an effort to standardize data communications codes, the United States adopted the Bell System model 33 teletype code as the United States of America Standard Code for Information Interchange (USASCII), better known simply as ASCII-63. Since its adoption, ASCII has generically progressed through the 1965, 1967, and 1977 versions, with the 1977 version being recommended by the CCITT as the International Alphabet No. 5. ASCII is a 7-bit character set which has 2^7 or 128 combinations. With ASCII, the least significant bit (LSB) is designated b_0 and the most significant bit (MSB) is designated b_6. b_7 is not part of the ASCII code but is generally reserved for the parity bit, which is explained later in this chapter. Actually, with any character set, all bits are equally

TABLE 13-1 BAUDOT CODE

Character shift		Binary code					
Letter	*Figure*	*Bit:*	*4*	*3*	*2*	*1*	*0*
A	—		1	1	0	0	0
B	?		1	0	0	1	1
C	:		0	1	1	1	0
D	$		1	0	0	1	0
E	3		1	0	0	0	0
F	!		1	0	1	1	0
G	&		0	1	0	1	1
H	#		0	0	1	0	1
I	8		0	1	1	0	0
J	'		1	1	0	1	0
K	(		1	1	1	1	0
L	)		0	1	0	0	1
M	.		0	0	1	1	1
N	,		0	0	1	1	0
O	9		0	0	0	1	1
P	0		0	1	1	0	1
Q	1		1	1	1	0	1
R	4		0	1	0	1	0
S	bel		1	0	1	0	0
T	5		0	0	0	0	1
U	7		1	1	1	0	0
V	;		0	1	1	1	1
W	2		1	1	0	0	1
X	/		1	0	1	1	1
Y	6		1	0	1	0	1
Z	"		1	0	0	0	1
Figure shift			1	1	1	1	1
Letter shift			1	1	0	1	1
Space			0	0	1	0	0
Line feed (LF)			0	1	0	0	0
Blank (null)			0	0	0	0	0

significant because the code does not represent a weighted binary number. It is common with character codes to refer to bits by their order; b_0 is the zero-order bit, b_1 is the first-order bit, b_7 is the seventh-order bit, and so on. With serial transmission, the bit transmitted first is called the LSB. With ASCII, the low-order bit (b_0) is the LSB and is transmitted first. ASCII is probably the code most often used today. The 1977 version of the ASCII code is shown in Table 13-2.

TABLE 13-2 ASCII-77 CODE—ODD PARITY

	Binary code								Hex		Binary code								Hex
Bit:	*7*	*6*	*5*	*4*	*3*	*2*	*1*	*0*		*Bit:*	*7*	*6*	*5*	*4*	*3*	*2*	*1*	*0*	*Hex*
NUL	1	0	0	0	0	0	0	0	00	@	0	1	0	0	0	0	0	0	40
SOH	0	0	0	0	0	0	0	1	01	A	1	1	0	0	0	0	0	1	41
STX	0	0	0	0	0	0	1	0	02	B	1	1	0	0	0	0	1	0	42
ETX	1	0	0	0	0	0	1	1	03	C	0	1	0	0	0	0	1	1	43
EOT	0	0	0	0	0	1	0	0	04	D	1	1	0	0	0	1	0	0	44
ENQ	1	0	0	0	0	1	0	1	05	E	0	1	0	0	0	1	0	1	45

TABLE 13-2 ASCII-77 CODE—ODD PARITY (CONTINUED)

Bit:	7	6	5	4	3	2	1	0	Hex	Bit:	7	6	5	4	3	2	1	0	Hex
				Binary code										Binary code					
ACK	1	0	0	0	0	1	1	0	06	F	0	1	0	0	0	1	1	0	46
BEL	0	0	0	0	0	1	1	1	07	G	1	1	0	0	0	1	1	1	47
BS	0	0	0	0	1	0	0	0	08	H	1	1	0	0	1	0	0	0	48
HT	1	0	0	0	1	0	0	1	09	I	0	1	0	0	1	0	0	1	49
NL	1	0	0	0	1	0	1	0	0A	J	0	1	0	0	1	0	1	0	4A
VT	0	0	0	0	1	0	1	1	0B	K	1	1	0	0	1	0	1	1	4B
FF	1	0	0	0	1	1	0	0	0C	L	0	1	0	0	1	1	0	0	4C
CR	0	0	0	0	1	1	0	1	0D	M	1	1	0	0	1	1	0	1	4D
SO	0	0	0	0	1	1	1	0	0E	N	1	1	0	0	1	1	1	0	4E
SI	1	0	0	0	1	1	1	1	0F	O	0	1	0	0	1	1	1	1	4F
DLE	0	0	0	1	0	0	0	0	10	P	1	1	0	1	0	0	0	0	50
DC1	0	0	0	1	0	0	0	1	11	Q	0	1	0	1	0	0	0	1	51
DC2	1	0	0	1	0	0	1	0	12	R	0	1	0	1	0	0	1	0	52
DC3	0	0	0	1	0	0	1	1	13	S	1	1	0	1	0	0	1	1	53
DC4	1	0	0	1	0	1	0	0	14	T	0	1	0	1	0	1	0	0	54
NAK	0	0	0	1	0	1	0	1	15	U	1	1	0	1	0	1	0	1	55
SYN	0	0	0	1	0	1	1	0	16	V	1	1	0	1	0	1	1	0	56
ETB	1	0	0	1	0	1	1	1	17	W	0	1	0	1	0	1	1	1	57
CAN	1	0	0	1	1	0	0	0	18	X	0	1	0	1	1	0	0	0	58
EM	0	0	0	1	1	0	0	1	19	Y	1	1	0	1	1	0	0	1	59
SUB	0	0	0	1	1	0	1	0	1A	Z	1	1	0	1	1	0	1	0	5A
ESC	1	0	0	1	1	0	1	1	1B	[	0	1	0	1	1	0	1	1	5B
FS	0	0	0	1	1	1	0	0	1C	\	1	1	0	1	1	1	0	0	5C
GS	1	0	0	1	1	1	0	1	1D	]	0	1	0	1	1	1	0	1	5D
RS	1	0	0	1	1	1	1	0	1E	∧	0	1	0	1	1	1	1	0	5E
US	0	0	0	1	1	1	1	1	1F	-	1	1	0	1	1	1	1	1	5F
SP	0	0	1	0	0	0	0	0	20	`	1	1	1	0	0	0	0	0	60
!	1	0	1	0	0	0	0	1	21	a	0	1	1	0	0	0	0	1	61
"	1	0	1	0	0	0	1	0	22	b	0	1	1	0	0	0	1	0	62
#	0	0	1	0	0	0	1	1	23	c	1	1	1	0	0	0	1	1	63
$	1	0	1	0	0	1	0	0	24	d	0	1	1	0	0	1	0	0	64
%	0	0	1	0	0	1	0	1	25	e	1	1	1	0	0	1	0	1	65
&	0	0	1	0	0	1	1	0	26	f	1	1	1	0	0	1	1	0	66
'	1	0	1	0	0	1	1	1	27	g	0	1	1	0	0	1	1	1	67
(	1	0	1	0	1	0	0	0	28	h	0	1	1	0	1	0	0	0	68
)	0	0	1	0	1	0	0	1	29	i	1	1	1	0	1	0	0	1	69
*	0	0	1	0	1	0	1	0	2A	j	1	1	1	0	1	0	1	0	6A
+	1	0	1	0	1	0	1	1	2B	k	0	1	1	0	1	0	1	1	6B
,	0	0	1	0	1	1	0	0	2C	l	1	1	1	0	1	1	0	0	6C
-	1	0	1	0	1	1	0	1	2D	m	0	1	1	0	1	1	0	1	6d
.	1	0	1	0	1	1	1	0	2E	n	0	1	1	0	1	1	1	0	6E
/	0	0	1	0	1	1	1	1	2F	o	1	1	1	0	1	1	1	1	6F
0	1	0	1	1	0	0	0	0	30	p	0	1	1	1	0	0	0	0	70
1	0	0	1	1	0	0	0	1	31	q	1	1	1	1	0	0	0	1	71
2	0	0	1	1	0	0	1	0	32	r	1	1	1	1	0	0	1	0	72
3	1	0	1	1	0	0	1	1	33	s	0	1	1	1	0	0	1	1	73
4	0	0	1	1	0	1	0	0	34	t	1	1	1	1	0	1	0	0	74
5	1	0	1	1	0	1	0	1	35	u	0	1	1	1	0	1	0	1	75
6	1	0	1	1	0	1	1	0	36	v	0	1	1	1	0	1	1	0	76
7	0	0	1	1	0	1	1	1	37	w	1	1	1	1	0	1	1	1	77
8	0	0	1	1	1	0	0	0	38	x	1	1	1	1	1	0	0	0	78
9	1	0	1	1	1	0	0	1	39	y	0	1	1	1	1	0	0	1	79
:	1	0	1	1	1	0	1	0	3A	z	0	1	1	1	1	0	1	0	7A

Continued

Data Communications Codes

TABLE 13-2 ASCII-77 CODE—ODD PARITY (CONTINUED)

	Binary code								Hex	Bit:	Binary code								Hex
Bit:	7	6	5	4	3	2	1	0	Hex	Bit:	7	6	5	4	3	2	1	0	Hex
;	0	0	1	1	1	0	1	1	3B	{	1	1	1	1	1	0	1	1	7B
<	1	0	1	1	1	1	0	0	3C	|	0	1	1	1	1	1	0	0	7C
=	0	0	1	1	1	1	0	1	3D	}	1	1	1	1	1	1	0	1	7D
>	0	0	1	1	1	1	1	0	3E	~	1	1	1	1	1	1	1	0	7E
?	1	0	1	1	1	1	1	1	3F	DEL	0	1	1	1	1	1	1	1	7F

NUL = null
SOH = start of heading
STX = start of text
ETX = end of text
EOT = end of transmission
ENQ = enquiry
ACK = acknowledge
BEL = bell
BS = back space
HT = horizontal tab
NL = new line
VT = vertical tab

FF = form feed
CR = carriage return
SO = shift-out
SI = shift-in
DLE = data link escape
DC1 = device control 1
DC2 = device control 2
DC3 = device control 3
DC4 = device control 4
NAK = negative acknowledge
SYN = synchronous

ETB = end of transmission block
CAN = cancel
SUB = substitute
ESC = escape
FS = field separator
GS = group separator
RS = record separator
US = unit separator
SP = space
DEL = delete

EBCDIC Code

EBCDIC is an 8-bit character code developed by IBM and used extensively in IBM and IBM-compatible equipment. With 8 bits, 2^8 or 256 combinations are possible, making EBCDIC the most powerful character set. Note that with EBCDIC the LSB is designated b_7 and the MSB is designated b_0. Therefore, with EBCDIC, the high-order bit (b_7) is transmitted first and the low-order bit (b_0) is transmitted last. The EBCDIC code does not facilitate the use of a parity bit. The EBCDIC code is shown in Table 13-3.

TABLE 13-3 EBCDIC CODE

Bit:	Binary code								Hex	Bit:	Binary code								Hex
	0	1	2	3	4	5	6	7	Hex		0	1	2	3	4	5	6	7	Hex
NUL	0	0	0	0	0	0	0	0	00		1	0	0	0	0	0	0	0	80
SOH	0	0	0	0	0	0	0	1	01	a	1	0	0	0	0	0	0	1	81
STX	0	0	0	0	0	0	1	0	02	b	1	0	0	0	0	0	1	0	82
ETX	0	0	0	0	0	0	1	1	03	c	1	0	0	0	0	0	1	1	83
	0	0	0	0	0	1	0	0	04	d	1	0	0	0	0	1	0	0	84
PT	0	0	0	0	0	1	0	1	05	e	1	0	0	0	0	1	0	1	85
	0	0	0	0	0	1	1	0	06	f	1	0	0	0	0	1	1	0	86
	0	0	0	0	0	1	1	1	07	g	1	0	0	0	0	1	1	1	87
	0	0	0	0	1	0	0	0	08	h	1	0	0	0	1	0	0	0	88
	0	0	0	0	1	0	0	1	09	i	1	0	0	0	1	0	0	1	89
	0	0	0	0	1	0	1	0	0A		1	0	0	0	1	0	1	0	8A
	0	0	0	0	1	0	1	1	0B		1	0	0	0	1	0	1	1	8B
FF	0	0	0	0	1	1	0	0	0C		1	0	0	0	1	1	0	0	8C
	0	0	0	0	1	1	0	1	0D		1	0	0	0	1	1	0	1	8D
	0	0	0	0	1	1	1	0	0E		1	0	0	0	1	1	1	0	8E
	0	0	0	0	1	1	1	1	0F		1	0	0	0	1	1	1	1	8F

TABLE 13-3 EBCDIC CODE (CONTINUED)

Bit:	0	1	2	3	4	5	6	7	Hex	Bit:	0	1	2	3	4	5	6	7	Hex
DLE	0	0	0	1	0	0	0	0	10		1	0	0	1	0	0	0	0	90
SBA	0	0	0	1	0	0	0	1	11	j	1	0	0	1	0	0	0	1	91
EUA	0	0	0	1	0	0	1	0	12	k	1	0	0	1	0	0	1	0	92
IC	0	0	0	1	0	0	1	1	13	l	1	0	0	1	0	0	1	1	93
	0	0	0	1	0	1	0	0	14	m	1	0	0	1	0	1	0	0	94
NL	0	0	0	1	0	1	0	1	15	n	1	0	0	1	0	1	0	1	95
	0	0	0	1	0	1	1	0	16	o	1	0	0	1	0	1	1	0	96
	0	0	0	1	0	1	1	1	17	p	1	0	0	1	0	1	1	1	97
	0	0	0	1	1	0	0	0	18	q	1	0	0	1	1	0	0	0	98
EM	0	0	0	1	1	0	0	1	19	r	1	0	0	1	1	0	0	1	99
	0	0	0	1	1	0	1	0	1A		1	0	0	1	1	0	1	0	9A
	0	0	0	1	1	0	1	1	1B		1	0	0	1	1	0	1	1	9B
DUP	0	0	0	1	1	1	0	0	1C		1	0	0	1	1	1	0	0	9C
SF	0	0	0	1	1	1	0	1	1D		1	0	0	1	1	1	0	1	9D
FM	0	0	0	1	1	1	1	0	1E		1	0	0	1	1	1	1	0	9E
ITB	0	0	0	1	1	1	1	1	1F		1	0	0	1	1	1	1	1	9F
	0	0	1	0	0	0	0	0	20		1	0	1	0	0	0	0	0	A0
	0	0	1	0	0	0	0	1	21	~	1	0	1	0	0	0	0	1	A1
	0	0	1	0	0	0	1	0	22	s	1	0	1	0	0	0	1	0	A2
	0	0	1	0	0	0	1	1	23	t	1	0	1	0	0	0	1	1	A3
	0	0	1	0	0	1	0	0	24	u	1	0	1	0	0	1	0	0	A4
	0	0	1	0	0	1	0	1	25	v	1	0	1	0	0	1	0	1	A5
ETB	0	0	1	0	0	1	1	0	26	w	1	0	1	0	0	1	1	0	A6
ESC	0	0	1	0	0	1	1	1	27	x	1	0	1	0	0	1	1	1	A7
	0	0	1	0	1	0	0	0	28	y	1	0	1	0	1	0	0	0	A8
	0	0	1	0	1	0	0	1	29	z	1	0	1	0	1	0	0	1	A9
	0	0	1	0	1	0	1	0	2A		1	0	1	0	1	0	1	0	AA
	0	0	1	0	1	0	1	1	2B		1	0	1	0	1	0	1	1	AB
	0	0	1	0	1	1	0	0	2C		1	0	1	0	1	1	0	0	AC
ENQ	0	0	1	0	1	1	0	1	2D		1	0	1	0	1	1	0	1	AD
	0	0	1	0	1	1	1	0	2E		1	0	1	0	1	1	1	0	AE
	0	0	1	0	1	1	1	1	2F		1	0	1	0	1	1	1	1	AF
	0	0	1	1	0	0	0	0	30		1	0	1	1	0	0	0	0	B0
	0	0	1	1	0	0	0	1	31		1	0	1	1	0	0	0	1	B1
SYN	0	0	1	1	0	0	1	0	32		1	0	1	1	0	0	1	0	B2
	0	0	1	1	0	0	1	1	33		1	0	1	1	0	0	1	1	B3
	0	0	1	1	0	1	0	0	34		1	0	1	1	0	1	0	0	B4
	0	0	1	1	0	1	0	1	35		1	0	1	1	0	1	0	1	B5
	0	0	1	1	0	1	1	0	36		1	0	1	1	0	1	1	0	B6
EOT	0	0	1	1	0	1	1	1	37		1	0	1	1	0	1	1	1	B7
	0	0	1	1	1	0	0	0	38		1	0	1	1	1	0	0	0	B8
	0	0	1	1	1	0	0	1	39		1	0	1	1	1	0	0	1	B9
	0	0	1	1	1	0	1	0	3A		1	0	1	1	1	0	1	0	BA
	0	0	1	1	1	0	1	1	3B		1	0	1	1	1	0	1	1	BB
RA	0	0	1	1	1	1	0	0	3C		1	0	1	1	1	1	0	0	BC
NAK	0	0	1	1	1	1	0	1	3D		1	0	1	1	1	1	0	1	BD
	0	0	1	1	1	1	1	0	3E		1	0	1	1	1	1	1	0	BE
SUB	0	0	1	1	1	1	1	1	3F		1	0	1	1	1	1	1	1	BF
SP	0	1	0	0	0	0	0	0	40	{	1	1	0	0	0	0	0	0	C0
	0	1	0	0	0	0	0	1	41	A	1	1	0	0	0	0	0	1	C1
	0	1	0	0	0	0	1	0	42	B	1	1	0	0	0	0	1	0	C2
	0	1	0	0	0	0	1	1	43	C	1	1	0	0	0	0	1	1	C3
	0	1	0	0	0	1	0	0	44	D	1	1	0	0	0	1	0	0	C4

Continued

TABLE 13-3 EBCDIC CODE (CONTINUED)

	0	1	2	3	4	5	6	7	Hex		0	1	2	3	4	5	6	7	Hex
	0	1	0	0	0	1	0	1	45	E	1	1	0	0	0	1	0	1	C5
	0	1	0	0	0	1	1	0	46	F	1	1	0	0	0	1	1	0	C6
	0	1	0	0	0	1	1	1	47	G	1	1	0	0	0	1	1	1	C7
	0	1	0	0	1	0	0	0	48	H	1	1	0	0	1	0	0	0	C8
	0	1	0	0	1	0	0	1	49	I	1	1	0	0	1	0	0	1	C9
¢	0	1	0	0	1	0	1	0	4A		1	1	0	0	1	0	1	0	CA
.	0	1	0	0	1	0	1	1	4B		1	1	0	0	1	0	1	1	CB
<	0	1	0	0	1	1	0	0	4C		1	1	0	0	1	1	0	0	CC
(	0	1	0	0	1	1	0	1	4D		1	1	0	0	1	1	0	1	CD
+	0	1	0	0	1	1	1	0	4E		1	1	0	0	1	1	1	0	CE
¦	0	1	0	0	1	1	1	1	4F		1	1	0	0	1	1	1	1	CF
&	0	1	0	1	0	0	0	0	50	}	1	1	0	1	0	0	0	0	D0
	0	1	0	1	0	0	0	1	51	J	1	1	0	1	0	0	0	1	D1
	0	1	0	1	0	0	1	0	52	K	1	1	0	1	0	0	1	0	D2
	0	1	0	1	0	0	1	1	53	L	1	1	0	1	0	0	1	1	D3
	0	1	0	1	0	1	0	0	54	M	1	1	0	1	0	1	0	0	D4
	0	1	0	1	0	1	0	1	55	N	1	1	0	1	0	1	0	1	D5
	0	1	0	1	0	1	1	0	56	O	1	1	0	1	0	1	1	0	D6
	0	1	0	1	0	1	1	1	57	P	1	1	0	1	0	1	1	1	D7
	0	1	0	1	1	0	0	0	58	Q	1	1	0	1	1	0	0	0	D8
	0	1	0	1	1	0	0	1	59	R	1	1	0	1	1	0	0	1	D9
!	0	1	0	1	1	0	1	0	5A		1	1	0	1	1	0	1	0	DA
$	0	1	0	1	1	0	1	1	5B		1	1	0	1	1	0	1	1	DB
*	0	1	0	1	1	1	0	0	5C		1	1	0	1	1	1	0	0	DC
)	0	1	0	1	1	1	0	1	5D		1	1	0	1	1	1	0	1	DD
;	0	1	0	1	1	1	1	0	5E		1	1	0	1	1	1	1	0	DE
¬	0	1	0	1	1	1	1	1	5F		1	1	0	1	1	1	1	1	DF
-	0	1	1	0	0	0	0	0	60	\	1	1	1	0	0	0	0	0	E0
/	0	1	1	0	0	0	0	1	61		1	1	1	0	0	0	0	1	E1
	0	1	1	0	0	0	1	0	62	S	1	1	1	0	0	0	1	0	E2
	0	1	1	0	0	0	1	1	63	T	1	1	1	0	0	0	1	1	E3
	0	1	1	0	0	1	0	0	64	U	1	1	1	0	0	1	0	0	E4
	0	1	1	0	0	1	0	1	65	V	1	1	1	0	0	1	0	1	E5
	0	1	1	0	0	1	1	0	66	W	1	1	1	0	0	1	1	0	E6
	0	1	1	0	0	1	1	1	67	X	1	1	1	0	0	1	1	1	E7
	0	1	1	0	1	0	0	0	68	Y	1	1	1	0	1	0	0	0	E8
	0	1	1	0	1	0	0	1	69	Z	1	1	1	0	1	0	0	1	E9
	0	1	1	0	1	0	1	0	6A		1	1	1	0	1	0	1	0	EA
,	0	1	1	0	1	0	1	1	6B		1	1	1	0	1	0	1	1	EB
%	0	1	1	0	1	1	0	0	6C		1	1	1	0	1	1	0	0	EC
	0	1	1	0	1	1	0	1	6D		1	1	1	0	1	1	0	1	ED
>	0	1	1	0	1	1	1	0	6E		1	1	1	0	1	1	1	0	EE
?	0	1	1	0	1	1	1	1	6F		1	1	1	0	1	1	1	1	EF
	0	1	1	1	0	0	0	0	70	0	1	1	1	1	0	0	0	0	F0
	0	1	1	1	0	0	0	1	71	1	1	1	1	1	0	0	0	1	F1
	0	1	1	1	0	0	1	0	72	2	1	1	1	1	0	0	1	0	F2
	0	1	1	1	0	0	1	1	73	3	1	1	1	1	0	0	1	1	F3
	0	1	1	1	0	1	0	0	74	4	1	1	1	1	0	1	0	0	F4
	0	1	1	1	0	1	0	1	75	5	1	1	1	1	0	1	0	1	F5
	0	1	1	1	0	1	1	0	76	6	1	1	1	1	0	1	1	0	F6
	0	1	1	1	0	1	1	1	77	7	1	1	1	1	0	1	1	1	F7
	0	1	1	1	1	0	0	0	78	8	1	1	1	1	1	0	0	0	F8
▲	0	1	1	1	1	0	0	1	79	9	1	1	1	1	1	0	0	1	F9

Continued

TABLE 13-3 EBCDIC CODE (CONTINUED)

Bit:	0	1	2	3	4	5	6	7	Hex	Bit:	0	1	2	3	4	5	6	7	Hex
				Binary code										Binary code					
:	0	1	1	1	1	0	1	0	7A		1	1	1	1	1	0	1	0	FA
#	0	1	1	1	1	0	1	1	7B		1	1	1	1	1	0	1	1	FB
@	0	1	1	1	1	1	0	0	7C		1	1	1	1	1	1	0	0	FC
,	0	1	1	1	1	1	0	1	7D		1	1	1	1	1	1	0	1	FD
=	0	1	1	1	1	1	1	0	7E		1	1	1	1	1	1	1	0	FE
"	0	1	1	1	1	1	1	1	7F		1	1	1	1	1	1	1	1	FF

DLE = data link escape
DUP = duplicate
EM = end of medium
ENQ = enquiry
EOT = end of transmission
ESC = escape
ETB = end of transmission
    block
ETX = end of text

EUA = erase unprotected
    to address
FF = form feed
FM = field mark
IC = insert cursor
ITB = end of intermediate
    transmission block
NUL = nul
PT = program tab

RA = repeat to address
SBA = set buffer address
SF = start field
SOH = start of heading
SP = space
STX = start of text
SUB = substitute
SYN = synchronous
NAK = negative acknowledge

ERROR CONTROL

A data communications circuit can be as short as a few feet or as long as several thousand miles, and the transmission medium can be as simple as a piece of wire or as complex as a microwave, satellite, or optical fiber system. Therefore, due to the nonideal transmission characteristics that are associated with any communications system, it is inevitable that errors will occur and that it is necessary to develop and implement procedures for error control. Error control can be divided into two general categories: error detection and error correction.

Error Detection

Error detection is simply the process of monitoring the received data and determining when a transmission error has occurred. Error detection techniques do not identify which bit (or bits) is in error, only that an error has occurred. The purpose of error detection is not to prevent errors from occurring but to prevent undetected errors from occurring. How a system reacts to transmission errors is system dependent and varies considerably. The most common error detection techniques used for data communications circuits are: redundancy, exact-count encoding, parity, vertical and longitudinal redundancy checking, and cyclic redundancy checking.

Redundancy. *Redundancy* involves transmitting each character twice. If the same character is not received twice in succession, a transmission error has occurred. The same concept can be used for messages. If the same sequence of characters is not received twice in succession, in exactly the same order, a transmission error has occurred.

Exact-count encoding. With *exact-count encoding,* the number of 1's in each character is the same. An example of an exact-count encoding scheme is the ARQ code

Error Control

TABLE 13-4 ARQ EXACT-COUNT CODE

	Binary code							Character	
Bit:	1	2	3	4	5	6	7	Letter	Figure
	0	0	0	1	1	1	0	Letter shift	
	0	1	0	0	1	1	0	Figure shift	
	0	0	1	1	0	1	0	A	-
	0	0	1	1	0	0	1	B	?
	1	0	0	1	1	0	0	C	:
	0	0	1	1	1	0	0	D	(WRU)
	0	1	1	1	0	0	0	E	3
	0	0	1	0	0	1	1	F	%
	1	1	0	0	0	0	1	G	@
	1	0	1	0	0	1	0	H	£
	1	1	1	0	0	0	0	I	8
	0	1	0	0	0	1	1	J	(bell)
	0	0	0	1	0	1	1	K	(
	1	1	0	0	0	1	0	L	)
	1	0	1	0	0	0	1	M	.
	1	0	1	0	1	0	0	N	,
	1	0	0	0	1	1	0	O	9
	1	0	0	1	0	1	0	P	0
	0	0	0	1	1	0	1	Q	1
	1	1	0	0	1	0	0	R	4
	0	1	0	1	0	1	0	S	'
	1	0	0	0	1	0	1	T	5
	0	1	1	0	0	1	0	U	7
	1	0	0	1	0	0	1	V	=
	0	1	0	0	1	0	1	W	2
	0	0	1	0	1	1	0	X	/
	0	0	1	0	1	0	1	Y	6
	0	1	1	0	0	0	1	Z	+
	0	0	0	0	1	1	1	(blank)	
	1	1	0	1	0	0	0	(space)	
	1	0	1	1	0	0	0	(line feed)	
	1	0	0	0	0	1	1	(carriage return)	

shown in Table 13-4. With the ARQ code, each character has three 1's in it, and therefore a simple count of the number of 1's received in each character can determine if a transmission error has occurred.

Parity. *Parity* is probably the simplest error detection scheme used for data communications systems and is used with both vertical and horizontal redundancy checking. With parity, a single bit (called a *parity bit*) is added to each character to force the total number of 1's in the character, including the parity bit, to be either an odd number (odd parity) or an even number (even parity). For example, the ASCII code for the letter "C" is 43 hex or P1000011 binary, with the P bit representing the parity bit. There are three 1's in the code, not counting the parity bit. If odd parity is used, the P bit is made a 0, keeping the total number of 1's at three, an odd number. If even parity is used, the P bit is made a 1 and the total number of 1's is four, an even number.

Taking a closer look at parity, it can be seen that the parity bit is independent of the number of 0's in the code and unaffected by pairs of 1's. For the letter "C," if all the 0 bits

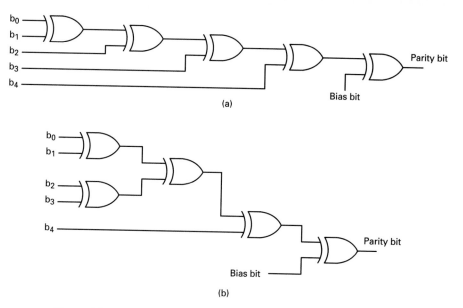

Figure 13-4 Parity generators: (a) serial; (b) parallel. 1, Odd parity; 2, even parity.

are dropped, the code is P1———11. For odd parity, the P bit is still a 0 and for even parity, the P bit is still a 1. If pairs of 1's are also excluded, the code is either P1———, P———1, or P———1—. Again, for odd parity the P bit is a 0, and for even parity the P bit is a 1.

The definition of parity is *equivalence or equality*. A logic gate that will determine when all its inputs are equal is the XOR gate. With an XOR gate, if all the inputs are equal (either all 0's or all 1's), the output is a 0. If all inputs are not equal, the output is a 1. Figure 13-4 shows two circuits that are commonly used to generate a parity bit. Essentially, both circuits go through a comparison process eliminating 0's and pairs of 1's. The circuit shown in Figure 13-4a uses *sequential (serial)* comparison, while the circuit shown in Figure 13-4(b) uses *combinational (parallel)* comparison. With the sequential parity generator b_0 is XORed with b_1, the result is XORed with b_2, and so on. The result of the last XOR operation is compared with a *bias* bit. If even parity is desired, the bias bit is made a logic 0. If odd parity is desired, the bias bit is made a logic 1. The output of the circuit is the parity bit, which is appended to the character code. With the parallel parity generator, comparisons are made in layers or levels. Pairs of bits (b_0 and b_1, b_2 and b_3, etc.) are XORed. The results of the first-level XOR gates are then XORed together. The process continues until only one bit is left, which is XORed with the bias bit. Again, if even parity is desired, the bias bit is made a logic 0 and if odd parity is desired, the bias bit is made a logic 1.

The circuits shown in Figure 13-4 can also be used for the parity checker in the receiver. A parity checker uses the same procedure as a parity generator except that the logic condition of the final comparison is used to determine if a parity violation has occurred (for odd parity a 1 indicates an error and a 0 indicates no error; for even parity, a 1 indicates an error and a 0 indicates no error).

The primary advantage of parity is its simplicity. The disadvantage is that when an even number of bits are received in error, the parity checker will not detect it (i.e., if the logic conditions of 2 bits are changed, the parity remains the same). Consequently, parity, over a long period of time, will detect only 50% of the transmission errors (this assumes an equal probability that an even or an odd number of bits could be in error).

Error Control

Vertical and horizontal redundancy checking. *Vertical redundancy checking* (VRC) is an error detection scheme that uses parity to determine if a transmission error has occurred within a character. Therefore, VRC is sometimes called *character parity*. With VRC, each character has a parity bit added to it prior to transmission. It may use even or odd parity. The example shown under the topic "parity" involving the ASCII character "C" is an example of how VRC is used.

Horizontal or longitudinal redundancy checking (HRC or LRC) is an error detection scheme that uses parity to determine if a transmission error has occurred in a message and is therefore sometimes called *message parity*. With LRC, each bit position has a parity bit. In other words, b_0 from each character in the message is XORed with b_0 from all of the other characters in the message. Similarly, b_1, b_2, and so on, are XORed with their respective bits from all the other characters in the message. Essentially, LRC is the result of XORing the "characters" that make up a message, whereas VRC is the XORing of the bits within a single character. With LRC, only even parity is used.

The LRC bit sequence is computed in the transmitter prior to sending the data, then transmitted as though it were the last character of the message. At the receiver, the LRC is recomputed from the data and the recomputed LRC is compared with the LRC transmitted with the message. If they are the same, it is assumed that no transmission errors have occurred. If they are different, a transmission error must have occurred.

Example 13-1 shows how VRC and LRC are determined.

EXAMPLE 13-1

Determine the VRC and LRC for the following ASCII-encoded message: THE CAT. Use odd parity for VRC and even parity for LRC.

Solution

Character		T	H	E	sp	C	A	T	LRC
Hex		*54*	*48*	*45*	*20*	*43*	*41*	*54*	*2F*
LSB	b_0	0	0	1	0	1	1	0	1
	b_1	0	0	0	0	1	0	0	1
ASCII	b_2	1	0	1	0	0	0	1	1
code	b_3	0	1	0	0	0	0	0	1
	b_4	1	0	0	0	0	0	1	0
	b_5	0	0	0	1	0	0	0	1
MSB	b_6	1	1	1	0	1	1	1	0
VRC	b_7	0	1	0	0	0	1	0	0

The LRC is 2FH or 00101111 binary. In ASCII, this is the character /.

The VRC bit for each character is computed in the vertical direction, and the LRC bits are computed in the horizontal direction. This is the same scheme that was used with early teletype paper tapes and keypunch cards and has subsequently been carried over to present-day data communications applications.

The group of characters that make up the message (i.e., THE CAT) is often called a *block* of data. Therefore, the bit sequence for the LRC is often called a *block check character* (BCC) or a *block check sequence* (BCS). BCS is more appropriate because the LRC has no function as a character (i.e., it is not an alpha/numeric, graphic, or data link control character); the LRC is simply a *sequence of bits* used for error detection.

Historically, LRC detects between 95 and 98% of all transmission errors. LRC will not detect transmission errors when an even number of characters have an error in the same bit position. For example, if b_4 in two different characters is in error, the LRC is still valid even though multiple transmission errors have occurred.

If VRC and LRC are used simultaneously, the only time an error would go undetected is when an even number of bits in an even number of characters were in error and the same bit positions in each character are in error, which is highly unlikely to happen. VRC does not identify which bit is in error in a character, and LRC does not identify which character has an error in it. However, for single bit errors, VRC used together with LRC will identify which bit is in error. Otherwise, VRC and LRC only identify that an error has occurred.

Cyclic redundancy checking. Probably the most reliable scheme for error detection is *cyclic redundancy checking* (CRC). With CRC, approximately 99.95% of all transmission errors are detected. CRC is generally used with 8-bit codes such as EBCDIC or 7-bit codes when parity is not used.

In the United States, the most common CRC code is CRC-16, which is identical to the international standard, CCITT V.41. With CRC-16, 16 bits are used for the BCS. Essentially, the CRC character is the remainder of a division process. A data message polynomial $G(x)$ is divided by a generator polynomial function $P(x)$, the quotient is discarded, and the remainder is truncated to 16 bits and added to the message as the BCS. With CRC generation, the division is not accomplished with a standard arithmetic division process. Instead of using straight subtraction, the remainder is derived from an XOR operation. At the receiver, the data stream and the BCS are divided by the same generating function $P(x)$. If no transmission errors have occurred, the remainder will be zero.

The generating polynomial for CRC-16 is

$$P(x) = x^{16} + x^{12} + x^5 + x^0$$

where $x^0 = 1$.

The number of bits in the CRC code is equal to the highest exponent of the generating polynomial. The exponents identify the bit positions that contain a 1. Therefore, b_{16}, b_{12}, b_5, and b_0 are 1's and all of the other bit positions are 0's.

Figure 13-5 shows the block diagram for a circuit that will generate a CRC-16 BCS for the CCITT V.41 standard. Note that for each bit position of the generating polynomial where there is a 1 an XOR gate is placed except for x^0.

EXAMPLE 13-2

Determine the BSC for the following data and CRC generating polynomials:

$$\text{data } G(x) = x^7 + x^5 + x^4 + x^2 + x^1 + x^0 \quad \text{or} \quad 10110111$$

$$\text{CRC } P(x) = x^5 + x^4 + x^1 + x^0 \quad \text{or} \quad 110011$$

Solution First $G(x)$ is multiplied by the number of bits in the CRC code, 5.

$$x^5(x^7 + x^5 + x^4 + x^2 + x^1 + x^0) = x^{12} + x^{10} + x^9 + x^7 + x^6 + x^5$$

$$= 1011011100000$$

Error Control

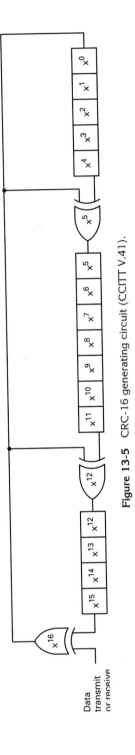

Figure 13-5 CRC-16 generating circuit (CCITT V.41).

Then divide the result by $P(x)$.

```
                          11010111
              110011 ⟌ 1011011100000
                       110011
                        111101
                        110011
                         111010
                         110011
                          100100
                          110011
                           101110
                           110011
                            111010
                            110011
                             01001 = CRC
```

The CRC is appended to the data to give the following transmitted data stream:

```
        G(x)              CRC
      10110111           01001
```

At the receiver, the transmitted data are again divided by $P(x)$.

```
                          11010111
              110011 ⟌ 1011011101001
                       110011
                        111101
                        110011
                         111010
                         110011
                          100110
                          110011
                           101010
                           110011
                            110011
                            110011
                             000000  remainder = 0
                                     no error occurred
```

Error Correction

Essentially, there are three methods of error correction: symbol substitution, retransmission, and forward error correction.

Symbol substitution. *Symbol substitution* was designed to be used in a human environment: when there is a human being at the receive terminal to analyze the received data and make decisions on its integrity. With symbol substitution, if a character is received in error, rather than revert to a higher level of error correction or display the incorrect character, a unique character that is undefined by the character code, such as a reverse question mark (⸮), is substituted for the bad character. If the character in error cannot be discerned by the operator, retransmission is called for (i.e., symbol substitution is a form of selective retransmission). For example, if the message "Name" had an error in the first character, it would be displayed as "? ame." An operator can discern the correct message by inspection and retransmission is unnecessary. However, if the message "$? ,000.00" were received, an operator could not determine the correct character, and retransmission is required.

Retransmission. *Retransmission,* as the name implies, is resending a message when it is received in error and the receive terminal automatically calls for retransmission of the entire message. Retransmission is often called ARQ, which is an old radio communications term that means *automatic request for retransmission.* ARQ is probably the most reliable method of error correction, although it is not always the most efficient. Impairments on transmission media occur in bursts. If short messages are used, the likelihood that an impairment will occur during a transmission is small. However, short messages require more acknowledgments and line turnarounds than do long messages. Acknowledgments and line turnarounds for error control are forms of *overhead* (characters other than data that must be transmitted). With long messages, less turnaround time is needed, although the likelihood that a transmission error will occur is higher than for short messages. It can be shown statistically that message blocks between 256 and 512 characters are of optimum size when using ARQ for error correction.

Forward error correction. *Forward error correction* (FEC) is the only error correction scheme that actually detects and corrects transmission errors at the receive end without calling for retransmission.

With FEC, bits are added to the message prior to transmission. A popular error-correcting code is the *Hamming code,* developed by R. W. Hamming at Bell Laboratories. The number of bits in the Hamming code is dependent on the number of bits in the data character. The number of Hamming bits that must be added to a character is determined from the following expression:

$$2^n \geq m + n + 1 \tag{13-1}$$

where n = number of Hamming bits
m = number of bits in the data character

EXAMPLE 13-3

For a 12-bit data string of 101100010010, determine the number of Hamming bits required, arbitrarily place the Hamming bits into the data string, determine the condition of each Hamming bit, assume an arbitrary single-bit transmission error, and prove that the Hamming code will detect the error.

Solution   Substituting into Equation 13-1, the number of Hamming bits is

$$2^n \geq m + n + 1$$

for $n = 4$:

$$2^4 = 16 \geq m + n + 1 = 12 + 4 + 1 = 17$$

$16 < 17$; therefore, 4 Hamming bits are insufficient.
For $n = 5$:

$$2^5 = 32 \geq m + n + 1 = 12 + 5 + 1 = 18$$

$32 > 18$; therefore, 5 Hamming bits are sufficient to meet the criterion of Equation 13-1. Therefore, a total of $12 + 5 = 17$ bits make up the data stream.

Arbitrarily place 5 Hamming bits into the data stream:

```
17 16 15 14 13 12 11 10 9 8 7 6 5 4 3 2 1
H  1  0  1  H  1  0  0  H H 0 1 0 H 0 1 0
```

To determine the logic condition of the Hamming bits, express all bit positions that contain a 1 as a 5-bit binary number and XOR them together.

Bit position	Binary number
2	00010
6	00110
XOR	00100
12	01100
XOR	01000
14	01110
XOR	00110
16	10000
XOR	10110 = Hamming code

$$b_{17} = 1, \quad b_{13} = 0, \quad b_9 = 1, \quad b_8 = 1, \quad b_4 = 0$$

The 17-bit encoded data stream becomes

```
H        H       H H       H
1 1 0 1 0 1 0 0 1 1 0 1 0 0 0 1 0
```

Assume that during transmission, an error occurs in bit position 14. The received data stream is

```
1 1 0 0 0 1 0 0 1 1 0 1 0 0 0 1 0
```

At the receiver to determine the bit position in error, extract the Hamming bits and XOR them with the binary code for each data bit position that contains a 1.

Bit position	Binary number
Hamming code	10110
2	00010
XOR	10100
6	00110
XOR	10010
12	01100
XOR	11110
16	10000
XOR	01110 = binary 14

Bit position 14 was received in error. To fix the error, simply complement bit 14.

The Hamming code described here will detect only single-bit errors. It cannot be used to identify multiple-bit errors or errors in the Hamming bits themselves. The Hamming code, like all FEC codes, requires the addition of bits to the data, consequently lengthening the transmitted message. The purpose of FEC codes is to reduce or eliminate the wasted time of retransmissions. However, the addition of the FEC bits to each message wastes transmission time in itself. Obviously, a trade-off is made between ARQ and FEC and system requirements determine which method is best suited to a particular system. FEC is often used for simplex transmissions to many receivers when acknowledgements are impractical.

Error Control

SYNCHRONIZATION

Synchronize means to coincide or agree in time. In data communications, there are four types of synchronization that must be achieved: bit or clock synchronization, modem or carrier synchronization, character synchronization, and message synchronization. The clock and carrier recovery circuits discussed in Chapter 12 accomplish bit and carrier synchronization, and message synchronization is discussed in Chapter 14.

Character Synchronization

Clock synchronization ensures that the transmitter and receiver agree on a precise time slot for the occurrence of a bit. When a continuous string of data is received, it is necessary to identify which bits belong to which characters and which bit is the least significant data bit, the parity bit, and the stop bit. In essence, this is character synchronization: identifying the beginning and the end of a character code. In data communications circuits, there are two formats used to achieve character synchronization: asynchronous and synchronous.

Asynchronous data format. With *asynchronous data*, each character is framed between a *start* and a *stop* bit. Figure 13-6 shows the format used to frame a character for asynchronous data transmission. The first bit transmitted is the start bit and is always a logic 0. The character code bits are transmitted next beginning with the LSB and continuing through the MSB. The parity bit (if used) is transmitted directly after the MSB of the character. The last bit transmitted is the stop bit, which is always a logic 1. There can be either 1, 1.5, or 2 stop bits.

A logic 0 is used for the start bit because an idle condition (no data transmission) on a data communications circuit is identified by the transmission of continuous 1's (these are often called *idle line 1's*). Therefore, the start bit of the first character is identified by a high-to-low transition in the received data, and the bit that immediately follows the start bit is the LSB of the character code. All stop bits are logic 1's, which guarantees a high-to-low transition at the beginning of each character. After the start bit is detected, the data and parity bits are clocked into the receiver. If data are transmitted in real time (i.e., as an operator types data into their computer terminal), the number of idle line 1's between each character will vary. During this *dead time,* the receiver will simply wait for the occurrence of another start bit before clocking in the next character.

EXAMPLE 13-4

For the following string of asynchronous ASCII-encoded data, identify each character (assume even parity and 2 stop bits).

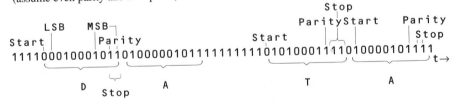

Figure 13-6 Asynchronous data format.

Synchronous data format. With *synchronous data,* rather than frame each character independently with start and stop bits, a unique synchronizing character called a SYN character is transmitted at the beginning of each message. For example with ASCII code, the SYN character is 16H. The receiver disregards incoming data until it receives the SYN character, then it clocks in the next 8 bits and interprets them as a character. The character that is used to signify the end of a transmission varies with the type of protocol used and what kind of transmission it is. Message-terminating characters are discussed in Chapter 14.

With asynchronous data, it is not necessary that the transmit and receive clocks be continuously synchronized. It is only necessary that they operate at approximately the same rate and be synchronized at the beginning of each character. This was the purpose of the start bit, to establish a time reference for character synchronization. With synchronous data, the transmit and receive clocks must be synchronized because character synchronization occurs only once at the beginning of the message.

EXAMPLE 13-5

For the following string of synchronous ASCII-encoded data, identity each character (assume odd parity).

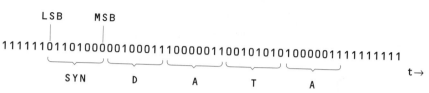

With asynchronous data, each character has 2 or 3 bits added to each character (1 start and 1 or 2 stop bits). These bits are additional overhead and thus reduce the efficiency of the transmission (i.e., the ratio of information bits to total transmitted bits). Synchronous data have two SYN characters (16 bits of overhead) added to each message. Therefore, asynchronous data are more efficient for short messages, and synchronous data are more efficient for long messages.

DATA COMMUNICATIONS HARDWARE

Figure 13-7 shows the block diagram of a multipoint data communications circuit that uses a bus topology. This arrangement is one of the most common configurations used for data communications circuits. At one station there is a mainframe computer and at each of the other two stations there is a *cluster* of computer terminals. The hardware and associated circuitry that connect the host computer to the remote computer terminals is called a *data communications link.* The station with the mainframe is called the *host* or *primary* and the other stations are called *secondaries* or simply *remotes.* An arrangement such as this is called a *centralized network;* there is one centrally located station (the host) with the responsibility of ensuring an orderly flow of data between the remote stations and itself. Data flow is controlled by an applications program which is stored at the primary station.

At the primary station there is a mainframe computer, a *line control unit* (LCU), and a *data modem* (a data modem is commonly referred to simply as a *modem*). At each secondary station there is a modem, an LCU, and terminal equipment, such as computer terminals, printers, and so on. The mainframe is the host of the network and is where the applications program is stored for each circuit it serves. For simplicity, Figure 13-7 shows

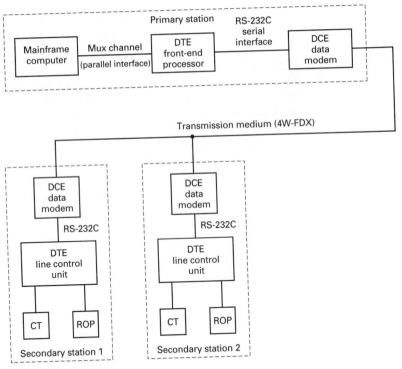

Figure 13-7 Multipoint data communications circuit block diagram.

only one circuit served by the primary, although there can be many different circuits served by one mainframe computer. The primary station has the capability of storing, processing, or retransmitting the data it receives from the secondary stations. The primary also stores software for data base management.

The LCU at the primary station is more complicated than the LCUs at the secondary stations. The LCU at the primary station directs data traffic to and from many different circuits, which could all have different characteristics (i.e., different bit rates, character codes, data formats, etc.). The LCU at a secondary station directs data traffic between one data link and a few terminal devices which all operate at the same speed and use the same character code. Generally speaking, if the LCU has software associated with it, it is called a *front-end processor* (FEP). The LCU at the primary station is usually an FEP.

Line Control Unit

The LCU has several important functions. The LCU at the primary station serves as an interface between the host computer and the circuits that it serves. Each circuit served is connected to a different port on the LCU. The LCU directs the flow of input and output data between the different data communications links and their respective applications program. The LCU performs parallel-to-serial and serial-to-parallel conversion of data. The mux interface channel between the mainframe computer and the LCU transfers data in parallel. Data transfers between the modem and the LCU are done serially. The LCU also houses the circuitry that performs error detection and correction. Also, data link control (DLC) characters are inserted and deleted in the LCU. Data link control characters are explained in Chapter 14.

Chap. 13   Data Communications

The LCU operates on the data when it is in digital form and is therefore called *data terminal equipment* (DTE). Within the LCU, there is a single integrated circuit that performs several of the LCU's functions. This circuit is called a UART when asynchronous transmission is used and a USRT when synchronous transmission is used.

Universal asynchronous receiver/transmitter (UART). The UART is used for asynchronous transmission of data between the DTE and the DCE. Asynchronous transmission means that an asynchronous data format is used and there is no clocking information transferred between the DTE and the DCE. The primary functions of the UART are:

1. To perform serial-to-parallel and parallel-to-serial conversion of data
2. To perform error detection by inserting and checking parity bits
3. To insert and detect start and stop bits

Functionally, the UART is divided into two sections: the transmitter and the receiver. Figure 13-8a shows a simplified block diagram of a UART transmitter.

Prior to transferring data in either direction, a *control* word must be programmed into the UART control register to indicate the nature of the data, such as the number of data bits; if parity is used, and if so, whether it is even or odd; and the number of stop bits. Essentially, the start bit is the only bit that is not optional; there is always only one start bit and it must be a logic 0. Figure 13-8b shows how to program the control word for the various functions. In the UART, the control word is used to set up the data-, parity-, and stop-bit steering logic circuit.

UART transmitter. The operation of the UART transmitter section is really quite simple. The UART sends a transmit buffer empty (TBMT) signal to the DTE to indicate that it is ready to receive data. When the DTE senses an active condition on TBMT, it sends a parallel data character to the transmit data lines (TD_0–TD_7) and strobes them into the transmit buffer register with the transmit data strobe signal ($\overline{TDS}$). The contents of the transmit buffer register are transferred to the transit shift register when the transmit-end-of-character (TEOC) signal goes active (the TEOC signal simply tells the buffer register when the shift register is empty and available to receive data). The data pass through the steering logic circuit, where they pick up the appropriate start, stop, and parity bits. After data have been loaded into the transmit shift register, they are serially outputted on the transmit serial output (TSO) pin with a bit rate equal to the transmit clock (TCP) frequency. While the data in the transmit shift register are sequentially clocked out, the DTE loads the next character into the buffer register. The process continues until the DTE has transferred all its data. The preceding sequence is shown in Figure 13-9.

UART receiver. A simplified block diagram of a UART receiver is shown in Figure 13-10. The number of stop bits, data bits, and the parity-bit information for the UART receiver are determined by the same control word that is used by the transmitter (i.e., the type of parity, the number of stop bits, and the number of data bits used for the UART receiver must be the same as that used for the UART transmitter).

The UART receiver ignores idle line 1's. When a valid start bit is detected by the start bit verification circuit, the data character is serially clocked into the receive shift register. If parity is used, the parity bit is checked in the parity check circuit. After one complete data character is loaded into the shift register, the character is transferred in parallel into the buffer register and the receive data available (RDA) flag is set in the status word register. To read the status register, the DTE monitors status word enable ($\overline{SWE}$) and if it is active, reads the character from the buffer register by placing an active condition on the

Data Communications Hardware

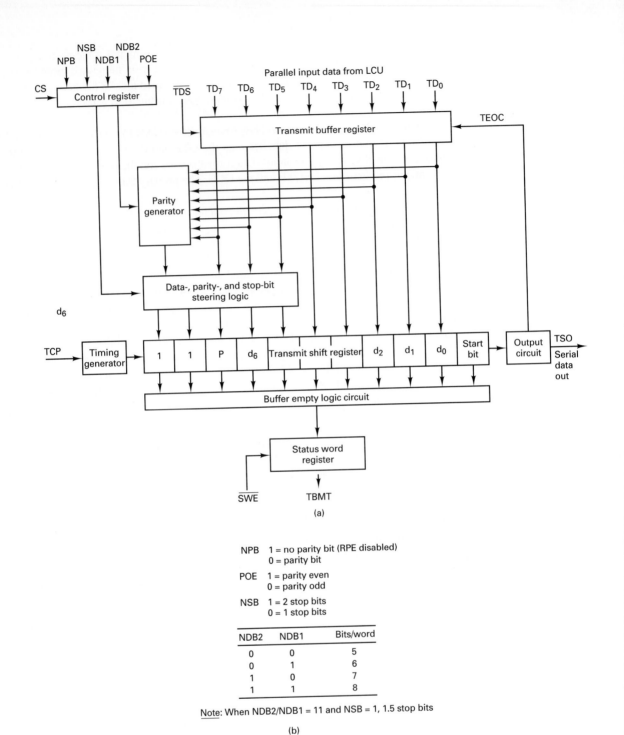

NPB   1 = no parity bit (RPE disabled)
         0 = parity bit

POE   1 = parity even
         0 = parity odd

NSB   1 = 2 stop bits
         0 = 1 stop bits

NDB2	NDB1	Bits/word
0	0	5
0	1	6
1	0	7
1	1	8

Note: When NDB2/NDB1 = 11 and NSB = 1, 1.5 stop bits

(b)

Figure 13-8   UART transmitter: (a) simplified block diagram; (b) control word.

receive data enable (RDE) pin. After reading the data, the DTE places an active signal on the receive data available reset ($\overline{\text{RDAR}}$) pin, which resets the RDA pin. Meanwhile, the next character is received and clocked into the receive shift register and the process

           Chap. 13   Data Communications

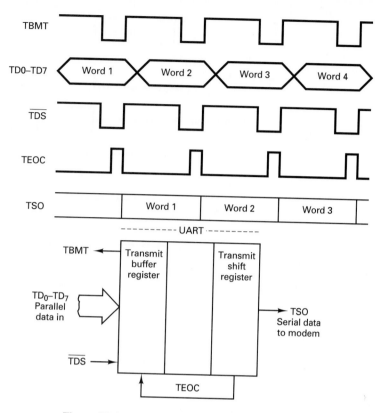

Figure 13-9 Timing diagram: UART transmitter.

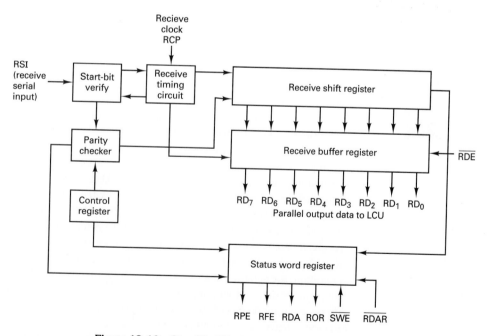

Figure 13-10 Simplified block diagram of a UART receiver.

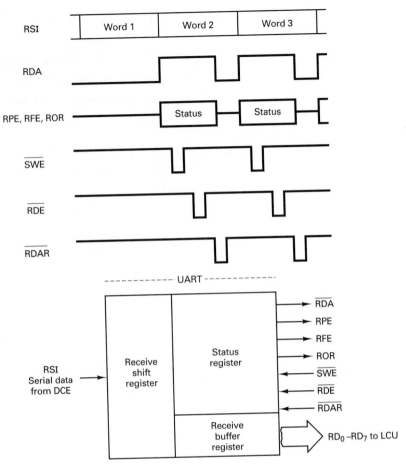

Figure 13-11 Timing diagram: UART receiver.

is repeated until all the data have been received. The preceding sequence is shown in Figure 13-11.

The status word register is also used for diagnostic information. The receive parity error (RPE) flag is set when a received character has a parity error in it. The receive framing error (RFE) flag is set when a character is received without any or an improper number of stop bits. The receive overrun (ROR) flag is set when a character in the buffer register is written over with another character (i.e., the DTE failed to service an active condition on RDA before the next character was received by the shift register).

The receive clock for the UART (RCP) is 16 times higher than the receive data rate. This allows the start-bit verification circuit to determine if a high-to-low transition in the received data is actually a valid start bit and not simply a negative-going noise spike. Figure 13-12 shows how this is accomplished. The incoming idle line 1's (continuous high condition) are sampled at a rate 16 times the actual bit rate. This assures that a high-to-low transition is detected within $\frac{1}{16}$ of a bit time after it occurs. Once a low is detected, the verification circuit counts off seven clock pulses, then resamples the data. If it is still low, it is assumed that a valid start bit has been detected. If it has reverted to the high condition, it is assumed that the high-to-low transition was simply a noise pulse and is therefore ignored. Once a valid start bit has been detected and verified, the verification circuit samples the

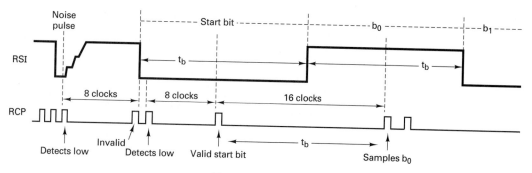

Figure 13-12 Start-bit verification.

incoming data once every 16 clock cycles, which is equal to the data rate. Sampling at 16 times the bit rate also establishes the sample time to within $\frac{1}{16}$ of a bit time from the center of a bit.

Universal synchronous receiver/transmitter (USRT). The USRT is used for synchronous data transmission between the DTE and the DCE. Synchronous transmission means that there is clocking information transferred between the USRT and the modem and each transmission begins with a unique SYN character. The primary functions of the USRT are:

1. To perform serial-to-parallel and parallel-to-serial conversion of data
2. To perform error detection by inserting and checking parity bits
3. To insert and detect SYN characters

The block diagram of the USRT is shown in Figure 13-13a. The USRT operates very similarly to the UART, and therefore only the differences are explained. With the USRT, start and stop bits are not allowed. Instead, unique SYN characters are loaded into the transmit and receive SYN registers prior to transferring data. The programming information for the control word is shown in Figure 13-13.

USRT transmitter. The transmit clock signal (TCP) is set at the desired bit rate and the desired SYN character is loaded from the parallel input pins (DB$_0$–DB$_7$) into the transmit SYN register by pulsing transmit SYN strobe (TSS). Data are loaded into the transmit data register from DB$_0$–DB$_7$ by pulsing the transmit data strobe (TDS). The next character transmitted is extracted from the transmit data register provided that the TDS pulse occurs during the presently transmitted character. If TDS is not pulsed, the next transmitted character is extracted from the transmit SYN register and the SYN character transmitted (SCT) signal is set. The transmit buffer empty (TBMT) signal is used to request the next character from the DTE. The serial output data appears on the transmit serial output (TSO) pin.

USRT receiver. The receive clock signal (RCP) is set at the desired bit rate and the desired SYN character is loaded into the receive SYN register from DB$_0$–DB$_7$ by pulsing receive SYN strobe (RSS). On a high-to-low transition of the receiver rest input (RR), the receiver is placed in the search (bit phase) mode. In the search mode, serially received data are examined on a bit-by-bit basis until a SYN character is found. After each bit is clocked into the receive shift register, its contents are compared to the contents of the receive SYN register. If they are identical, a SYN character has been found and the SYN character receive (SCR) output is set. This character is transferred into the receive buffer register

Data Communications Hardware

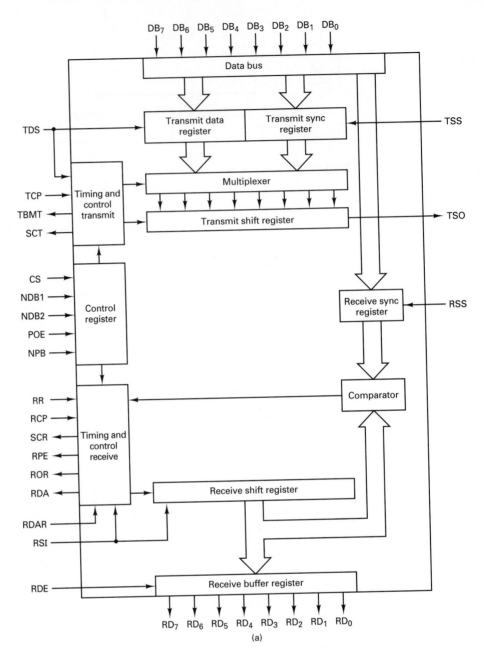

NPB 1 = no parity bit (RPE disabled)
     0 = parity bit

POE 1 = parity even
     0 = parity odd

NDB2	NDB1	Bits/word
0	0	5
0	1	6
1	0	7
1	1	8

(b)

Figure 13-13 USRT transceiver: (a) block diagram; (b) control word.

Chap. 13   **Data Communications**

and the receiver is placed into the character mode. In the character mode, receive data are examined on a character-by-character basis and receiver flags for receive data available (RDA), receiver overrun (ROR), receive parity error (RPE), and SYN character received are provided to the status word register. Parallel receive data are outputted to the DTE on RB_0–RB_7.

SERIAL INTERFACES

To ensure an orderly flow of data between the line control unit and the modem, a *serial interface* is placed between them. This interface coordinates the flow of data, control signals, and timing information between the DTE and the DCE.

Before serial interfaces were standardized, every company that manufactured data communications equipment used a different interface configuration. More specifically, the cabling arrangement between the DTE and the DCE, the type and size of the connectors used, and the voltage levels varied considerably from vender to vender. To interconnect equipment manufactured by different companies, special level converters, cables, and connectors had to be built. The Electronic Industries Association (EIA), in an effort to standardize interface equipment between the data terminal equipment and data communications equipment, agreed on a set of standards which are called the RS-232C specifications. The RS-232C specifications identify the mechanical, electrical, and functional description for the interface between the DTE and the DCE. The RS-232C interface is similar to the combined CCITT standards V.28 (electrical specifications) and V.24 (functional description) and is designed for serial transmission of data up to 20,000 bps for a distance of approximately 50 ft. The EIA has adopted a new set of standards called the RS-449A, which when used in conjunction with the RS-422A or RS-423A standard, can operate at data rates up to 10 Mbps and span distances up to 1200 m.

RS-232C Interface

The RS-232C interface specifies a 25-wire cable with a DB25P/DB25S-compatible connector. Figure 13-14 shows the electrical characteristics of the RS-232C interface. The terminal load capacitance of the cable is specified as 2500 pF, which includes cable

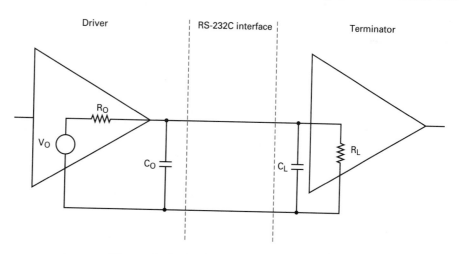

Figure 13-14   RS-232C electrical specifications.

capacitance. The impedance at the terminating end must be between 3000 and 7000 Ω, and the output impedance is specified as greater than 300 Ω. With these electrical specifications and for a maximum bit rate of 20,000 bps, the nominal maximum length of the RS-232C interface is approximately 50 ft.

Although the RS-232C interface is simply a cable and two connectors, the standard also specifies limitations on the voltage levels that the DTE and DCE can output onto or receive from the cable. In both the DTE and DCE, there are circuits that convert their internal logic levels to RS-232C values. For example, a DTE uses TTL logic and is interfaced to a DCE which uses ECL logic; they are not compatible. Voltage-leveling circuits convert the internal voltage values of the DTE and DCE to RS-232C values. If both the DCE and DTE output and input RS-232C levels, they are electrically compatible regardless of which logic family they use internally. A leveler is called a *driver* if it outputs a signal voltage to the cable and a *terminator* if it accepts a signal voltage from the cable. Table 13-5 lists the voltage limits for both drivers and terminators. Note that the data lines use negative logic and the control lines use positive logic.

From Table 13-5 it can be seen that the limits for a driver are more inclusive than those for a terminator. The driver can output any voltage between +5 and +15 or −5 and −15 V dc, and a terminator will accept any voltage between +3 and +25 and −3 and −25 V dc. The difference in the voltage levels between a driver and a terminator is called *noise margin*. The noise margin reduces the susceptibility of the interface to noise transients on the cable. Typical voltages used for data and control signals are ±7 V dc and ±10 V dc.

The pins on the RS-232C interface cable are functionally categorized as either ground, data, control (handshaking), or timing pins. All the pins are unidirectional (signals are propagated only from the DTE to the DCE, or vice versa). Table 13-6 lists the 25 pins of the RS-232C interface, their designations, and the direction of signal propagation (i.e., either toward the DTE or toward the DCE). The RS-232C specifications designate the ground, data, control, and timing pins as A, B, C, and D, respectively. These are nondescriptive designations. It is more practical and useful to use acronyms to designate the pins that reflect the pin functions. Table 13-6 lists the CCITT and EIA designations and the nomenclature more commonly used by industry in the United States.

EIA RS-232C pin functions. Twenty of the 25 pins of the RS-232C interface are designated for specific purposes or functions. Pins 9, 10, 11, 18, and 25 are unassigned; pins 1 and 7 are grounds; pins 2, 3, 14 and 16 are data pins; pins 15, 17, and 24 are timing pins; and all the other assigned pins are reserved for control or handshaking signals. There

TABLE 13-5 RS232C VOLTAGE SPECIFICATIONS (V DC)

	Data pins	
	Logic 1	Logic 0
Driver	−5 to −15	+5 to +15
Terminator	−3 + −25	+3 to +25
	Control pins	
	Enable "on"	Disable "off"
Driver	+5 to +15	−5 to −15
Terminator	+3 to +25	−3 to −25

Chap. 13   Data Communications

TABLE 13-6 EIA RS-232C PIN DESIGNATIONS

Pin Number	EIA nomenclature	Common acronyms	Direction
1	Protective ground (AA)	GWG	None
2	Transmitted data (BA)	TD, SD	DTE to DCE
3	Received data (BB)	RD	DCE to DTE
4	Request to send (CA)	RS, RTS	DTE to DCE
5	Clear to send (CB)	CS, CTS	DCE to DTE
6	Data set ready (CC)	DSR, MR	DCE to DTE
7	Signal ground (AB)	GND	None
8	Received line signal detect (CF)	RLSD, CD	DCE to DTE
9	Unassigned		
10	Unassigned		
11	Unassigned		
12	Secondary received line signal detect (SCF)	SRLSD	DCE to DTE
13	Secondary clear to send (SCB)	SCS	DCE to DTE
14	Secondary transmitted data (SBA)	STD	DTE to DCE
15	Transmission signal element timing (DB)	SCT	DCE to DTE
16	Secondary received data (SBB)	SRD	DCE to DTE
17	Receiver signal element timing (DD)	SCR	DCE to DTE
18	Unassigned		
19	Secondary request to send (SCA)	SRS	DTE to DCE
20	Data terminal ready (CD)	DTR	DTE to DCE
21	Signal quality detector (CG)	SQD	DCE to DTE
22	Ring indicator (CE)	RI	DCE to DTE
23	Data signal rate selector (CH)	DSRS	DTE to DCE
24	Transmit signal element timing (DA)	SCTE	DTE to DCE
25	Unassigned		

are two full-duplex data channels available with the RS-232C interface; one channel is for primary data (actual information) and the second channel is for secondary data (diagnostic information and handshaking signals). The functions of the 20 assigned pins are summarized below.

Pin 1—protective ground. This pin is frame ground and is used for protection against electrical shock. Pin 1 should be connected to the third-wire ground of the ac electrical system at one end of the cable (either at the DTE or the DCE, but not at both ends).

Pin 2—transmit data (TD). Serial data on the primary channel from the DTE to the DCE are transmitted on this pin. TD is enabled by an active condition on the CS pin.

Pin 3—received data (RD). Serial data on the primary channel are transferred from the DCE to the DTE on this pin. RD is enabled by an active condition on the RLSD pin.

Serial Interfaces

Pin 4—request to send (RS).   The DTE bids for the primary communications channel from the DCE on this pin. An active condition on RS turns on the modem's analog carrier. The analog carrier is modulated by a unique bit pattern called a training sequence which is used to initialize the communications channel and synchronize the receive modem. RS cannot go active unless pin 6 (DSR) is active.

Pin 5—clear to send (CS).   This signal is a handshake from the DCE to the DTE in response to an active condition on request to send. CS enables the TD pin.

Pin 6—data set ready (DSR).   On this pin the DCE indicates the availability of the communications channel. DSR is active as long as the DCE is connected to the communications channel (i.e., the modem or the communications channel is not being tested or is not in the voice mode).

Pin 7—signal ground.   This pin is the signal reference for all the data, control, and timing pins. Usually, this pin is strapped to frame ground (pin 1).

Pin 8—receive line signal detect (RLSD).   The DCE uses this pin to signal the DTE when the DCE is receiving an analog carrier on the primary data channel. RSLD enables the RD pin.

Pin 9.   Unassigned.

Pin 10.   Unassigned.

Pin 11.   Unassigned.

Pin 12—secondary receive line signal detect (SRLSD).   This pin is active when the DCE is receiving an analog carrier on the secondary channel. SRLSD enables the SRD pin.

Pin 13—secondary clear to send (SCS).   This pin is used by the DCE to send a handshake to the DTE in response to an active condition on the secondary request to send pin. SCS enables the STD pin.

Pin 14—secondary transmit data (STD).   Diagnostic data are transferred from the DTE to the DCE on this pin. STD is enabled by an active condition on the SCS pin.

Pin 15—transmission signal element timing (SCT).   Transmit clocking signals are sent from the DCE to the DTE on this pin.

Pin 16—secondary received data (SRD).   Diagnostic data are transferred from the DCE to the DTE on this pin. SRD is enabled by an active condition on the SCS pin.

Pin 17—receive signal element timing (SCR).   Receive clocking signals are sent from the DCE to the DTE on this pin. The clock frequency is equal to the bit rate of the primary data channel.

Pin 18.   Unassigned.

Pin 19—secondary request to send (SRS).   The DTE bids for the secondary communications channel from the DCE on this pin.

Pin 20—data terminal ready (DTR).   The DTE sends information to the DCE on this pin concerning the availability of the data terminal equipment (i.e., access to the mainframe at the primary station or status of the computer terminal at the secondary station). DTR is used primarily with dial-up data communications circuits to handshake with RI.

Pin 21—signal quality detector (SQD). The DCE sends signals to the DTE on this pin that reflect the quality of the received analog carrier.

Pin 22—ring indicator (RI). This pin is used with dial-up lines for the DCE to signal the DTE that there is an incoming call.

Pin 23—data signal rate selector (DSRS). The DTE uses this pin to select the transmission bit rate (clock frequency) of the DCE.

Pin 24—transmit signal element timing (SCTE). Transmit clocking signals are sent from the DTE to the DCE on this pin when the master clock oscillator is located in the DTE.

Pin 25. Unassigned.

Pins 1 through 8 are used with both asynchronous and synchronous modems. Pins 15, 17, and 24 are used for only synchronous modems. Pins 12, 13, 14, 16, and 19 are used only when the DCE is equipped with a secondary channel. Pins 19 and 22 are used exclusively for dial-up telephone connections.

The basic operation of the RS-232C interface is shown in Figure 13-15 and described as follows. When the DTE has primary data to send, it enables request to send ($t = 0$ ms). After a predetermined time delay (50 ms), CS goes active. During the RS/CS delay the modem is outputting an analog carrier that is modulated by a unique bit pattern called a *training sequence.* The training sequence is used to initialize the communications line and synchronize the carrier and clock recovery circuits in the receive modem. After the RS/CS delay, TD is enabled and the DTE begins to transmit data. After the receive DTE detects an analog carrier, RD is enabled. When the transmission is complete ($t = 150$ ms), RS

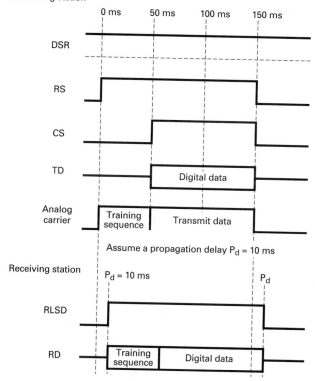

Figure 13-15 Timing diagram: basic operation of the RS-232C interface.

Serial Interfaces

goes low turning off the analog carrier and shutting off CS. For a more detailed explanation, timing diagrams, and illustrative examples, see V. Alisouskas and W. Tomasi, *Digital and Data Communications* (Englewood Cliffs, N.J.: Prentice Hall, 1985).

RS-449A Interface

Contemporary data rates have exceeded the capabilities of the RS-232C interface. Therefore, it was necessary to adopt and implement a new standard that allows higher bit rates to be transmitted for longer distances. The RS-232C has a maximum bit rate of 20,000 bps and a maximum distance of approximately 50 ft. Consequently, the EIA has adopted a new standard: the RS-449A interface. The RS-449A is essentially an updated version of the RS-232C except that the RS-449A outlines only the mechanical and functional specifications of the cable and connectors.

The RS-449A specifies two cables: one with 37 wires that is used for serial data transmission and one with 9 wires that is used for secondary diagnostic information. Table 13-7 lists the 37 pins of the RS-449A primary cable and their designations, and Table 13-8 lists the 9 pins of the diagnostic cable and their designations. Note that the acronyms used with the RS-449A are more descriptive than those recommended by the EIA for the RS-232C. The functions specified by the RS-449A are very similar to the RS-232C. The major difference between the two standards is the separation of the primary data and secondary diagnostic channels onto two cables.

TABLE 13-7   EIA RS-449A PRIMARY CHANNEL PIN DESIGNATIONS

Pin number	Mnemonic	Circuit name
1	None	Shield
2	SI	Signaling rate indicator
3,21	None	Spare
4,22	SD	Send data
5,23	ST	Send timing
6,24	RD	Receive data
7,25	RS	Request to send
8,26	RT	Receive timing
9,27	CS	Clear to send
10	LL	Local loopback
11,29	DM	Data mode
12,30	TR	Terminal ready
13,31	RR	Receiver ready
14	RL	Remote loopback
15	IC	Incoming call
16	SF/SR	Select frequency/signaling rate
17,23	TT	Terminal timing
18	TM	Test mode
19	SG	Signal ground
20	RC	Receive common
28	IS	Terminal in service
32	SS	Select standby
33	SQ	Signal quality
34	NS	New signal
36	SB	Standby indicator
37	SC	Send common

TABLE 13-8 EIA RS-449A SECONDARY
DIAGNOSTIC CHANNEL PIN DESIGNATIONS

Pin number	Mnemonic	Circuit name
1	None	Shield
2	SRR	Secondary receiver ready
3	SSD	Secondary send data
4	SRD	Secondary receive data
5	SG	Signal ground
6	RC	Receive common
7	SRS	Secondary request to send
8	SCS	Secondary clear to send
9	SC	Send common

The RS-232C and RS-449A standards provide specifications for answering calls, but not for dialing. The EIA has a different standard, RS-366, for automatic calling units. The principal use of RS-366 is for dial backup of private-line data circuits and for automatic dialing of remote terminals.

The electrical specifications used with the RS-449A are specified by either the RS-422A or the RS-423A standard. The RS-422A standard specifies a balanced interface cable that will operate at bit rates up to 10 Mbps and span distances up to 1200 m. This does not mean that 10 Mbps can be transmitted 1200 m. At 10 Mbps the maximum distance is 15 m, and 90 kbps is the maximum bit rate that can be transmitted 1200 m. The RS-423A standard specifies an unbalanced interface cable that will operate at a maximum line speed of 100 kbps and span a maximum distance of 90 m.

Figure 13-16 shows the *balanced* digital interface circuit for the RS-422A, and Figure 13-17 shows the *unbalanced* digital interface circuit for the RS-423A.

A balanced interface, such as the RS-422A, transfers information to a *balanced transmission line.* With a balanced transmission line, both conductors carry current except the current in the two wires travel in opposite directions. With a bidirectional *unbalanced* line, one wire is at ground potential and the currents in the two wires may be different.

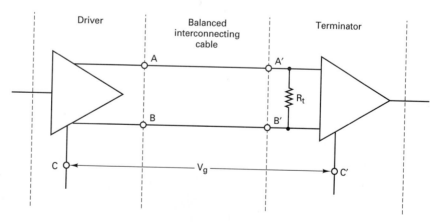

Figure 13-16 RS-422A interface circuit, R_t, optional cable termination resistance; V_g, ground potential difference; A, B, driver interface points; A′, B′, terminator interface points; C, driver circuit ground; C′, terminator circuit ground; A-B, balanced driver output; A′-B′, balanced terminator input.

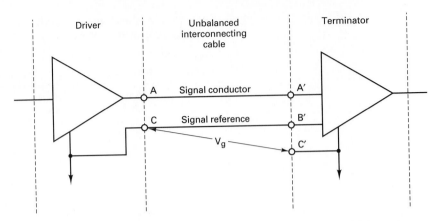

Figure 13-17 RS-423A interface circuit. A, C, driver interface; A', B', terminator interface; V_g, ground potential difference; C, driver circuit ground;
C', terminator circuit ground.

Currents that flow in opposite directions in a balanced wire pair are called *metallic circuit*
currents. Currents that flow in the same direction are called *longitudinal* currents. A balanced pair has the advantage that most noise interference is induced equally in both wires,
producing longitudinal currents that cancel in the load. Figure 13-18 shows the results of
metallic and longitudinal currents on a balanced transmission line. It can be seen that longitudinal currents (generally produced by static interference) cancel in the load. Balanced
transmission lines can be connected to unbalanced loads and vice versa, with special transformers called *baluns* (*balanced* to *unbalanced*).

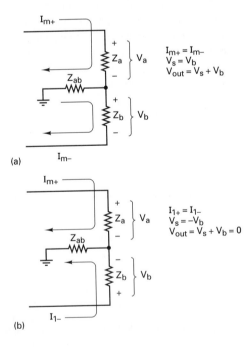

Figure 13-18 Results of metallic and
longitudinal currents on a balanced
transmission line; (a) metallic currents
due to signal voltages; (b) longitudinal
currents due to noise voltages.

CCITT X.21

In 1976, the CCITT introduced the X.21 recommendation, which includes the specifications for placing and receiving calls and for sending and receiving data using full-duplex synchronous transmission. The X.21 recommendation presumes a direct digital connection to a digital telephone network. Thus all data transmissions must be synchronous, and the data communications equipment will need to provide both bit and character synchronization. The minimum data rate for X.21 will probably be 64 kbps because this is the bit rate currently used to encode voice in digital form on the telephone network.

The X.21 specifies only six signals, which are listed in Table 13-9. Data are transmitted toward the modem on the Transmit line, and the modem returns data on the Receive line. The Control and Indication lines are control channels for the two transmission directions. The Signal Element Timing line carries the bit timing signal (clock) and the Byte Timing line carries the character synchronization information. The electrical specifications for X.21 are listed either in recommendation X.26 (balanced) or recommendation X.27 (unbalanced).

The major advantage of the X.21 standard over the RS-232C and RS-449A standards is that X.21 signals are encoded in serial digital form, which sets the stage for providing special new services in computer communications.

TRANSMISSION MEDIA AND DATA MODEMS

In its simplest form, data communication is the transmittal of digital information between two DTEs. The DTEs may be separated by a few feet or several thousand miles. At the present time, there is an insufficient number of transmission media to carry digital information from source to destination in digital form. Therefore, the most convenient alternative is to use the existing public telephone network (PTN) as the transmission medium for data communications circuits. Unfortunately, the PTN was designed (and most of it constructed) long before the advent of large-scale data communications. The PTN was intended to be used for transferring voice telephone communications signals, not digital data. Therefore, to use the PTN for data communications, the data must be converted to a form more suitable for transmission over analog carrier systems.

TABLE 13-9   CCITT X.21 PIN DESIGNATIONS

Interchange circuit	Name	Direction
G	Signal ground	[a]
GA	DTE common return	DTE to DCE
T	Transmit	DTE to DCE
R	Receive	DCE to DTE
C	Control	DTE to DCE
I	Indication	DCE to DTE
S	Signal element timing	DCE to DTE
B	Byte timing	DCE to DTE

[a]See X.24 Recommendations

Transmission Media

As stated previously, the public telephone network is a convenient alternative to constructing alternate digital facilities (at a tremendous cost) for carrying only digital data. The public telephone network comprises over 2000 local telephone companies and several long-distance common carriers such as Microwave Communications Incorporated (MCI), GTE Sprint, and the American Telephone and Telegraph Company (AT&T). Local telephone companies provide voice and data services for relatively small geographic areas, whereas long-distance common carriers provide voice and data services for relatively large geographic areas.

Essentially, there are two types of circuits available from the public telephone network: *direct distance dialing* (DDD) and *private line*. The DDD network is commonly called the *dial-up network*. Anyone who has a telephone number subscribes to the DDD network. With the DDD network, data links are established and disconnected in the same manner as normal voice calls are established and disconnected—with a standard telephone or some kind of an automatic dial/answer machine. Data links that are established through the DDD network use *common usage* equipment and facilities. Common usage means that a subscriber uses the equipment and transmission medium for the duration of the call, then they are relinquished to the network for other subscribers to use. With private-line circuits, a subscriber has a permanent dedicated communications link 24 hours a day.

Figure 13-19 shows a simplified block diagram of a telephone communications link. Each subscriber has a dedicated cable facility between his station and the nearest telephone office called a *local loop*. The local loop is used by the subscriber to access the PTN. The facilities used to interconnect telephone offices are called *trunk* circuits and can be a metallic cable, a digital carrier system, a microwave radio, an optical fiber link, or a satellite radio system, depending on the distance between the two offices. For temporary connections using the DDD network, telephone offices are interconnected through sophisticated electronic switching systems (ESS) and use intricate switching arrangements. With private-line circuits, data links are permanently hardwired through telephone offices without going through a switch. Dial-up data links are preferred when there are a large number

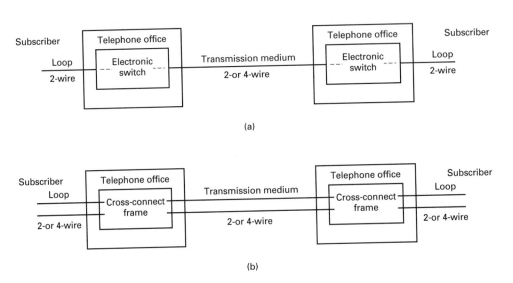

Figure 13-19 Telephone communications link: (a) direct distance dialing; (b) dedicated private line.

of subscribers in a network or if there is a small volume of data traffic. Private-line circuits are preferred for limited-access networks when there is a large volume of data throughput.

The quality of a dial-up circuit is guaranteed to meet the minimum requirements for a *voice band* (VB) communications circuit. With a private-line circuit, the communications link can be improved by adding amplifiers and equalizers to the circuit. This is called *conditioning* the line. A voice-grade circuit using the PTN has an ideal passband from 0 to 4 kHz, although the usable passband is limited to approximately 300 to 3000 Hz. The minimum-quality circuit available using the PTN is called a basic voice grade (VG) circuit. The quality of a dial-up circuit is guaranteed to meet basic requirements and can be as good as a private-line circuit. However, with the DDD network, the transmission characteristics of the data link vary from call to call, while in a private line circuit they remain relatively constant. With the DDD network, *contention* can be a problem; each subscriber must contend for a connection through the network with every other subscriber in the network. With private-line circuits, there is no contention because each circuit has only one subscriber. Consequently, there are several advantages that private-line circuits have over dial-up networks: increased availability, more consistent performance, greater reliability, and lower costs for moderate to high volumes of data. Dial-up circuits are limited to two-wire operation, whereas private-line circuits can operate either two- or four-wire.

Data Modems

The primary purpose of the data modem is to interface the digital terminal equipment to an analog communications channel. The data modem is also called a DCE, a *dataset,* a *dataphone,* or simply a *modem.* At the transmit end, the modem converts digital pulses from the serial interface to analog signals, and at the receive end, the modem converts analog signals to digital pulses.

Modems are generally classified as either asynchronous or synchronous and use either FSK, PSK, or QAM modulation. With synchronous modems, clocking information is recovered in the receive modem; with asynchronous modems, it is not. Asynchronous modems use FSK modulation and are restricted to low-speed applications (below 2000 bps). Synchronous modems use PSK and QAM modulation and are used for medium-speed (2400 to 4800 bps) and high-speed (9600 bps) applications.

Asynchronous modems.   Asynchronous modems are used primarily for low-speed dial-up circuits. There are several standard modem designs commonly used for asynchronous data transmission. For half-duplex operation using the two-wire DDD network or full-duplex operation with four-wire private line circuit, the Western Electric 202T/S or equivalent is a popular modem. The 202T is a four-wire full-duplex modem and the 202S is a two-wire, half-duplex modem.

The 202T modem is an asynchronous transceiver utilizing frequency shift keying. It uses a 1700-Hz carrier that can be shifted at a maximum rate of 1200 times a second. When a logic 1 (mark) is applied to the modulator, the carrier is shifted down 500 Hz, to 1200 Hz. When a logic 0 (space) is applied, the carrier is shifted up 500 Hz, to 2200 Hz. Consequently, as the data input signal alternates between 1 and 0, the carrier is shifted back and forth between 1200 and 2200 Hz, respectively. This process can be related to conventional frequency modulation. The difference between the mark and space frequencies (1200 to 2200 Hz) is the peak-to-peak frequency deviation, and the rate of change of the digital input signal (bit rate) is equal to twice the frequency of the modulating signal. Therefore, for the worst-case situation, the 1700-Hz carrier is frequency modulated by a 1200-Hz square wave.

A figure of merit often used to express the degree of modulation achieved in an FSK modulator is the *h factor*, which is defined as

$$h = \frac{|f_m - f_s|}{\text{bps}} \qquad (13\text{-}2)$$

where f_m = mark (logic 1) frequency (Hz)
   f_s = space (logic 0) frequency (Hz)
   bps = input bit rate (bps)

For the 202T modem,

$$h = \frac{|1200 - 2200|}{1200} = \frac{1000}{1200} = 0.83$$

As a general rule and for best performance, the *h* factor is limited to a value less than 1. The *h* factor is equivalent to the modulation index for conventional FM. Consequently, with FSK the number of side frequencies generated is directly related to the *h* factor. The separation between adjacent side frequencies is equal to one-half the input bit rate. The frequency spectrum for the 202T modem is shown in Figure 13-20. As the figure shows, for an *h* factor of 0.83, only two sets of significant side frequencies are generated, resulting in a worst-case bandwidth of 2400 Hz.

To operate full duplex with a two-wire dial-up circuit, it is necessary to divide the usable bandwidth of a voice band circuit in half, creating two equal-capacity data channels. A popular modem that does this is the Western Electric 103 or equivalent. The 103 modem is capable of full-duplex operation over a two-wire line at bit rates up to 300 bps. With the 103 modem, there are two data channels each with separate mark and space frequencies. One channel is the *low-band channel* and occupies a passband from 300 to 1650 Hz. The second channel is the *high-band channel* and occupies a passband from 1650 to 3000 Hz. The mark and space frequencies for the low-band channel are 1270 and 1070 Hz, respectively. The mark and space frequencies for the high-band channel are 2225 and 2025 Hz, respectively. For a bit rate of 300 bps, the modulation index for the 103 modem is 0.67. The output spectrum for the 103 modem is shown in Figure 13-21. The high- and low-band data channels occupy different frequency bands and can therefore use the same two-wire facility without interfering with each other. This is called *frequency-division multiplexing* and is explained in detail in Chapter 17.

The low-band channel is commonly called the *originate channel* and the high-band channel is called the *answer channel*. It is standard procedure on a dial-up circuit for the station that originates the call to transmit on the low-band frequencies and receive on the high-band frequencies, and the station that answers the call to transmit on the high-band frequencies and receive on the low-band frequencies.

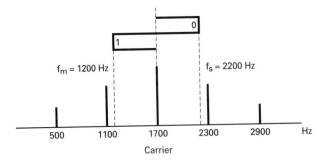

Figure 13-20 Output spectrum for a 202 T/S modem. Carrier frequency = 1700 Hz, input data = 1200 bps alternating 1/0 pattern, modulation index = 0.83.

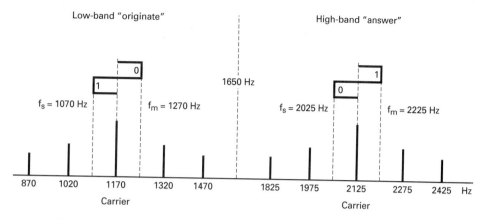

Figure 13-21 Output spectrum for a 103 modem. Carrier frequecy: low band = 1170, high band = 2125; input data = 300 bps alternating 1/0 sequence; modulation index = 0.67.

Synchronous modems. Synchronous modems are used for medium- and high-speed data transmission and use either PSK or QAM modulation. With synchronous modems the transmit clock, together with the data, digitally modulate an analog carrier. The modulated carrier is transmitted to the receive modem, where a coherent carrier is recovered and used to demodulate the data. The transmit clock is recovered from the data and used to clock the received data into the DTE. Because of the clock and carrier recovery circuits, a synchronous modem is more complicated and thus more expensive than its asynchronous counterpart.

PSK modulation is used for medium-speed (2400 to 4800 bps) synchronous modems. More specifically, QPSK is used with 2400-bps modems and 8-PSK is used with 4800-bps modems. QPSK has a bandwidth efficiency of 2 bps/Hz; therefore, the baud rate and minimum bandwidth for a 2400-bps synchronous modem are 1200 baud and 1200 Hz. The standard 2400-bps synchronous modem is the Western Electric 201C or equivalent. The 201C uses a 1600-Hz carrier and has an output spectrum that extends from 1000 to 2200 Hz. 8-PSK has a bandwidth efficiency of 3 bps/Hz; therefore, the baud rate and minimum bandwidth for 4800-bps synchronous modems are 1600 baud and 1600 Hz. The standard 4800-bps synchronous modem is the Western Electric 208A or equivalent. The 208A also uses a 1600-Hz carrier but has an output spectrum that extends from 800 to 2400 Hz. Both the 201C and 208A are full-duplex modems designed to be used with four-wire private line circuits. The 201C and 208A can operate over two-wire dial-up circuits but only in the simplex mode. There are half-duplex two-wire versions of both models: the 201B and 208B.

High-speed synchronous modems operate at 9600 bps and use 16-QAM modulation. 16-QAM has a bandwidth efficiency of 4 bps/Hz; therefore, the baud rate and minimum bandwidth for 9600-bps synchronous modems are 2400 baud and 2400 Hz. The standard 9600-bps modem is the Western Electric 209A or equivalent. The 209A uses a 1650-Hz carrier and has an output spectrum that extends from 450 to 2850 Hz. The Western Electric 209A is a four-wire synchronous modem designed to be used on full-duplex private-line circuits. The 209B is the two-wire version designed for half-duplex dial-up circuits.

Normally, an asynchronous data format is used with asynchronous modems and a synchronous data format is used with synchronous modems. However, asynchronous data

Transmission Media and Data Modems

549

are occasionally used with synchronous modems; this is called *isochronous transmission*. Synchronous data are never used with asynchronous modems.

Table 13-10 summarizes the standard Western Electric modems.

MODEM SYNCHRONIZATION

During the RTS/CTS delay, the transmit modem outputs a special, internally generated bit pattern called the *training sequence*. This bit pattern is used to synchronize (train) the receive modem. Depending on the type of modulation, transmission bit rate, and the complexity of the modem, the training sequence accomplishes one or more of the following functions in the receive modem:

1. Verify continuity (activate RLSD).
2. Initialize the descrambler circuits. (These circuits are used for clock recovery—explained later in this chapter.)
3. Initialize the automatic equalizer. (These circuits compensate for telephone line impairments—explained later in this chapter.)
4. Synchronize the transmitter and receiver carrier oscillators.
5. Synchronize the transmitter and receiver clock oscillators.
6. Disable any echo suppressors in the circuit.
7. Establish the gain of any AGC amplifiers in the circuit.

Low-Speed Modems

Since these modems are generally asynchronous and use noncoherent FSK, the transmit carrier and clock frequencies need not be recovered by the receive modem. Therefore, scrambler and descrambler circuits are unnecessary. The pre- and post-equalization circuits, if used, are generally manual and do not require initialization. The special bit pattern transmitted during the RTS/CTS delay is usually a constant string of 1's (idle line 1's) and is used to verify continuity, set the gain of the AGC amplifiers, and disable any echo suppressors in dial-up applications.

TABLE 13-10 MODEM SUMMARY

Western Electric designation	Line facility	Operating mode	Synchronization	Type of modulation	Maximum data rate (bps)
103	Dial-up	FDX	Asynchronous	FSK	300
113A	Dial-up	Simplex	Asynchronous	FSK	300
113B	Dial-up	Simplex	Asynchronous	FSK	300
201B	Dial-up	HDX	Synchronous	QPSK	2400
201C	Private	HDX/FDX	Synchronous	QPSK	2400
202S	Dial-up	HDX	Asynchronous	FSK	1200
202T	Private	HDX/FDX	Asynchronous	FSK	1200 (basic) 1800 (CI conditioning)
208A	Private	HDX/FDX	Synchronous	8-PSK	4800
208B	Dial-up	HDX	Synchronous	8-PSK	4800
209A	Private	HDX/FDX	Synchronous	16-QAM	9600 (DI conditioning)
209B	Dial-up	HDX	Synchronous	16-QAM	9600

Medium- and High-Speed Modems

These modems are used where transmission rates of 2400 bps or more are required. In order to transmit at these higher bit rates, PSK or QAM modulation is used which requires the receive carrier oscillators to be at least frequency coherent (and possibly phase coherent). Since these modems are synchronous, clock timing recovery by the receive modem must be achieved. These modems contain *scrambler* and *descrambler circuits* and *adaptive (automatic) equalizers.*

Training. The type of modulation and encoding technique used determines the number of bits required and therefore the duration of the training sequence. The 208 modem is a synchronous, 4800-bps modem which uses 8-DPSK. The training sequence for this modem is shown in Figure 13-22. Each symbol represents 3 bits (1 tribit) and is 0.625 ms in duration. The four-phase idle code sequences through four of the eight possible phase shifts. This allows the receiver to recover the carrier and the clock timing information rapidly. The four-phase test word allows the adaptive equalizer in the receive modem to adjust to its final setting. The eight-phase initialization period prepares the descrambler circuits for eight-phase operation. The entire training sequence (234 bits) requires 48.75 ms for transmission.

Clock recovery. Although timing (clock) synchronization is first established during the training sequence, it must be maintained for the duration of the transmission. The clocking information can be extracted from either the I or the Q channel, or from the output of the bit combiner. If an alternating 1/0 pattern is assumed at the output of the LPF (Figure 13-23) a clock frequency at the bit rate of the I (or Q) channel can be recovered. The waveforms associated with Figure 13-23 are shown in Figure 13-24.

This clocking information is used to phase-lock loop the receive clock oscillator onto the transmitter clock frequency. To recover clocking information by this method successfully, there must be sufficient transitions in the received data stream. That these transitions will automatically occur cannot be assumed. In a QPSK system, an alternating 1/0 pattern applied to the transmit modulator produces a sequence of all 1's in the I or Q channel, and a sequence of all 0's in the opposite channel. A prolonged sequence of all 1's or all 0's applied to the transmit modulator would not provide any transitions in either the I, Q, or the composite received data stream. Restrictions could be placed on the customer's protocol and message format to prevent an undesirable bit sequence from occurring, but this is a poor solution to the problem.

Scramblers and descramblers. A better method is to scramble the customer's data before it modulates the carrier. The receiver circuitry must contain the corresponding

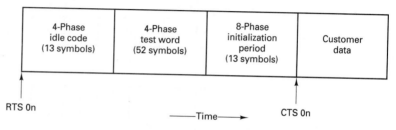

Figure 13-22 Training sequence for a 208 modem.

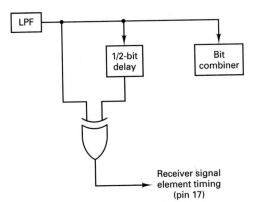

Figure 13-23 Clock recovery circuit for QPSK demodulator.

descrambling algorithm to recover the original bit sequence before data are sent to the DTE. The purpose of a scrambler is not simply to randomize the transmitted bit sequence, but to detect the occurrence of an undesirable bit sequence and convert it to a more acceptable pattern.

A block diagram of a scrambler and descrambler circuit is shown in Figure 13-25. These circuits are incomplete since an additional gate would be required to detect a varying sequence that would create an all 1 or all 0 sequence in a modulator channel after the bits were split.

EXAMPLE 13-6

For QPSK or 4-QAM:

```
0  1  0  1  0  1  0  1  0  1  0  1
I  Q  I  Q  I  Q  I  Q  I  Q  I  Q
```

For 8-PSK or 8-QAM:

```
0  1  1  0  1  0  0  1  1  0  1  1
I  Q     I  Q     I  Q     I  Q
```

The scrambler circuit is inserted prior to the bit splitter in the QPSK modulator and the descrambler is inserted after the bit combiner in the QPSK demodulator. In general, the output of the scrambler or descrambler OR gate is $A \ B \ C \ D + A' \ B' \ C' \ D'$.

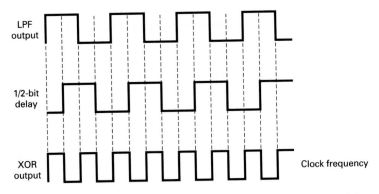

Figure 13-24 Clock recovery from I (or Q) channel of a QPSK demodulator.

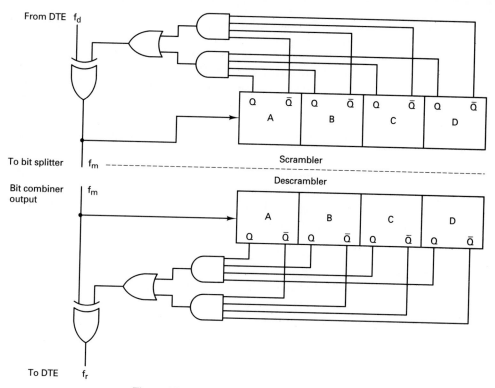

Figure 13-25 Scrambler and descrambler circuits.

$$f_m = f_d \oplus (A\,B\,C\,D + A'\,B'\,C'\,D') \quad \text{top XOR gate}$$

$$f_r = f_m \oplus (A\,B\,C\,D + A'\,B'\,C'\,D') \text{ bottom XOR gate}$$

Substituting for f_m in the second equation, we have

$$f_r = f_d \oplus (A\,B\,C\,D + A'\,B'\,C'\,D') \oplus (A\,B\,C\,D + A'\,B'\,C'\,D')$$

Since any identity XORed with itself yields 0,

$$f_r = f_d \oplus 0$$

$$f_r = f_d$$

This simply shows that the original transmitted data (f_d) will be fully recovered by the receiver.

The output of either OR gate will be a 1 if the 4-bit register contains either all 1's or all 0's. Neither of these is a desirable sequence. If the OR gate output is a 1, f_m will be the complement (opposite) of f_d, or f_r will be complement of f_m. The intent is to create transitions in a prolonged bit stream of either all 1's or all 0's. If the output of the OR gate is a 0, neither of these undesired conditions exists and $f_m = f_d$ or $f_r = f_m$: the data pass through the XOR gate unchanged. If the other logic gates (AND, OR, NAND, NOR) were used either alone or in combination in place of the XOR gates, the necessary transitions could be created in the scrambler circuit, but the original data could not be recovered in the descrambler circuit. If a long string of all 1's or all 0's is applied to the scrambler circuit, this circuit will introduce transitions. However, there may be times when the scrambler

Modem Synchronization

creates an undesired sequence. The XOR output is always either a 1 or a 0. No matter what the output of the OR gate, a value of f_d may be found to produce a 1 or a 0 at the XOR output. If either value for f_d was equiprobable, the scrambler circuit would be unnecessary. If the 4-bit register contains all 1's, if $f_d = 1$, we would like to see it inverted. However, if $f_d = 0$, we'd prefer to pass it through the XOR gate unchanged. The scrambler circuit for this situation inverts the 0 and extends the output string of 1's. It is beyond the intended scope of this book to delve deeply into all parameters involved in scrambler design. Let it be enough to say that scramblers will cure more problems than they create.

Equalizers. *Equalization* is the compensation for the phase delay distortion and amplitude distortion of a telephone line. One form of equalization is C-type conditioning. Additional equalization may be performed by the modems. *Compromise equalizers* are contained in the transmit section of the modem and they provide *pre-equalization*. They shape the transmitted signal by altering its delay and gain characteristics before it reaches the telephone line. It is an attempt to compensate for impairments anticipated in the bandwidth parameters of the line. When a modem is installed, the compromise equalizers are manually adjusted to provide the best *bit error rate* (BER). Typically, compromise equalizer settings affect:

1. Amplitude only
2. Delay only
3. Amplitude and delay
4. Neither amplitude nor delay

The setting above may be applied to either the high or low voice band frequencies or symmetrically to both at the same time. Once a compromise equalizer setting has been selected, it can only be changed manually. The setting that achieves the best BER is dependent on the electrical length of the circuit and the type of facilities that make it up. *Adaptive equalizers* are located in the receiver section of the modem and provide *post-equalization* to the received analog signal. Adaptive equalizers automatically adjust their gain and delay characteristics to compensate for telephone-line impairments. An adaptive equalizer may determine the quality of the received signal within its own circuitry or it may acquire this information from the demodulator or descrambler circuits. Whichever the case, the adaptive equalizer may continuously vary its settings to achieve the best overall bandwidth characteristics for the circuit.

QUESTIONS

13-1. Define *data communications*.

13-2. What was the significance of the Carterfone decision?

13-3. Explain the difference between a two-point and a multipoint circuit.

13-4. What is a data communications topology?

13-5. Define the four transmission modes for data communications circuits.

13-6. Which of the four transmission modes can be used only with multipoint circuits?

13-7. Explain the differences between two-wire and four-wire circuits.

13-8. What is a data communications code? What are some of the other names for data communications codes?

13-9. What are the three types of characters used in data communications codes?

13-10. Which data communications code is the most powerful? Why?

13-11. What are the two general categories of error control? What is the difference between them?

13-12. Explain the following error detection techniques: redundancy, exact-count encoding, parity, vertical redundancy checking, longitudinal redundancy checking, and cyclic redundancy checking.

13-13. Which error detection technique is the simplest?

13-14. Which error detection technique is the most reliable?

13-15. Explain the following error correction techniques: symbol substitution, retransmission, and forward error correction.

13-16. Which error correction technique is designed to be used in a human environment?

13-17. Which error correction technique is the most reliable?

13-18. Define *character synchronization.*

13-19. Describe the asynchronous data format.

13-20. Describe the synchronous data format.

13-21. Which data format is best suited to long messages? Why?

13-22. What is a cluster?

13-23. Describe the functions of a control unit.

13-24. What is the purpose of the data modem?

13-25. What are the primary functions of the UART?

13-26. What is the maximum number of bits that can make up a single character with a UART?

13-27. What do the status signals RPE, RFE, and ROR indicate?

13-28. Why does the receive clock for a UART operate 16 times faster than the receive bit rate?

13-29. What are the major differences between a UART and a USRT?

13-30. What is the purpose of the serial interface?

13-31. What is the most prominent serial interface in the United States?

13-32. Why did the EIA establish the RS-232C interface?

13-33. What is the nominal maximum length for the RS-232C interface?

13-34. What are the four general classifications of pins on the RS-232C interface?

13-35. What is the maximum positive voltage that a driver will output?

13-36. Which classification of pins uses negative logic?

13-37. What is the primary difference between the RS-449A interface and the RS-232C interface?

13-38. Higher bit rates are possible with a (balanced, unbalanced) interface cable.

13-39. Who provides the most commonly used transmission medium for data communications circuits? Why?

13-40. Explain the differences between DDD circuits and private line circuits.

13-41. Define the following terms: local loop, trunk, common usage, and dial switch.

13-42. What is a DCE?

13-43. What is the primary difference between a synchronous and an asynchronous modem?

13-44. What is necessary for full-duplex operation using a two-wire circuit?

13-45. What do *originate* and *answer mode* mean?

13-46. What modulation scheme is used for low-speed applications? For medium-speed applications? For high-speed applications?

13-47. Why are synchronous modems required for medium- and high-speed applications?

PROBLEMS

13-1. Determine the LRC and VRC for the following message (use even parity for LRC and odd parity for VRC).

D A T A sp C O M M U N I C A T I O N S

13-2. Determine the BCS for the following data- and CRC-generating polynomials.

$$G(x) = x^7 + x^4 + x^2 + x^0 = 1\ 0\ 0\ 1\ 0\ 1\ 0\ 1$$
$$P(x) = x^5 + x^4 + x^1 + x^0 = 1\ 1\ 0\ 0\ 1\ 1$$

13-3. How many Hamming bits are required for a single ASCII character?

13-4. Determine the Hamming bits for the ASCII character "B." Insert the Hamming bits into every other location starting at the left.

14
Data Communications Protocols

INTRODUCTION

The primary goal of *network architecture* is to give the users of the network the tools necessary for setting up the network and for performing flow control. A network architecture outlines the way in which a data communications network is arranged or structured and generally includes the concept of *levels* or *layers* within the architecture. Each layer within the network consists of specific *protocols* or rules for communicating that perform a given set of functions.

Protocols are arrangements between people or processes. In essence, a protocol is a set of customs or regulations dealing with formality or precedence, such as diplomatic or military protocol. A *data communications network protocol* is a set of rules governing the orderly exchange of data.

As stated previously, the function of a line control unit is to control the flow of data between the applications program and the remote terminals. Therefore, there must be a set of rules that govern how an LCU reacts to or initiates different types of transmissions. This set of rules is called a *data link protocol*. Essentially, a data link protocol is a set of procedures, including precise character sequences, that ensure an orderly exchange of data between two LCUs.

In a data communications circuit, the station that is presently transmitting is called the *master* and the receiving station is called the *slave*. In a centralized network, the primary station controls when each secondary station can transmit. When a secondary station is transmitting, it is the master and the primary station is now the slave. The role of master is temporary and which station is master is delegated by the primary. Initially, the primary is master. The primary station solicits each secondary station, in turn, by *polling* it. A poll is an invitation from the primary to a secondary to transmit a message. Secondaries cannot poll a primary. When a primary polls a secondary, the primary is initiating a *line turn-around;* the polled secondary has been designated the master and must respond. If the primary *selects* a secondary, the secondary is identified as a receiver. A selection is an interrogation by the primary of a secondary to determine the secondary's status (i.e., ready to receive or not ready to receive a message). Secondary stations cannot select the primary. Transmissions from the primary go to all the secondaries; it is up to the secondary stations to individually decode each transmission and determine if it is intended for them. When a secondary transmits, it sends only to the primary.

Data link protocols are generally categorized as either asynchronous or synchronous. As a rule, asynchronous protocols use an asynchronous data format and asynchronous modems, whereas synchronous protocols use a synchronous data format and synchronous modems.

OPEN SYSTEMS INTERCONNECTION

The term *open systems interconnection* (OSI) is the name for a set of standards for communications among computers. The primary purpose of OSI standards is to serve as a structural guideline for exchanging information between computers, terminals, and networks. The OSI is endorsed by both the ISO and CCITT, which have worked together to establish a set of ISO standards and CCITT recommendations that are essentially identical. In 1983, the ISO and CCITT adopted a seven-layer communication architecture reference model. Each layer consists of specific protocols for communicating.

The ISO Protocol Hierarchy

The ISO–Open Systems Interconnection Seven-Layer Model is shown in Figure 14-1. This hierarchy was developed to facilitate the intercommunications of data processing equipment by separating network responsibilities into seven distinct layers. The basic concept of layering responsibilities is that each layer adds value to services provided by the sets of lower layers. In this way, the highest level is offered the full set of services needed to run a distributed data application.

There are several advantages to using a layered architecture for the OSI model. The different layers allow different computers to communicate at different levels. In addition, as technological advances occur, it is easier to modify one layer's protocol without having to modify all of the other layers. Each layer is essentially independent of every other layer. Therefore, many of the functions found in the lower layers have been removed entirely from software tasks and replaced with hardware. Some examples of these functions are shown in Figure 14-1. The primary disadvantage of the seven-layer architecture is the tremendous amount of overhead required in adding headers to the information being transmitted through the various layers. In fact, if all seven levels are addressed, less than 15% of the transmitted message is source information; the rest is overhead. The result of adding headers to each layer is illustrated in Figure 14-1.

Levels 4, 5, 6, and 7 allow for two host computers to communicate directly. The three bottom layers are concerned with the actual mechanics of moving data (at the bit level) from one machine to another. The basic services provided by each layer of the hierarchy are summarized below.

1. Physical layer. The physical layer is the lowest level of the hierarchy and specifies the physical, electrical, functional, and procedural standards for accessing the data communications network. Definitions such as maximum and minimum voltage levels and

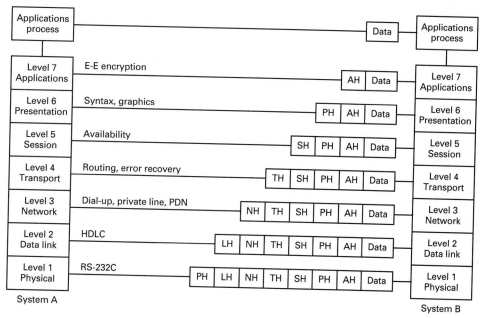

Figure 14-1 ISO international protocol hierarchy. AH, applications header; PH, presentation header; SH, session header; TH, transport header; NH, network header; LH, link header; PH, physical header.

circuit impedances are made at the physical layer. The specifications outlined by the physical layer are similar to those specified by the EIA RS-232C serial interface standard.

2. Data link layer. The data link layer is responsible for communications between primary and secondary nodes within the network. The data link layer provides a means to activate, maintain, and deactivate the data link. The data link layer provides the final framing of the information envelope, facilitates the orderly flow of data between nodes, and allows for error detection and correction. Examples of data link protocols are IBM's bisynchronous communications (Bisync) and synchronous data link control (SDLC).

3. Network layer. The network layer determines which network configuration (dial-up, leased, or packet) is most appropriate for the function provided by the network. The network layer also defines the mechanism in which messages are broken into data packets and routed from a sending node to a receiving node within a communications network.

4. Transport layer. The transport layer controls the end-to-end integrity of the message, which includes message routing, segmenting, and error recovery. The transport layer is the highest layer in terms of communications. Layers above the transport layer are not concerned with technological aspects of the network. The upper three layers address the applications aspects of the network, where the lower three layers address the message transfer. Thus the transport layer acts as the interface between the network and the session layers.

5. Session layer. The session layer is responsible for network availability (i.e., buffer storage and processor capacity). Session responsibilities include network log-on and log-off procedures and user authentication. A session is a temporary condition that exists when data are actually in the process of being transferred and does not include procedures such as call establishment, setup, or disconnect procedures. The session layer determines the type of dialogue available (i.e., simplex, half-duplex, or full duplex).

Open Systems Interconnection

6. *Presentation layer.* The presentation layer addresses any code or syntax conversion necessary to present the data to the network in a common format for communications. Presentation functions include data file formatting, encoding (ASCII, EBCDIC, etc.), encryption and decryption of messages, dialogue procedures, data compression, synchronization, interruption, and termination. The presentation layer performs code and character set translation and determines the display mechanism for messages.

7. *Applications layer.* The application layer is the highest layer in the hierarchy and is analogous to the general manager of the network. The application layer controls the sequence of activities within an application and also the sequence of events between the computer application and the user of another application. The application layer communicates directly with the user's application program.

ASYNCHRONOUS PROTOCOLS

Two of the most commonly used asynchronous data protocols are Western Electric's *selective calling system* (8A1/8B1) and IBM's *asynchronous data link protocol* (83B). In essence, these two protocols are the same set of procedures.

Asynchronous protocols are *character oriented.* That is, unique data link control characters such as end of transmission (EOT) and start of text (STX), no matter where they occur in a transmission, warrant the same action or perform the same function. For example, the end-of-transmission character used with ASCII is 04H. No matter when 04H is received by a secondary, the LCU is cleared and placed in the line monitor mode. Consequently, care must be taken to ensure that the bit sequences for data link control characters do not occur within a message unless they are intended to perform their designated data link functions. Vertical redundancy checking (parity) is the only type of error detection used with asynchronous protocols, and symbol substitution and ARQ (retransmission) are used for error correction. With asynchronous protocols, each secondary station is generally limited to a single terminal/printer pair. This station arrangement is called a *stand alone.* With the stand-alone configuration, all messages transmitted from or received on the terminal CRT are also written on the printer. Thus the printer simply generates a hard copy of all transmissions.

In addition to the line monitoring mode, a remote station can be in any one of three operating modes: *transmit, receive,* and *local.* A secondary station is in the transmit mode whenever it has been designated master. In the transmit mode, the secondary can send formatted messages or acknowledgments. A secondary is in the receive mode whenever it has been selected by the primary. In the receive mode, the secondary can receive formatted messages from the primary. For a terminal operator to enter information into his or her computer terminal, the terminal must be in the local mode. A terminal can be placed in the local mode through software commands sent from the primary or the operator can do it manually from the keyboard.

The polling sequence for most asynchronous protocols is quite simple and usually encompasses sending one or two data link control characters, then a *station polling address.* A typical polling sequence is

```
E D
O C A
T 3
```

The EOT character is the *clearing* character and always precedes the polling sequence. EOT places all the secondaries in the line monitor mode. When in the line monitor mode, a secondary station listens to the line for its polling or selection address. When DC3 immediately follows EOT, it indicates that the next character is a station polling address. For this example, the station polling address is the single ASCII character "A." Station "A" has been designated the master and must respond with either a formatted message or an acknowledgment. There are two acknowledgment sequences that may be transmitted in response to a poll. They are listed below together with their functions.

Acknowledgment	Function
A \ C K	No message to transmit, ready to receive
\\	No message to transmit, not ready to receive

The selection sequence, which is very similar to the polling sequence, is

```
E
O  X  Y
T
```

Again, the EOT character is transmitted first to ensure that all the secondary stations are in the line monitor mode. Following the EOT is a two-character selection address "XY." Station XY has been selected by the primary and designated as a receiver. Once selected, a secondary station must respond with one of three acknowledgment sequences indicating its status. They are listed below together with their functions.

Acknowledgment	Function
A \ C K	Ready to receive
\\	Not ready to receive, terminal in local, or printer out of paper
**	Not ready to receive, have a formatted message to transmit

More than one station can be selected simultaneously with *group* or *broadcast* addresses. Group addresses are used when the primary desires to select more than one but not all of the remote stations. There is a single broadcast address that is used to select simultaneously all the remote stations. With asynchronous protocols, acknowledgment procedures for group and broadcast selections are somewhat involved and for this reason are seldom used.

Messages transmitted from the primary and secondary use exactly the same data format. The format is as follows:

```
S                    E
T  message data  0
X                    T
```

The preceding format is used by the secondary to transmit data to the primary in response to a poll. The STX and EOT characters frame the message. STX precedes the data and indicates that the message begins with the character that immediately follows it. The EOT character signals the end of the message and relinquishes the role of master to the primary. The same format is used when the primary transmits a message except that the STX and EOT characters have an additional function. The STX is a *blinding* character. Upon receipt of the STX character, all previously unselected stations are "blinded," which means that they ignore all transmissions except EOT. Consequently, the subsequent message transmitted by the primary is received only by the previously selected station. The unselected secondaries remain blinded until they receive an EOT character, at which time they will return to the line monitor mode and again listen to the line for their polling or selection addresses. STX and EOT are not part of the message; they are data link control characters and are inserted and deleted by the LCU.

Sometimes it is necessary or desirable to transmit coded data in addition to the message that are used only for data link management, such as date, time of message, message number, message priority, routing information, and so on. This bookkeeping information is not part of the message; it is overhead and is transmitted as *heading* information. To identify the heading, the message begins with a start-of-heading character (SOH). SOH is transmitted first, followed by the heading information, STX, then the message. The entire sequence is terminated with an EOT character. When a heading is included, STX terminates the heading and also indicates the beginning of the message. The format for transmitting heading information together with message data is

```
S            S              E
0 heading    T message data 0
H            X              T
```

SYNCHRONOUS PROTOCOLS

With synchronous protocols, a secondary station can have more than a single terminal/printer pair. The group of devices is commonly called a *cluster*. A single LCU can serve a cluster with as many as 50 devices (terminals and printers). Synchronous protocols can be either character or bit oriented. The most commonly used character-oriented synchronous protocol is IBM's 3270 binary synchronous communications (BSC or bisync), and the most popular bit-oriented protocol (BOP) is IBM's synchronous data link control (SDLC).

IBM's Bisync Protocol

With bisync, each transmission is preceded by a unique SYN character: 16H for ASCII and 32H for EBCDIC. The SYN character places the receive USRT in the character or byte mode and prepares it to receive data in 8-bit groupings. With bisync, SYN characters are always transmitted in pairs (hence the name "bisync"). Therefore, if 8 successive bits are received in the middle of a message that are equivalent to a SYN character, they are ignored. For example, the characters "A" and "b" have the following hex and binary codes:

```
A = 41H = 0 1 0 0 0 0 0 1
b = 62H = 0 1 1 0 0 0 1 0
```

If the ASCII characters A and b occur successively during a message or heading, the following bit sequence occurs:

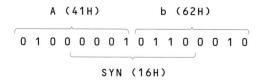

As you can see, it appears that a SYN character has been transmitted when actually it has not. To avoid this situation, SYN characters are always transmitted in pairs, and consequently, if only one is received, it is ignored. The likelihood of two false SYN characters occurring one immediately after the other is remote.

With synchronous protocols, the concepts of polling, selecting, and acknowledging are identical to those used with asynchronous protocols except, with bisync, group, and broadcast selections are not allowed. There are two polling formats used with bisync: general and specific. The format for a general poll is

```
P  S  S  E  P  S  S  S        E  P
A  Y  Y  O  A  Y  Y  P  "  "   N  A
D  N  N  T  D  N  N  A  A      Q  D
```

The PAD character at the beginning of the sequence is called a *leading* pad and is either a 55H or an AAH (01010101 or 10101010 binary). As you can see, a leading pad is simply a string of alternating 1's and 0's. The purpose of the leading pad is to ensure that transitions occur in the data prior to the actual message. The transitions are needed for clock recovery in the receive modem to maintain bit synchronization. Next, there are two SYN characters to establish character synchronization. The EOT character is again used as a clearing character and places all the secondary stations into the line monitor mode. The PAD character immediately following the second SYN character is simply a string of successive logic 1's that is used for a time fill, giving each of the secondary stations time to clear. The number of 1's transmitted during this time fill may not be a multiple of 8 bits. Consequently, the two SYN characters are repeated to reestablish character synchronization. The SPA is not an ASCII or EBCDIC character. The letters SPA stand for *station polling address*. Each secondary station has a unique SPA. Two SPAs are transmitted for the purpose of error detection (redundancy). A secondary will not respond to a poll unless its SPA appears twice. The two quotation marks signify that the poll is for any device at that station that is in the send mode. If two or more devices are in the send mode when a general poll is received, the LCU determines which device's message is transmitted. The enquiry (ENQ) character is sometimes called a *format* or *line turnaround character* because it completes the polling format and initiates a line turnaround (i.e., the secondary station identified by the SPA is designated master and must respond).

The PAD character at the end of the polling sequence is called a *trailing* pad and is simply a 7FH (DEL or delete character). The purpose of the trailing pad is to ensure that the RLSD signal in the receive modem is held active long enough for the entire received message to be demodulated. If the carrier were shut off immediately at the end of the message, RLSD would go inactive and disable the receive data pin. If the last character of the message were not completely demodulated, the end of it would be cut off.

The format for a specific poll is

Synchronous Protocols

```
P S S E P S S S     E P
A Y Y O A Y Y P P D D N A
D N N T D N N A A A A Q D
```

The character sequence for a specific poll is similar to that of a general poll except that the two DAs (*device addresses*) are substituted for the two quotation marks. With a specific poll, both the station and device addresses are included. Therefore, a specific poll is an invitation for a specific device at a given station to transmit its message. Again, two DAs are transmitted for redundancy error detection.

The character sequence for a selection is

```
P S S E P S S S     E P
A Y Y O A Y Y S S D D N A
D N N T D N N A A A A Q D
```

The sequence for a selection is similar to that of a specific poll except that two SSA characters are substituted for the two SPAs. SSA stands for "station select address." All selections are specific; they are for a specific device (device DA). Table 14-1 lists the SPAs, SSAs, and DAs for a network that can have a maximum of 32 stations and the LCU at each station can serve a 32-device cluster.

EXAMPLE 14-1

Determine the character sequences for (a) a general poll for station 8, (b) a specific poll for device 6 at station 8, and (c) a selection of device 6 at station 8.

Solution (a) From Table 14-1 the SPA for station 8 is H; therefore the sequence for a general poll is

```
P S S E P S S         E P
A Y Y O A Y Y H H " " N A
D N N T D N N         Q D
```

TABLE 14-1 STATION AND DEVICE ADDRESSES

Station or device number	SPA	SSA	DA	Station or device number	SPA	SSA	DA
0	sp	-	sp	16	&	Ø	&
1	A	/	A	17	J	1	J
2	B	S	B	18	K	2	K
3	C	T	C	19	L	3	L
4	D	U	D	20	M	4	M
5	E	V	E	21	N	5	N
6	F	W	F	22	O	6	O
7	G	X	G	23	P	7	P
8	H	Y	H	24	Q	8	Q
9	I	Z	I	25	R	9	R
10	[	¦	[	26	]	:	]
11	.	,	.	27	$	#	$
12	<	%	<	28	*	@	*
13	(	—	(	29	)	'	)
14	+	>	+	30	;	=	;
15	!	?	!	31	∧	"	∧

Chap. 14　Data Communications Protocols

(b) From Table 14-1 the DA for device 6 is F; therefore, the sequence for a specific poll is

```
P S S E P S S       E P
A Y Y O A Y Y H H F F N A
D N N T D N N       Q D
```

(c) From Table 14-1 the SSA for station 8 is Y; therefore, the sequence for a selection is

```
P S S E P S S       E P
A Y Y O A Y Y Y Y F F N A
D N N T D N N       Q D
```

With bisync, there are only two ways in which a secondary can respond to a poll: with a formatted message or with a *handshake*. A handshake is simply a response from the secondary that indicates it has no formatted messages to transmit (i.e., a handshake is a negative acknowledgment to a poll). The character sequence for a handshake is

```
P S S E P
A Y Y O A
D N N T D
```

A secondary can respond to a selection with either a positive or a negative acknowledgment. A positive acknowledgment to a selection indicates that the device selected is ready to receive. The character sequence for a positive acknowledgment is

```
P S S D   P
A Y Y L O A
D N N E   D
```

A negative acknowledgment to a selection indicates that the device selected is not ready to receive. A negative acknowledgment is called a *reverse interrupt* (RVI). The character sequence for an RVI is

```
P S S D   P
A Y Y L < A
D N N E   D
```

With bisync, formatted messages are sent from a secondary to the primary in response to a poll and sent from the primary to a secondary after the secondary has been selected. Formatted messages use the following format:

```
P S S S          S         E B P
A Y Y O heading  T message T C A
D N N H          X         X C D
```

Note: If CRC-16 is used for error detection, there are two block check characters.

Longitudinal redundancy checking (LRC) is used for error detection with ASCII-coded messages, and cyclic redundancy checking (CRC) is used for EBCDIC. The BCC is computed beginning with the first character after SOH and continues through and includes ETX. (If there is no heading, the BCC is computed beginning with the first character after

STX.) With synchronous protocols, data are transmitted in blocks. Blocks of data are generally limited to 256 characters. ETX is used to terminate the last block of a message. ETB is used for multiple block messages to terminate all message blocks except the last one. The last block of a message is always terminated with ETX. All BCCs must be acknowledged by the receiving station. A positive acknowledgment indicates that the BCC was good and a negative acknowledgment means that the BCC was bad. A negative acknowledgment is an automatic request for retransmission. The character sequences for positive and negative acknowledgments are as follows:

Positive acknowledgment:

```
P S S D   P        P S S D   P
A Y Y L 0 A  or  A Y Y L 1 A
D N N E   D        D N N E   D

even-numbered      odd-numbered
   blocks             blocks
```

Negative acknowledgment:

```
P S S N P
A Y Y A A
D N N K D
```

Examples of dialogue using bisync protocol.

```
P S S E P S S       E P
A Y Y 0 A Y Y A A " " N A  ———→
D N N T D N N       Q D
```

Primary station sends a general poll for station 1.

```
        P S S E P
 ←——— A Y Y 0 A ———
        D N N T D
```

Station 1 responds with a negative acknowledgment—no messages to transmit.

```
 ┌── P S S E P S S       E P
 │   A Y Y 0 A Y Y B B " " N A  ———→
 │   D N N T D N N       Q D
```

Primary station sends a general poll for station 2.

```
 P S S S         S  message E B P
 A Y Y 0 heading T  block 1 T C A  ———
 D Y Y H         X          B C D
```

Station 2 responds with the first block of a multiblock message.

```
        P S S D   P
 ←——— A Y Y L 1 A ———→
        D N N E   D
```

Chap. 14   **Data Communications Protocols**

Primary sends a positive acknowledgment indicating that block
1 was received without any errors—because block 1 is an odd-
numbered block, DLE 1 is used.

```
P S S         E B P
A Y Y T message T C A
D N N X block 2 X C D
```

Station 2 sends the second and final block of the message—note
that there is no heading at the beginning of the second block—a
heading is transmitted only with the first block of a message.

```
P S S N P
A Y Y A A
D N N K D
```

Primary sends a negative acknowledgment to station 2 indicating
that block 2 was received with an error and must be transmitted.

```
P S S         E B P
A Y Y T message T C A
D N N X block 2 X C D
```

Station 2 resends block 2.

```
P S S D   P
A Y Y L O A
D N N E   D
```

Primary sends a positive acknowledgment to station 2 indicating
that block 2 was received without any errors—because block 2 is
an even-numbered block, DLE 0 is used.

```
P S S E P
A Y Y O A
D N N T D
```

Secondary responds with a handshake—a secondary sends a handshake whenever it is its turn to transmit but it has nothing to say.

```
P S S E P S S       E P
A Y Y O A Y Y T T E E N A
D N N T D N N       Q D
```

Primary selects station 3, device 5.

```
P S S D   P
A Y Y L O A
D N N E   D
```

Station 3 sends a positive acknowledgment to the selection; device
5 is ready to receive.

Synchronous Protocols

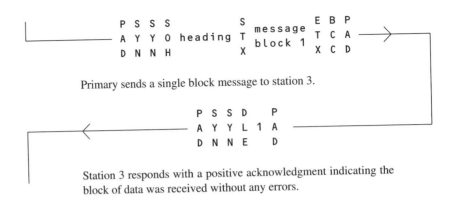

```
          P S S         S           E B P
        - A Y Y 0  heading  T  message  T C A  ----->
          D N N H         X  block 1   X C D
```

Primary sends a single block message to station 3.

```
                    P S S D   P
      <---------   - A Y Y L 1 A  ---------
                    D N N E   D
```

Station 3 responds with a positive acknowledgment indicating the
block of data was received without any errors.

Transparency. It is possible that a device that is attached to one of the ports of a station LCU is not a computer terminal or a printer. For example, a microprocessor-controlled monitor system that is used to monitor environmental conditions (temperature, humidity, etc.) or a security alarm system. If so, the data transferred between it and the applications program are not ASCII- or EBCDIC-encoded characters; they are microprocessor op-codes or binary-encoded data. Consequently, it is possible that an 8-bit sequence could occur in the message that is equivalent to a data link control character. For example, if the binary code 00000011 (03H) occurred in a message, the LCU would misinterpret it as the ASCII code for ETX. Consequently, the receive LCU would prematurely terminate the message and interpret the next 8-bit sequence as a BCC. To prevent this from occurring, the LCU is made *transparent* to the data. With bisync, a *data link escape* character (DLE) is used to achieve transparency. To place an LCU in the transparent mode, STX is preceded by a DLE. This causes the LCU to transfer the data to the selected device without searching through the message for data link control characters. To come out of the transparent mode, DLE ETX is transmitted. To transmit a DLE as part of the text, it must be preceded by DLE (i.e., DLE DLE). Actually, there are only five characters that it is necessary to precede with DLE:

1. *DLE STX:* places the receive LCU into the transparent mode.
2. *DLE ETX:* used to terminate the last block of transparent text and take the LCU out of the transparent mode.
3. *DLE ETB:* used to terminate blocks of transparent text other than the final block.
4. *DLE ITB:* used to terminate blocks of transparent text other than the final block when ITB is used for a block terminating character.
5. *DLE SYN:* used only with transparent messages that are more than 1 s long. With bisync, two SYN characters are inserted in the text every 1 s to ensure that the receive LCU does not lose character synchronization. In a multipoint circuit with a polling environment, it is highly unlikely that any blocks of data would exceed 1 s in duration. SYN character insertion is used almost exclusively for two-point circuits.

Synchronous Data Link Communications

Synchronous data link communications (SDLC) is a synchronous *bit-oriented* protocol developed by IBM. A bit-oriented protocol (BOP) is a discipline for serial-by-bit information transfer over a data communication channel. With a BOP, data link control information is transferred and interpreted on a bit-by-bit basis rather than with unique data link control characters. SDLC can transfer data either simplex, half duplex, or full duplex.

With a BOP, there is a single control field that performs essentially all the data link control functions. The character language used with SDLC is EBCDIC and data are transferred in groups called *frames*. Frames are generally limited to 256 characters in length. There are two types of stations in SDLC: primary stations and secondary stations. The *primary station* controls data exchange on the communications channel and issues *commands*. The *secondary station* receives commands and returns *responses* to the primary.

There are three transmission states with SDLC: transient, idle, and active. The *transient state* exists before and after the initial transmission and after each line turnaround. An *idle state* is presumed after 15 or more consecutive 1's have been received. The *active state* exists whenever either the primary or a secondary station is transmitting information or control signals.

Figure 14-2 shows the frame format used with SDLC. The frames sent from the primary and the frames sent from a secondary use exactly the same format. There are five fields used with SDLC: the flag field, the address field, the control field, the text or information field, and the frame check field.

Information field. All information transmitted in an SDLC frame must be in the information field (I field), and the number of bits in the I field must be a multiple of 8. An I field is not allowed with all SDLC frames. The types of frames that allow an I field are discussed later.

Flag field. There are two flag fields per frame: the beginning flag and the ending flag. The flags are used for the *delimiting sequence* and to achieve character synchronization. The delimiting sequence sets the limits of the frame (i.e., when the frame begins and when it ends). The flag is used with SDLC in the same manner that SYN characters are used with bisync, to achieve character synchronization. The sequence for a flag is 7EH, 01111110 binary, or the EBCDIC character "=." There are several variations of how flags are used. They are:

1. One beginning and one ending flag for each frame.

```
      beginning flag                                   ending flag
 . . . 01111110 address control text FCC 01111110 . . .
```

2. The ending flag from one frame can be used for the beginning flag for the next frame.

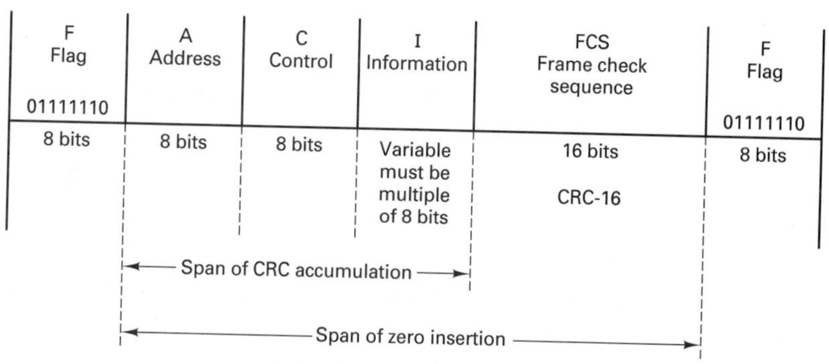

Figure 14-2 SDLC frame format.

```
                    ←─────────────────────────── frame N + 1 ───────────────→
        ←──────── frame N ────────→
. . . text FCC 01111110 address control text FCC 01111110 . . .
                          /     \
                ending flag beginning flag
                  frame N      frame N + 1
```

3. The last zero of an ending flag is also the first zero of the beginning flag of the next frame.

```
                              ←──────────────── frame N + 1 ───────
                    shared 0
          frame N ────→  /
. . . text FCC 011111101111110 address control text FCC . . .
                    /      \
          ending flag beginning flag
            frame N       frame N + 1
```

4. Flags are transmitted in lieu of idle line 1's.

```
01111110111111011111101111110 address control text . . .
        /       /       /      _____⌄_____/
idle line flags  beginning flag
```

Address field.   The address field has 8 bits; thus 256 addresses are possible with SDLC. The address 00H (00000000) is called the *null* or *void address* and is never assigned to a secondary. The null address is used for network testing. The address FFH (11111111) is the *broadcast address* and is common to all secondaries. The remaining 254 addresses can be used as *unique* station addresses or as *group* addresses. In frames sent from the primary, the address field contains the address of the destination station (a secondary). In frames sent from a secondary, the address field contains the address of that secondary. Therefore, the address is always that of a secondary. The primary station has no address because all transmissions from secondary stations go to the primary.

Control field.   The control field is an 8-bit field that identifies the type of frame it is. The control field is used for polling, confirming previously received information frames, and several other data link management functions. There are three frame formats used with SDLC: *information, supervisory,* and *unnumbered.*

Information frame.   With an information frame there must be an information field. Information frames are used for transmitting sequenced information. The bit pattern for the control field of an information frame is

Bit:	b_0	b_1	b_2	b_3	b_4	b_5	b_6	b_7
Function:	←	nr	→	P or F	←	ns	→	0
				$\overline{\text{P or F}}$				

An information frame is identified by a 0 in the least significant bit position (b_7 with EBCDIC code). Bits b_4, b_5, and b_6 are used for numbering transmitted frames (ns = number sent). With 3 bits, the binary numbers 000 through 111 (0–7) can be represented. The first frame transmitted is designated frame 000, the second frame 001, and so on up to frame 111 (the eighth frame); then the count cycles back to 000 and repeats.

Bits b_0, b_1, and b_2 are used to confirm correctly received information frames (nr = number received) and to automatically request retransmission of incorrectly received information frames. The nr is the number of the next frame that the transmitting station expects to receive, or the number of the next frame that the receiving station will transmit. The nr confirms received frames through nr-1. Frame nr-1 is the last frame received without a transmission error. Any transmitted I frame not confirmed must be retransmitted. Together, the ns and nr bits are used for error correction (ARQ). The primary must keep track of an ns and nr for each secondary. Each secondary must keep track of only its ns and nr. After all frames have been confirmed, the primary's ns must agree with the secondary's nr, and vice versa. For the example shown next, the primary and secondary stations begin with their ns and nr counters reset to 000. The primary sends three numbered information frames (ns = 0, 1, and 2). At the same time the primary sends nr = 0 because the next frame it expects to receive is frame 0, which is the secondary's present ns. The secondary responds with two information frames (ns = 0 and 1). The secondary received all three frames from the primary without any errors, so the nr transmitted in the secondary's control field is 3 (which is the number of the next frame that the primary will send). The primary now sends information frames 3 and 4 with an nr = 2. The nr = 2 confirms the correct reception of frames 0 and 1. The secondary responds with frames ns = 2, 3, and 4 with an nr = 4. The nr = 4 confirms reception of only frame 3 from the primary (nr-1). Consequently, the primary must retransmit frame 4. Frame 4 is retransmitted together with four additional frames (ns = 5, 6, 7, and 0). The primary's nr = 5, which confirms frames 2, 3, and 4 from the secondary. Finally, the secondary sends information frame 5 with an nr = 1. The nr = 1 confirms frames 4, 5, 6, 7 and 0 from the primary. At this point, all the frames transmitted have been confirmed except frame 5 from the secondary.

Primary's ns:	0 1 2		3 4		4 5 6 7 0
Primary's nr:	0 0 0		2 2		5 5 5 5 5
Secondary's ns:		0 1		2 3 4	5
Secondary's nr:		3 3		4 4 4	1

With SDLC, a station can never send more than seven numbered frames without receiving a confirmation. For example, if the primary sent eight frames (ns = 0, 1, 2, 3, 4, 5, 6, and 7) and the secondary responded with an nr = 0, it is ambiguous which frames are being confirmed. Does nr = 0 mean that all eight frames were received correctly, or that frame 0 had an error in it and all eight frames must be retransmitted? (With SDLC, all previously transmitted frames beginning with frame nr-1 must be retransmitted.)

Bit b_3 is the *poll* (P) or *not-a-poll* ($\overline{P}$) bit when sent from the primary and the *final* (F) or *not-a-final* ($\overline{F}$) bit when sent by a secondary. In a frame sent by the primary, if the primary desires to poll the secondary, the P bit is set (1). If the primary does not wish to poll the secondary, the P bit is reset (0). A secondary cannot transmit unless it receives a frame addressed to it with the P bit set. In a frame sent from a secondary, if it is the last (final) frame of the message, the F bit is set (1). If it is not the final frame, the F bit is reset (0). With I frames, the primary can select a secondary station, send formatted information, confirm previously received I frames, and poll with a single transmission.

EXAMPLE 14-2

Determine the bit pattern for the control field of a frame sent from the primary to a secondary station for the following conditions: primary is sending information frame 3, it is a poll, and the primary is confirming the correct reception of frames 2, 3, and 4 from the secondary.

Solution

$b_7 = 0$ because it is an information frame.
b_4, b_5, and b_6 are 011 (binary 3 for ns = 3).
$b_3 = 1$, it is a polling frame.
b_0, b_1, and b_2 are 101 (binary 5 for nr = 5).
control field = B6H.

b_0	b_1	b_2	b_3	b_4	b_5	b_6	b_7
1	0	1	1	0	1	1	0

Supervisory frame. An information field is not allowed with a supervisory frame. Consequently, supervisory frames cannot be used to transfer information; they are used to assist in the transfer of information. Supervisory frames are used to confirm previously received information frames, convey ready or busy conditions, and to report frame numbering errors. The bit pattern for the control field of a supervisory frame is

Bit:	b_0	b_1	b_2	b_3	b_4	b_5	b_6	b_7
Function:	←	nr	→	P or F / $\overline{\text{P or F}}$	X	X	0	1

A supervisory frame is identified by a 01 in bit positions b_6 and b_7, respectively, of the control field. With the supervisory format, bit b_3 is again the poll/not-a-poll or final/not-a-final bit and b_0, b_1, and b_2 are the nr bits. However, with a supervisory format, b_4 and b_5 are used to indicate either the receive status of the station transmitting the frame or to request transmission or retransmission of sequenced information frames. With two bits, there are four combinations possible. The four combinations and their functions are as follows:

b_4	b_5	Receiver status
0	0	Ready to receive (RR)
0	1	Ready not to receive (RNR)
1	0	Reject (REJ)
1	1	Not used with SDLC

When the primary sends a supervisory frame with the P bit set and a status of ready to receive, it is equivalent to a general poll with bisync. Supervisory frames are used by the primary for polling and for confirming previously received information frames when there is no information to send. A secondary uses the supervisory format for confirming previously received information frames and for reporting its receive status to the primary. If a secondary sends a supervisory frame with RNR status, the primary cannot send it numbered information frames until that status is cleared. RNR is cleared when a secondary sends an information frame with the F bit = 1 or a RR or REJ frame with the F bit = 0. The REJ command/response is used to confirm information frames through nr-1 and to request retransmission of numbered information frames beginning with the frame number identified in the REJ frame. An information field is prohibited with a supervisory frame and the REJ command/response is used only with full-duplex operation.

Chap. 14 **Data Communications Protocols**

EXAMPLE 14-3

Determine the bit pattern for the control field of a supervisory frame sent from a secondary station to the primary for the following conditions: the secondary is ready to receive, it is the final frame, and the secondary station is confirming frames 3, 4, and 5.

Solution

b_6 and $b_7 = 01$ because it is a supervisory frame.
b_4 and $b_5 = 00$ (ready to receive).
$b_3 = 1$ (it is the final frame).
b_0, b_1, and $b_2 = 110$ (binary 6 for nr = 6).
control field = D1H

$$
\begin{array}{cccccccc}
b_0 & b_1 & b_2 & b_3 & b_4 & b_5 & b_6 & b_7 \\
1 & 1 & 0 & 1 & 0 & 0 & 0 & 1
\end{array}
$$

Unnumbered frame. An unnumbered frame is identified by making bits b_6 and b_7 in the control field 11. The bit pattern for the control field of an unnumbered frame is

Bit:	b_0	b_1	b_2	b_3	b_4	b_5	b_6	b_7
Function:	X	X	X	$\frac{P \text{ or } F}{\overline{P} \text{ or } \overline{F}}$	X	X	1	1

With an unnumbered frame, bit b_3 is again either the $P/\overline{P}$ or $F/\overline{F}$ bit. Bits b_0, b_1, b_2, b_4, and b_5 are used for various unnumbered commands and responses. With 5 bits available, 32 unnumbered commands/responses are possible. The control field in an unnumbered frame sent by the primary is a command. The control field in an unnumbered frame sent by a secondary is a response. With unnumbered frames, there are no ns or nr bits. Therefore, numbered information frames cannot be sent or confirmed with the unnumbered format. Unnumbered frames are used to send network control and status information. Two examples of control functions are (1) placing secondary stations on-line and off-line and (2) LCU initialization. Table 14-2 lists several of the more commonly used unnumbered commands and responses. An information field is prohibited with all the unnumbered commands/responses except UI, FRMR, CFGR, TEST, and XID.

TABLE 14-2 UNNUMBERED COMMANDS AND RESPONSES

Binary configuration b_0		b_7	Acronym	Command	Response	I field prohibited	Resets ns and nr
000	P/F	0011	UI	Yes	Yes	No	No
000	F	0111	RIM	No	Yes	Yes	No
000	P	0111	SIM	Yes	No	Yes	Yes
100	P	0011	SNRM	Yes	No	Yes	Yes
000	F	1111	DM	No	Yes	Yes	No
010	P	0011	DISC	Yes	No	Yes	No
011	F	0011	UA	No	Yes	Yes	No
100	F	0111	FRMR	No	Yes	No	No
111	F	1111	BCN	No	Yes	Yes	No
110	P/F	0111	CFGR	Yes	Yes	No	No
010	F	0011	RD	No	Yes	Yes	No
101	P/F	1111	XID	Yes	Yes	No	No
001	P	0011	UP	Yes	No	Yes	No
111	P/F	0011	TEST	Yes	Yes	No	No

A secondary station must be in one of three modes: the initialization mode, the normal response mode, or the normal disconnect mode. The procedures for the *initialization mode* are system specified and vary considerably. A secondary in the *normal response mode* cannot initiate unsolicited transmissions; it can transmit only in response to a frame received with the P bit set. When in the *normal disconnect mode,* a secondary is off-line. In this mode, a secondary can receive only a TEST, XID, CFGR, SNRM, or SIM command from the primary and can respond only if the P bit is set.

The unnumbered commands and responses are summarized below.

Unnumbered information (UI).　UI is a command/response that is used to send unnumbered information. Unnumbered information transmitted in the I field is not confirmed.

Set initialization mode (SIM).　SIM is a command that places the secondary station into the initialization mode. The initialization procedure is system specified and varies from a simple self-test of the station controller to executing a complete IPL (initial program logic) program. SIM resets the ns and nr counters at the primary and secondary stations. A secondary is expected to respond to a SIM command with a UA response.

Request initialization mode (RIM).　RIM is a response sent by a secondary station to request the primary to send an SIM command.

Set normal response mode (SNRM).　SNRM is a command that places a secondary station in the normal response mode (NRM). A secondary station cannot send or receive numbered information frames unless it is in the normal response mode. Essentially, SNRM places a secondary station on-line. SNRM resets the ns and nr counters at the primary and secondary stations. UA is the normal response to an SNRM command. Unsolicited responses are not allowed when the secondary is in the NRM. A secondary remains in the NRM until it receives a DISC or SIM command.

Disconnect mode (DM).　DM is a response that is sent from a secondary station if the primary attempts to send numbered information frames to it when the secondary is in the normal disconnect mode.

Request disconnect (RD).　RD is a response sent when a secondary wishes to be placed in the disconnect mode.

Disconnect (DISC).　DISC is a command that places a secondary station in the normal disconnect mode (NDM). A secondary cannot send or receive numbered information frames when it is in the normal disconnect mode. When in the normal disconnect mode, a secondary can receive only an SIM or SNRM command and can transmit only a DM response. The expected response to a DISC command is UA.

Unnumbered acknowledgment (UA).　UA is an affirmative response that indicates compliance to a SIM, SNRM, or DISC command. UA is also used to acknowledge unnumbered information frames.

Frame reject (FRMR).　FRMR is for reporting procedural errors. The FRMR sequence is a response transmitted when the secondary has received an invalid frame from the primary. A received frame may be invalid for any one of the following reasons:

1. The control field contains an invalid or unassigned command.
2. The amount of data in the information field exceeds the buffer space at the secondary.

3. An information field is received in a frame that does not allow information.

4. The nr received is incongruous with the secondary's ns. For example, if the secondary transmitted ns frames 2, 3, and 4 and then the primary responded with an nr of 7.

A secondary cannot release itself from the FRMR condition, nor does it act on the frame that caused the condition. The secondary repeats the FRMR response until it receives one of the following *mode-setting* commands: SNRM, DISC, or SIM. The information field for a FRMR response always contains three bytes (24 bits) and has the following format:

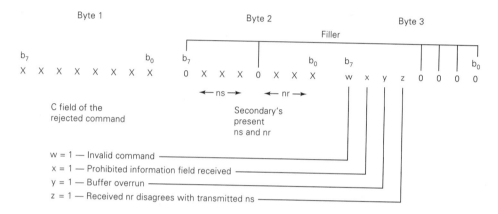

TEST. Test is a command that can be sent in any mode to solicit a TEST response. If an information field is included with the command, the secondary returns it with the response. The TEST command/response is exchanged for link testing purposes.

Exchange station identification (XID). As a command, XID solicits the identification of the secondary station. An information field can be included in the frame to convey the identification data of either the primary or secondary station. For dial-up circuits, it is often necessary that the secondary station identify itself before the primary will exchange information frames with it, although XID is not restricted to only dial-up data circuits.

Frame check sequence field. The FCS field contains the error detection mechanism for SDLC. The FCS is equivalent to the BCC used with bisync. SDLC uses CRC-16 and the following generating polynomial: $x^{16} + x^{12} + x^5 + x^1$.

SDLC Loop Operation

An SDLC *loop* is operated in the half-duplex mode. The primary difference between the loop and bus configurations is that in a loop, all transmissions travel in the same direction on the communications channel. In a loop configuration, only one station transmits at a time. The primary transmits first, then each secondary station responds sequentially. In an SDLC loop, the transmit port of the primary station controller is connected to one or more secondary stations in a serial fashion; then the loop is terminated back at the receive port of the primary. Figure 14-3 shows an SDLC loop configuration.

In an SDLC loop, the primary transmits frames that are addressed to any or all of the secondary stations. Each frame transmitted by the primary contains an address of the secondary station to which that frame is directed. Each secondary station, in turn, decodes the address field of every frame, then serves as a repeater for all stations that are down-loop

Synchronous Protocols

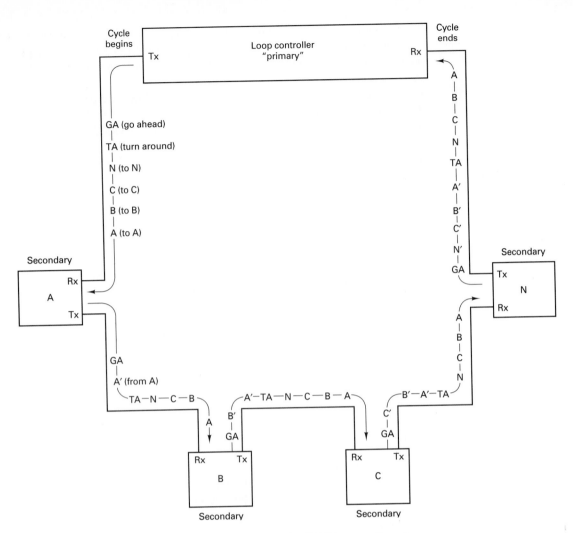

Figure 14-3 SDLC loop configuration.

from it. If a secondary detects a frame with its address, it accepts the frame, then passes it on to the next down-loop station. All frames transmitted by the primary are returned to the primary. When the primary has completed transmitting, it follows the last flag with eight consecutive 0's. A flag followed by eight consecutive 0's is called a *turnaround* sequence which signals the end of the primary's transmission. Immediately following the turn-around sequence, the primary transmits continuous 1's, which generates a *go-ahead* sequence (01111111). A secondary cannot transmit until it has received a frame addressed to it with the P bit set, a turnaround sequence, and then a go-ahead sequence. Once the primary has begun transmitting 1's, it goes into the receive mode.

The first down-loop secondary station that has received a frame addressed to it with the P bit set, changes the seventh 1 bit in the go-ahead sequence to a 0, thus creating a flag. That flag becomes the beginning flag of the secondary's response frame or frames. After the secondary has transmitted its last frame, it again becomes a repeater for the idle line 1's from the primary. These idle line 1's again become the go-ahead sequence for the next secondary station. The next down-loop station that has received a frame addressed to it with the P bit set detects the turnaround sequence, any frames transmitted from up-loop secondaries, and then the go-ahead sequence. Each secondary station inserts its response frames immediately after the last repeated frame. The cycle is completed when the primary receives its own turnaround sequence, a series of response frames, and then the go-ahead sequence.

Configure command/response.   The configure command/response (CFGR) is an unnumbered command/response that is used only in a loop configuration. CFGR contains a one-byte *function descriptor* (essentially a subcommand) in the information field. A CFGR command is acknowledged with a CFGR response. If the low-order bit of the function descriptor is set, a specified function is initiated. If it is reset, the specified function is cleared. There are six subcommands that can appear in the configure command's function field.

1. Clear—00000000.   A clear subcommand causes all previously set functions to be cleared by the secondary. The secondary's response to a clear subcommand is another clear subcommand, 00000000.

2. Beacon test (BCN)—0000000X.   The beacon test causes the secondary receiving it to turn on or turn off its carrier. If the X bit is set, the secondary suppresses transmission of the carrier. If the X bit is reset, the secondary resumes transmission of the carrier. The beacon test is used to isolate an open-loop problem. Also, whenever a secondary detects the loss of a receive carrier, it automatically begins to transmit its beacon response. The secondary will continue transmitting the beacon until the loop resumes normal status.

3. Monitor mode—0000010X.   The monitor command causes the addressed secondary to place itself into a monitor (receive only) mode. Once in the monitor mode, a secondary cannot transmit until it receives a monitor mode clear (00000100) or a clear (00000000) subcommand.

4. Wrap—0000100X.   The wrap command causes the secondary station to loop its transmissions directly to its receiver input. The wrap command places the secondary effectively off-line for the duration of the test. A secondary station does not send the results of a wrap test to the primary.

5. Self-test—0000101X.   The self-test subcommand causes the addressed secondary to initiate a series of internal diagnostic tests. When the tests are completed, the secondary will respond. If the P bit in the configure command is set, the secondary will respond following completion of the self-test at its earliest opportunity. If the P bit is reset, the secondary will respond following completion of the test to the next poll-type frame it receives. All other transmissions are ignored by the secondary while it is performing the self-tests. The secondary indicates the results of the self-test by setting or resetting the low-order bit (X) of its self-test response. A 1 indicates that the tests were unsuccessful, and a 0 indicates that they were successful.

6. Modified link test—0000110X.   If the modified link test function is set (X bit set), the secondary station will respond to a TEST command with a TEST response that has an information field containing the first byte of the TEST command information field repeated *n* times. The number *n* is system implementation dependent. If the X bit is reset, the secondary station will respond to a TEST command, with or without an information field, with a TEST response with a zero-length information field. The modified link test is an optional subcommand and is only used to provide an alternative form of link test to that previously described for the TEST command.

Synchronous Protocols

Transparency

The transparency mechanism used with SDLC is called *zero-bit insertion* or *zero stuffing*. The flag bit sequence (01111110) can occur in a frame where this pattern is not intended to be a flag. For example, any time that 7EH occurs in the address, control, information, or FCS field it would be interpreted as a flag and disrupt character synchronization. Therefore, 7EH must be prohibited from occurring except when it is intended to be a flag. To prevent a 7EH sequence from occurring, a zero is automatically inserted after any occurrence of five consecutive 1's except in a designated flag sequence (i.e., flags are not zero inserted). When five consecutive 1's are received and the next bit is a 0, the 0 is deleted or removed. If the next bit is a 1, it must be a valid flag. An example of zero insertion/deletion is shown below.

Original frame bits at the transmit station:

```
01111110   01101111   11010011   1110001100110101   01111110
  flag      address    control           FCS            flag
```

After zero insertion but prior to transmission:

```
01111110   01101111   101010011   11100001100110101   01111110
  flag      address    ↑ control          ↑  FCS          flag
                       inserted zeros
```

After zero deletion at the receive end:

```
01111110   01101111   11010011   1110001100110101   01111110
  flag      address    control          FCS             flag
```

Message Abort

Message abort is used to prematurely terminate a frame. Generally, this is only done to accommodate high-priority messages such as emergency link recovery procedures, and so on. A message abort is any occurrence of 7 to 14 consecutive 1's. Zeros are not inserted in an abort sequence. A message abort terminates an existing frame and immediately begins the higher-priority frame. If more than 14 consecutive 1's occur in succession, it is considered an idle line condition. Therefore, 15 or more successive 1's place the circuit into the idle state.

Invert-on-Zero Encoding

A binary synchronous transmission such as SDLC is time synchronized to enable identification of sequential binary digits. Synchronous data communications assumes that bit or clock synchronization is provided by either the DCE or the DTE. With synchronous transmissions, a receiver samples incoming data at the same rate that they were transmitted. Although minor variations in timing can exist, synchronous modems provide received data clock recovery and dynamically adjusted sample timing to keep sample times midway between bits. For a DTE or a DCE to recover the clock, it is necessary that transitions occur in the data. *Invert-on-zero coding* is an encoding scheme that guarantees at least one transition in the data for every 7 bits transmitted. Invert-on-zero coding is also called NRZI (*nonreturn-to-zero inverted*).

With NRZI encoding, the data are encoded in the transmitter, then decoded in the receiver. Figure 14-4 shows an example of NRZI encoding. The encoded waveform is unchanged by 1's in the NRZI encoder. However, 0's invert the encoded transmission level. Consequently, consecutive 0's generate an alternating high/low sequence. With SDLC, there can never be more than six 1's in succession (a flag). Therefore, a high-to-low transition is guaranteed to occur at least once for every 7 bits transmitted except during a message abort or an idle line condition. In a NRZI decoder, whenever a high/low transition occurs in the received data, a 0 is generated. The absence of a transition simply generates a 1. In Figure 14-4, a high level is assumed prior to encoding the incoming data.

NRZI encoding was intended to be used with asynchronous modems which do not have clock recovery capabilities. Consequently, the DTE must provide time synchronization which is aided by using NRZI-encoded data. Synchronous modems have built in scramblers and descramblers which ensure that transitions occur in the data, and thus NRZI encoding is unnecessary. The NRZI encoder/decoder is placed between the DTE and the DCE.

High-Level Data Link Control

In 1975, the International Standards Organization (ISO) defined several sets of substandards that, when combined, are called *high-level data link control* (HDLC). Since HDLC is a superset of SDLC, only the added capabilities are explained.

HDLC comprises three standards (subdivisions) that, when combined, outline the frame structure, control standards, and class of operation for a bit-oriented data link control (DLC).

ISO 3309–1976(E). This standard defines the frame structure, delimiting, sequence, and transparency mechanism used with HDLC. These are essentially the same as with SDLC except that HDLC has extended addressing capabilities and checks the FCS in a slightly different manner. The delimiting sequence used with HDLC is identical to SDLC: a 01111110 sequence.

HDLC can use either the *basic* 8-bit address field or an *extended* addressing format. With extended addressing the address field may be extended recursively. If b_0 in the address byte is a logic 1, the 7 remaining bits are the secondary's address (the ISO defines the low-order bit as b_0, whereas SDLC designates the high-order bit as b_0). If b_0 is a logic 0, the next byte is also part of the address. If b_0 of the second byte is a 0, a third address byte follows, and so on, until an address byte with a logic 1 for the low-order bit is encountered. Essentially, there are 7 bits available in each address byte for address encoding. An example of a three-byte extended addressing scheme is shown, b_0 in the first two

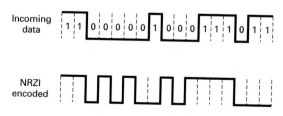

Figure 14-4 NRZI encoding.

bytes of the address field are 0's, indicating that additional address bytes follow and b_0 in the third address byte is a logic 1, which terminates the address field.

```
          b₀ = 0      b₀ = 0      b₀ = 1
          |           |           |
01111110  0XXXXXXX    0XXXXXXX    1XXXXXXX    . . .
  flag          three-byte address field   control field, etc.
```

HDLC uses CRC-16 with a generating polynomial specified by CCITT V.41. At the transmit station, the CRC is computed such that if it is included in the FCS computation at the receive end, the remainder for an errorless transmission is always F0BBH.

ISO 4335–1979(E). This standard defines the elements of procedure for HDLC. The control field, information field, and supervisory format have increased capabilities over SDLC.

Control field. With HDLC, the control field can be extended to 16 bits. Seven bits are for the ns and 7 bits are for the nr. Therefore, with the extended control format, there can be a maximum of 127 outstanding (unconfirmed) frames at any give time.

Information field. HDLC permits any number of bits in the information field of an information command or response (SDLC is limited to 8-bit bytes). With HDLC any number of bits may be used for a character in the I field as long as all characters have the same number of bits.

Supervisory format. With HDLC, the supervisory format includes a fourth status condition: selective reject (SREJ). SREJ is identified by a 11 in bit position b_4 and b_5 of a supervisory control field. With a SREJ, a single frame can be rejected. A SREJ calls for the retransmission of only the frame identified by nr, whereas a REJ calls for the retransmission of all frames beginning with nr. For example, the primary sends I frames ns = 2, 3, 4, and 5. Frame 3 was received in error. A REJ would call for a retransmission of frames 3, 4, and 5; a SREJ would call for the retransmission of only frame 3. SREJ can be used to call for the retransmission of any number of frames except that only one is identified at a time.

Operational modes. HDLC has two operational modes not specified in SDLC: asynchronous response mode and asynchronous disconnect mode.

1. *Asynchronous response mode (ARM).* With the ARM, secondary stations are allowed to send unsolicited responses. To transmit, a secondary does not need to have received a frame from the primary with the P bit set. However, if a secondary receives a frame with the P bit set, it must respond with a frame with the F bit set.

2. *Asynchronous disconnect mode (ADM).* An ADM is identical to the normal disconnect mode except that the secondary can initiate a DM or RIM response at any time.

ISO 7809–1985(E). This standard combines previous standards 6159(E) (unbalanced) and 6256(E) (balanced) and outlines the class of operation necessary to establish the link-level protocol.

Unbalanced operation. This class of operation is logically equivalent to a multipoint private line circuit with a polling environment. There is a single primary station responsible for central control of the network. Data transmission may be either half- or full-duplex.

Balanced operation.   This class of operation is logically equivalent to a two-point private line circuit. Each station has equal data link responsibilities, and channel access is through contention using the asynchronous response mode. Data transmission may be half- or full-duplex.

PUBLIC DATA NETWORK

A *public data network* (PDN) is a switched data communications network similar to the public telephone network except that a PDN is designed for transferring data only. Public data networks combine the concepts of both *value-added networks* (VANs) and *packet-switching networks.*

Value-Added Network

A value-added network "*adds value*" to the services or facilities provided by a common carrier to provide new types of communication services. Examples of added values are error control, enhanced connection reliability, dynamic routing, failure protection, logical multiplexing, and data format conversions. A VAN comprises an organization that leases communications lines from common carriers such as AT&T and MCI and adds new types of communications services to those lines. Examples of value-added networks are GTE Telnet, DATAPAC, TRANSPAC, and Tymnet Inc.

Packet-Switching Network

Packet switching involves dividing data messages into small bundles of information and transmitting them through communications networks to their intended destinations using computer-controlled switches. Three common switching techniques are used with public data networks: *circuit switching, message switching,* and *packet switching.*

Circuit switching.   Circuit switching is used for making a standard telephone call on the public telephone network. The call is established, information is transferred, and then the call is disconnected. The time required to establish the call is called the *setup* time. Once the call has been established, the circuits interconnected by the network switches are allocated to a single user for the duration of the call. After a call has been established, information is transferred in *real time.* When a call is terminated, the circuits and switches are once again available for another user. Because there are a limited number of circuits and switching paths available, *blocking* can occur. Blocking is the inability to complete a call because there are no facilities or switching paths available between the source and destination locations. When circuit switching is used for data transfer, the terminal equipment at the source and destination must be compatible; they must use compatible modems and the same bit rate, character set, and protocol.

A circuit switch is a *transparent* switch. The switch is transparent to the data; it does nothing more than interconnect the source and destination terminal equipment. A circuit switch adds no value to the circuit.

Message switching.   Message switching is a form of *store-and-forward* network. Data, including source and destination identification codes, are transmitted into the network and stored in a switch. Each switch within the network has message storage capabilities. The network transfers the data from switch to switch when it is convenient to do so. Consequently, data are not transferred in real time; there can be a delay at each switch. With message switching, blocking cannot occur. However, the delay time from message transmission to reception varies from call to call and can be quite long (possibly as long as 24 hours). With message switching, once the information has entered the network, it is converted to a more suitable format for transmission through the network. At the receive end, the data are converted to a format compatible with the receiving data terminal equipment. Therefore, with message switching, the source and destination data terminal equipment do not need to be compatible. Message switching is more efficient than circuit switching because data that enter the network during busy times can be held and transmitted later when the load has decreased.

A message switch is a *transactional* switch because it does more than simply transfer the data from the source to the destination. A message switch can store data or change its format and bit rate, then convert the data back to their original form or an entirely different form at the receive end. Message switching multiplexes data from different sources onto a common facility.

Packet switching.   With packet switching, data are divided into smaller segments called *packets* prior to transmission through the network. Because a packet can be held in memory at a switch for a short period of time, packet switching is sometimes called a *hold-and-forward* network. With packet switching, a message is divided into packets and each packet can take a different path through the network. Consequently, all packets do not necessarily arrive at the receive end at the same time or in the same order in which they were transmitted. Because packets are small, the hold time is generally quite short and message transfer is near real time and blocking cannot occur. However, packet-switching networks require complex and expensive switching arrangements and complicated protocols. A packet switch is also a transactional switch. Circuit, message, and packet switching techniques are summarized in Table 14-3.

CCITT X.1 International User Class of Service

The CCITT X.1 standard divides the various classes of service into three basic modes of transmission for a public data network. The three modes are: *start/stop, synchronous,* and *packet.*

Start/stop mode.   With the start/stop mode, data are transferred from the source to the network and from the network to the destination in an asynchronous data format (i.e., each character is framed within a start and stop bit). Call control signaling is done in International Alphabet No. 5 (ASCII-77). Two common protocols used for start/stop transmission are IBM's 83B protocol and AT&T's 8A1/B1 selective calling arrangement.

Synchronous mode.   With the synchronous mode, data are transferred from the source to the network and from the network to the destination in a synchronous data format (i.e., each message is preceded by a unique synchronizing character). Call control signaling is identical to that used with private line data circuits and common protocols used for synchronous transmission are IBM's 3270 bisync, Burrough's BASIC, and UNIVAC's UNISCOPE.

TABLE 14-3 SWITCHING TECHNIQUE SUMMARY

Circuit switching	Message switching	Packet switching
Dedicated transmission path	No dedicated transmission path	No dedicated transmission path
Continuous transmission of data	Transmission of messages	Transmission of packets
Operates in real time	Not real time	Near real time
Messages not stored	Messages stored	Messages held for short time
Path established for entire message	Route established for each message	Route established for each packet
Call setup delay	Message transmission delay	Packet transmission delay
Busy signal if called party busy	No busy signal	No busy signal
Blocking may occur	Blocking cannot occur	Blocking cannot occur
User responsible for message-loss protection	Network responsible for lost messages	Network may be responsible for each packet but not for entire message
No speed or code conversion	Speed and code conversion	Speed and code conversion
Fixed bandwidth transmission (i.e., fixed information capacity)	Dynamic use of bandwidth	Dynamic use of bandwidth
No overhead bits after initial setup delay	Overhead bits in each message	Overhead bits in each packet

Packet mode.   With the packet mode, data are transferred from the source to the network and from the network to the destination in a frame format. The ISO HDLC frame format is the standard data link protocol used with the packet mode. Within the network, data are divided into smaller packets and transferred in accordance with the CCITT X.25 user to network interface protocol.

Figure 14-5 illustrates a typical layout for a public data network showing each of the three modes of operation. The packet assembler/disassembler (PAD) interfaces user data to X.25 format when the user's data are in either the asynchronous or synchronous mode of operation. A PAD is unnecessary when the user is operating in the packet mode. X.75 is

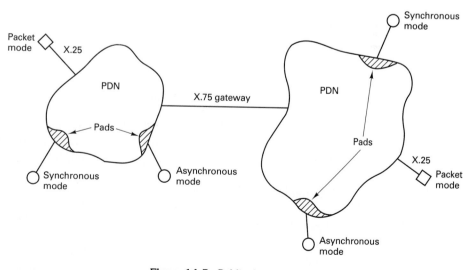

Figure 14-5 Public data network.

Public Data Network

recommended by the CCITT for the gateway protocol. A gateway is used to interface two public data networks.

CCITT X.25 USER-TO-NETWORK INTERFACE PROTOCOL

In 1976, the CCITT designated the X.25 user interface as the international standard for packet network access. Keep in mind that X.25 is strictly a *user-to-network* interface and addresses only the physical, data link, and network layers in the ISO seven layer model. X.25 uses existing standards whenever possible. For example, X.25 specifies X.21, X.26, and X.27 standards as the physical interface, which correspond to EIA RS-232C, RS-423A, and RS-422A standards, respectively. X.25 defines HDLC as the international standard for the data link layer and the American National Standards Institute (ANSI) 3.66 *advanced data communications control procedures* (ADCCP) as the U.S. standard. ANSI 3.66 and ISO HDLC specify exactly the same set of data link control procedures. However, ANSI 3.66 and HDLC were designed for private line data circuits with a polling environment. Consequently, the addressing and control procedures outlined by them are not appropriate for packet data networks. ANSI 3.66 and HDLC were selected for the data link layer because of their frame format, delimiting sequence, transparency mechanism, and error detection method.

At the link level, the protocol specified by X.25 is a subset of HDLC, referred to as *Link Access Procedure Balanced* (LAPB). LAPB provides for two-way, full-duplex communications between DTE and DCE at the packet network gateway. Only the address of the DTE or DCE may appear in the address field of a LAPB frame. The address field refers to a link address, not a network address. The network address of the destination terminal is embedded in the packet header, which is part of the information field.

Tables 14-4 and 14-5 show the commands and responses, respectively, for an LAPB frame. During LAPB operation, most frames are commands. A response frame is compelled only when a command frame is received containing a poll (P-bit) = 1. SABM/UA is a command/response pair used to initialize all counters and timers at the beginning of a session. Similarly, DISC/DM is a command/response pair used at the end of a session. FRMR is a response to any illegal command for which there is no indication of transmission errors according to the frame check sequence field.

Information (I) commands are used to transmit packets. Packets are never sent as responses. Packets are acknowledged using ns and nr just as they were in SDLC. RR is sent by a station when it needs to respond (acknowledge) something, but has no information packets to send. A response to an information command could be RR with $F = 1$. This procedure is called *checkpointing*.

REJ is another way of requesting transmission of frames. RNR is used for the flow control to indicate a busy condition and prevents further transmissions until cleared with an RR.

The network layer of X.25 specifies three switching services offered in a switched data network: permanent virtual circuit, virtual call, and datagram.

Permanent Virtual Circuit

A *permanent virtual circuit* (PVC) is logically equivalent to a two-point dedicated private line circuit except slower. A PVC is slower because a hardwired end to end connection is not provided. The first time a connection is requested, the appropriate switches and circuits must be established through the network to provide the interconnection. A PVC

TABLE 14-4 LAPB COMMANDS

| Command name | Bit number | | | |
	8 7 6	5	4 3 2	1
I (information)	nr	P	ns	0
RR (receiver ready)	nr	P	0 0 0	1
RNR (receiver not ready)	nr	P	0 1 0	1
REJ (reject)	nr	P	1 0 0	1
SABM (set asynchronous balanced mode)	0 0 1	P	1 1 1	1
DISC (disconnect)	0 1 0	P	0 0 1	1

TABLE 14-5 LAPB RESPONSES

| Command name | Bit number | | | |
	8 7 6	5	4 3 2	1
RR (receiver ready)	nr	F	0 0 0	1
RNR (receiver not ready)	nr	F	0 1 0	1
REJ (reject)	nr	F	1 0 0	1
UA (unnumbered acknowledgment)	0 1 1	F	0 0 1	1
DM (disconnect mode)	0 0 0	F	1 1 1	1
FRMR (frame rejected)	1 0 0	F	0 1 1	1

identifies the routing between two predetermined subscribers of the network that is used for all subsequent messages. With a PVC, a source and destination address are unnecessary because the two users are fixed.

Virtual Call

A *virtual call* (VC) is logically equivalent to making a telephone call through the DDD network except no direct end to end connection is made. A VC is a one-to-many arrangement. Any VC subscriber can access any other VC subscriber through a network of switches and communication channels. Virtual calls are temporary virtual connections that use common usage equipment and circuits. The source must provide its address and the address of the destination before a VC can be completed.

Datagram

A *datagram* (DG) is, at best, vaguely defined by X.25 and, until it is completely outlined, has very limited usefulness. With a DG, users send small packets of data into the network. Each packet is self-contained and travels through the network independent of other packets of the same message by whatever means available. The network does not acknowledge packets nor does it guarantee successful transmission. However, if a message will fit into a single packet, a DG is somewhat reliable. This is called a *single-packet-per-segment* protocol.

X.25 Packet Format

A virtual call is the most efficient service offered for a packet network. There are two packet formats used with virtual calls: a call request packet and a data transfer packet.

Call request packet. Figure 14-6 shows the field format for a call request packet. The delimiting sequence is 01111110 (an HDLC flag), and the error detection/correction mechanism is CRC-16 with ARQ. The link address field and the control field have little use and are therefore seldom used with packet networks. The rest of the fields are defined in sequence.

Format identifier. The format identifier identifies whether the packet is a new call request or a previously established call. The format identifier also identifies the packet numbering sequence (either 0–7 or 0–127).

Logical channel identifier (LCI). The LCI is a 12-bit binary number that identifies the source and destination users for a given virtual call. After a source user has gained access to the network and has identified the destination user, they are assigned an LCI. In subsequent packets, the source and destination addresses are unnecessary; only the LCI is needed. When two users disconnect, the LCI is relinquished and can be reassigned to new users. There are 4096 LCIs available. Therefore, there may be as many as 4096 virtual calls established at any given time.

Packet type. This field is used to identify the function and the content of the packet (i.e., new request, call clear, call reset, etc.).

Calling address length. This 4-bit field gives the number of digits (in binary) that appear in the calling address field. With 4 bits, up to 15 digits can be specified.

Called address length. This field is the same as the calling address field except that it identifies the number of digits that appear in the called address field.

Called address. This field contains the destination address. Up to 15 BCD digits (60 bits) can be assigned to a destination user.

Calling address. This field is the same as the called address field except that it contains up to 15 BCD digits that can be assigned to a source user.

Facilities length field. This field identifies (in binary) the number of 8-bit octets present in the facilities field.

Facilities field. This field contains up to 512 bits of optional network facility information, such as reverse billing information, closed user groups, and whether it is a simplex transmit or simplex receive connection.

Protocol identifier. This 32-bit field is reserved for the subscriber to insert user-level protocol functions such as log-on procedures and user identification practices.

User data field. Up to 96 bits of user data can be transmitted with a call request packet. These are unnumbered data which are not confirmed. This field is generally used for user passwords.

Data transfer packet. Figure 14-7 shows the field format for a data transfer packet. A data transfer packet is similar to a call request packet except that a data transfer packet has considerably less overhead and can accommodate a much larger user data field. The data transfer packet contains a send and receive packet sequence field that were not included with the call request format.

The flag, link address, link control, format identifier, LCI, and FCS fields are identical to those used with the call request packet. The send and receive packet sequence fields are described as follows.

Flag	Link address field	Link control field	Format identifier	Logical channel identifier	Packet type	Calling address length	Called address length	Called address	Calling address	0	Facilities field length	Facilities field	Protocol ID	User data	Frame check sequence	Flag
8	8	8	4	12	8	4	4	To 60	To 60	2	6	To 512	32	To 96	16	8

Figure 14-6 Call request packet format.

Flag	Link address field	Link control field	Format identifier	Logical channel identifier	Send packet sequence number P(s)	0	Receive packet sequence number P(r)	0	User data	Frame check sequence	Flag
Bits: 8	8	8	4	12	3/7	5/1	3/7	5/1	To 1024	16	8

Figure 14-7 Data transfer packet format.

Send packet sequence field. This field is used in the same manner that the ns and nr sequences are used with SDLC and HDLC. P(s) is analogous to ns, and P(r) is analogous to nr. Each successive data transfer packet is assigned the next P(s) number in sequence. The P(s) can be a 14- or a 7-bit binary number and thus number packets from either 0–7 or 0–127. The numbering sequence is identified in the format identifier. The send packet field always contains 8 bits and the unused bits are reset.

Receive packet sequence field. P(r) is used to confirm received packets and call for retransmission of packets received in error (ARQ). The I field in a data transfer packet can have considerably more source information than an I field in a call request packet.

The X Series of Recommended Standards

X.25 is part of the X series of CCITT-recommended standards for public data networks. The X series is classified into two categories: X.1 through X.39, which deal with services and facilities, terminals, and interfaces; and X.40 through X.199, which deal with network architecture, transmission, signaling, switching, maintenance, and administrative arrangements. Table 14-6 lists the most important X standards with their titles and descriptions.

TABLE 14-6 CCITT X SERIES STANDARDS

X.1	International user classes of service in public data networks. Assigns numerical class designations to different terminal speeds and types.
X.2	International user services and facilities in public data networks. Specifies essential and additional services and facilities.
X.3	Packet assembly/disassembly facility (PAD) in a public data network. Describes the packet assembler/dissassembler, which normally is used at a network gateway to allow connection of a start/stop terminal to a packet network.
X.20-bis	Use on public data networks of DTE designed for interfacing to asynchronous full-duplex V-series modems. Allows use of V.24/V.28 (essentially the same as EIA RS-232C).
X.21-bis	Use on public data networks of DTE designed for interfacing to synchronous full-duplex V-series modems. Allows use of V.24/V.28 (essentially the same as EIA RS-232C) or V.35.
X.25	Interface between DTE and DCE for terminals operating in the packet mode on public data networks. Defines the architecture of three levels of protocols existing in the serial interface cable between a packet-mode terminal and a gateway to a packet network.
X.28	DTE/DCE interface for a start/stop mode DTE accessing the PAD in a public data network situated in the same country. Defines the architecture of protocols existing in a serial interface cable between a start/stop terminal and an X.3 PAD.
X.29	Procedures for the exchange of control information and user data between a PAD and a packet mode DTE or another PAD. Defines the architecture of protocols behind the X.3 PAD, either between two PADs or between a PAD and a packet-mode terminal on the other side of the network.
X.75	Terminal and transit call control procedures and data transfer system on international circuits between packet-switched data networks. Defines the architecture of protocols between two public packet networks.
X.121	International numbering plan for public data networks. Defines a numbering plan including code assignments for each nation.

LOCAL AREA NETWORKS

A *local area network* (LAN) is a data communications network that is designed to provide two-way communications between a large variety of data communications terminal equipment within a relatively small geographic area. LANs are privately owned and operated and are used to interconnect data terminal equipment in the same building, building complex, or geographical area.

Local Area Network System Considerations

Topology. The topology or physical architecture of a LAN identifies how the stations are interconnected. The most common configurations used with LANs are the star, bus, ring, and mesh topologies.

Connecting medium. Presently, most LANs use coaxial cable as the transmission medium, although in the near future it is likely that optical fiber cables will predominate. Fiber cables can operate at higher bit rates and, consequently, have a larger capacity to transfer information than coaxial cables. LANs that use a coaxial cable are limited to an overall length of approximately 1500 m. Fiber links are expected to far exceed this distance.

Transmission format. There are two basic approaches to transmission format for LANs: *baseband* and *broadband*. Baseband transmission uses the connecting medium as a single-channel device. Only one station can transmit at a time and all stations must transmit and receive the same types of signals (encoding schemes and bit rates). Essentially, a baseband format time division multiplexes signals onto the transmission medium. Broadband transmission uses the connecting medium as a multi-channel device. Each channel occupies a different frequency band (i.e., frequency-division multiplexing). Consequently, each channel can contain different encoding schemes and operate at different bit rates. A broadband network permits voice, digital data, and video to be transmitted simultaneously over the same transmission medium. However, broadband systems require RF modems, amplifiers, and more complicated transceivers than baseband systems. For this reason, baseband systems are more prevalent. Table 14-7 summarizes baseband and broadband transmission techniques.

Channel Accessing

Channel accessing describes the mechanism used by a station to gain access to a local area network. There are essentially two methods used for channel accessing with LANs: carrier sense, multiple access with collision detection (CSMA/CD) and token passing.

Carrier sense, multiple access with collision detection. With CSMA/CD, a station monitors (listens to) the line to determine if the line is busy. If a station has a message to transmit but the line is busy, it waits for an idle condition before it transmits its message. If two stations begin transmitting at the same time, a *collision* occurs. When this happens, both stations cease transmitting (*back off*) and each station waits a random period of time before attempting a retransmission. The random delay time for each station is different and therefore allows for prioritizing the stations on the network. With CSMA/CD,

Local Area Networks

TABLE 14-7 TRANSMISSION FORMAT SUMMARY

Baseband	Broadband
Characteristics	
Digital signaling	Analog signaling (requires RF modem)
Entire bandwidth used by signal	FDM possible (i.e., multiple data channels)
Bidirectional	Unidirectional
Bus topology	Bus topology
Maximum length approximately 1500 meters	Maximum length up to tens of kilometers
Advantages	
Less expensive	High capacity
Simpler technology	Multiple traffic types
Easy and quick to install	More flexible circuit configurations
	Larger area covered
Disadvantages	
Single channel	Modem required
Limited capacity	Complex installation and maintenance
Grounding problems	Double propagation delay
Limited distance	

stations must contend for the network. A station is not guaranteed access to the network. To detect the occurrence of a collision, a station must be capable of transmitting and receiving simultaneously. CSMA/CD is used by most baseband LANs in the bus configuration. Ethernet is a popular local area network that uses baseband transmission with CSMA/CD. The transmission rate with Ethernet is 10 Mbps over a coaxial cable. Collision detection is accomplished by monitoring the line for phase violations in a Manchester-encoded (biphase) digital encoding scheme.

Token passing. Token passing is a channel-accessing arrangement that is best suited for a ring topology with either a baseband or a broadband network. With token passing, an electrical *token* (*code*) is circulated around the ring from station to station. Each station, in turn, acquires the token. In order to transmit, a station must first possess the token; then the station removes the token and places its message on the line. After a station transmits, it passes the token on to the next sequential station. With token passing, each station has equal access to the transmission medium. The Cambridge ring is a popular local area network that uses baseband transmission with token passing. The transmission rate with a Cambridge ring is 10 Mbps. Table 14-8 lists several local area networks and some of their characteristics.

In 1980, the IEEE local area network committee was established to standardize the means of connecting digital computer equipment and peripherals with the local area network environment. In 1983, the committee established IEEE standards 802.3 (CSMA/CD) and 802.4 (token passing) for a bus topology. In 1987, IEEE standard 802.5 (token passing) with a ring topology and Fiber Distributed Data Interface (FDDI) were adopted.

TABLE 14-8 LOCAL AREA NETWORK SUMMARY

Ethernet	Developed by Xerox Corporation in conjunction with Digital Equipment Corporation and Intel Corporation; baseband system using CSMA/CD; 10 Mbps
Wangnet	Developed by Wang Computer Corporation; broadband system using CSMA/CD
Localnet	Developed by Sytek Corporation; broadband system using CSMA/CD
Domain	Developed by Apollo Computer Corporation; broadband network using token passing
Cambridge ring	Developed by the University of Cambridge; baseband system using CSMA/CD; 10 Mbps

ETHERNET

Ethernet is a baseband system which uses a bus topology and is designed for use in a local area network. Since there are many computer manufacturers, the problem is to design a system in which different types of computers can communicate with each other. In a demonstration in 1982, the computers of 10 major companies were linked on a 1500-ft Ethernet cable for electronic mail, word processing, and so on.

Before explaining the operation of Ethernet, a description of this system's hardware and software is provided.

Software

Information is transmitted from one station (computer) to another in the form of packets. The format for a packet is shown in Figure 14-8. The preamble is used for bit synchronization. The source and destination addresses and the field type make up the header information. The CRC is computed on the header information and the data field. The packet information is converted to Manchester code and transmitted at a 10-Mbps rate. In the Manchester encoding, each bit cell is divided into two parts. The first half contains the complement of the bit value and the second contains the actual bit value. This is illustrated in Figure 14-9. This code ensures that a signal transition occurs in every bit.

The destination of a packet may be a single station or a group of stations. A *physical address* is a unique address of a single station and is denoted by a 0 as the first bit (LSB) of the destination address. A *multicast address* has two forms. If only a partial group of the total stations is addressed, the first bit is a 1. The destination address is all 1's if all of the stations in the network are to receive the transmitted packet. A delay of 9.6 μs is required between the transmission of packets.

Hardware

Coaxial cable is the transmission medium for Ethernet. A cable segment may have a maximum distance of 500 m (1640 ft). Each segment may have up to 100 transceivers attached by way of pressure taps. Since these taps do not require cable cutting for installation, additional stations may be added without interfering with normal system operations. Interlan has developed the NT10 transceiver, in which the taps consist of two probes. One probe

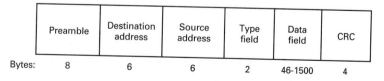

Preamble	Destination address	Source address	Type field	Data field	CRC
Bytes: 8	6	6	2	46-1500	4

Figure 14-8 Packet format for Ethernet.

Ethernet

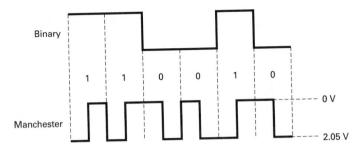

Figure 14-9 Manchester code.

contacts the center conductor while the other contacts the shield. Color bands on the segment cable identify where the taps may be placed. A maximum of three segments may be connected end to end by way of repeaters to extend the system length to 1500 m (4921 ft). In Figure 14-10, this is the path between A and C. Remote repeaters may be used with a maximum distance of 1000 m between them. If these are used, the maximum end-to-end distance would be extended to 2.5 km (8202 ft). In Figure 14-10, if the center and right segments were interchanged, there would be a 2.5-km separation between A and B. There must always be only one signal path between any two stations. A station is connected to a controller, which is then interfaced with the transceiver to the transmission system. This is shown in Figure 14-11.

System Operation

Transmission. The data link control for Ethernet is CSMA/CD. Stations acquire access to the transmission system through contention. They are not polled, nor do they have specific time slots for transmission. A station wishing to transmit first determines if another station is currently using the transmission system. The controller accomplishes this through the transceiver by sensing the presence of a carrier on the line. The controller may be a hardware or a software function, depending on the complexity of the station. In 1983, Xerox Corporation, in conjunction with Digital Equipment Corporation and Intel Corporation, developed and introduced a single-chip controller for Ethernet. Later, Mostek Corporation, working with Digital Equipment Corporation and Advanced Micro Devices, announced a two-chip set for this purpose. One chip is the local area network controller for Ethernet (LANCE) and the second is a Serial Interface Adapter.

The presence of a carrier is denoted by the signal transitions on the line produced by the Manchester code. If a carrier is detected, the station defers transmission until the line is quiet. After the required delay, the station sends digital data to the controller. The controller converts these data to Manchester code, inserts the CRC, adds the preamble, and places the packet on-line. The transmission of the entire packet is not yet assured. A different station may also have detected the quiet line and started to transmit its own packet. The first station monitors the line for a period called the *collision window* or the *collision interval*. This interval is a function of the end-to-end propagation delay of the line. This delay, including the delay caused by any repeaters, measured in distance, cannot exceed 2.5 km. If a data collision has not occurred in this interval, the station is said to have line acquisition and will continue to transmit the entire packet. Should a collision be detected, both of the transmitting stations will immediately abort their transmissions for a random period of time and then attempt retransmission. A data collision may be detected by the transceiver by comparing the received signal with the transmitted signal. To make this comparison, a station must still be transmitting its packet while a previously transmitted

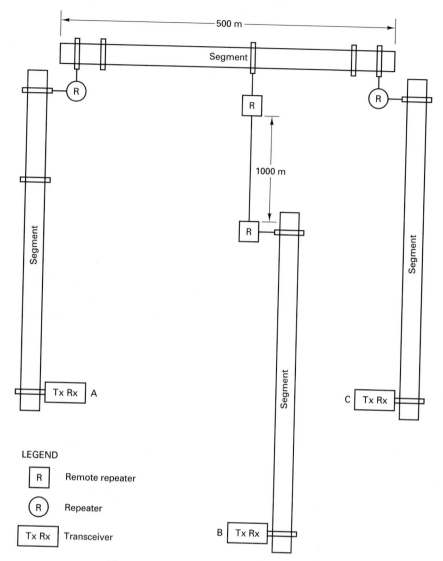

Figure 14-10 Ethernet transmission system.

signal has propagated to the end of the line and back. This dictates that a packet be of some minimum size. If a collision is detected, it is the controller that must take the necessary action. The controller-transceiver interface contains a line for notification of Collision Presence (10-MHz square wave). The remaining three lines of this interface are the Transmit Data, Receive Data, and power for the transceiver. Since a collision is manifested in some form of phase violation, the controller alone may detect a collision. In Ethernet, data collision is detected mainly by the transceiver. When feasible, this is supplemented by a collision detection facility in the controller. To ensure that all stations are aware of the collision, a *collision enforcement consensus procedure* is invoked. When the controller

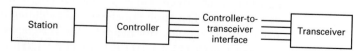

Figure 14-11 Ethernet station connection to the transmission system.

Ethernet

detects a collision, it transmits four to six bytes of random data. These bytes are called the *jam sequence*. If a collision has occurred, the station's random waiting time before retransmission is determined from a *binary exponential back-off algorithm*. The transmission time slot is usually set to be slightly longer than the round-trip time of the channel. The delay time is randomly selected from this interval.

EXAMPLE 14-4

Transmission time slot = time of 512 bits

$$\text{Maximum time of the interval} = \frac{512 \text{ bits}}{10 \text{ Mbps}} = 51.2 \ \mu s$$

$$\text{Time interval} = 0 \text{ to } 51.2 \ \mu s$$

For each succeeding collision encountered by the same packet, the time interval is doubled until a maximum interval is reached. The maximum interval is given as $2^{10} \times$ transmission time slot. After 15 unsuccessful attempts at transmitting a packet have been made, no further attempts are made and the error is reported to the station. This is the major drawback of Ethernet—it cannot guarantee packet delivery at a time of heavy transmission load.

Reception.   The line is monitored until the station's address is detected. The controller strips the preamble, checks the CRC, and converts the Manchester code back to digital format. If the packet contains any errors, it is discarded. The end of the packet is recognized by the absence of a carrier on the transmission line. This means that no transitions were detected for the period of 75 to 125 ns since the center of the last bit cell. The decoding is accomplished through a phase-locked loop. The phase-locked loop is initialized by the known pattern of the preamble.

INTEGRATED SERVICES DIGITAL NETWORK (ISDN)

Introduction to ISDN

The data and telephone communications industry is continually changing to meet the demands of contemporary telephone, video, and computer communications systems. Today more and more people have a need to communicate with each other than ever before. In order to meet these needs, old standards are being updated and new standards are being developed and implemented on almost a daily basis.

The *integrated services digital network* (ISDN) is a proposed network designed by the major telephone companies in conjunction with the CCITT with the intent of providing worldwide telecommunications support of voice, data, video, and facsimile information within the same network (in essence, ISDN is the integrating of a wide range of services into a single multipurpose network). ISDN is a network that proposes to interconnect an unlimited number of independent users through a common communications network.

However, to date only a small number of ISDN facilities have been developed. However the telephone industry is presently implementing ISDN system so that in the near future, subscribers will access the ISDN system using existing public telephone and data networks. The basic principles and evolution of ISDN have been outlined by the International Consultative Committeé for Telegraphy and Telephony (CCITT) in its recommendation CCITT 1.120 (1984). CCITT 1.120 lists the following principles and evolution of ISDN:

Principles of ISDN

1. The main feature of the ISDN concept is to support a wide range of voice (telephone) and non voice (digital data) applications in the same network using a limited number of standardized facilities.

2. ISDN's support a wide variety of applications including both switched and non-switched (dedicated) connections. Switched connections include both circuit- and packet-switched connections and their concatenations.

3. Whenever practical, new services introduced into an ISDN should be compatible with 64 kbps switched digital connections. The 64 kbps digital connection is the basic building block of ISDN.

4. An ISDN will contain intelligence for the purpose of providing service features, maintenance, and network management functions. In other words, ISDN is expected to provide services beyond the simple setting up of switched circuit calls.

5. A layered protocol structure should be used to specify the access procedures to an ISDN and can be mapped into the open system interconnection (OSI) model. Standards already developed for OSI-related applications can be used for ISDN, such as X.25 level 3 for access to packet-switching services.

6. It is recognized that ISDNs may be implemented in a variety of configurations according to specific national situations. This accommodates both single-source or competitive national policy.

Evolution of ISDNs

1. ISDNs will be based on the concepts developed for telephone ISDNs and may evolve by progressively incorporating additional functions and network features including those of any other dedicated networks such as circuit and packet switching for data so as to provide for existing and new services.

2. The transition from an existing network to a comprehensive ISDN may require a period of time extending over one or more decades. During this period, arrangements must be developed for the internetworking of services on ISDNs and services on other networks.

3. In the evolution towards an ISDN, digital end-to-end connectivity will be obtained via plant and equipment used in existing networks, such as digital transmission, time-division multiplex and/or space-division multiplex switching. Existing relevant recommendations for these constituent elements of an ISDN are contained in the appropriate series of Recommendations of CCITT and CCIR.

4. In the early stages of the evolution of ISDNs, some interim user-network arrangements may need to be adopted in certain countries to facilitate early penetration of digital service capabilities.

5. An evolving ISDN may also include at later stages switched connections at bit rates higher and lower than 64 kbps.

ISDN Architecture

A block diagram indicating the proposed architecture for ISDN functions is shown in Figure 14-12. The ISDN is designed to support an entirely new physical connection for the user, a digital subscriber loop, and a variety of transmission services.

A *common physical interface* will be defined to provide a DTE-DCE interface connection. A single interface will be used for telephones, computer terminals, and video

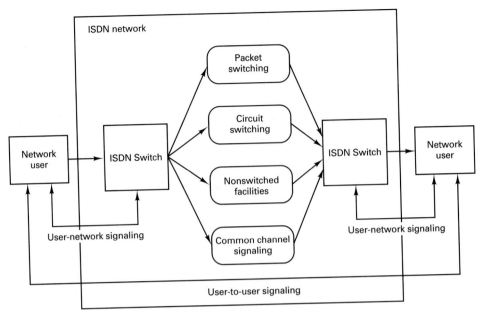

Figure 14-12 Architecture for ISDN functions.

equipment. Therefore, various protocols will be required to allow control information to be exchanged between the user's device and the ISDN. There are three basic types of channels available with ISDN. They are:

B channel: 64 kbps

D channel: 16 or 64 kbps

H channel: 384, 1536, or 1920 kbps.

ISDN standards specify that residential users of the network (that is, the subscribers) be provided a *basic access* consisting of three full-duplex time-division multiplexed digital channels, two operating at 64 kbps (designated the B channels, for *bearer*) and one at 16 kbps (designated the D channel, for *data*). The B and D bit rates were selected to be compatible with existing DS1–DS4 digital carrier systems. The D channel is used for carrying signaling information and for exchanging network control information. One B channel is used for digitally encoded voice and the other for applications such as data transmission, PCM-encoded digitized voice, and videotex. The 2B + D service is sometimes called the *basic rate interface* (BRI). BRI systems require bandwidths that can accommodate two 64-kbps B channels and one 16-kbps D channel plus framing, synchronization, and other overhead bits for a total bit rate of 192 kbps. The H channels are used to provide higher bit rates for special services such as fast facsimile, video, high-speed data, and high-quality audio.

There is another service called the *primary service, primary access*, or *primary rate interface* (PRI) that will provide multiple 64 kbps channels intended to be used by the higher volume subscribers to the network. In the United States, Canada, Japan, and Korea, the primary rate interface consists of twenty-three 64-kbps B channels and one 64-kbps D channel (23B + D) for a combined bit rate of 1.544 Mbps. In Europe, the primary rate interface uses thirty 64-kbps B channels and one 64-kbps D channel for a combined bit rate of 2.048 Mbps.

It is intended that ISDN provide a circuit-switched B channel with the existing telephone system, however, packet-switched B channels for data transmission at nonstandard rates would have to be created.

The subscriber's loop, as with the twisted pair cable used with a common telephone, provides the physical signal path from the subscriber's equipment to the ISDN central office. The subscriber loop must be capable of supporting full-duplex digital transmission for both basic and primary data rates. Ideally, as the network grows, optical fiber cables will replace the metallic cables.

Table 14-9 lists the services proposed to be used by ISDN subscribers. BC designates a circuit switched B channel, BP designates a packed-switched B channel, and D designates a D channel.

ISDN System Connections and Interface Units

ISDN subscriber units and interfaces are defined by their function and reference within the network. Figure 14-13 shows how users may be connected to an ISDN. As the figure shows, subscribers must access the network through one of two different types of entry devices, *Terminal equipment type 1* (TE1) and *terminal equipment type 2* (TE2). TE1 equipment supports standard ISDN interfaces and, therefore, requires no protocol translation. Data enters the network and are immediately configured into ISDN protocol format. TE2 equipment are classified as non-ISDN, thus, computer terminals are connected to the system through physical interfaces such as the RS-232C and host computers with X.25. Translation between non-ISDN data protocol and ISDN protocol is performed in a device called a *terminal adapter* (TA). Terminal adapters convert the user's data into the 64 kbps ISDN channel B or the 16 kbps channel D format, and X.25 packets are converted to ISDN packet formats. If any additional signaling is required, it is added by the terminal adapter. The terminal adapters can also support traditional analog telephones and facsimile signals by using a 3.1 kHz audio service channel. The analog signals are digitized and put into ISDN format before entering the network.

User data at points designated as *reference point S (system)* are presently in ISDN format and provide the 2B + D data at 192 kbps. These reference points separate user terminal equipment from network-related system functions. *Reference point T (terminal)* locations correspond to a minimal ISDN network termination at the user's location. These reference points separate the network provider's equipment from the user's equipment. *Reference point R (rate)* provides an interface between non-ISDN compatible user equipment and the terminal adapters. *Network termination 1* (NT1) provides the functions associated with the physical interface between the user and the common carrier and are designated by the letter T (these functions corresponds to OSI layer 1). The NT1 is a bound-

TABLE 14-9 PROJECTED ISDN SERVICES

Service	Transmission rate	Channel
Telephone	64 kbps	BC
System Alarms	100 kbps	D
Utility company metering	100 kbps	D
Energy management	100 kbps	D
Video	2.4–64 kbps	BP
Electronic mail	4.8–64 kbps	BP
Facsimile	4.8–64 kbps	BC
Slow scan television	64 kbps	BC

Integrated Services Digital Network (ISDN)

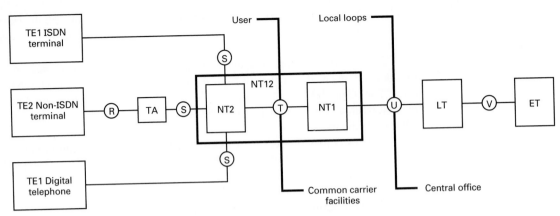

Figure 14-13 ISDN connections and reference points.

ary to the network and may be controlled by the ISDN provider. The NTI performs line maintenance functions and supports multiple channels at the physical level (e.g., 2B + D). Data from these channels are time-division multiplexed together. Network terminal 2 devices are intelligent and can perform *concentration* and switching functions (functionally up through OSI level 3). NT2 terminations can also be used to terminate several S point connections and provide local switching functions and two-wire to four-wire and four-wire to two-wire conversions. *U-reference points* refer to interfaces between the common carrier subscriber loop and the *central office switch*. A *U loop* is the media interface point between an NT1 and the central office. Network termination 1,2 (NT12) constitutes one piece of equipment that combines the functions of NT1 and NT2. U loops are terminated at the central office by a *line termination* (LT) unit, which provides physical layer interface functions between the central office and the loop lines. The LT unit is connected to an *exchange termination* (ET) at *reference point* V. An ET routes data to an outgoing channel or central office user.

There are several types of transmission channels in addition to the B and D types described in the previous section. They include the following:

HO channel—this interface supports multiple 384-kbps HO channels. These structures are 3HO + D and 4HO + D for the 1.544-Mbps interface and 5HO + D for the 2.048 Mbps interface.

H11 channel—this interface consists of one 1.536-Mbps H11 channel (twenty-four 64-kbps channels).

H12 channel—European version of H11 that uses 30 channels for a combined data rate of 1.92 Mbps.

E channel—Packet switched using 64 kbps (similar to the standard D channel)

ISDN Protocols

Standards developed for ISDN include protocols that allow interaction between ISDN users and the network itself and also for interaction between one ISDN user and another. In addition, it is desirable to fit the new ISDN protocols into the OSI framework. Figure 14-14 shows the relationships between OSI and ISDN. In essence, ISDN is not concerned with OSI layers 4–7. These layers are for end-to-end exchange of information between the network users.

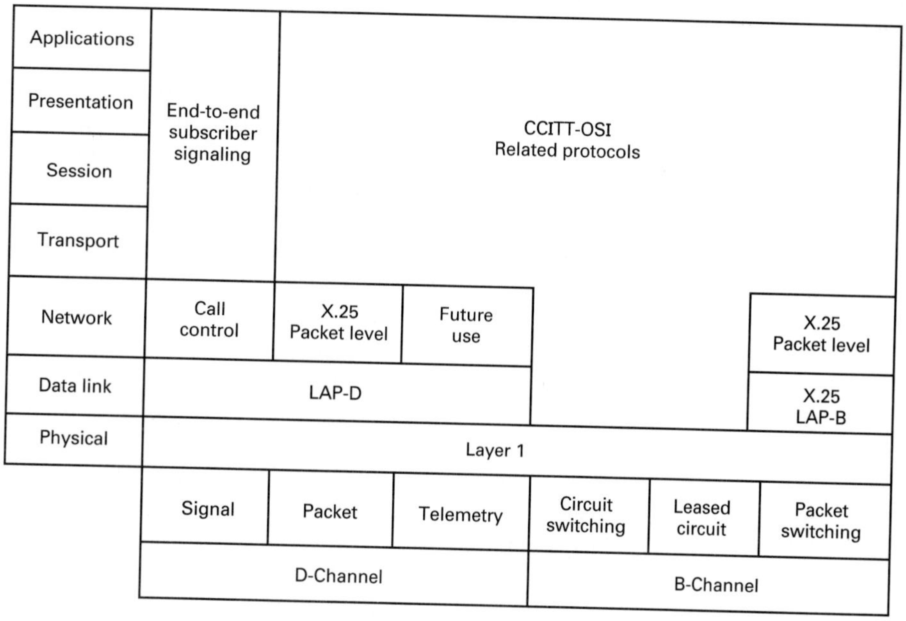

Figure 14-14 Structured relationships between OSI and ISDN user network.

Layer 1 specifies the physical interface for both basic and primary access to the network. B and D channels are time-division multiplexed onto the same interface, consequently, the same standards apply to both types of channels. However, the protocols of layers 2 and 3 differ for the two channels. The protocol used by ISDN for the data link layer is very similar to the HDLC format discussed earlier in this chapter and is called *Link Access Protocol for D-Channels* (LAP-D) and *Link Access Protocol for B-Channels* (LAP-B). CCITT standards Q920 and Q921 give the details of these specifications.

LAP-D services.　All data transmissions on the LAP-D channel are between the subscriber equipment and an ISDN switching element. LAP-D provides two types of services: *unacknowledged* and *acknowledged information transfer*. The unacknowledged information transfer provides for the transfer of data frames with no acknowledgement. This service supports both point-to-point or broadcast transmission but does not guarantee successful transmission of data, nor does it inform the sender if the transmission fails. Unacknowledged information transfer does not provide any type of data flow control or error control mechanism. Error detection is used to detect and discard any damaged frames. Transmission of unacknowledged information transfer simply provides a means of transferring data quickly and is useful for services such as sending alarm messages.

The acknowledged information transfer service is much more commonly used. This service is similar to the services offered by LAP-B. With acknowledged information transfer, a logical connection is established between two subscribers prior to the transfer of any data. Then data are transferred in sequentially numbered frames that are acknowledged either individually or in groups. Both error and flow control are included in this service. This type of service is sometimes referred to as multiple-frame operation.

LAP-D format.　With LAP-D protocol, subscriber information as well as protocol control information and parameters are transmitted in frames. The basic LAP-D frame is identical to that of HDLC except for the address field. The LAP-D frame format is shown

Integrated Services Digital Network (ISDN)

in Figure 14-15. As shown, the frame begins with the transmission of a flag field, followed by the address field. The flag field delimits the frame at both ends with a hex 7E (binary 01111110). LAP-D has to contend with two types of multiplexing: subscriber site mulit-plexing where there may be multiple devices sharing the same physical interface, and within each user device, there may be multiple forms of traffic multiplexed together (such as packet-switched data and control signalling).

To accommodate the two forms of multiplexing, LAP-D uses a two-part address field that consists of a *terminal endpoint identifier* (TEI) and a *service point identifier* (SAPI). Typically each subscriber is assigned a unique TEI. It is also possible for a single device to have more than one TEI, such as for a terminal concentrator. TEI assignments can be made automatically when the equipment is first turned up, or manually by the sub-scriber. The SAPI identifies a layer 3 subscriber of LAP-D. Only four SAPI identifiers are currently in use. A SAPI of 0 is used for call control procedures for managing B channel circuits, a SAPI of 1 is used for packet-mode transmissions using I.451 control procedures (such as subscriber-to-subscriber signaling), a SAPI of 16 is reserved for packet-mode transmissions on the D channel using X.25 level 3, and a SAPI of 63 is used for the exchange of layer 2 management and control information. The TEI and SAPI assignments are given in Table 14-10 and 14-11, respectively. The TEI and SAPI can be used together to uniquely identify a logical connection. When used in this manner, the combination of TEI and SAPI is referred to as a *data link connection identifier* (DLCI).

An additional bit, called the *command/response* (C/R) bit, is used to identify the frame as a command (logic 0 for traffic from terminals and logic 1 for traffic from the net-work) or a response message (opposite logic condition).

The control field identifies the type of frame and keeps track of the frame sequence the same as with SDLC and HDLC. Information, supervisory, and unnumbered control fields are also the same as with SDLC and HDLC with these exceptions. The set normal response unnumbered frame is replaced with a set asynchronous balanced mode extended frame, which functions in a similar manner and establishes a data link for acknowledged data transfers of information frames. An additional unnumbered frame, *Transfer ID* (XID), is also included to allow stations to identify themselves for line management purposes.

The information field is allowed only with information frames and certain special unnumbered frames. The information field can contain any sequence of bits so long as they consist of an integral multiple of eight (octets). The information field is variable within system-defined specifications, however, for both control signaling and packet information, the maximum length is 260 octets.

The frame check sequence (FCS) is a CRC-CCITT code used for error-detection of all bits within the frame except the flags.

Beginning flag	Address field		Control field	Information field	Frame check sequence	Ending flag
7E 01111110	C/R SAPI 0 X XXXXXX	TEI 1 XXXXXXX	Control word	Data	CRC-16	7E 01111110
1-byte	1-byte	1-byte	1-2 bytes	1-128 or 0-260 bytes	2-bytes	1-byte

SAPI = Service point identifier
TEI = Terminal endpoint identifier
CRC = Cyclic redundancy check
C/R = Command/response

Figure 14-15 ISDN LAP-D format.

TABLE 14-10 TEI ASSIGNMENTS

TEI	User type
0–63	Nonautomatic TEI assignment subscriber equipment
64–126	Automatic TEI assignment subscriber equipment
127	Used during automatic TEI assignment

TABLE 14-11 SAPI ASSIGNMENTS

SAPI	Related layer 3 or management function
0	Call control procedures
1	Reserved for packet-mode transmissions using I.451 call control procedures
16	Packet-mode transmissions using X.25 level 3
63	Layer 2 management and control procedures
all others	Reserved for future uses

LAP-B frames are similar to LAP-D although XID and unnumbered information (UI) frames are not used. Also, with LAP-B the frames are limited to 14-bit modulo 8 numbers (0–7) where LAP-D uses 7-bit modulo 128 numbers (0–127).

Broadband ISDN (BISDN)

Broadband ISDN (BISDN) is defined by the CCITT as a service that provide transmission channels capable of supporting transmission rates greater than the primary data rate. With BISDN, services requiring data rates of a magnitude beyond those provided by ISDN, such as video transmission, will become available. With the advent of BISDN, the original concept of ISDN is being referred to as *narrowband* ISDN.

In 1988, the CCITT first recommended as part of its I-series recommendations relating to BISDN: I.113, *Vocabulary of terms for broadband aspects of ISDN*, and I.121, *Broadband aspects of ISDN*. These two documents are a concensus concerning the aspects of the future of BISDN. They outline preliminary descriptions of future standards and development work.

The new BISDN standards are based upon the concept of an *asynchronous transfer mode* (ATM), which will incorporate optical fiber cable as the transmission medium for data transmission. The BISDN specifications set a maximum length of 1 km per cable length but is making provisions for repeated interface extensions. The expected data rates on the optical fiber cables will be either 11, 155, or 600 Mbps, depending on the specific application and the location of the fiber cable within the network.

CCITT classifies the services that could be provided by BISDN as interactive and distribution services. *Interactive services* include those in which there is a two-way exchange of information (excluding control signaling) between two subscribers or between a subscriber and a service provider. *Distribution services* are those in which information transfer is primarily from service provider to subscriber. On the other hand, *conversational services* will provide a means for bidirectional end-to-end data transmission, in real time, between two subscribers or between a subscriber and a service provider.

The authors of BISDN composed specifications that require the new services meet both existing ISDN interface specifications and the new BISDN needs. A standard ISDN terminal and a *broadband terminal interface* (BTI) will be serviced by the *subscriber's premise network* (SPN), which will multiplex incoming data and transfer them to the

Virtual channel identifier	Header error detection character	Undefined	Channel data

Figure 14-16 ATM cell header format.

broadband node. The broadband node is called a *broadband network termination* (BNT), which codes the data information into smaller packets used by the BISDN network. Data transmissions within the BISDN network can be asymmetric (that is, access onto and off of the network may be accomplished at different transmission rates, depending on system requirements).

Asynchronous transfer mode. The asynchronous transfer mode (ATM) is a means by which data can enter and exit the BISDN network in an asynchronous (time independent) fashion. ATM uses labeled channels that are transferable at fixed data rates. The data rates can be anywhere from 16 kbps up to the maximum rate of the system. Once data has entered the network, they are transferred into fixed time slots called *cells*. A cell is identified by a *label* in the *cell header*. The format for a cell header is shown in Figure 14-16. The *virtual channel identifier* indicates the node source and packet destination. The channel is virtual, rather than specific, which allows the actual physical routing of the packet and network entry and exit times to be determined by network availability and access rights. Immediately after the virtual channel identifier, is the *header error detection character* which may be a CRC character or any other form of error detection. The CRC is for the header label only and separate error detection methods are used for the actual data field. The next section of the header cell is unidentified and is reserved for future use.

BISDN configuration. Figure 14-17 shows how access to the BISDN network is accomplished. Each peripheral device is interfaced to the *access node* of a BISDN network through a *broadband distant terminal* (BDT). The BDT is responsible for the electrical to optical conversion, multiplexing of peripherals, and maintenance of the subscriber's local system. Access nodes concentrate several BDTs into high-speed optical fiber lines directed through a *feeder point* into a *service node*. Most of the control functions for system

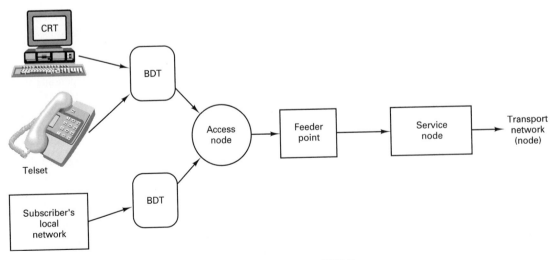

Figure 14-17 BISDN access.

Chap. 14 **Data Communications Protocols**

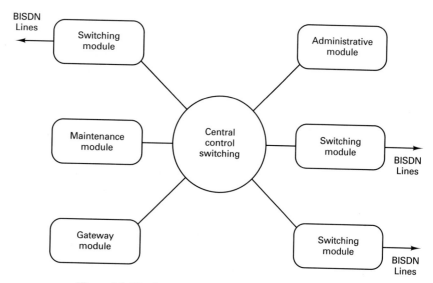

Figure 14-18 BISDN functional module interconnections.

access are managed by the service node, such as call processing, administrative functions, and switching and maintenance functions. The functional modules are interconnected in a star configuration and include switching, administrative, gateway, and maintenance modules. The interconnection of the functional modules is shown in Figure 14-18. The central control hub acts as the end user interface for control signaling and data traffic maintenance. In essence, it oversees the operation of the modules.

Subscriber terminals near the central office may bypass the access nodes entirely and be directly connected to the BISDN network through a service node. BISDN networks that use optical fiber cables can utilize much wider bandwidths and, consequently, have higher transmission rates and offer more channel-handling capacity than ISDN systems.

Broadband channel rates. The CCITT has published preliminary definitions of new broadband channel rates that will be added to the existing ISDN narrowband channel rates. The new channel rates are:

1. H21: 32.768 Mbps
2. H22: 43 to 45 Mbps
3. H4: 132 to 138.24 Mbps

The H21 and H22 data rates are intended to be used for full-motion video transmission for video conferencing, video telephone, and video messaging. The H4 data rate is intended for bulk data transfer of text, facsimile, and enhanced video information. The H21 data rate is equivalent to 512 64-kbps channels. The H22 and H4 data rates must be multiples of the basic 64 kbps transmission rate.

QUESTIONS

14-1. Define data communications protocol.

14-2. What is a master station? A slave station?

14-3. Define *polling* and *selecting*.

14-4. What is the difference between a synchronous and an asynchronous protocol?

14-5. What is the difference between a character-oriented protocol and a bit-oriented protocol?

14-6. Define the three operating modes used with data communications circuits.

14-7. What is the function of the clearing character?

14-8. What is a unique address? A group address? A broadcast address?

14-9. What does a negative acknowledgment to a poll indicate?

14-10. What is the purpose of a heading?

14-11. Why is IBM's 3270 synchronous protocol called bisync?

14-12. Why are SYN characters always transmitted in pairs?

14-13. What is an SPA? An SSA? A DA?

14-14. What is the purpose of a leading pad? A trailing pad?

14-15. What is the difference between a general poll and a specific poll?

14-16. What is a handshake?

14-17. (Primary, secondary) stations transmit polls.

14-18. What does a negative acknowledgment to a poll indicate?

14-19. What is a positive acknowledgment to a poll?

14-20. What is the difference between ETX, ETB, and ITB?

14-21. What character is used to terminate a heading and begin a block of text?

14-22. What is transparency? When is it necessary? Why?

14-23. What is the difference between a command and a response with SDLC?

14-24. What are the three transmission states used with SDLC? Explain them.

14-25. What are the five fields used with an SDLC frame? Briefly explain each.

14-26. What is the delimiting sequence used with SDLC?

14-27. What is the null address in SDLC? When is it used?

14-28. What are the three frame formats used with SDLC? Explain what each format is used for.

14-29. How is an information frame identified in SDLC? A supervisory frame? An unnumbered frame?

14-30. What are the purposes of the nr and ns sequences in SDLC?

14-31. With SDLC, when is the P bit set? The F bit?

14-32. What is the maximum number of unconfirmed frames that can be outstanding at any one time with SDLC? Why?

14-33. With SDLC, which frame formats can have an information field?

14-34. With SLDC, which frame formats can be used to confirm previously received frames?

14-35. What command/response is used for reporting procedural errors with SDLC?

14-36. Explain the three modes in SDLC that a secondary station can be in.

14-37. When is the configure command/response used with SDLC?

14-38. What is a go-ahead sequence? A turnaround sequence?

14-39. What is the transparency mechanism used with SDLC?

14-40. What is a message abort? When is it transmitted?

14-41. Explain invert-on-zero encoding. Why is it used?

14-42. What supervisory condition exists with HDLC that is not included with SDLC?

14-43. What is the delimiting sequence used with HDLC? The transparency mechanism?

14-44. Explain extended addressing as it is used with HDLC.

14-45. What is the difference between the basic control format and the extended control format with HDLC?

14-46. What is the difference in the information fields used with SDLC and HDLC?

14-47. What operational modes are included with HDLC that are not included with SDLC?

14-48. What is a public data network?

14-49. Describe a value-added network.

14-50. Explain the differences in circuit-, message-, and packet-switching techniques.

14-51. What is blocking? With which switching techniques is blocking possible?

14-52. What is a transparent switch? A transactional switch?

14-53. What is a packet?

14-54. What is the difference between a store-and-forward and a hold-and-forward network?

14-55. Explain the three modes of transmission for public data networks.

14-56. What is the user-to-network protocol designated by CCITT?

14-57. What is the user-to-network protocol designated by ANSI?

14-58. Which layers of the ISO protocol hierarchy are addressed by X.25?

14-59. Explain the following terms: permanent virtual circuit, virtual call, and datagram.

14-60. Why was HDLC selected as the link-level protocol for X.25?

14-61. Briefly explain the fields that make up an X.25 call request packet.

14-62. Describe a local area network.

14-63. What is the connecting medium used with local area networks?

14-64. Explain the two transmission formats used with local area networks.

14-65. Explain CSMA/CD.

14-66. Explain token passing.

14-67. Describe what an ISDN is and who proposed its concept.

14-68. What are the primary principles of ISDN?

14-69. What are the evolutions of ISDN?

14-70. Describe the proposed architecture for ISDN.

14-71. Describe an ISDN D-channel. An ISDN B-Channel.

14-72. Describe the following ISDN terms: terminal equipment 1, terminal equipment 2, terminal adapter, reference point S, reference point T, reference point R, network termination 1, network termination 2, network termination 1,2, line termination unit, exchange termination, and reference point V.

14-73. Describe the ISDN LAP-D format.

14-74. What are the differences between the ISDN LAP-D and LAP-B formats?

14-75. What is an ISDN service point identifier and when is it used?

14-76. What is an ISDN terminal endpoint identifier and when is it used?

14-77. Describe the basic concepts of BISDN and how it differs from narrowband ISDN.

14-78. What is meant by the asynchronous transfer mode?

14-79. What are the proposed new BISDN channel data rates?

PROBLEMS

14-1. Determine the hex code for the control field in an SDLC frame for the following conditions: information frame, poll, transmitting frame 4, and confirming reception of frames 2, 3, and 4.

14-2. Determine the hex code for the control field in an SDLC frame for the following conditions: supervisory frame, ready to receive, final, confirming reception of frames 6, 7, and 0.

14-3. Insert 0's into the following SDLC data stream.

111 001 000 011 111 111 100 111 110 100 111 101 011 111 111 111 001 011

14-4. Delete 0's from the following SDLC data stream.

010 111 110 100 011 011 111 011 101 110 101 111 101 011 100 011 111 00

14-5. Sketch the NRZI waveform for the following data stream (start with a high condition).

1 0 0 1 1 1 0 0 1 0 1 0

Chap. 14   Data Communications Protocols

Digital Transmission

INTRODUCTION

As stated previously, digital transmission is the transmittal of digital pulses between two points in a communications system. The original source information may already be in digital form or it may be analog signals that must be converted to digital pulses prior to transmission and converted back to analog form at the receive end. With digital transmission systems, a physical facility such as a metallic wire pair, a coaxial cable, or an optical fiber link is required to interconnect the two points in the system. The pulses are contained in and propagate down the facility.

Advantages of Digital Transmission

1. The primary advantage of digital transmission is noise immunity. Analog signals are more susceptible than digital pulses to undesired amplitude, frequency, and phase variations. This is because with digital transmission, it is not necessary to evaluate these parameters as precisely as with analog transmission. Instead, the received pulses are evaluated during a sample interval, and a simple determination is made whether the pulse is above or below a certain threshold.

2. Digital pulses are better suited to processing and multiplexing than analog signals. Digital pulses can be stored easily, whereas analog signals cannot. Also, the transmission rate of a digital system can easily be changed to adapt to different environments and to interface with different types of equipment. Multiplexing is explained in detail in Chapter 16.

3. Digital systems use signal regeneration rather than signal amplification, thus producing a more noise resistant system than their analog counterpart.

4. Digital signals are simpler to measure and evaluate. Therefore, it is easier to compare the performance of digital systems with different signaling and information capacities than it is with comparable analog systems.

5. Digital systems are better suited to evaluate error performance (i.e., error detection and correction) than analog systems.

Disadvantages of Digital Transmission

1. The transmission of digitally encoded analog signals requires more bandwidth than simply transmitting the analog signal.

2. Analog signals must be converted to digital codes prior to transmission and converted back to analog at the receiver.

3. Digital transmission requires precise time synchronization between transmitter and receiver clocks.

4. Digital transmission systems are incompatible with existing analog facilities.

PULSE MODULATION

Pulse modulation includes many different methods of converting information into pulse form for transferring pulses from a source to a destination. The four predominant methods are *pulse width modulation* (PWM), *pulse position modulation* (PPM), *pulse amplitude modulation* (PAM), and *pulse code modulation* (PCM). The four most common methods of pulse modulation are summarized below and shown in Figure 15-1.

1. *PWM.* This method is sometimes called pulse duration modulation (PDM) or pulse length modulation (PLM). The pulse width (active portion of the duty cycle) is proportional to the amplitude of the analog signal.

2. *PPM.* The position of a constant-width pulse within a prescribed time slot is varied according to the amplitude of the analog signal.

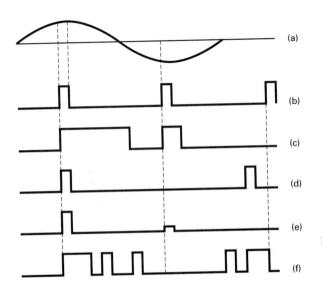

Figure 15-1 Pulse modulation:
(a) analog signal; (b) sample pulse;
(c) PWM; (d) PPM; (e) PAM; (f) PCM.

Chap. 15   Digital Transmission

3. *PAM.* The amplitude of a constant-width, constant-position pulse is varied according to the amplitude of the analog signal.

4. *PCM.* The analog signal is sampled and converted to a fixed-length, serial binary number for transmission. The binary number varies according to the amplitude of the analog signal.

PAM is used as an intermediate form of modulation with PSK, QAM, and PCM, although it is seldom used by itself. PWM and PPM are used in special-purpose communications systems (usually for the military) but are seldom used for commercial systems. PCM is by far the most prevalent method of pulse modulation and consequently, will be the topic of discussion for the remainder of this chapter.

PULSE CODE MODULATION

Pulse code modulation (PCM) is the only one of the digitally encoded pulse modulation techniques previously mentioned that is used in a digital transmission system. With PCM, the pulses are of fixed length and fixed amplitude. PCM is a binary system; a pulse or lack of a pulse within a prescribed time slot represents either a logic 1 or a logic 0 condition. With PWM, PPM, or PAM, a single pulse does not represent a single binary digit (bit).

Figure 15-2 shows a simplified block diagram of a single-channel, *simplex (one-way-only)* PCM system. The bandpass filter limits the input analog signal to the standard voice band frequency range 300 to 3000 Hz. The *sample-and-hold* circuit periodically samples the analog input and converts those samples to a multilevel PAM signal. The *analog-to-digital converter* (ADC) converts the PAM samples to a serial binary data stream for transmission. The transmission medium is a metallic wire or optical fiber.

At the receive end, the *digital-to-analog converter* (DAC) converts the serial binary data stream to a multilevel PAM signal. The hold circuit and low-pass filter convert the PAM signal back to its original analog form. An integrated circuit that performs the PCM encoding and decoding is called a *codec* (*co*der/*dec*oder). The codec is explained in detail in Chapter 16.

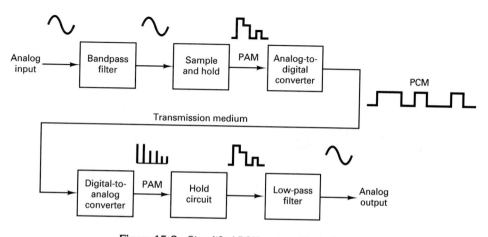

Figure 15-2 Simplified PCM system block diagram.

Pulse Code Modulation

Sample-and-Hold Circuit

The purpose of the sample-and-hold circuit is to sample periodically the continually changing analog input signal and convert the samples to a series of constant-amplitude PAM levels. For the ADC to accurately convert a signal to a digital code, the signal must be relatively constant. If not, before the ADC can complete the conversion, the input would change. Therefore, the ADC would continually be attempting to following the analog changes and never stabilize on any PCM code.

Figure 15-3 shows the schematic diagram of a sample-and-hold circuit. The FET acts like a simple switch. When turned "on," it provides a low-impedance path to deposit the analog sample voltage on capacitor C1. The time that Q1 is "on" is called the *aperture* or *acquisition time*. Essentially, C1 is the hold circuit. When Q1 is "off," the capacitor does not have a complete path to discharge through and therefore stores the sampled voltage. The *storage time* of the capacitor is also called the A/D *conversion time* because it is during this time that the ADC converts the sample voltage to a digital code. The acquisition time should be very short. This assures that a minimum change occurs in the analog signal while it is being deposited across C1. If the input to the ADC is changing while it is performing the conversion, distortion results. This distortion is called *aperture distortion.* Thus, by having a short aperture time and keeping the input to the ADC relatively constant, the sample-and-hold circuit reduces aperture distortion. If the analog signal is sampled for a short period of time and the sample voltage is held at a constant amplitude during the A/D conversion time, this is called *flat-top sampling.* If the sample time is made longer and the analog-to-digital conversion takes place with a changing analog signal, this is called *natural sampling.* Natural sampling introduces more aperture distortion than flat-top sampling and requires a faster A/D converter.

Figure 15-4 shows the input analog signal, the sampling pulse, and the waveform developed across C1. It is important that the output impedance of voltage follower Z1 and the "on" resistance of Q1 be as small as possible. This assures that the *RC* charging time constant of the capacitor is kept very short, allowing the capacitor to charge or discharge rapidly during the short acquisition time. The rapid drop in the capacitor voltage immediately following each sample pulse is due to the redistribution of the charge across C1. The interelectrode capacitance between the gate and drain of the FET is placed in series with C1 when the FET is "off," thus acting like a capacitive voltage-divider network. Also, note the gradual discharge across the capacitor during the conversion time. This is called *droop* and is caused by the capacitor discharging through its own leakage resistance and the input impedance of voltage follower Z2. Therefore, it is important that the input impedance of Z2 and the leakage resistance of C1 be as high as possible. Essentially, voltage

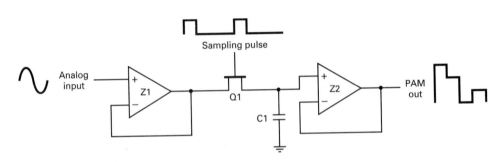

Figure 15-3 Sample-and-hold circuit.

Chap. 15   Digital Transmission

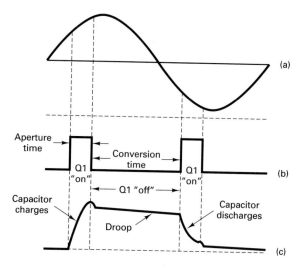

Aperture time

Conversion time

Q1 "on"

Q1 "off"

Q1 "on"

Capacitor charges

Droop

Capacitor discharges

(a)

(b)

(c)

Figure 15-4 Sample-and-hold waveforms: (a) analog in; (b) sample pulse; (c) capacitor voltage.

followers Z1 and Z2 isolate the sample-and-hold circuit (Q1 and C1) from the input and output circuitry.

EXAMPLE 15-1

For the sample-and-hold circuit shown in Figure 15-3, determine the largest-value capacitor that can be used. Use an output impedance for Z1 of 10 Ω, an "on" resistance for Q1 of 10 Ω, an acquisition time of 10 μs, a maximum peak-to-peak input voltage of 10 V, a maximum output current from Z1 of 10 mA, and an accuracy of 1%.

Solution The expression for the current through a capacitor is

$$i = C\frac{dv}{dt}$$

Rearranging and solving for C yields

$$C = i\frac{dt}{dv}$$

where C = maximum capacitance (farads)
i = maximum output current from Z1, 10 mA
dv = maximum change in voltage across C1, which equals 10 V
dt = charge time, which equals the aperture time, 10 μs

Therefore,

$$C_{max} = \frac{(10 \text{ mA})(10 \text{ } \mu\text{s})}{10 \text{ V}} = 10 \text{ nF}$$

The charge time constant for C when Q1 is "on" is

$$\tau = RC$$

where τ = one charge time constant (seconds)
R = output impedance of Z1 plus the "on" resistance of Q1 (ohms)
C = capacitance value of C1 (farads)

Rearranging and solving for C gives us

$$C_{max} = \frac{\tau}{R}$$

Pulse Code Modulation

The charge time of capacitor C1 is also dependent on the accuracy desired from the device. The percent accuracy and its required RC time constant are summarized as follows:

Accuracy (%)	Charge time
10	3τ
1	4τ
0.1	7τ
0.01	9τ

For an accuracy of 1%,

$$C = \frac{10\ \mu s}{4(20)} = 125\ nF$$

To satisfy the output current limitations of Z1, a maximum capacitance of 10 nF was required. To satisfy the accuracy requirements, 125 nF was required. To satisfy both requirements, the smaller-value capacitor must be used. Therefore, C1 can be no larger than 10 nF.

Sampling Rate

The Nyquist sampling theorem establishes the *minimum sampling rate* (f_s) that can be used for a given PCM system. For a sample to be reproduced accurately at the receiver, each cycle of the analog input signal (f_a) must be sampled at least twice. Consequently, the minimum sampling rate is equal to twice the highest audio input frequency. If f_s is less than two times f_a, distortion will result. This distortion is called *aliasing* or *foldover distortion*. Mathematically, the minimum Nyquist sample rate is

$$f_s \geq 2f_a \qquad (15\text{-}1)$$

where f_s = minimum Nyquist sample rate (hertz)
f_a = highest frequency to be sampled (hertz)

Essentially, a sample-and-hold circuit is an AM modulator. The switch is a nonlinear device that has two inputs: the sampling pulse and the input analog signal. Consequently, *nonlinear mixing* (*heterodyning*) occurs between these two signals. Figure 15-5a shows the frequency-domain representation of the output spectrum from a sample-and-hold circuit. The output includes the two original inputs (the audio and the fundamental frequency of the sampling pulse), their sum and difference frequencies ($f_s \pm f_a$), all the harmonics of f_s and f_a ($2f_s$, $2f_a$, $3f_s$, $3f_a$, etc.), and their associated cross products ($2f_s \pm f_a$, $3f_s \pm f_a$, etc.).

Because the sampling pulse is a repetitive waveform, it is made up of a series of harmonically related sine waves. Each of these sine waves is amplitude modulated by the analog signal and produces sum and difference frequencies symmetrical around each of the harmonics of f_s. Each sum and difference frequency generated is separated from its respective center frequency by f_a. As long as f_s is at least twice f_a, none of the side frequencies from one harmonic will spill into the sidebands of another harmonic and aliasing does not occur. Figure 15-5b shows the results when an analog input frequency greater than $f_s/2$ modulates f_s. The side frequencies from one harmonic fold over into the sideband of another harmonic. The frequency that folds over is an alias of the input signal (hence the names "aliasing" or "foldover distortion"). If an alias side frequency from the first harmonic

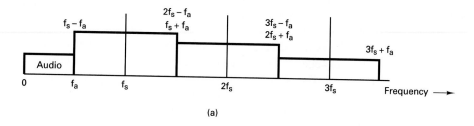

(a)

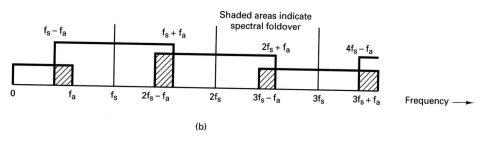

(b)

Figure 15-5 Output spectrum for a sample-and-hold circuit; (a) no aliasing; (b) aliasing distortion.

folds over into the input audio spectrum, it cannot be removed through filtering or any other technique.

EXAMPLE 15-2

For a PCM system with a maximum audio input frequency of 4 kHz, determine the minimum sample rate and the alias frequency produced if a 5-kHz audio signal were allowed to enter the sample-and-hold circuit.

Solution Using Nyquist's sampling theorem (Equation 15-1), we have

$$f_s \geq 2f_a \quad \text{therefore,} \, f_s \geq 8 \text{ kHz}$$

If a 5-kHz audio frequency entered the sample-and-hold circuit, the output spectrum shown in Figure 15-6 is produced. It can be seen that the 5-kHz signal produces an alias frequency of 3 kHz that has been introduced into the original audio spectrum.

The input bandpass filter shown in Figure 15-2 is called an *antialiasing* or *antifoldover filter*. Its upper cutoff frequency is chosen such that no frequency greater than one-half of the sampling rate is allowed to enter the sample-and-hold circuit, thus eliminating the possibility of foldover distortion occurring.

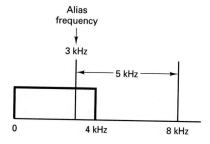

Figure 15-6 Output spectrum for Example 15-2.

Pulse Code Modulation

With PCM, the analog input signal is sampled, then converted to a serial binary code. The binary code is transmitted to the receiver, where it is converted back to the original analog signal. The binary codes used for PCM are *n*-bit codes, where *n* may be any positive integer greater than 1. The codes currently used for PCM are *sign-magnitude codes,* where the *most significant bit* (MSB) is the sign bit and the remaining bits are used for magnitude. Table 15-1 shows an *n*-bit PCM code where *n* equals 3. The most significant bit is used to represent the sign of the sample (logic 1 = positive and logic 0 = negative). The two remaining bits represent the magnitude. With 2 magnitude bits, there are four codes possible for positive numbers and four possible for negative numbers. Consequently, there is a total of eight possible codes ($2^3 = 8$).

Folded Binary Code

The PCM code shown in Table 15-1 is called a *folded binary code.* Except for the sign bit, the codes on the bottom half of the table are a mirror image of the codes on the top half. (If the negative codes were folded over on top of the positive codes, they would match perfectly.) Also, with folded binary there are two codes assigned to zero volts: 100 (+0) and 000 (−0). For this example, the magnitude of the minimum step size is 1 V. Therefore, the maximum voltage that may be encoded with this scheme is +3 V (111) or −3 V (011). If the magnitude of a sample exceeds the highest quantization interval, *overload distortion* (also called *peak limiting*) occurs. Assigning PCM codes to absolute magnitudes is called *quantizing.* The magnitude of the minimum step size is called *resolution,* which is equal in magnitude to the voltage of the least significant bit (V_{lsb} or the magnitude of the minimum step size of the DAC). The resolution is the minimum voltage other than 0 V that can be decoded by the DAC at the receiver. The smaller the magnitude of the minimum step size, the better (smaller) the resolution and the more accurately the quantization interval will resemble the actual analog sample.

In Table 15-1, each 3-bit code has a range of input voltages that will be converted to that code. For example, any voltage between +0.5 and +1.5 will be converted to the code 101. Any voltage between +1.5 and +2.5 will be encoded as 110. Each code has a *quantization range* equal to + or − one-half the resolution except the codes for +0 V and −0 V. The 0-V codes each have an input range equal to only one-half the resolution, but because there are two 0-V codes, the range for 0 V is also + or − one-half the resolution. Consequently, the maximum input voltage to the system is equal to the voltage of the highest magnitude code plus one-half of the voltage of the least significant bit.

TABLE 15-1 3-BIT PCM CODE

Sign	Magnitude		Level	Decimal
1	1	1		+3
1	1	0		+2
1	0	1		+1
1	0	0		+0
0	0	0		−0
0	0	1		−1
0	1	0		−2
0	1	1		−3

Figure 15-7 shows an analog input signal, the sampling pulse, the corresponding PAM signal, and the PCM code. The analog signal is sampled three times. The first sample occurs at time t_1 when the analog voltage is +2 V. The PCM code that corresponds to sample 1 is 110. Sample 2 occurs at time t_2 when the analog voltage is −1 V. The corresponding PCM code is 001. To determine the PCM code for a particular sample, simply divide the voltage of the sample by the resolution, convert it to an *n*-bit binary code, and add the sign bit to it. For sample 1, the sign bit is 1, indicating a positive voltage. The magnitude code (10) corresponds to a binary 2. Two times 1 V equals 2 V, the magnitude of the sample.

Sample 3 occurs at time t_3. The voltage at this time is +2.6 V. The folded PCM code for +2.6 V is 2.6/1 = 2.6. There is no code for this magnitude. If successive approximation ADCs are used, the magnitude of the sample is rounded off to the nearest valid code (111 or +3 V for this example). This results in an error when the code is converted back to analog by the DAC at the receive end. This error is called *quantization error* (Qe). The quantization error is equivalent to additive noise (it alters the signal amplitude). Like noise, the quantization error may add to or subtract from the actual signal. Consequently, quantization error is also called *quantization noise* (Qn) and its maximum magnitude is one-half the voltage of the minimum step size ($V_{lsb}/2$). For this example, Qe = 1 V/2 or 0.5 V.

Figure 15-8 shows the input-versus-output transfer function for a linear analog-to-digital converter (sometimes called a linear quantizer). As the figure shows for a linear analog input signal (i.e., a ramp), the quantized signal is a staircase. Thus, as shown in Figure 15-8c, the maximum quantization error is the same for any magnitude input signal.

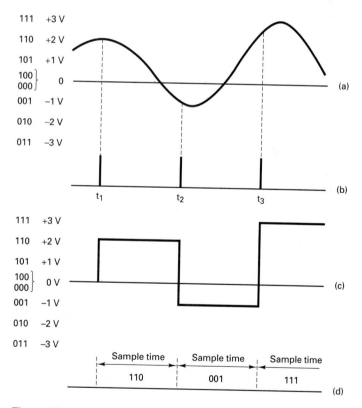

Figure 15-7 (a) Analog input signal; (b) sample pulse; (c) PAM signal; (d) PCM code.

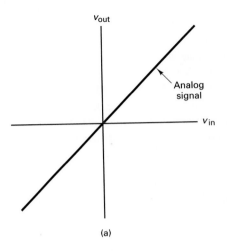

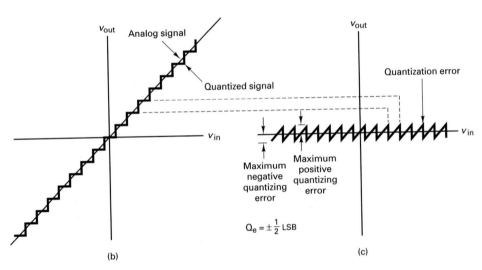

Figure 15-8 Linear input-versus-output transfer curve: (a) linear transfer function; (b) quantization; (c) Q_e.

Figure 15-9 shows the same analog input signal used in Figure 15-7 being sampled at a faster rate. As the figure shows, reducing the time between samples (i.e., increasing the sample rate) produces a PAM signal that more closely resembles the original analog input signal. However, it should also be noted that increasing the sample rate does not reduce the quantization error of the samples.

Dynamic Range

The number of PCM bits transmitted per sample is determined by several variables, which include maximum allowable input amplitude, resolution, and dynamic range. *Dynamic range* (DR) is the ratio of the largest possible magnitude to the smallest possible magnitude that can be decoded by the DAC. Mathematically, dynamic range is

$$DR = \frac{V_{max}}{V_{min}} \tag{15-2a}$$

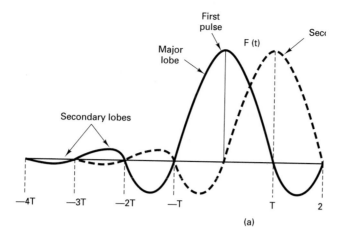

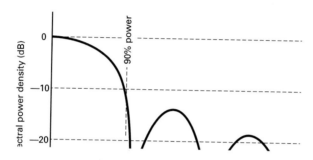

Figure 15-9 PAM: (a) input signal; (b) sample pulse; (c) PAM signal.

where V_{min} is equal to the resolution and V_{max} is the maximum voltage magnitude that can be decoded by the DACs. Thus

$$DR = \frac{V_{max}}{\text{resolution}}$$

For the system shown in Table 15-1,

$$DR = \frac{3 \text{ V}}{1 \text{ V}} = 3$$

It is common to represent dynamic range in decibels; therefore,

$$DR = 20 \log\frac{V_{max}}{V_{min}} = 20 \log\frac{3}{1} = 9.54 \text{ dB} \qquad (15\text{-}2b)$$

A dynamic range of 3 indicates that the ratio of the largest to the smallest decoded signal voltage is 3.

　　If a smaller resolution is desired, such as 0.5 V, to maintain a dynamic range of 3, the maximum allowable input voltage must be reduced by the same factor, one-half.

$$DR = \frac{1.5}{0.5} = 3$$

Therefore, V_{max} is reduced by a factor of 2 and the dynamic range is independent of resolution. If the resolution were reduced by a factor of 2 (0.25 V), to maintain the same maximum input amplitude, the dynamic range must double:

PCM Codes

$$DR = \frac{1.5}{0.25} = 6$$

The number of bits used for a PCM code depends on the dynamic range. With a 2-bit PCM code, the minimum decodable magnitude has a binary code of 01. The maximum magnitude is 11. The ratio of the maximum binary code to the minimum binary code is 3, the same as the dynamic range. Because the minimum binary code is always 1, DR is simply the maximum binary number for a system. Consequently, to determine the number of bits required for a PCM code the following mathematical relationship is used:

$$2^n - 1 \geq DR$$

and for a minimum value of n,

$$2^n - 1 = DR \qquad (15\text{-}3a)$$

where n = number of PCM bits, excluding sign bit
   DR = absolute value of Dynamic Range

Why $2^n - 1$? One PCM code is used for 0 V, which is not considered for dynamic range. Therefore,

$$2^n = DR + 1 \qquad (15\text{-}3b)$$

To solve for n, convert to logs:

$$\log 2^n = \log (DR + 1) \qquad (15\text{-}3c)$$

$$n \log 2 = \log (DR + 1)$$

$$n = \frac{\log (3 + 1)}{\log 2} = \frac{0.602}{0.301} = 2$$

For a dynamic range of 3, a PCM code with 2 bits is required.

EXAMPLE 15-3

A PCM system has the following parameters: a maximum analog input frequency of 4 kHz, a maximum decoded voltage at the receiver of $\pm 2.55V$, and a minimum dynamic range of 46 dB. Determine the following: minimum sample rate, minimum number of bits used in the PCM code, resolution, and quantization error.

Solution   Substituting into Equation 15-1, the minimum sample rate is

$$f_s = 2f_a = 2 \ (4 \ \text{kHz}) = 8 \ \text{kHz}$$

To determine the absolute value of the dynamic range, substitute into Equation 15-2b:

$$46 \ \text{dB} = 20 \log \frac{V_{max}}{V_{min}}$$

$$2.3 = \log \frac{V_{max}}{V_{min}}$$

$$10^{2.3} = \frac{V_{max}}{V_{min}} = DR$$

$$199.5 = DR$$

Substitute into Equation 15-3b and solve for n:

$$n = \frac{\log(199.5 + 1)}{\log 2} = 7.63$$

The closest whole number greater than 7.63 is 8; therefore, 8 bits must be used for the magnitude.

Because the input amplitude range is $\pm 2.55V$, one additional bit, the sign bit, is required. Therefore, the total number of PCM bits is 9 and the total number of PCM codes is 2^9 or 512. (There are 255 positive codes, 255 negative codes, and 2 zero codes.)

To determine the actual dynamic range, substitute into Equation 15-3c:

$$DR = 20 \log 255 = 48.13 \text{ dB}$$

To determine the resolution, divide the maximum + or − magnitude by the number of positive or negative nonzero PCM codes.

$$\text{resolution} = \frac{V_{max}}{2^n - 1}$$

$$= \frac{2.55}{2^8 - 1} = \frac{2.55}{256 - 1} = 0.01 \text{ V}$$

The maximum quantization error is

$$Qe = \frac{\text{resolution}}{2} = \frac{0.01}{2} = 0.005 \text{ V}$$

Coding Efficiency

Coding efficiency is a numerical indication of how efficiently a PCM code is utilized. Coding efficiency is the ratio of the minimum number of bits required to achieve a certain dynamic range to the actual number of PCM bits used. Mathematically, coding efficiency is

$$\text{coding efficiency} = \frac{\text{minimum number of bits}}{\text{actual number of bits}} \times 100 \qquad (15\text{-}4)$$
$$\text{(including sign bit)}$$

The coding efficiency for Example 15-3 is

$$\text{coding efficiency} = \frac{8.63}{9} \times 100 = 95.89\%$$

Signal-to-Quantization Noise Ratio

The 3-bit PCM coding scheme described in the preceding section is a linear code. That is, the magnitude change between any two successive codes is the same. Consequently, the magnitude of their quantization error is also the same. The maximum quantization noise is the voltage of the least significant bit divided by 2. Therefore, the worst possible *signal voltage-to-quantization noise voltage ratio* (SQR) occurs when the input signal is at its minimum amplitude (101 or 001). Mathematically, the worst-case voltage SQR is

$$SQR = \frac{\text{minimum voltage}}{\text{quantization noise voltage}} = \frac{V_{lsb}}{V_{lsb}/2} = 2$$

For a maximum amplitude input signal of 3 V (either 111 or 011), the maximum quantization noise is also the voltage of the least significant bit divided by 2. Therefore, the voltage SQR for a maximum input signal condition is

$$SQR = \frac{\text{maximum voltage}}{\text{quantization noise voltage}} = \frac{V_{max}}{V_{lsb}/2} = \frac{3}{0.5} = 6$$

From the preceding example it can be seen that even though the magnitude of error remains constant throughout the entire PCM code, the percentage of error does not; it decreases as the magnitude or the input signal increases. As a result, the SQR is not constant.

The preceding expression for SQR is for voltage and presumes the maximum quantization error and a constant-amplitude analog signal; therefore, it is of little practical use and is shown only for comparison purposes. In reality and as shown in Figure 15-7, the difference between the PAM waveform and the analog input waveform varies in magnitude. Therefore, the signal-to-quantization noise ratio is not constant. Generally, the quantization error or distortion caused by digitizing an analog sample is expressed as an average signal power-to-average noise power ratio. For linear PCM codes (all quantization intervals have equal magnitudes), the signal power-to-quantizing noise power ratio (also called *signal-to-distortion ratio* or *signal-to-noise ratio*) is determined as follows:

$$SQR \text{ (dB)} = 10 \log \frac{v^2/R}{(q^2/12)/R}$$

where R = resistance

v = rms signal voltage

q = quantization interval

$\dfrac{v^2}{R}$ = average signal power

$\dfrac{q^2/12}{R}$ = average quantization noise power

If the resistances are assumed to be equal,

$$SQR \text{ (dB)} = 10 \log \frac{v^2}{q^2/12} \qquad (15\text{-}5a)$$

$$= 10.8 + 20 \log \frac{v}{q} \qquad (15\text{-}5b)$$

Linear versus Nonlinear PCM Codes

Early PCM systems used *linear codes* (i.e., the magnitude change between any two successive steps is uniform). With linear encoding, the accuracy (resolution) for the higher-amplitude analog signals is the same as for the lower-amplitude signals, and the SQR for the lower-amplitude signals is less than for the higher-amplitude signals. With voice transmission, low-amplitude signals are more likely to occur than large-amplitude signals. Therefore, if there were more codes for the lower amplitudes, it would increase the accuracy where the accuracy is needed. As a result, there would be fewer codes available for the higher amplitudes, which would increase the quantization error for the larger-amplitude signals (thus decreasing the SQR). Such a coding technique is called *nonlinear* or *nonuniform encoding.* With nonlinear encoding, the step size increases with the amplitude of the input signal. Figure 15-10 shows the step outputs from a linear and a nonlinear ADC.

Note, with nonlinear encoding, there are more codes at the bottom of the scale than there are at the top, thus increasing the accuracy for the smaller signals. Also note that the distance between successive codes is greater for the higher-amplitude signals, thus increasing the quantization error and reducing the SQR. Also, because the ratio of V_{max} to V_{min} is increased with nonlinear encoding, the dynamic range is larger than with a uniform code. It is evident that nonlinear encoding is a compromise; SQR is sacrificed for the high-amplitude signals to achieve more accuracy for the low-amplitude signals and to achieve a larger dynamic range. It is difficult to fabricate nonlinear ADCs; consequently, alternative methods of achieving the same results have been devised and are discussed later in this chapter.

Idle Channel Noise

During times when there is no analog input signal, the only input to the PAM sampler is random, thermal noise. This noise is called *idle channel noise* and is converted to a PAM sample just as if it were a signal. Consequently, even input noise is quantized by the ADC. Figure 15-11 shows a way to reduce idle channel noise by a method called *midtread quantization.* With midtread quantizing, the first quantization interval is made larger in amplitude than the rest of the steps. Consequently, input noise can be quite large and still be quantized as a positive or negative zero code. As a result, the noise is suppressed during the encoding process.

In the PCM codes described thus far, the lowest-magnitude positive and negative codes have the same voltage range as all the other codes (+ or − one-half the resolution). This is called *midrise quantization.* Figure 15-11 contrasts the idle channel noise transmitted with a midrise PCM code to the idle channel noise transmitted when midtread quantization is used. The advantage of midtread quantization is less idle channel noise. The disadvantage is a larger possible magnitude for Qe in the lowest quantization interval.

With a folded binary PCM code, residual noise that fluctuates slightly above and below 0 V is converted to either a + or − zero PCM code and is consequently eliminated. In systems that do not use the two 0-V assignments, the residual noise could cause the PCM encoder to alternate between the zero code and the minimum + or − code. Consequently, the decoder would reproduce the encoded noise. With a folded binary code, most of the residual noise is inherently eliminated by the encoder.

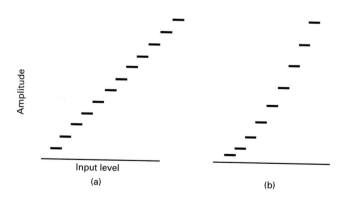

Figure 15-10 (a) Linear versus (b) nonlinear encoding.

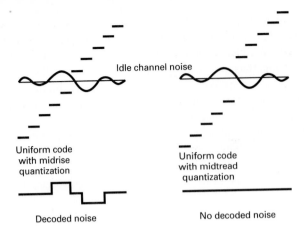

Uniform code
with midrise
quantization

Uniform code
with midtread
quantization

Decoded noise

No decoded noise

Figure 15-11 Idle channel noise.

Coding Methods

There are several coding methods used to quantize PAM signals into 2^n levels. These methods are classified according to whether the coding operation proceeds a level at a time, a digit at a time, or a word at a time.

Level-at-a-time coding. This type of coding compares the PAM signal to a ramp waveform while a binary counter is being advanced at a uniform rate. When the ramp waveform equals or exceeds the PAM sample, the counter contains the PCM code. This type of coding requires a very fast clock if the number of bits in the PCM code is large. Level-at-a-time coding also requires that 2^n sequential decisions be made for each PCM code generated. Therefore, level-at-a-time coding is generally limited to low-speed applications. Nonuniform coding is achieved by using a nonlinear function as the reference ramp.

Digit-at-a-time coding. This type of coding determines each digit of the PCM code sequentially. Digit-at-a-time coding is analogous to a balance where known reference weights are used to determine an unknown weight. Digit-at-a-time coders provide a compromise between speed and complexity. One common kind of digit-at-a-time coder, called a *feedback coder,* uses a successive approximation register (SAR). With this type of coder, the entire PCM code word is determined simultaneously.

Word-at-a-time coding. Word-at-a-time coders are flash encoders and are more complex; however, they are more suitable for high-speed applications. One common type of word-at-a-time coder uses multiple threshold circuits. Logic circuits sense the highest threshold circuit sensed by the PAM input signal and produce the approximate PCM code. This method is again impractical for large values of *n.*

Companding

Companding is the process of *compressing,* then *expanding.* With companded systems, the higher-amplitude analog signals are compressed (amplified less than the lower-amplitude signals) prior to transmission, then expanded (amplified more than the smaller-amplitude signals) at the receiver.

Figure 15-12 illustrates the process of companding. An input signal with a dynamic

Chap. 15   Digital Transmission

range of 120 dB is compressed to 60 dB for transmission, then expanded to 120 dB at the receiver. With PCM, companding may be accomplished through analog or digital techniques. Early PCM systems used analog companding, whereas more modern systems use digital companding.

Analog Companding

Historically, analog compression was implemented using specially designed diodes inserted in the analog signal path in the PCM transmitter prior to the sample-and-hold circuit. Analog expansion was also implemented with diodes that were placed just after the receive low-pass filter. Figure 15-13 shows the basic process of analog companding. In the transmitter, the analog signal is compressed, sampled, then converted to a linear PCM code. In the receiver, the PCM code is converted to a PAM signal, filtered, then expanded back to its original input amplitude characteristics.

Different signal distributions require different companding characteristics. For instance, voice signals require relatively constant SQR performance over a wide dynamic range, which means that the distortion must be proportional to signal amplitude for any input signal level. This requires a logarithmic compression ratio. A truly logarithmic assignment code requires an infinite dynamic range and an infinite number of PCM codes, which is impossible. There are two methods of analog companding currently being used that closely approximate a logarithmic function and are often called *log-PCM* codes. They are μ-*law* and *A-law companding.*

μ-Law companding. In the United States and Japan, μ-law companding is used. The compression characteristic for μ-law is

$$V_{\text{out}} = \frac{V_{\text{max}} \times \ln\left(1 + \mu V_{\text{in}}/V_{\text{max}}\right)}{\ln\left(1 + \mu\right)} \tag{15-6}$$

where V_{max} = maximum uncompressed analog input amplitude
          V_{in} = amplitude of the input signal at a particular instant of time
          μ = parameter used to define the amount of compression
          V_{out} = compressed output amplitude

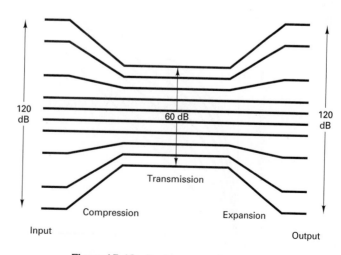

Figure 15-12 Basic companding process.

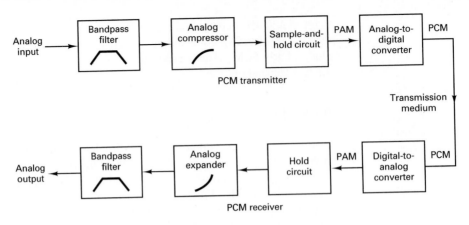

PCM transmitter

Transmission
medium

PCM receiver

Figure 15-13 PCM system with analog companding.

Figure 15-14 shows the compression for several values of μ. Note that the higher the μ, the more compression. Also note that for a μ = 0, the curve is linear (no compression).

The parameter μ determines the range of signal power in which the SQR is relatively constant. Voice transmission requires a minimum dynamic range of 40 dB and a 7-bit PCM code. For a relatively constant SQR and a 40-dB dynamic range, μ = 100 or larger is required. The early Bell System digital transmission systems used a 7-bit PCM code with μ = 100. The most recent digital transmission systems use 8-bit PCM codes and μ = 255.

EXAMPLE 15-4

For a compressor with μ = 255, determine the gain for the following values of V_{in}: V_{max}, 0.75 V_{max}, 0.5 V_{max}, and 0.25 V_{max}.

Solution Substituting into Equation 15-6, the following gains are achieved for various input magnitudes:

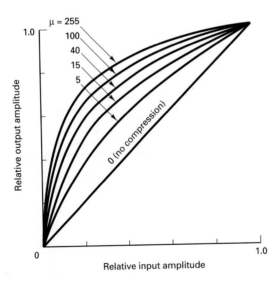

Figure 15-14 μ-Law compression characteristics.

V_{in}	Gain
V_{max}	1
$0.75V_{max}$	1.26
$0.5V_{max}$	1.75
$0.25V_{max}$	3

It can be seen that as the input signal amplitude increases, the gain decreases or is compressed.

A-Law companding. In Europe, the CCITT has established A-law companding to be used to approximate true logarithmic companding. For an intended dynamic range, A-law companding has a slightly flatter SQR than μ-law. A-law companding, however, is inferior to μ-law in terms of small-signal quality (idle channel noise). The compression characteristic for A-law companding is

$$V_{out} = V_{max} \frac{A V_{in}/V_{max}}{1 + \ln A} \qquad 0 \le \frac{V_{in}}{V_{max}} \le \frac{1}{A} \qquad (15\text{-}7a)$$

$$= V_{max} \frac{1 + \ln (A V_{in}/V_{max})}{1 + \ln A} \qquad \frac{1}{A} \le \frac{V_{in}}{V_{max}} \le 1 \qquad (15\text{-}7b)$$

Digital Companding

Digital companding involves compression at the transmit end after the input sample has been converted to a linear PCM code and expansion at the receive end prior to PCM decoding. Figure 15-15 shows the block diagram of a digitally companded PCM system.

With digital companding, the analog signal is first sampled and converted to a linear code, then the linear code is digitally compressed. At the receive end, the compressed PCM code is received, expanded, then decoded. The most recent digitally compressed PCM systems use a 12-bit linear code and an 8-bit compressed code. This companding process closely resembles a μ = 255 analog compression curve by approximating the curve with a set of eight straight line *segments* (segments 0 through 7). The slope of each successive segment is exactly one-half that of the previous segment. Figure 15-16 shows the 12-bit-to-8-bit digital compression curve for positive values only. The curve for negative values is identical except the inverse. Although there are 16 segments (eight positive and eight negative) this scheme is often called *13-segment compression*. This is because

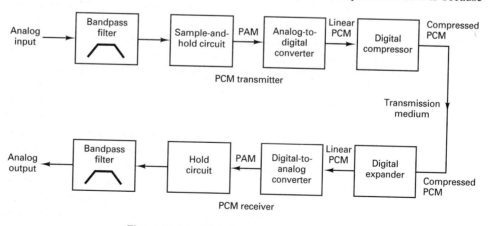

Figure 15-15 Digitally companded PCM system.

PCM Codes

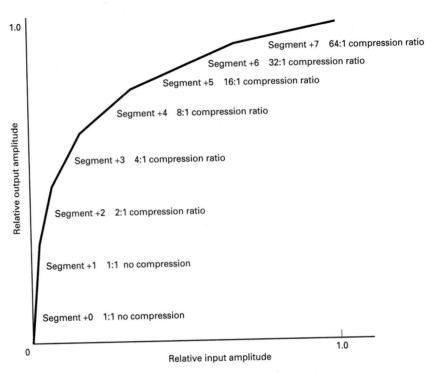

Figure 15-16 μ255 compression characteristics (positive values only).

the curve for segments +0, +1, −0, and −1 is a straight line with a constant slope and is often considered as one segment.

The digital companding algorithm for a 12-bit-linear-to-8-bit-compressed code is actually quite simple. The 8-bit compressed code is comprised of a sign bit, a 3-bit segment identifier, and a 15-bit magnitude code which identifies the *quantization interval* within the specified segment (see Figure 15-17a).

In the μ255 encoding table shown in Figure 15-17b, the bit positions designated with an X are truncated during compression and are consequently lost. Bits designated A, B, C, and D are transmitted as is. The sign bit (s) is also transmitted as is. Note that for segments 0 and 1, the original 12 bits are duplicated exactly at the output of the decoder (Figure 15-17c), whereas for segment 7, only the most significant 6 bits are recovered. With 11 magnitude bits, there are 2048 possible codes. There are 16 codes in segment 0 and in segment 1. In segment 2, there are 32 codes; segment 3 has 64. Each successive segment beginning with segment 3 has twice as many codes as the previous segment. In each of the eight segments, only sixteen 12-bit codes can be recovered. Consequently, in segments 0 and 1, there is no compression (of the 16 possible codes, all 16 can be recovered). In segment 2, there is a compression ratio of 2:1 (32 possible transmit codes and 16 possible recovered codes). In segment 3, there is a 4:1 compression ratio (64 possible transmit codes and 16 possible recovered codes). The compression ratio doubles with each successive segment. The compression ratio in segment 7 is 1024/16 or 64:1.

The compression process is as follows. The analog signal is sampled and converted to a linear 12-bit sign-magnitude code. The sign bit is transferred directly to the 8-bit code. The segment is determined by counting the number of leading 0's in the 11-bit magnitude portion of the code beginning with the MSB. Subtract the number of leading 0's (not to exceed 7) from 7. The result is the segment number, which is converted to a 3-bit binary

Sign bit 1 = + 0 = −	3-Bit segment identifier	4-Bit quantization interval A B C D
	000 to 111	0000 to 1111

(a)

Segment	12-Bit linear code	8-Bit compressed code	8-Bit compressed code	12-Bit recovered code	Segment
0	s0000000ABCD	s000ABCD	s000ABCD	s0000000ABCD	0
1	s0000001ABCD	s001ABCD	s001ABCD	s0000001ABCD	1
2	s000001ABCDX	s010ABCD	s010ABCD	s000001ABCD1	2
3	s00001ABCDXX	s011ABCD	s011ABCD	s00001ABCD10	3
4	s0001ABCDXXX	s100ABCD	s100ABCD	s0001ABCD100	4
5	s001ABCDXXXX	s101ABCD	s101ABCD	s001ABCD1000	5
6	s01ABCDXXXXX	s110ABCD	s110ABCD	s01ABCD10000	6
7	s1ABCDXXXXXX	s111ABCD	s111ABCD	s1ABCD100000	7

(b)   (c)

Figure 15-17 12-bit to 8-bit digital companding: (a) 8-bit µ255 compressed code format; (b) µ255 encoding table; (c) µ255 decoding table.

number and substituted into the 8-bit code as the segment identifier. The four magnitude bits (A, B, C, and D) are the quantization interval and are substituted into the least significant 4 bits of the 8-bit compressed code.

Essentially, segments 2 through 7 are subdivided into smaller subsegments. Each segment has 16 subsegments, which correspond to the 16 conditions possible for the bits A, B, C, and D (0000–1111). In segment 2 there are two codes per subsegment. In segment 3 there are four. The number of codes per subsegment doubles with each subsequent segment. Consequently, in segment 7, each subsegment has 64 codes. Figure 15-18 shows the breakdown of segments versus subsegments for segments 2, 5, and 7. Note that in each subsegment, all 12-bit codes, once compressed and expanded, yield a single 12-bit code. This is shown in Figure 15-18.

From Figures 15-17 and 15-18, it can be seen that the most significant of the truncated bits is reinserted at the decoder as a 1. The remaining truncated bits are reinserted as 0's. This ensures that the maximum magnitude of error introduced by the compression and expansion process is minimized. Essentially, the decoder guesses what the truncated bits were prior to encoding. The most logical guess is halfway between the minimum- and maximum-magnitude codes. For example, in segment 5, the 5 least significant bits are truncated during compression. At the receiver, the decoder must determine what those bits were. The possibilities are any code between 00000 and 11111. The logical guess is 10000, approximately half the maximum magnitude. Consequently, the maximum compression error is slightly more than one-half the magnitude of that segment.

EXAMPLE 15-5

For a resolution of 0.01 V and analog sample voltages of (a) 0.05 V, (b) 0.32 V, and (c) 10.23 V, determine the 12-bit linear code, the 8-bit compressed code, and the recovered 12-bit code.

Solution (a) To determine the 12-bit linear code for 0.05 V, Simply divide the sample voltage by the resolution and convert the result to a 12-bit sign-magnitude binary number.

12-bit linear code:

$$\frac{0.05 \text{ V}}{0.01 \text{ V}} = 5 = \begin{array}{l} 1\ 0\ 0\ 0\ 0\ 0\ 0\ 0\ 0\ 1\ 0\ 1 \\ \texttt{s------magnitude-----} \end{array}$$

```
(11-bit binary number)
```

8-bit compressed code:

```
1   0   0   0   0   0   0   0     0   1   0   1
s       (7 - 7 = 0 or 000)        A   B   C   D
1             0 0 0               0   1   0   1
↑
sign bit          unit            quanti-
  (+)          identifier          zation
              (segment 0)         interval
```

12-bit recovered code:

```
1                    0   0   0         0   1   0   1
s        (7 - 0 = 7 leading 0's)       A   B   C   D
1   0    0   0   0   0   0   0         0   1   0   1
↑
sign bit      segment identifier     quantization
              determines the            interval
              number of leading
                    0's
```

As you can see, the recovered 12-bit code is exactly the same as the original 12-bit linear code. This is true for all codes in segment 0 and 1. Consequently, there is no compression error in these two segments.

(b) For the 0.32-V sample:

12-bit linear code:

$$\frac{0.32 \text{ V}}{0.01 \text{ V}} = 32 = \begin{array}{l} 1\ 0\ 0\ 0\ 0\ 0\ 1\ 0\ 0\ 0\ 0\ 0 \\ \texttt{s ------magnitude-----} \end{array}$$

8-bit compressed code:

```
1   0   0   0   0   0   1     0   0   0   0   0
s   (7 - 5 = 2 or 010)        A   B   C   D   X
1       0   1   0             0   0   0   0   ↑
(+)     (segment 2)                      truncated
```

12-bit recovered code:

```
1                    0   1   0        0   0   0   0
s    (7 - 2 = 5 leading 0's)          A   B   C   D   X
1   0   0   0   0   0   1             0   0   0   0   1
                         ↑                            ↑
                     inserted                     inserted
```

Segment	12-Bit linear code		12-Bit expanded code	Subsegment
7	s11111111111 ⋯ s11111000000	64 : 1	s11111100000	15
7	s11110111111 ⋯ s11110000000	64 : 1	s11110100000	14
7	s11101111111 ⋯ s11101000000	64 : 1	s11101100000	13
7	s11100111111 ⋯ s11100000000	64 : 1	s11100100000	12
7	s11011111111 ⋯ s11011000000	64 : 1	s11011100000	11
7	s11010111111 ⋯ s11010000000	64 : 1	s11010100000	10
7	s11001111111 ⋯ s11001000000	64 : 1	s11001100000	9
7	s11000111111 ⋯ s11000000000	64 : 1	s11000100000	8
7	s10111111111 ⋯ s10111000000	64 : 1	s10111100000	7
7	s10110111111 ⋯ s10110000000	64 : 1	s10110100000	6
7	s10101111111 ⋯ s10101000000	64 : 1	s10101100000	5
7	s10100111111 ⋯ s10100000000	64 : 1	s10100100000	4
7	s10011111111 ⋯ s10011000000	64 : 1	s10011100000	3
7	s10010111111 ⋯ s10010000000	64 : 1	s10010100000	2
7	s10001111111 ⋯ s10001000000	64 : 1	s10001100000	1
7	s10000111111 ⋯ s10000000000	64 : 1	s10000100000	0
	s1ABCD-------			

(a)

Figure 15-18 12-bit segments divided into subsegments: (a) segment 7;

Note the two inserted 1's in the decoded 12-bit code. The least significant bit is determined from the decoding table in Figure 15-17. The stuffed 1 in bit position 6 was dropped during the 12-bit-to-8-bit conversion. Transmission of this bit is redundant because if it were not a 1, the sample would not be in segment 3. Consequently, in all segments except 0, a 1 is automatically inserted after the reinserted zeros. For this sample, there is an error in the received voltage equal to the resolution, 0.01 V. In segment 2, for every two 12-bit codes possible, there is only one recovered 12-bit code. Thus a coding compression of 2:1 is realized.

(c) To determine the codes for 10.23 V, the process is the same:

Segment	12-Bit linear code		12-Bit expanded code	Subsegment
5	s00111111111 .... s00111110000	} 16 : 1	s00111111000	15
5	s00111101111 .... s00111100000	} 16 : 1	s00111101000	14
5	s00111011111 .... s00111010000	} 16 : 1	s00111011000	13
5	s00111001111 .... s00111000000	} 16 : 1	s001110010000	12
5	s00110111111 .... s00110110000	} 16 : 1	s00110111000	11
5	s00110101111 .... s00110100000	} 16 : 1	s00110101000	10
5	s00110011111 .... s00110010000	} 16 : 1	s00110011000	9
5	s00110001111 .... s00110000000	} 16 : 1	s00110001000	8
5	s00101111111 .... s00101110000	} 16 : 1	s00101111000	7
5	s00101101111 .... s00101100000	} 16 : 1	s00101101000	6
5	s00101011111 .... s00101010000	} 16 : 1	s00101011000	5
5	s00101001111 .... s00101000000	} 16 : 1	s00101001000	4
5	s00100111111 .... s00100110000	} 16 : 1	s00100111000	3
5	s00100101111 .... s00100100000	} 16 : 1	s00100101000	2
5	s00100011111 .... s00100010000	} 16 : 1	s00100011000	1
5	s00100001111 .... s00100000000	} 16 : 1	s00100001000	0
	s001ABCD----			

(b)

(b) segment 5;

12-bit linear code:

```
1    1  1111    111111
↑
s       ABCD    truncated
```

8-bit compressed code:

```
↑       111    1111
s    segment    ABCD
```

Chap. 15   Digital Transmission

Segment	12-Bit linear code		12-Bit expanded code	Subsegment
2	s00000111111 .... s00000111110	} 2 : 1	s00000111111	15
2	s00000111101 .... s00000111100	} 2 : 1	s00000111101	14
2	s00000111011 .... s00000111010	} 2 : 1	s00000111011	13
2	s00000111001 .... s00000111000	} 2 : 1	s00000111001	12
2	s00000110111 .... s00000110110	} 2 : 1	s00000110111	11
2	s00000110101 .... s00000110100	} 2 : 1	s00000110101	10
2	s00000110011 .... s00000110010	} 2 : 1	s00000110011	9
2	s00000110001 .... s00000110000	} 2 : 1	s00000110001	8
2	s00000101111 .... s00000101110	} 2 : 1	s00000101111	7
2	s00000101101 .... s00000101100	} 2 : 1	s00000101101	6
2	s00000101011 .... s00000101010	} 2 : 1	s00000101011	5
2	s00000101001 .... s00000101000	} 2 : 1	s00000101001	4
2	s000000100111 .... s00000100110	} 2 : 1	s00000100111	3
2	s00000100101 .... s00000100100	} 2 : 1	s00000100101	2
2	s000000100011 .... s00000100010	} 2 : 1	s00000100011	1
2	s00000100001 .... s00000100000	} 2 : 1	s00000100001	0
	s000001ABCD-			

(c)

(c) segment 2.

12-bit recovered code:

```
    1   1   1111    100000
    ↑   ↑   └──┬──┘  └──┬──┘
    s   │    ABCD   inserted
        │
    inserted
```

The difference in the original 12-bit linear code and the recovered 12-bit code is

$$
\begin{array}{r}
111111111111 \\
-111111100000 \\
\hline
000000011111
\end{array}
= 31 \ (0.01 \ V) = 0.31 \ V
$$

PCM Codes

Percentage Error

For comparison purposes, the following formula is used for computing the *percentage of error* introduced by digital compression:

$$\% \text{ error} = \frac{|\text{Tx voltage} - \text{Rx voltage}|}{\text{Rx voltage}} \times 100 \qquad (15\text{-}8)$$

EXAMPLE 15-6

The maximum percentage of error will occur for the smallest number in the lowest subsegment within any given segment. Because there is no compression error in segments 0 and 1, for segment 3 the maximum % error is computed as follows:

```
Transmit 12-bit code:   s00001000000
Receive  12-bit code:   s00001000010
Magnitude of error:      00000000010
```

$$\% \text{ error} = \frac{|1000000 - 1000010|}{1000010} \times 100$$

$$= \frac{|64 - 66|}{66} \times 100 = 3.03\%$$

For segment 7:

```
Transmit 12-bit code:   s10000000000
Receive  12-bit code:   s10000100000
Magnitude of error:      00000100000
```

$$\% \text{ error} = \frac{|10000000000 - 10000100000|}{10000100000} \times 100$$

$$= \frac{|1024 - 1056|}{1056} \times 100 = 3.03\%$$

Although the magnitude of error is higher for segment 7, the percentage of error is the same. The maximum percentage of error is the same for segments 3 through 7, and consequently, the SQR degradation is the same for each segment.

Although there are several ways in which the 12-bit-to-8-bit compression and the 8-bit-to-12-bit expansion can be accomplished with hardware, the simplest and most economical method is with a look-up table in ROM (read-only memory).

Essentially every function performed by a PCM encoder and decoder is now accomplished with a single integrated-circuit chip called a *codec*. Most of the more recently developed codecs include an antialiasing (bandpass) filter, a sample-and-hold circuit, and an analog-to-digital converter in the transmit section and a digital-to-analog converter, a sample-and-hold circuit, and a bandpass filter in the receive section. The operation of a codec is explained in detail in Chapter 16.

Vocoders

The PCM coding and decoding processes described in the preceding sections were concerned primarily with reproducing waveforms as accurately as possible. The precise nature of the waveform was unimportant as long as it occupied the voice band frequency range. When digitizing speech signals only, special voice encoders/decoders called *vocoders* are often used. To achieve acceptable speech communications, the short-term power spectrum of the speech information is all that must be preserved. The human ear is relatively insensitive to the phase relationship between individual frequency components within a voice waveform. Therefore, vocoders are designed to reproduce only the short-term power spectrum, and the decoded time waveforms often only vaguely resemble the original input signal. Vocoders cannot be used in applications where analog signals other than voice are present, such as output signals from voice band data modems. Vocoders typically produce *unnatural* sounding speech and are therefore generally used for recorded information such as "wrong number" messages, encrypted voice for transmission over analog telephone circuits, computer output signals, and educational games.

The purpose of a vocoder is to encode the minimum amount of speech information necessary to reproduce a perceptible message with fewer bits than those needed by a conventional encoder/decoders. Vocoders are used primarily in limited bandwidth applications. Essentially, there are three vocoding techniques available: the *channel vocodor*, the *formant vocoder*, and the *linear predictive coder*.

Channel vocoders. The first channel vocoder was developed by Homer Dudley in 1928. Dudley's vocoder compressed conventional speech waveforms into an analog signal with a total bandwidth of approximately 300 Hz. Present-day digital vocoders operate at less than 2 kbps. Digital channel vocoders use bandpass filters to separate the speech waveform into narrower *subbands*. Each subband is full-wave rectified, filtered, then digitally encoded. The encoded signal is transmitted to the destination receiver, where it is decoded. Generally speaking, the quality of the signal at the output of a vocoder is quite poor. However, some of the more advanced channel vocoders operate at 2400 bps and can produce a highly intelligible, although slightly synthetic sounding speech.

Formant vocoders. A formant vocoder takes advantage of the fact that the short-term spectral density of typical speech signals seldom distributes uniformly across the entire voice band spectrum (300 to 3000 Hz). Instead, the spectral power of most speech energy concentrates at three or four peak frequencies called *formants*. A formant vocoder simply determines the location of these peaks and encodes and transmits only the information with the most significant short-term components. Therefore, formant vocoders can operate at lower bit rates and thus require narrower bandwidths. Formant vocoders sometimes have trouble tracking changes in the formants. However, once the formants have been identified, a formant vocoder can transfer intelligible speech at less than 1000 bps.

Linear predictive coders. A linear predictive coder extracts the most significant portions of speech information directly from the time waveform rather than from the frequency spectrum as with the channel and formant vocoders. A linear predictive coder produces a time-varying model of the *vocal tract excitation* and transfer function directly from the speech waveform. At the receive end, a *synthesizer* reproduces the speech by passing the specified excitation through a mathematical model of the vocal tract. Linear predictive coders provide more-natural-sounding speech than does either the channel or formant vocoder. Linear predictive coders typically encode and transmit speech at between 1.2 and 2.4 kbps.

DELTA MODULATION PCM

Delta modulation uses a single-bit PCM code to achieve digital transmission of analog signals. With conventional PCM, each code is a binary representation of both the sign and magnitude of a particular sample. Therefore, multiple-bit codes are required to represent the many values that the sample can be. With delta modulation, rather than transmit a coded representation of the sample, only a single bit is transmitted which simply indicates whether that sample is larger or smaller than the previous sample. The algorithm for a delta modulation system is quite simple. If the current sample is smaller than the previous sample, a logic 0 is transmitted. If the current sample is larger than the previous sample, a logic 1 is transmitted.

Delta Modulation Transmitter

Figure 15-19 shows a block diagram of a delta modulation transmitter. The input analog is sampled and converted to a PAM signal which is compared to the output of the DAC. The output of the DAC is a voltage equal to the regenerated magnitude of the previous sample, which was stored in the up-down counter as a binary number. The up-down counter is incremented or decremented depending on whether the previous sample is larger or smaller than the current sample. The up-down counter is clocked at a rate equal to the sample rate. Therefore, the up-down counter is updated after each comparison.

Figure 15-20 shows the ideal operation of a delta modulation encoder. Initially, the up-down counter is zeroed and the DAC is outputting 0 V. The first sample is taken, converted to a PAM signal, and compared to zero volts. The output of the comparator is a logic 1 condition (+V), indicating that the current sample is larger in amplitude than the previous sample. On the next clock pulse, the up-down counter is incremented to a count of 1. The DAC now outputs a voltage equal to the magnitude of the minimum step size (resolution). The steps change value at a rate equal to the clock frequency (sample rate). Consequently, with the input signal shown, the up-down counter follows the input analog

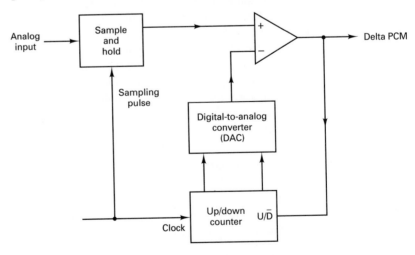

Figure 15-19 Delta modulation transmitter.

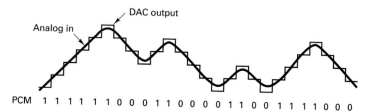

Figure 15-20 Ideal operation of a delta modulation encoder.

signal up until the output of the DAC exceeds the analog sample; then the up-down counter will begin counting down until the output of the DAC drops below the sample amplitude. In the idealized situation (shown in Figure 15-20), the DAC output follows the input signal. Each time the up-down counter is incremented, a logic 1 is transmitted, and each time the up-down counter is decremented, a logic 0 is transmitted.

Delta Modulation Receiver

Figure 15-21 shows the block diagram of a delta modulation receiver. As you can see, the receiver is almost identical to the transmitter except for the comparator. As the logic 1's and 0's are received, the up-down counter is incremented or decremented accordingly. Consequently, the output of the DAC in the decoder is identical to the output of the DAC in the transmitter.

With delta modulation, each sample requires the transmission of only one bit; therefore, the bit rates associated with delta modulation are lower than conventional PCM systems. However, there are two problems associated with delta modulation that do not occur with conventional PCM: slope overload and granular noise.

Slope overload.   Figure 15-22 shows what happens when the analog input signal changes at a faster rate than the DAC can keep up with. The slope of the analog signal is greater than the delta modulator can maintain. This is called *slope overload.* Increasing the clock frequency reduces the probability of slope overload occurring. Another way is to increase the magnitude of the minimum step size.

Granular noise.   Figure 15-23 contrasts the original and reconstructed signals associated with a delta modulation system. It can be seen that when the original analog input signal has a relatively constant amplitude, the reconstructed signal has variations that were not present in the original signal. This is called *granular noise.* Granular noise in delta modulation is analogous to quantization noise in conventional PCM.

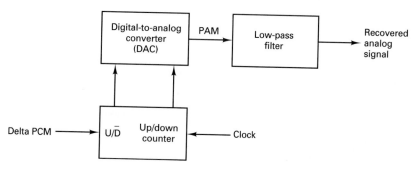

Figure 15-21 Delta modulation receiver.

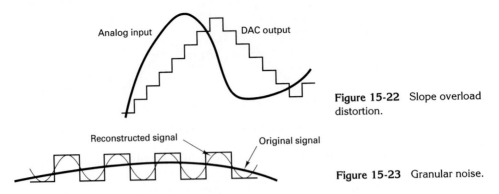

Figure 15-22 Slope overload distortion.

Figure 15-23 Granular noise.

Granular noise can be reduced by decreasing the step size. Therefore, to reduce the granular noise, a small resolution is needed, and to reduce the possibility of slope overload occurring, a large resolution is required. Obviously, a compromise is necessary.

Granular noise is more prevalent in analog signals that have gradual slopes and whose amplitudes vary only a small amount. Slope overload is more prevalent in analog signals that have steep slopes or whose amplitudes vary rapidly.

ADAPTIVE DELTA MODULATION PCM

Adaptive delta modulation is a delta modulation system where the step size of the DAC is automatically varied depending on the amplitude characteristics of the analog input signal. Figure 15-24 shows how an adaptive delta modulator works. When the output of the transmitter is a string of consecutive 1's or 0's, this indicates that the slope of the DAC output is less than the slope of the analog signal in either the positive or negative direction. Essentially, the DAC has lost track of exactly where the analog samples are and the possibility of slope overload occurring is high. With an adaptive delta modulator, after a predetermined number of consecutive 1's or 0's, the step size is automatically increased. After the next sample, if the DAC output amplitude is still below the sample amplitude, the next step is increased even further until eventually the DAC catches up with the analog signal. When an alternating sequence of 1's and 0's is occurring, this indicates that the possibility of granular noise occurring is high. Consequently, the DAC will automatically revert to its minimum step size and thus reduce the magnitude of the noise error.

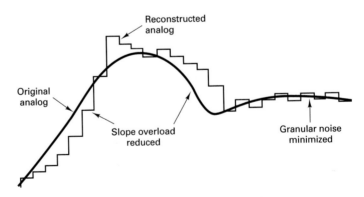

Figure 15-24 Adaptive delta modulation.

Chap. 15   **Digital Transmission**

A common algorithm for an adaptive delta modulator is when three consecutive 1's or 0's occur, the step size of the DAC is increased or decreased by a factor of 1.5. Various other algorithms may be used for adaptive delta modulators, depending on particular system requirements.

DIFFERENTIAL PULSE CODE MODULATION

In a typical PCM-encoded speech waveform, there are often successive samples taken in which there is little difference between the amplitudes of the two samples. This necessitates transmitting several identical PCM codes, which is redundant. Differential pulse code modulation (DPCM) is designed specifically to take advantage of the sample-to-sample redundancies in typical speech waveforms. With DPCM, the difference in the amplitude of two successive samples is transmitted rather than the actual sample. Since the range of sample differences is typically less than the range of individual samples, fewer bits are required for DPCM than conventional PCM.

Figure 15-25 shows a simplified block diagram of a DPCM transmitter. The analog input signal is bandlimited to one-half of the sample rate, then compared to the preceding accumulated signal level in the differentiator. The output of the differentiator is the difference between the two signals. The difference is PCM encoded and transmitted. The A/D converter operates the same as in a conventional PCM system except that it typically uses fewer bits per sample.

Figure 15-26 shows a simplified block diagram of a DPCM receiver. Each received sample is converted back to analog, stored, and then summed with the next sample received. In the receiver shown in Figure 15-26 the integration is performed on the analog signals, although it could also be performed digitally.

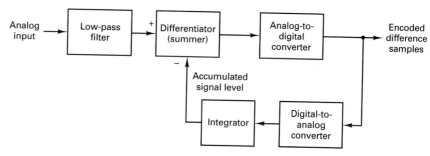

Figure 15-25 DPCM transmitter.

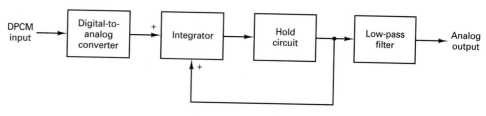

Figure 15-26 DPCM receiver.

Differential Pulse Code Modulation

PULSE TRANSMISSION

All digital carrier systems involve the transmission of pulses through a medium with a finite bandwidth. A highly selective system would require a large number of filter sections, which is impractical. Therefore, practical digital systems generally utilize filters with bandwidths that are approximately 30% or more in excess of the ideal Nyquist bandwidth. Figure 15-27a shows the typical output waveform from a *bandlimited* communications channel when a narrow pulse is applied to its input. The figure shows that bandlimiting a pulse causes the energy from the pulse to be spread over a significantly longer time in the form of *secondary lobes*. The secondary lobes are called *ringing tails*. The output frequency spectrum corresponding to a rectangular pulse is referred to as a (sin x)/x response and is given as

$$f(\omega) = (T)\frac{\sin(\omega T/2)}{\omega T/2} \qquad \text{where} \quad \begin{array}{l} \omega = 2\pi f \text{ (rad)} \\ T = \text{pulse width (sec)} \end{array} \qquad (15\text{-}9)$$

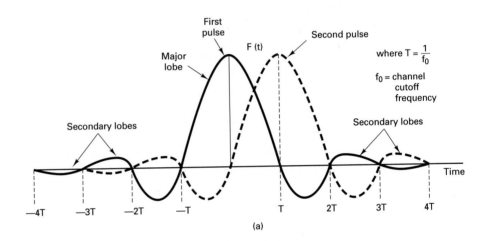

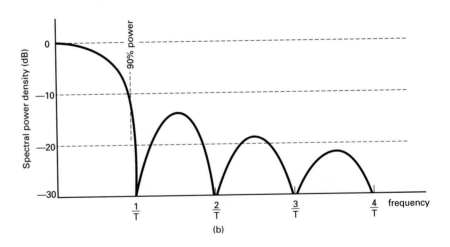

Figure 15-27 Pulse response: (a) typical pulse response of a bandlimited filter; (b) spectrum of square pulse with duration $1/T$.

Chap. 15   Digital Transmission

Figure 15-27b shows the distribution of the total spectrum power. It can be seen that approximately 90% of the signal power is contained within the first *spectral null* (i.e., $f = 1/T$). Therefore, the signal can be confined to a bandwidth $B = 1/T$ and still pass most of the energy from the original waveform. In theory, only the amplitude at the middle of each pulse interval needs to be preserved. Therefore, if the bandwidth is confined to $B = 1/2T$, the maximum signaling rate achievable through a low-pass filter with a specified bandwidth without causing excessive distortion is given as the Nyquist rate and is equal to twice the bandwidth. Mathematically, the Nyquist rate is

$$R = 2B \qquad (15\text{-}10)$$

where R = signaling rate = $1/T$
   B = specified bandwidth

Intersymbol Interference

Figure 15-28 shows the input signal to an ideal minimum bandwidth lowpass filter. The input signal is a random, binary non-return-to-zero (NRZ) sequence. Figure 15-28b shows the output of a lowpass filter that does not introduce any phase or amplitude distortion.

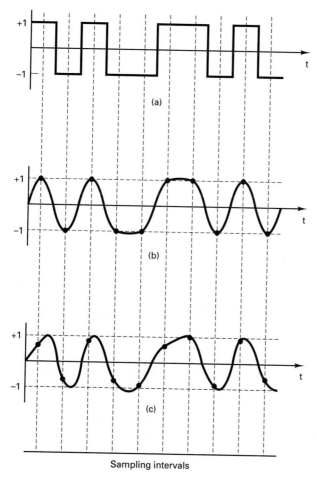

Sampling intervals

Figure 15-28 Pulse response: (a) NRZ input signal; (b) output from a perfect filter; (c) output from an imperfect filter.

Pulse Transmission

Note that the output signal reaches its full value for each transmitted pulse at precisely the center of each sampling interval. However, if the lowpass filter is imperfect (which in reality it will be), the output response will more closely resemble that shown in Figure 15-28c. At the sampling instants (i.e., the center of the pulses), the signal does not always attain the maximum value. The ringing tails of several pulses have *overlapped,* thus interfering with the *major pulse lobe.* Assuming no time delays through the system, energy in the form of spurious responses from the third and fourth impulses from one pulse appears during the sampling instant ($T = 0$) of another pulse. This interference is commonly called *intersymbol interference* or simply *ISI.* ISI is an important consideration in the transmission of pulses over circuits with a limited bandwidth and a nonlinear phase response. Simply stated, rectangular pulses will not remain rectangular in less than an infinite bandwidth. The narrower the bandwidth, the more rounded the pulses. If the phase distortion is excessive, the pulse will *tilt* and, consequently, affect the next pulse. When pulses from more than one source are multiplexed together, the amplitude, frequency, and phase responses become even more critical. ISI causes *crosstalk* between channels that occupy adjacent time slots in a time-division-multiplexed carrier system. Special filters called *equalizers* are inserted in the transmission path to "*equalize*" the distortion for all frequencies, creating a uniform transmission medium and reducing transmission impairments. The four primary causes of ISI are:

1. Timing inaccuracies. In digital transmission systems, transmitter timing inaccuracies cause intersymbol interference if the rate of transmission does not conform to the *ringing frequency* designed into the communications channel. Generally, timing inaccuracies of this type are insignificant. Since receiver clocking information is derived from the received signals, which are contaminated with noise, inaccurate sample timing is more likely to occur in receivers than in transmitters.

2. Insufficient bandwidth. Timing errors are less likely to occur if the transmission rate is well below the channel bandwidth (i.e., the Nyquist bandwidth is significantly below the channel bandwidth). As the bandwidth of a communications channel is reduced, the ringing frequency is reduced and intersymbol interference is more likely to occur.

3. Amplitude distortion. Filters are placed in a communications channel to bandlimit signals and reduce or eliminate predicted noise and interference. Filters are also used to produce a specific pulse response. However, the frequency response of a channel cannot always be predicted absolutely. When the frequency characteristics of a communications channel depart from the normal or expected values, *pulse distortion* results. Pulse distortion occurs when the peaks of pulses are reduced, causing improper ringing frequencies in the time domain. Compensation for such impairments is called amplitude equalization.

4. Phase distortion. A pulse is simply the superposition of a series of harmonically related sine waves with specific amplitude and phase relationships. Therefore, if the relative phase relations of the individual sine waves are altered, phase distortion occurs. Phase distortion occurs when frequency components undergo different amounts of time delay while propagating through the transmission medium. Special delay equalizers are placed in the transmission path to compensate for the varying delays, thus reducing the phase distortion. Phase equalizers can be manually adjusted or designed to automatically adjust themselves to varying transmission characteristics.

Eye Patterns

The performance of a digital transmission system depends, in part, on the ability of a repeater to regenerate the original pulses. Similarly, the quality of the regeneration process depends on the decision circuit within the repeater and the quality of the signal at the input to the decision circuit. Therefore, the performance of a digital transmission system can be measured by displaying the received signal on an oscilloscope and triggering the time base at the data rate. Thus all waveform combinations are superimposed over adjacent signaling intervals. Such a display is called an *eye pattern* or *eye diagram*. An eye pattern is a convenient technique for determining the effects of the degradations introduced into the pulses as they travel to the regenerator. The test setup to display an eye pattern is shown in Figure 15-29. The received pulse stream is fed to the vertical input of the oscilloscope, and the symbol clock is fed to the external trigger input, while the sweep rate is set approximately equal to the symbol rate.

Figure 15-30 shows an eye pattern generated by a symmetrical waveform for *ternary* signals in which the individual pulses at the input to the regenerator have a cosine-squared shape. In an *m*-level system, there will be $m - 1$ separate eyes. The horizontal lines labeled +1, 0, and −1 correspond to the ideal received amplitudes. The vertical lines, separated by the signaling interval, *T*, correspond to the ideal *decision times*. The decision levels for the regenerator are represented by *crosshairs*. The vertical hairs represents the decision time, while the horizontal hairs represents the decision level. The eye pattern shows the quality of shaping and timing and discloses any noise and errors that might be present in the line equalization. The eye opening (the area in the middle of the eye pattern) defines a boundary within which no waveform *trajectories* can exist under any code-pattern condition. The eye opening is a function of the number of code levels and the intersymbol interference caused by the ringing tails of any preceding or succeeding pulses. To regenerate the pulse sequence without error, the eye must be open (i.e., a decision area must exist), and the decision crosshairs must be within the open area. The effect of pulse degradation is a reduction in the size of the ideal eye. In Figure 15-30 it can be seen that at the center of the eye (i.e., the sampling instant) the opening is about 90%, indicating only minor ISI degradation due to filtering imperfections. The small degradation is due to the nonideal Nyquist amplitude and phase characteristics of the transmission system. Mathematically, the ISI degradation is

$$20 \log \frac{h}{H} \tag{15-11}$$

where H = ideal vertical opening
        h = degraded vertical opening

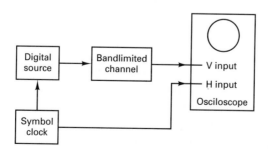

Figure 15-29 Eye diagram measurement setup.

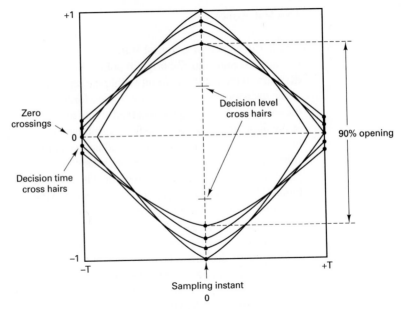

Figure 15-30 Eye diagram.

For the eye diagram shown in Figure 15-30,

$$20 \log \frac{90}{100} = 0.915 \text{ dB ISI degradation}$$

In Figure 15-30 it can also be seen that the overlapping signal pattern does not cross the horizontal zero line at exact integer multiples of the symbol clock. This is an impairment known as data transition jitter. This jitter has an effect on the symbol timing (clock) recovery circuit and, if excessive, may significantly degrade the performance of cascaded regenerative sections.

QUESTIONS

15-1. Contrast the advantages and disadvantages of digital transmission.

15-2. What are the four most common methods of pulse modulation?

15-3. Which method listed in Question 15-2 is the only form of pulse modulation that is used in a digital transmission system? Explain.

15-4. What is the purpose of the sample-and-hold circuit?

15-5. Define *aperture* and *acquisition time*.

15-6. What is the difference between natural and flat-top sampling?

15-7. Define *droop*. What causes it?

15-8. What is the Nyquist sampling rate?

15-9. Define and state the causes of foldover distortion.

15-10. Explain the difference between a magnitude-only code and a sign-magnitude code.

15-11. Explain overload distortion.

15-12. Explain quantizing.

15-13. What is quantization range? Quantization error?

15-14. Define *dynamic range*.

15-15. Explain the relationship between dynamic range, resolution, and the number of bits in a PCM code.

15-16. Explain coding efficiency.

15-17. What is SQR? What is the relationship between SQR, resolution, dynamic range, and the number of bits in a PCM code?

15-18. Contrast linear and nonlinear PCM codes.

15-19. Explain idle channel noise.

15-20. Contrast midtread and midrise quantization.

15-21. Define *companding*.

15-22. What does the parameter μ determine?

15-23. Briefly explain the process of digital companding.

15-24. What is the effect of digital compression on SQR, resolution, quantization interval, and quantization noise?

15-25. Contrast delta modulation PCM and standard PCM.

15-26. Define *slope overload* and *granular noise*.

15-27. What is the difference between adaptive delta modulation and conventional delta modulation?

15-28. Contrast differential and conventional PCM.

PROBLEMS

15-1. Determine the Nyquist sample rate for a maximum analog input frequency of **(a)** 4 kHz, and **(b)** 10 kHz.

15-2. For the sample-and-hold circuit shown in Figure 15-3, determine the largest-value capacitor that can be used. Use the following parameters: an output impedance for Z1 = 20 Ω, an "on" resistance of Q1 of 20 Ω, an acquisition time of 10 μs, a maximum output current from Z1 of 20 mA, and an accuracy of 1%.

15-3. For a sample rate of 20 kHz, determine the maximum analog input frequency.

15-4. Determine the alias frequency for a 15-kHz sample rate and an analog input frequency of 1.5 kHz.

15-5. Determine the dynamic range for a 10-bit sign-magnitude PCM code.

15-6. Determine the minimum number of bits required in a PCM code for a dynamic range of 80 dB. What is the coding efficiency?

15-7. For a resolution of 0.04 V, determine the voltages for the following linear 7-bit sign-magnitude PCM codes.
   (a) 0110101
   (b) 0000011
   (c) 1000001
   (d) 0111111
   (e) 1000000

15-8. Determine the SQR for a 2-V rms signal and a quantization interval of 0.2 V.

15-9. Determine the resolution and quantization noise for an 8-bit linear sign-magnitude PCM code for a maximum decoded voltage of 1.27V.

15-10. A 12 bit linear PCM code is digitally compressed into 8 bits. The resolution = 0.03 V. Determine the following for an analog input voltage of 1.465 V.
   (a) 12-bit linear PCM code
   (b) 8-bit compressed code
   (c) Decoded 12-bit code

(d) Decoded voltage

(e) Percentage error

15-11. For a 12-bit linear PCM code with a resolution of 0.02 V, determine the voltage range that would be converted to the following PCM codes.

(a) 1 0 0 0 0 0 0 0 0 0 0 1

(b) 0 0 0 0 0 0 0 0 0 0 0 0

(c) 1 1 0 0 0 0 0 0 0 0 0 0

(d) 0 1 0 0 0 0 0 0 0 0 0 0

(e) 1 0 0 1 0 0 0 0 0 0 0 1

(f) 1 0 1 0 1 0 1 0 1 0 1 0

15-12. For each of the following 12-bit linear PCM codes, determine the 8-bit compressed code to which they would be converted.

(a) 1 0 0 0 0 0 0 0 1 0 0 0

(b) 1 0 0 0 0 0 0 0 1 0 0 1

(c) 1 0 0 0 0 0 0 1 0 0 0 0

(d) 0 0 0 0 0 0 1 0 0 0 0 0

(e) 0 1 0 0 0 0 0 0 0 0 0 0

(f) 0 1 0 0 0 0 1 0 0 0 0 0

16
Multiplexing

INTRODUCTION

Multiplexing is the transmission of information (either voice or data) from more than one source to more than one destination on the same transmission medium *(facility)*. Transmissions occur on the same facility but not necessarily at the same time. The transmission medium may be a metallic wire pair, a coaxial cable, a terresteral microwave radio system, a satellite microwave radio, or an optical fiber cable. There are several ways in which multiplexing can be achieved, although the two most common methods are *frequency-division multiplexing* (FDM) and *time-division multiplexing* (TDM).

TIME-DIVISION MULTIPLEXING

With TDM, transmissions from multiple sources occur on the same facility but not at the same time. Transmissions from various sources are *interleaved* in the time domain. The most common type of modulation used with TDM systems is PCM. With a PCM-TDM system, two or more voice-band channels are sampled, converted to PCM codes, and then time division muliplexed onto a single metallic cable pair or an optical fiber cable.

Figure 16-1a shows a simplified block diagram of a two-channel PCM-TDM carrier system. Each channel is alternately sampled and converted to a PCM code. While the PCM code for channel 1 is being transmitted, channel 2 is sampled and converted to a PCM code. While the PCM code from channel 2 is being transmitted, the next sample is taken from channel 1 and converted to a PCM code. This process continues and samples are taken alternately from each channel, converted to PCM codes, and transmitted. The multiplexer is simply an electronic switch with two inputs and one output. Channel 1 and

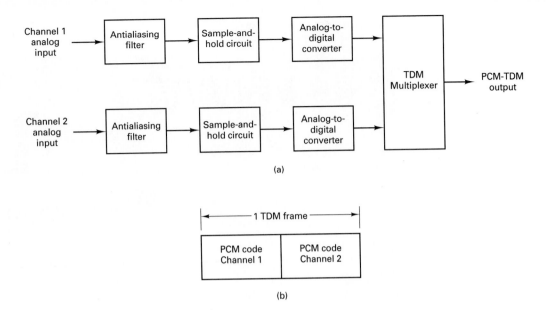

Figure 16-1 Two-channel PCM-TDM system: (a) block diagram; (b) TDM frame.

channel 2 are alternately selected and connected to the multiplexer output. The time it takes to transmit one sample from each channel is called the *frame time.*

The PCM code for each channel occupies a fixed time slot (epoch) within the total TDM frame. With a two-channel system, the time allocated for each channel is equal to one-half of the total frame time. A sample from each channel is taken once during each frame. Therefore, the total frame time is equal to the reciprocal of the sample rate ($1/f_s$). Figure 16-1b shows the TDM frame allocation for a two-channel system.

T1 DIGITAL CARRIER SYSTEM

A digital carrier is a communications system that uses digital pulses to encode information rather than analog signals. Figure 16-2 shows the block diagram of the Bell System T1 digital carrier system. This system is the North American telephone standard. A T1 carrier time-division multiplexes 24 PCM encoded samples for transmission over a single metallic wire pair or optical fiber. Again, the multiplexer is simply a switch except that now it has 24 inputs and 1 output. The 24 voice-band channels are sequentially selected and connected to the multiplexer output. Each voice-band channel occupies a 300- to 3000-Hz bandwidth.

Simply time-division multiplexing 24 voice-band channels does not in itself constitute a T1 carrier. At this point, the output of the multiplexer is simply a multiplexed digital signal (DS-1). It does not actually become a T1 carrier until it is line encoded and placed on special conditioned wire pairs called *T1 lines.* This is explained in more detail later in this chapter under the heading "North American Digital Hierarchy."

With the Bell System T1 carrier system, D-type (digital) channel banks perform the sampling, encoding, and multiplexing of 24 voice-band channels. Each channel contains an 8-bit PCM code and is sampled 8000 times per second. (Each channel is sampled at the same rate but not necessarily at the same time; see Figure 16-3.) Therefore, a 64-kbps PCM encoded sample is transmitted for each voice band channel during each frame.

$$\frac{8 \text{ bits}}{\text{sample}} \times \frac{8000 \text{ samples}}{\text{second}} = 64 \text{ kbps}$$

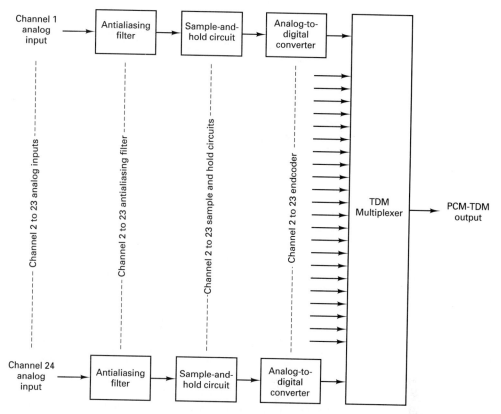

Figure 16-2 Bell system T1 PCM-TDM digital carrier system block diagram.

Within each frame an additional bit called a *framing bit* is added. The framing bit occurs at an 8000-bps rate and is recovered in the receiver circuitry and used to maintain frame and sample synchronization between the TDM transmitter and receiver. As a result, each TDM frame contains 193 bits.

$$\frac{8 \text{ bits}}{\text{channel}} \times \frac{24 \text{ channels}}{\text{frame}} = \frac{192 \text{ bits}}{\text{frame}} + \frac{1 \text{ framing bit}}{\text{frame}} = \frac{193 \text{ bits}}{\text{frame}}$$

As a result, the line speed (bps) for the T1 carrier is

$$\text{line speed} = \frac{193 \text{ bits}}{\text{frame}} \times \frac{8000 \text{ frames}}{\text{second}} = 1.544 \text{ Mbps}$$

D-Type Channel Banks

The early T1 carrier systems were equipped with D1A channel banks that use a 7-bit magnitude-only PCM code with analog companding and $\mu = 100$. A later version of the D1 channel bank (D1D) used an 8-bit sign-magnitude PCM code. With D1A channel banks an eighth bit (the s bit) is added to each PCM code word for the purpose of *signaling* (supervision: on-hook, off-hook, dial pulsing, and so on). Consequently, the signaling rate for D1 channel banks is 8 kbps. Also, with D1 channel banks, the framing bit sequence is simply an alternating 1/0 pattern. Figure 16-4 shows the frame and sample alignment for the T1 carrier system using D1A channel banks.

T1 Digital Carrier System

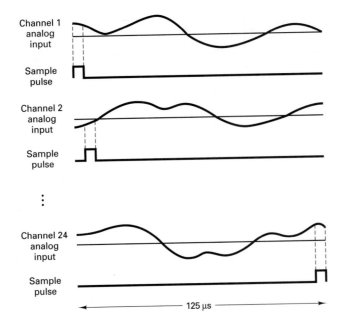

Figure 16-3 T1 sampling sequence.

Generically, the T1 carrier system has progressed through the D2, D3, D4, D5, and D6 channel banks. D4, D5, and D6 use a digitally companded, 8-bit sign-magnitude compressed PCM code with $\mu = 255$. In the D1 channel bank, the compression and expansion characteristics were implemented in circuitry separate from the encoder and decoder. The D2, D3, D4, and D5 channel banks incorporate the companding functions directly in the encoders and decoders. Although the D2 and D3 channel banks are functionally similar, the D3 channel banks were the first to incorporate separate customized LSI integrated circuits (codecs) for each voice-band channel. With D1, D2, and D3 channel banks, common equipment performs the encoding and decoding functions. Consequently, a single equipment malfunction constitutes a total system failure.

D1A channel banks use a magnitude-only code; consequently, an error in the most significant bit (MSB) of a channel sample always produces a decoded error equal to one-half the total quantization range (V_{max}). Because D1D, D2, D3, D4, and D5 channel banks use a sign-magnitude code, an error in the MSB (sign bit) causes a decoded error equal to twice the sample magnitude (from $+V$ to $-V$, or vice versa). The worst-case error is

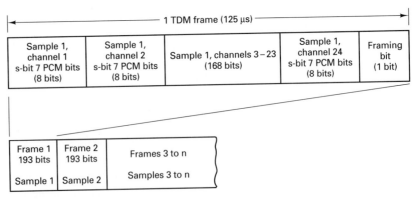

Figure 16-4 T1 carrier system frame and sample alignment using D1 channel banks.

equal to twice the total quantization range. However, maximum amplitude samples occur rarely, and most errors with D1D, D2, D3, D4, and D5 coding are less than one-half the coding range. On the average, the error performance with a sign-magnitude code is better than with a magnitude-only code.

Superframe Format

The 8-kbps signaling rate used with D1 channel banks is excessive for voice transmission. Therefore, with D2 and D3 channel banks, a signaling bit is substituted only into the least significant bit (LSB) of every sixth frame. Therefore, five out of every six frames have 8-bit resolution, while one out of every six frames (the signaling frame) has only 7-bit resolution. Consequently, the signaling rate on each channel is 1.333 kbps (8000 bps/6), and the effective number of bits per sample is actually $7\frac{5}{6}$ bits and not 8.

Because only every sixth frame includes a signaling bit, it is necessary that all the frames be numbered so that the receiver knows when to extract the signaling information. Also, because the signaling is accomplished with a 2-bit binary word, it is necessary to identify the MSB and LSB of the signaling word. Consequently, the *superframe* format shown in Figure 16-5 was devised. Within each superframe, there are 12 consecutively numbered frames (1–12). The signaling bits are substituted in frames 6 and 12, the MSB into frame 6 and the LSB into frame 12. Frames 1–6 are called the A-highway, with frame 6 designated as the A-channel signaling frame. Frames 7–12 are called the B-highway,

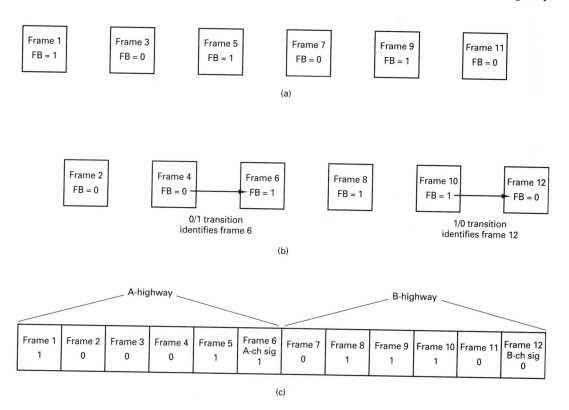

Figure 16-5 Framing bit sequence or the T1 superframe format using D2 or D3 channel banks: (a) frame synchronizing bits (odd-numbered frames); (b) signaling frame alignment bits (even-numbered frames); (c) composite frame alignment.

T1 Digital Carrier System

with frame 12 designated as the B-channel signaling frame. Therefore, in addition to identifying the signaling frames, the sixth and twelfth frames must be positively identified.

To identify frames 6 and 12, a different framing bit sequence is used for the odd- and even-numbered frames. The odd frames (frames 1, 3, 5, 7, 9, and 11) having an alternating 1/0 pattern, and the even frames (frames 2, 4, 6, 8, 10, and 12) have a 0 0 1 1 1 0 repetitive pattern. As a result, the combined bit pattern for the framing bits is a 1 0 0 0 1 1 0 1 1 1 0 0 repetitive pattern. The odd-numbered frames are used for frame and sample synchronization, while the even-numbered frames are used to identify the A- and B-channel signaling frames (6 and 12). Frame 6 is identified by a 0/1 transition in the framing bit between frames 4 and 6. Frame 12 is identified by a 1/0 transition in the framing bit between frames 10 and 12.

Figure 16-6 shows the frame, sample, and signaling alignment for the T1 carrier system using D2 or D3 channel banks.

In addition to *multiframe alignment* bits and PCM sample bits, certain time slots are used to indicate alarm conditions. For example, in the case of a transmit power supply failure, a common equipment failure, or loss of multiframe alignment, the second bit in each channel is made a 0 until the alarm condition has cleared. Also, the framing bit in frame 12 is complemented whenever multiframe alignment is lost (this is assumed whenever frame alignment is lost). In addition, there are special framing conditions that must be avoided in order to maintain clock and bit synchronization at the receive demultiplexing equipment. These special conditions are explained later in this chapter.

D4 channel banks time division multiplex 48 voice-band channels and operate at a transmission rate of 3.152 Mbps. This is slightly more than twice the line speed for 24-channel D1, D2, or D3 channel banks. This is because with D4 channel banks, rather than transmit a single framing bit with each frame, a 10-bit frame synchronization pattern is used. Consequently, the total number of bits in a D4 (DS-1C) TDM frame is

$$\frac{8 \text{ bits}}{\text{channel}} \times \frac{48 \text{ channels}}{\text{frame}} = \frac{384 \text{ bits}}{\text{frame}} + \frac{10 \text{ syn bits}}{\text{frame}} = \frac{394 \text{ bits}}{\text{frame}}$$

and the line speed is

$$\text{line speed} = \frac{394 \text{ bits}}{\text{frame}} \times \frac{8000 \text{ frames}}{\text{second}} = 3.152 \text{ Mbps}$$

The framing for the DS-1 (T1) system or the framing pattern for the DS-1C (T1C) time division multiplexed carrier systems is added to the multiplexed digital signal at the output of the multiplexer. Figure 16-7 shows the framing bit circuitry for the 24-channel T1 carrier system using either D1, D2, or D3 channel banks (DS-1). Note that the bit rate at the output of the TDM multiplex is 1.536 Mbps and the bit rate at the output of the 193-bit shift register is 1.544 Mbps. The difference (8 kbps) is due to the addition of the framing bit in the shift register.

CCITT TIME DIVISION MULTIPLEXED CARRIER SYSTEM

Figure 16-8 shows the frame alignment for the CCITT (Comité Consultatif International Téléphonique et Télégraphique) European standard PCM-TDM system. With the CCITT system, a 125-μs frame is divided into 32 equal time slots. Time slot 0 is used for a frame

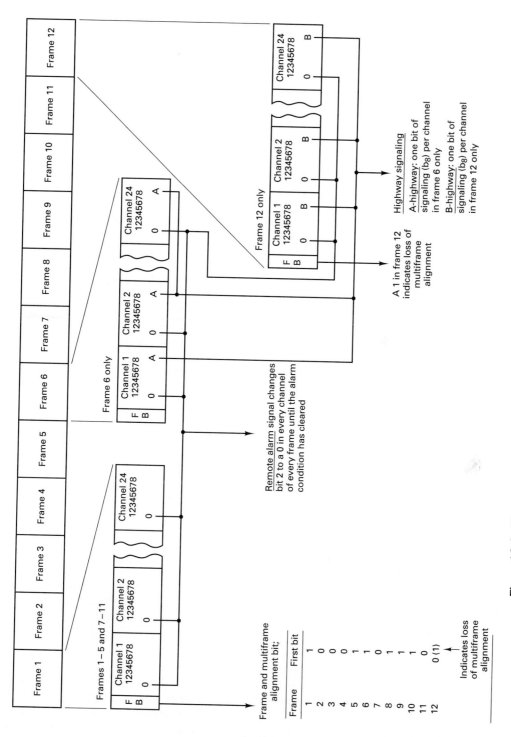

Figure 16-6 T1 carrier frame, sample, and signaling alignment for D2 and D3 channel banks.

651

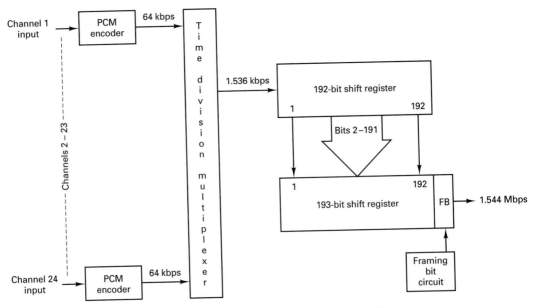

Figure 16-7 Framing bit circuitry for the DS-1 T1 carrier system.

alignment pattern and for an alarm channel. Time slot 17 is used for a common signaling channel. The signaling for all the voice-band channels is accomplished on the common signaling channel. Consequently, 30 voice-band channels are time division multiplexed into each CCITT frame.

With the CCITT standard, each time slot has 8 bits. Consequently, the total number of bits per frame is

$$\frac{8 \text{ bits}}{\text{time slot}} \times \frac{32 \text{ time slots}}{\text{frame}} = \frac{256 \text{ bits}}{\text{frame}}$$

and the line speed is

$$\text{line speed} = \frac{256 \text{ bits}}{\text{frame}} \times \frac{8000 \text{ frames}}{\text{second}} = 2.048 \text{ Mbps}$$

CODECS

A *codec* is a large-scale-integration (LSI) chip designed for use in the telecommunications industry for *private branch exchanges* (PBXs), central office switches, digital handsets, voice store-and-forward systems, and digital echo suppressors. Essentially, the codec is applicable for any purpose that requires the digitizing of analog signals, such as in a PCM-TDM carrier system.

Codec is a generic term that refers to the *co*ding functions performed by a device that converts analog signals to digital codes and digital codes to analog signals. Recently developed codecs are called *combo* chips because they combine codec and filter functions in the same LSI package. The input/output filter performs the following functions: band-limiting, noise rejection, antialiasing, and reconstruction of analog audio waveforms after decoding. The codec performs the following functions: analog sampling, encoding/decoding (analog-to-digital and digital-to-analog conversions), and digital companding.

Time slot 0	Time slot 1	Time slots 2–16	Time slot 17	Time slots 18–30	Time slot 31
Framing and alarm channel	Voice channel 1	Voice channels 2–15	Common signaling channel	Voice channels 16–29	Voice channel 30
8 bits	8 bits	112 bits	8 bits	112 bits	8 bits

(a)

Time slot 17

16 frames equal one multiframe; 500 multiframes are transmitted each second

Frame	Bits 1234	5678
0	0000	xyxx
1	ch 1	ch 16
2	ch 2	ch 17
3	ch 3	ch 18
4	ch 4	ch 19
5	ch 5	ch 20
6	ch 6	ch 21
7	ch 7	ch 22
8	ch 8	ch 23
9	ch 9	ch 24
10	ch 10	ch 25
11	ch 11	ch 26
12	ch 12	ch 27
13	ch 13	ch 28
14	ch 14	ch 29
15	ch 15	ch 30

x = spare
y = loss of multiframe alignment if a 1

4 bits per channel are transmitted once every 16 frames, resulting in a 500-bps signaling rate for each channel

(b)

Figure 16-8 CCITT TDM frame alignment and common signaling channel alignment: (a) CCITT TDM frame (125 μs, 256 bits, 2.048 Mbps); (b) common signaling channel.

COMBO CHIPS

A combo chip can provide the analog-to-digital and the digital-to-analog conversions and the transmit and receive filtering necessary to interface a full-duplex (four-wire) voice telephone circuit to the PCM highway of a TDM carrier system. Essentially, a combo chip replaces the older codec and filter chip combination.

Table 16-1 lists several of the combo chips available and their prominent features.

General Operation

The following major functions are provided by a combo chip:

1. Bandpass filtering of the analog signals prior to encoding and after decoding
2. Encoding and decoding of voice and call progress signals
3. Encoding and decoding of signaling and supervision information
4. Digital companding

Figure 16-9 shows the block diagram of a typical combo chip.

Combo Chips

TABLE 16-1 FEATURES OF SEVERAL CODEC/FILTER COMBO CHIPS

2916 (16-pin)	2917 (16-pin)	2913 (20-pin)	2914 (24-pin)
μ-Law companding only	A-law companding only	μ/A-law companding	μ/A-law companding
Master clock, 2.048 MHz only	Master clock, 2.048 MHz only	Master clock, 1.536 MHz, 1.544 MHz, or 2.048 MHz	Master clock, 1.536 MHz, 1.544 MHz, or 2.048 MHz
Fixed data rate	Fixed data rate	Fixed data rate	Fixed data rate
Variable data rate, 64 kbps–2.048 Mbps	Variable data rate, 64 kbps–4.096 Mbps	Variable data rate, 64 kbps–4.096 Mbps	Variable data rate, 64 kbps–4.096 Mbps
78-dB dynamic	78-dB dynamic range	78-dB dynamic range	78-dB dynamic range
ATT D3/4 compatible	ATT D3/4 compatible	ATT D3/4 compatible	ATT D3/4 compatible
Single-ended input	Single-ended input	Differential input	Differential input
Single-ended output	Single-ended output	Differential output	Differential output
Gain adjust transmit only	Gain adjust transmit only	Gain adjust transmit and receive	Gain adjust transmit and receive
Synchronous clocks	Synchronous clocks	Synchronous clocks	Synchronous clocks Asynchronous clocks
			Analog loopback
			Signaling

Fixed-data-rate Mode

In the *fixed-data-rate mode*, the master *transmit* and *receive clocks* on a combo chip (CLKX and CLKR) perform the following functions:

1. Provide the master clock for the on-board switched capacitor filter
2. Provide the clock for the analog-to-digital and digital-to-analog converters
3. Determine the input and output data rates between the codec and the PCM highway

Therefore, in the fixed-data-rate mode, the transmit and receive data rates must be either 1.536, 1.544, or 2.048 Mbps, the same as the master clock rate.

Transmit and receive frame synchronizing pulses (FSX and FSR) are 8-kHz inputs that set the transmit and receive sampling rates and distinguish between *signaling* and *nonsignaling* frames. $\overline{TSX}$ is a *time-slot strobe buffer enable* output that is used to gate the PCM word onto the PCM highway when an external buffer is used to drive the line. $\overline{TSX}$ is also used as an external gating pulse for a time-division multiplexer (see Figure 16-10).

Data are transmitted to the PCM highway from DX on the first eight positive transitions of CLKX following the rising edge of FSX. On the receive channel, data are received from the PCM highway from DR on the first eight falling edges of CLKR after the occurrence of FSR. Therefore, the occurrence of FSX and FSR must be synchronized between codecs in a multiple-channel system to ensure that only one codec is transmitting to or receiving from the PCM highway at any given time.

Figure 16-10 shows the block diagram and timing sequence for a single-channel PCM system using a combo chip in the fixed-data-rate mode and operating with a master clock frequency of 1.536 MHz. In the fixed-data-rate mode, data are input and output for a single channel in short bursts. (This mode of operation is sometimes called the *burst mode*.) With only a single channel, the PCM highway is active only $\frac{1}{24}$ of the total frame time. Additional channels can be added to the system provided that their transmissions are synchronized so that they do not occur at the same time as transmissions from any other channel.

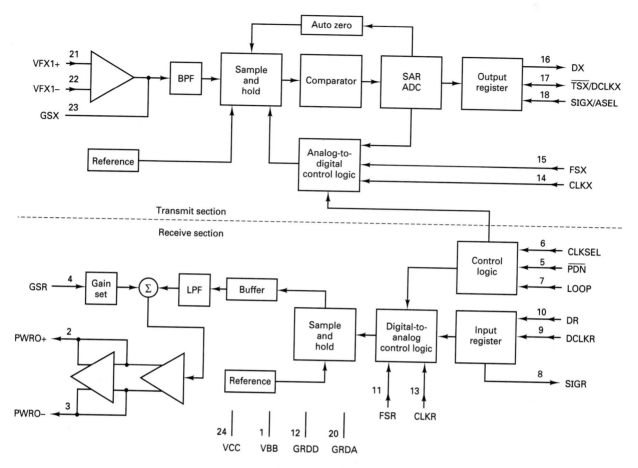

Figure 16-9 Block diagram of a combo chip.

From Figure 16-10 the following observations can be made:

1. The input and output bit rates from the codec are equal to the master clock frequency, 1.536 Mbps.

2. The codec inputs and outputs 64,000 PCM bits per second.

3. The data output (DX) and data input (DR) are enabled only $\frac{1}{24}$ of the total frame time (125 μs).

To add channels to the system shown in Figure 16-10, the occurrence of the FSX, FSR, and $\overline{TSX}$ signals for each additional channel must be synchronized so that they follow a timely sequence and do not allow more than one codec to transmit or receive at the same time. Figure 16-11 shows the block diagram and timing sequence for a 24-channel PCM-TDM system operating with a master clock frequency of 1.536 MHz.

Variable-data-rate Mode

The *variable-data-rate mode* allows for a flexible data input and output clock frequency. It provides the ability to vary the frequency of the transmit and receive bit clocks. In the variable-data-rate mode, a master clock frequency of 1.536, 1.544, or 2.048 MHz is still

Combo Chips

655

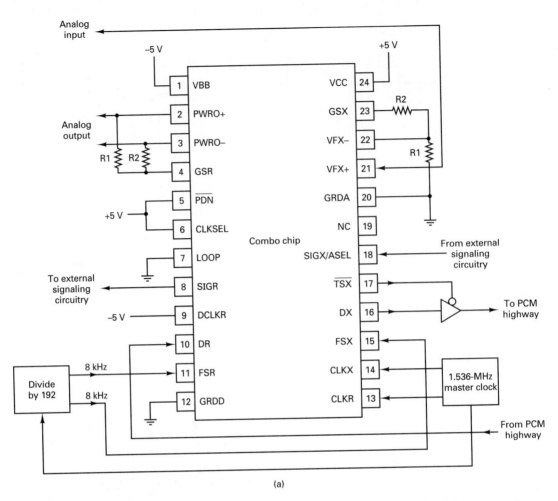

Figure 16-10 Single-channel PCM system using a combo chip in the fixed-data-rate mode: (a) block diagram; *(Continued next page.)*

required for proper operation of the onboard bandpass filters and the analog-to-digital and digital-to-analog converters. However, in the variable-data-rate mode, DCLKR and DCLKX become the data clocks for the receive and transmit PCM highways, respectively. When FSX is high, data are transmitted onto the PCM highway on the next eight consecutive positive transitions of DCLKX. Similarly, while FSR is high, data from the PCM highway are clocked into the codec on the next eight consecutive negative transitions of DCLKR. This mode of operation is sometimes called the *shift register mode.*

On the transmit channel, the last transmitted PCM word is repeated in all remaining time slots in the 125-μs frame as long as DCLKX is pulsed and FSX is held active high. This feature allows the PCM word to be transmitted to the PCM highway more than once per frame. Signaling is not allowed in the variable-data-rate mode because this mode provides no means to specify a signaling frame.

Figure 16-12 shows the block diagram and timing sequence for a two-channel PCM-TDM system using a combo chip in the variable-data-rate mode with a master clock frequency of 1.536 MHz, a sample rate of 8 kHz, and a transmit and receive data rate of 128 kbps.

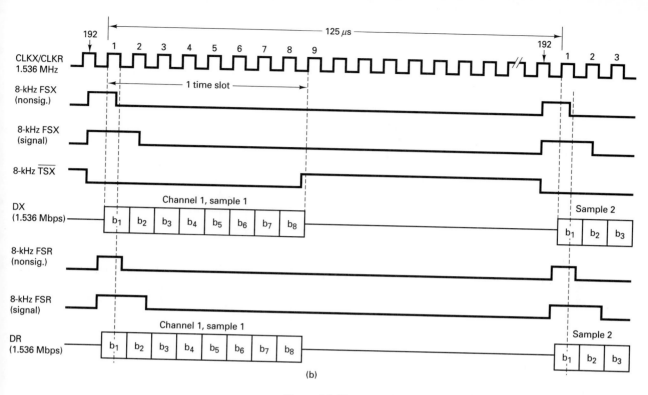

Figure 16-10 (continued) (b) timing sequence.

With a sample rate of 8 kHz, the frame time is 125 μs. Therefore, one 8-bit PCM word from each channel is transmitted and/or received during each 125-μs frame. For 16 bits to occur in 125 μs, a 128-kHz transmit and receive data clock is required.

$$t_b = \frac{1 \text{ channel}}{8 \text{ bits}} \times \frac{1 \text{ frame}}{2 \text{ channels}} \times \frac{125 \text{ μs}}{\text{frame}} = \frac{125 \text{ μs}}{16 \text{ bits}} = \frac{7.8125 \text{ μs}}{\text{bit}}$$

$$\text{bit rate} = \frac{1}{t_b} = \frac{1}{7.8125 \text{ μs}} = 128 \text{ kbps}$$

The transmit and receive enable signals (FSX and FSR) for each codec are active for one-half of the total frame time. Consequently, 8-kHz, 50% duty cycle transmit and receive data enable signals (FSX and FXR) are fed directly to one codec and fed to the other codec 180° out of phase (inverted), thereby enabling only one codec at a time.

To expand to a four-channel system, simply increase the transmit and receive data clock rates to 256 kHz and change the enable signals to 8-kHz, 25% duty cycle pulses.

Supervisory Signaling

With a combo chip, *supervisory signaling* can be used only in the fixed-data-rate mode. A transmit signaling frame is identified by making the FSX and FSR pulses twice their normal width. During a transmit signaling frame, the signal present on input SIGX is substituted into the least significant bit position (b_1) of the encoded PCM word. At the receive end, the signaling bit is extracted from the PCM word prior to decoding and placed on output SIGR until updated by reception of another signaling frame.

Combo Chips

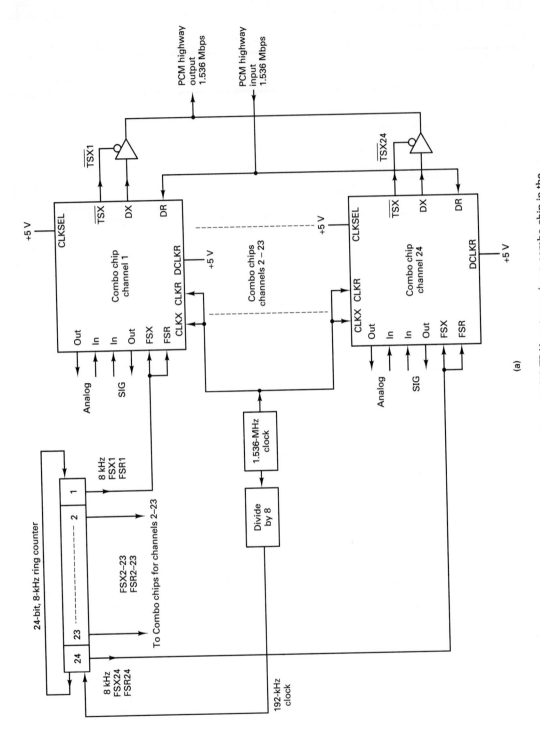

Figure 16-11 Twenty-four-channel PCM-TDM system using a combo chip in the fixed-data-rate mode and operating with a masterclock frequency of 1.536 MHz: (a) block diagram; *(continued on next page.)*

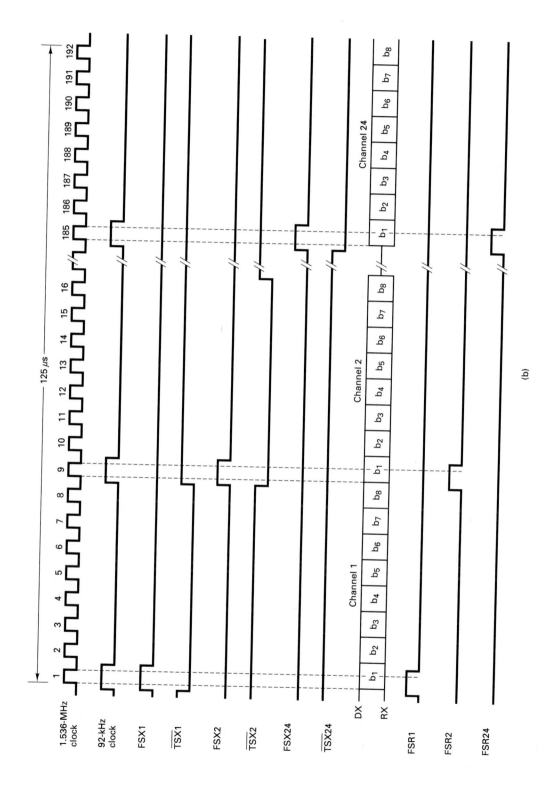

Figure 16-11 *(continued)* (b) timing diagram.

(b)

659

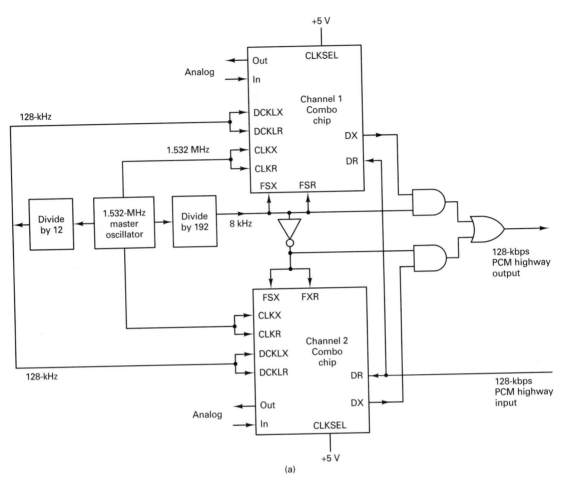

Figure 16-12 Two-channel PCM-TDM system using a combo chip in the variable-data-rate mode with a master clock frequency of 1.536 MHz: (a) block diagram; *(Continued on next page.)*

Asynchronous operation occurs when the master transmit and receive clocks are derived from separate independent sources. A combo chip can be operated in either the synchronous or asynchronous mode using separate digital-to-analog converters and voltage references in the transmit and receive channels, which allows them to be operated completely independent of each other. With either synchronous or asynchronous operation, the master clock, data clock, and time-slot strobe must be synchronized at the beginning of each frame. In the variable-data-rate mode, CLKX and DCLKX must be synchronized once per frame, but may be different frequencies.

NORTH AMERICAN DIGITAL HIERARCHY

Multiplexing signals in digital form lends itself easily to interconnecting digital transmission facilities with different transmission bit rates. Figure 16-13 shows the American Telephone and Telegraph Company's (AT&T) North American Digital Hierarchy for multiplexing digital signals with the same bit rates into a single pulse stream suitable for transmission on the next higher level of the hierarchy. To upgrade from one level in the hierarchy to the next higher level, special devices called *muldems* (*mul*tiplexers/

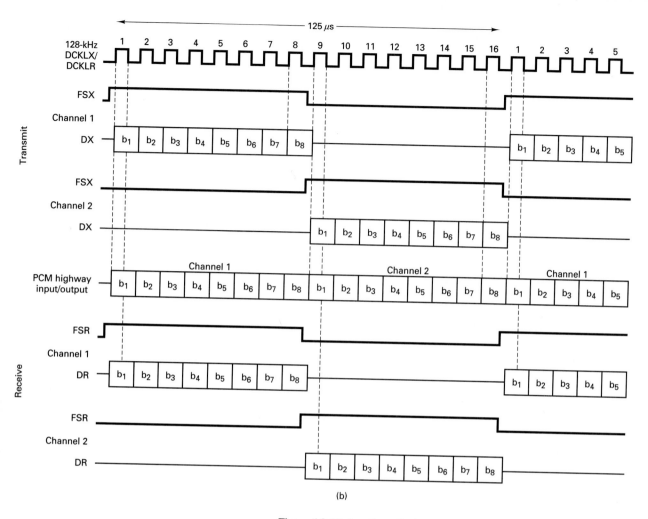

Figure 16-12 *(continued)* (b) timing diagram.

*dem*ultiplexers) are used. Muldems can handle bit-rate conversions in both directions. The muldem designations (M12, M23, and so on) identify the input and output digital signals associated with that muldem. For instance, an M12 muldem interfaces DS-1 and DS-2 *digital signals*. An M23 muldem interfaces DS-2 and DS-3 signals. DS-1 signals may be further multiplexed or line encoded and placed on specially conditioned lines called T1 lines. DS-2, DS-3, DS-4 and DS-5 signals may be placed on T2, T3, T4M, and T5 lines, respectively.

Digital signals are routed at central locations called *digital cross-connects*. A digital cross-connect (DSX) provides a convenient place to make patchable interconnects and to perform routine maintenance and troubleshooting. Each type of digital signal (DS-1, DS-2, and so on) has its own digital switch (DSX-1, DSX-2, and so on). The output from a digital switch may be upgraded to the next higher level or line encoded and placed on its respective T lines (T1, T2, and so on).

Table 16-2 lists the digital signals, their bit rates, channel capacities, and services offered for the line types included in the North American Digital Hierarchy.

When the bandwidth of the signals to be transmitted is such that after digital conversion it occupies the entire capacity of a digital transmission line, a single-channel terminal

North American Digital Hierarchy

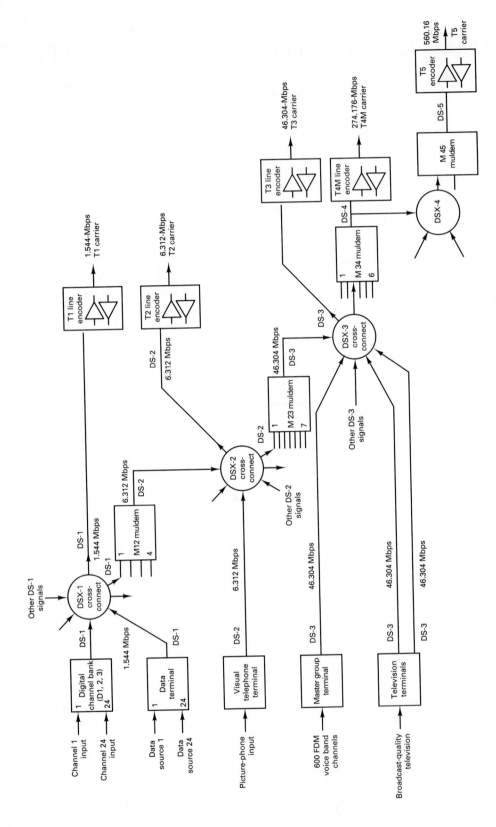

Figure 16-13 North American Digital Hierarchy.

TABLE 16-2 SUMMARY OF THE NORTH AMERICAN DIGITAL HIERARCHY

Line type	Digital signal	Bit rate (Mbps)	Channel capacities	Services offered
T1	DS-1	1.544	24	Voice-band telephone
T1C	DS-1C	3.152	48	Voice-band telephone
T2	DS-2	6.312	96	Voice-band telephone and picturephone
T3	DS-3	46.304	672	Voice-band telephone, picturephone, and broadcast-quality television
T4M	DS-4	274.176	4032	Same as T3 except more capacity
T5	DS-5	560.160	8064	Same as T4 except more capacity

is provided. Examples of such single-channel terminals are picturephone, mastergroup, and commercial television terminals.

Mastergroup and Commercial Television Terminals

Figure 16-14 shows the block diagram of a mastergroup and commercial television terminal. The mastergroup terminal receives voice-band channels that have already been frequency division multiplexed (a topic covered later in this chapter) without requiring that each voice-band channel be demultiplexed to voice frequencies. The signal processor provides frequency shifting for the mastergroup signals (shifts it from a 564- to 3084-kHz bandwidth to a 0- to 2520-kHz bandwidth) and dc restoration for the television signal. By shifting the mastergroup band, it is possible to sample at a 5.1-MHz rate. Sampling of the commercial television signal is at twice that rate or 10.2 MHz.

To meet the transmission requirements, a 9-bit PCM code is used to digitize each sample of the mastergroup or television signal. The digital output from the terminal is therefore approximately 46 Mbps for the mastergroup and twice that much (92 Mbps) for the television signal.

The digital terminal shown in Figure 16-14 has three specific functions; it converts

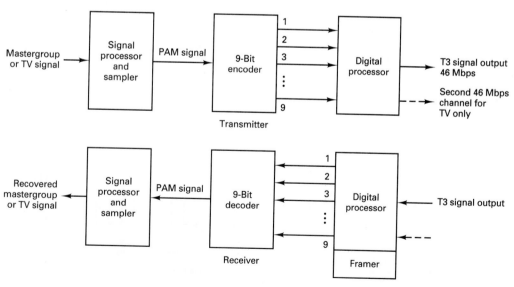

Figure 16-14 Block diagram of a mastergroup or commercial television digital terminal.

the parallel data from the output of the encoder to serial data, it inserts frame synchronizing bits, and it converts the serial binary signal to a form more suitable for transmission. In addition, for the commercial television terminal, the 92-Mbps digital signal must be split into two 46-Mbps digital signals because there is no 92-Mbps line speed in the digital hierarchy.

Picturephone Terminal

Essentially, *picturephone* is a low-quality video transmission for use between nondedicated subscribers. For economic reasons it is desirable to encode a picturephone signal into the T2 capacity of 6.312 Mbps, which is substantially less than that for commercial network broadcast signals. This substantially reduces the cost and makes the service affordable. At the same time, it permits the transmission of adequate detail and contrast resolution to satisfy the average picturephone subscriber. Picturephone service is ideally suited to a differential PCM code. Differential PCM is similar to conventional PCM except that the exact magnitude of a sample is not transmitted. Instead, only the difference between that sample and the previous sample is encoded and transmitted. To encode the difference between samples requires substantially fewer bits than encoding the actual sample.

Data Terminal

The portion of communications traffic that involves data (signals other than voice) is increasing exponentially. Also, in most cases the data rates generated by each individual subscriber are substantially less than the data rate capacities of digital lines. Therefore, it seems only logical that terminals be designed that transmit data signals from several sources over the same digital line.

Data signals could be sampled directly; however, this would require excessively high sample rates, resulting in excessively high transmission bit rates, especially for sequences of data with few or no transitions. A more efficient method is one that codes the transition times. Such a method is shown in Figure 16-15. With the coding format shown, a 3-bit code is used to identify when transitions occur in the data and whether that transition is from a 1 to a 0, or vice versa. The first bit of the code is called the address bit. When this

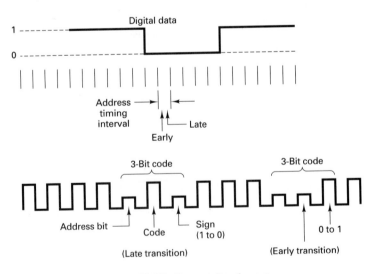

Figure 16-15 Data coding format.

Chap. 16   Multiplexing

bit is a logic 1, this indicates that no transition occurred; a logic 0 indicates that a transition did occur. The second bit indicates whether the transition occurred during the first half (0) or during the second half (1) of the sample interval. The third bit indicates the sign or direction of the transition; a 1 for this bit indicates a 0-to-1 transition and a 0 indicates a 1-to-0 transition. Consequently, when there are no transitions in the data, a signal of all 1's is transmitted. Transmission of only the address bit would be sufficient; however, the sign bit provides a degree of error protection and limits error propagation (when one error leads to a second error, and so on). The efficiency of this format is approximately 33%; there are 3 code bits for each data bit. The advantage of using a coded format rather than the original data is that coded data are more efficiently substituted for voice in analog systems. Without this coding format, transmitting a 250-kbps data signal, requires the same bandwidth as would be required to transmit 60 voice channels with analog multiplexing. With this coded format, a 50-kbps data signal displaces three 64-kbps PCM encoded channels, and a 250-kbps data stream displaces only 12 voice-band channels.

LINE ENCODING

Line encoding involves converting standard logic levels (TTL, CMOS, and the like) to a form more suitable to telephone line transmission. Essentially, five primary factors must be considered when selecting a line-encoding format:

1. Transmission voltages and dc component
2. Timing (clock) recovery
3. Transmission bandwidth
4. Ease of detection and decoding
5. Error detection

Transmission Voltages and DC Component

Transmission voltages or levels can be categorized as either *unipolar* (UP) or *bipolar* (BP). Unipolar transmission of binary data involves the transmission of only a single nonzero voltage level (for example, $+V$ for logic 1 and 0 V or ground for a logic 0). In bipolar transmission, two nonzero voltage levels are involved (for example, $+V$ for a logic 1 and $-V$ for a logic 0).

Over a digital transmission line, it is more power efficient to encode binary data with voltages that are equal in magnitude but opposite in polarity and symmetrically balanced about 0 V. For example, assuming a 1-Ω resistance and a logic 1 level of $+5$ V and a logic 0 level of 0 V, the average power required is 12.5 W (assuming an equal probability of the occurrence of a 1 or a 0). With a logic 1 level of $+2.5$ V and a logic 0 level of -2.5 V, the average power is only 6.25 W. Thus, by using bipolar symmetrical voltages, the average power is reduced by a factor of 50%.

Duty Cycle

The *duty cycle* of a binary pulse can also be used to categorize the type of transmission. If the binary pulse is maintained for the entire bit time, this is called *nonreturn to zero* (NRZ). If the active time of the binary pulse is less than 100% of the bit time, this is called *return to zero* (RZ).

Unipolar and bipolar transmission voltages and return-to-zero and nonreturn-to-zero encoding can be combined in several ways to achieve a particular line encoding scheme. Figure 16-16 shows five line-encoding possibilities.

In Figure 16-16a, there is only one nonzero voltage level (+V = logic 1); a zero voltage simply implies a binary 0. Also, each logic 1 maintains the positive voltage for the entire bit time (100% duty cycle). Consequently, Figure 16-16a represents a unipolar non-return-to-zero signal (UPNRZ). In Figure 16-16b, there are two nonzero voltages (+V = logic 1 and −V = logic 0) and a 100% duty cycle is used. Figure 16-16b represents a bipolar nonreturn-to-zero signal (BPNRZ). In Figure 16-16c, only one nonzero voltage is used, but each pulse is active for only 50% of the bit time. Consequently, Figure 16-16c represents a unipolar return-to-zero signal (UPRZ). In Figure 16-16d, there are two nonzero voltages (+V = logic 1 and −V = logic 0). Also, each pulse is active only 50% of the total bit time. Consequently, Figure 16-16d represents a bipolar return-to-zero (BPRZ) signal. In Figure 16-16e, there are again two nonzero voltage levels (−V and +V), but here both polarities represent a logic 1 and 0 V represents a logic 0. This method of encoding is called *alternate mark inversion* (AMI). With AMI transmissions, each successive logic 1 is inverted in polarity from the previous logic 1. Because return to zero is used, this encoding technique is called *bipolar-return-to-zero alternate mark inversion* (BPRZ-AMI).

With NRZ encoding, a long string of either 1's or 0's produces a condition in which

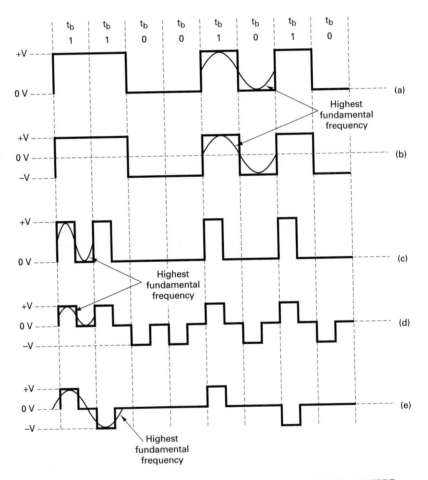

Figure 16-16 Line-encoding formats: (a) UPNRZ; (b) BPNRZ; (c) UPRZ; (d) BPRZ; (e) BPRZ-AMI.

Chap. 16   Multiplexing

a receiver may lose its amplitude reference for optimum discrimination between received 1's and 0's. This condition is called *dc wandering*. The problem may also arise when there is a significant imbalance in the number of ones and zeros transmitted. Figure 16-17 shows how dc wandering is produced by a long string of successive logic 1's. It can be seen that after a long string of 1's, 1-to-0 errors are more likely than 0-to-1 errors. Similarly, long strings of 0's increase the probability of a 0-to-1 error.

The method of line encoding used determines the minimum bandwidth required for transmission, how easily a clock may be extracted from it, how easily it may be decoded, the average dc level, and whether it offers a convenient means of detecting errors.

Bandwidth Considerations

To determine the minimum bandwidth required a propagate a line-encoded signal, you must determine the highest fundamental frequency associated with it (see Figure 16-16). The highest fundamental frequency is determined from the worst-case (fastest transition) binary bit sequence. With UPNRZ, the worst-case condition is an alternating 1/0 sequence; the period of the highest fundamental frequency takes the time of 2 bits and is therefore equal to one-half the bit rate. With BPNRZ, again the worst-case condition is an alternating 1/0 sequence, and the highest fundamental frequency is one-half of the bit rate. With UPRZ, the worst-case condition is two successive 1's. The minimum bandwidth is therefore equal to the bit rate. With BPRZ, the worst-case condition is either successive 1's or 0's, and the minimum bandwidth is again equal to the bit rate. With BPRZ-AMI, the worst-case condition is two or more consecutive 1's, and the minimum bandwidth is equal to one-half of the bit rate.

Clock Recovery

To recover and maintain clocking information from received data, there must be a sufficient number of transitions in the data signal. With UPNRZ and BPNRZ, a long string of consecutive 1's or 0's generates a data signal void of transitions and is therefore inadequate for clock synchronization. With UPRZ and BPRZ-AMI, a long string of 0's also generates a data signal void of transitions. With BPRZ, a transition occurs in each bit position regardless of whether the bit is a 1 or a 0. In the clock recovery circuit, the data are simply full wave rectified to produce a data-independent clock equal to the receive bit rate. Therefore, BPRZ encoding is best suited for clock recovery. If long sequences of 0's are prevented from occurring. BPRZ-AMI encoding is sufficient to ensure clock synchronization.

Error Detection

When UPNRZ, BPNRZ, UPRZ, and BPRZ transmissions, there is no way to determine if the received data have errors. With BPRZ-AMI transmissions, an error in any bit will cause a bipolar violation (the reception of two or more consecutive 1's with the same polarity). Therefore, BPRZ-AMI has a built-in error-detection mechanism.

Figure 16-17   DC wandering.

TABLE 16-3   LINE-ENCODING SUMMARY

Encoding format	Minimum BW	Average DC	Clock recovery	Error detection
UPNRZ	$f_b/2^a$	$+V/2$	Poor	No
BPNRZ	$f_b/2^a$	$0\ V^a$	Poor	No
UPRZ	f_b	$+V/2$	Good	No
BPRZ	f_b	$0\ V^a$	$Best^a$	No
BPRZ-AMI	$f_b/2^a$	$0\ V^a$	Good	Yes^a

[a]Denotes best performance or quality.

Ease of Detection and Decoding

Because unipolar transmission involves the transmission of only one polarity voltage, an average dc voltage is associated with the signal equal to $+V/2$. Assuming an equal probability of 1's and 0's occurring, bipolar transmissions have an average dc component of 0 V. A dc component is undesirable because it biases the input to a conventional threshold detector (a biased comparator) and could cause a misinterpretation of the logic condition of the received pulses. Therefore, bipolar transmission is better suited to data detection.

Table 16-3 summarizes the minimum bandwidth, average dc voltage, clock recovery, and error-detection capabilities of the line-encoding formats shown in Figure 16-16. From Table 16-3 it can be seen that BPRZ-AMI encoding has the best overal characteristics; it is therefore the most commonly used method.

Digital Biphase

Digital *biphase* (sometimes called the *Manchester code* or *diphase*) is a popular type of line encoding that produces a strong timing component for clock recovery and does not cause dc wandering. Biphase is a form of BPRZ transmission that uses one cycle of a square wave at 0° phase to represent a logic 1 and one cycle of a square wave at 180° phase to represent a logic 0. Digital biphase encoding is shown in Figure 16-18. Notice that a transition occurs in the center of every signaling element, regardless of its phase. Thus biphase produces a strong timing component for clock recovery. In addition, assuming an equal ability of 1's and 0's, the average dc voltage is 0 V and there is no dc wandering. A disadvantage of biphase is that it contains no means of error detection.

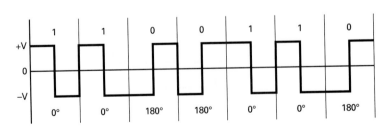

Figure 16-18   Digital biphase.

Chap. 16   Multiplexing

T carriers are used for the transmission of PCM-encoded time division multiplexed digital signals. In addition, T carriers utilize special line-encoded signals and metallic cables that have been conditioned to meet the relatively high bandwidths required for high-speed digital transmissions. Digital signals deteriorate as they propagate along a cable due to power loss in the metallic conductors and the low-pass filtering inherent in parallel wire transmission lines. Consequently, *regenerative repeaters* must be placed at periodic intervals. The distance between repeaters depends on the transmission bit rate and the line-encoding technique used.

Figure 16-19 shows the block diagram of a regenerative repeater. Essentially there are three functional blocks: an amplifier/equalizer, a timing circuit, and the regenerator. The amplifier/equalizer shapes the incoming digital signal and raises its power level so that a pulse/no pulse decision can be made by the regenerator circuit. The timing circuit recovers the clocking information from the received data and provides the proper timing information to the regenerator so that decisions can be made at the optimum time that minimizes the chance of an error occurring. Spacing of the repeaters is designed to maintain an adequate signal-to-noise ratio for error-free performance. The signal-to-noise ratio (S/N) at the output of a regenerator is exactly what it was at the output of the transmit terminal or at the output of the previous regenerator (that is, the S/N does not deteriorate as a digital signal propagates through a regenerator; in fact, a regenerator reconstructs the original pulses with the original S/N ratio).

T1 and T1C Carrier Systems

The T1 carrier system utilizes PCM and TDM techniques to provide short-haul transmission of 24 voice-band signals. The lengths of T1 carrier systems range from about 5 to 50 miles. T1 carriers use BPRZ-AMI encoding with regenerative repeaters placed every 6000 ft; 6000 ft was chosen because telephone company manholes are located at approximately 6000-ft intervals and these same manholes are used for placement of the repeaters, facilitating convenient installation, maintenance, and repair. The transmission medium for T1 carriers is either a 19- or 22-gauge wire pair.

Because T1 carriers use BPRZ-AMI encoding, they are susceptible to losing synchronization on a long string of consecutive 0's. With a folded binary PCM code, the pos-

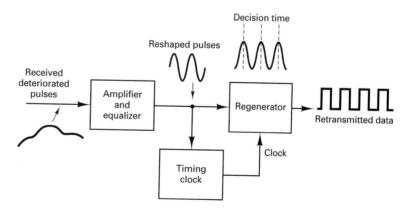

Figure 16-19 Regenerative repeater block diagram.

sibility of generating a long string of consecutive 0's is high (whenever a channel is idle it generates a ±0-V code, which is either seven or eight consecutive 0's). If two or more adjacent voice channels are idle, there is a high probability that a long string of consecutive 0's will be transmitted. To reduce this possibility, the PCM code is inverted prior to transmission and inverted again at the receiver prior to decoding. Consequently, the only time a long string of consecutive 0's is transmitted is when two or more adjacent voice-band channels each encode the maximum possible positive sample voltage, which is unlikely to happen.

With T1 and T1C carrier systems, provisions are taken to prevent more than 14 consecutive 0's from occurring. The transmissions from each frame are monitored for the presence of either 15 consecutive 0's or any one PCM sample (8 bits) without at least one nonzero bit. If either of these conditions occurs, a 1 is substituted into the appropriate bit position. The worst-case conditions are as follows:

```
                 MSB    LSB MSB      LSB
Original         1000  0000  0000   0001    14 consecutive 0's
DS-1 signal                                 (no substitution)


                 MSB    LSB MSB      LSB
Original         1000  0000  0000   0000    15 consecutive 0's
DS-1 signal


Substituted      1000  0000  0000   0010
DS-1 signal                           ↑
                                Substituted
                                bit
```

A 1 is substituted into the second least significant bit. This introduces an encoding error equal to twice the amplitude resolution. This bit is selected rather than the least significance bit because, with the superframe format, during every sixth frame the LSB is the signaling bit and to alter it would alter the signaling word.

```
                 MSB    LSB MSB       LSB MSB    LSB
Original         1010  1000  0000   0000  0000  0001
DS-1 signal


Substituted      1010  1000  0000   0010  0000  0001
DS-1 signal                           ↑
                                Substituted
                                bit
```

The process shown is used for T1 and T1C carrier systems. Also, if at any time 32 consecutive 0's are received, it is assumed that the system is not generating pulses and is therefore out of service; this is because the occurrence of 32 consecutive 0's is prohibited.

T2 Carrier System

The T2 carrier utilizes PCM to time division multiplex 96 voice-band channels into a single 6.312-Mbps data signal for transmission up to 500 miles over a special LOCAP (low capacitance) cable. A T2 carrier is also used to carry a single picturephone signal. T2 carriers also use BPRZ-AMI encoding. However, because of the higher transmission rate, clock synchronization becomes more critical. A sequence of six consecutive 0's could be sufficient to cause loss of clock synchronization. Therefore, T2 carrier systems use an alternative method of ensuring that ample transitions occur in the data. This method is called *binary six zero substitution* (B6ZS).

With B6ZS, whenever six consecutives 0's occur, one of the following codes is substituted in its place: $0 - + 0 + -$ or $0 + - 0 - +$. The + and − represent positive and negative logic 1's. A zero simply indicates a logic 0 condition. The 6-bit code substituted for the six 0's is selected to purposely cause a bipolar violation. If the violation is caught at the receiver and the B6ZS code is detected, the original six 0's can be substituted back into the data signal. The substituted patterns cause a bipolar violation in the second and fifth bits of the substituted pattern. If DS-2 signals are multiplexed to form DS-3 signals, the B6ZS code must be detected and stripped from the DS-2 signal prior to DS-3 multiplexing. An example of B6ZS is as follows:

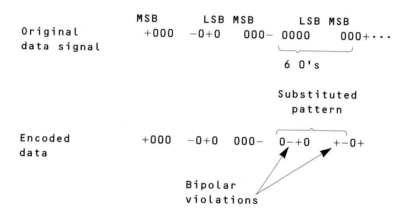

T3 Carrier System

A T3 carrier time division multiplexes 672 PCM-encoded voice channels for transmission over a single metallic cable. The transmission rate for T3 signals is 46.304 Mbps. The encoding technique used with T3 carriers is *binary three zero substitution* (B3ZS). Substitutions are made for any occurrence of three consecutive 0's. There are four substitution patterns used: $00-$, $-0-$, $00+$, and $+0+$. The pattern chosen should cause a bipolar error in the third substitute bit. An example is as follows:

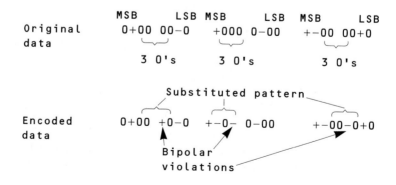

T4M Carrier System

A T4M carrier time division multiplexes 4032 PCM-encoded voice-band channels for transmission over a single coaxial cable up to 500 miles. The transmission rate is sufficiently high that substitute patterns are impractical. Instead, T4M carriers transmit scrambled unipolar NRZ digital signals; the scrambling and descrambling functions are performed in the subscriber's terminal equipment.

T5 Carrier System

A T5 carrier system time division multiplexes 8064 PCM-encoded voice-band channels and transmits them at a 560.16-Mbps rate over a single coaxial cable.

FRAME SYNCHRONIZATION

With TDM systems it is imperative that a frame be identified and that individual time slots (samples) within the frame also be identified. To acquire frame synchronization, a certain amount of overhead must be added to the transmission. Five methods are commonly used to establish frame synchronization: added digit framing, robbed digit framing, added channel framing, statistical framing, and unique line signal framing.

Added Digit Framing

T1 carriers using D1, D2, or D3 channel banks use *added digit framing*. A special *framing digit* (framing pulse) is added to each frame. Consequently, for an 8-kHz sample rate (125-μs frame), 8000 digits are added per second. With T1 carriers, an alternating 1/0 frame synchronizing pattern is used.

To acquire frame synchronization, the receive terminal searches through the incoming data until it finds the alternating 1/0 sequence used for the framing bit pattern. This encompasses testing a bit, counting off 193 bits and then testing again for the opposite condition. This process continues until an alternating 1/0 sequence is found. Initial frame synchronization depends on the total frame time, the number of bits per frame, and the period of each bit. Searching through all possible bit positions requires N tests, where N is the number of bit positions in the frame. On average, the receiving terminal dwells at a false framing position for two frame periods during a search; therefore, the maximum average synchronization time is

$$\text{synchronization time} = 2NT = 2N^2t$$

where T = frame period of Nt
   N = number of bits per frame
   t = bit time

For the T1 carrier, $N = 193$, $T = 125$ μs, and $t = 0.648$ μs; therefore, a maximum of 74,498 bits must be tested and the maximum average synchronization time is 48.25 ms.

Robbed Digit Framing

When a short frame is used, added digit framing is very inefficient. This occurs in single-channel PCM systems such as those used in television terminals. An alternative solution is to replace the least significant bit of every nth frame with a framing bit. The parameter n is chosen as a compromise between reframe time and signal impairment. For $n = 10$, the SQR is impaired by only 1 dB. *Robbed digit framing* does not interrupt transmission, but instead periodically replaces information bits with forced data errors to maintain clock synchronization. B6ZS and B3ZS are examples of systems that use robbed digit techniques.

Added Channel Framing

Essentially, *added channel framing* is the same as added digit framing except that digits are added in groups or words instead of as individual bits. The CCITT multiplexing scheme previously discussed uses added channel framing. One of the 32 time slots in each frame is dedicated to a unique synchronizing sequence. The average frame synchronization time for added channel framing is

$$\text{synchronization time (bits)} = \frac{N^2}{2(2^L - 1)}$$

where N = number of bits per frame
L = number of bits in the frame code

For the CCITT 32-channel system, $N = 256$ and $L = 8$. Therefore, the average number of bits needed to acquire frame synchronization is 128.5. At 2.048 Mbps, the synchronization time is approximately 62.7 μs.

Statistical Framing

With *statistical framing*, it is not necessary to either rob or add digits. With the Gray code, the second bit is a 1 in the central half of the code range and 0 at the extremes. Therefore, a signal that has a centrally peaked amplitude distribution generates a high probability of a 1 in the second digit. A mastergroup signal has such a distribution. With a mastergroup encoder, the probability that the second bit will be a 1 is 95%. For any other bit, it is less than 50%. Therefore, the second bit can be used for a framing bit.

Unique Line Code Framing

With *unique line code framing*, the framing bit is different from the information bits. It is either made higher or lower in amplitude or of a different time duration. The earliest PCM/TDM systems used unique line code framing. D1 channel banks used framing pulses that were twice the amplitude of normal data bits. With unique line code framing, added digit or added word framing can be used or data bits can be used to simultaneously convey information and carry synchronizing signals. The advantage of unique line code framing is that synchronization is immediate and automatic. The disadvantage is the additional processing requirements required to generate and recognize the unique framing bit.

BIT INTERLEAVING VERSUS WORD INTERLEAVING

When time division multiplexing two or more PCM systems, it is necessary to interleave the transmissions from the various terminals in the time domain. Figure 16-20 shows two methods of interleaving PCM transmissions: *bit interleaving* and *word interleaving*.

T1 carrier systems use word interleaving; 8-bit samples from each channel are interleaved into a single 24-channel TDM frame. Higher-speed TDM systems and delta modulation systems use bit interleaving. The decision as to which type of interleaving to use is usually determined by the nature of the signals to be multiplexed.

FREQUENCY-DIVISION MULTIPLEXING

In *frequency-division multiplexing* (FDM), multiple sources that originally occupied the same frequency spectrum are each converted to a different frequency band and transmitted simultaneously over a single transmission medium. Thus many relatively narrowband channels can be transmitted over a single wideband transmission system.

FDM is an analog multiplexing scheme; the information entering an FDM system is analog and it remains analog throughout transmission. An example of FDM is the AM commercial broadcast band, which occupies a frequency spectrum from 535 to 1605 kHz. Each station carries an intelligence signal with a bandwidth of 0 to 5 kHz. If the audio from each station were transmitted with the original frequency spectrum, it would be impossible to separate one station from another. Instead, each station amplitude modulates a different carrier frequency and produces a 10-kHz double-sideband signal. Because adjacent stations' carrier frequencies are separated by 10 kHz, the total commerical AM band is divided into 107 10-kHz frequency slots stacked next to each other in the frequency domain. To receive a particular station, a receiver is simply tuned to the frequency

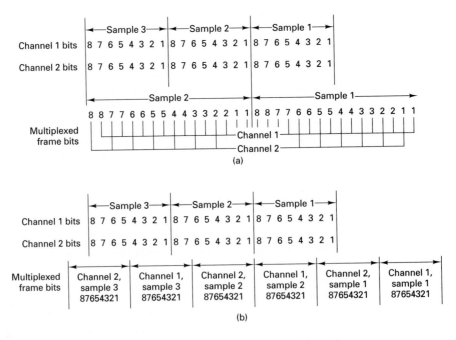

Figure 16-20 Interleaving: (a) bit; (b) word.

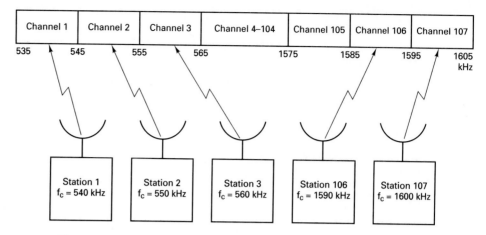

Figure 16-21 Frequency division multiplexing commercial AM broadcast band stations.

band associated with that station's transmissions. Figure 16-21 shows how commercial AM broadcast station signals are frequency division multiplexed and transmitted over a single transmission medium (free space).

There are many other applications for FDM, such as commercial FM and television broadcasting and high-volume telecommunications systems. Within any of the commercial broadcast bands, each station's transmissions are independent of all the other stations' transmissions. Consequently, the multiplexing (stacking) process is accomplished without any synchronization between stations. With a high-volume telephone communication system, many voice-band telephone channels may originate from a common source and terminate in a common destination. The source and destination terminal equipment is most likely a high-capacity *electronic switching system* (ESS). Because of the possibility of a large number of narrowband channels originating and terminating at the same location, all multiplexing and demultiplexing operations must be synchronized.

AT&T's FDM HIERARCHY

Although AT&T is no longer the only long-distance common carrier in the United States, it still provides a vast majority of the long-distance services and, if for no other reason than its overwhelming size, has essentially become the standards organization for the telephone industry in North America.

AT&T's nationwide communications network is subdivided into two classifications: *short haul* (short distance) and *long haul* (long distance). The T1 carrier explained earlier in this chapter is an example of a short-haul communications system.

Long-haul Communications with FDM

Figure 16-22 shows AT&T's North American FDM hierarchy for long-haul communications. Only a transmit terminal is shown, although a complete set of inverse functions must be performed at the receiving terminal.

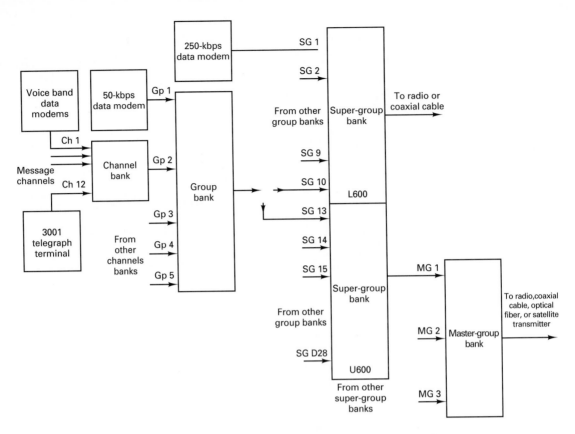

Figure 16-22 AT&T's long-haul FDM hierarchy.

Message Channel

The *message channel* is the basic building block of the FDM hierarchy. The basic message channel was originally intended for voice transmission, although it now includes any transmissions that utilize voice-band frequencies (0 to 4 kHz), such as voice-band data circuits. The basic voice-band (VB) circuit is called a 3002 channel and is actually bandlimited to a 300- to 3000-Hz band, although for practical considerations it is considered a 4-kHz channel. The basic 3002 channel can be subdivided into 24 narrower 3001 (telegraph) channels that have been frequency division multiplexed to form a single 3002 channel.

Basic Group

A *group* is the next higher level in the FDM hierarchy above the basic message channel and is, consequently, the first multiplexing step for the message channels. A basic group is comprised of 12 voice-band channels stacked next to each other in the frequency domain. The 12-channel modulating block is called an *A-type* (analog) channel bank. The 12-channel *group* output of the A-type channel bank is the standard building block for most long-haul *broadband* communications systems. Additions and deletions in total system capacity are accomplished with a minimum of one group (12 VB channels). The A-type channel bank has generically progressed from the early A1 channel bank to the most recent A6 channel bank.

Basic Supergroup

The next higher level in the FDM hierarchy shown in Figure 16-22 is the combination of five groups into a *supergroup*. The multiplexing of five groups is accomplished in a group bank. A single supergroup can carry information from 60 VB channels or handle high-speed data up to 250 kbps.

Basic Mastergroup

The next higher level in the FDM hierarchy is the basic *mastergroup*. A mastergroup is comprised of 10 supergroups (10 supergroups of five groups each = 600 VB channels). Supergroups are combined in supergroup banks to form mastergroups. There are two categories of mastergroups (U600 and L600), which occupy different frequency bands. The type of mastergroup used depends on the system capacity and whether the transmission medium is a coaxial cable, a microwave radio, an optical fiber, or a satellite link.

Larger Groupings

Master groups can be further multiplexed in mastergroup banks to form *jumbogroups, multijumbogroups*, and *superjumbogroups*. A basic FDM/FM microwave radio channel carries three mastergroups (1800 VB channels), a jumbogroup has 3600 VB channels, and a superjumbogroup has three jumbogroups (10,800 VB channels).

COMPOSITE BASEBAND SIGNAL

Baseband describes the modulating signal (intelligence) in a communications system. A single message channel is baseband. A group, supergroup, or mastergroup is also baseband. The composite baseband signal is the total intelligence signal prior to modulation of the final carrier. In Figure 16-22 the output of a channel bank is baseband. Also, the output of a group or supergroup bank is baseband. The final output of the FDM multiplexer is the *composite* (total) baseband. The formation of the composite baseband signal can include channel, group, supergroup, and mastergroup banks, depending on the capacity of the system.

Formation of a Group

Figure 16-23a shows how a group is formed with an A-type channel bank. Each voice-band channel is bandlimited with an antialiasing filter prior to modulating the channel carrier. FDM uses single-sideband suppressed carrier (SSBSC) modulation. The combination of the balanced modulator and the bandpass filter makes up the SSBSC modulator. A balanced modulator is a double-sideband suppressed carrier modulator, and the bandpass filter is tuned to the difference between the carrier and the input voice-band frequencies (LSB). The ideal input frequency range for a single voice-band channel is 0 to 4 kHz. The carrier frequencies for the channel banks are determined from the following expression:

$$f_c = 112 - 4n \quad \text{kHz}$$

where *n* is the channel number. Table 16-4 lists the carrier frequencies for channels 1

Composite Baseband Signal

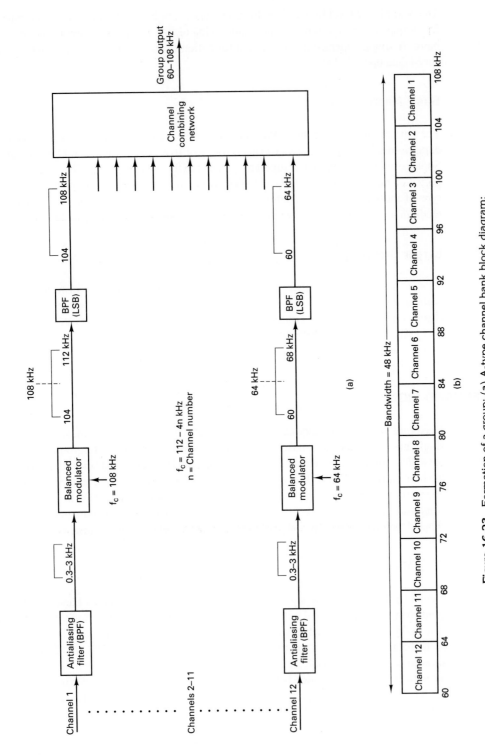

Figure 16-23 Formation of a group: (a) A-type channel bank block diagram; (b) output spectrum.

TABLE 16-4 CHANNEL
CARRIER FREQUENCIES

Channel	Carrier frequency (kHz)
1	108
2	104
3	100
4	96
5	92
6	88
7	84
8	80
9	76
10	72
11	68
12	64

through 12. Therefore, for channel 1, a 0- to 4-kHz band of frequencies modulates a 108-kHz carrier. Mathematically, the output of a channel bandpass filter is

$$f_{\text{out}} = f_c - f_m \text{ to } f_c$$

where f_c = channel carrier frequency ($112 - 4n$ kHz)
f_m = channel frequency spectrum (0 to 4 kHz)

For channel 1,

$$f_{\text{out}} = 108 \text{ kHz} - (0 \text{ to } 4 \text{ kHz}) = 104 \text{ to } 108 \text{ kHz}$$

For channel 2,

$$f_{\text{out}} = 104 \text{ kHz} - (0 \text{ to } 4 \text{ kHz}) = 100 \text{ to } 104 \text{ kHz}$$

For channel 12:

$$f_{\text{out}} = 64 \text{ kHz} - (0 \text{ to } 4 \text{ kHz}) = 60 \text{ to } 64 \text{ kHz}$$

The outputs from the 12 A-type channel modulators are summed in the *linear* combiner to produce the total group spectrum shown in Figure 16-23b (60 to 108 kHz). Note that the total group bandwidth is equal to 48 kHz (12 channels × 4 kHz).

Figure 16-24a shows how a supergroup is formed with a group bank and combining network. Five groups are combined to form a supergroup. The frequency spectrum for each group is 60 to 108 kHz. Each group is mixed with a different group carrier frequency in a balanced modulator and then bandlimited with a bandpass filter tuned to the difference frequency band (LSB) to produce a SSBSC signal. The group carrier frequencies are derived from the following expression:

$$f_c = 372 + 48n \quad \text{kHz}$$

where n is the group number. Table 16-5 lists the carrier frequencies for groups 1 through 5. For group 1, a 60- to 80-kHz group signal modulates a 420-kHz group carrier frequency. Mathematically, the output of a group bandpass filter is

$$f_{\text{out}} = f_c - f_g \text{ to } f_c$$

Composite Baseband Signal

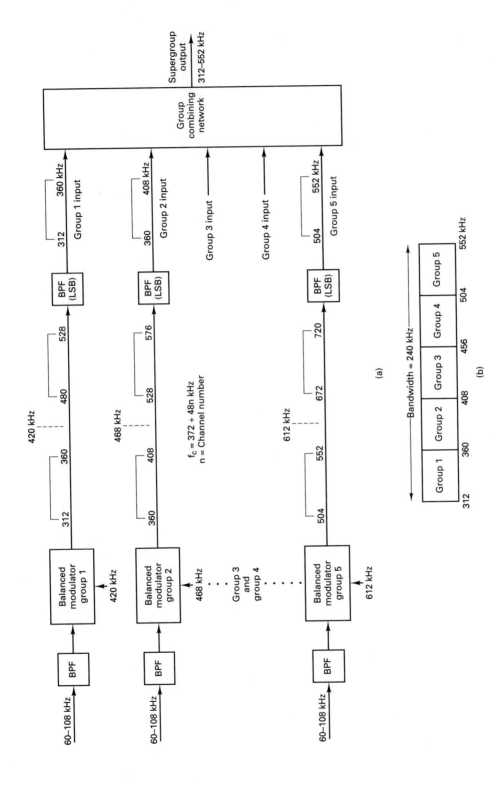

Figure 16-24 Formation of a supergroup: (a) group bank and combining network block diagram; (b) output spectrum.

TABLE 16-5   GROUP CARRIER FREQUENCIES

Group	Carrier frequency (kHz)
1	420
2	468
3	516
4	564
5	612

where f_c = group carrier frequency $(372 + 48n$ kHz)
f_g = group frequency spectrum (60 to 108 kHz)

For group 1,

$$f_{out} = 420 \text{ kHz} - (60 \text{ to } 108 \text{ kHz}) = 312 \text{ to } 360 \text{ kHz}$$

For group 2,

$$f_{out} = 468 \text{ kHz} - (60 \text{ to } 108 \text{ kHz}) = 360 \text{ to } 408 \text{ kHz}$$

For group 5,

$$f_{out} = 612 \text{ kHz} - (60 \text{ to } 108 \text{ kHz}) = 504 \text{ to } 552 \text{ kHz}$$

The outputs from the five group modulators are summed in the linear combiner to produce the total supergroup spectrum shown in Figure 16-24b (312 to 552 kHz). Note that the total supergroup bandwidth is equal to 240 kHz (60 channels × 4 kHz).

FORMATION OF A MASTERGROUP

There are two types of mastergroups: L600 and U600 type. The L600 mastergroup is used for low-capacity microwave systems, while the U600 mastergroup may be further multiplexed and used for higher-capacity microwave radio systems.

U600 Mastergroup.   Figure 16-25a shows how a U600 mastergroup is formed with a supergroup bank and combining network. Ten supergroups are combined to form a mastergroup. The frequency spectrum for each supergroup is 312 to 552 kHz. Each supergroup is mixed with a different supergroup carrier frequency in a balanced modulator. The output is then bandlimited to the difference frequency band (LSB) to form a SSBSC signal. The 10 supergroup carrier frequencies are listed in Table 16-6. For supergroup 13, a 312- to 552-kHz supergroup band of frequencies modulates a 1116-kHz carrier frequency. Mathematically, the output from a supergroup bandpass filter is

$$f_{out} = f_c - f_s + f_c$$

where f_c = supergroup carrier frequency
f_s = supergroup frequency spectrum (312 to 552 kHz)

For supergroup 13,

$$f_{out} = 1116 \text{ kHz} - (312 \text{ to } 552 \text{ kHz}) = 564 \text{ to } 804 \text{ kHz}$$

Formation of a Mastergroup

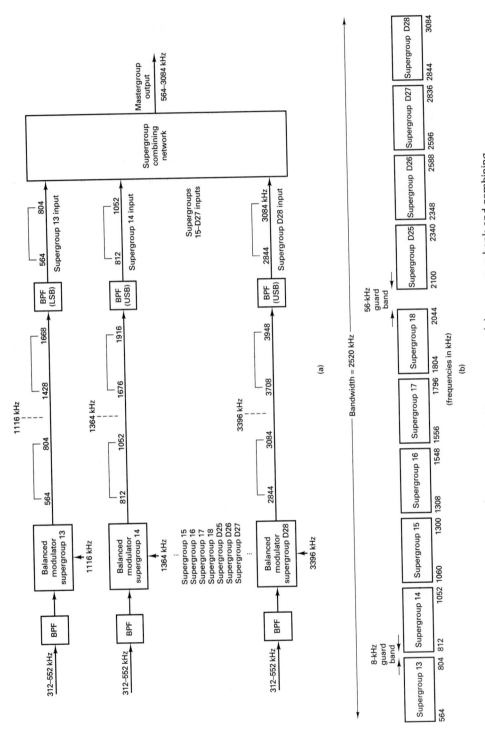

Figure 16-25 Formation of a U600 mastergroup: (a) supergroup bank and combining network block diagram; (b) output spectrum.

TABLE 16-6 SUPERGROUP
CARRIER FREQUENCIES
FOR A U600
MASTERGROUP

Supergroup	Carrier frequency (kHz)
13	1116
14	1364
15	1612
16	1860
17	2108
18	2356
D25	2652
D26	2900
D27	3148
D28	3396

For supergroup 14,

$$f_{out} = 1364 \text{ kHz} - (312 \text{ to } 552 \text{ kHz}) = 812 \text{ to } 1052 \text{ kHz}$$

For supergroup D28,

$$f_{out} = 3396 \text{ kHz} - (312 \text{ to } 552 \text{ kHz}) = 2844 \text{ to } 3084 \text{ kHz}$$

The outputs from the 10 supergroup modulators are summed in the linear summer to produce the total mastergroup spectrum shown in Figure 16-25b (564 to 3084 kHz). Note that between any two adjacent supergroups there is a void band of frequencies that is not included within any supergroup band. These voids are called *guard bands*. The guard bands are necessary because the demultiplexing process is accomplished through filtering and down-converting. Without the guard bands, it would be difficult to separate one supergroup from an adjacent supergroup. The guard bands reduce the *quality factor (Q)* required to perform the necessary filtering. The guard band is 8 kHz between all supergroups except 18 and D25, where it is 56 kHz. Consequently, the bandwidth of a U600 mastergroup is 2520 kHz (564 to 3084 kHz), which is greater than is necessary to stack 600 voice-band channels (600 × 4 kHz = 2400 kHz).

Guard bands were not necessary between adjacent groups because the group frequencies are sufficiently low and it is relatively easy to build bandpass filters to separate one group from another.

In the channel bank, the antialiasing filter at the channel input passes a 0.3- to 3-kHz band. The separation between adjacent channel carrier frequencies is 4 kHz. Therefore, there is a 1300-Hz guard band between adjacent channels. This is shown in Figure 16-26.

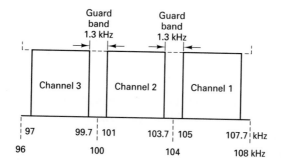

Figure 16-26 Channel guard bands.

TABLE 16-7 SUPERGROUP
CARRIER FREQUENCIES
FOR A L600
MASTERGROUP

Supergroup	Carrier frequency (kHz)
1	612
2	Direct
3	1116
4	1364
5	1612
6	1860
7	2108
8	2356
9	2724
10	3100

L600 Mastergroup. With an L600 mastergroup, 10 supergroups are combined as with the U600 mastergroup, except that the supergroup carrier frequencies are lower. Table 16-7 lists the supergroup carrier frequencies for a L600 mastergroup. With an L600 mastergroup, the composite baseband spectrum occupies a lower-frequency band than the U-type mastergroup (Figure 16-27). An L600 mastergroup is not further multiplexed. Therefore, the maximum channel capacity for a microwave or coaxial cable system using a single L600 mastergroup is 600 voice-band channels.

Formation of a Radio Channel

A *radio channel* comprises either a single L600 mastergroup or up to three U600 master-groups (1800 voice-band channels). Figure 16-28 shows how an 1800-channel composite FDM baseband signal is formed for transmission over a single microwave radio channel. Mastergroup 1 is transmitted directly as is, while mastergroups 2 and 3 undergo an additional multiplexing step. The three mastergroups are summed in a mastergroup combining network to produce the output spectrum shown in Figure 16-28b. Note the 80-kHz guard band between adjacent mastergroups.

The system shown in Figure 16-28 can be increased from 1800 voice-band channels to 1860 by adding an additional supergroup (supergroup 12) directly to mastergroup 1. The additional 312- to 552-kHz supergroup extends the composite output spectrum to 312 to 8284 kHz.

HYBRID DATA

With *hybrid* data it is possible to combine digitally encoded signals with FDM signals and transmit them as one composite baseband signal. There are four primary types of hybrid data: data under voice (DUV), data above voice (DAV), data above video (DAVID), and data in voice (DIV).

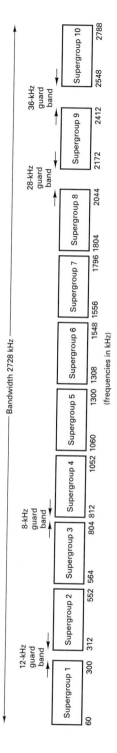

Figure 16-27 L600 mastergroup.

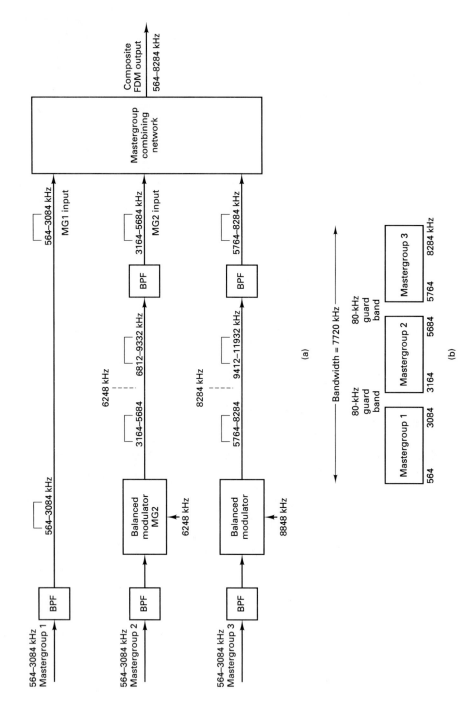

Figure 16-28 Three-mastergroup radio channel: (a) block diagram; (b) output spectrum.

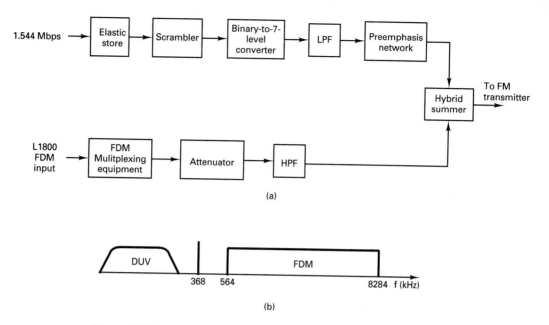

Figure 16-29 Data under voice (DUV): (a) block diagram; (b) frequency spectrum.

Data Under Voice

Figure 16-29a shows the block diagram of AT&T's 1.544-Mbps *data under FDM voice* system. With the L1800 FDM system explained earlier in this chapter, the 0 to 564-kHz frequency spectrum is void of baseband signals. With FM transmission, the lower baseband frequencies realize the highest signal-to-noise ratios. Consequently, the best portion of the baseband spectrum was unused. DUV is a means of utilizing this spectrum for the transmission of digitally encoded signals. A T1 carrier system can be converted to a quasi-analog signal and then frequency division multiplexed onto the lower portion of the FDM spectrum.

In Figure 16-29a, the *elastic store* removes timing jitter from the incoming data stream. The data are then *scrambled* to suppress the discrete high-power spectral components. The advantage of scrambling is that the randomized data output spectrum is continuous and has a predictable effect on the FDM radio system. In other words, the data present a load to the system equivalent to adding additional FDM voice channels. The serial seven-level *partial response* encoder (correlative coder) compresses the data bandwidth and allows a 1.544-Mbps signal to be transmitted in a bandwidth less than 400 kHz. The low-pass filter performs the final spectral shaping of the digital information and suppresses the spectral power above 386 kHz. This prevents the DUV information from interfering with the 386-kHz pilot control tone. The DUV signal is preemphasized and combined with the L1800 baseband signal. The output spectrum is shown in Figure 16-29b.

AT&T uses DUV for *digital data service* (DDS). DDS is intended to provide a communications medium for the transfer of digital data from station to station without the use of a data modem. DDS circuits are guaranteed to average 99.5% error-free seconds at 56 kbps.

Data Above Voice

Figure 16-30 shows the block diagram and frequency spectrum for a *data above voice* system. The advantage of DAV is for FDM systems that extend into the low end of the baseband spectrum; the low-frequency baseband does not have to be vacated for data

Hybrid Data

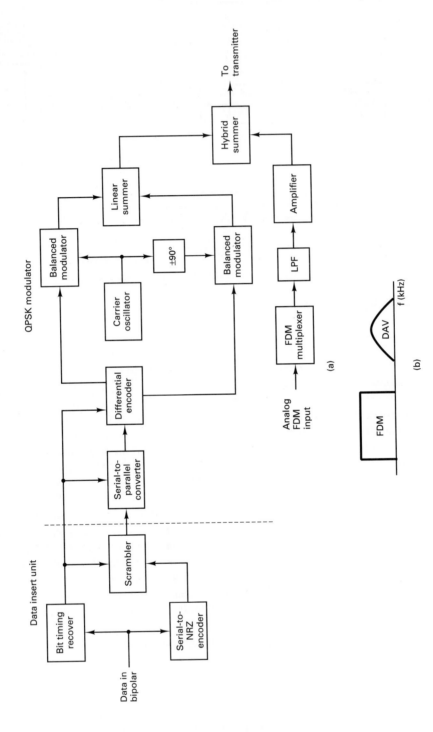

Figure 16-30 Data above voice (DAV): (a) block diagram; (b) frequency spectrum.

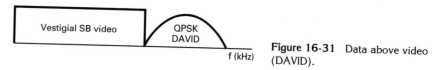

Figure 16-31 Data above video (DAVID).

transmission. With DAV, data PSK modulates a carrier that is then upconverted to a frequency above the FDM message. With DAV, up to 3.152 Mbps can be cost effectively transmitted using existing FDM/FM microwave systems (Chapter 17).

Data Above Video

Essentially, *data above video* is the same as DAV except that the lower baseband spectrum is a *vestigial sideband* video signal rather than a composite FDM signal. Figure 16-31 shows the frequency spectrum for a DAVID system.

Data In Voice

Data in voice, developed by Fujitsu of Japan, uses an eight-level PAM-VSB modulation technique with steep filtering. It uses a highly compressed partial response encoding technique, which gives it a high bandwidth efficiency of nearly 5 bps/Hz (1.544-Mbps data are transmitted in a 344-kHz bandwidth).

QUESTIONS

16-1. Define *multiplexing*.

16-2. Describe time-division mutliplexing.

16-3. Describe the Bell System T1 carrier system.

16-4. What is the purpose of the signaling bit?

16-5. What is frame synchronization? How is it achieved in a PCM/TDM system?

16-6. Describe the superframe format. Why is it used?

16-7. What is a codec? A combo chip?

16-8. What is a fixed-data-rate mode?

16-9. What is a variable-data-rate mode?

16-10. What is a DSX? What is it used for?

16-11. Explain *line coding*.

16-12. Briefly explain unipolar and bipolar transmission.

16-13. Briefly explain return-to-zero and nonreturn-to-zero transmission.

16-14. Contrast the bandwidth considerations of return-to-zero and nonreturn-to-zero transmission.

16-15. Contrast the clock recovery capabilities with return-to-zero and nonreturn-to-zero transmission.

16-16. Contrast the error detection and decoding capabilities of return-to-zero and nonreturn-to-zero transmission.

16-17. What is a regenerative repeater?

16-18. Explain B6ZS and B3ZS. When or why would you use one rather than the other?

16-19. Briefly explain the following framing techniques: added digit framing, robbed digit framing, added channel framing, statistical framing, and unique line-code framing.

16-20. Contrast bit and word interleaving.

16-21. Describe frequency division multiplexing.

16-22. Describe a message channel.

16-23. Describe the formation of a group, a supergroup, and a mastergroup.

16-24. Define *baseband* and *composite baseband*.

16-25. What is a guard band? When is a guard band used?

16-26. What are the four primary types of hybrid data network?

16-27. What is the difference between a DUV and a DAV network?

PROBLEMS

16-1. A PCM/TDM system multiplexes 24 voice-band channels. Each sample is encoded into 7 bits and a framing bit is added to each frame. The sampling rate if 9000 samples/second. BPRZ-AMI encoding is the line format. Determine:
   (a) Line speed in bits per second.
   (b) Minimum Nyquist bandwidth.

16-2. A PCM/TDM system multiplexes 32 voice-band channels each with a bandwidth of 0 to 4 kHz. Each sample is encoded with an 8-bit PCM code. UPNRZ encoding is used. Determine:
   (a) Minimum sample rate.
   (b) Line speed in bits per second.
   (c) Minimum Nyquist bandwidth.

16-3. For the following bit sequence, draw the timing diagram for UPRZ, UPNRZ, BPRZ, BPNRZ, and BPRZ-AMI encoding:

bit stream: 1 1 1 0 0 1 0 1 0 1 1 0 0

16-4. Encode the following BPRZ-AMI data stream with B6ZS and B3ZS.

+ − 0 0 0 0 + − + 0 − 0 0 0 0 0 + − 0 0 +

16-5. Calculate the 12 channel carrier frequencies for the U600 FDM system.

16-6. Calculate the five group carrier frequencies for the U600 FDM system.

17

Microwave Radio Communications and System Gain

INTRODUCTION

Presently, terrestrial (earth) *microwave radio relay systems* provide less than half of the total message circuit mileage in the United States. However, at one time microwave systems carried the bulk of long-distance communications for the public telephone network, military and governmental agencies, and specialized private communications networks. There are many different types of microwave systems that operate over distances varying from 15 to 4000 miles in length. *Intrastate* or *feeder service* systems are generally categorized as *short haul* because they are used for relatively short distances. *Long-haul* radio systems are those used for relatively long distances, such as interstate and backbone route applications. Microwave system capacities range from less than 12 voice band channels to more than 22,000. Early microwave radio systems carried frequency-division-multiplexed voice band circuits and used conventional, noncoherent frequency modulation techniques. More recently developed microwave systems carry pulse-code-modulated time-division-multiplexed voice band circuits and use more modern digital modulation techniques, such as phase shift keying and quadrature amplitude modulation. This chapter deals primarily with conventional FDM/FM microwave systems, and Chapter 19 deals with the more modern PCM/PSK techniques.

FREQUENCY VERSUS AMPLITUDE MODULATION

Frequency modulation (FM) is used in microwave radio systems rather than amplitude modulation (AM) because amplitude-modulated signals are more sensitive to amplitude nonlinearities inherent in *wideband microwave amplifiers.* Frequency-modulated signals are relatively insensitive to this type of nonlinear distortion and can be transmitted through amplifiers that have compression or amplitude nonlinearity with little penalty. In addition, FM signals are less sensitive to random noise and can be propagated with lower transmit powers.

Intermodulation noise is a major factor when designing FM radio systems. In AM systems, intermodulation noise is caused by repeater amplitude nonlinearity. In FM systems, intermodulation noise is caused primarily by transmission gain and delay distortion. Consequently, in AM systems, intermodulation noise is a function of signal amplitude, but in FM systems it is a function of signal amplitude and the magnitude of the frequency deviation. Thus the characteristics of frequency-modulated signals are more suitable than amplitude-modulated signals for microwave transmission.

SIMPLIFIED FM MICROWAVE RADIO SYSTEM

A simplified block diagram of an FM microwave radio system is shown in Figure 17-1. The *baseband* is the composite signal that modulates the FM carrier and may comprise one or more of the following:

1. Frequency-division-multiplexed voice band channels
2. Time-division-multiplexed voice band channels
3. Broadcast-quality composite video or picturephone
4. Wideband data

FM Microwave Radio Transmitter

In the FM *microwave transmitter* shown in Figure 17-1a, a *preemphasis* network precedes the FM deviator. The preemphasis network provides an artificial boost in amplitude to the higher baseband frequencies. This allows the lower baseband frequencies to frequency modulate the IF carrier and the higher baseband frequencies to phase modulate it. This scheme assures a more uniform signal-to-noise ratio throughout the entire baseband spectrum. An FM deviator provides the modulation of the IF carrier which eventually becomes the main microwave carrier. Typically, IF carrier frequencies are between 60 and 80 MHz, with 70 MHz the most common. *Low-index* frequency modulation is used in the FM deviator. Typically, modulation indices are kept between 0.5 and 1. This produces a *narrowband* FM signal at the output of the deviator. Consequently, the IF bandwidth resembles conventional AM and is approximately equal to twice the highest baseband frequency.

The IF and its associated sidebands are up-converted to the microwave region by the AM mixer, microwave oscillator, and bandpass filter. Mixing, rather than multiplying, is used to translate the IF frequencies to RF frequencies because the modulation index is unchanged by the heterodyning process. Multiplying the IF carrier would also multiply the frequency deviation and the modulation index, thus increasing the bandwidth. Typically, frequencies above 1000 MHz (1 GHz) are considered microwave frequencies. Presently, there are microwave systems operating with carrier frequencies up to approximately 18 GHz. The most common microwave frequencies currently being used are the 2-, 4-, 6-, 12-, and 14-GHz bands. The channel-combining network provides a means of connecting more than one microwave transmitter to a single transmission line feeding the antenna.

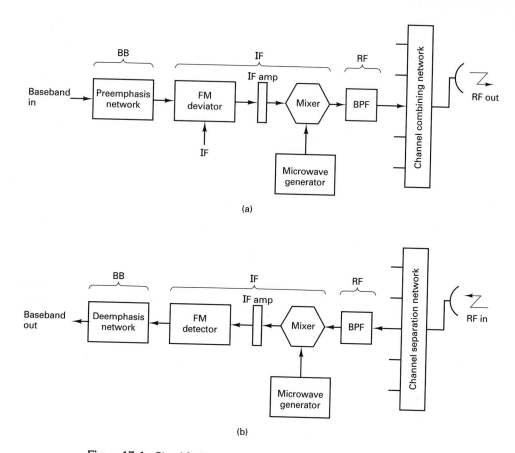

Figure 17-1 Simplified block diagram of an FM microwave radio system: (a) transmitter; (b) receiver.

FM Microwave Radio Receiver

In the FM microwave receiver shown in Figure 17-1b, the channel separation network provides the isolation and filtering necessary to separate individual microwave channels and direct them to their respective receivers. The bandpass filter, AM mixer, and microwave oscillator down-convert the RF microwave frequencies to IF frequencies and pass them on to the FM demodulator. The FM demodulator is a conventional, *noncoherent* FM detector (i.e., a discriminator or a PLL demodulator). At the output of the FM detector, a deemphasis network restores the baseband signal to its original amplitude versus frequency characteristics.

FM MICROWAVE RADIO REPEATERS

The permissible distance between an FM microwave transmitter and its associated microwave receiver depends on several system variables, such as transmitter output power, receiver noise threshold, terrain, atmospheric conditions, system capacity, reliability objectives, and performance expectations. Typically, this distance is between 15 and 40 miles. Longhaul microwave systems span distances considerably longer than this. Consequently, a single-hop microwave system, such as the one shown in Figure 17-1, is inadequate for most practical system applications. With systems that are longer than 40 miles

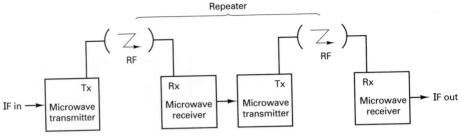

Figure 17-2 Microwave repeater.

or when geographical obstructions, such as a mountain, block the transmission path, *repeaters* are needed. A microwave repeater is a receiver and a transmitter placed back to back or in tandem with the system. A block diagram of a microwave repeater is shown in Figure 17-2. The repeater station receives a signal, amplifies and reshapes it, then retransmits the signal to the next repeater or terminal station downline from it.

Basically, there are two types of microwave repeaters: *baseband* and *IF* (Figure 17-3). IF repeaters are also called *heterodyne* repeaters. With an IF repeater (Figure 17-3a), the received RF carrier is down-converted to an IF frequency, amplified, reshaped, up-converted to an RF frequency, and then retransmitted. The signal is never demodulated below IF. Consequently, the baseband intelligence is unmodified by the repeater. With a baseband repeater (Figure 17-3b), the received RF carrier is down-converted to an IF frequency, amplified, filtered, and then further demodulated to baseband. The baseband signal, which is typically frequency-division-multiplexed voice band channels, is further demodulated to a mastergroup, supergroup, group, or even channel level. This allows the baseband signal to be reconfigured to meet the routing needs of the overall communications network. Once the baseband signal has been reconfigured, it FM modulates an IF carrier which is up-converted to an RF carrier and then retransmitted.

Figure 17-3c shows another baseband repeater configuration. The repeater demodulates the RF to baseband, amplifies and reshapes it, then modulates the FM carrier. With this technique, the baseband is not reconfigured. Essentially, this configuration accomplishes the same thing that an IF repeater accomplishes. The difference is that in a baseband configuration, the amplifier and equalizer act on baseband frequencies rather than IF frequencies. The baseband frequencies are generally less than 9 MHz, whereas the IF frequencies are in the range 60 to 80 MHz. Consequently, the filters and amplifiers necessary for baseband repeaters are simpler to design and less expensive than the ones required for IF repeaters. The disadvantage of a baseband configuration is the addition of the FM terminal equipment.

DIVERSITY

Microwave systems use *line-of-sight* transmission. There must be a direct, line-of-sight signal path between the transmit and the receive antennas. Consequently, if that signal path undergoes a severe degradation, a service interruption will occur. *Diversity* suggests that there is more than one transmission path or method of transmission available between a transmitter and a receiver. In a microwave system, the purpose of using diversity is to increase the reliability of the system by increasing its availability. When there is more than one transmission path or method of transmission available, the system can select the path or method that produces the highest-quality received signal. Generally, the highest quality is determined by evaluating the carrier-to-noise (*C/N*) ratio at the receiver input or by

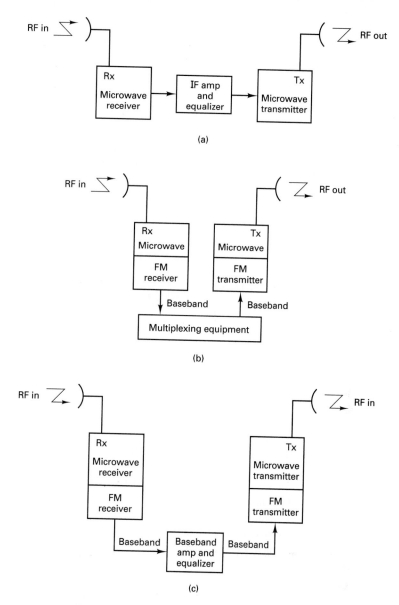

Figure 17-3 Microwave repeaters: (a) IF; (b) and (c) baseband.

simply measuring the received carrier power. Although there are many ways of achieving diversity, the most common methods used are *frequency, space,* and *polarization.*

Frequency Diversity

Frequency diversity is simply modulating two different RF carrier frequencies with the same IF intelligence, then transmitting both RF signals to a given destination. At the destination, both carriers are demodulated, and the one that yields the better-quality IF signal is selected. Figure 17-4 shows a single-channel frequency-diversity microwave system.

In Figure 17-4a, the IF input signal is fed to a power splitter, which directs it to microwave transmitters A and B. The RF outputs from the two transmitters are combined

Diversity

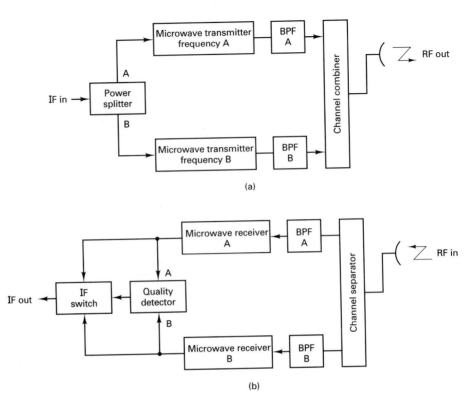

Figure 17-4 Frequency diversity microwave system: (a) transmitter; (b) receiver.

in the channel-combining network and fed to the transmit antenna. At the receive end (Figure 17-4b), the channel separator directs the A and B RF carriers to their respective microwave receivers, where they are down-converted to IF. The quality detector circuit determines which channel, A or B, is the higher quality and directs that channel through the IF switch to be further demodulated to baseband. Many of the temporary, adverse atmospheric conditions that degrade an RF signal are frequency selective; they may degrade one frequency more than another. Therefore, over a given period of time, the IF switch may switch back and forth from receiver A to receiver B, and vice versa many times.

Space Diversity

With space diversity, the output of a transmitter is fed to two or more antennas that are physically separated by an appreciable number of wavelengths. Similarly, at the receiving end, there may be more than one antenna providing the input signal to the receiver. If multiple receiving antennas are used, they must also be separated by an appreciable number of wavelengths. Figure 17-5 shows a single-channel space-diversity microwave system.

When space diversity is used, it is important that the electrical distance from a transmitter to each of its antennas and to a receiver from each of its antennas is an equal multiple of wavelengths long. This is to ensure that when two or more signals of the same frequency arrive at the input to a receiver, they are in phase and additive. If received out of phase, they will cancel and, consequently, result in less received signal power than if simply one antenna system were used. Adverse atmospheric conditions are often isolated to a very small geographical area. With space diversity, there is more than one transmission

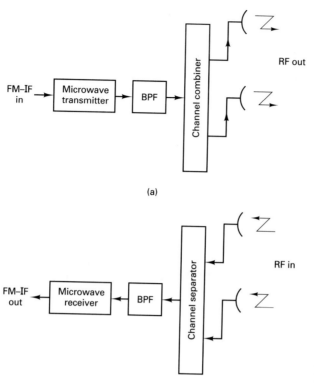

Figure 17-5 Space-diversity microwave system: (a) transmitter; (b) receiver.

path between a transmitter and a receiver. When adverse atmospheric conditions exist in one of the paths, it is unlikely that the alternate path is experiencing the same degradation. Consequently, the probability of receiving an acceptable signal is higher when space diversity is used than when no diversity is used. An alternate method of space diversity uses a single transmitting antenna and two receiving antennas separated vertically. Depending on the atmospheric conditions at a particular time, one of the receiving antennas should be receiving an adequate signal. Again, there are two transmission paths that are unlikely to be affected simultaneously by fading.

Polarization Diversity

With polarization diversity, a single RF carrier is propagated with two different electromagnetic polarizations (vertical and horizontal). Electromagnetic waves of different polarizations do not necessarily experience the same transmission impairments. Polarization diversity is generally used in conjunction with space diversity. One transmit/receive antenna pair is vertically polarized and the other is horizontally polarized. It is also possible to use frequency, space, and polarization diversity simultaneously.

PROTECTION SWITCHING

Radio path losses vary with atmospheric conditions. Over a period of time, the atmospheric conditions between transmitting and receiving antenna can vary significantly, causing a corresponding reduction in the received signal strength of 20, 30, 40, or more dB.

This reduction in signal strength is referred to as a *radio fade. Automatic gain control circuits,* built into radio receivers, can compensate for fades of 25 to 40 dB, depending on the system design. However, fades in excess of 40 dB cause a total loss of the received signal. When this happens, service continuity is lost. To avoid a service interruption during periods of deep fades or equipment failures, alternate facilities are temporarily made available in what is called a *protection switching* arrangement. Essentially, there are two types of protection switching arrangements: *hot standby* and *diversity.* With hot standby protection, each working radio channel has a dedicated backup or spare channel. With diversity protection, a single backup channel is made available to as many as 11 working channels. Hot standby systems offer 100% protection for each working radio channel. A diversity system offers 100% protection only to the first working channel that fails. If two radio channels fail at the same time, a service interruption will occur.

Hot Standby

Figure 17-6a shows a single-channel hot standby protection switching arrangement. At the transmitting end, the IF goes into a *head-end bridge,* which splits the signal power and directs it to the working and the spare (standby) microwave channels simultaneously. Consequently, both the working and standby channels are carrying the same baseband information. At the receiving end, the IF switch passes the IF signal from the working channel to the FM terminal equipment. The IF switch continuously monitors the received signal power on the working channel and if it fails, switches to the standby channel. When the IF signal on the working channel is restored, the IF switch resumes its normal position.

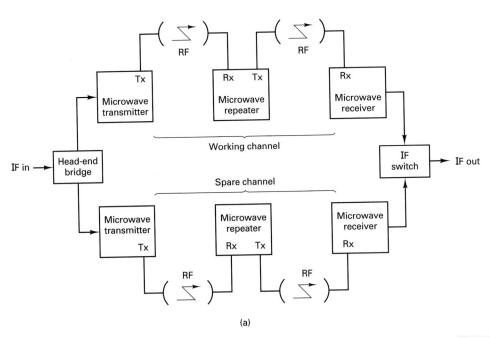

(a)

Figure 17-6 Microwave protection switching arrangements: (a) hot standby; *(Continued on next page.)*

Diversity

Figure 17-6b shows a diversity protection switching arrangement. This system has two working channels (channel 1 and channel 2), one spare channel, and an *auxiliary* channel. The IF switch at the receive end continuously monitors the receive signal strength of both working channels. If either one should fail, the IF switch detects a loss of carrier and sends back to the transmitting station IF switch a VF (*voice frequency*) tone-encoded signal that directs it to switch the IF signal from the failed channel onto the spare microwave channel. When the failed channel is restored, the IF switches resume their normal positions. The auxiliary channel simply provides a transmission path between the two IF switches. Typically, the auxiliary channel is a low-capacity low-power microwave radio that is designed to be used for a maintenance channel only.

Reliability

The number of repeater stations between protection switches depends on the *reliability objectives* of the system. Typically, there are between two and six repeaters between switching stations.

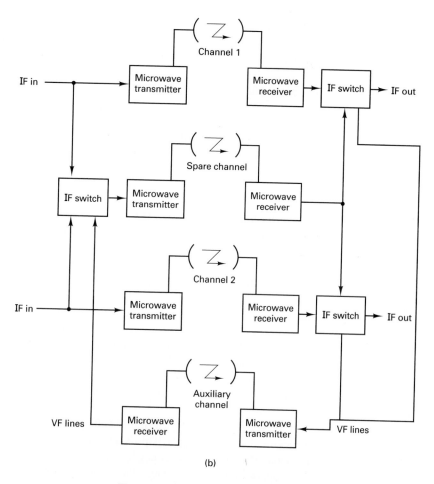

(b)

Figure 17-6 *(continued)* (b) diversity.

As you can see, diversity systems and protection switching arrangements are quite similar. The primary difference between the two is that diversity systems are permanent arrangements and are intended only to compensate for temporary, abnormal atmospheric conditions between only two selected stations in a system. Protection switching arrangements, on the other hand, compensate for both radio fades and equipment failures and may include from six to eight repeater stations between switches. Protection channels may also be used as temporary communication facilities, while routine maintenance is performed on a regular working channel. With a protection switching arrangement, all signal paths and radio equipment are protected. Diversity is used selectively, that is, only between stations that historically experience severe fading a high percentage of the time.

A statistical study of outage time (i.e., service interruptions) caused by radio fades, equipment failures, and maintenance is important in the design of a microwave radio system. From such a study, engineering decisions can be made on which type of diversity system and protection switching arrangement is best suited for a particular application.

FM MICROWAVE RADIO STATIONS

Basically, there are two types of FM microwave stations: terminals and repeaters. *Terminal stations* are points in the system where baseband signals either originate or terminate. *Repeater stations* are points in a system where baseband signals may be reconfigured or where RF carriers are simply "repeated" or amplified.

Terminal Station

Essentially, a terminal station consists of four major sections: the baseband, wire line entrance link (WLEL), FM-IF, and RF sections. Figure 17-7 shows the block diagram of the baseband, WLEL, and FM-IF sections. As mentioned previously, the baseband may be one of several different types of signals. For our example, frequency-division-multiplexed voice band channels are used.

Wire line entrance link (WLEL).   Very often in large communications networks such as the American Telephone and Telegraph Company (AT&T), the building that houses the radio station is quite large. Consequently, it is desirable that similar equipment be physically placed at a common location (i.e., all FDM equipment in the same room). This simplifies alarm systems, providing dc power to the equipment, maintenance, and other general cabling requirements. Dissimilar equipment may be separated by a considerable distance. For example, the distance between the FDM multiplexing equipment and the FM-IF section is typically several hundred feet and in some cases several miles. For this reason a WLEL is required. A WLEL serves as the interface between the multiplex terminal equipment and the FM-IF equipment. A WLEL generally consists of an amplifier and an equalizer (which together compensate for cable transmission losses) and level-shaping devices commonly called pre- and deemphasis networks.

IF section.   The FM terminal equipment shown in Figure 17-7 generates a frequency-modulated IF carrier. This is accomplished by mixing the outputs of two deviated oscillators that differ in frequency by the desired IF carrier. The oscillators are deviated in phase opposition, which reduces the magnitude of phase deviation required of a single deviator by a factor of 2. This technique also reduces the deviation linearity requirements

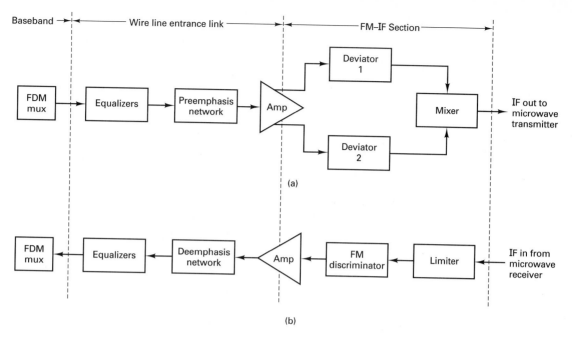

Figure 17-7 Microwave terminal station, baseband, wire line entrance link, and FM-IF:
(a) transmitter; (b) receiver.

for the oscillators and provides for the partial cancellation of unwanted modulation products. Again, the receiver is a conventional noncoherent FM detector.

RF section. A block diagram of the RF section of a microwave terminal station is shown in Figure 17-8. The IF signal enters the transmitter (Figure 17-8a) through a protection switch. The IF and compression amplifiers help keep the IF signal power constant and at approximately the required input level to the transmit modulator (*transmod*). A transmod is a balanced modulator that when used in conjunction with a microwave generator, power amplifier, and bandpass filter, up-converts the IF carrier to an RF carrier and amplifies the RF to the desired output power. Power amplifiers for microwave radios must be capable of amplifying very high frequencies and passing very wide bandwidth signals. *Klystron tubes, traveling-wave tubes* (TWTs), and *IMPATT* (*imp*act/*a*valanche and *T*ransit *T*ime) diodes are several of the devices currently being used in microwave power amplifiers. Because high-gain antennas are used and the distance between microwave stations is relatively short, it is not necessary to develop a high output power from the transmitter output amplifiers. Typical gains for microwave antennas range from 10 to 40 dB, and typical transmitter output powers are between 0.5 and 10 W.

A *microwave generator* provides the RF carrier input to the up-converter. It is called a microwave generator rather than an oscillator because it is difficult to construct a stable circuit that will oscillate in the gigahertz range. Instead, a crystal-controlled oscillator operating in the range 5 to 25 MHz is used to provide a base frequency that is multiplied up to the desired RF carrier frequency.

An *isolator* is a unidirectional device often made from a ferrite material. The isolator is used in conjunction with a channel-combining network to prevent the output of one transmitter from interfering with the output of another transmitter.

The RF receiver (Figure 17-8b) is essentially the same as the transmitter except that it works in the opposite direction. However, one difference is the presence of an IF amplifier

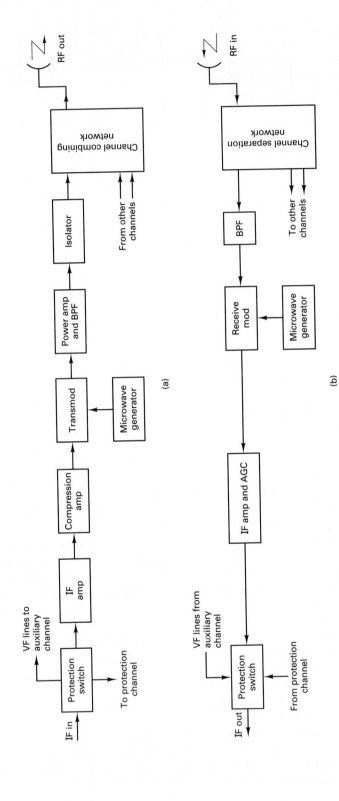

Figure 17-8 Microwave terminal station: (a) transmitter; (b) receiver.

in the receiver. This IF amplifier has an *automatic gain control* (AGC) circuit. Also, very often, there are no RF amplifiers in the receiver. Typically, a very sensitive, low-noise-balanced demodulator is used for the receive demodulator (receive mod). This eliminates the need for an RF amplifier and improves the overall signal-to-noise ratio. When RF amplifiers are required, high-quality, *low-noise amplifiers* (LNAs) are used. Examples of commonly used LNAs are tunnel diodes and parametric amplifiers.

Repeater Station

Figure 17-9 shows the block diagram of a microwave IF repeater station. The received RF signal enters the receiver through the channel separation network and bandpass filter. The receive mod down-converts the RF carrier to IF. The IF AMP/AGC and equalizer circuits amplify and reshape the IF. The equalizer compensates for *gain versus frequency nonlinearities* and *envelope delay distortion* introduced in the system. Again, the transmod up-converts the IF to RF for retransmission. However, in a repeater station, the method used to generate the RF microwave carrier frequencies is slightly different from the method used in a terminal station. In the IF repeater, only one microwave generator is required to supply both the transmod and the receive mod with an RF carrier signal. The microwave generator, shift oscillator, and shift modulator allow the repeater to receive one RF carrier frequency, down-convert it to IF, and then up-convert the IF to a different RF carrier frequency. It is possible for station C to receive the transmissions from both station A and station B simultaneously (this is called *multihop interference* and is shown in Figure 17-10a). This can occur only when three stations are placed in a geographical straight line in the system. To prevent this from occurring, the allocated bandwidth for the system is divided in half, creating a low-frequency and a high-frequency band. Each station, in turn, alternates from a low-band to a high-band transmit carrier frequency (Figure 17-10b). If a transmission from station A is received by station C, it will be rejected in the channel separation network and cause no interference. This arrangement is called a high/low microwave repeater system. The rules are simple: If a repeater station receives a low-band RF carrier, it retransmits a high-band RF carrier, and vice versa. The only time that multiple carriers of the same frequency can be received is when a transmission from one station is received from another station that is three hops away. This is unlikely to happen.

Another reason for using a high/low-frequency scheme is to prevent the power that "leaks" out the back and sides of a transmit antenna from interfering with the signal entering the input of a nearby receive antenna. This is called *ringaround*. All antennas, no matter how high their gain or how directive their radiation pattern, radiate a small percentage of their power out the back and sides; giving a finite *front-to-back* ratio for the antenna. Although the front-to-back ratio of a typical microwave antenna is quite high, the relatively small amount of power that is radiated out the back of the antenna may be quite substantial compared to the normal received carrier power in the system. If the transmit and receive carrier frequencies are different, filters in the receiver separation network will prevent ringaround from occurring.

A high/low microwave repeater station (Figure 17-10b) needs two microwave carrier supplies for the down- and up-converting process. Rather than use two microwave generators, a single generator together with a shift oscillator, a shift modulator, and a bandpass filter can generate the two required signals. One output from the microwave generator is fed directly into the transmod and another output (from the same microwave generator) is mixed with the shift oscillator signal in the shift modulator to produce a second microwave carrier frequency. The second microwave carrier frequency is offset from the

FM Microwave Radio Stations

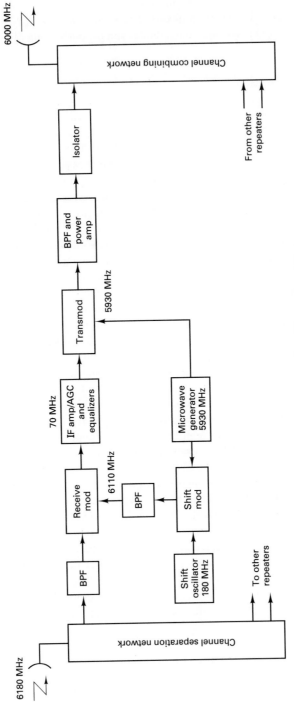

Figure 17-9 Microwave IF repeater station.

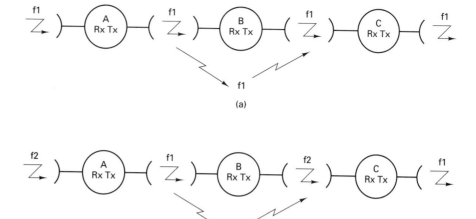

Figure 17-10 (a) Multihop interference and (b) high/low microwave system.

first by the shift oscillator frequency. The second microwave carrier frequency is fed into the receive modulator.

EXAMPLE 17-1

In Figure 17-9 the received RF carrier frequency is 6180 MHz, and the transmitted RF carrier frequency is 6000 MHz. With a 70-MHz IF frequency, a 5930-MHz microwave generator frequency, and a 180-MHz shift oscillator frequency, the output filter of the shift mod must be tuned to 6110 MHz. This is the sum of the microwave generator and the shift oscillator frequencies (5930 MHz + 180 MHz = 6110 MHz).

This process does not reduce the number of oscillators required, but it is simpler and cheaper to build one microwave generator and one relatively low-frequency shift oscillator than to build two microwave generators. This arrangement also provides a certain degree of synchronization between repeaters. The obvious disadvantage of the high/low scheme is that the number of channels available in a given bandwidth is cut in half.

Figure 17-11 shows a high/low-frequency plan with eight channels (four high-band and four low-band). Each channel occupies a 29.7-MHz bandwidth. The west terminal transmits the low-band frequencies and receives the high-band frequencies. Channel 1 and 3 (Figure 17-11a) are designated as *V channels*. This means that they are propagated with vertical polarization. Channels 2 and 4 are designated as H or horizontally polarized channels. This is not a polarization diversity system. Channels 1 through 4 are totally independent of each other; they carry different baseband information. The transmission of *orthogonally* polarized carriers (90° out of phase) further enhances the isolation between the transmit and receive signals. In the west-to-east direction, the repeater receives the low-band and transmits the high-band frequencies. After channel 1 is received and down-converted to IF, it is up-converted to a different RF frequency and a different polarization for retransmission. The low-band channel 1 corresponds to the high-band channel 11, channel 2 to channel 12, and so on. The east-to-west direction (Figure 17-11b) propagates the high- and low-band carriers in the sequence opposite to the west-to-east system. The polarizations are also reversed. If some of the power from channel 1 of the west terminal were to propagate directly to the east terminal receiver, it has a different frequency and polarization than channel 11's transmissions. Consequently, it would not interfere with the

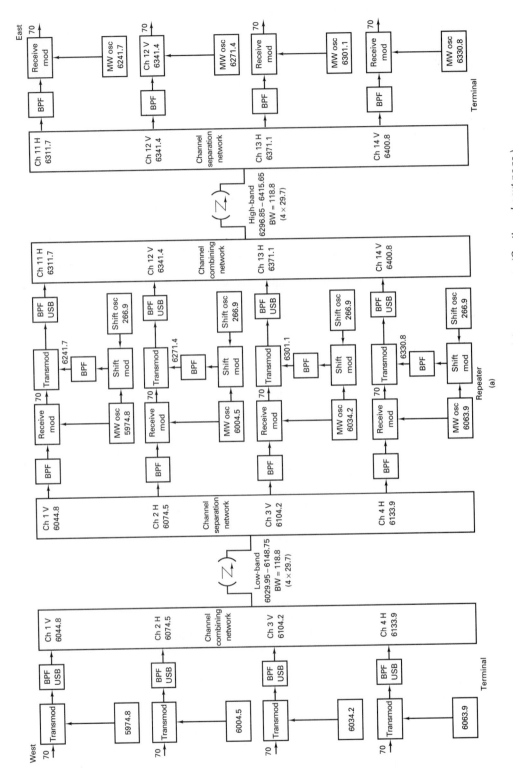

Figure 17-11 Eight-channel high/low frequency plan: (a) west to east; *(Continued next page.)*

706

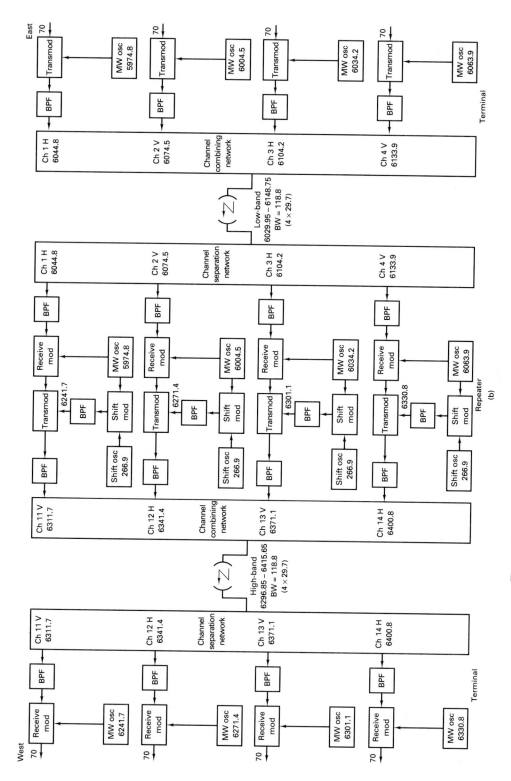

Figure 17-11 (continued) (b) east to west. All frequencies in megahertz.

reception of channel 11 (no multihop interference). Also, note that none of the transmit or receive channels at the repeater station has both the same frequency and polarization. Consequently, the interference from the transmitters to the receivers due to ringaround is insignificant.

PATH CHARACTERISTICS

The normal *propagation paths* between two radio antennas in a microwave radio system is shown in Figure 17-12. The *free-space path* is the *line-of-sight path* directly between the transmit and receive antennas (this is also called the *direct wave*). The *ground-reflected wave* is the portion of the transmit signal that is reflected off Earth's surface and captured by the receive antenna. The *surface wave* consists of the electric and magnetic fields associated with the currents induced in Earth's surface. The magnitude of the surface wave depends on the characteristics of Earth's surface and the electromagnetic polarization of the wave. The sum of these three paths (taking into account their amplitude and phase) is called the *ground wave*. The *sky wave* is the portion of the transmit signal that is returned (reflected) back to Earth's surface by the ionized layers of Earth's atmosphere.

All of the paths shown in Figure 17-12 exist in any microwave radio system, but some are negligible in certain frequency ranges. At frequencies below 1.5 MHz, the surface wave provides the primary coverage, and the sky wave helps to extend this coverage at night when the absorption of the ionosphere is at a minimum. For frequencies above about 30 to 50 MHz, the free-space and ground-reflected paths are generally the only paths of importance. The surface wave can also be neglected at these frequencies, provided that the antenna heights are not too low. The sky wave is only a source of occasional long-distance interference and not a reliable signal for microwave communications purposes. In this chapter the surface and sky-wave propagations are neglected, and attention is focused on those phenomena that affect the direct and reflected waves.

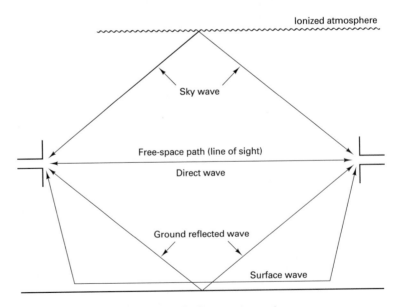

Figure 17-12 Propagation paths.

In its simplest form, *system gain* is the difference between the nominal output power of a transmitter and the minimum input power required by a receiver. System gain must be greater than or equal to the sum of all the gains and losses incurred by a signal as it propagates from a transmitter to a receiver. In essence, it represents the net loss of a radio system. System gain is used to predict the reliability of a system for given system parameters. Mathematically, system gain is

$$G_s = P_t - C_{min}$$

where: G_s = system gain (dB)
P_t = transmitter output power (dBm)
C_{min} = minimum receiver input power for a given quality objective (dBm)

and where:

$$P_t - C_{min} \geq losses - gains$$

Gains:

A_t = transmit antenna gain (dB) relative to an isotropic radiator
A_r = receive antenna gain (dB) relative to an isotropic radiator

Losses:

L_p = free-space path loss between antennas (dB)
L_f = waveguide feeder loss (dB) between the distribution network (channel combining network or channel separation network) and its respective antenna (see Table 17-1)
L_b = total coupling or branching loss (dB) in the circulators, filters, and distribution network between the output of a transmitter or the input to a receiver and its respective waveguide feed (see Table 17-1)
F_m = fade margin for a given reliability objective

TABLE 17-1 SYSTEM GAIN PARAMETERS

Frequency (GHz)	Feeder loss, L_f		Branching loss (dB) Diversity:		Antenna gain, A_t or A_r	
	Type	Loss (dB/100 m)	Frequency	Space	Size (m)	Gain (dB)
1.8	Air-filled coaxial cable	5.4	5	2	1.2	25.2
					2.4	31.2
					3.0	33.2
					3.7	34.7
7.4	EWP 64 eliptical waveguide	4.7	3	2	1.5	38.8
					2.4	43.1
					3.0	44.8
					3.7	46.5
8.0	EWP 69 eliptical waveguide	6.5	3	2	2.4	43.8
					3.0	45.6
					3.7	47.3
					4.8	49.8

Mathematically, system gain is

$$G_s = P_t - C_{\min} \geq F_m + L_p + L_f + L_b - A_t - A_r \qquad (17\text{-}1)$$

where all values are expressed in dB or dBm. Because system gain is indicative of a net loss, the losses are represented with positive dB values and the gains are represented with negative dB values. Figure 17-13 shows an overall microwave system diagram and indicates where the respective losses and gains are incurred.

Free-Space Path Loss

Free-space path loss is defined as the loss incurred by an electromagnetic wave as it propagates in a straight line through a vacuum with no absorption or reflection of energy from nearby objects. The expression for free-space path loss is given as

$$L_p = \left(\frac{4\pi D}{\lambda} \right)^2 = \left(\frac{4\pi f D}{c} \right)^2$$

where: L_p = free-space path loss
$\quad\quad D$ = distance
$\quad\quad f$ = frequency
$\quad\quad \lambda$ = wavelength
$\quad\quad c$ = velocity of light in free space (3×10^8 m/s)

Converting to dB yields

$$L_p \text{ (dB)} = 20 \log \frac{4\pi f D}{c} = 20 \log \frac{4\pi}{c} + 20 \log f + 20 \log D$$

When the frequency is given in MHz and the distance in km,

$$L_p \text{ (dB)} = 20 \log \frac{4\pi (10)^6 (10)^3}{3 \times 10^8} + 20 \log f \text{ (MHz)} + 20 \log D \text{ (km)} \qquad (17\text{-}2)$$

$$= 32.4 + 20 \log f \text{ (MHz)} + 20 \log D \text{ (km)}$$

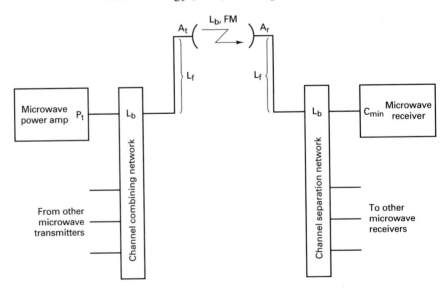

Figure 17-13 System gains and losses.

When the frequency is given in GHz and the distance in km,

$$L_p \text{ (dB)} = 92.4 + 20 \log f \text{ (GHz)} + 20 \log D \text{ (km)} \qquad (17\text{-}3)$$

Similar conversions can be made using distance in miles, frequency in kHz, and so on.

EXAMPLE 17-2

For a carrier frequency of 6 GHz and a distance of 50 km, determine the free-space path loss.

Solution

$$L_p \text{ (dB)} = 32.4 + 20 \log 6000 + 20 \log 50$$
$$= 32.4 + 75.6 + 34$$
$$= 142 \text{ dB}$$

or

$$L_p \text{ (dB)} = 92.4 + 20 \log 6 + 20 \log 50$$
$$= 92.4 + 15.6 + 34$$
$$= 142 \text{ dB}$$

Fade Margin

Essentially, *fade margin* is a "fudge factor" included in the system gain equation that considers the nonideal and less predictable characteristics of radio-wave propagation, such as *multipath propagation* (*multipath loss*) and *terrain sensitivity*. These characteristics cause temporary, abnormal atmospheric conditions that alter the free-space path loss and are usually detrimental to the overall system performance. Fade margin also considers system reliability objectives. Thus fade margin is included in the system gain equation as a loss.

Solving the Barnett–Vignant reliability equations for a specified annual system availability for an unprotected, nondiversity system yields the following expression:

$$F_m = \underbrace{30 \log D}_{\substack{\text{multipath} \\ \text{effect}}} + \underbrace{10 \log (6ABf)}_{\substack{\text{terrain} \\ \text{sensitivity}}} - \underbrace{10 \log (1 - R)}_{\substack{\text{reliability} \\ \text{objectives}}} - \underbrace{70}_{\text{constant}} \qquad (17\text{-}4)$$

where: F_m = fade margin (dB)
D = distance (km)
f = frequency (GHz)
R = reliability expressed as a decimal (i.e., 99.99% = 0.9999 reliability)
$1 - R$ = reliability objective for a one-way 400-km route
A = roughness factor
= 4 over water or a very smooth terrain
= 1 over an average terrain
= 0.25 over a very rough, mountainous terrain
B = factor to convert a worst-month probability to an annual probability
= 1 to convert an annual availability to a worst-month basis
= 0.5 for hot humid areas
= 0.25 for average inland areas
= 0.125 for very dry or mountainous areas

System Gain

711

EXAMPLE 17-3

Consider a space-diversity microwave radio system operating at an RF carrier frequency of 1.8 GHz. Each station has a 2.4-m-diameter parabolic antenna that is fed by 100 m of air-filled coaxial cable. The terrain is smooth and the area has a humid climate. The distance between stations is 40 km. A reliability objective of 99.99% is desired. Determine the system gain.

Solution Substituting into Equation 17-4, we find that the fade margin is

$$F_m = 30 \log 40 + 10 \log (6) (4) (0.5) (1.8) - 10 \log (1 - 0.9999) - 70$$

$$= 48.06 + 13.34 - (-40) - 70$$

$$= 48.06 + 13.34 + 40 - 70$$

$$= 31.4 \text{ dB}$$

Substituting into Equation 17-3, we obtain path loss

$$L_p = 92.4 + 20 \log 1.8 + 20 \log 40$$

$$= 92.4 + 5.11 + 32.04$$

$$= 129.55 \text{ dB}$$

From Table 17-1,

$$L_b = 4 \text{ dB} (2 + 2 = 4)$$

$$L_f = 10.8 \text{ dB} (100 \text{ m} + 100 \text{ m} = 200 \text{ m})$$

$$A_t = A_r = 31.2 \text{ dB}$$

Substituting into Equation 17-1 gives us system gain

$$G_s = 31.4 + 129.55 + 10.8 + 4 - 31.2 - 31.2 = 113.35 \text{ dB}$$

The results indicate that for this system to perform at 99.99% reliability with the given terrain, distribution networks, transmission lines, and antennas, the transmitter output power must be at least 113.35 dB more than the minimum receive signal level.

Receiver Threshold

Carrier-to-noise (*C/N*) is probably the most important parameter considered when evaluating the performance of a microwave communications system. The minimum wideband carrier power (C_{min}) at the input to a receiver that will provide a usable baseband output is called the receiver *threshold* or, sometimes, receiver *sensitivity*. The receiver threshold is dependent on the wideband noise power present at the input of a receiver, the noise introduced within the receiver, and the noise sensitivity of the baseband detector. Before C_{min} can be calculated, the input noise power must be determined. The input noise power is expressed mathematically as

$$N = KTB$$

where: N = noise power (watts)
  K = Boltzmann's constant (1.38×10^{-23} J/K)
  T = equivalent noise temperature of the receiver (Kelvin)
    (room temperature = 290 K)
  B = noise bandwidth (hertz)

Expressed in dBm,

$$N \text{ (dBm)} = 10 \log \frac{KTB}{0.001} = 10 \log \frac{KT}{0.001} + 10 \log B$$

For a 1-Hz bandwidth at room temperature,

$$N = 10 \log \frac{(1.38 \times 10^{-23})(290)}{0.001} + 10 \log 1$$

$$= -174 \text{ dBm}$$

Thus

$$N \text{ (dBm)} = -174 \text{ dBm} + 10 \log B \qquad (17\text{-}5)$$

EXAMPLE 17-4

For an equivalent noise bandwidth of 10 MHz, determine the noise power.

Solution Substituting into equation 17-5 yields

$$N = -174 \text{ dBm} + 10 \log (10 \times 10^6)$$

$$= -174 \text{ dBm} + 70 \text{ dB} = -104 \text{ dBm}$$

If the minimum C/N requirement for a receiver with a 10-MHz noise bandwidth is 24 dB, the minimum receive carrier power is

$$C_{\min} = \frac{C}{N} \text{ (dB)} + N \text{ (dB)}$$

$$= 24 \text{ dB} + (-104 \text{ dBm}) = -80 \text{ dBm}$$

For a system gain of 113.35 dB, it would require a minimum transmit carrier power (P_t) of

$$P_t = G_s + C_{\min}$$

$$= 113.35 \text{ dB} + (-80 \text{ dBm}) = 33.35 \text{ dBm}$$

This indicates that a minimum transmit power of 33.35 dBm (2.16 W) is required to achieve a carrier-to-noise ratio of 24 dB with a system gain of 113.35 dB and a bandwidth of 10 MHz.

Carrier-to-Noise versus Signal-to-Noise

Carrier-to-noise (C/N) is the ratio of the wideband "carrier" (actually, not just the carrier, but rather the carrier and its associated sidebands) to the wideband noise power (the noise bandwidth of the receiver). C/N can be determined at an RF or an IF point in the receiver. Essentially, C/N is a *predetection* (before the FM demodulator) signal-to-noise ratio. Signal-to-noise (S/N) is a *postdetection* (after the FM demodulator) ratio. At a baseband point in the receiver, a single voice band channel can be separated from the rest of the baseband and measured independently. At an RF or IF point in the receiver, it is impossible to separate a single voice band channel from the composite FM signal. For example, a typical bandwidth for a single microwave channel is 30 MHz. The bandwidth of a voice band channel is 4 kHz. C/N is the ratio of the power of the composite RF signal to the total noise power in the 30-MHz bandwidth. S/N is the ratio of the signal power of a single voice band channel to the noise power in a 4-kHz bandwidth.

System Gain

Noise Figure

In its simplest form, *noise figure* (NF) is the signal-to-noise ratio of an ideal noiseless device divided by the *S/N* ratio at the output of an amplifier or a receiver. In a more practical sense, noise figure is defined as the ratio of the *S/N* ratio at the input to a device divided by the *S/N* ratio at the output. Mathematically, noise figure is

$$NF(dB) = 10 \log \frac{(S/N)_{in}}{(S/N)_{out}}$$

Thus noise figure is a ratio of ratios. The noise figure of a totally noiseless device is unity or 0 dB. Remember, the noise present at the input to an amplifier is amplified by the same gain as the signal. Consequently, only noise added within the amplifier can decrease the signal-to-noise ratio at the output and increase the noise figure. (Keep in mind, the higher the noise figure, the worse the *S/N* ratio at its output).

Essentially, noise figure indicates the relative increase of the noise power to the increase in signal power. A noise figure of 10 means that the device added sufficient noise to reduce the *S/N* ratio by a factor of 10, or the noise power increased tenfold in respect to the increase in signal power.

When two or more amplifiers or devices are cascaded together (Figure 17-14), the total noise figure (NF$_T$) is an accumulation of the individual noise figures. Mathematically, the total noise figure is

$$NF_T = NF_1 + \frac{NF_2 - 1}{A_1} + \frac{NF_3 - 1}{A_1 A_2} + \frac{NF_4 - 1}{A_1 A_2 A_3} \quad \text{etc.} \qquad (17\text{-}6)$$

where: NF$_T$ = total noise figure
 NF$_1$ = noise figure of amplifier 1
 NF$_2$ = noise figure of amplifier 2
 NF$_3$ = noise figure of amplifier 3
 A_1 = power gain of amplifier 1
 A_2 = power gain of amplifier 2

Note: In equation 17-6 noise figures and gains are expressed as absolute values rather than in dB.

It can be seen that the noise figure of the first amplifier (NF$_1$) contributes the most toward the overall noise figure. The noise introduced in the first stage is amplified by each of the succeeding amplifiers. Therefore, when compared to the noise introduced in the first stage, the noise added by each succeeding amplifier is effectively reduced by a factor equal to the product of the power gains of the preceding amplifiers.

When precise noise calculations (0.1 dB or less) are necessary, it is generally more convenient to express noise figure in terms of noise temperature or equivalent noise temperature rather than as an absolute power (Chapter 19). Because noise power (*N*) is proportional to temperature, the noise present at the input to a device can be expressed as a function of the device's environmental temperature (*T*) and its equivalent noise temperature

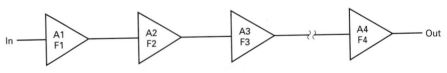

Figure 17-14 Total noise figure.

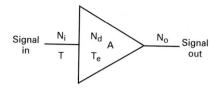

Figure 17-15 Noise figure as a function of temperature.

(T_e). Noise figure can be converted to a term dependent on temperature only as follows (refer to Figure 17-15).

Let

$$N_d = \text{noise power added by a single amplifier, referred to its input}$$

Then

$$N_d = KT_eB$$

where T_e is the equivalent noise temperature. Let

$$N_o = \text{total output noise power of an amplifier}$$
$$N_i = \text{total input noise power of an amplifier}$$
$$A = \text{power gain of an amplifier}$$

Therefore,

$$N_o \quad \text{may be expressed as}$$
$$N_o = AN_i + AN_d$$

and

$$N_o = AKTB + AKT_eB$$

Simplifying yields

$$N_o = AKB(T + T_e)$$

and the overall noise figure (NF_T) equals

$$NF_T = \frac{(S/N)_{\text{in}}}{(S/N)_{\text{out}}} = \frac{S/N_i}{AS/N_o} = \frac{N_o}{AN_i} = \frac{AKB(T + T_e)}{AKTB} \tag{17-7}$$

$$= \frac{T + T_e}{T} = 1 + \frac{T_e}{T}$$

EXAMPLE 17-5

In Figure 17-14, let $NF_1 = NF_2 = NF_3 = 3$ dB and $A_1 = A_2 = A_3 = 10$ dB. Solve for the total noise figure.

Solution Substituting into equation 17-6 (*Note:* All gains and noise figures have been converted to absolute values) yields

$$NF_T = NF_1 + \frac{NF_2 - 1}{A_1} + \frac{NF_3 - 1}{A_1A_2}$$

$$= 2 + \frac{2 - 1}{10} + \frac{2 - 1}{100}$$

$$= 2.11 \text{ or } 10 \log 2.11 = 3.24 \text{ dB}$$

An overall noise figure of 3.24 dB indicates that the *S/N* ratio at the output of A3 is 3.24 dB less than the *S/N* ratio at the input to A1.

System Gain

The noise figure of a receiver must be considered when determining C_{min}. The noise figure is included in the system gain equation as an equivalent loss. (Essentially, a gain in the total noise power is equivalent to a corresponding loss in the signal power.)

EXAMPLE 17-6

Refer to Figure 17-16. For a system gain of 112 dB, a total noise figure of 6.5 dB, an input noise power of -104 dBm, and a minimum $(S/N)_{out}$ of the FM demodulator of 32 dB, determine the minimum receive carrier power and the minimum transmit power.

Solution To achieve a S/N ratio of 32 dB out of the FM demodulator, an input C/N of 15 dB is required (17 dB of improvement due to FM quieting). Solving for the receiver input carrier-to-noise ratio gives

$$\frac{C_{min}}{N} = \frac{C}{N} + NF_T = 15 \text{ dB} + 6.5 \text{ dB} = 21.5 \text{ dB}$$

Thus

$$C_{min} = \frac{C_{min}}{N} + N$$

$$= 21.5 \text{ dB} + (-104 \text{ dBm}) = -82.5 \text{ dBm}$$

$$P_t = G_s + C_{min}$$

$$= 112 \text{ dB} + (-82.5 \text{ dBm}) = 29.5 \text{ dBm}$$

EXAMPLE 17-7

For the system shown in Figure 17-17, determine the following: G_s, C_{min}/N, C_{min}, N, G_s, and P_t.

Solution The minimum C/N at the input to the FM receiver is 23 dB.

$$\frac{C_{min}}{N} = \frac{C}{N} + NF_T$$

$$= 23 \text{ dB} + 4.24 \text{ dB} = 27.24 \text{ dB}$$

Substituting into equation 17-5 yields

$$N = -174 \text{ dBm} + 10 \log B$$

$$= -174 \text{ dBm} + 68 \text{ dB} = -106 \text{ dBm}$$

$$C_{min} = \frac{C_{min}}{N} + N$$

$$= 27.24 \text{ dB} + (-106 \text{ dBm}) = -78.76 \text{ dBm}$$

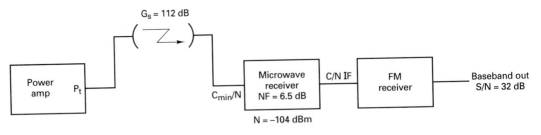

Figure 17-16 System gain example.

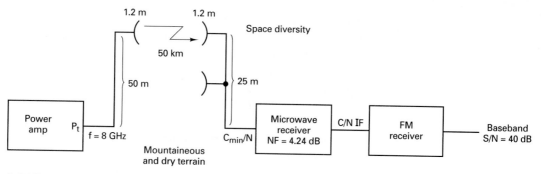

Reliability objective = 99.999%
Bandwidth = 6.3 MHz

Figure 17-17 System gain example.

Substituting into equation 17-4 gives us

$$F_m = 30 \log 50 + 10 \log [(6)\,(0.25)\,(0.125)\,(8)]$$

$$= -10 \log (1 - 0.99999) - 70$$

$$= 32.76 \text{ dB}$$

Substituting into equation 17-3, we have

$$L_p = 92.4 \text{ dB} + 20 \log 8 + 20 \log 50$$

$$= 92.4 \text{ dB} + 18.06 \text{ dB} + 33.98 \text{ dB} = 144.44 \text{ dB}$$

From Table 17-1,

$$L_b = 4 \text{ dB}$$

$$L_f = 0.75\,(6.5 \text{ dB}) = 4.875 \text{ dB}$$

$$A_t = A_r = 37.8 \text{ dB}$$

Note: The gain of an antenna increases or decreases proportional to the square of its diameter (i.e., if its diameter changes by a factor of 2, its gain changes by a factor of 4 which is 6 dB).
 Substituting into equation 17-1 yields

$$G_s = 32.76 + 144.44 + 4.875 + 4 - 37.8 - 37.8 = 110.475 \text{ dB}$$

$$P_t = G_s + C_{min}$$

$$= 110.475 \text{ dB} + (-78.76 \text{ dBm}) = 31.715 \text{ dBm}$$

QUESTIONS

17-1. What constitutes a short-haul microwave system? A long-haul microwave system?

17-2. Describe the baseband signal for a microwave system.

17-3. Why do FDM/FM microwave systems use low-index FM?

17-4. Describe a microwave repeater. Contrast baseband and IF repeaters.

17-5. Define *diversity*. Describe the three most commonly used diversity schemes.

17-6. Describe a protection switching arrangement. Contrast the two types of protection switching arrangements.

17-7. Briefly describe the four major sections of a microwave terminal station.

Questions

717

17-8. Define *ringaround.*

17-9. Briefly describe a high/low microwave system.

17-10. Define *system gain.*

17-11. Define the following terms: free-space path loss, branching loss, and feeder loss.

17-12. Define *fade margin.* Describe multipath losses, terrain sensitivity, and reliability objectives and how they affect fade margin.

17-13. Define *receiver threshold.*

17-14. Contrast carrier-to-noise ratio and signal-to-noise ratio.

17-15. Define *noise figure.*

PROBLEMS

17-1. Calculate the noise power at the input to a receiver that has a radio carrier frequency of 4 GHz and a bandwidth of 30 MHz (assume room temperature).

17-2. Determine the path loss for a 3.4-GHz signal propagating 20,000 m.

17-3. Determine the fade margin for a 60-km microwave hop. The RF carrier frequency is 6 GHz, the terrain is very smooth and dry, and the reliability objective is 99.95%.

17-4. Determine the noise power for a 20-MHz bandwidth at the input to a receiver with an input noise temperature of 290°C.

17-5. For a system gain of 120 dB, a minimum input C/N of 30 dB, and an input noise power of -115 dBm, determine the minimum transmit power (P_t).

17-6. Determine the amount of loss attributed to a reliability objective of 99.98%.

17-7. Determine the terrain sensitivity loss for a 4-GHz carrier that is propagating over a very dry, mountainous area.

17-8. A frequency-diversity microwave system operates at an RF carrier frequency of 7.4 GHz. The IF is a low-index frequency-modulated subcarrier. The baseband signal is the 1800-channel FDM system described in Chapter 16 (564 to 8284 kHz). The antennas are 4.8-m-diameter parabolic dishes. The feeder lengths are 150 m at one station and 50 m at the other station. The reliability objective is 99.999%. The system propagates over an average terrain that has a very dry climate. The distance between stations is 50 km. The minimum carrier-to-noise ratio at the receiver input is 30 dB. Determine the following: fade margin, antenna gain, free-space path loss, total branching and feeder losses, receiver input noise power, C_{min}, minimum transmit power, and system gain.

17-9. Determine the overall noise figure for a receiver that has two RF amplifiers each with a noise figure of 6 dB and a gain of 10 dB, a mixer down-converter with a noise figure of 10 dB, and a conversion gain of -6 dB, and 40 dB of IF gain with a noise figure of 6 dB.

17-10. A microwave receiver has a total input noise power of -102 dBm and an overall noise figure of 4 dB. For a minimum C/N ratio of 20 dB at the input to the FM detector, determine the minimum receive carrier power.

INTRODUCTION

In the early 1960s, the American Telephone and Telegraph Company (AT&T) released studies indicating that a few powerful satellites of advanced design could handle more traffic than the entire AT&T long-distance communications network. The cost of these satellites was estimated to be only a fraction of the cost of equivalent terrestrial microwave facilities. Unfortunately, because AT&T was a utility, government regulations prevented them from developing the satellite systems. Smaller and much less lucrative corporations were left to develop the satellite systems, and AT&T continued to invest billions of dollars each year in conventional terrestrial microwave systems. Because of this, early developments in satellite technology were slow in coming.

Throughout the years the prices of most goods and services have increased substantially; however, satellite communications services have become more affordable each year. In most instances, satellite systems offer more flexibility than submarine cables, buried underground cables, line-of-sight microwave radio, tropospheric scatter radio, or optical fiber systems.

Essentially, a communications satellite is a radio repeater in the sky (*transponder*). A satellite system consists of a transponder, a ground-based station to control its operation, and a user network of earth stations that provide the facilities for transmission and reception of communications traffic through the satellite system. Satellite transmissions are categorized as either *bus* or *payload*. The bus includes control mechanisms that support the payload operation. The payload is the actual user information that is conveyed through the system. Although in recent years new data services and television broadcasting are more and more in demand, the transmission of conventional speech telephone signals (in analog or digital form) is still the bulk of the satellite payload.

HISTORY OF SATELLITES

The simplest type of satellite is a *passive reflector,* a device that simply "bounces" a signal from one place to another. The moon is a natural satellite of the earth and, consequently, in the late 1940s and early 1950s, became the first passive satellite. In 1954, the U.S. Navy successfully transmitted the first messages over this earth-to-moon-to-earth relay. In 1956, a relay service was established between Washington, D.C. and Hawaii and, until 1962, offered reliable long-distance communications. Service was limited only by the availability of the moon.

In 1957, Russia launched *Sputnik I,* the first *active* earth satellite. An active satellite is capable of receiving, amplifying, and retransmitting information to and from earth stations. *Sputnik I* transmitted telemetry information for 21 days. Later in the same year, the United States launched *Explorer I,* which transmitted telemetry information for nearly 5 months.

In 1958, NASA launched *Score,* a 150-pound conical-shaped satellite. With an on-board tape recording, *Score* rebroadcast President Eisenhower's 1958 Christmas message. *Score* was the first artificial satellite used for relaying terrestrial communications. *Score* was a *delayed repeater satellite*; it received transmissions from earth stations, stored them on magnetic tape, and rebroadcast them to ground stations farther along in its orbit.

In 1960, NASA in conjunction with Bell Telephone Laboratories and the Jet Propulsion Laboratory launched *Echo,* a 100-ft-diameter plastic balloon with an aluminum coating. *Echo* passively reflected radio signals from a large earth antenna. *Echo* was simple and reliable but required extremely high power transmitters at the earth stations. The first transatlantic transmission using a satellite was accomplished using *Echo.* Also in 1960, the Department of Defense launched *Courier. Courier* transmitted 3 W of power and lasted only 17 days.

In 1962, AT&T launched *Telstar I,* the first satellite to receive and transmit simultaneously. The electronic equipment in *Telstar I* was damaged by radiation from the newly discovered Van Allen belts and, consequently, lasted only a few weeks. *Telstar II* was electronically identical to *Telstar I,* but it was made more radiation resistant. *Telstar II* was successfully launched in 1963. It was used for telephone, television, facsimile, and data transmissions. The first successful transatlantic transmission of video was accomplished with *Telstar II.*

Early satellites were both of the passive and active type. Again, a passive satellite is one that simply reflects a signal back to earth; there are no gain devices on board to amplify or repeat the signal. An active satellite is one that electronically repeats a signal back to earth (i.e., receives, amplifies, and retransmits the signal). An advantage of passive satellites is that they do not require sophisticated electronic equipment on board, although they are not necessarily void of power. Some passive satellites require a *radio beacon transmitter* for tracking and ranging purposes. A beacon is a continuously transmitted unmodulated carrier that an earth station can lock onto and use to align its antennas or to determine the exact location of the satellite. A disadvantage of passive satellites is their inefficient use of transmitted power. With *Echo,* for example, only 1 part in every 10^{18} of the earth station transmitted power was actually returned to the earth station receiving antenna.

ORBITAL SATELLITES

The satellites mentioned thus far are called *orbital* or *nonsynchronous* satellites. Nonsynchronous satellites rotate around the earth in a low-altitude elliptical or circular pattern. If the satellite is orbiting in the same direction as Earth's rotation and at an angular velocity greater than that of Earth, the orbit is called a *prograde orbit.* If the satellite is orbiting in the opposite direction as Earth's rotation or in the same direction but at an angular velocity less than that of Earth, the orbit is called a *retrograde orbit.* Consequently, nonsynchronous satellites are continuously either gaining or falling back on earth and do not remain stationary relative to any particular point on earth. Thus nonsynchronous satellites have to be used when available, which may be as short a period of time as 15 minutes per orbit. Another disadvantage of orbital satellites is the need for complicated and expensive tracking equipment at the earth stations. Each earth station must locate the satellite as it comes into view on each orbit and then lock its antenna onto the satellite and track it as it passes overhead. A major advantage of orbital satellites is that propulsion rockets are not required on board the satellites to keep them in their respective orbits.

One of the more interesting orbital satellite systems is the Soviet *Molniya* system. This is also spelled *Molnya* and *Molnia,* which means "lightning" in Russian (in colloquial Russian it means "news flash"). The Molniya satellites are used for television broadcasting and are presently the only nonsynchronous-orbit commercial satellite system in use. Molniya uses a highly elliptical orbit with *apogee* at about 40,000 km and *perigee* at about 1000 km (see Figure 18-1). The apogee is the farthest distance from earth a satellite orbit reaches, the perigee is the minimum distance, and the *line of apsides* is the line joining the perigee and apogee through the center of the earth. With the *Molniya* system, the apogee is reached while over the northern hemisphere and the perigee while over the southern hemisphere. The size of the ellipse was chosen to make its period exactly one-half of a sidereal day (the time it takes the earth to rotate back to the same constellation). Because of its unique orbital pattern, the *Molniya* satellite is synchronous with the rotation of the earth. During its 12-h orbit, it spends about 11 h over the north hemisphere.

GEOSTATIONARY SATELLITES

Geostationary or *geosynchronous* satellites are satellites that orbit in a circular pattern with an angular velocity equal to that of earth. Consequently, they remain in a fixed position in respect to a given point on earth. An obvious advantage is they are available to all the earth stations within their *shadow* 100% of the time. The shadow of a satellite includes all earth stations that have a line-of-sight path to it and lie within the radiation pattern of the satellite's antennas. An obvious disadvantage is they require sophisticated and heavy propulsion devices on board to keep them in a fixed orbit. The orbital time of a geosynchronous satellite is 24 h, the same as earth.

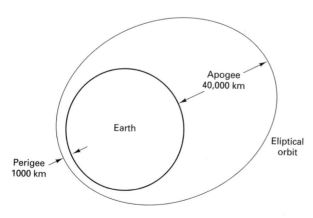

Figure 18-1 Soviet *Molniya* satellite orbit.

Syncom I, launched in February 1963, was the first attempt to place a geosynchronous satellite into orbit. *Syncom I* was lost during orbit injection. *Syncom II* and *Syncom III* were successfully launched in February 1963 and August 1964, respectively. The *Syncom III* satellite was used to broadcast the 1964 Olympic Games from Tokyo. The *Syncom* projects demonstrated the feasibility of using geosynchronous satellites.

Since the *Syncom* projects, a number of nations and private corporations have successfully launched satellites that are currently being used to provide national as well as regional and international global communications. There are more than 200 satellite communications systems operating in the world today. They provide worldwide fixed common-carrier telephone and data circuits; point-to-point cable television (CATV); network television distribution; music broadcasting; mobile telephone service; and private networks for corporations, governmental agencies, and military applications.

In 1964 a commercial global satellite network known as *Intelsat* (International Telecommunications Satellite Organization) was established. Intelsat is owned and operated by a consortium of more than 100 countries. Intelsat is managed by the designated communications entities in their respective countries. The first Intelsat satellite was *Early Bird 1*, which was launched in 1965 and provided 480 voice channels. From 1966 to 1987, a series of satellites designated *Intelsat II, III, IV, V,* and *VI* were launched. *Intelsat VI* has a capacity of 80,000 voice channels.

Domestic satellites (*domsats*) are used to provide satellite services within a single country. In the United States, all domsats are situated in geostationary orbit. Table 18-1 is a partial list of current international and domestic satellite systems and their primary payload.

ORBITAL PATTERNS

Once projected, a satellite remains in orbit because the centrifugal force caused by its rotation around the earth is counterbalanced by the earth's gravitational pull. The closer to earth the satellite rotates, the greater the gravitational pull and the greater the velocity required to keep it from being pulled to earth. Low-altitude satellites that orbit close to earth (100 to 300 miles in height) travel at approximately 17,500 miles per hour. At this speed, it takes approximately $1\frac{1}{2}$ h to rotate around the entire earth. Consequently, the time that the satellite is in line of sight of a particular earth station is only $\frac{1}{4}$ h or less per orbit. Medium-altitude satellites (6000 to 12,000 miles in height) have a rotation period of 5 to 12 h and remain in line of sight of a particular earth station for 2 to 4 h per orbit. High-altitude, geosynchronous satellites (19,000 to 25,000 miles in height) travel at approximately 6879 miles per hour and have a rotation period of 24 h, exactly the same as the earth. Consequently, they remain in a *fixed* position in respect to a given earth station and have a 24-h availability time. Figure 18-2 shows a low-, medium-, and high-altitude satellite orbit. It can be seen that three equally spaced, high-altitude geosynchronous satellites rotating around the earth above the equator can cover the entire earth except for the unpopulated areas of the north and south poles.

Figure 18-3 shows the three paths that a satellite may take as it rotates around the earth. When the satellite rotates in an orbit above the equator, it is called an *equatorial orbit*. When the satellite rotates in an orbit that takes it over the north and south poles, it is called a *polar orbit*. Any other orbital path is called an *inclined orbit*. An *ascending node* is the point where the orbit crosses the equatorial plane going from south to north, and a *descending node* is the point where the orbit crosses the equatorial plane going from north to south. The line joining the ascending and descending nodes through the center of earth is called the *line of nodes*.

TABLE 18-1 CURRENT SATELLITE COMMUNICATIONS SYSTEMS

	Characteristic system				
	Westar	*Intelsat V*	*SBS*	*Fleet-satcom*	*ANIK-D*
Operator	Western Union Telegraph	Intelsat	Satellite Business Systems	U.S. Dept. of Defense	Telsat Canada
Frequency band	C	C and Ku	Ku	UHF, X	C, Ku
Coverage	Conus	Global, zonal, spot	Conus	Global	Canada, northern U.S.
Number of transponders	12	21	10	12	24
Transponder BW (MHz)	36	36–77	43	0.005–0.5	36
EIRP (dBW)	33	23.5–29	40–43.7	26–28	36
Multiple Access	FDMA, TDMA	FDMA, TDMA, reuse	TDMA	FDMA	FDMA
Modulation	FM, QPSK	FDM/FM, QPSK	QPSK	FM, QPSK	FDM, FM, FM/TVD, SCPC
Service	Fixed tele, TTY	Fixed tele, TVD	Fixed tele, TVD	Mobile military	Fixed tele

C-band: 3.4–6.425 GHz
Ku-band: 10.95–14.5 GHz
X-band: 7.25–8.4 GHz

TTY	Teletype
TVD	TV distribution
FDMA	Frequency-division multiple access
TDMA	Time-division multiple access
Conus	Continental United States

It is interesting to note that 100% of the earth's surface can be covered with a single satellite in a polar orbit. The satellite is rotating around the earth in a longitudinal orbit while the earth is rotating on a latitudinal axis. Consequently, the satellite's radiation pattern is a diagonal spiral around the earth which somewhat resembles a barber pole. As a result, every location on earth lies within the radiation pattern of the satellite twice each day.

SUMMARY

Advantages of Geosynchronous Orbits

1. The satellite remains almost stationary in respect to a given earth station. Consequently, expensive tracking equipment is not required at the earth stations.

2. There is no need to switch from one satellite to another as they orbit overhead. Consequently, there are no breaks in transmission because of the switching times.

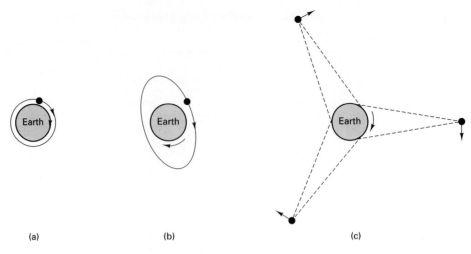

Figure 18-2 Satellite orbits: (a) low altitude (circular orbit, 100-300 mi);
(b) medium altitude (elliptical orbit, 6000 to 12,000 mi); (c) high altitude
(geosynchronous orbit, 19,000 to 25,000 mi).

3. High-altitude geosynchronous satellites can cover a much larger area of the earth
than their low-altitude orbital counterparts.

4. The effects of Doppler shift are negligible.

Disadvantages of Geosynchronous Orbits

1. The higher altitudes of geosynchronous satellites introduce much longer propagation times. The round-trip propagation delay between two earth stations through a geosynchronous satellite is 500 to 600 ms.

2. Geosynchronous satellites require higher transmit powers and more sensitive
receivers because of the longer distances and greater path losses.

3. High-precision spacemanship is required to place a geosynchronous satellite into
orbit and to keep it there. Also, propulsion engines are required on board the satellites to
keep them in their respective orbits.

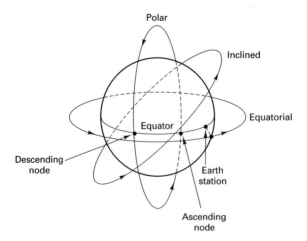

Figure 18-3 Satellite orbits.

To orient an earth station antenna toward a satellite, it is necessary to know the *elevation angle* and *azimuth* (Figure 18-4). These are called the *look angles.*

Angle of Elevation

The angle of elevation is the angle formed between the direction of travel of a wave radiated from an earth station antenna and the horizontal, or the angle subtended at the earth station antenna between the satellite and the horizontal. The smaller the angle of elevation, the greater the distance a propagated wave must pass through Earth's atmosphere. As with any wave propagated through Earth's atmosphere, it suffers absorption and may also be severely contaminated by noise. Consequently, if the angle of elevation is too small and the distance the wave is within Earth's atmosphere is too long, the wave may deteriorate to a degree that it provides inadequate transmission. Generally, 5° is considered as the minimum acceptable angle of elevation. Figure 18-5 shows how the angle of elevation affects the signal strength of a propagated wave due to normal atmospheric absorption, absorption due to thick fog, and absorption due to a heavy rain. It can be seen that the 14/12-GHz

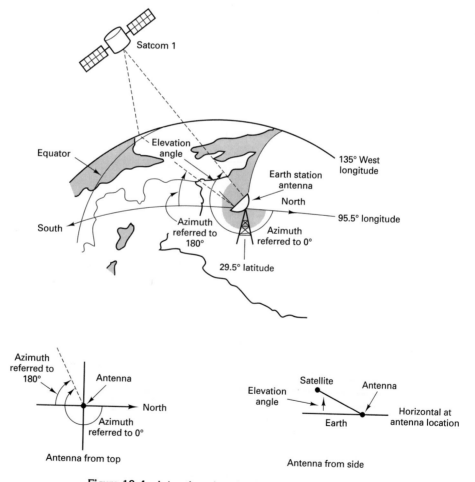

Figure 18-4 Azimuth and angle of elevation "look angles."

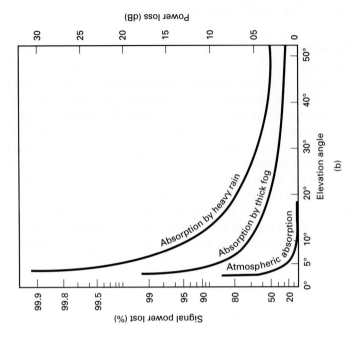

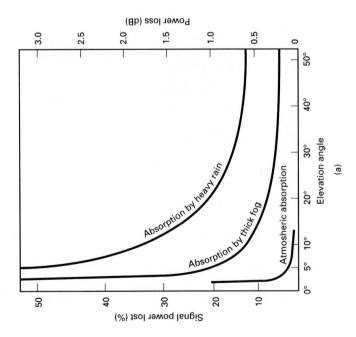

Figure 18-5 Attenuation due to atmospheric absorption: (a) 6/4-GHz band; (b) 14/12-GHz band.

726

band (Figure 18-5b) is more severely affected than the 6/4-GHz band (Figure 18-5a). This is due to the smaller wavelengths associated with the higher frequencies. Also, at elevation angles less than 5°, the attenuation increases rapidly.

Azimuth

Azimuth is defined as the horizontal pointing angle of an antenna. It is usually measured in a clockwise direction in degrees from true north. The angle of elevation and the azimuth both depend on the latitude of the earth station and the longitude of both the earth station and the orbiting satellite. For a geosynchronous satellite in an equatorial orbit, the procedure is as follows: From a good map, determine the longitude and latitude of the earth station. From Table 18-2, determine the longitude of the satellite of interest. Calculate the difference, in degrees (ΔL), between the longitude of the satellite and the longitude of the earth station. Then, from Figure 18-6, determine the azimuth and elevation angle for the antenna. Figure 18-6 is for a geosynchronous satellite in an equatorial orbit.

EXAMPLE 18-1

An earth station is located at Houston, Texas, which has a longitude of 95.5°W and a latitude of 29.5°N. The satellite of interest is RCA's *Satcom 1*, which has a longitude of 135°W. Determine the azimuth and elevation angle for the earth station antenna.

Solution First determine the difference between the longitude of the earth station and the satellite.

$$\Delta L = 135° - 95.5° = 39.5°$$

Locate the intersection of ΔL and the latitude of the earth station on Figure 18-6. From the figure the angle of elevation is approximately 35°, and the azimuth is approximately 59° west of south.

TABLE 18-2
LONGITUDINAL
POSITION OF
SEVERAL CURRENT
SYNCHRONOUS
SATELLITES
PARKED IN AN
EQUATORIAL ARC[a]

Satellite	Longitude (°W)
Satcom I	135
Satcom V	143
ANIK I	104
Westar I	99
Westar II	123.5
Westar III	91
Westar IV	98.5
Westar V	119.5
RCA	126
Mexico	116.5
Galaxy	74
Telstar	96

[a]0° Latitude.

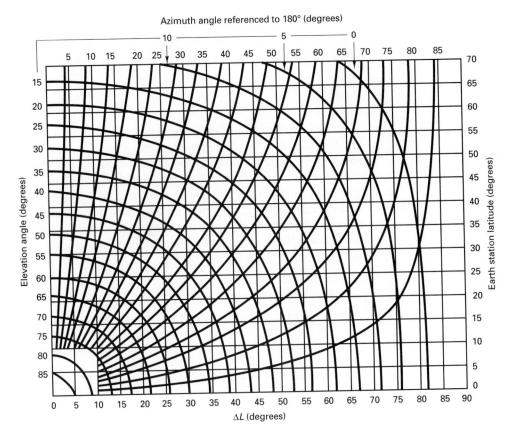

Azimuth angle referenced to 180° (degrees)

Figure 18-6 Azimuth and elevation angle for earth stations located in the northern hemisphere (referred to 180°).

ORBITAL CLASSIFICATIONS, SPACING, AND FREQUENCY ALLOCATION

There are two primary classifications for communications satellites: *spinners* and *three-axis stabilizer satellites*. Spinner satellites use the angular momentum of its spinning body to provide roll and yaw stabilization. With a three-axis stabilizer, the body remains fixed relative to Earth's surface while an internal subsystem provides roll and yaw stabilization. Figure 18-7 shows the two main classifications of communications satellites.

Geosynchronous satellites must share a limited space and frequency spectrum within a given arc of a geostationary orbit. Each communications satellite is assigned a longitude in the geostationary arc approximately 22,300 miles above the equator. The position in the slot depends on the communications frequency band used. Satellites operating at or near the same frequency must be sufficiently separated in space to avoid interfering with each other (Figure 18-8). There is a realistic limit to the number of satellite structures that can be stationed (*parked*) within a given area in space. The required *spatial separation* is dependent on the following variables:

1. Beamwidths and sidelobe radiation of both the earth station and satellite antennas
2. RF carrier frequency
3. Encoding or modulation technique used

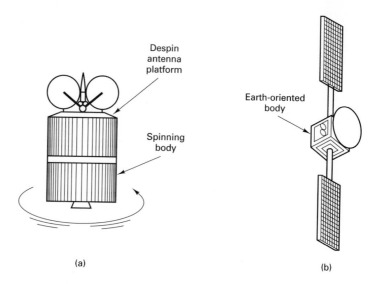

Figure 18-7 Satellite classes: (a) spinner; (b) three-axis stabilized.

4. Acceptable limits of interference
5. Transmit carrier power

Generally, 3 to 6° of spatial separation is required depending on the variables stated above.

The most common carrier frequencies used for satellite communications are the 6/4- and 14/12-GHz bands. The first number is the up-link (earth station-to-transponder) frequency, and the second number is the down-link (transponder-to-earth station) frequency. Different up-link and down-link frequencies are used to prevent ringaround from occurring (Chapter 17). The higher the carrier frequency, the smaller the diameter required of an antenna for a given gain. Most domestic satellites use the 6/4-GHz band. Unfortunately, this band is also used extensively for terrestrial microwave systems. Care must be

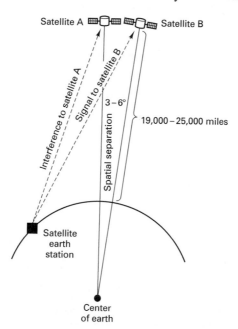

Figure 18-8 Spatial separation of satellites in geosynchronous orbit.

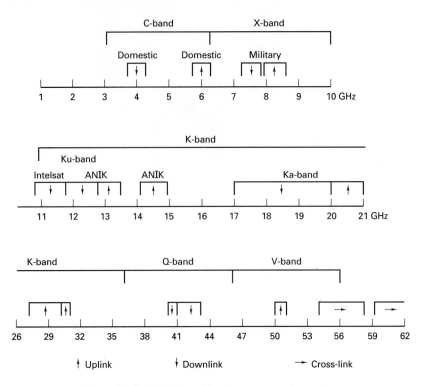

Figure 18-9 WARC satellite frequency assignments.

taken when designing a satellite network to avoid interference from or interference with established microwave links.

Certain positions in the geosynchronous orbit are in higher demand than the others. For example, the mid-Atlantic position which is used to interconnect North America and Europe is in exceptionally high demand. The mid-Pacific position is another.

The frequencies allocated by WARC (World Administrative Radio Conference) are summarized in Figure 18-9. Table 18-3 shows the bandwidths available for various services in the United States. These services include *fixed-point* (between earth stations located at

TABLE 18-3 SATELLITE BANDWIDTHS AVAILABLE IN THE UNITED STATES

Band	Up-link	Down-link	Bandwidth (MHz)
	Frequency band (GHz)		
C	5.9–6.4	3.7–4.2	500
X	7.9–8.4	7.25–7.75	500
Ku	14–14.5	11.7–12.2	500
Ka	27–30	17–20	—
	30–31	20–21	—
V	50–51	40–41	1000
Q	—	41–43	2000
V	54–58		3900
(ISL)	59–64		5000

fixed geographical points on earth), *broadcast* (wide-area coverage), *mobile* (ground-to-aircraft, ships, or land vehicles), and *intersatellite* (satellite-to-satellite cross-links).

RADIATION PATTERNS: FOOTPRINTS

The area of the earth covered by a satellite depends on the location of the satellite in its geosynchronous orbit, its carrier frequency, and the gain of its antennas. Satellite engineers select the antenna and carrier frequency for a particular spacecraft to concentrate the limited transmitted power on a specific area of Earth's surface. The geographical representation of a satellite antenna's radiation pattern is called a *footprint* (Figure 18-10). The contour lines represent limits of equal receive power density.

The radiation pattern from a satellite antenna may be categorized as either *spot, zonal,* or *earth* (Figure 18-11). The radiation patterns of earth coverage antennas have a beamwidth of approximately 17° and include coverage of approximately one-third of the earth's surface. Zonal coverage includes an area less than one-third of the earth's surface. Spot beams concentrate the radiated power in a very small geographic area.

Reuse

When an allocated frequency band is filled, additional capacity can be achieved by *reuse* of the frequency spectrum. By increasing the size of an antenna (i.e., increasing the antenna gain) the beamwidth of the antenna is also reduced. Thus different beams of the same frequency can be directed to different geographical areas of the earth. This is called frequency reuse. Another method of frequency reuse is to use dual polarization. Different information signals can be transmitted to different earth station receivers using the same band of frequencies simply by orienting their electromagnetic polarizations in an orthogonal manner (90° out of phase). Dual polarization is less effective because Earth's atmosphere has a tendency to reorient or repolarize an electromagnetic wave as it passes through. Reuse is simply another way to increase the capacity of a limited bandwidth.

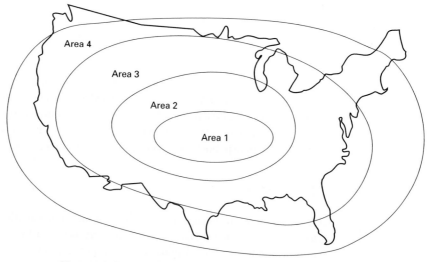

Figure 18-10 Satellite antenna radiation patterns ("footprints").

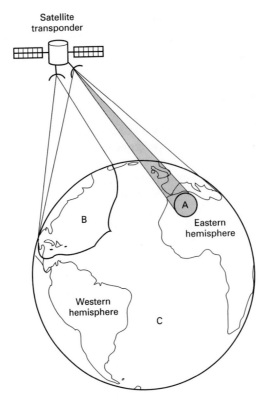

Figure 18-11 Beams: A, spot; B, zonal; C, earth.

SATELLITE SYSTEM LINK MODELS

Essentially, a satellite system consists of three basic sections: an uplink, a satellite transponder, and a downlink.

Uplink Model

The primary component within the *uplink* section of a satellite system is the earth station transmitter. A typical earth station transmitter consists of an IF modulator, an IF-to-RF microwave up-converter, a high-power amplifier (HPA), and some means of bandlimiting the final output spectrum (i.e., an output bandpass filter). Figure 18-12 shows the block diagram of a satellite earth station transmitter. The IF modulator converts the input baseband signals to either an FM, a PSK, or a QAM modulated intermediate frequency. The up-converter (mixer and bandpass filter) converts the IF to an appropriate RF carrier frequency. The HPA provides adequate input sensitivity and output power to propagate the signal to the satellite transponder. HPAs commonly used are klystons and traveling-wave tubes.

Transponder

A typical *satellite transponder* consists of an input bandlimiting device (BPF), an input *low-noise amplifier* (LNA), a *frequency translator,* a low-level power amplifier, and an output bandpass filter. Figure 18-13 shows a simplified block diagram of a satellite transponder. This transponder is an RF-to-RF repeater. Other transponder configurations

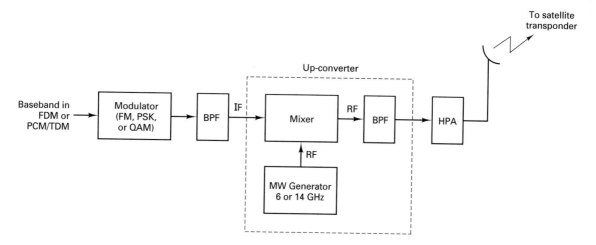

Figure 18-12   Satellite uplink model.

are IF and baseband repeaters similar to those used in microwave repeaters. In Figure 18-13, the input BPF limits the total noise applied to the input of the LNA. (A common device used as an LNA is a tunnel diode.) The output of the LNA is fed to a frequency translator (a shift oscillator and a BPF) which converts the high-band uplink frequency to the low-band downlink frequency. The low-level power amplifier, which is commonly a traveling-wave tube, amplifies the RF signal for transmission through the downlink to Earth station receivers. Each RF satellite channel requires a separate transponder.

Downlink Model

An earth station receiver includes an input BPF, an LNA, and an RF-to-IF down-converter. Figure 18-14 shows a block diagram of a typical earth station receiver. Again, the BPF limits the input noise power to the LNA. The LNA is a highly sensitive, low-noise device such as a tunnel diode amplifier or a parametric amplifier. The RF-to-IF down-converter is a mixer/bandpass filter combination which converts the received RF signal to an IF frequency.

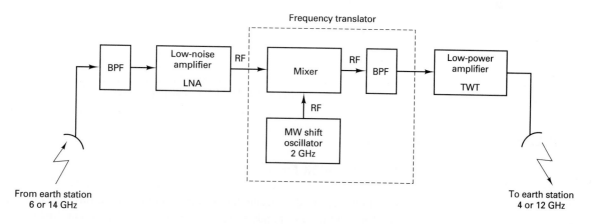

Figure 18-13   Satellite transponder.

Satellite System Link Models

733

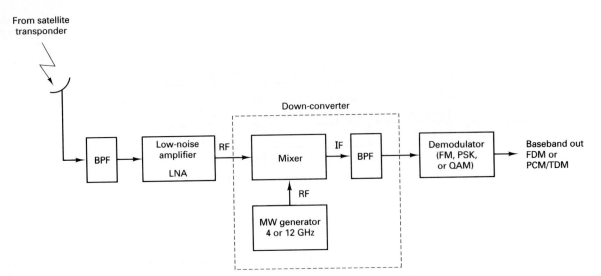

Figure 18-14 Satellite downlink model.

Cross-Links

Occasionally, there is an application where it is necessary to communicate between satellites. This is done using *satellite cross-links* or *intersatellite links* (ISLs), shown in Figure 18-15. A disadvantage of using an ISL is that both the transmitter and receiver are *space-bound*. Consequently, both the transmitter's output power and the receiver's input sensitivity are limited.

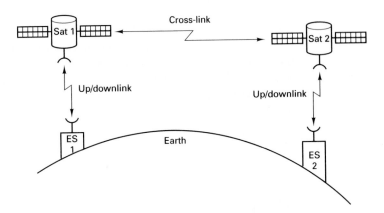

Figure 18-15 Intersatellite link.

SATELLITE SYSTEM PARAMETERS

Transmit Power and Bit Energy

High-power amplifiers used in earth station transmitters and the traveling-wave tubes typically used in satellite transponders are *nonlinear devices*; their gain (output power-versus-input power) is dependent on input signal level. A typical input/output power characteristic curve is shown in Figure 18-16. It can be seen that as the input power is reduced by 5 dB, the output power is reduced by only 2 dB. There is an obvious *power compression*. To reduce the amount of intermodulation distortion caused by the nonlinear amplification of the HPA, the input power must be reduced (*backed off*) by several dB. This allows the HPA to operate in a more *linear* region. The amount the output level is backed off from rated levels is equivalent to a loss and is appropriately called *back-off loss* (L_{bo}).

To operate as efficiently as possible, a power amplifier should be operated as close as possible to saturation. The *saturated output power* is designated P_o (sat) or simply P_t. The output power of a typical satellite earth station transmitter is much higher than the output power from a terrestrial microwave power amplifier. Consequently, when dealing with satellite systems, P_t is generally expressed in dBW (decibels in respect to 1 W) rather than in dBm (decibels in respect to 1 mW).

Most modern satellite systems use either phase shift keying (PSK) or quadrature amplitude modulation (QAM) rather than conventional frequency modulation (FM). With PSK and QAM, the input baseband is generally a PCM-encoded, time-division-multiplexed signal which is digital in nature. Also, with PSK and QAM, several bits may be encoded in a single transmit signaling element. Consequently, a parameter more meaningful than carrier power is *energy per bit* (E_b). Mathematically, E_b is

$$E_b = P_t T_b \qquad (18\text{-}1a)$$

where E_b = energy of a single bit (joules per bit)
   P_t = total carrier power (watts)
   T_b = time of a single bit (seconds)

or because $T_b = 1/f_b$, where f_b is the bit rate in bits per second.

$$E_b = \frac{P_t}{f_b} \qquad (18\text{-}1b)$$

EXAMPLE 18-2

For a total transmit power (P_t) of 1000 W, determine the energy per bit (E_b) for a transmission rate of 50 Mbps.

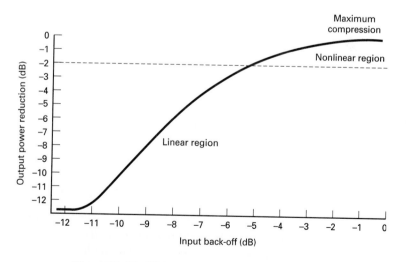

Figure 18-16 HPA input/output characteristic curve.

Solution

$$T_b = \frac{1}{f_b} = \frac{1}{50 \times 10^6 \text{ bps}} = 0.02 \times 10^{-6} \text{ s}$$

(It appears that the units for T_b should be s/bit but the per bit is implied in the definition of T_b, time of bit.)

Substituting into Equation 18-1a yields

$$E_b = 1000 \text{ J/s } (0.02 \times 10^{-6} \text{ s/bit}) = 20 \text{ μJ/bit}$$

(Again the units appear to be J/bit, but the per bit is implied in the definition of E_b, energy per bit.)

$$E_b = \frac{1000 \text{ J/s}}{50 \times 10^6 \text{ bps}} = 20 \text{ μJ}$$

Expressed as a log,

$$E_b = 10 \log (20 \times 10^{-6}) = -47 \text{ dBJ}$$

It is common to express P_t in dBW and E_b in dBW/bps. Thus

$$P_t = 10 \log 1000 = 30 \text{ dBW}$$

$$E_b = P_t - 10 \log f_b$$

$$= P_t - 10 \log (50 \times 10^6)$$

$$= 30 \text{ dBW} - 77 \text{ dB} = -47 \text{ dBW/bps}$$

or simply -47 dBW.

Effective Isotropic Radiated Power

Effective isotropic radiated power (EIRP) is defined as an equivalent transmit power and is expressed mathematically as

$$\text{EIRP} = P_r A_t$$

where   EIRP = effective isotropic radiated power (watts)
   P_r = total power radiated from an antenna (watts)
   A_t = transmit antenna gain (unitless ratio)

Expressed as a log,

$$\text{EIRP (dBW)} = P_r \text{ (dBW)} + A_t \text{ (dB)}$$

In respect to the transmitter output,

$$P_r = P_t - L_{bo} - L_{bf}$$

Thus

$$\text{EIRP} = P_t - L_{bo} - L_{bf} + A_t \tag{18-2}$$

where P_t = actual power output of the transmitter (dBW)
   L_{bo} = back-off losses of HPA (dB)
   L_{bf} = total branching and feeder loss (dB)
   A_t = transmit antenna gain (dB)

EXAMPLE 18-3

For an earth station transmitter with an output power of 40 dBW (10,000 W), a back-off loss of 3 dB, a total branching and feeder loss of 3 dB, and a transmit antenna gain of 40 dB, determine the EIRP.

Solution Substituting into Equation 18-2 yields

$$EIRP = P_t - L_{bo} - L_{bf} + A_t$$

$$= 40 \text{ dBW} - 3 \text{ dB} - 3 \text{ dB} + 40 \text{ dB} = 74 \text{ dBW}$$

Equivalent Noise Temperature

With terrestrial microwave systems, the noise introduced in a receiver or a component within a receiver was commonly specified by the parameter noise figure. In satellite communications systems, it is often necessary to differentiate or measure noise in increments as small as a tenth or a hundredth of a decibel. Noise figure, in its standard form, is inadequate for such precise calculations. Consequently, it is common to use *environmental temperature* (T) and *equivalent noise temperature* (T_e) when evaluating the performance of a satellite system. In Chapter 17 total noise power was expressed mathematically as

$$N = KTB$$

Rearranging and solving for T gives us

$$T = \frac{N}{KB}$$

where N = total noise power (watts)
K = Boltzmann's constant (joules per degree Kelvin)
B = bandwidth (hertz)
T = temperature of the environment (degree Kelvin)

Again from Chapter 17 (Equation 17-7),

$$NF = 1 + \frac{T_e}{T}$$

where T_e = equivalent noise temperature (degree Kelvin)
NF = noise figure expressed as an absolute value
T = temperature of the environment (degree Kelvin)

Rearranging Equation 17-7, we have

$$T_e = T(NF - 1)$$

Typically, equivalent noise temperatures of the receivers used in satellite transponders are about 1000 K. For earth station receivers T_e values are between 20 and 1000 K. Equivalent noise temperature is generally more useful when expressed logarithmically with the unit of dBK, as follows:

$$T_e \text{ (dBK)} = 10 \log T_e$$

For an equivalent noise temperature of 100 K, T_e (dBK) is

$$T_e \text{ (dBK)} = 10 \log 100 \text{ or } 20 \text{ dBK}$$

Satellite System Parameters

Equivalent noise temperature is a hypothetical value that can be calculated but cannot be measured. Equivalent noise temperature is often used rather than noise figure because it is a more accurate method of expressing the noise contributed by a device or a receiver when evaluating its performance. Essentially, equivalent noise temperature (T_e) represents the noise power present at the input to a device plus the noise added internally by that device. This allows us to analyze the noise characteristics of a device by simply evaluating an equivalent input noise temperature. As you will see in subsequent discussions, T_e is a very useful parameter when evaluating the performance of a satellite system.

EXAMPLE 18-4

Convert noise figures of 4 and 4.01 to equivalent noise temperatures. Use 300 K for the environmental temperature.

Solution Substituting into Equation 18-7 yields

$$T_e = T(\text{NF} - 1)$$

For NF = 4:

$$T_e = 300(4 - 1) = 900 \text{ K}$$

For NF = 4.01:

$$T_e = 300(4.01 - 1) = 903 \text{ K}$$

It can be seen that the 3° difference in the equivalent temperatures is 300 times as large as the difference between the two noise figures. Consequently, equivalent noise temperature is a more accurate way of comparing the noise performances of two receivers or devices.

Noise Density

Simply stated, *noise density* (N_0) is the total noise power normalized to a 1-Hz bandwidth, or the noise power present in a 1-Hz bandwidth. Mathematically, noise density is

$$N_0 = \frac{N}{B} \quad \text{or} \quad KT_e \tag{18-3a}$$

where N_0 = noise density (W/Hz) (N_0 is generally expressed as simply watts; the per hertz is implied in the definition of N_0)
N = total noise power (watts)
B = bandwidth (herts)
K = Boltzmann's constant (joules per degree Kelvin)
T_e = equivalent noise temperature (degree Kelvin)

Expressed as a log,

$$N_0 \text{ (dBW/Hz)} = 10 \log N - 10 \log B \tag{18-3b}$$

$$= 10 \log K + 10 \log T_e \tag{18-3c}$$

EXAMPLE 18-5

For an equivalent noise bandwidth of 10 MHz and a total noise power of 0.0276 pW, determine the noise density and equivalent noise temperature.

Solution Substituting into Equation 18-3a, we have

$$N_0 = \frac{N}{B} = \frac{276 \times 10^{-16} \text{ W}}{10 \times 10^6 \text{ Hz}} = 276 \times 10^{-23} \frac{\text{W}}{\text{Hz}}$$

or simply, 276×10^{-23} W.

$$N_0 = 10 \log (276 \times 10^{-23}) = -205.6 \text{ dBW/Hz}$$

or simply -205.6 dBW. Substituting into Equation 18-3b gives us

$$N_0 = N \text{ (dBW)} - B \text{ (dB/Hz)}$$

$$= -135.6 \text{ dBW} - 70 \text{ (dB/Hz)} = -205.6 \text{ dBW}$$

Rearranging Equation 18-3a and solving for equivalent noise temperature yields

$$T_e = \frac{N_0}{K}$$

$$= \frac{276 \times 10^{-23} \text{ J/cycle}}{1.38 \times 10^{-23} \text{ J/K}} = 200 \text{ K/cycle}$$

$$= 10 \log 200 = 23 \text{ dBK}$$

$$= N_0 \text{ (dBW)} - 10 \log K$$

$$= -205.6 \text{ dBW} - (-228.6 \text{ dBWK}) = 23 \text{ dBK}$$

Carrier-to-Noise Density Ratio

C/N_0 is the average wideband carrier power-to-noise density ratio. The *wideband carrier power* is the combined power of the carrier and its associated sidebands. The noise is the thermal noise present in a normalized 1-Hz bandwidth. The carrier-to-noise density ratio may also be written as a function of noise temperature. Mathematically, C/N_0 is

$$\frac{C}{N_0} = \frac{C}{KT_e} \qquad (18\text{-}4a)$$

Expressed as a log,

$$\frac{C}{N_0} \text{ (dB)} = C \text{ (dBW)} - N_0 \text{ (dBW)} \qquad (18\text{-}4b)$$

Energy of Bit-to-Noise Density Ratio

E_b/N_0 is one of the most important and most often used parameters when evaluating a digital radio system. The E_b/N_0 ratio is a convenient way to compare digital systems that use different transmission rates, modulation schemes, or encoding techniques. Mathematically, E_b/N_0 is

$$\frac{E_b}{N_0} = \frac{C/f_b}{N/B} = \frac{CB}{Nf_b} \qquad (18\text{-}5)$$

E_b/N_0 is a convenient term used for digital system calculations and performance comparisons, but in the real world, it is more convenient to measure the wideband carrier power-to-noise density ratio and convert it to E_b/N_0. Rearranging Equation 18-5 yields the following expression:

$$\frac{E_b}{N_0} = \frac{C}{N} \times \frac{B}{f_b}$$

The E_b/N_0 ratio is the product of the carrier-to-noise ratio (C/N) and the noise bandwidth-to-bit rate ratio (B/f_b). Expressed as a log,

$$\frac{E_b}{N_0} \text{ (dB)} = \frac{C}{N} \text{ (dB)} + \frac{B}{f_b} \text{ (dB)} \qquad (18\text{-}6)$$

The energy per bit (E_b) will remain constant as long as the total wideband carrier power (C) and the transmission rate (bps) remain unchanged. Also, the noise density (N_0) will remain constant as long as the noise temperature remains constant. The following conclusion can be made: For a given carrier power, bit rate, and noise temperature, the E_b/N_0 ratio will remain constant regardless of the encoding technique, modulation scheme, or bandwidth used.

Figure 18-17 graphically illustrates the relationship between an expected probability of error $P(e)$ and the minimum C/N ratio required to achieve the $P(e)$. The C/N specified is for the minimum double-sided Nyquist bandwidth. Figure 18-18 graphically illustrates the relationship between an expected $P(e)$ and the minimum E_b/N_0 ratio required to achieve that $P(e)$.

A $P(e)$ of 10^{-5} ($1/10^5$) indicates a probability that 1 bit will be in error for every 100,000 bits transmitted. $P(e)$ is analogous to the bit error rate (BER).

EXAMPLE 18-6

A coherent binary phase-shift-keyed (BPSK) transmitter operates at a bit rate of 20 Mbps. For a probability of error $P(e)$ of 10^{-4}:

(a) Determine the minimum theoretical C/N and E_b/N_0 ratios for a receiver bandwidth equal to the minimum double-sided Nyquist bandwidth.

(b) Determine the C/N if the noise is measured at a point prior to the bandpass filter, where the bandwidth is equal to twice the Nyquist bandwidth.

(c) Determine the C/N if the noise is measured at a point prior to the bandpass filter where the bandwidth is equal to three times the Nyquist bandwidth.

Solution (a) With BPSK, the minimum bandwidth is equal to the bit rate, 20 MHz. From Figure 18-17, the minimum C/N is 8.8 dB. Substituting into Equation 18-6 gives us

$$\frac{E_b}{N_0} \text{ (dB)} = \frac{C}{N} \text{ (dB)} + \frac{B}{f_b} \text{ (dB)}$$

$$= 8.8 \text{ dB} + 10 \log \frac{20 \times 10^6}{20 \times 10^6}$$

$$= 8.8 \text{ dB} + 0 \text{ dB} = 8.8 \text{ dB}$$

Note: The minimum E_b/N_0 equals the minimum C/N when the receiver noise bandwidth equals the minimum Nyquist bandwidth. The minimum E_b/N_0 of 8.8 can be verified from Figure 18-18.

What effect does increasing the noise bandwidth have on the minimum C/N and E_b/N_0 ratios? The wideband carrier power is totally independent of the noise bandwidth. Similarly, an increase in the bandwidth causes a corresponding increase in the noise power. Consequently, a decrease in C/N is realized that is directly proportional to the increase in the noise bandwidth. E_b is dependent on the wideband carrier power and the bit rate only. Therefore, E_b is unaffected by an increase in the noise bandwidth. N_0 is the noise power normalized to a 1-Hz bandwidth and, consequently, is also unaffected by an increase in the noise bandwidth.

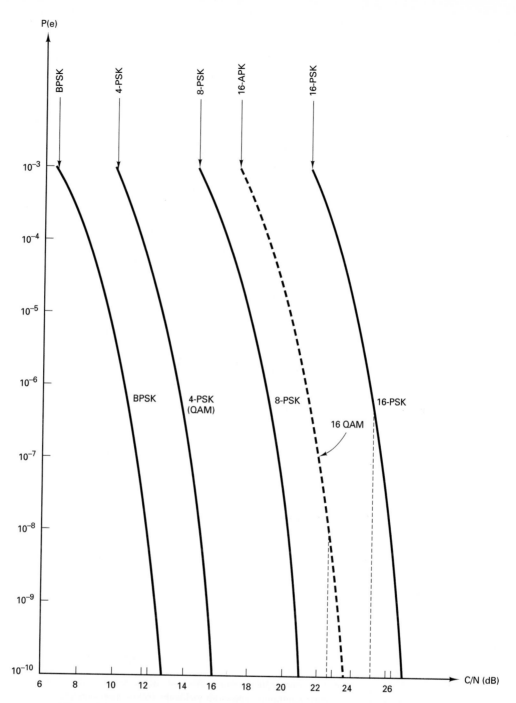

Figure 18-17 *P(e)* performance of *M*-ary PSK, QAM, QPR, and *M*-ary APK coherent systems. The rms *C/N* is specified in the double-sided Nyquist bandwidth.

(b) Since E_b/N_0 is independent of bandwidth, measuring the *C/N* at a point in the receiver where the bandwidth is equal to twice the minimum Nyquist bandwidth has absolutely no effect on E_b/N_0. Therefore, E_b/N_0 becomes the constant in Equation 18-6 and is used to solve for the new value of *C/N*. Rearranging Equation 18-6 and using the calculated E_b/N_0 ratio, we have

Satellite System Parameters

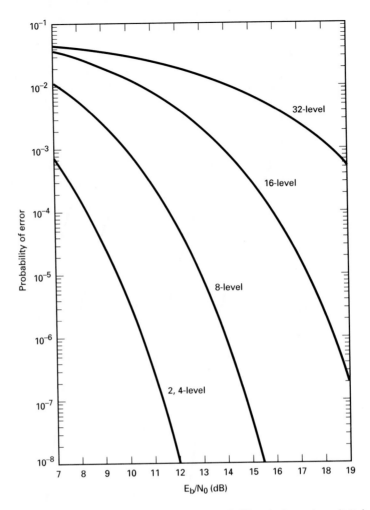

Figure 18-18 Probability of error $P(e)$ versus E_b/N_0 ratio for various digital modulation schemes.

$$\frac{C}{N} \text{ (dB)} = \frac{E_b}{N_0} \text{ (dB)} - \frac{B}{f_b} \text{ (dB)}$$

$$= 8.8 \text{ dB} - 10 \log \frac{40 \times 10^6}{20 \times 10^6}$$

$$= 8.8 \text{ dB} - 10 \log 2$$

$$= 8.8 \text{ dB} - 3 \text{ dB} = 5.8 \text{ dB}$$

(c) Measuring the *C/N* ratio at a point in the receiver where the bandwidth equals three times the minimum bandwidth yields the following results for *C/N*.

$$\frac{C}{N} = \frac{E_b}{N_0} - 10 \log \frac{60 \times 10^6}{20 \times 10^6}$$

$$= 8.8 \text{ dB} - 10 \log 3$$

$$= 4.03 \text{ dB}$$

The *C/N* ratios of 8.8, 5.8, and 4.03 dB indicate the *C/N* ratios that would be measured at the three specified points in the receiver to achieve the desired minimum E_b/N_0 and $P(e)$.

Because E_b/N_0 cannot be directly measured to determine the E_b/N_0 ratio, the wideband carrier-to-noise ratio is measured and then substituted into Equation 18-6. Consequently, to accurately determine the E_b/N_0 ratio, the noise bandwidth of the receiver must be known.

EXAMPLE 18-7

A coherent 8-PSK transmitter operates at a bit rate of 90 Mbps. For a probability of error of 10^{-5}:

(a) Determine the minimum theoretical C/N and E_b/N_0 ratios for a receiver bandwidth equal to the minimum double-sided Nyquist bandwidth.

(b) Determine the C/N if the noise is measured at a point prior to the bandpass filter where the bandwidth is equal to twice the Nyquist bandwidth.

(c) Determine the C/N if the noise is measured at a point prior to the bandpass filter where the bandwidth is equal to three times the Nyquist bandwidth.

Solution (a) 8-PSK has a bandwidth efficiency of 3 bps/Hz and, consequently, requires a minimum bandwidth of one-third the bit rate or 30 MHz. From Figure 18-17, the minimum C/N is 18.5 dB. Substituting into Equation 18-6, we obtain

$$\frac{E_b}{N_0} \text{ (dB)} = 18.5 \text{ dB} + 10 \log \frac{30 \text{ MHz}}{90 \text{ Mbps}}$$

$$= 18.5 \text{ dB} + (-4.8 \text{ dB}) = 13.7 \text{ dB}$$

(b) Rearranging Equation 18-6 and substituting for E_b/N_0 yields

$$\frac{C}{N} \text{ (dB)} = 13.7 \text{ (dB)} - 10 \log \frac{60 \text{ MHz}}{90 \text{ Mbps}}$$

$$= 13.7 \text{ dB} - (-1.77 \text{ dB}) = 15.47 \text{ dB}$$

(c) Again, rearranging Equation 18-6 and substituting for E_b/N_0 gives us

$$\frac{C}{N} \text{ (dB} = 13.7 \text{ (dB)} - 10 \log \frac{90 \text{ MHz}}{90 \text{ Mbps}}$$

$$= 13.7 \text{ dB} - 0 \text{ dB} = 13.7 \text{ dB}$$

It should be evident from Examples 18-6 and 18-7 that the E_b/N_0 and C/N ratios are equal only when the noise bandwidth is equal to the bit rate. Also, as the bandwidth at the point of measurement increases, the C/N decreases.

When the modulation scheme, bit rate, bandwidth, and C/N ratios of two digital radio systems are different it is often difficult to determine which system has the lower probability of error. Because E_b/N_0 is independent of bit rate, bandwidth, and modulation scheme; it is a convenient common denominator to use for comparing the probability of error performance of two digital radio systems.

EXAMPLE 18-8

Compare the performance characteristics of the two digital systems listed below, and determine which system has the lower probability of error.

	QPSK	8-PSK
Bit rate	40 Mbps	60 Mbps
Bandwidth	1.5 × minimum	2 × minimum
C/N	10.75 dB	13.76 dB

Solution Substituting into Equation 18-6 for the QPSK system gives us

$$\frac{E_b}{N_0} \text{ (dB)} = \frac{C}{N} \text{ (dB)} + 10 \log \frac{B}{f_b}$$

$$= 10.75 \text{ dB} + 10 \log \frac{1.5 \times 20 \text{ MHz}}{40 \text{ Mbps}}$$

$$= 10.75 \text{ dB} + (-1.25 \text{ dB})$$

$$= 9.5 \text{ dB}$$

From Figure 18-18, the $P(e)$ is 10^{-4}.

Substituting into Equation 18-6 for the 8-PSK system gives us

$$\frac{E_b}{N_0} \text{ (dB)} = 13.76 \text{ dB} + 10 \log \frac{2 \times 20 \text{ MHz}}{60 \text{ Mbps}}$$

$$= 13.76 \text{ dB} + (-1.76 \text{ dB})$$

$$= 12 \text{ dB}$$

From Figure 18-18, the $P(e)$ is 10^{-3}.

Although the QPSK system has a lower C/N and E_b/N_0 ratio, the $P(e)$ of the QPSK system is 10 times lower (better) than the 8-PSK system.

Gain-to-Equivalent Noise Temperature Ratio

Essentially, *gain-to-equivalent noise temperature ratio* (G/T_e) is a figure of merit used to represent the quality of a satellite or an earth station receiver. The G/T_e of a receiver is the ratio of the receive antenna gain to the equivalent noise temperature (T_e) of the receiver. Because of the extremely small receive carrier powers typically experienced with satellite systems, very often an LNA is physically located at the feedpoint of the antenna. When this is the case, G/T_e is a ratio of the gain of the receiving antenna plus the gain of the LNA to the equivalent noise temperature. Mathematically, gain-to-equivalent noise temperature ratio is

$$\frac{G}{T_e} = \frac{A_r + A(\text{LNA})}{T_e} \tag{18-7}$$

Expressed in logs, we have

$$\frac{G}{T_e} \text{ (dBK}^{-1}) = A_r \text{ (dB)} + A(\text{LNA})(\text{dB}) - T_e \text{ (dBK)} \tag{18-8}$$

G/T_e is a very useful parameter for determining the E_b/N_0 and C/N ratios at the satellite transponder and earth station receivers. G/T_e is essentially the only parameter required at a satellite or an earth station receiver when completing a link budget.

EXAMPLE 18-9

For a satellite transponder with a receiver antenna gain of 22 dB, an LNA gain of 10 dB, and an equivalent noise temperature of 22 dBK; determine the G/T_e figure of merit.

Solution Substituting into Equation 18-8 yields

$$\frac{G}{T_e} \text{ (dBK}^{-1}) = 22 \text{ dB} + 10 \text{ dB} - 22 \text{ dBK}$$

$$= 10 \text{ dBK}^{-1}$$

SATELLITE SYSTEM LINK EQUATIONS

The error performance of a digital satellite system is quite predictable. Figure 18-19 shows a simplified block diagram of a digital satellite system and identifies the various gains and losses that may affect the system performance. When evaluating the performance of a digital satellite system, the uplink and downlink parameters are first considered separately, then the overall performance is determined by combining them in the appropriate manner. Keep in mind, a digital microwave or satellite radio simply means the original and demodulated baseband signals are digital in nature. The RF portion of the radio is analog; that is, FSK, PSK, QAM, or some other higher-level modulation riding on an analog microwave carrier.

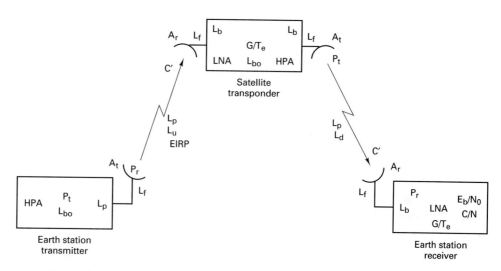

Figure 18-19 Overall satellite system showing the gains and losses incurred in both the uplink and downlink sections. HPA, High-power amplifier; P_t, HPA output power; L_{bo}, back-off loss; L_f, feeder loss; L_b, branching loss; A_t, transmit antenna gain; P_r, total radiated power $= P_t - L_{bo} - L_b - L_f$; EIRP, effective isotropic radiated power $= P_r A_t$; L_u, additional uplink losses due to atmosphere; L_p, path loss; A_r, receive antenna gain; G/T_e, gain-to-equivalent noise ratio; L_d, additional downlink losses due to atmosphere; LNA, low-noise amplifier; C/T_e, carrier-to-equivalent noise ratio; C/N_0, carrier-to-noise density ratio; E_b/N_0, energy of bit-to-noise density ratio; C/N, carrier-to-noise ratio.

LINK EQUATIONS

The following *link equations* are used to separately analyze the uplink and the downlink sections of a single radio-frequency carrier satellite system. These equations consider only the ideal gains and losses and effects of thermal noise associated with the earth station transmitter, earth station receiver, and the satellite transponder. The nonideal aspects of the system are discussed later in this chapter.

Uplink Equation

$$\frac{C}{N_0} = \frac{A_t P_r (L_p L_u) A_r}{K T_e} = \frac{A_t P_r (L_p L_u)}{K} \times \frac{G}{T_e}$$

Link Equations

where L_d and L_u are the additional uplink and downlink atmospheric losses, respectively. The uplink and downlink signals must pass through Earth's atmosphere, where they are partially absorbed by the moisture, oxygen, and particulates in the air. Depending on the elevation angle, the distance the RF signal travels through the atmosphere varies from one earth station to another. Because L_p, L_u, and L_d represent losses, they are decimal values less than 1. G/T_e is the receive antenna gain plus the gain of the LNA divided by the equivalent input noise temperature.

Expressed as a log,

$$\frac{C}{N_0} = \underbrace{10 \log A_t P_r}_{\substack{\text{EIRP} \\ \text{earth} \\ \text{station}}} - \underbrace{20 \log \left(\frac{4\pi D}{\lambda} \right)}_{\substack{\text{free-space} \\ \text{path loss}}} + \underbrace{10 \log \left(\frac{G}{T_e} \right)}_{\substack{\text{satellite} \\ G/T_e}} - \underbrace{10 \log L_u}_{\substack{\text{additional} \\ \text{atmospheric} \\ \text{losses}}} - \underbrace{10 \log K}_{\substack{\text{Boltzmann's} \\ \text{constant}}}$$

$$= \text{EIRP (dBW)} - L_p \text{ (dB)} + \frac{G}{T_e} \text{ (dBK}^{-1}\text{)} - L_u \text{ (dB)} - K \text{ (dBWK)}$$

Downlink Equation

$$\frac{C}{N_0} = \frac{A_t P_r (L_p L_d) A_r}{K T_e} = \frac{A_t P_r (L_p L_d)}{K} \times \frac{G}{T_e}$$

Expressed as a log

$$\frac{C}{N_0} = \underbrace{10 \log A_t P_r}_{\substack{\text{EIRP} \\ \text{satellite}}} - \underbrace{20 \log \left(\frac{4\pi D}{\lambda} \right)}_{\substack{\text{free-space} \\ \text{path loss}}} + \underbrace{10 \log \left(\frac{G}{T_e} \right)}_{\substack{\text{satellite} \\ G/T_e}} - \underbrace{10 \log L_d}_{\substack{\text{additional} \\ \text{atmospheric} \\ \text{losses}}} - \underbrace{10 \log K}_{\substack{\text{Boltzmann's} \\ \text{constant}}}$$

$$= \text{EIRP (dBW)} - L_p \text{ (dB)} + \frac{G}{T_e} \text{ (dBK}^{-1}\text{)} - L_d \text{ (dB)} - K \text{ (dBWK)}$$

LINK BUDGET

Table 18-4 lists the system parameters for three typical satellite communication systems. The systems and their parameters are not necessarily for an existing or future system; they are hypothetical examples only. The system parameters are used to construct a *link budget*. A link budget identifies the system parameters and is used to determine the projected C/N and E_b/N_0 ratios at both the satellite and earth station receivers for a given modulation scheme and desired $P(e)$.

EXAMPLE 18-10

Complete the link budget for a satellite system with the following parameters.

Uplink

1. Earth station transmitter output power   33 dBW
   at saturation, 2000 W

**TABLE 18-4 SYSTEM PARAMETERS FOR THREE HYPOTHETICAL
SATELLITE SYSTEMS**

	System A: 6/4 GHz, earth coverage QPSK modulation, 60 Mbps	System B: 14/12 GHz, earth coverage 8-PSK modulation, 90 Mbps	System C: 14/12 GHz, earth coverage 8-PSK modulation, 120 Mbps
Uplink			
Transmitter output power (saturation, dBW)	35	25	33
Earth station back-off loss (dB)	2	2	3
Earth station branching and feeder loss (dB)	3	3	4
Additional atmospheric (dB)	0.6	0.4	0.6
Earth station antenna gain (dB)	55	45	64
Free-space path loss (dB)	200	208	206.5
Satellite receive antenna gain (dB)	20	45	23.7
Satellite branching and feeder loss (dB)	1	1	0
Satellite equivalent noise temperature (K)	1000	800	800
Satellite G/T_e (dBK^{-1})	−10	16	−5.3
Downlink			
Transmitter output power (saturation, dBW)	18	20	10
Satellite back-off loss (dB)	0.5	0.2	0.1
Satellite branching and feeder loss (dB)	1	1	0.5
Additional atmospheric loss (dB)	0.8	1.4	0.4
Satellite antenna gain (dB)	16	44	30.8
Free-space path loss (dB)	197	206	205.6
Earth station receive antenna gain (dB)	51	44	62
Earth station branching and feeder loss (dB)	3	3	0
Earth station equivalent noise temperature (K)	250	1000	270
Earth station G/T_e (dBK^{-1})	27	14	37.7

2. Earth station back-off loss   3 dB

3. Earth station branching and feeder losses   4 dB

4. Earth station transmit antenna gain (from Figure 18-20, 15 m at 14 GHz)   64 dB

5. Additional uplink atmospheric losses   0.6 dB

6. Free-space path loss (from Figure 18-21, at 14 GHz)   206.5 dB

7. Satellite receiver G/T_e ratio   −5.3 dBK^{-1}

8. Satellite branching and feeder losses   0 dB

9. Bit rate   120 Mbps

10. Modulation scheme   8-PSK

Downlink

1. Satellite transmitter output power at saturation 10 W   10 dBW

2. Satellite back-off loss   0.1 dB

3. Satellite branching and feeder losses   0.5 dB

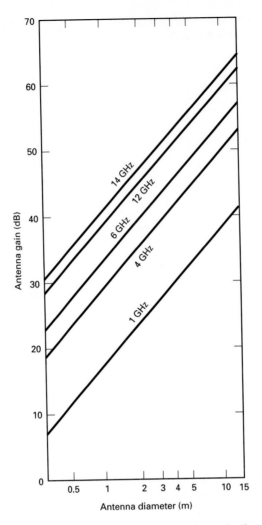

Figure 18-20 Antenna gain based on the gain equation for a parabolic antenna:

$$A \ (\text{db}) = 10 \log \eta \ (\pi \ D/\lambda)^2$$

where D is the antenna diameter, $\lambda =$ the wavelength, and $\eta =$ the antenna efficiency. Here $\eta = 0.55$. To correct for a 100% efficient antenna, add 2.66 dB to the value.

4. Satellite transmit antenna gain (from Figure 18-20, 0.37 m at 12 GHz)	30.8 dB
5. Additional downlink atmospheric losses	0.4 dB
6. Free-space path loss (from Figure 18-21, at 12 GHz)	205.6 dB
7. Earth station receive antenna gain (15 m, 12 GHz)	62 dB
8. Earth station branching and feeder losses	0 dB
9. Earth station equivalent noise temperature	270 K
10. Earth station G/T_e ratio	37.7 dBK^{-1}
11. Bit rate	120 Mbps
12. Modulation scheme	8-PSK

Solution *Uplink budget:* Expressed as a log,

$$\text{EIRP (earth station)} = P_t + A_t - L_{\text{bo}} - L_{bf}$$
$$= 33 \text{ dBW} + 64 \text{ dB} - 3 \text{ dB} - 4 \text{ dB} = 90 \text{ dBW}$$

Carrier power density at the satellite antenna:

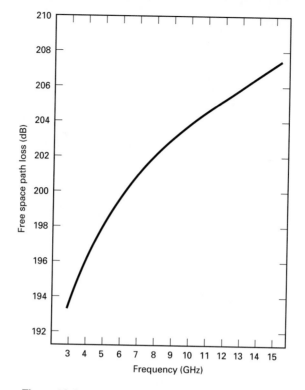

Elevation angle correction:	
Angle	+dB
90°	0
45°	0.44
0°	1.33

Figure 18-21 Free-space path loss (L_p) determined from $L_p = 183.5 + 20 \log f$ (GHz), elevation angle = 90°, and distance = 35,930 km.

$$C' = \text{EIRP (earth station)} - L_p - L_u$$

$$= 90 \text{ dBw} - 206.5 \text{ dB} - 0.6 \text{ dB} = -117.1 \text{ dBW}$$

C/N_0 at the satellite:

$$\frac{C}{N_0} = \frac{C}{KT_e} = \frac{C}{T_e} \times \frac{1}{K} \qquad \text{where } \frac{C}{T_e} = C' \times \frac{G}{T_e}$$

Thus

$$\frac{C}{N_0} = C' \times \frac{G}{T_e} \times \frac{1}{K}$$

Expressed as a log,

$$\frac{C}{N_0} \text{ (dB)} = C' \text{ (dBW)} + \frac{G}{T_e} \text{ (dBK}^{-1}) - 10 \log (1.38 \times 10^{-23})$$

$$\frac{C}{N_0} = -117.1 \text{ dBW} + (-5.3 \text{ dBK}^{-1}) - (-228.6 \text{ dBWK}) = 106.2 \text{ dB}$$

Thus

$$\frac{E_b}{N_0} \text{ (dB)} = \frac{C/f_b}{N_0} \text{ (dB)} = \frac{C}{N_0} \text{ (dB)} - 10 \log f_b$$

$$\frac{E_b}{N_0} = 106.2 \text{ dB} - 10 (\log 120 \times 10^6) = 25.4 \text{ dB}$$

Link Budget

and for a minimum bandwidth system,

$$\frac{C}{N} = \frac{E_b}{N_0} - \frac{B}{f_b} = 25.4 - 10 \log \frac{40 \times 10^6}{120 \times 10^6} = 30.2 \text{ dB}$$

Downlink budget: Expressed as a log,

$$\text{EIRP (satellite transponder)} = P_t + A_t - L_{bo} - L_{bf}$$

$$= 10 \text{ dBW} + 30.8 \text{ dB} - 0.1 \text{ dB} - 0.5 \text{ dB}$$

$$= 40.2 \text{ dBW}$$

Carrier power density at earth station antenna:

$$C' = \text{EIRP (dBW)} - L_p \text{ (dB)} - L_d \text{ (dB)}$$

$$= 40.2 \text{ dBW} - 205.6 \text{ dB} - 0.4 \text{ dB} = -165.8 \text{ dBW}$$

C/N_0 at the earth station receiver:

$$\frac{C}{N_0} = \frac{C}{KT_e} = \frac{C}{T_e} \times \frac{1}{K} \qquad \text{where } \frac{C}{T_e} = C' \times \frac{G}{T_e}$$

Thus

$$\frac{C}{N_0} = C' \times \frac{G}{T_e} \times \frac{1}{K}$$

Expressed as a log,

$$\frac{C}{N_0} \text{ (dB)} = C' \text{ (dBW)} + \frac{G}{T_e} \text{ (dBK}^{-1}) - 10 \log (1.38 \times 10^{-23})$$

$$= -165.8 \text{ dBW} + (37.7 \text{ dBK}^{-1}) - (-228.6 \text{ dBWK}) = 100.5 \text{ dB}$$

An alternative method of solving for C/N_0 is

$$\frac{C}{N_0} \text{ (dB)} = C' \text{ (dBW)} + A_r \text{ (dB)} - T_e \text{ (dBK}^{-1}) - K \text{ (dBWK)}$$

$$= -165.8 \text{ dBW} + 62 \text{ dB} - 10 \log 270 - (-228.6 \text{ dBWK})$$

$$\frac{C}{N_0} = -165.8 \text{ dBW} + 62 \text{ dB} - 24.3 \text{ dBK}^{-1} + 228.6 \text{ dBWK} = 100.5 \text{ dB}$$

$$\frac{E_b}{N_0} \text{ (dB)} = \frac{C}{N_0} \text{ (dB)} - 10 \log f_b$$

$$= 100.5 \text{ dB} - 10 \log (120 \times 10^6)$$

$$= 100.5 \text{ dB} - 80.8 \text{ dB} = 19.7 \text{ dB}$$

and for a minimum bandwidth system,

$$\frac{C}{N} = \frac{E_b}{N_0} - \frac{B}{f_b} = 19.7 - 10 \log \frac{40 \times 10^6}{120 \times 10^6} = 24.5 \text{ dB}$$

With careful analysis and a little algebra, it can be shown that the overall energy of bit-to-noise density ratio (E_b/N_0), which includes the combined effects of the uplink ratio $(E_b/N_0)_u$ and the downlink ratio $(E_b/N_0)_d$, is a standard product over the sum relationship and is expressed mathematically as

$$\frac{E_b}{N_0} \text{ (overall)} = \frac{(E_b/N_0)_u \, (E_b/N_0)_d}{(E_b/N_0)_u + (E_b/N_0)_d} \tag{18-9}$$

where all E_b/N_0 ratios are in absolute values. For Example 18-10, the overall E_b/N_0 ratio is

$$\frac{E_b}{N_0} \text{ (overall)} = \frac{(346.7)(93.3)}{346.7 + 93.3} = 73.5$$

$$= 10 \log 73.5 = 18.7 \text{ dB}$$

As with all product-over-sum relationships, the smaller of the two numbers dominates. If one number is substantially smaller than the other, the overall result is approximately equal to the smaller of the two numbers.

The system parameters used for Example 18-10 were taken from system C in Table 18-4. A complete link budget for the system is shown in Table 18-5.

TABLE 18-5   LINK BUDGET FOR EXAMPLE 18-10

Uplink

1. Earth station transmitter output power at saturation, 2000 W	33 dBW
2. Earth station back-off loss	3 dB
3. Earth station branching and feeder losses	4 dB
4. Earth station transmit antenna gain	64 dB
5. Earth station EIRP	90 dBW
6. Additional uplink atmospheric losses	0.6 dB
7. Free-space path loss	206.5 dB
8. Carrier power density at satellite	−117.1 dBW
9. Satellite branching and feeder losses	0 dB
10. Satellite G/T_e ratio	−5.3 dBK^{-1}
11. Satellite C/T_e ratio	−122.4 dBWK^{-1}
12. Satellite C/N_0 ratio	106.2 dB
13. Satellite C/N ratio	30.2 dB
14. Satellite E_b/N_0 ratio	25.4 dB
15. Bit rate	120 Mbps
16. Modulation scheme	8-PSK

Downlink

1. Satellite transmitter output power at saturation, 10 W	10 dBW
2. Satellite back-off loss	0.1 dB
3. Satellite branching and feeder losses	0.5 dB
4. Satellite transmit antenna gain	30.8 dB
5. Satellite EIRP	40.2 dBW
6. Additional downlink atmospheric losses	0.4 dB
7. Free-space path loss	205.6 dB
8. Earth station receive antenna gain	62 dB
9. Earth station equivalent noise temperature	270 K
10. Earth station branching and feeder losses	0 dB
11. Earth station G/T_e ratio	37.7 dBK^{-1}
12. Carrier power density at earth station	−165.8 dBW
13. Earth station C/T_e ratio	−128.1 dBWK^{-1}
14. Earth station C/N_0 ratio	100.5 dB
15. Earth station C/N ratio	24.5 dB
16. Earth station E_b/N_0 ratio	19.7 dB
17. Bit rate	120 Mbps
18. Modulation scheme	8-PSK

Link Budget

NONIDEAL SYSTEM PARAMETERS

Additional *nonideal parameters* include the following impairments: AM/AM conversion and AM/PM conversion, which result from nonlinear amplification in HPAs and limiters; *pointing error,* which occurs when the earth station and satellite antennas are not exactly aligned; *phase jitter,* which results from imperfect carrier recovery in receivers; *nonideal filtering,* due to the imperfections introduced in bandpass filters; *timing error,* due to imperfect clock recovery in receivers; and *frequency translation errors* introduced in the satellite transponders. The degradation caused by the preceding impairments effectively reduces the E_b/N_0 ratios determined in the link budget calculations. Consequently, they have to be included in the link budget as equivalent losses. An in-depth coverage of the nonideal parameters is beyond the intent of this text.

QUESTIONS

18-1. Briefly describe a satellite.

18-2. What is a passive satellite? An active satellite?

18-3. Contrast nonsynchronous and synchronous satellites.

18-4. Define *prograde* and *retrograde.*

18-5. Define *apogee* and *perigee.*

18-6. Briefly explain the characteristics of low-, medium-, and high-altitude satellite orbits.

18-7. Explain equatorial, polar, and inclined orbits.

18-8. Contrast the advantages and disadvantages of geosynchronous satellites.

18-9. Define *look angles, angle of elevation,* and *azimuth.*

18-10. Define *satellite spatial separation* and list its restrictions.

18-11. Describe a "footprint."

18-12. Describe spot, zonal, and earth coverage radiation patterns.

18-13. Explain *reuse.*

18-14. Briefly describe the functional characteristics of an uplink, a transponder, and a downlink model for a satellite system.

18-15. Define *back-off loss* and its relationship to saturated and transmit power.

18-16. Define *bit energy.*

18-17. Define *effective isotropic radiated power.*

18-18. Define *equivalent noise temperature.*

18-19. Define *noise density.*

18-20. Define *carrier-to-noise density ratio* and *energy of bit-to-noise density ratio.*

18-21. Define *gain-to-equivalent noise temperature ratio.*

18-22. Describe what a satellite link budget is and how it is used.

PROBLEMS

18-1. An earth station is located at Houston, Texas, which has a longitude of 99.5° and a latitude of 29.5° north. The satellite of interest is Satcom 2. Determine the look angles for the earth station antenna.

18-2. A satellite system operates at 14-GHz uplink and 11-GHz downlink and has a projected $P(e)$ of 10^{-7}. The modulation scheme is 8-PSK, and the system will carry 120 Mbps. The equivalent noise temperature of the receiver is 400 K, and the receiver noise bandwidth is equal to the minimum Nyquist frequency. Determine the following parameters: minimum theoretical C/N ratio, minimum theoretical E_b/N_0 ratio, noise density, total receiver input noise, minimum receive carrier power, and the minimum energy per bit at the receiver input.

18-3. A satellite system operates at 6-GHz uplink and 4-GHz downlink and has a projected $P(e)$ of 10^{-6}. The modulation scheme is QPSK and the system will carry 100 Mbps. The equivalent receiver noise temperature is 290 K, and the receiver noise bandwidth is equal to the minimum Nyquist frequency. Determine the C/N ratio that would be measured at a point in the receiver prior to the BPF where the bandwidth is equal to (a) $1\frac{1}{2}$ times the minimum Nyquist frequency, and (b) 3 times the minimum Nyquist frequency.

18-4. Which system has the best projected BER?
  (a) 8-QAM, $C/N = 15$ dB, $B = 2f_N$, $f_b = 60$ Mbps.
  (b) QPSK, $C/N = 16$ dB, $B = f_N$, $f_b = 40$ Mbps.

18-5. An earth station satellite transmitter has an HPA with a rated saturated output power of 10,000 W. The back-off ratio is 6 dB, the branching loss is 2 dB, the feeder loss is 4 dB, and the antenna gain is 40 dB. Determine the actual radiated power and the EIRP.

18-6. Determine the total noise power for a receiver with an input bandwidth of 20 MHz and an equivalent noise temperature of 600 K.

18-7. Determine the noise density for Problem 18-6.

18-8. Determine the minimum C/N ratio required to achieve a $P(e)$ of 10^{-5} for an 8-PSK receiver with a bandwidth equal to f_N.

18-9. Determine the energy per bit-to-noise density ratio when the receiver input carrier power is -100 dBW, the receiver input noise temperature is 290 K, and a 60-Mbps transmission rate is used.

18-10. Determine the carrier-to-noise density ratio for a receiver with a -70-dBW input carrier power, an equivalent noise temperature of 180 K, and a bandwidth of 20 MHz.

18-11. Determine the minimum C/N ratio for an 8-PSK system when the transmission rate is 60 Mbps, the minimum energy of bit-to-noise density ratio is 15 dB, and the receiver bandwidth is equal to the minimum Nyquist frequency.

18-12. For an earth station receiver with an equivalent input temperature of 200 K, a noise bandwidth of 20 MHz, a receive antenna gain of 50 dB, and a carrier frequency of 12 GHz, determine the following: G/T_e, N_0, and N.

18-13. For a satellite with an uplink E_b/N_0 of 14 dB and a downlink E_b/N_0 of 18 dB, determine the overall E_b/N_0 ratio.

18-14. Complete the following link budget:

Uplink Parameters

1. Earth station transmitter output power at saturation, 1 kW
2. Earth station back-off loss, 3 dB
3. Earth station total branching and feeder losses, 3 dB
4. Earth station transmit antenna gain for a 10-m parabolic dish at 14 GHz
5. Free-space path loss for 14 GHz
6. Additional uplink losses due to the earth's atmosphere, 0.8 dB
7. Satellite transponder G/Te, -4.6 dBK^{-1}
8. Transmission bit rate, 90 Mbps, 8-PSK

Downlink Parameters

1. Satellite transmitter output power at saturation, 10 W
2. Satellite transmit antenna gain for a 0.5-m parabolic dish at 12 GHz
3. Satellite modulation back-off loss, 0.8 dB
4. Free-space path loss for 12 GHz
5. Additional downlink losses due to earth's atmosphere, 0.6 dB
6. Earth station receive antenna gain for a 10-m parabolic dish at 12 GHz
7. Earth station equivalent noise temperature, 200 K
8. Earth station branching and feeder losses, 0 dB
9. Transmission bit rate, 90 Mbps, 8-PSK

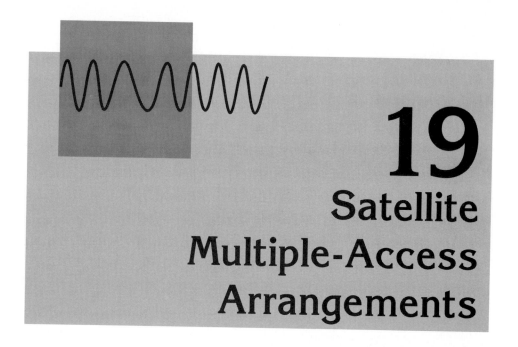

19

Satellite
Multiple-Access
Arrangements

INTRODUCTION

In Chapter 18 we analyzed the link parameters of *single-channel satellite transponders*. In this chapter, we will extend the discussion of satellite communications to systems designed for *multiple carriers*. Whenever multiple carriers are utilized in satellite communications, it is necessary that a *multiple-accessing format* be established over the system. This format allows for a distinct separation between the uplink and downlink transmissions to and from a multitude of different earth stations. Each format has its own specific characteristics, advantages, and disadvantages.

FDM/FM SATELLITE SYSTEMS

Figure 19-1a shows a single-link (two earth stations) *fixed-frequency* FDM/FM system using a single satellite transponder. With earth coverage antennas and for full-duplex operation, each link requires two RF satellite channels (i.e., four RF carrier frequencies, two uplink and two downlink). In Figure 19-1a, earth station 1 transmits on a high-band carrier (f11, f12, f13, etc.) and receives on a low-band carrier (f1, f2, f3, etc.). To avoid interfering with earth station 1, earth station 2 must transmit and receive on different RF carrier frequencies. The RF carrier frequencies are fixed and the satellite transponder is simply an RF-to-RF repeater that provides the uplink/downlink frequency translation. This arrangement is economically impractical and extremely inefficient as well. Additional earth stations can communicate through different transponders within the same satellite structure (Figure 19-1b) but each additional link requires four more RF carrier

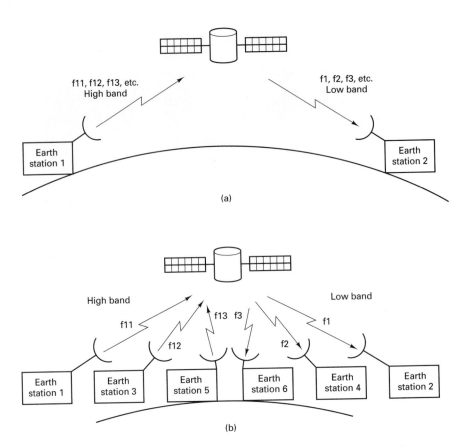

Figure 19-1 Fixed-frequency earth station satellite system: (a) single link; (b) multiple link.

frequencies. It is unlikely that any two-point link would require the capacity available in an entire RF satellite channel. Consequently, most of the available bandwidth is wasted. Also, with this arrangement, each earth station can communicate with only one other earth station. The RF satellite channels are fixed between any two earth stations; thus the voice band channels from each earth station are committed to a single destination.

In a system where three or more earth stations wish to communicate with each other, fixed-frequency or *dedicated channel* systems such as those shown in Figure 19-1 are inadequate; a method of *multiple accessing* is required. That is, each earth station using the satellite system has a means of communicating with each of the other earth stations in the system through a common satellite transponder. Multiple accessing is sometimes called *multiple destination* because the transmissions from each earth station are received by all the other earth stations in the system. The voice band channels between any two earth stations may be *preassigned* (*dedicated*) or *demand-assigned* (*switched*). When pre-assignment is used, a given number of the available voice band channels from each earth station are assigned a dedicated destination. With demand assignment, voice band channels are assigned on an as-needed basis. Demand assignment provides more versatility and more efficient use of the available frequency spectrum. On the other hand, demand assignment requires a control mechanism that is common to all the earth stations to keep track of channel routing and the availability of each voice band channel.

Remember, in an FDM/FM satellite system, each RF channel requires a separate transponder. Also, with FDM/FM transmissions, it is impossible to differentiate (separate) multiple transmissions that occupy the same bandwidth. Fixed-frequency systems may be used in a multiple-access configuration by switching the RF carriers at the satellite, reconfiguring the baseband signals with multiplexing/demultiplexing equipment on board the satellite, or by using multiple spot beam antennas (reuse). All three of these methods require relatively complicated, expensive, and heavy hardware on the spacecraft.

ANIK-D COMMUNICATIONS SATELLITE

The *ANIK-D* communications satellite is a domsat satellite that was launched in August 1982. *ANIK-D* is operated by Telsat Canada. Figure 19-2 shows the frequency and polarization plan for the *ANIK-D* satellite system. There are 12 transponder channels (each approximately 36 MHz wide). However, by using horizontal polarization for one group of 12 channels (group A) and vertical polarization for another group of 12 channels (group B), a total of 24 channels can occupy a bandwidth of approximately 500 MHz. This

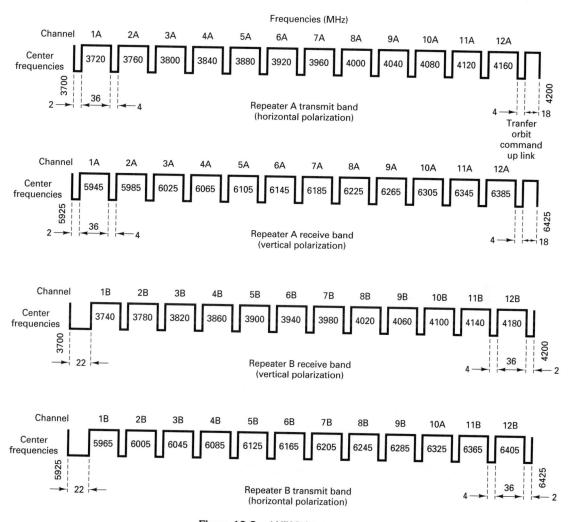

Figure 19-2 *ANIK-D* frequency and polarization plan.

ANIK-D Communications Satellite

scheme is called *frequency reuse* and is possible by using *orthogonal polarization* and spacing adjacent channels 20 MHz apart. There are 12 primary channels and 12 spare or preemptible channels.

MULTIPLE ACCESSING

Figure 19-3 shows the three most commonly used multiple accessing arrangements: frequency-division multiple accessing (FDMA), time-division multiple accessing (TDMA), and code-division multiple accessing (CDMA). With FDMA, each earth station's transmissions are assigned specific uplink and downlink frequency bands within an allotted satellite channel bandwidth; they may be preassigned or demand assigned. Consequently, transmissions from different earth stations are separated in the frequency domain. With TDMA, each earth station transmits a short burst of information during a specific time slot (*epoch*) within a TDMA frame. The bursts must be synchronized so that each station's *burst* arrives at the satellite at a different time. Consequently, transmissions from different earth stations are separated in the time domain. With CDMA, all earth stations transmit within the same frequency band and, for all practical purposes, have no limitation on when they may transmit or on which carrier frequency. Signal separation is accomplished with *envelope encryption/decryption* techniques.

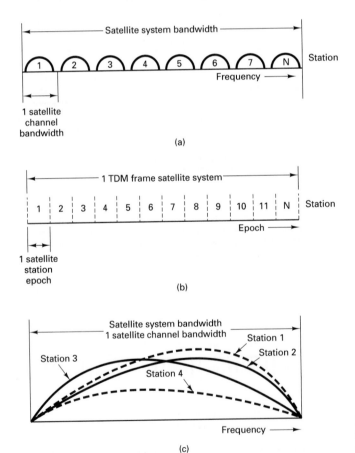

Figure 19-3 Multiple-accessing arrangements: (a) FDMA; (b) TDMA; (c) CDMA.

Frequency-Division Multiple Access

Frequency-division multiple access (FDMA) is a method of multiple accessing where a given RF channel bandwidth is divided into smaller frequency bands called *subdivisions*. Each subdivision is used to carry one voice band channel. A control mechanism is used to ensure that no two earth stations transmit in the same subdivision at the same time. Essentially, the control mechanism designates a receive station for each of the subdivisions. In demand-assignment systems, the control mechanism is also used to establish or terminate the voice band links between the source and destination earth stations. Consequently, any of the subdivisions may be used by any of the participating earth stations at any given time. Typically, each subdivision is used to carry a single 4-kHz voice band channel, but occasionally, groups, supergroups, or even mastergroups are assigned a larger subdivision.

SPADE system. The first FDMA demand-assignment system for satellites was developed by Comsat for use on the *Intelsat IV* satellite. This system was called *SPADE* (single-channel-per-carrier PCM multiple-access demand assignment equipment). Figures 19-4 and 19-5 show the block diagram and IF frequency assignments for SPADE, respectively.

With SPADE, 800 PCM-encoded voice band channels separately QPSK modulate an IF carrier signal (hence the name *single carrier per channel*, SCPC). Each 4-kHz voice band channel is sampled at an 8-kHz rate and converted to an 8-bit PCM code. This produces a 64-kbps PCM code for each voice band channel. The PCM code from each voice band channel QPSK modulates a different IF carrier frequency. With QPSK, the minimum required bandwidth is equal to one-half the input bit rate. Consequently, the output of each QPSK modulator requires a minimum bandwidth of 32 kHz. Each channel is allocated a 45-kHz bandwidth, allowing for a 13-kHz guard band between pairs of frequency-

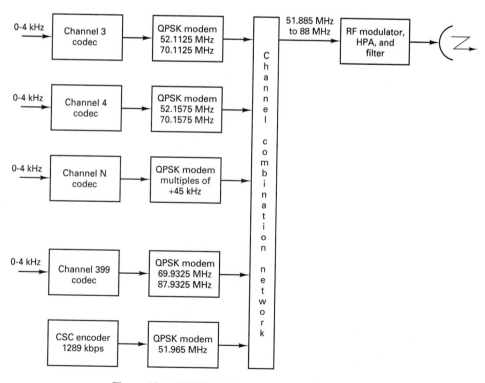

Figure 19-4 FDMA, SPADE earth station transmitter.

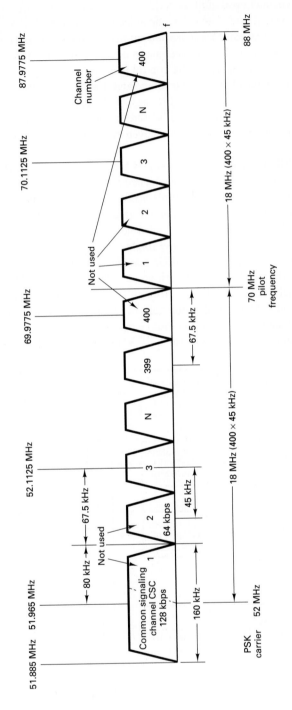

Figure 19-5 Carrier frequency assignments for the Intelsat single channel-per-carrier PCM multiple access demand assignments equipment (SPADE).

division-multiplexed channels. The IF carrier frequencies begin at 52.0225 MHz (low-band channel 1) and increase in 45-kHz steps to 87.9775 MHz (high-band channel 400). The entire 36-MHz band (52 to 88 MHz) is divided in half, producing two 400-channel bands (a low-band and a high-band). For full-duplex operation, four hundred 45-kHz channels are used for one direction of transmission and 400 are used for the opposite direction. Also, channels 1, 2, and 400 from each band are left permanently vacant. This reduces the number of usable full-duplex voice band channels to 397. The 6-GHz C-band extends from 5.725 to 6.425 GHz (700 MHz). This allows for approximately nineteen 36-MHz RF channels per system. Each RF channel has a capacity of 397 full-duplex voice band channels.

Each RF channel (Figure 19-5) has a 160-kHz *common signaling channel* (CSC). The CSC is a time-division-multiplexed transmission that is frequency-division multiplexed into the IF spectrum below the QPSK-encoded voice band channels. Figure 19-6 shows the TDM frame structure for the CSC. The total frame time is 50 ms, which is subdivided into fifty 1-ms epochs. Each earth station transmits on the CSC channel only during its preassigned 1-ms time slot. The CSC signal is a 128-bit binary code. To transmit a 128-bit code in 1 ms, a transmission rate of 128 kbps is required. The CSC code is used for establishing and disconnecting voice band links between two earth station users when demand-assignment channel allocation is used.

EXAMPLE 19-1

For the system shown in Figure 19-7, a user earth station in New York wishes to establish a voice band link between itself and London. New York randomly selects an idle voice band channel. It then transmits a binary-coded message to London on the CSC channel during its respective time slot, requesting that a link be established on the randomly selected channel. London responds on the CSC channel during its time slot with a binary code, either confirming or denying the establishment of the voice band link. The link is disconnected in a similar manner when the users are finished.

The CSC channel occupies a 160-kHz bandwidth, which includes the 45 kHz for low-band channel 1. Consequently, the CSC channel extends from 51.885 MHz to 52.045 MHz. The 128-kbps CSC binary code QPSK modulates a 51.965-MHz carrier. The minimum bandwidth required for the CSC channel is 64 kHz; this results in a 48-kHz guard band on either side of the CSC signal.

With FDMA, each earth station may transmit simultaneously within the same 36-MHz RF spectrum, but on different voice band channels. Consequently, simultaneous transmissions of voice band channels from all earth stations within the satellite network are inter-

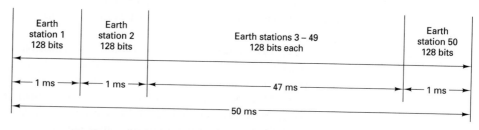

128 bits/1ms × 1000 ms/1s = 128 kbps or 6400 bits/frame × 1 frame/50 ms = 128 kbps

Figure 19-6 FDMA, SPADE common signaling channel (CSC).

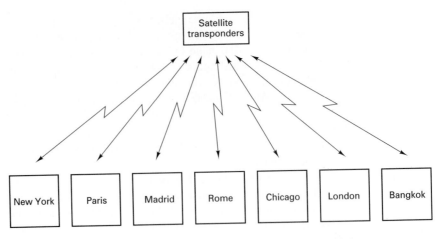

Figure 19-7 Diagram of the system for Example 19-1.

leaved in the frequency domain in the satellite transponder. Transmissions of CSC signals are interleaved in the time domain.

An obvious disadvantage of FDMA is that carriers from multiple earth stations may be present in a satellite transponder at the same time. This results in cross-modulation distortion between the various earth station transmissions. This is alleviated somewhat by shutting off the IF subcarriers on all unused 45-kHz voice band channels. Because balanced modulators are used in the generation of QPSK, carrier suppression is inherent. This also reduces the power load on a system and increases its capacity by reducing the idle channel power.

Time-Division Multiple Access

Time-division multiple access (TDMA) is the predominant multiple-access method used today. It provides the most efficient method of transmitting digitally modulated carriers (PSK). TDMA is a method of time-division multiplexing digitally modulated carriers between participating earth stations within a satellite network through a common satellite transponder. With TDMA, each earth station transmits a short *burst* of a digitally modulated carrier during a precise time slot (epoch) within a TDMA frame. Each station's burst is synchronized so that it arrives at the satellite transponder at a different time. Consequently, only one earth station's carrier is present in the transponder at any given time, thus avoiding a collision with another station's carrier. The transponder is an RF-to-RF repeater that simply receives the earth station transmissions, amplifies them, and then retransmits them in a down-link beam which is received by all the participating earth stations. Each earth station receives the bursts from all other earth stations and must select from them the traffic destined only for itself.

Figure 19-8 shows a basic TDMA frame. Transmissions from all earth stations are synchronized to a *reference burst*. Figure 19-8 shows the reference burst as a separate transmission, but it may be the *preamble* which precedes a reference station's transmission of data. Also, there may be more than one synchronizing reference burst.

The reference burst contains a *carrier recovery sequence* (CRS) from which all receiving stations recover a frequency and phase coherent carrier for PSK demodulation. Also included in the reference burst is a binary sequence for *bit timing recovery* (BTR, i.e., clock recovery). At the end of each reference burst, a *unique word* (UW) is transmitted.

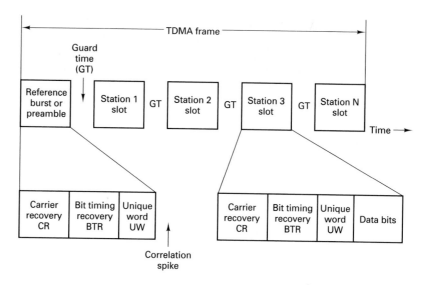

Figure 19-8 Basic time-division-multiple accessing (TDMA) frame.

The UW sequence is used to establish a precise time reference that each of the earth stations uses to synchronize the transmission of its burst. The UW is typically a string of successive binary 1's terminated with a binary 0. Each earth station receiver demodulates and integrates the UW sequence. Figure 19-9 shows the result of the integration process. The integrator and threshold detector are designed so that the threshold voltage is reached precisely when the last bit of the UW sequence is integrated. This generates a *correlation spike* at the output of the threshold detector at the exact time the UW sequence ends.

Each earth station synchronizes the transmission of its carrier to the occurrence of the UW correlation spike. Each station waits a different length of time before it begins transmitting. Consequently, no two stations will transmit the carrier at the same time. Note

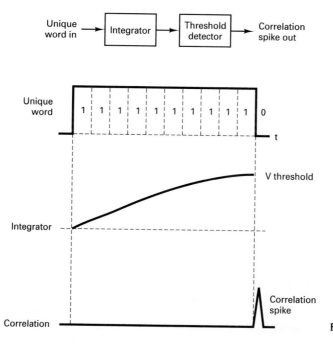

Figure 19-9 Unique word correlator.

Multiple Accessing   763

the *guard time* (GT) between transmissions from successive stations. This is analogous to a guard band in a frequency-division-multiplexed system. Each station precedes the transmission of data with a *preamble*. The preamble is logically equivalent to the reference burst. Because each station's transmissions must be received by all other earth stations, all stations must recover carrier and clocking information prior to demodulating the data. If demand assignment is used, a common signaling channel must also be included in the preamble.

CEPT primary multiplex frame. Figures 19-10 and 19-11 show the block diagram and timing sequence for the CEPT primary multiplex frame respectively (CEPT—Conference of European Postal and Telecommunications Administrations; the CEPT sets many of the European telecommunications standards). This is a commonly used TDMA frame format for digital satellite systems.

Essentially, TDMA is a *store-and-forward* system. Earth stations can transmit only during their specified time slot, although the incoming voice band signals are continuous. Consequently, it is necessary to sample and store the voice band signals prior to transmission. The CEPT frame is made up of 8-bit PCM encoded samples from 16 independent voice band channels. Each channel has a separate codec that samples the incoming voice signals at a 16-kHz rate and converts those samples to 8-bit binary codes. This results in 128-kbps transmitted at a 2.048 MHz rate from each voice channel codec. The sixteen 128-kbps transmissions are time-division multiplexed into a subframe that contains one 8-bit sample from each of the 16 channels (128 bits). It requires only 62.5 μs to accumulate the 128 bits (2.048-Mbps transmission rate). The CEPT multiplex format specifies a

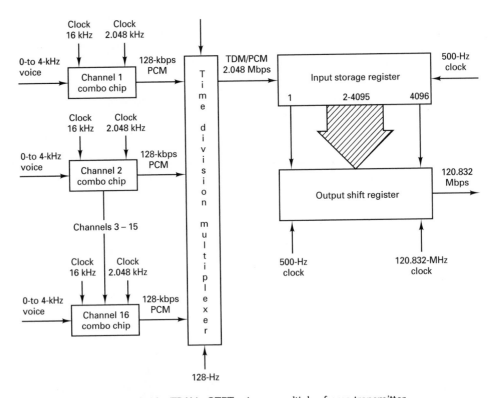

Figure 19-10 TDMA, CEPT primary multiplex frame transmitter.

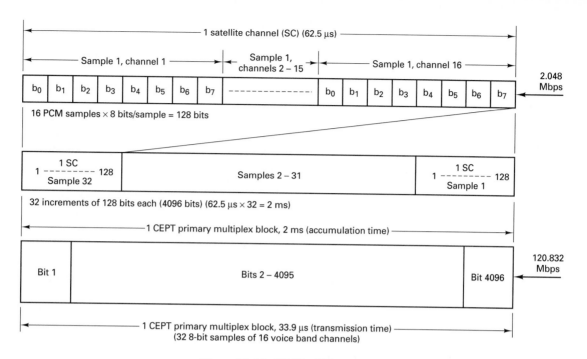

Figure 19-11 TDMA, CEPT primary multiplex frame.

2-ms frame time. Consequently, each earth station can transmit only once every 2 ms and therefore must store the PCM-encoded samples. The 128 bits accumulated during the first sample of each voice band channel are stored in a holding register while a second sample is taken from each channel and converted into another 128-bit *subframe*. This 128-bit sequence is stored in the holding register behind the first 128 bits. The process continues for 32 subframes (32×62.5 μs = 2 ms). After 2 ms, thirty-two 8-bit samples have been taken from each of 16 voice band channels for a total of 4096 bits ($32 \times 8 \times 16 = 4096$). At this time, the 4096 bits are transferred to an output shift register for transmission. Because the total TDMA frame is 2 ms long and during this 2-ms period each of the participating earth stations must transmit at different times, the individual transmissions from each station must occur in a significantly shorter time period. In the CEPT frame, a transmission rate of 120.832 Mbps is used. This rate is the fifty-ninth multiple of 2.048 Mbps. Consequently, the actual transmission of the 4096 accumulated bits takes approximately 33.9 μs. At the earth station receivers, the 4096 bits are stored in a holding register and shifted at a 2.048-Mbps rate. Because all the clock rates (500 Hz, 16 kHz, 128 kHz, 2.048 MHz, and 120.832 MHz) are synchronized, the PCM codes are accumulated, stored, transmitted, received, and then decoded in perfect synchronization. To the users, the voice transmission appears to be a continuous process.

There are several advantages of TDMA over FDMA. The first, and probably the most significant, is that with TDMA only the carrier from one earth station is present in the satellite transponder at any given time, thus reducing intermodulation distortion. Second, with FDMA, each earth station must be capable of transmitting and receiving on a multitude of carrier frequencies to achieve multiple accessing capabilities. Third, TDMA is much better suited to the transmission of digital information than FDMA. Digital signals are more naturally acclimated to storage, rate conversions, and time-domain processing than their analog counterparts.

Multiple Accessing

The primary disadvantage of TDMA as compared to FDMA is that in TDMA precise synchronization is required. Each earth station's transmissions must occur during an exact time slot. Also, bit and frame timing must be achieved and maintained with TDMA.

Code-Division Multiple Access (Spread-Spectrum Multiple Accessing)

With FDMA, earth stations are limited to a specific bandwidth within a satellite channel or system but have no restriction on when they can transmit. With TDMA, earth station's transmissions are restricted to a precise time slot but have no restriction on what frequency or bandwidth they may use within a specified satellite system or channel allocation. With *code-division multiple access* (CDMA), there are no restrictions on time or bandwidth. Each earth station transmitter may transmit whenever it wishes and can use any or all of the bandwidth allocated a particular satellite system or channel. Because there is no limitation on the bandwidth, CDMA is sometimes referred to as *spread-spectrum multiple access*; transmissions can spread throughout the entire allocated bandwidth. Transmissions are separated through envelope encryption/decryption techniques. That is, each earth station's transmissions are encoded with a unique binary word called a *chip code*. Each station has a unique chip code. To receive a particular earth station's transmission, a receive station must know the chip code for that station.

Figure 19-12 shows the block diagram of a CDMA encoder and decoder. In the encoder (Figure 19-12a), the input data (which may be PCM-encoded voice band signals or raw digital data) is multiplied by a unique chip code. The product code PSK modulates an IF carrier which is up-converted to RF for transmission. At the receiver (Figure 19-12b), the RF is down-converted to IF. From the IF, a coherent PSK carrier is recovered. Also, the chip code is acquired and used to synchronize the receive station's code generator. Keep in mind, the receiving station knows the chip code but must generate a chip code that is synchronous in time with the receive code. The recovered synchronous chip code multiplies the recovered PSK carrier and generates a PSK modulated signal that contains the PSK carrier plus the chip code. The received IF signal that contains the chip code, the PSK carrier, and the data information is compared to the received IF signal in the *correlator*. The function of the correlator is to compare the two signals and recover the original data. Essentially, the correlator subtracts the recovered PSK carrier + chip code from the received PSK carrier + chip code + data. The resultant is the data.

The correlation is accomplished on the analog signals. Figure 19-13 shows how the encoding and decoding is accomplished. Figure 19-13a shows the correlation of the correctly received chip code. A $+1$ indicates an in-phase carrier and a -1 indicates an out-of-phase carrier. The chip code is multiplied by the data (either $+1$ or -1). The product is either an in-phase code or one that is 180° out of phase with the chip code. In the receiver, the recovered synchronous chip code is compared in the correlator to the received signaling elements. If the phases are the same, a $+1$ is produced; if they are 180° out of phase, a -1 is produced. It can be seen that if all the recovered chips correlate favorably with the incoming chip code, the output of the correlator will be a $+6$ (which is the case when a logic 1 is received). If all the code chips correlate 180° out of phase, a -6 is generated (which is the case when a logic 0 is received). The bit decision circuit is simply a threshold detector. Depending on whether a $+6$ or -6 is generated, the threshold detector will output a logic 1 or a logic 0, respectively.

As the name implies, the correlator looks for a correlation (similarity) between the incoming coded signal and the recovered chip code. When a correlation occurs, the bit decision circuit generates the corresponding logic condition.

With CDMA, all earth stations within the system may transmit on the same frequency

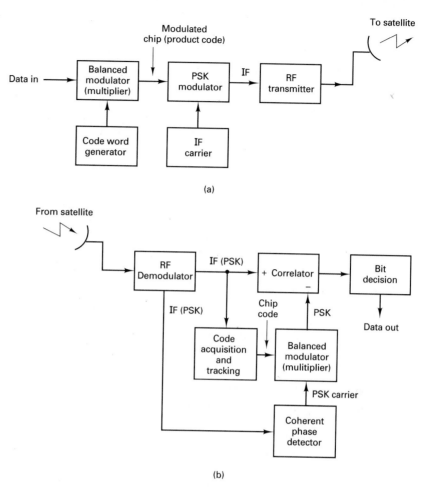

Figure 19-12 Code-division multiple access (CDMA): (a) encoder;
(b) decoder.

at the same time. Consequently, an earth station receiver may be receiving coded PSK signals simultaneously from more than one transmitter. When this is the case, the job of the correlator becomes considerably more difficult. The correlator must compare the recovered chip code with the entire received spectrum and separate from it only the chip code from the desired earth station transmitter. Consequently, the chip code from one earth station must not correlate with the chip codes from any of the other earth stations.

Figure 19-13b shows how such a coding scheme is achieved. If half of the bits within a code were made the same and half were made exactly the opposite, the resultant would be zero cross correlation between chip codes. Such a code is called an *orthogonal code*. In Figure 19-13b it can be seen that when the orthogonal code is compared with the original chip code, there is no correlation (i.e., the sum of the comparison is zero). Consequently, the orthogonal code, although received simultaneously with the desired chip code, had absolutely no effect on the correlation process. For this example, the orthogonal code is received in exact time synchronization with the desired chip code; this is not always the case. For systems that do not have time synchronous transmissions, codes must be developed where there is no correlation between one station's code and any phase of another station's code. For more than two participating earth stations, this is impossible to do. A

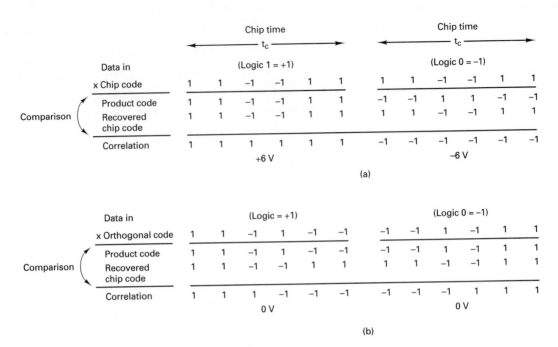

Figure 19-13 CDMA code/data alignment: (a) correct code; (b) orthogonal code.

code set has been developed called the *Gold code*. With the Gold code, there is a minimum correlation between different chips' codes. For a reasonable number of users, it is impossible to achieve perfect orthogonal codes. You can design only for a minimum *cross correlation* between chips.

One of the advantages of CDMA was that the entire bandwidth of a satellite channel or system may be used for each transmission from every earth station. For our example, the chip rate was six times the original bit rate. Consequently, the actual transmission rate of information was one-sixth of the PSK modulation rate, and the bandwidth required is six times that required to simply transmit the original data as binary. Because of the coding inefficiency resulting from transmitting chips for bits, the advantage of more bandwidth is partially offset and is thus less of an advantage. Also, if the transmission of chips from the various earth stations must be synchronized, precise timing is required for the system to work. Therefore, the disadvantage of requiring time synchronization in TDMA systems is also present with CDMA. In short, CDMA is not all that it is cracked up to be. The most significant advantage of CDMA is immunity to interference (jamming), which makes CDMA ideally suited for military applications.

FREQUENCY HOPPING

Frequency hopping is a form of CDMA where a digital code is used to continually change the frequency of the carrier. With frequency hopping, the total available bandwidth is partitioned into smaller frequency bands and the total transmission time is subdivided into smaller time slots. The idea is to transmit within a limited frequency band for only a short period of time, then switch to another frequency band, and so on. This process continues indefinitely. The frequency hopping pattern is determined by a binary code. Each station uses a different code sequence. A typical *hopping pattern* (*frequency-time matrix*) is shown in Figure 19-14.

With frequency hopping, each earth station within a CDMA network is assigned a different frequency hopping pattern. Each transmitter switches (hops) from one frequency band to the next according to their assigned pattern. With frequency hopping, each station uses the entire RF spectrum but never occupies more than a small portion of that spectrum at any one time.

FSK is the modulation scheme most commonly used with frequency hopping. When it is a given station's turn to transmit, it sends one of the two frequencies (either mark or space) for the particular band in which it is transmitting. The number of stations in a given frequency hopping system is limited by the number of unique hopping patterns that can be generated.

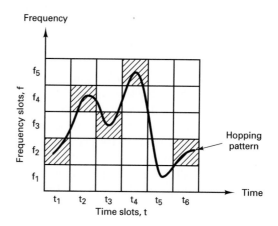

(a)

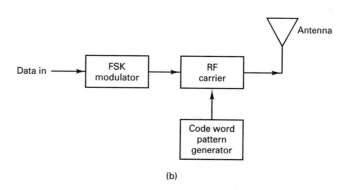

(b)

Figure 19-14 Frequency hopping: (a) frequency time-hopping matrix; (b) frequency hopping transmitter.

Frequency Hopping

769

Essentially, there are two methods used to interface terrestrial voice band channels with satellite channels: digital noninterpolated interfaces (DNI) and digital speech interpolated interfaces (DSI).

Digital Noninterpolated Interfaces

A *digital noninterpolated interface* assigns an individual terrestrial channel (TC) to a particular satellite channel (SC) for the duration of the call. A DNI system can carry no more traffic than the number of satellite channels it has. Once a TC has been assigned an SC, the SC is unavailable to the other TCs for the duration of the call. DNI is a form of preassignment; each TC has a permanent dedicated SC.

Digital Speech Interpolated Interfaces

A *digital speech interpolated interface* assigns a terrestrial channel to a satellite channel only when speech energy is present on the TC. DSI interfaces have *speech detectors* that are similar to *echo suppressors*; they sense speech energy, then seize an SC. Whenever a speech detector senses energy on a TC, the TC is assigned to an SC. The SC assigned is randomly selected from the idle SCs. On a given TC, each time speech energy is detected, the TC could be assigned to a different SC. Therefore, a single TC can use several SCs for a single call. For demultiplexing purposes, the TC/SC assignment information must be conveyed to the receive terminal. This is done on a common signaling channel similar to the one used on the SPADE system. DSI is a form of demand assignment; SCs are randomly assigned on an as-needed basis.

With DSI it is apparent that there is a *channel compression*; there can be more TCs assigned than there are SCs. Generally, a TC:SC ratio of 2:1 is used. For a full-duplex (two-way simultaneous) communication circuit, there is speech in each direction 40% of the time, and for 20% of the time the circuit is idle in both directions. Therefore, a DSI gain slightly more than 2 is realized. The DSI gain is affected by a phenomenon called *competitive clipping*. Competitive clipping is when speech energy is detected on a TC and there is no SC to assign it to. During the *wait* time, speech information is lost. Competitive clipping is not noticed by a subscriber if its duration is less than 50 ms.

To further enhance the channel capacity, a technique called *bit stealing* is used. With bit stealing, channels can be added to fully loaded systems by stealing bits from the in-use channels. Generally, an overload channel is generated by stealing the least significant bit from seven other satellite channels. Bit stealing results in eight channels with 7-bit resolution for the time that the *overload channel* is in use. Consequently, bit stealing results in a lower SQR than normal.

Time-Assignment Speech Interpolation

Time-assignment speech interpolation (TASI) is a form of analog channel compression that has been used for suboceanic cables for many years. TASI is very similar to DSI except that the signals interpolated are analog rather than digital. TASI also uses a 2:1 compression ratio. TASI was also the first means used to scramble voice for military security. TASI is similar to a packet data network; the voice message is chopped up into smaller segments comprised of sounds or portions of sounds. The sounds are sent through the network as separate bundles of energy, then put back together at the receive end to reform the original voice message.

QUESTIONS

19-1. Discuss the drawbacks of using FDM/FM modulation for statellite multiple-accessing systems.

19-2. Contrast *preassignment* and *demand assignment.*

19-3. What are the three most common multiple-accessing arrangements used with satellite systems?

19-4. Briefly describe the multiple-accessing arrangements listed in Question 19-3.

19-5. Briefly describe the operation of Comsat's *Spade* system.

19-6. What is meant by *single carrier per channel*?

19-7. What is a common signaling channel, and how is it used?

19-8. Describe what a reference burst is for TDMA and explain the following terms: preamble, carrier recovery sequence, bit timing recovery, unique word, and correlation spike.

19-9. Describe guard time.

19-10. Briefly describe the operation of the CEPT primary multiplex frame.

19-11. What is a store-and-forward system?

19-12. What is the primary advantage of TDMA as compared to FDMA?

19-13. What is the primary advantage of FDMA as compared to TDMA?

19-14. Briefly describe the operation of a CDMA multiple-accessing system.

19-15. Describe a chip code.

19-16. Describe what is meant by an orthogonal code.

19-17. Describe cross correlation.

19-18. What are the advantages of CDMA as compared to TDMA and FDMA?

19-19. What are the disadvantages of CDMA?

19-20. What is a Gold code?

19-21. Describe frequency hopping.

19-22. What is a frequency-time matrix?

19-23. Describe digital noninterpolated interfaces.

19-24. Describe digital speech interpolated interfaces.

19-25. What is channel compression, and how is it accomplished with a DSI system?

19-26. Describe competitive clipping.

19-27. What is meant by *bit stealing*?

19-28. Describe time-assignment speech interpolation.

PROBLEMS

19-1. How many satellite transponders are required to interlink six earth stations with FDM/FM modulation?

19-2. For the *Spade* system, what are the carrier frequencies for channel 7? What are the allocated passbands for channel 7? What are the actual passband frequencies (excluding guard bands) required?

19-3. If a 512-bit preamble precedes each CEPT station's transmission, what is the maximum number of earth stations that can be linked together with a single satellite transponder?

19-4. Determine an orthogonal code for the following chip code (101010). Prove that your selection will not produce any cross correlation for an in-phase comparison. Determine the cross correlation for each out-of-phase condition that is possible.

20

Optical Fiber Communications

INTRODUCTION

During the past 10 years, the electronic communications industry has experienced many remarkable and dramatic changes. A phenomenal increase in voice, data, and video communications has caused a corresponding increase in the demand for more economical and larger capacity communications systems. This has caused a technical revolution in the electronic communications industry. Terrestrial microwave systems have long since reached their capacity, and satellite systems can provide, at best, only a temporary relief to the ever-increasing demand. It is obvious that economical communications systems that can handle large capacities and provide high-quality service are needed.

Communications systems that use light as the carrier of information have recently received a great deal of attention. As we shall see later in this chapter, propagating light waves through Earth's atmosphere is difficult and impractical. Consequently, systems that use glass or plastic fiber cables to "contain" a light wave and guide it from a source to a destination are presently being investigated at several prominent research and development laboratories. Communications systems that carry information through a *guided fiber cable* are called *fiber optic* systems.

The *information-carrying capacity* of a communications system is directly proportional to its bandwidth; the wider the bandwidth, the greater its information-carrying capacity. For comparison purposes, it is common to express the bandwidth of a system as a percentage of its carrier frequency. For instance, a VHF radio system operating at 100 MHz could have a bandwidth equal to 10 MHz (i.e., 10% of the carrier frequency). A microwave radio system operating at 6 GHz with a bandwidth equal to 10% of its carrier frequency would have a bandwidth equal to 600 MHz. Thus the higher the carrier frequency, the wider the bandwidth possible and consequently, the greater the information-carrying capacity. Light frequencies used in fiber optic systems are between 10^{14} and 4×10^{14} Hz (100,000 to 400,000 GHz). Ten percent of 100,000 GHz is 10,000 GHz. To meet today's communications needs or the needs of the foreseeable future, 10,000 GHz is an excessive bandwidth. However, it does illustrate the capabilities of optical fiber systems.

HISTORY OF FIBER OPTICS

In 1880, Alexander Graham Bell experimented with an apparatus he called a *photophone.* The photophone was a device constructed from mirrors and selenium detectors that transmitted sound waves over a beam of light. The photophone was awkward, unreliable, and had no real practical application. Actually, visual light was a primary means of communicating long before electronic communications came about. Smoke signals and mirrors were used ages ago to convey short, simple messages. Bell's contraption, however, was the first attempt at using a beam of light for carrying information.

Transmission of light waves for any useful distance through Earth's atmosphere is impractical because water vapor, oxygen, and particulates in the air absorb and attenuate the the signals at light frequencies. Consequently, the only practical type of optical communications system is one that uses a fiber guide. In 1930, J. L. Baird, an English scientist, and C. W. Hansell, a scientist from the United States, were granted patents for scanning and transmitting television images through uncoated fiber cables. A few years later a German scientist named H. Lamm successfully transmitted images through a single glass fiber. At that time, most people considered fiber optics more of a toy or a laboratory stunt and consequently, it was not until the early 1950s that any substantial breakthrough was made in the field of fiber optics.

In 1951, A. C. S. van Heel of Holland and H. H. Hopkins and N. S. Kapany of England experimented with light transmission through *bundles* of fibers. Their studies led to the development of the *flexible fiberscope,* which is used extensively in the medical field. It was Kapany who coined the term "fiber optics" in 1956.

In 1958, Charles H. Townes, an American, and Arthur L. Schawlow, a Canadian, wrote a paper describing how it was possible to use stimulated emission for amplifying light waves (laser) as well as microwaves (maser). Two years later, Theodore H. Maiman, a scientist with Hughes Aircraft Company, built the first optical maser.

The *laser* (*l*ight *a*mplification by *s*timulated *e*mission of *r*adiation) was invented in 1960. The laser's relatively high output power, high frequency of operation, and capability of carrying an extremely wide bandwidth signal make it ideally suited for high-capacity communications systems. The invention of the laser greatly accelerated research efforts in fiber optic communications, although it was not until 1967 that K. C. Kao and G. A. Bockham of the Standard Telecommunications Laboratory in England proposed a new communications medium using *cladded* fiber cables.

The fiber cables available in the 1960s were extremely *lossy* (more than 1000 dB/km), which limited optical transmissions to short distances. In 1970, Kapron, Keck, and Maurer of Corning Glass Works in Corning, New York, developed an optical fiber with losses less than 2 dB/km. That was the "big" breakthrough needed to permit practical fiber optics communications systems. Since 1970, fiber optics technology has grown exponentially. Recently, Bell Laboratories successfully transmitted 1 billion bps through a fiber cable for 600 miles without a regenerator.

In the late 1970s and early 1980s, the refinement of optical cables and the development of high-quality, affordable light sources and detectors opened the door to the development of high-quality, high-capacity, and efficient fiber optics communications systems. The branch of electronics that deals with light is called *optoelectronics.*

OPTICAL FIBERS VERSUS METALLIC CABLE FACILITIES

Communications through glass or plastic fiber cables has several overwhelming advantages over communications using conventional *metallic* or *coaxial* cable facilities.

Advantages of Fiber Systems

1. Fiber systems have a greater capacity due to the inherently larger bandwidths available with optical frequencies. Metallic cables exhibit capacitance between and inductance along their conductors. These properties cause them to act like low-pass filters which limit their transmission frequencies and bandwidths.

2. Fiber systems are immune to crosstalk between cables caused by *magnetic induction.* Glass or plastic fibers are nonconductors of electricity and therefore do not have a magnetic field associated with them. In metallic cables, the primary cause of crosstalk is magnetic induction between conductors located near each other.

3. Fiber cables are immune to *static* interference caused by lightning, electric motors, fluorescent lights, and other electrical noise sources. This immunity is also attributable to the fact that optical fibers are nonconductors of electricity. Also, fiber cables do not radiate RF energy and therefore cannot cause interference with other communications systems. This characteristic makes fiber systems ideally suited to military applications, where the effects of nuclear weapons (EMP—electromagnetic pulse interference) has a devastating effect on conventional communications systems.

4. Fiber cables are more resistive to environmental extremes. They operate over a larger temperature variation than their metallic counterparts, and fiber cables are affected less by corrosive liquids and gases.

5. Fiber cables are safer and easier to install and maintain. Because glass and plastic fibers are nonconductors, there are no electrical currents or voltages associated with them. Fibers can be used around volatile liquids and gases without worrying about their causing explosions or fires. Fibers are smaller and much more lightweight than their metallic counterparts. Consequently, they are easier to work with. Also, fiber cables require less storage space and are cheaper to transport.

6. Fiber cables are more secure than their copper counterparts. It is virtually impossible to tap into a fiber cable without the user knowing about it. This is another quality attractive for military applications.

7. Although it has not yet been proven, it is projected that fiber systems will last longer than metallic facilities. This assumption is based on the higher tolerances that fiber cables have to changes in the environment.

8. The long-term cost of a fiber optic system is projected to be less than that of its metallic counterpart.

Disadvantages of Fiber Systems

At the present time, there are few disadvantages of fiber systems. One significant disadvantage is the higher initial cost of installing a fiber system, although in the future it is believed that the cost of installing a fiber system will be reduced dramatically. Another disadvantage of fiber systems is the fact that they are unproven; there are no systems that have been in operation for an extended period of time. Maintenance and repair of fiber systems is also more difficult and expensive than metallic systems.

ELECTROMAGNETIC SPECTRUM

The total electromagnetic frequency spectrum is shown in Figure 20-1. It can be seen that the frequency spectrum extends from the *subsonic* frequencies (a few hertz) to *cosmic rays* (10^{22} Hz). The light frequency spectrum can be divided into three general bands:

1. *Infrared:* band of light wavelengths that are too long to be seen by the human eye
2. *Visible:* band of light wavelengths that the human eye will respond to
3. *Ultraviolet:* band of light wavelengths that are too short to be seen by the human eye

When dealing with higher-frequency electromagnetic waves, such as light, it is common to use units of *wavelength* rather than frequency. Wavelength is the length of the wave that one cycle of an electromagnetic wave occupies in space. The length of a wavelength depends on the frequency of the wave and the velocity of light. Mathematically, wavelength is

$$\lambda = \frac{c}{f} \tag{20-1}$$

where λ = wavelength (meters per cycle)
   c = velocity of light (300,000,000 m/s)
   f = frequency (hertz)

With light frequencies, wavelength is often stated in *microns* (1 micron = 1 micrometer) or *nanometers* (1 nanometer = 10^{-9} meter or 0.001 micron). However, when describing the optical spectrum, the unit *angstrom* (Å) has been often used to express wavelength (1 Å = 10^{-10} meter or 0.0001 micron). Figure 20-2 shows the total electromagnetic wave length spectrum.

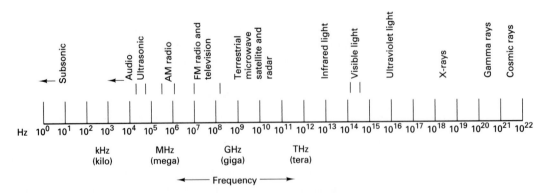

Figure 20-1   Electromagnetic frequency spectrum.

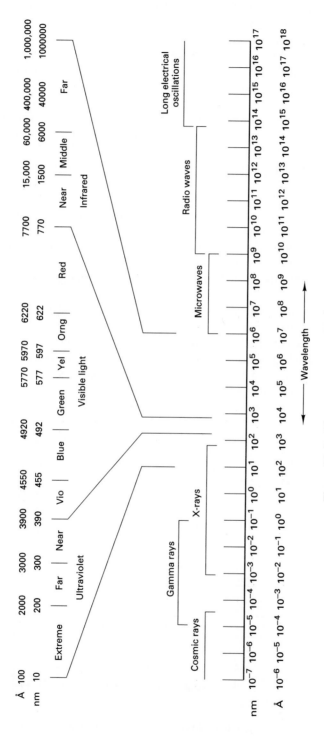

Figure 20-2 Electromagnetic wavelength spectrum.

Figure 20-3 shows a simplified block diagram of a optical fiber communications link. The three primary building blocks of the link are the *transmitter,* the *receiver,* and the *fiber guide.* The transmitter consists of an analog or digital interface, a voltage-to-current converter, a light source, and a source-to-fiber light coupler. The fiber guide is either an ultra-pure glass or plastic cable. The receiver includes a fiber-to-light detector coupling device, a photo detector, a current-to-voltage converter, an amplifier, and an analog or digital interface.

In an optical fiber transmitter, the light source can be modulated by a digital or an analog signal. For analog modulation, the input interface matches impedances and limits the input signal amplitude. For digital modulation, the original source may already be in digital form or, if in analog form, it must be converted to a digital pulse stream. For the latter case, an analog-to-digital converter must be included in the interface.

The voltage-to-current converter serves as an electrical interface between the input circuitry and the light source. The light source is either a light-emitting diode (LED) or an injection laser diode (ILD). The amount of light emitted by either an LED or an ILD is proportional to the amount of drive current. Thus the voltage-to-current converter converts an input signal voltage to a current which is used to drive the light source.

The source-to-fiber coupler (such as a lens) is a mechanical interface. Its function is to couple the light emitted by the source into the optical fiber cable. The optical fiber consists of a glass or plastic fiber core, a cladding, and a protective jacket. The fiber-to-light detector coupling device is also a mechanical coupler. Its function is to couple as much light as possible from the fiber cable into the light detector.

The light detector is very often either a PIN (*p*-type-*i*ntrinsic-*n*-type) diode or an APD (*a*valanche *p*hoto*d*iode). Both the APD and the PIN diode convert light energy to current. Consequently, a current-to-voltage converter is required. The current-to-voltage converter transforms changes in detector current to changes in output signal voltage.

The analog or digital interface at the receiver output is also an electrical interface. If analog modulation is used, the interface matches impedances and signal levels to the output circuitry. If digital modulation is used, the interface must include a digital-to-analog converter.

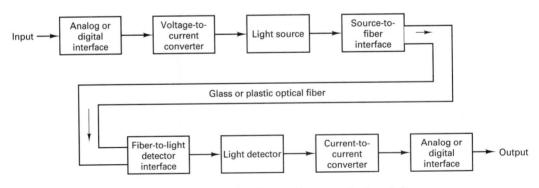

Figure 20-3 Fiber optic communications link.

OPTICAL FIBERS

Fiber Types

Essentially, there are three varieties of optical fibers available today. All three varieties are constructed of either glass, plastic, or a combination of glass and plastic. The three varieties are:

1. Plastic core and cladding
2. Glass core with plastic cladding (often called PCS fiber, plastic-clad silica)
3. Glass core and glass cladding (often called SCS, silica-clad silica)

Presently, Bell Laboratories is investigating the possibility of using a fourth variety that uses a *nonsilicate* substance, *zinc chloride*. Preliminary experiments have indicated that fibers made of this substance will be as much as 1000 times as efficient as glass, their silica-based counterpart.

Plastic fibers have several advantages over glass fibers. First, plastic fibers are more flexible and, consequently, more rugged than glass. They are easy to install, can better withstand stress, are less expensive, and weigh approximately 60% less than glass. The disadvantage of plastic fibers is their high attenuation characteristic; they do not propagate light as efficiently as glass. Consequently, plastic fibers are limited to relatively short runs, such as within a single building or a building complex.

Fibers with glass cores exhibit low attenuation characteristics. However, PCS fibers are slightly better than SCS fibers. Also, PCS fibers are less affected by radiation and are therefore more attractive to military applications. SCS fibers have the best propagation characteristics and they are easier to terminate than PCS fibers. Unfortunately, SCS cables are the least rugged, and they are more susceptible to increases in attenuation when exposed to radiation.

The selection of a fiber for a given application is a function of specific system requirements. There are always trade-offs based on the economics and logistics of a particular application.

Fiber Construction

There are many different cable designs available today. Figure 20-4 shows examples of several fiber optic cable configurations. Depending on the configuration, the cable may include a *core,* a *cladding,* a *protective tube, buffers, strength members,* and one or more *protective jackets.*

With the *loose* tube construction (shown in Figure 20-4a) each fiber is contained in a protective tube. Inside the protective tube, a polyurethane compound encapsules the fiber and prevents the intrusion of water.

Figure 20-4b shows the construction of a *constrained* optical fiber cable. Surrounding the fiber cable are a primary and a secondary buffer. The buffer jackets provide protection for the fiber from external mechanical influences which could cause fiber breakage or excessive optical attenuation. Kelvar is a yarn-type material that increases the tensile strength of the cable. Again, an outer protective tube is filled with polyurethane, which prevents moisture from coming into contact with the fiber core.

Figure 20-4c shows a *multiple-strand* configuration. To increase the tensile strength, a steel central member and a layer of Mylar tape wrap are included in the package. Figure

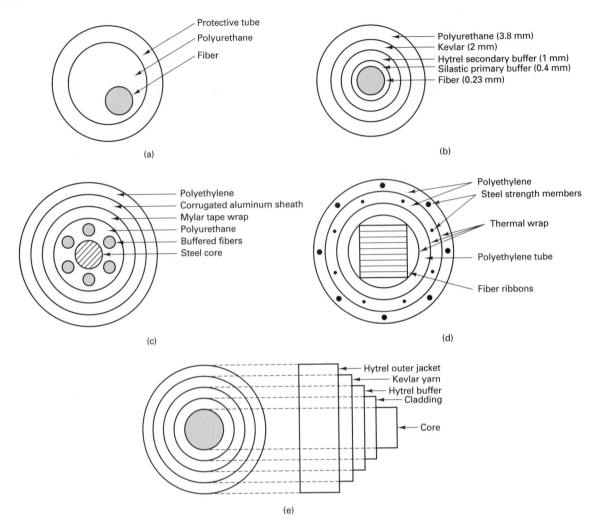

Figure 20-4 Fiber optic cable configurations: (a) loose tube construction; (b) constrained fiber; (c) multiple strands; (d) telephone cable; (e) plastic-clad silica cable.

20-4d shows a *ribbon* configuration, which is frequently seen in telephone systems using fiber optics. Figure 20-4e shows both the end and side views of a plastic-clad silica cable.

The type of cable construction used depends on the performance requirements of the system and both the economic and environmental constraints.

LIGHT PROPAGATION

The Physics of Light

Although the performance of optical fibers can be analyzed completely by application of Maxwell's equations, this is necessarily complex. For most practical applications, *geometric wave tracing* may be used instead of Maxwell's equations; ray tracing will yield sufficiently accurate results.

Chap. 20 **Optical Fiber Communications**

An atom has several energy levels or states, the lowest of which is the ground state. Any energy level above the ground state is called an *excited state*. If an atom in one energy level decays to a lower energy level, the loss of energy (in electron volts) is emitted as a photon. The energy of the photon is equal to the difference between the energy of the two energy levels. The process of decay from one energy level to another energy level is called *spontaneous decay* or *spontaneous emission.*

Atoms can be irradiated by a light source whose energy is equal to the difference between the ground level and an energy level. This can cause an electron to change from one energy level to another by absorbing light energy. The process of moving from one energy level to another is called *absorption.* When making the transition from one energy level to another, the atom absorbs a packet of energy called a *photon.* This process is similar to that of emission.

The energy absorbed or emitted (photon) is equal to the difference between the two energy levels. Mathematically,

$$E_2 - E_1 = E_p \qquad (20\text{-}2)$$

where E_p is the energy of the photon. Also,

$$E_p = hf \qquad (20\text{-}3)$$

where
$$h = \text{Planck's constant}$$

$$= 6.625 \times 10^{34} \text{ J-s}$$

$$f = \text{frequency of light emitted (Hz)}$$

Photon energy may also be expressed in terms of wavelength. Substituting Equation 20-1 into Equation 20-3 yields

$$E_p = hf \qquad (20\text{-}4)$$

$$= \frac{hc}{\lambda}$$

Velocity of Propagation

Electromagnetic energy, such as light, travels at approximately 300,000,000 m/s (186,000 miles per second) in free space. Also, the velocity of propagation is the same for all light frequencies in free space. However, it has been demonstrated that in materials more dense than free space, the velocity is reduced. When the velocity of an electromagnetic wave is reduced as it passes from one medium to another medium of a denser material, the light ray is *refracted* (bent) toward the normal. Also, in materials more dense than free space, all light frequencies do not propagate at the same velocity.

Refraction

Figure 20-5a shows how a light ray is refracted as it passes from a material of a given density into a less dense material. (Actually, the light ray is not bent, but rather, it changes direction at the interface.) Figure 20-5b shows how sunlight, which contains all light frequencies, is affected as it passes through a material more dense than free space. Refraction occurs at both air/glass interfaces. The violet wavelengths are refracted the most, and the red wavelengths are refracted the least. The spectral separation of white light in this manner is called *prismatic refraction.* It is this phenomenon that causes rainbows; water

Light Propagation

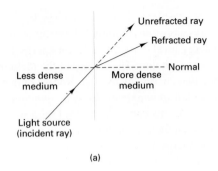

(a)

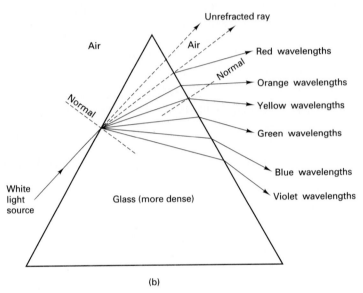

(b)

Figure 20-5 Refraction of light: (a) light refraction; (b) prismatic refraction.

droplets in the atmosphere act like small prisms that split the white sunlight into the various wavelengths, creating a visible spectrum of color.

Refractive Index

The amount of bending or refraction that occurs at the interface of two materials of different densities is quite predictable and depends on the *refractive index* (also called *index of refraction*) of the two materials. The refractive index is simply the ratio of the velocity of propagation of a light ray in free space to the velocity of propagation of a light ray in a given material. Mathematically, the refractive index is

$$ n = \frac{c}{v} $$

where c = speed of light in free space (300,000,000 m/s)
        v = speed of light in a given material

Although the refractive index is also a function of frequency, the variation in most applications is insignificant and therefore omitted from this discussion. The indexes of refraction of several common materials are given in Table 20-1.

TABLE 20-1 TYPICAL INDEXES
OF REFRACTION

Medium	Index of refraction[a]
Vacuum	1.0
Air	1.0003 ($\approx$1.0)
Water	1.33
Ethyl alcohol	1.36
Fused quartz	1.46
Glass fiber	1.5–1.9
Diamond	2.0–2.42
Silicon	3.4
Gallium-arsenide	3.6

[a]Index of refraction is based on a wavelength of light emitted from a sodium flame (5890 Å).

How a light ray reacts when it meets the interface of two transmissive materials that have different indexes of refraction can be explained with *Snell's law*. Snell's law simply states:

$$n_1 \sin \theta_1 = n_2 \sin \theta_2 \tag{20-5}$$

where n_1 = refractive index of material 1 (unitless)
n_2 = refractive index of material 2 (unitless)
θ_1 = angle of incidence (degrees)
θ_2 = angle of refraction (degrees)

A refractive index model for Snell's law is shown in Figure 20-6. At the interface, the incident ray may be refracted toward the normal or away from it, depending on whether n_1 is less than or greater than n_2.

Figure 20-7 shows how a light ray is refracted as it travels from a more dense (higher refractive index) material into a less dense (lower refractive index) material. It can be seen that the light ray changes direction at the interface, and the angle of refraction is greater than the angle of incidence. Consequently, when a light ray enters a less dense material,

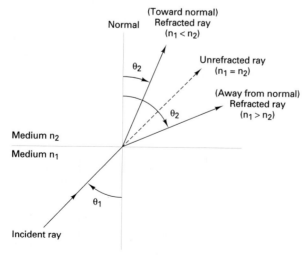

Figure 20-6   Refractive model for Snell's law.

Light Propagation

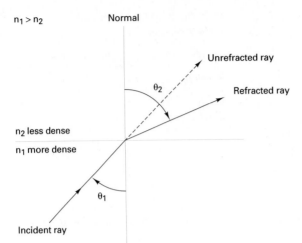

$n_1 > n_2$

Normal

Unrefracted ray

θ_2

Refracted ray

n_2 less dense

n_1 more dense

θ_1

Incident ray

Figure 20-7 Light ray refracted away from the normal.

the ray bends away from the normal. The normal is simply a line drawn perpendicular to the interface at the point where the incident ray strikes the interface. Similarly, when a light ray enters a more dense material, the ray bends toward the normal.

EXAMPLE 20-1

In Figure 20-7, let medium 1 be glass and medium 2 be ethyl alcohol. For an angle of incidence of 30°, determine the angle of refraction.

Solution From Table 20-1,

$$n_1 \text{ (glass)} = 1.5$$

$$n_2 \text{ (ethyl alcohol)} = 1.36$$

Rearranging Equation 20-5 and substituting for n_1, n_2, and θ_1 gives us

$$\frac{n_1}{n_2} \sin \theta_1 = \sin \theta_2$$

$$\frac{1.5}{1.36} \sin 30 = 0.5514 = \sin \theta_2$$

$$\theta_2 = \sin^{-1} 0.5514 = 33.47°$$

The result indicates that the light ray refracted (bent) or changed direction by 3.47° at the interface. Because the light was traveling from a more dense material into a less dense material, the ray bent away from the normal.

Critical Angle

Figure 20-8 shows a condition in which an *incident ray* is at an angle such that the angle of refraction is 90° and the refracted ray is along the interface. (It is important to note that the light ray is traveling from a medium of higher refractive index to a medium with a lower refractive index.) Again, using Snell's law,

$$\sin \theta_1 = \frac{n_2}{n_1} \sin \theta_2$$

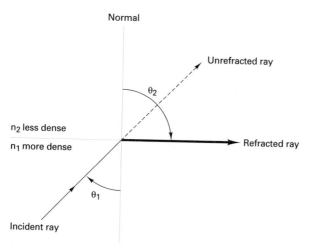

Figure 20-8 Critical angle refraction.

With $\theta_2 = 90°$,

$$\sin \theta_1 = \frac{n_2}{n_1} \ (1) \qquad \text{or} \qquad \sin \theta_1 = \frac{n_2}{n_1}$$

and

$$\sin^{-1} \frac{n_2}{n_1} = \theta_1 = \theta_c \tag{20-6}$$

where θ_c is the critical angle.

The *critical angle* is defined as the minimum angle of incidence at which a light ray may strike the interface of two media and result in an angle of refraction of 90° or greater. (This definition pertains only when the light ray is traveling from a more dense medium into a less dense medium.) If the angle of refraction is 90° or greater, the light ray is not allowed to penetrate the less dense material. Consequently, total reflection takes place at

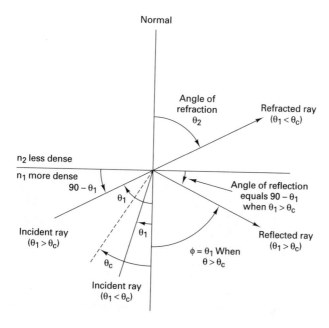

Figure 20-9 Angle of reflection and refraction.

Light Propagation

the interface, and the angle of reflection is equal to the angle of incidence. Figure 20-9 shows a comparison of the angle of refraction and the angle of reflection when the angle of incidence is less than or more than the critical angle.

PROPAGATION OF LIGHT THROUGH AN OPTICAL FIBER

Light can be propagated down an optical fiber cable by either reflection or refraction. How the light is propagated depends on the *mode of propagation* and the *index profile* of the fiber.

Mode of Propagation

In fiber optics terminology, the word *mode* simply means path. If there is only one path for light to take down the cable, it is called *single mode*. If there is more than one path, it is called *multimode*. Figure 20-10 shows single and multimode propagation of light down an optical fiber.

Index Profile

The index profile of an optical fiber is a graphical representation of the value of the refractive index across the fiber. The refractive index is plotted on the horizontal axis and the radial distance from the core axis is plotted on the vertical axis. Figure 20-11 shows the core index profiles of three types of fiber cables.

There are two basic types of index profiles: step and graded. A *step-index fiber* has a central core with a uniform refractive index. The core is surrounded by an outside cladding with a uniform refractive index less than that of the central core. From Figure 20-11 it can be seen that in a step-index fiber there is an abrupt change in the refractive index at the core/cladding interface. In a *graded-index fiber* there is no cladding, and the refractive index of the core is nonuniform; it is highest at the center and decreases gradually with distance toward the outer edge.

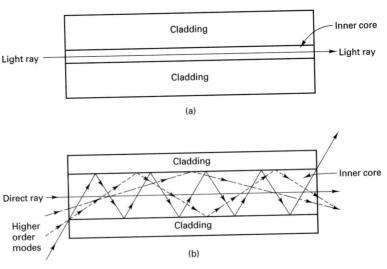

Figure 20-10 Modes of propagation: (a) single mode; (b) multimode.

Chap. 20   **Optical Fiber Communications**

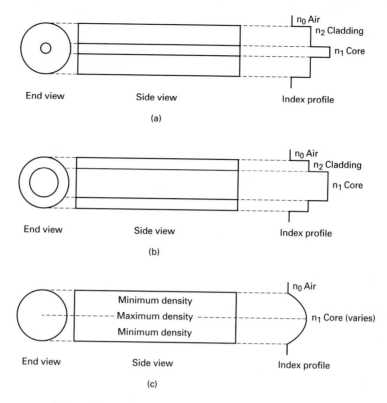

Figure 20-11 Core index profiles: (a) single-mode step index;
(b) multimode step index; (c) multimode graded index.

OPTICAL FIBER CONFIGURATIONS

Essentially, there are three types of optical fiber configurations: single-mode step-index, multimode step-index, and multimode graded-index.

Single-Mode Step-Index Fiber

A *single-mode step-index fiber* has a central core that is sufficiently small so that there is essentially only one path that light may take as it propagates down the cable. This type of fiber is shown in Figure 20-12. In the simplest form of single-mode step-index fiber, the outside cladding is simply air (Figure 20-12a). The refractive index of the glass core (n_1) is approximately 1.5, and the refractive index of the air cladding (n_0) is 1. The large difference in the refractive indexes results in a small critical angle (approximately 42°) at the glass/air interface. Consequently, the fiber will accept light from a wide aperture. This makes it relatively easy to couple light from a source into the cable. However, this type of fiber is typically very weak and of limited practical use.

A more practical type of single-mode step-index fiber is one that has a cladding other than air (Figure 20-12b). The refractive index of the cladding (n_2) is slightly less than that of the central core (n_1) and is uniform throughout the cladding. This type of cable is physically stronger than the air-clad fiber, but the critical angle is also much higher (approximately 77°). This results in a small acceptance angle and a narrow source-to-fiber aperture, making it much more difficult to couple light into the fiber from a light source.

Optical Fiber Configurations

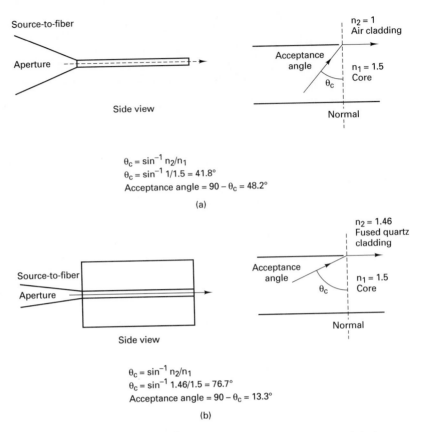

$$\theta_c = \sin^{-1} n_2/n_1$$
$$\theta_c = \sin^{-1} 1/1.5 = 41.8°$$
$$\text{Acceptance angle} = 90 - \theta_c = 48.2°$$

(a)

$$\theta_c = \sin^{-1} n_2/n_1$$
$$\theta_c = \sin^{-1} 1.46/1.5 = 76.7°$$
$$\text{Acceptance angle} = 90 - \theta_c = 13.3°$$

(b)

Figure 20-12 Single-mode step-index fibers: (a) air cladding (b) glass cladding.

With both types of single-mode step-index fibers, light is propagated down the fiber through reflection. Light rays that enter the fiber propagate straight down the core or, perhaps, are reflected once. Consequently, all light rays follow approximately the same path down the cable and take approximately the same amount of time to travel the length of the cable. This is one overwhelming advantage of single-mode step-index fibers and will be explained in more detail later in this chapter.

Multimode Step-Index Fiber

A *multimode step-index fiber* is shown in Figure 20-13. It is similar to the single-mode configuration except that the center core is much larger. This type of fiber has a large light-to-fiber aperture and, consequently, allows more light to enter the cable. The light rays that strike the core/cladding interface at an angle greater than the critical angle (ray A) are propagated down the core in a zigzag fashion, continuously reflecting off the interface boundary. Light rays that strike the core/cladding interface at an angle less than the critical angle (ray B) enter the cladding and are lost. It can be seen that there are many paths that a light ray may follow as it propagates down the fiber. As a result, all light rays do not follow the same path and, consequently, do not take the same amount of time to travel the length of the fiber.

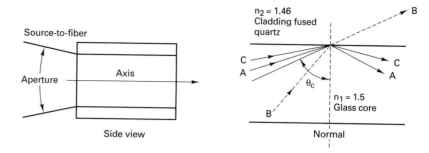

Figure 20-13 Multimode step-index fiber.

Multimode Graded-Index Fiber

A *multimode graded-index fiber* is shown in Figure 20-14. A multimode graded-index fiber is characterized by a central core that has a refractive index that is nonuniform; it is maximum at the center and decreases gradually toward the outer edge. Light is propagated down this type of fiber through refraction. As a light ray propagates diagonally across the core toward the center it is continually intersecting a less-dense-to-more-dense interface. Consequently, the light rays are constantly being refracted, which results in a continuous bending of the light rays. Light enters the fiber at many different angles. As they propagate down the fiber, the light rays that travel in the outermost area of the fiber travel a greater distance than the rays traveling near the center. Because the refractive index decreases with distance from the center and the velocity is inversely proportional to the refractive index, the light rays traveling farthest from the center propagate at a higher velocity. Consequently, they take approximately the same amount of time to travel the length of the fiber.

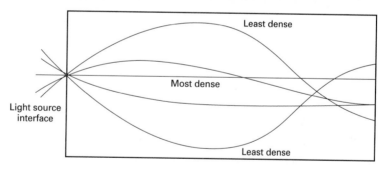

Figure 20-14 Multimode graded-index fiber.

COMPARISON OF THE THREE TYPES OF OPTICAL FIBERS

Single-Mode Step-Index Fiber

Advantages

1. There is minimum dispersion. Because all rays propagating down the fiber take approximately the same path, they take approximately the same amount of time to travel down the cable. Consequently, a pulse of light entering the cable can be reproduced at the receiving end very accurately.

2. Because of the high accuracy in reproducing transmitted pulses at the receive end, larger bandwidths and higher information transmission rates are possible with single-mode step-index fibers than with the other types of fibers.

Disadvantages

1. Because the central core is very small, it is difficult to couple light into and out of this type of fiber. The source-to-fiber aperture is the smallest of all the fiber types.

2. Again, because of the small central core, a highly directive light source such as a laser is required to couple light into a single-mode step-index fiber.

3. Single-mode step-index fibers are expensive and difficult to manufacture.

Multimode Step-Index Fiber

Advantages

1. Multimode step-index fibers are inexpensive and simple to manufacture.

2. It is easy to couple light into and out of multimode step-index fibers; they have a relatively large source-to-fiber aperture.

Disadvantages

1. Light rays take many different paths down the fiber, which results in large differences in their propagation times. Because of this, rays traveling down this type of fiber have a tendency to spread out. Consequently, a pulse of light propagating down a multimode step-index fiber is distorted more than with the other types of fibers.

2. The bandwidth and rate of information transfer possible with this type of cable are less than the other types.

Multimode Graded-Index Fiber

Essentially, there are no outstanding advantages or disadvantages of this type of fiber. Multimode graded-index fibers are easier to couple light into and out of than single-mode step-index fibers but more difficult than multimode step-index fibers. Distortion due to multiple propagation paths is greater than in single-mode step-index fibers but less than in multimode step-index fibers. Graded-index fibers are easier to manufacture than single-mode step-index fibers but more difficult than multimode step-index fibers. The multimode graded-index fiber is considered an intermediate fiber compared to the other types.

ACCEPTANCE ANGLE AND ACCEPTANCE CONE

In previous discussions, the *source-to-fiber aperture* was mentioned several times, and the *critical* and *acceptance* angles at the point where a light ray strikes the core/cladding interface were explained. The following discussion deals with the light-gathering ability of the fiber, the ability to couple light from the source into the fiber cable.

Figure 20-15 shows the source end of a fiber cable. When light rays enter the fiber, they strike the air/glass interface at normal A. The refractive index of air is 1 and the refractive index of the glass core is 1.5. Consequently, the light entering at the air/glass interface propagates from a less dense medium into a more dense medium. Under these

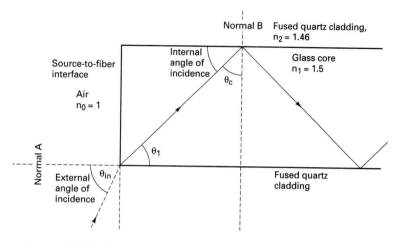

Figure 20-15 Ray propagation into and down an optical fiber cable.

conditions and according to Snell's law, the light rays will refract toward the normal. This causes the light rays to change direction and propagate diagonally down the core at an angle (θ_c) which is different than the external angle of incidence at the air/glass interface (θ_{in}). In order for a ray of light to propagate down the cable, it must strike the internal core/cladding interface at an angle that is greater than the critical angle (θ_c).

Applying Snell's law to the external angle of incidence yields the following expression:

$$n_0 \sin \theta_{in} = n_1 \sin \theta_1 \qquad (20\text{-}7)$$

and

$$\theta_1 = 90 - \theta_c$$

Thus

$$\sin \theta_1 = \sin (90 - \theta_c) = \cos \theta_c \qquad (20\text{-}8)$$

Substituting Equation 20-8 into Equation 20-7 yields the following expression:

$$n_0 \sin \theta_{in} = n_1 \cos \theta_c$$

Rearranging and solving for $\sin \theta_{in}$ gives us

$$\sin \theta_{in} = \frac{n_1}{n_0} \cos \theta_c \qquad (20\text{-}9)$$

Figure 20-16 shows the geometric relationship of Equation 20-9.

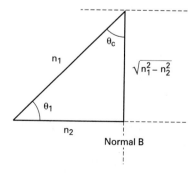

Figure 20-16 Geometric relationship of Equation 20-9.

Acceptance Angle and Acceptance Cone

From Figure 20-16 and using the Pythagorean theorem, we obtain

$$\cos \theta_c = \frac{\sqrt{n_1^2 - n_2^2}}{n_1} \tag{20-10}$$

Substituting Equation 20-10 into Equation 20-9 yields

$$\sin \theta_{in} = \frac{n_1}{n_0} \frac{\sqrt{n_1^2 - n_2^2}}{n_1}$$

Reducing the equation gives

$$\sin \theta_{in} = \frac{\sqrt{n_1^2 - n_2^2}}{n_0} \tag{20-11}$$

and

$$\theta_{in} = \sin^{-1} \frac{\sqrt{n_1^2 - n_2^2}}{n_0} \tag{20-12}$$

Because light rays generally enter the fiber from an air medium, n_0 equals 1. This simplifies Equation 20-12 to

$$\theta_{in(max)} = \sin^{-1} \sqrt{n_1^2 - n_2^2} \tag{20-13}$$

θ_{in} is called the *acceptance angle* or *acceptance cone* half-angle. It defines the maximum angle in which external light rays may strike the air/fiber interface and still propagate down the fiber with a response that is no greater than 10 dB down from the peak value. Rotating the acceptance angle around the fiber axis describes the acceptance cone of the fiber input. This is shown in Figure 20-17.

Numerical Aperture

Numerical aperture (NA) is a figure of merit that is used to describe the light-gathering or light-collecting ability of an optical fiber. The larger the magnitude of NA, the greater the amount of light accepted by the fiber from the external light source. For a step-index fiber, numerical aperture is mathematically defined as the sine of the acceptance half-angle. Thus

$$NA = \sin \theta_{in}$$

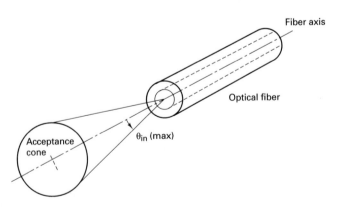

Figure 20-17 Acceptance cone of a fiber cable.

Chap. 20   Optical Fiber Communications

and

$$NA = \sqrt{n_1^2 - n_2^2} \qquad (20\text{-}14)$$

Also,

$$\sin^{-1} NA = \theta_{in}$$

For a graded index, NA is simply the sin of the critical angle:

$$NA = \sin \theta_c$$

EXAMPLE 20-2

For this example refer to Figure 20-15. For a multimode step-index fiber with a glass core ($n_1 = 1.5$) and a fused quartz cladding ($n_2 = 1.46$), determine the critical angle (θ_c), acceptance angle (θ_{in}), and numerical aperture. The source-to-fiber media is air.

Solution Substituting into Equation 20-6, we have

$$\theta_c = \sin^{-1} \frac{n_2}{n_1} = \sin^{-1} \frac{1.46}{1.5} = 76.7°$$

Substituting into Equation 20-13 yields

$$\theta_{in} = \sin^{-1} \sqrt{n_1^2 - n_2^2} = \sin^{-1} \sqrt{1.5^2 - 1.46^2}$$

$$= 20.2°$$

Substituting into Equation 20-14 gives us

$$NA = \sin \theta_{in} = \sin 20.2$$

$$= 0.344$$

LOSSES IN OPTICAL FIBER CABLES

Transmission losses in optical fiber cables are one of the most important characteristics of the fiber. Losses in the fiber result in a reduction in the light power and thus reduce the system bandwidth, information transmission rate, efficiency, and overall system capacity. The predominant fiber losses are as follows:

1. Absorption losses
2. Material or Rayleigh scattering losses
3. Chromatic or wavelength dispersion
4. Radiation losses
5. Modal dispersion
6. Coupling losses

Absorption Losses

Absorption loss in optical fibers is analogous to power dissipation in copper cables; impurities in the fiber absorb the light and convert it to heat. The ultrapure glass used to manufacture optical fibers is approximately 99.9999% pure. Still, absorption losses between 1 and 1000 dB/km are typical. Essentially, there are three factors that contribute to the absorption losses in optical fibers: ultraviolet absorption, infrared absorption, and ion resonance absorption.

Ultraviolet absorption. Ultraviolet absorption is caused by valence electrons in the silica material from which fibers are manufactured. Light *ionizes* the valence electrons into conduction. The ionization is equivalent to a loss in the total light field and, consequently, contributes to the transmission losses of the fiber.

Infrared absorption. Infrared absorption is a result of *photons* of light that are absorbed by the atoms of the glass core molecules. The absorbed photons are converted to random mechanical vibrations typical of heating.

Ion resonance absorption. Ion resonance absorption is caused by OH^- ions in the material. The source of the OH^- ions is water molecules that have been trapped in the glass during the manufacturing process. Ion absorption is also caused by iron, copper, and chromium molecules.

Figure 20-18 shows typical losses in optical fiber cables due to ultraviolet, infrared, and ion resonance absorption.

Material or Rayleigh Scattering Losses

During the manufacturing process, glass is drawn into long fibers of very small diameter. During this process, the glass is in a plastic state (not liquid and not solid). The tension applied to the glass during this process causes the cooling glass to develop submicroscopic irregularities that are permanently formed in the fiber. When light rays that are propagating down a fiber strike one of these impurities, they are *diffracted*. Diffraction causes the light to disperse or spread out in many directions. Some of the diffracted light continues down the fiber and some of it escapes through the cladding. The light rays that escape

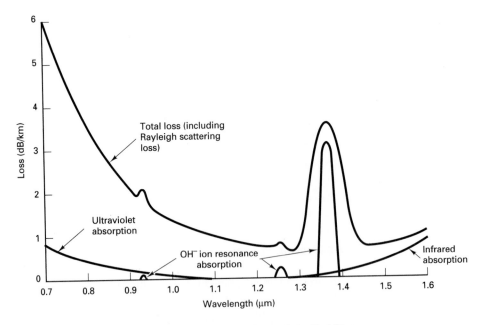

Figure 20-18 Absorption losses in optical fibers.

Chap. 20   **Optical Fiber Communications**

represent a loss in light power. This is called *Rayleigh scattering loss.* Figure 20-19 graphically shows the relationship between wavelength and Rayleigh scattering loss.

Chromatic or Wavelength Dispersion

As stated previously, the refractive index of a material is wavelength dependent. Light-emitting diodes (LEDs) emit light that contains a combination of wavelengths. Each wavelength within the composite light signal travels at a different velocity. Consequently, light rays that are simultaneously emitted from an LED and propagated down an optical fiber do not arrive at the far end of the fiber at the same time. This results in a distorted receive signal; the distortion is called *chromatic distortion.* Chromatic distortion can be eliminated by using a monochromatic source such as an injection laser diode (ILD).

Radiation Losses

Radiation losses are caused by small bends and kinks in the fiber. Essentially, there are two types of bends: microbends and constant-radius bends. *Microbending* occurs as a result of differences in the thermal contraction rates between the core and cladding material. A microbend represents a discontinuity in the fiber where Rayleigh scattering can occur. *Constant-radius bends* occur when fibers are bent during handling or installation.

Modal Dispersion

Modal dispersion or *pulse spreading,* is caused by the difference in the propagation times of light rays that take different paths down a fiber. Obviously, modal dispersion can occur only in multimode fibers. It can be reduced considerably by using graded-index fibers and almost entirely eliminated by using single-mode step-index fibers.

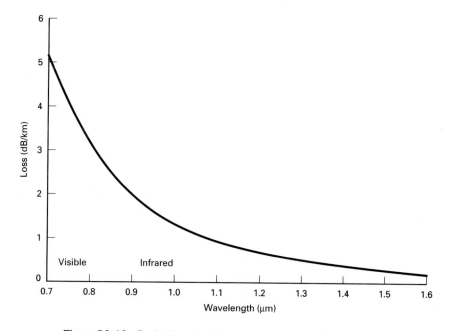

Figure 20-19 Rayleigh scattering loss as a function of wavelength.

Losses in Optical Fiber Cables

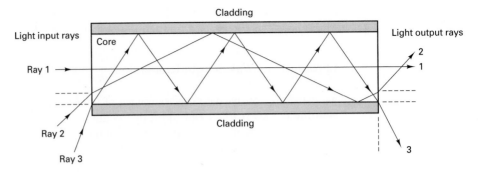

Figure 20-20 Light propagation down a mutimode step-index fiber.

Modal dispersion can cause a pulse of light energy to spread out as it propagates down a fiber. If the pulse spreading is sufficiently severe, one pulse may fall back on top of the next pulse (this is an example of intersymbol interference). In a multimode step-index fiber, a light ray that propagates straight down the axis of the fiber takes the least amount of time to travel the length of the fiber. A light ray that strikes the core/cladding interface at the critical angle will undergo the largest number of internal reflections and, consequently, take the longest time to travel the length of the fiber.

Figure 20-20 shows three rays of light propagating down a multimode step-index fiber. The lowest-order mode (ray 1) travels in a path parallel to the axis of the fiber. The middle-order mode (ray 2) bounces several times at the interface before traveling the length of the fiber. The highest-order mode (ray 3) makes many trips back and forth across the fiber as it propagates the entire length. It can be seen that ray 3 travels a considerably longer distance than ray 1 as it propagates down the fiber. Consequently, if the three rays of light were emitted into the fiber at the same time and represented a pulse of light energy, the three rays would reach the far end of the fiber at different times and result in a spreading out of the light energy in respect to time. This is called modal dispersion and results in a stretched pulse which is also reduced in amplitude at the output of the fiber. All three rays of light propagate through the same material at the same velocity, but ray 3 must travel a longer distance and, consequently, takes a longer period of time to propagate down the fiber.

Figure 20-21 shows light rays propagating down a single-mode step-index fiber. Because the radial dimension of the fiber is sufficiently small, there is only a single path for each of the rays to follow as they propagate down the length of the fiber. Consequently, each ray of light travels the same distance in a given period of time and the light rays have exactly the same time relationship at the far end of the fiber as they had when they entered the cable. The result is no *modal dispersion* or *pulse stretching*.

Figure 20-22 shows light propagating down a multimode graded-index fiber. Three rays are shown traveling in three different modes. Each ray travels a different path but they all take approximately the same amount of time to propagate the length of fiber. This is because the refractive index of the fiber decreases with distance from the center, and the

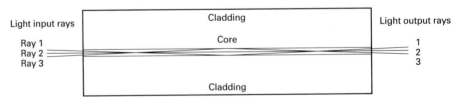

Figure 20-21 Light propagation down a single-mode step-index fiber.

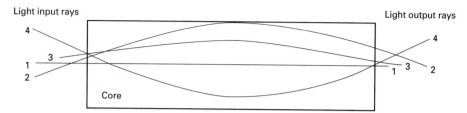

Figure 20-22 Light propagation down a multimode graded-index fiber.

velocity at which a ray travels is inversely proportional to the refractive index. Consequently, the farther rays 2 and 3 travel from the center of the fiber, the faster they propagate.

Figure 20-23 shows the relative time/energy relationship of a pulse of light as it propagates down a fiber cable. It can be seen that as the pulse propagates down the fiber, the light rays that make up the pulse spread out in time, which causes a corresponding reduction in the pulse amplitude and stretching of the pulse width. This is called *pulse spreading* or *pulse-width dispersion* and causes errors in digital transmission. It can also be seen that as light energy from one pulse falls back in time, it will interfere with the next pulse causing intersymbol interference.

Figure 20-24a shows a unipolar return-to-zero (UPRZ) digital transmission. With UPRZ transmission (assuming a very narrow pulse) if light energy from pulse A were to fall back (*spread*) one bit time (T_b), it would interfere with pulse B and change what was a logic 0 to a logic 1. Figure 20-24b shows a unipolar nonreturn-to-zero (UPNRZ) digital transmission where each pulse is equal to the bit time. With UPNRZ transmission, if energy from pulse A were to fall back one-half of a bit time, it would interfere with pulse B. Consequently, UPRZ transmissions can tolerate twice as much delay or spread as UPNRZ transmissions.

The difference between the absolute delay times of the fastest and slowest rays of

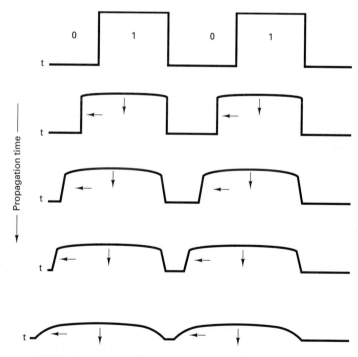

Figure 20-23 Pulse-width dispersion in an optical fiber cable.

Losses in Optical Fiber Cables

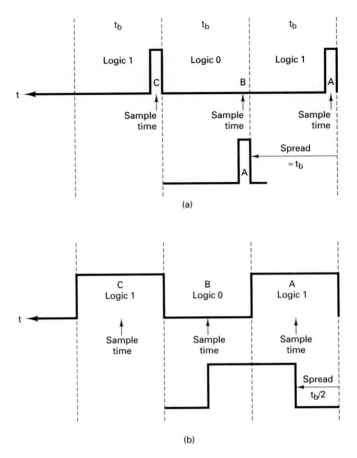

Figure 20-24 Pulse spreading of digital transmissions: (a) UPRZ; (b) UPNRZ.

light propagating down a fiber of unit length is called the *pulse-spreading constant* (Δt) and is generally expressed in nanoseconds per kilometer (ns/km). The total pulse spread (ΔT) is then equal to the pulse spreading constant (Δt) times the total fiber length (L). Mathematically, ΔT is

$$\Delta T \text{ (ns)} = \Delta t \left(\frac{\text{ns}}{\text{km}} \right) \times L \text{ (km)} \tag{20-15}$$

For UPRZ transmissions, the maximum data transmission rate in bits per second (bps) is expressed as

$$f_b \text{ (bps)} = \frac{1}{\Delta t \times L} \tag{20-16}$$

and for UPNRZ transmissions, the maximum transmission rate is

$$f_b \text{ (bps)} = \frac{1}{2 \, \Delta t \times L} \tag{20-17}$$

EXAMPLE 20-3

For an optical fiber 10 km long with a pulse-spreading constant of 5 ns/km, determine the maximum digital transmission rates for (a) return-to-zero, and (b) nonreturn-to-zero transmissions.

Solution (a) Substituting into Equation 20-16 yields

$$f_b = \frac{1}{5 \text{ ns/km} \times 10 \text{ km}} = 20 \text{ Mbps}$$

(b) Substituting into Equation 20-17 yields

$$f_b = \frac{1}{(2 \times 5 \text{ ns/km}) \times 10 \text{ km}} = 10 \text{ Mbps}$$

The results indicate that the digital transmission rate possible for this optical fiber is twice as high (20 Mbps versus 10 Mbps) for UPRZ as for UPNRZ transmission.

Coupling Losses

In fiber cables coupling losses can occur at any of the following three types of optical junctions: light source-to-fiber connections, fiber-to-fiber connections, and fiber-to-photodetector connections. Junction losses are most often caused by one of the following alignment problems: lateral misalignment, gap misalignment, angular misalignment, and imperfect surface finishes. These impairments are shown in Figure 20-25.

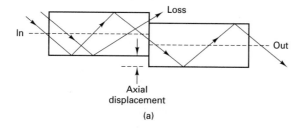

(a)

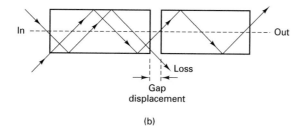

(b)

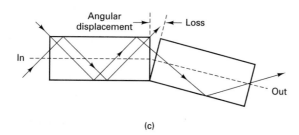

(c)

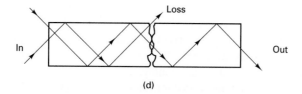

(d)

Figure 20-25 Fiber alignment impairments: (a) lateral misalignment; (b) gap displacement; (c) angular misalignment; (d) surface finish.

Losses in Optical Fiber Cables

Lateral misalignment.   This is shown in Figure 20-25a and is the lateral or axial displacement between two pieces of adjoining fiber cables. The amount of loss can be from a couple of tenths of a decibel to several decibels. This loss is generally negligible if the fiber axes are aligned to within 5% of the smaller fiber's diameter.

Gap misalignment.   This is shown in Figure 20-25b and is sometimes called *end separation.* When *splices* are made in optical fibers, the fibers should actually touch. The farther apart the fibers are, the greater the loss of light. If two fibers are joined with a connector, the ends should not touch. This is because the two ends rubbing against each other in the connector could cause damage to either or both fibers.

Angular misalignment.   This is shown in Figure 20-25c and is sometimes called *angular displacement.* If the angular displacement is less than 2°, the loss will be less than 0.5 dB.

Imperfect surface finish.   This is shown in Figure 20-25d. The ends of the two adjoining fibers should be highly polished and fit together squarely. If the fiber ends are less than 3° off from perpendicular, the losses will be less than 0.5 dB.

LIGHT SOURCES

Essentially, there are two devices commonly used to generate light for fiber optic communications systems: light-emitting diodes (LEDs) and injection laser diodes (ILDs). Both devices have advantages and disadvantages and selection of one device over the other is determined by system economic and performance requirements.

Light-Emitting Diodes

Essentially, a *light-emitting diode* (LED) is simply a P-N junction diode. It is usually made from a semiconductor material such as aluminum-gallium-arsenide (AlGaAs) or gallium-arsenide-phosphide (GaAsP). LEDs emit light by spontaneous emission; light is emitted as a result of the recombination of electrons and holes. When forward biased, minority carriers are injected across the *p-n* junction. Once across the junction, these minority carriers recombine with majority carriers and give up energy in the form of light. This process is essentially the same as in a conventional diode except that in LEDs certain semiconductor materials and dopants are chosen such that the process is radiative; a photon is produced. A photon is a quantum of electromagnetic wave energy. Photons are particles that travel at the speed of light but at rest have no mass. In conventional semiconductor diodes (germanium and silicon, for example), the process is primarily nonradiative and no photons are generated. The energy gap of the material used to construct an LED determines whether the light emitted by it is invisible or visible and of what color.

The simplest LED structures are homojunction, epitaxially grown, or single-diffused devices and are shown in Figure 20-26. *Epitaxially grown LEDs* are generally constructed of silicon-doped gallium-arsenide (Figure 20-26a). A typical wavelength of light emitted from this construction is 940 nm, and a typical output power is approximately 3 mW at 100 mA of forward current. *Planar diffused (homojunction) LEDs* (Figure 20-26b) output approximately 500 μW at a wavelength of 900 nm. The primary disadvantage of homojunction LEDs is the nondirectionality of their light emission, which makes them a poor choice as a light source for fiber optic systems.

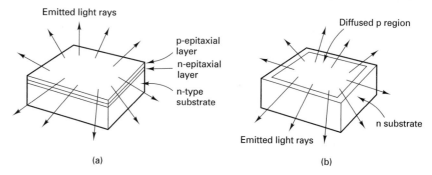

Figure 20-26 Homojunction LED structures: (a) silicon-doped gallium arsenide; (b) planar diffused.

The *planar heterojunction LED* (Figure 20-27) is quite similar to the epitaxially grown LED except that the geometry is designed such that the forward current is concentrated to a very small area of the active layer. Because of this the planar heterojunction LED has several advantages over the homojunction type. They are:

1. The increase in current density generates a more brilliant light spot.
2. The smaller emitting area makes it easier to couple its emitted light into a fiber.
3. The small effective area has a smaller capacitance, which allows the planar heterojunction LED to be used at higher speeds.

Burrus etched-well surface-emitting LED. For the more practical applications, such as telecommunications, data rates in excess of 100 Mbps are required. For these applications, the etched-well LED was developed. Burrus and Dawson of Bell Laboratories developed the etched-well LED. It is a surface-emitting LED and is shown in Figure 20-28. The Burrus etched-well LED emits light in many directions. The etched well helps concentrate the emitted light to a very small area. Also, domed lenses can be placed over the emitting surface to direct the light into a smaller area. These devices are more efficient than the standard surface emitters and they allow more power to be coupled into the optical fiber, but they are also more difficult and expensive to manufacture.

Edge-emitting LED. The edge-emitting LED, which was developed by RCA, is shown in Figure 20-29. These LEDs emit a more directional light pattern than do the surface-emitting LEDs. The construction is similar to the planar and Burrus diodes except that the emitting surface is a stripe rather than a confined circular area. The light is emitted from an active stripe and forms an elliptical beam. Surface-emitting LEDs are more commonly used than edge emitters because they emit more light. However, the coupling losses with surface emitters are greater and they have narrower bandwidths.

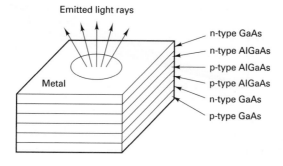

Figure 20-27 Planar heterojunction LED.

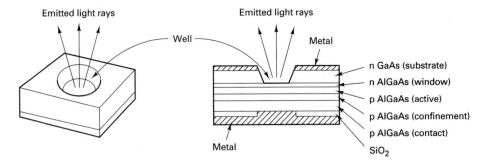

Figure 20-28 Burrus etched-well surface-emitting LED.

The *radiant* light power emitted from an LED is a linear function of the forward current passing through the device (Figure 20-30). It can also be seen that the optical output power of an LED is, in part, a function of the operating temperature.

Injection Laser Diode

The word *laser* is an acronym for *l*ight *a*mplification by *s*timulated *e*mission of *r*adiation. Lasers are constructed from many different materials, including gases, liquids, and solids, although the type of laser used most often for fiber optic communications is the semiconductor laser.

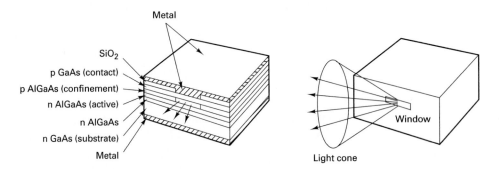

Figure 20-29 Edge-emitting LED.

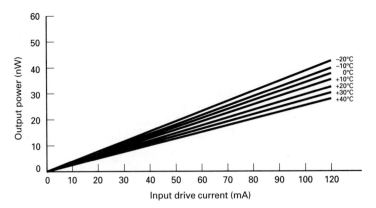

Figure 20-30 Output power versus forward current and operating temperature for an LED.

Chap. 20 **Optical Fiber Communications**

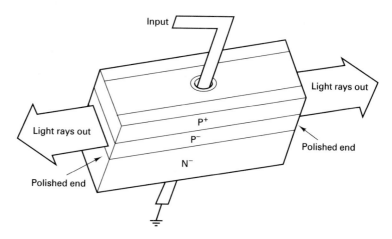

Figure 20-31 Injection laser diode construction.

The *injection laser diode* (ILD) is similar to the LED. In fact, below a certain threshold current, an ILD acts like an LED. Above the threshold current, an ILD oscillates; lasing occurs. As current passes through a forward-biased *p-n* junction diode, light is emitted by spontaneous emission at a frequency determined by the energy gap of the semiconductor material. When a particular current level is reached, the number of minority carriers and photons produced on either side of the *p-n* junction reaches a level where they begin to collide with already excited minority carriers. This causes an increase in the ionization energy level and makes the carriers unstable. When this happens, a typical carrier recombines with an opposite type of carrier at an energy level that is above its normal before-collision value. In the process, two photons are created; one is stimulated by another. Essentially, a gain in the number of photons is realized. For this to happen, a large forward current that can provide many carriers (holes and electrons) is required.

The construction of an ILD is similar to that of an LED (Figure 20-31) except that the ends are highly polished. The mirror-like ends trap the photons in the active region and, as they reflect back and forth, stimulate free electrons to recombine with holes at a higher-than-normal energy level. This process is called *lasing*.

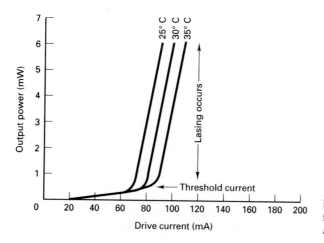

Figure 20-32 Output power versus forward current and temperature for an ILD.

Light Sources

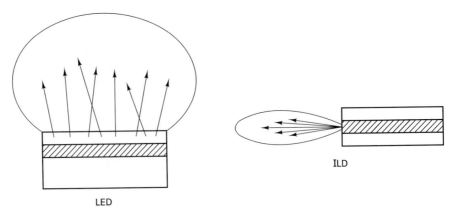

Figure 20-33 LED and ILD radiation patterns.

The radiant output light power of a typical ILD is shown in Figure 20-32. It can be seen that very little output power is realized until the threshold current is reached; then lasing occurs. After lasing begins, the optical output power increases dramatically, with small increases in drive current. It can also be seen that the magnitude of the optical output power of the ILD is more dependent on operating temperature than is the LED.

Figure 20-33 shows the light radiation patterns typical of an LED and an ILD. Because light is radiated out the end of an ILD in a narrow concentrated beam, it has a more direct radiation pattern.

Advantages of ILDs

1. Because ILDs have a more direct radiation pattern, it is easier to couple their light into an optical fiber. This reduces the coupling losses and allows smaller fibers to be used.

2. The radiant output power from an ILD is greater than that for an LED. A typical output power for an ILD is 5 mW (7 dBm) and 0.5 mW (−3 dBm) for LEDs. This allows ILDs to provide a higher drive power and to be used for systems that operate over longer distances.

3. ILDs can be used at higher bit rates than can LEDs.

4. ILDs generate monochromatic light, which reduces chromatic or wavelength dispersion.

Disadvantages of ILDs

1. ILDs are typically on the order of 10 times more expensive than LEDs.

2. Because ILDs operate at higher powers, they typically have a much shorter lifetime than LEDs.

3. IDLs are more temperature dependent than LEDs.

LIGHT DETECTORS

There are two devices that are commonly used to detect light energy in fiber optic communications receivers; PIN (p-type-intrinsic-n-type) diodes and APD (avalanche photodiodes).

PIN Diodes

A *PIN diode* is a *depletion-layer photodiode* and is probably the most common device used as a light detector in fiber optic communications systems. Figure 20-34 shows the basic construction of a PIN diode. A very lightly doped (almost pure or intrinsic) layer of *n*-type semiconductor material is sandwiched between the junction of the two heavily doped *n*- and *p*-type contact areas. Light enters the device through a very small window and falls on the carrier-void intrinsic material. The intrinsic material is made thick enough so that most of the photons that enter the device are absorbed by this layer. Essentially, the PIN photodiode operates just the opposite of an LED. Most of the photons are absorbed by electrons in the valence band of the intrinsic material. When the photons are absorbed, they add sufficient energy to generate carriers in the depletion region and allow current to flow through the device.

Photoelectric effect. Light entering through the window of a PIN diode is absorbed by the intrinsic material and adds enough energy to cause electrons to move from the valence band into the conduction band. The increase in the number of electrons that move into the conduction band is matched by an increase in the number of holes in the valence band. To cause current to flow in a photodiode, light of sufficient energy must be absorbed to give valence electrons enough energy to jump the energy gap. The energy gap for silicon is 1.12 eV (electron volts). Mathematically, the operation is as follows.

For silicon, the energy gap (E_g) equals 1.12 eV:

$$1 \text{ eV} = 1.6 \times 10^{-19} \text{ J}$$

Thus the energy gap for silicon is

$$E_g = (1.12 \text{ eV})\left(1.6 \times 10^{-19} \, \frac{\text{J}}{\text{eV}}\right) = 1.792 \times 10^{-19} \text{ J}$$

and

$$\text{energy } (E) = hf$$

where h = Planck's constant = 6.6256×10^{-34} J/Hz
     f = frequency (Hz)

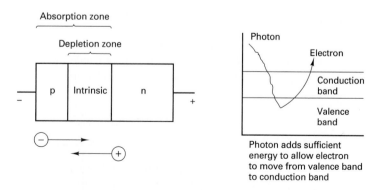

Figure 20-34 PIN photodiode construction.

Rearranging and solving for f yields

$$f = \frac{E}{h}$$

For a silicon photodiode,

$$f = \frac{1.792 \times 10^{-19} \text{ J}}{6.6256 \times 10^{-34} \text{ J/Hz}}$$

$$= 2.705 \times 10^{14} \text{ Hz}$$

Converting to wavelength yields

$$\lambda = \frac{c}{f} = \frac{3 \times 10^{8} \text{ m/s}}{2.705 \times 10^{14} \text{ Hz}} = 1109 \text{ nm/cycle}$$

Consequently, light wavelengths of 1109 nm or shorter, or light frequencies of 2.705×10^{14} Hz or higher, are required to cause enough electrons to jump the energy gap of a silicon photodiode.

Avalanche Photodiodes

Figure 20-35 shows the basic construction of an *avalanche photodiode* (APD). An APD is a *pipn* structure. Light enters the diode and is absorbed by the thin, heavily doped *n*-layer. A high electric field intensity developed across the *i-p-n* junction by reverse-bias causes impact ionization to occur. During impact ionization, a carrier can gain sufficient energy to ionize other bound electrons. These ionized carriers, in turn, cause more ionizations to occur. The process continues like an avalanche and is, effectively, equivalent to an internal gain or carrier multiplication. Consequently, APDs are more sensitive than PIN diodes and require less additional amplification. The disadvantages of APDs are relatively long transit times and additional internally generated noise due to the avalanche multiplication factor.

Characteristics of Light Detectors

The most important characteristics of light detectors are:

Responsivity.   A measure of the conversion efficiency of a photodetector. It is the ratio of the output current of a photodiode to the input optical power and has the unit of amperes/watt. Responsivity is generally given for a particular wavelength or frequency.

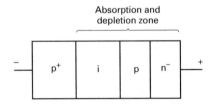

Absorption and
depletion zone

Figure 20-35   Avalanche photodiode construction.

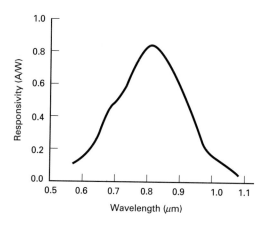

Figure 20-36 Spectral response curve.

Dark current. The leakage current that flows through a photodiode with no light input. Dark current is caused by thermally generated carriers in the diode.

Transit time. The time it takes a light-induced carrier to travel across the depletion region. This parameter determines the maximum bit rate possible with a particular photodiode.

Spectral response. The range of wavelength values that can be used for a given photodiode. Generally, relative spectral response is graphed as a function of wavelength or frequency. Figure 20-36 is an illustrative example of a spectral response curve. It can be seen that this particular photodiode more efficiently absorbs energy in the range 800 to 820 nm.

LASERS

Laser technology deals with the concentration of light into very small, powerful beams. The word *laser* is an acronym derived from the initials of the term "*l*ight *a*mplification by *s*timulated *e*mission of *r*adiation." The acronym was chosen when technology shifted from microwaves to light waves.

The first laser was developed by Theodore H. Maiman, a scientist who worked for Hughes Aircraft Company in California. Maiman directed a beam of light into ruby crystals with a xenon flashlamp and measured emitted radiation from the ruby. He discovered that when the emitted radiation increased beyond threshold it caused emitted radiation to become extremely intense and highly directional. Uranium lasers were developed in 1960 along with other rare-earth materials. Also in 1960, A. Javin of Bell Laboratories developed the helium laser. Semiconductor lasers (injection laser diodes) were manufactured in 1962 by General Electric, IBM, and Lincoln Laboratories.

Laser Types

Basically, there are four types of lasers: gas, liquid, solid, and semiconductor.

1. Gas lasers. Gas lasers use a mixture of helium and neon enclosed in a glass tube. A flow of coherent (one frequency) light waves is emitted through the output coupler when an electric current is discharged into the gas. The continuous light-wave output is monochromatic (one color).

Lasers

2. Liquid lasers. Liquid lasers use organic dyes enclosed in a glass tube for an active medium. Dye is circulated into the tube with a pump. A powerful pulse of light excites the organic dye.

3. Solid lasers. Solid lasers use a solid, cylindrical crystal such as ruby for the active medium. Each end of the ruby is polished and parallel. The ruby is excited by a tungsten lamp tied to an alternating-current power supply. The output from the laser is a continuous wave.

4. Semiconductor lasers. Semiconductor lasers are made from semiconductor *p-n* junctions and are commonly called *injection laser diodes* (ILDs). The excitation mechanism is a direct-current power supply which controls the amount of current to the active medium. The output light from an ILD is easily modulated, making it very useful in many electronic communications applications.

Laser Characteristics

All types of lasers have several common characteristics: (1) they all use an active material to convert energy into laser light, (2) a pumping source to provide power or energy, (3) optics to direct the beam through the active material to be amplified, (4) optics to direct the beam into a narrow powerful cone of divergence, (5) a feedback mechanism to provide continuous operation, and (6) an output coupler to transmit power out of the laser.

The radiation of a laser is extremely intense and directional. When focused into a fine hairlike beam, it can concentrate all its power into the narrow beam. If the beam of light were allowed to diverge, it would lose most of its power.

Laser Construction

Figure 20-37 shows the construction of a basic laser. A power source is connected to a flashtube that is coiled around a glass tube that holds the active medium. One end of the glass tube is a polished mirror face for 100% internal reflection. The flashtube is energized by a trigger pulse and produces a high-level burst of light (similar to a flashbulb). The flash causes the chromium atoms within the active crystalline structure to become excited. The process of pumping raises the level of the chromium atoms from ground state to an excited energy state. The ions then decay, falling to an intermediate energy level. When the population of ions in the intermediate level is greater than the ground state, a population inversion occurs. The population inversion causes laser action (lasing) to occur. After a period of time, the excited chromium atoms will fall to the ground energy level. At this time, photons are emitted. A photon is a packet of radiant energy. The emitted photons strike atoms and two other photons are emitted (hence the term "stimulated emission"). The frequency of the energy determines the strength of the photons; higher frequencies cause greater strength photons.

Laser Applications

Since its inception, lasers have become commonly used devices for both commercial and industrial applications. Lasers are used in electronics communications, holography, medicine, direction finding, and manufacturing.

In electronics communications, lasers are used in audio, radio, and television transmission. Laser beams have a very narrow bandwidth and they are also highly directional.

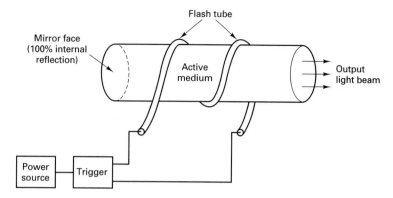

Figure 20-37 Laser construction.

Modulated light is a necessity for optical fiber applications. In medicine, ruby lasers are used for precise applications such as eye surgery. Argon ion lasers are replacing scalpels. The military uses lasers for distance measuring and surveying. In manufacturing, the laser is used for holography to detect stains and measure irregular objects. High-power lasers are used to cut reams of cloth and drill fine holes. Because of its narrow beamwidth, laser can be used to cut fabric within the accuracy of a single thread. There is really no end in sight for the application for lasers.

QUESTIONS

20-1. Define a fiber optic system.

20-2. What is the relationship between information capacity and bandwidth?

20-3. What development in 1951 was a substantial breakthrough in the field of fiber optics? In 1960? In 1970?

20-4. Contrast the advantages and disadvantages of fiber optic cables and metallic cables.

20-5. Outline the primary building blocks of a fiber optic system.

20-6. Contrast glass and plastic fiber cables.

20-7. Briefly describe the construction of a fiber optic cable.

20-8. Define the following terms: velocity of propagation, refraction, and refractive index.

20-9. State Snell's law for refraction and outline its significance in fiber optic cables.

20-10. Define *critical angle*.

20-11. Describe what is meant by mode of operation; by index profile.

20-12. Describe a step-index fiber cable; a graded-index cable.

20-13. Contrast the advantages and disadvantages of step-index, graded-index, single-mode propagation, and multimode propagation.

20-14. Why is single-mode propagation impossible with graded-index fibers?

20-15. Describe the source-to-fiber aperture.

20-16. What are the acceptance angle and the acceptance cone for a fiber cable?

20-17. Define *numerical aperture*.

20-18. List and briefly describe the losses associated with fiber cables.

20-19. What is *pulse spreading*?

20-20. Define *pulse spreading constant*.

20-21. List and briefly describe the various coupling losses.

20-22. Briefly describe the operation of a light-emitting diode.

20-23. What are the two primary types of LEDs?

20-24. Briefly describe the operation of an injection laser diode.

20-25. What is lasing?

20-26. Contrast the advantages and disadvantages of ILDs and LEDs.

20-27. Briefly describe the function of a photodiode.

20-28. Describe the photoelectric effect.

20-29. Explain the difference between a PIN diode and an APD.

20-30. List and describe the primary characteristics of light detectors.

PROBLEMS

20-1. Determine the wavelengths in nanometers and angstroms for the following light frequencies.
   (a) 3.45×10^{14} Hz
   (b) 3.62×10^{14} Hz
   (c) 3.21×10^{14} Hz

20-2. Determine the light frequency for the following wavelengths.
   (a) 670 nm
   (b) 7800 Å
   (c) 710 nm

20-3. For a glass ($n = 1.5$)/quartz ($n = 1.38$) interface and an angle of incidence of 35°, determine the angle of refraction.

20-4. Determine the critical angle for the fiber described in Problem 20-3.

20-5. Determine the acceptance angle for the cable described in Problem 20-3.

20-6. Determine the numerical aperture for the cable described in Problem 20-3.

20-7. Determine the maximum bit rate for RZ and NRZ encoding for the following pulse-spreading constants and cable lengths.
   (a) $\Delta t = 10$ ns/m, $L = 100$ m
   (b) $\Delta t = 20$ ns/m, $L = 1000$ m
   (c) $\Delta t = 2000$ ns/km, $L = 2$ km

20-8. Determine the lowest light frequency that can be detected by a photodiode with an energy gap $= 1.2$ eV.

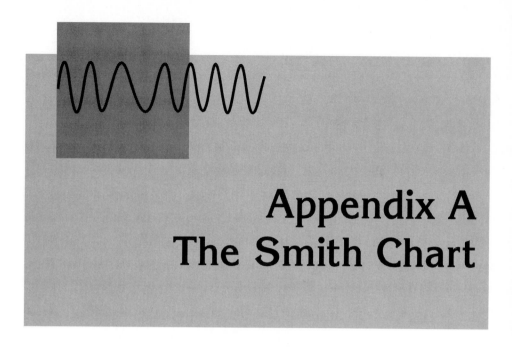

Appendix A
The Smith Chart

INTRODUCTION

Mathematical solutions for transmission-line impedances are laborious. Therefore, it is common practice to use charts to graphically solve transmission-line impedance problems. Equation A-1 is the formula for determining the impedance at a given point on a transmission line.

$$Z = Z_O \left[\frac{Z_L + jZ_O \tan \beta S}{Z_O + jZ_L \tan \beta S} \right] \tag{A-1}$$

where Z = line impedance at a given point
  Z_L = load impedance
  Z_O = line characteristic impedance
  βS = distance from the load to the point where the impedance value is to be calculated

Several charts are available on which the properties of transmission lines are graphically presented. However, the most useful graphical representations are those that give the impedance relations that exist along a lossless transmission line for varying load conditions. The *Smith chart* is the most widely used transmission-line calculator of this type. The Smith chart is a special kind of impedance coordinate system that portrays the relationship of impedance at any point along a uniform transmission line to the impedance at any other point on the line.

The Smith chart was developed by Philip H. Smith at Bell Telephone Laboratories and was originally described in an article entitled "Transmission Line Calculator"

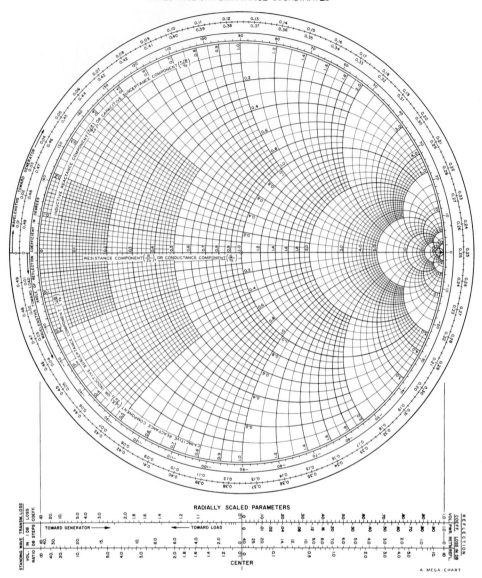

Figure A-1 Smith chart, transmission-line calculator.

(*Electronics*, January 1939). A Smith chart is shown in Figure A-1. This chart is based on two sets of *orthogonal* circles. One set represents the ratio of the resistive component of the line impedance (R) to the characteristic impedance of the line (Z_O), which for a lossless line is also purely resistive. The second set of circles represents the ratio of the reactive component of the line impedance ($\pm jX$) to the characteristic impedance of the line (Z_O). Parameters plotted on the Smith chart include the following:

1. Impedance (or admittance) at any point along a transmission line
   a. Reflection coefficient magnitude (Γ)
   b. Reflection coefficient angle in degrees

2. Length of transmission line between any two points in wavelengths
3. Attenuation between any two points
   a. Standing-wave loss coefficient
   b. Reflection loss
4. Voltage or current standing-wave ratio
   a. Standing-wave ratio
   b. Limits of voltage and current due to standing waves

SMITH CHART DERIVATION

The impedance of a transmission line, Z, is made up of both *real* and *imaginary* components of either sign (that is, $Z = R \pm jX$). Figure A-2(a) shows three typical circuit elements, and Figure A-2(b) shows their impedances graphed on a *rectangular* coordinate plane. All values of Z that correspond to passive networks must be plotted on or to the right of the imaginary axis of the Z plane (this is because a negative real component implies that the network is capable of supplying energy). To display the impedance of all possible passive networks on a rectangular plot, the plot must extend to infinity in three directions ($+R$, $+jX$, and $-jX$). The Smith chart overcomes this limitation by plotting the complex *reflection coefficient*,

$$\Gamma = \frac{z - 1}{z + 1} \tag{A-2}$$

where z = the impedance normalized to the characteristic impedance (that is, $z = Z/Z_O$).

Equation A-2 shows that for all passive impedance values, z, the magnitude of Γ is between 0 and 1. Also, since $|\Gamma| \leq 1$, the entire right side of the z plane can be mapped

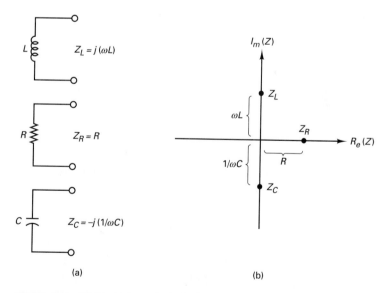

(a)   (b)

Figure A-2 (a) Typical circuit elements; (b) impedances graphed on rectangular coordinate plane. (Note: ω is the angular frequency at which Z is measured.)

Smith Chart Derivation

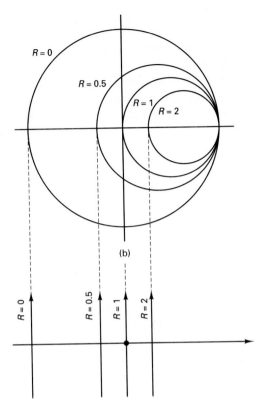

(b)

Figure A-3 (a) Rectangular plot; (b) I plane.

onto a circular area on the Γ plane. The resulting circle has a radius $r = 1$ and a center at $\Gamma = 0$, which corresponds to $z = 1$ or $Z = Z_O$.

Lines of Constant $R_e(z)$

Figure A-3(a) shows the rectangular plot of four lines of constant resistance $R_e(z) = 0, 0.5,$ 1, and 2. For example, any impedance with a real part $R_e = 1$ will lie on the $R = 1$ line. Impedances with a positive reactive component (X_L) will fall above the real axis, while impedances with a negative reactive component (X_C) will fall below the real axis. Figure A-3(b) shows the same four values of R mapped onto the Γ plane. $R_e(z)$ are now circles of $R_e(\Gamma)$. However, inductive impedances are still transferred to the area above the horizontal axis, and capacitive impedances are transferred to the area below the horizontal axis. The primary difference between the two graphs is that with the circular plot the lines no longer extend to infinity. The infinity points all meet on the plane at a distance of 1 to the right of the origin. This implies that for $z = \infty$ (whether real, inductive, or capacitive) $\Gamma = 1$.

Lines of Constant $X(z)$

Figure A-4(a) shows the rectangular plot of three lines of constant inductive reactance $(X = 0.5, 1,$ and 2), three lines of constant capacitive reactance $(X = -0.5, -1,$ and $-2)$, and a line of zero reactance $(X = 0)$. Figure A-4(b) shows the same seven values of jX plotted onto the Γ plane. It can be seen that all values of infinite magnitude again meet at $\Gamma = 1$. The entire rectangular z plane curls to the right, and its three axes (which

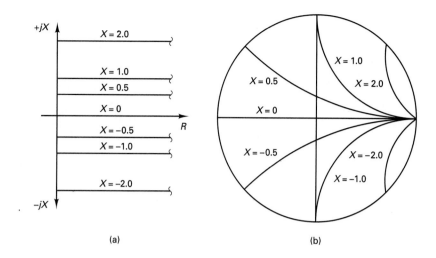

(a)

(b)

Figure A-4 (a) Rectangular plot; (b) I plane.

previously extended infinitely) meet at the intersection of the $\Gamma = 1$ circle and the horizontal axis.

Impedance Inversion (Admittance)

Admittance (Y) is the mathematical inverse of Z (that is, $Y = 1/Z$). Y, or for that matter any complex number, can be found graphically using the Smith chart by simply plotting z on the complex Γ plane and then rotating this point 180° about $\Gamma = 0$. By rotating every point on the chart by 180°, a second set of coordinates (the y coordinates) can be developed that is an inverted mirror image of the original chart. See Figure A-5(a). Occasionally, the

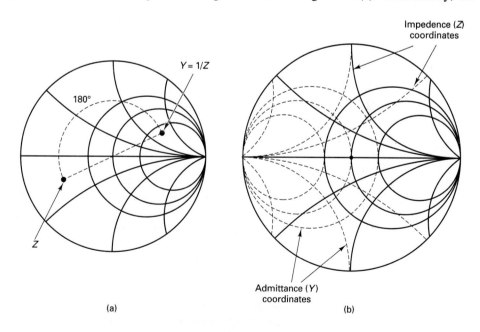

(a)

(b)

Figure A-5 Impedance inversion.

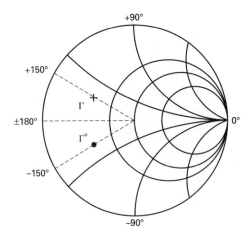

admittance coordinates are superimposed on the same chart as the impedance coordinates. See Figure A-5(b). Using the combination chart, both impedance and admittance values can be read directly by using the proper set of coordinates.

Complex Conjugate

The *complex conjugate* can easily be determined using the Smith chart by simply reversing the sign of the angle of Γ. On the Smith chart, Γ is usually written in polar form and angles become more negative (phase lagging) when rotated in a clockwise direction around the chart. Hence, $0°$ is on the right end of the real axis and $\pm180°$ is on the left end. For example, let $\Gamma = 0.5\ \underline{/+150°}$. The complex conjugate, Γ^*, is $0.5\ \underline{/-150°}$. In Figure A-6, it is shown that Γ^* is found by mirroring Γ about the real axis.

PLOTTING IMPEDANCE, ADMITTANCE, AND SWR ON THE SMITH CHART

Any impedance Z can be plotted on the Smith chart by simply *normalizing* the impedance value to the characteristic impedance (that is, $z = Z/Z_O$) and plotting the real and imaginary parts. For example, for a characteristic impedance $Z_O = 50\ \Omega$ and an impedance $Z = 25\ \Omega$ resistive, the normalized impedance z is determined as follows:

$$z = \frac{Z}{Z_O} = \frac{25}{50} = 0.5$$

Because z is purely resistive, its plot must fall directly on the horizontal axis ($\pm jX = 0$). $Z = 25$ is plotted on Figure A-7 at point A (that is, $z = 0.5$). Rotating $180°$ around the chart gives a normalized admittance value $y = 2$ (where $y = Y/Y_O$). y is plotted on Figure A-7 at point B.

As previously stated, a very important characteristic of the Smith chart is that any lossless line can be represented by a circle having its origin at $1 \pm j0$ (the center of the chart) and radius equal to the distance between the origin and the impedance plot. Therefore, the *standing-wave ratio, SWR,* corresponding to any particular circle is equal to the value of Z/Z_O at which the circle crosses the horizontal axis on the right side of the chart. Therefore, for this example, SWR $= 0.5$ ($Z/Z_O = 25/50 = 0.5$). It should also be noted

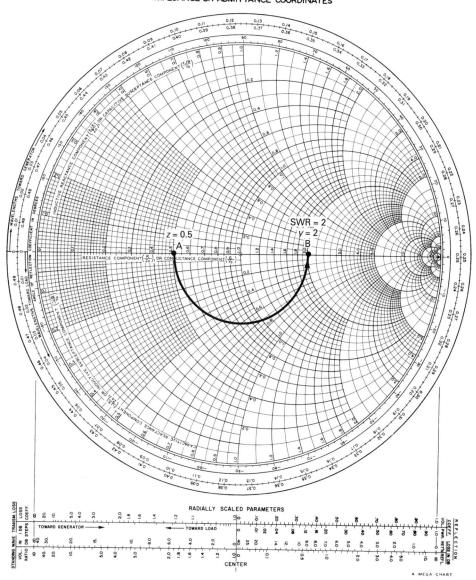

Figure A-7 Resistive impedance.

that any impedance or admittance point can be rotated 180° by simply drawing a straight line from the point through the center of the chart to where the line intersects the circle on the opposite side.

For a characteristic impedance $Z_O = 50$ and an inductive load $Z = +j25$, the normalized impedance z is determined as follows:

$$z = \frac{Z}{Z_O} = \frac{+jX}{Z_O} = \frac{+j25}{50} = +j0.5$$

Because z is purely inductive, its plot must fall on the $R = 0$ axis, which is the outer circle on the chart. $z = +j0.5$ is plotted on Figure A-8 at point A, and its admittance

Plotting Impedance, Admittance, and SWR on the Smith Chart   817

$y = -j2$ is graphically found by simply rotating 180° around the chart (point *B*). SWR for this example must lie on the far right end of the horizontal axis, which is plotted at point *C* and corresponds to SWR = ∞, which is inevitable for a purely reactive load. SWR is plotted at point *C*.

For a complex impedance $Z = 25 + j25$, z is determined as follows:

$$z = \frac{25 + j25}{50} = 0.5 + j0.5$$

$$= 0.707 \underline{/45°}$$

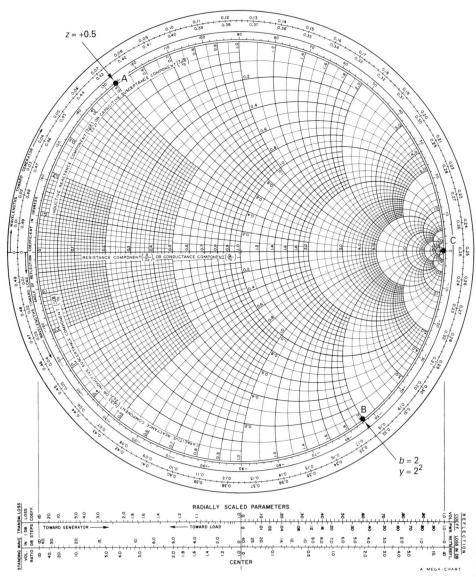

Figure A-8   Inductive load.

Appendix A   The Smith Chart

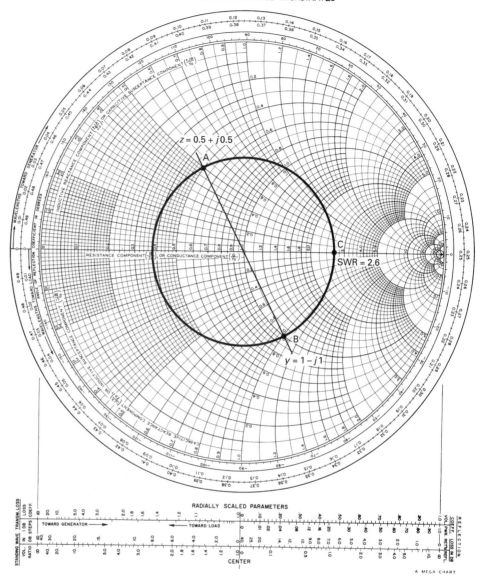

Figure A-9 Complex impedance.

Therefore,

$$Z = 0.707 \,\underline{/45°} \times 50 = 35.35 \,\underline{/45°}$$

and

$$Y = \frac{1}{35.35 \,\underline{/45°}} = 0.02829 \,\underline{/-45°}$$

Thus

$$y = \frac{Y}{Y_O} = \frac{0.02829}{0.02} = 1.414$$

and

$$y = 1 - j1$$

z is plotted on the Smith chart by locating the point where the $R = 0.5$ arc intersects

Plotting Impedance, Admittance, and SWR on the Smith Chart   **819**

the $X = 0.5$ arc on the top half of the chart. $z = 0.5 + j0.5$ is plotted on Figure A-9 at point A, and y is plotted at point B $(1 - j1)$. From the chart, SWR is approximately 2.6 (point C).

INPUT IMPEDANCE AND THE SMITH CHART

The Smith chart can be used to determine the input impedance of a transmission line at any distance from the load. The two outermost scales on the Smith chart indicate distance in *wavelengths* (see Figure A-1). The outside scale gives distance from the load toward the generator and increases in a clockwise direction, and the second scale gives distance from the source toward the load and increases in a counterclockwise direction. However, neither scale necessarily indicates the position of either the source or the load. One complete revolution (360°) represents a distance of one-half wavelength (0.5λ), half of a revolution (180°) represents a distance of one-quarter wavelength (0.25λ), and so on.

A transmission line that is terminated in an open circuit has an impedance at the open end that is purely resistive and equal to infinity (Chapter 8). On the Smith chart, this point is plotted on the right end of the $X = 0$ line (point A on Figure A-10). As you move toward the source (generator), the input impedance is found by rotating around the chart in a clockwise direction. It can be seen that input impedance immediately becomes capacitive and maximum. As you rotate farther around the circle (move toward the generator), the capacitance decreases to a normalized value of unity (that is, $z = -j1$) at a distance of one-eighth wavelength from the load (point C on Figure A-10) and a minimum value just short of one-quarter wavelength. At a distance of one-quarter wavelength, the input impedance is purely resistive and equal to 0 Ω (point B on Figure A-10). As described in Chapter 8, there is an impedance inversion every one-quarter wavelength on a transmission line. Moving just past one-quarter wavelength, the impedance becomes inductive and minimum; then the inductance increases to a normalized value of unity (that is, $z = +j1$) at a distance of three-eighths wavelength from the load (point D on Figure A-10) and a maximum value just short of one-half wavelength. At a distance of one-half wavelength, the input impedance is again purely resistive and equal to infinity (return to point A on Figure A-10). The results of the preceding analysis are identical to those achieved with phasor analysis in Chapter 8 and plotted in Figure 8-20.

A similar analysis can be done with a transmission line that is terminated in a short circuit, although the opposite impedance variations are achieved as with an open load. At the load, the input impedance is purely resistive and equal to 0. Therefore, the load is located at point B on Figure A-10, and point A represents a distance one-quarter wavelength from the load. Point D is a distance of one-eighth wavelength from the load and point C, a distance of three-eighths wavelength. The results of such an analysis are identical to those achieved with phasors in Chapter 8 and plotted in Figure 8-21.

For a transmission line terminated in a purely resistive load not equal to Z_o, Smith chart analysis is very similar to the process described in the preceding section. For example, for a load impedance $Z_L = 37.5$ Ω resistive and a transmission-line characteristic impedance $Z_O = 75$ Ω, the input impedance at various distances from the load is determined as follows:

1. The normalized load impedance z is

$$z = \frac{Z_L}{Z_O} = \frac{37.5}{75} = 0.5$$

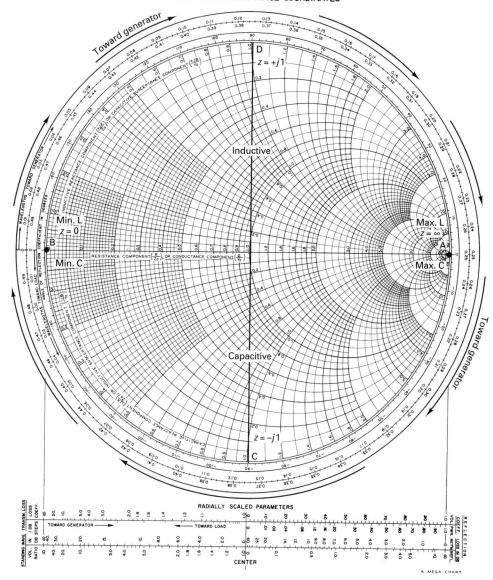

Figure A-10 Transmission-line input impedance for shorted and open line.

2. $z = 0.5$ is plotted on the Smith chart (point A on Figure A-11). A circle is drawn that passes through point A with its center located at the intersection of the $R = 1$ circle and the $x = 0$ arc.

3. SWR is read directly from the intersection of the $z = 0.5$ circle and the $X = 0$ line on the right side (point F), SWR $= 2$. The impedance circle can be used to describe all impedances along the transmission line. Therefore, the input impedance (Z_i) at a distance of 0.125λ from the load is determined by extending the z circle to the outside of the chart, moving point A to a similar position on the outside scale (point B on Figure A-11) and moving around the scale in a clockwise direction a distance of 0.125λ.

4. Rotate from point B a distance equal to the length of the transmission line (point

Input Impedance and The Smith Chart   **821**

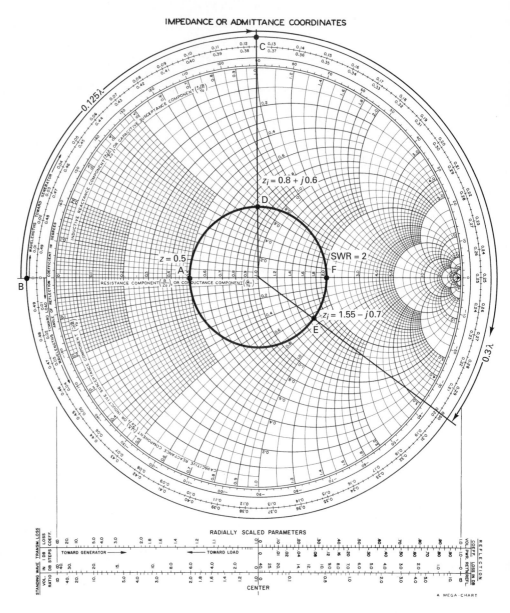

Figure A-11 Input impedance calculations.

C on Figure A-11). Transfer this point to a similar position on the $z = 0.5$ circle (point D on Figure A-11). The normalized input impedance is located at point D (0.8 + j0.6). The actual input impedance is found by multiplying the normalized impedance by the characteristic impedance of the line. Therefore, the input impedance Z_i is

$$Z_i = (0.8 + j0.6)75 = 60 + j45$$

Input impedances for other distances from the load are determined in the same way. Simply rotate in a clockwise direction from the initial point a distance equal to the length of the transmission line. At a distance of 0.3λ from the load, the normalized input imped-ance is found at point E ($z = 1.55 - j0.7$ and $Z_i = 116.25 - j52.5$). For distances greater than 0.5λ, simply continue rotating around the circle, with each complete rotation accounting for 0.5λ. A length of 1.125λ is found by rotating around the circle two complete revolutions and an additional 0.125λ.

EXAMPLE A-1

Determine the input impedance and SWR for a transmission line 1.25λ long with a characteristic impedance $Z_O = 50$ Ω and a load impedance $Z_L = 30 + j40$ Ω.

Solution The normalized load impedance z is

$$z = \frac{30 + j40}{50} = 0.6 + j0.8$$

z is plotted on Figure A-12 at point A and the impedance circle is drawn. SWR is read off the Smith chart from point B.

$$SWR = 2.9$$

The input impedance 1.25λ from the load is determined by rotating from point C 1.25λ in a clockwise direction. Two complete revolutions account for 1λ. Therefore, the additional 0.25λ is simply added to point C.

$$0.12λ + 0.25λ = 0.37λ \quad \text{point } D$$

Point D is moved to a similar position on the $z = 0.6 + j0.8$ circle (point E), and the input impedance is read directly from the chart.

$$z_i = 0.63 - j0.77$$

$$Z_i = 50(0.63 - j0.77) = 31.5 - j38.5$$

Quarter-wave Transformer Matching with the Smith Chart

As described in Chapter 8, a length of transmission line acts like a transformer (that is, there is an impedance inversion every one-quarter wavelength). Therefore, a transmission line with the proper length located the correct distance from the load can be used to match a load to the impedance of the transmission line. The procedure for matching a load to a transmission line with a quarter-wave transformer using the Smith chart is outlined in the following steps.

1. A load $Z_L = 75 + j50$ Ω can be matched to a 50-Ω source with a quarter-wave transformer. The normalized load impedance z is

$$z = \frac{75 + j50}{50} = 1.5 + j1$$

2. $z = 1.5 + j1$ is plotted on the Smith chart (point A, Figure A-13) and the impedance circle is drawn.

3. Extended point A to the outermost scale (point B). The characteristic impedance of an ideal transmission line is purely resistive. Therefore, if a quarter-wave transformer is located at a distance from the load where the input impedance is purely resistive, the transformer can match the transmission line to the load. There are two points on the impedance circle where the input impedance is purely resistive: where the circle intersects the $X = 0$ line (points C and D on Figure A-13). Therefore, the distance from the load to a point where the input impedance is purely resistive is determined by simply calculating the dis-

Input Impedance and The Smith Chart

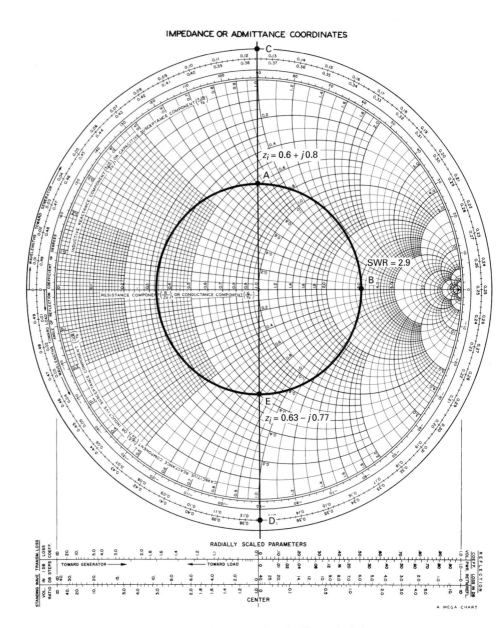

Figure A-12 Smith chart for Example A-1.

tance in wavelengths from point B on Figure A-13 to either point C or D, whichever is the shortest. The distance from point B to point C is

$$
\begin{array}{lr}
\text{point } C & 0.250\lambda \\
-\text{point } D & -0.192\lambda \\
\hline
\text{distance} & 0.058\lambda
\end{array}
$$

If a quarter-wave transformer is placed 0.058λ from the load, the input impedance is read directly from Figure A-13, $z_i = 2.4$ (point C).

4. Note that 2.4 is also the SWR of the mismatched line and is read directly from the chart.

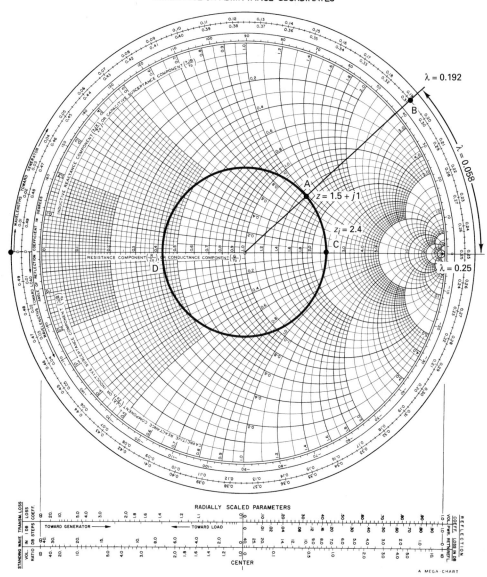

IMPEDANCE OR ADMITTANCE COORDINATES

Figure A-13 Smith chart, quarter-wave transformer.

5. The actual input impedance $Z_i = 50(2.4) = 120 \ \Omega$. The characteristic impedance of the quarter-wave transformer is determined from Equation 8-32.

$$Z_O' = \sqrt{Z_O Z_1} = \sqrt{50 \times 120} = 77.5 \ \Omega$$

Thus, if a quarter-wavelength of a 77.5-Ω transmission line is inserted 0.058λ from the load, the line is matched. It should be noted that a quarter-wave transformer does not totally eliminate standing waves on the transmission line. It simply eliminates them from the transformer back to the source. Standing waves are still present on the line between the transformer and the load.

Input Impedance and The Smith Chart   825

EXAMPLE A-2

Determine the SWR, characteristic impedance of a quarter-wave transformer, and the distance the transformer must be placed from the load to match a 75-Ω transmission line to a load $Z_L = 25 - j50$.

Solution The normalized load impedance z is

$$z = \frac{25 - j50}{75} = 0.33 - j0.66$$

z is plotted at point A on Figure A-14 and the corresponding impedance circle is drawn. The SWR is read directly from point F:

$$\text{SWR} = 4.6$$

The closest point on the Smith chart where z_i is purely resistive is point D. Therefore, the distance that the quarter-wave transformer must be placed from the load is

$$\begin{array}{ll} \text{point } D & 0.5\lambda \\ -\text{point } B & 0.4\lambda \\ \hline \text{distance} & 0.1\lambda \end{array}$$

The normalized input impedance is found by moving point D to a similar point on the z circle (point E), $z_i = 0.22$. The actual input impedance is

$$Z_i = 0.22(75) = 16.5 \ \Omega$$

The characteristic impedance of the quarter-wavelength transformer is again found from Equation 8-32.

$$Z_O' = \sqrt{75 \times 165} = 35.2 \ \Omega$$

Stub Matching with the Smith Chart

As described in Chapter 8, shorted and open stubs can be used to cancel the reactive portion of a complex load impedance and thus match the load to the transmission line. Shorted stubs are preferred because open stubs have a greater tendency to radiate.

Matching a complex load $Z_L = 50 - j100$ to a 75-Ω transmission line using a shorted stub is accomplished quite simply with the aid of a Smith chart. The procedure is outlined in the following steps:

1. The normalized load impedance z is

$$z = \frac{50 - j100}{75} = 0.67 - j1.33$$

2. $z = 0.67 - j1.33$ is plotted on the Smith chart shown in Figure A-15 at point A and the impedance circle is drawn in. Because stubs are shunted across the load (that is, placed in parallel with the load), admittances are used rather than impedances to simplify the calculations, and the circles and arcs on the Smith chart are now used for conductance and susceptance.

3. The normalized admittance y is determined from the Smith chart by simply rotating the impedance plot, z, 180°. This is done on the Smith chart by simply drawing a line from point A through the center of the chart to the opposite side of the circle (point B).

4. Rotate the admittance point clockwise to a point on the impedance circle where it intersects the $R = 1$ circle (point C). The real component of the input impedance at this

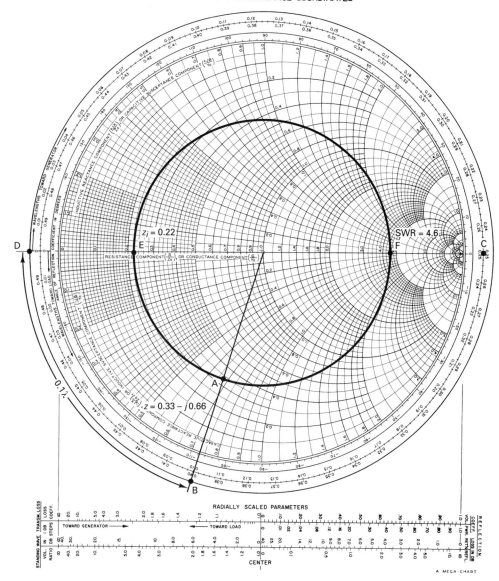

Figure A-14 Smith chart for Example A-2.

point is equal to the characteristic impedance Z_O, $Z_{in} = R \pm jX$, where $R = Z_O$). At point C, the admittance $y = 1 + j1.7$.

5. The distance from point B to point C is how far from the load the stub must be placed. For this example, the distance is $0.18\lambda - 0.09\lambda = 0.09\lambda$. The stub must have an impedance with a zero resistive component and a susceptance that has the opposite polarity (that is, $y_s = 0 - j1.7$).

6. To find the length of the stub with an admittance $y_s = 0 - j1.7$, move around the outside circle of the Smith chart (the circle where $R = 0$), having a wavelength identified at point D, until an admittance $y = 1.7$ is found (wavelength value identified at point E). You begin at point D because a shorted stub has minimum resistance ($R = 0$) and, conse-

Input Impedance and The Smith Chart 827

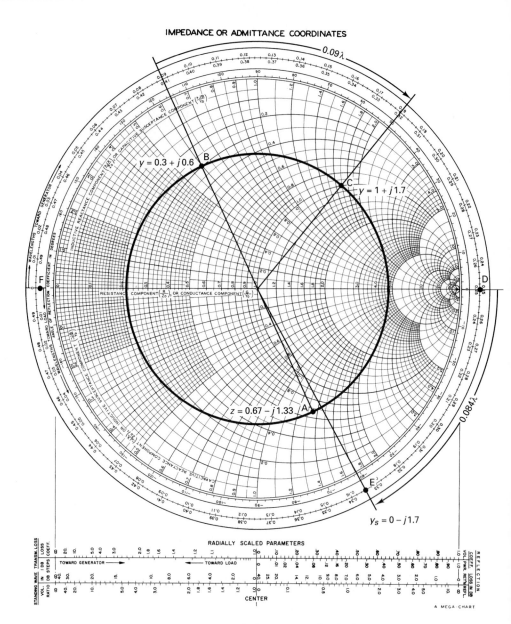

Figure A-15 Stub matching, Smith chart.

quently, a susceptance $B = \infty$. Point D is such a point. (If an open stub were used, you would begin your rotation at the opposite side of the $X = 0$ line, point F).

7. The distance from point D to point E is the length of the stub. For this example, the length of the stub is $0.334\lambda - 0.25\lambda = 0.084\lambda$.

EXAMPLE A-3

For a transmission line with a characteristic impedance $Z_O = 300 \ \Omega$ and a load with a complex impedance $Z_L = 450 + j600$, determine SWR, the distance a shorted stub must be placed from the load to match the load to the line, and the length of the stub.

Solution   The normalized load impedance z is

$$z = \frac{450 + j600}{300} = 1.5 + j2$$

$z = 1.5 + j2$ is plotted on Figure A-16 at point A and the corresponding impedance circle is drawn.

SWR is read directly from the chart at point B.

$$\text{SWR} = 4.7$$

Rotate point A 180° around the impedance circle to determine the normalized admittance.

IMPEDANCE OR ADMITTANCE COORDINATES

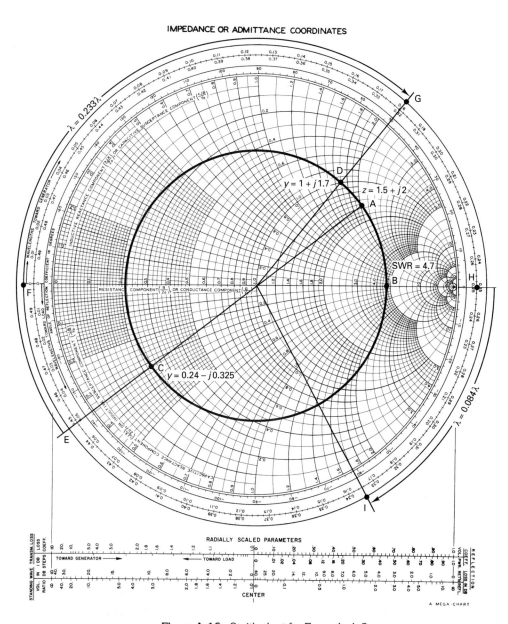

Figure A-16   Smith chart for Example A-3.

Input Impedance and The Smith Chart

$$y = 0.24 - j0.325 \text{ (point } C)$$

To determine the distance from the load to the stub, rotate clockwise around the outer scale beginning from point C until the circle intersects the $R = 1$ circle (point D).

$$y = 1 + j1.7$$

The distance from point C to point D is the sum of the distances from point E to point F and point F to point G.

$$\begin{aligned} E \text{ to } F &= 0.5\lambda - 0.449\lambda = 0.051\lambda \\ +F \text{ to } G &= \underline{0.18\lambda - 0\lambda} \quad\ \ = \underline{0.18\lambda} \\ &\quad\ \text{total distance} = 0.231\lambda \end{aligned}$$

To determine the length of the shorted stub, calculate the distance from the $y = \infty$ point (point H) to the $y_s = 0 - j1.7$ point (point I).

$$\text{stub length} = 0.334\lambda - 0.25\lambda = 0.084\lambda$$

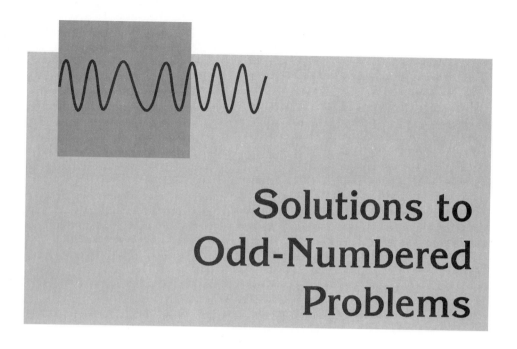

Solutions to Odd-Numbered Problems

CHAPTER 1

1-1. (a) $f_1 = 500$ Hz, $f_2 = 1000$ Hz, $f_3 = 1500$ Hz, $f_4 = 2000$ Hz, $f_5 = 2500$ Hz $V_1 = 10.19$ V$_p$, $V_2 = 0$ V, $V_3 = 3.4$ V$_p$, $V_4 = 0$ V, $V_5 = 2.04$ V$_p$

(b)

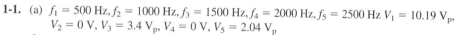

(c)

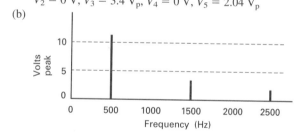

1-3. The spectrum shows five harmonically related frequency components beginning with a fundamental frequency of 1 kHz. This output spectrum would be typical for a nonlinear amplifier with a single input frequency of 1 kHz and four significant higher harmonics.

1-5. (a) 7 kHz, 14 kHz, and 21 kHz
4 kHz, 8 kHz, and 12 kHz

(b)

m	n	cross products
1	1	11 and 3 kHz
1	2	15 and 1 kHz
2	1	18 and 10 kHz
2	2	22 and 6 kHz

(c)

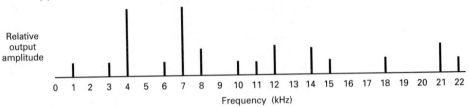

1-7. 6.63%

1-9. (a) 4.14×10^{-21} watts, -173.8 dBm
(b) 8.28×10^{-17} watts, -130.8 dBm
(c) $0.128\ \mu V$

1-11. 2.0101 or 3.03 dB

1-13. 10%

1-15. (a) 1×10^{-21} watts
(b) 4×10^{-17} watts
(c) The noise density is independent of bandwidth and thus remains at 1×10^{-21} watts.

CHAPTER 2

2-1. (a) $f_o = 19.9984$ MHz
(b) 19.9968 MHz
(c) 20.0032 MHz

2-3. 159.2 kHz

2-5. 0.0022 μF

2-7. (a) $V_{out} = 1.5$ V
(b) $V_{out} = 0.5$ V
(c) $V_{out} = 1.85$ V
(d) 0.5 V/rad

2-9.

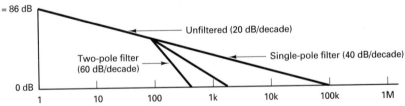

2-11. 0.373 V

2-13. 0.25 rad

2-15. 647,126 Hz.

CHAPTER 3

3-1. $m = 0.25$, %M $= 25$%

3-3. (a) 25 V_p

(b) 15 V_p

(c) $m = 0.6$, %M $= 60$%

3-5. An AM DSBFC envelope with a maximum positive voltage of 22.4 V, a maximum negative voltage of -22.4 V, a minimum positive voltage of 9.6 V, and a minimum negative voltage of -9.6 V.

3-7. (a) $f_{usf} = 825$ kHz

$f_{lsf} = 775$ kHz

(b) m $= 0.25$, %M $= 25$%

(c) $V_{max} = 50$ V, $V_{min} = 30$ V

(d)

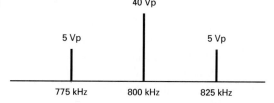

(e) An AM DSBFC envelope with a maximum positive voltage of 50 V, a maximum negative voltage of -50 V, a minimum positive voltage of 30 V, and a minimum negative voltage of -30 V.

3-9. (a) 6.25 watts

(b) The power of the modulated carrier is the same as the power of the unmodulated carrier, 6.25 watts. $P_{usf} = P_{lsf} = 0.5625$ watts

3-11. An AM DSBFC envelope with a maximum positive voltage of 510 mV, a maximum negative voltage of -510 mV, a minimum positive voltage of 150 mV, and a minimum negative voltage of -150 mV.

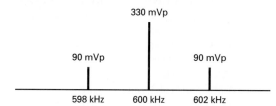

3-13. (a) 0.4

(b) 40%

(c) 20 Vp

(d) 4 Vp

CHAPTER 4

4-1. 12, 1200%

4-3. 900 K

4-5. $f_{if} = 600$ kHz and $f_{usf} = 596$ kHz

4-7. (a) 1810 kHz

(b) 122 or 20.9 dB

4-9. (a) 20 kHz

   (b) 14,276 Hz

CHAPTER 5

5-1. (a) 396 to 400 kHz and 400 to 404 kHz

   (b) 397.2 kHz and 402.8 kHz

5-3. (a) Balanced modulator 1 output: LSB = 96 to 100 kHz, USB = 100 to 104 kHz

   BPF1 output: USB = 100 to 104 kHz

   Balanced modulator 2 output: LSB = 3.896 to 3.9 MHz, USB = 4.1 to 4.104 MHz

   BPF2 output: USB = 4.1 to 4.104 MHz

   Balanced modulator 3 output: LSB = 25.896 to 25.9 MHz, USB = 34.1 to 34.104 MHz

   BPF3 output: 34.1 to 34.104 MHz

   (b) BPF1 output: 101.5 kHz, BPF2 = 4.1015 MHz, and BPF3 = 34.1015 MHz

5-5. (a) USB = 28 to 28.003 MHz

   (b) 28.0022 MHz

5-7. (a) USB = 496 to 500 kHz

   (b) 497 kHz

5-9. (a) Balanced modulator A output: LSB = 196 to 200 kHz, USB = 200 to 204 kHz

   BPFA output: LSB = 196 to 200 kHz

   Balanced modulator B output: LSB = 196 to 200 kHz, USB = 200 to 204 kHz

   BPFB output: USB = 200 to 204 kHz

   Hybrid output: LSB = 196 to 200 kHz, USB = 200 to 204 kHz

   Linear summer output: LSB = 196 to 200 kHz, USB = 200 to 204 kHz, and a 200 kHz carrier

   Balanced modulator 3 output: LSB = 3.796 to 3.804 MHz and a 3.8 MHz carrier, USB = 4.196 to 4.204 MHz and a 4.2 MHz carrier

   BPF3 output: 4.196 to 4.204 MHz and a 4.2 MHz carrier

   Balanced modulator 4 output: LSB = 27.796 to 27.804 MHz and a 27.8 MHz carrier, USB = 36.196 to 36.204 MHz and a 36.2 MHz carrier

   BPF4 output: 36.196 to 36.204 MHz and a 36.2 MHz carrier

   (b) BPFA = 197.5 kHz

   BPFB = 203 kHz

   BPF3 = 4.1975, 4.2, and 4.203 kHz

   BPF4 = 36.1975, 36.2, and 36.203 MHz

5-11. IF = 10.602 MHz

   BFO = 10.6 MHz

5-13. (a) USF1 = 202 kHz

   USF2 = 203 kHz

   (b) 5.57 watts

   2.278 watts

CHAPTER 6

6-1. 0.5 kHz/V, 1 kHz

6-3. (a) 40 kHz

   (b) 80 kHz

   (c) 20

6-5. 80%

6-7. (a) From the Bessel table, there are 4 sets of significant sidebands

(b)

J_n	V	f
0.22	1.76	carrier
0.58	4.64	1st set of sidebands
0.35	2.80	2nd set of sidebands
0.13	1.04	3rd set of sidebands
0.03	0.24	4th set of sidebands

(c)

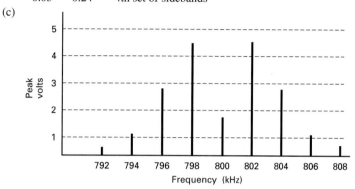

(d) 16 kHz

(e) 32 kHz

6-9. 50 kHz

6-11. DR = 2

6-13. P_c = 23.716 watts
P_{1st} = 7.744 watts
P_{2nd} = 0.484 watts
P_{3rd} = 0.016 watts

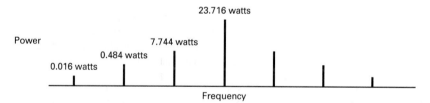

6-15. 0.0377 rad

6-17. (a) 0.0036 rad and 7.2 rad

(b) 7.2 kHz and 14.4 kHz

(c) 90 MHz

CHAPTER 7

7-1. S/N(predetection) = 18 dB and S/N(input) = 22 dB

7-3. f_{image} = 114.15 MHz
f_{lo} = 103.45 MHz

7-5. 66.5 kHz peak and 133 kHz peak-to-peak

$$f_{\text{upper}} = 20.4665 \text{ MHz}$$
$$f_{\text{lower}} = 20.3335 \text{ MHz}$$

7-7. 0.4 V

7-9. 3.3 dB, FM improvement = 3.3 dB, Predetection S/N ratio = 13.3 dB

CHAPTER 8

8-1.

f	l
1 kHz	300 km
100 kHz	3 km
1 MHz	300 m
1 GHz	0.3 m

8-3. 261 ohms

8-5. 111.8 ohms

8-7. 0.05

8-9. 12

8-11. 1.5

8-13. 252 meters

8.15. 0.8

CHAPTER 9

9-1. 0.2 μW/m^2

9-3. The power density is inversely proportional to the distance squared. Therefore, the power density would decrease by a factor of 3^2 or 9.

9-5. 14.14 MHz

9-7. 0.0057 V/m

9-9. The power density is inversely proportional to the square of the distance from the source. Therefore, if the distance increases by a factor of 4, the power density decreases by a factor of 4^2 or 16.

9-11. 20 dB

9-13. 8.94 miles

9-15. 17.89 miles

CHAPTER 10

10-1. (a) 25 ohms
(b) 92.6%
(c) 926 watts

10-3. 38.13 dB

10-5. 82 watts
EIRP = 163,611.5 watts or 82.1 dBm

10-7. 30 watts, 0.106μW/m^2

10-9. 16 dB

Solutions to Odd-Numbered Problems

10-11. 97.9%

10-13. 98.2%

10-15. 5.35 dB, 108.7°

10-17. (a) 6 GHz
(b) 5 cm
(c) 1.54×10^8 m/s
(d) 5.82×10^8 m/2

CHAPTER 11

11-1.

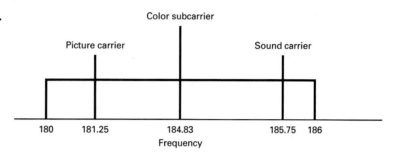

11-3. 31.5 scan lines

11-5. 1.27 ms

11-7. Channel 55 = 721.75 MHz
Channel 66 = 787.75 MHz

11-9. I = 0.248 V
Q = −0.082 V

CHAPTER 12

12-1. Baud = 10 Megabaud
B = 40 MHz

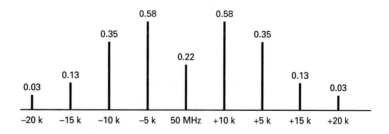

12-3.

Q	I	I	Q	Output phase
0	0	$-\sin \omega_c t$	$-\cos \omega_c t$	+135°
0	1	$+\sin \omega_c t$	$-\cos \omega_c t$	+45°
1	0	$-\sin \omega_c t$	$+\cos \omega_c t$	−135°
1	1	$+\sin \omega_c t$	$+\cos \omega_c t$	−45°

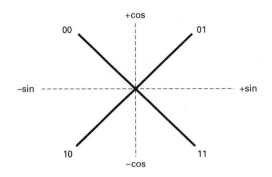

12-5. Baud = 6.667 Megabaud
B = 6.667 MHz

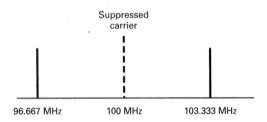

12-7. Baud = 5 Megabaud
f_N = 5 MHz

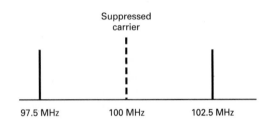

12-9. (a) QPSK = 2 bps/Hz
(b) 8-PSK = 3 bps/Hz
(c) 16-QAM = 4 bps/Hz

12-11. (a) −100 dBm
(b) −132.23 dBm
(c) −180 dBm
(d) −174.77 dBJ
(e) 32.2 dB
(f) 35.23 dB

CHAPTER 13

13-1.

	D	A	T	A	Sp	C	O	M	M	U	N	I	C	A	T	I	O	N	S	E	T	X	B	S	C
b_0	0	1	0	1	0	1	1	1	1	1	0	1	1	1	0	1	1	0	1	1		0
b_1	0	0	0	0	0	1	1	0	0	0	1	0	1	0	0	0	1	1	1	1		0
b_2	1	0	1	0	0	0	1	1	1	1	1	0	0	0	1	0	1	1	0	0		0
b_3	0	0	0	0	0	0	1	1	1	0	1	1	0	0	0	1	1	1	0	0		0
b_4	0	0	1	0	0	0	0	0	0	1	0	0	0	0	1	0	0	0	1	0		0
b_5	0	0	0	0	1	0	0	0	0	0	0	0	0	0	0	0	0	0	0	0		1
b_6	1	1	1	1	0	1	1	1	1	1	1	1	1	1	1	1	1	1	1	0		0
VCR	1	1	0	1	0	0	0	1	1	1	1	0	0	1	0	0	0	1	1	1		1

13-3. 4 Hamming bits.

CHAPTER 14

14-1.

	b_0	b_1	b_2	b_3	b_4	b_5	b_6	b_7
B8H	1	0	1	1	1	0	0	0

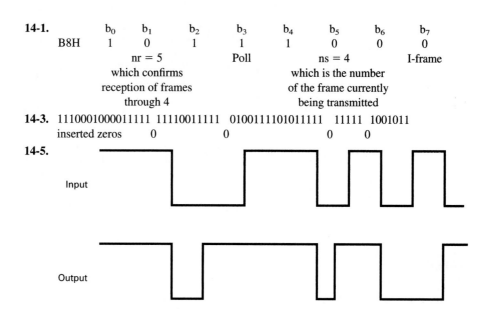

nr = 5   Poll   ns = 4   I-frame

which confirms   which is the number

reception of frames   of the frame currently

through 4   being transmitted

14-3. 1110001000011111 11110011111 0100111101011111 11111 1001011
inserted zeros   0   0   0   0

14-5.

Input

Output

CHAPTER 15

15-1. (a) $f_s \geq 8$ kHz
   (b) $f_s \geq 20$ kHz

15-3. 10 kHz or less.

15-5. DR = 511 or 20 log 511 = 54 dB

15-7. (a) -2.12 V
   (b) -0.12 V
   (c) $+0.04$ V
   (d) -2.52 V
   (e) $+0$ V

15-9. resolution = 0.01 V
   quantization error = 0.005 V

15-11. (a) $+0.01$ to $+0.03$ V
   (b) -0.01 to 0 V
   (c) $+10.23$ to $+10.25$ V
   (d) -10.23 to -10.25 V
   (e) $+5.13$ to $+5.15$ V
   (f) $+13.63$ to $+13.65$ V

CHAPTER 16

16-1. (a) 1.521 Mbps
   (b) 760.5 kHz

16-3.

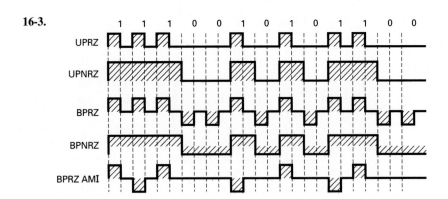

16-5. Ch1 = 108 kHz, Ch2 = 104 kHz, Ch3 = 100 kHz, Ch4 = 96 kHz, Ch5 = 92 kHz, Ch6 = 88 kHz, Ch7 = 84 kHz, Ch8 = 80 kHz, Ch9 = 76 kHz, Ch10 = 72 kHz, Ch11 = 68 kHz, Ch12 = 64 kHz

CHAPTER 17

17-1. −99.23 dBm
17-3. 28.9 dB
17-5. 35 dBm
17-7. −1.25 dB
17-9. 6.54 dB

CHAPTER 18

18-1. Elevation angle = 51°, azimuth = 33° west of south
18-3. (a) 11.74 dB
     (b) 8.75 dB
18-5. Radiated power = 28 dBW, EIRP = 68 dBW
18-7. −200.8 dBW
18-9. 26.18 dB
18-11. 19.77 dB
18-13. 12.5 dB

CHAPTER 19

19-1. 15 transponders
19-3. 44 stations

CHAPTER 20

20-1. (a) 869 nm, 8690 A
      (b) 828 nm, 8280 A
      (c) 935 nm, 9350 A

20-3. 38.57°

20-5. 36°

20-7. (a) RZ = 1 Mbps, NRZ = 500 kbps
      (b) RZ = 50 kbps, NRZ = 25 kbps
      (c) RZ = 250 kbps, NRZ = 125 kbps

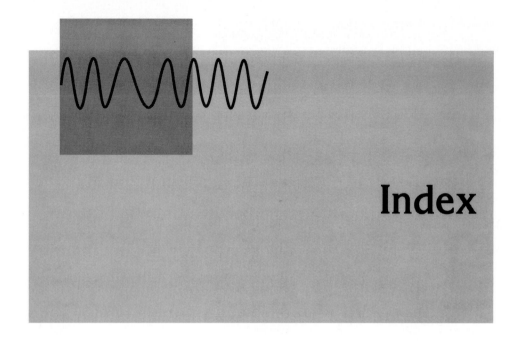

Index

American Radio Telephone Service (ARTS), 305
American Standard Code for Information Interchange (*see* ASCII)
American Telephone and Telegraph (*see* AT&T)
Amplitude:
  compandoring single sideband, 223
  distortion, 40, 148
  limiters, 281
Amplitude modulation (*see* AM)
Analog, 454
  communications system, 2, 454
  companding, 623
  multiplier, 71
Analog-to-digital converter (ADC), 316, 481
Angle of
  elevation, 725
  incidence, 362, 363, 783
  reflection, 363
  refraction, 362
Angle modulation, 229, 230
  average power, 245
  bandwidth requirements, 241
  phasor representation, 244
  receivers, 271
Angstrom, 776
Angular displacement, 230
Angular misalignment, 800
ANIK-D communications satellite, 757
ANSI, 509, 584
  3.66, 584
Answer channel, 548
Antenna, 3, 377
  arrays, 393
  bandwidth, 385
  basic operation, 377
  beamwidth, 385
  broadside array, 394
  dipole, 378
  directive gain, 382
  directivity, 382
  efficiency, 380, 381
  elementary doublet, 386
  elements, 393
  endfire array, 395, 400
  fat dipole, 396
  feed mechanisms, 408
  folded dipole, 396
  grounded, 390
  half-wave dipole, 378, 387, 389
  helical, 400
  Hertz, 378, 387, 391
  input impedance, 386
  loading, 392
  log periodic, 397
  loop, 399

Marconi, 378, 390
microwave, 402
monopole, 378, 390
nonresonant, 395
omnidirectional, 357, 380
parabolic reflector, 403
phased array, 400
polarization, 385
power gain, 382, 383
quarter-wave dipole, 37
radiation efficiency, 377
radiation pattern, 378
resonant, 387
Rhombic, 395
short dipole, 386
UHF, 402
Yagi-Uda, 396, 437
Antialiasing filter, 613, 683
Antifoldover distortion, 613
Antipodal signaling, 589
Aperture distortion, 700
Aperture ratio, 405
Aperture time, 610
Apogee, 721
Application layer, 560
Application program, 507, 529, 557, 560
Armstrong, Major Edwin Howard, 3, 229
Armstrong indirect FM transmitter, 263
ARQ code, 519
Ascending node, 722
ASCII code, 513
Associated Press (AP), 507, 513
Asynchronous
  data, 528
    format, 528
  disconnect mode, 580
  modems, 547
  protocols, 560
  receivers, 150
  response mode, 580
  transfer mode (ATM), 601, 602
Atmospheric noise, 35
AT&T (American Telephone and Telegraph), 12, 305, 508, 546, 581, 660, 687, 700, 719
Attenuation, 359
Attenuation factor, 145
A-type channel bank, 676
Audio:
  detector, 154
  frequency band, 5
  section, 154
  subcarrier, 217
Aural signal, 425
Automatic:
  call selection, 304
  desensing, 185

frequency and phase control (AFPC), 451
frequency control (*see* AFC)
gain control (*see* AGC)
request for retransmission (ARQ), 519, 526, 560
teller machine (ATM), 507
volume control (AVC), 184
Auxiliary channel, 699
Avalanche photodiodes, 778, 804, 806
Axes symmetry, 16
Azimuth, 727

B

Back off, 589
Back off loss, 735
Back-to-back coupling, 403
Backporch, 434
Baird, John L., 425, 774
Balanced:
  bridge modulator, 206
  digital interface, 543
  diode mixer, 170
  lattice modulator, 201
  modulator, 170, 201, 463, 467
  operation, 581
  ring modulator, 201, 463
  signal transmission, 321
  slope detector, 272
  transmission, 543
  transmission line, 321
Balun, 325, 544
  bazooka, 326
  choke, 326
  sleeve, 326
Bandlimiting, 28, 144
Bandpass limiter/amplifier (BPL), 282
Bandwidth, 7, 455, 459, 773
  efficiency, 489
  IF, 460
  improvement, 146
  minimum, 460
  Nyquist, 460
  reduction, 175
Bardeen, John, 3
Barkhausen criterion, 51
Barnett-Vignant reliability equations, 711
Base:
  frequency, 93
  station, 305, 308, 315
Base and collector modulation, 129
Baseband, 4, 588, 692
Baseband repeaters, 694
Baseband signal, 4
Basic:
  group, 686

Electronic:
   communications, 1, 3, 654
   communications systems, 1, 5, 654
   Industries Association (*see* EIA)
   Mobile Xchange (EMX), 310
   push to talk, 300
   switching center, 310
   switching machines, 305
   switching systems (ESS), 546, 675
Electrostatic deflection, 431
Elementary dipole, 387
Elementary doublet, 386
Elevation angle, 725
Elliptically polarized, 385
Emission classifications, 7
Emitter modulation, 121
End-fire array, 395
End-fire helical antenna, 400
Energy bit per (E_b), 496, 735
Energy bit-to-noise power density
   (E_b/N_o), 497, 739
Energy gap, 805
Energy spectra, 25
Envelope:
   AM, 103
   delay, 703
   detection, 150
   detection receiver, 222
   detector, 181
   encryption/decryption, 758
Epoch, 646, 758
Equalization:
   post, 550
   pre, 550
Equalizer, 554
   adaptive, 551
   compromise, 554
Equalizing pulses, 435, 443
Equatorial orbit, 722
Equipartition law, 36, 37
Equivalent noise resistance, 164
Equivalent noise temperature (T_e),
   149, 737
Ericcson, 310
Error:
   control, 519
   correction, 519
   detection, 460, 667
      block check character (BCC), 522
      block check sequence (BCS), 522
      cyclic redundancy checking
         (CRC), 523
      exact count encoding, 519
      frame check sequence (FCS),
         575, 580
      horizontal redundancy checking
         (HRC), 520, 522
      longitudinal redundancy check-
         ing (LRC), 522
      redundancy, 519

   parity, 520
   vertical redundancy checking
      (VRC), 522
forward error correction (FEC), 526
retransmission, 526
symbol substitution, 525
voltage, 462
Ethernet, 590, 591
Even function, 16
Even symmetry, 15
Exact count coding, 519
EXAR Corporation, 70
Excess noise, 42
Exchange station identification (XID),
   575
Exchange termination (ET), 598
Exhalted carrier, 196
Expanding, 622
Explorer 1, 720
Extended addressing, 579
Extended binary-coded decimal inter-
   change code (*see* EBCDIC)
Extended source, 291
External noise, 35
Extraterrestrial noise, 35
Eye diagram, 641
Eye patterns, 641

F

Facilities field, 586
Facilities length field, 586
Facility, 1, 455, 509
Fade margin, 709, 711
Fading, 697
Far field, 380
Fast Fourier transforms, 26
Fat dipole, 396
FDM (frequency division multiplex-
   ing), 292, 548, 674, 761
   AT&T FDM hierarchy, 675
FDM/FM satellite system, 755
Federal Communications Commission
   (FCC), 6, 59, 236, 291, 299,
   304, 305, 308
Feed mechanisms, 403, 408
   Cassegrain feed, 409
   center-feed, 408
   horn-feed, 409
Feedback:
   coder, 622
   degenerative, 51
   loop, 51
   negative, 51
   oscillator, 51
   oscillator circuits, 59
   positive, 51
   ratio, 51
   regenerative, 51

Feeder losses, 709
Feeder point, 602
Feeder services, 691
Feedpoint, 386
Feedthrough capacitor, 164
Ferrite core, 177
FET push-pull balanced modulator,
   203
Fiber distributed data interface
   (FDDI), 590
Fiber optic communications, 773
Fiber optics (*see* optical fiber)
Fidelity, 148
Field, 430
Field intensity, 356
Figure shift, 513
Filter:
   ceramic, 213
   crystal, 211
   mechanical, 214
   surface acoustical wave (SAW),
      214
First detector, 153
First IF, 188
Fixed data rate mode, 654
Flag, 569
Flag field, 569
Flat top sampling, 610
F-layer, 372
Flexible fiber scope, 774
Flexible waveguide, 421
Flux linkage, 173
Flyback time, 430
FM (Frequency modulation), 3, 4,
   229, 454, 459, 691
   bandwidth requirements, 241
   capture effect, 281
   demodulator, 237, 272
   deviation, 235
   deviation ratio, 243
   direct, 230
   direct modulators, 253
   direct transmitters, 258
   double conversion superheterodyne
      receivers, 272
   indirect, 230
   indirect modulator, 258
   indirect transmitter, 263
   integrated circuit modulators, 256
   low index, 692
   microwave radio:
      receiver, 693
      repeater, 693
      stations, 700
      system, 692
      transmitter, 692
   modulator, 237
   narrowband, 461
   noise triangle, 247
   percent modulation, 236

X

Y

Z